T0189706

IFIP – The International Federation for Information Processing

IFIP was founded in 1960 under the auspices of UNESCO, following the first World Computer Congress held in Paris the previous year. A federation for societies working in information processing, IFIP's aim is two-fold: to support information processing in the countries of its members and to encourage technology transfer to developing nations. As its mission statement clearly states:

> IFIP is the global non-profit federation of societies of ICT professionals that aims at achieving a worldwide professional and socially responsible development and application of information and communication technologies.

IFIP is a non-profit-making organization, run almost solely by 2500 volunteers. It operates through a number of technical committees and working groups, which organize events and publications. IFIP's events range from large international open conferences to working conferences and local seminars.

The flagship event is the IFIP World Computer Congress, at which both invited and contributed papers are presented. Contributed papers are rigorously refereed and the rejection rate is high.

As with the Congress, participation in the open conferences is open to all and papers may be invited or submitted. Again, submitted papers are stringently refereed.

The working conferences are structured differently. They are usually run by a working group and attendance is generally smaller and occasionally by invitation only. Their purpose is to create an atmosphere conducive to innovation and development. Refereeing is also rigorous and papers are subjected to extensive group discussion.

Publications arising from IFIP events vary. The papers presented at the IFIP World Computer Congress and at open conferences are published as conference proceedings, while the results of the working conferences are often published as collections of selected and edited papers.

IFIP distinguishes three types of institutional membership: Country Representative Members, Members at Large, and Associate Members. The type of organization that can apply for membership is a wide variety and includes national or international societies of individual computer scientists/ICT professionals, associations or federations of such societies, government institutions/government related organizations, national or international research institutes or consortia, universities, academies of sciences, companies, national or international associations or federations of companies.

More information about this series at http://www.springer.com/series/6102

IFIP Advances in Information and Communication Technology

475

Editor-in-Chief

Kai Rannenberg, Goethe University Frankfurt, Germany

Editorial Board

Lazaros Iliadis · Ilias Maglogiannis (Eds.)

Artificial Intelligence Applications and Innovations

12th IFIP WG 12.5 International Conference
and Workshops, AIAI 2016
Thessaloniki, Greece, September 16–18, 2016
Proceedings

 Springer

Editors
Lazaros Iliadis
Democritus University of Thrace
Orestiada
Greece

Ilias Maglogiannis
University of Piraeus
Piraeus
Greece

ISSN 1868-4238   ISSN 1868-422X (electronic)
IFIP Advances in Information and Communication Technology
ISBN 978-3-319-83168-8   ISBN 978-3-319-44944-9   (eBook)
DOI 10.1007/978-3-319-44944-9

Printed on acid-free paper

This Springer imprint is published by Springer Nature
The registered company is Springer International Publishing AG Switzerland

Preface

Artificial Intelligence (AI) is a rapidly evolving and growing research area. During the last few decades it has expanded from a field of promise to one of actual delivery, with good practical application in almost every scientific domain. More specifically, during the last 5 years, AI algorithms have been applied more and more by Google in Facebook, by Microsoft (e.g., the CNTK that is an open source deep learning toolkit on GitHub) by Amazon, and by Baidu in China. The common core of all these recent research efforts is deep learning. Joaquin Candela, head of Facebook's Applied Machine Learning group, stated: "We're trying to build more than 1.5 billion AI agents, one for every person who uses Facebook or any of its products." Facebook is using a machine-learning platform known as the FBLearner Flow. In fact, Facebook is already building AI that builds AI! Deep learning, deep neural networks, multi-agent systems and autonomous agents, image processing, biologically inspired neural networks (spiking ANN) are already a reality. Deep neural networks are changing the Internet.

The International Federation for Information Processing (IFIP) was founded in 1960 under the auspices of UNESCO, following the first historical World Computer Congress held in Paris in 1959. The First AIAI conference (Artificial Intelligence Applications and Innovations) was organized in Toulouse, France in 2004 by the IFIP. Since then, it has always been technically supported by the Working Group 12.5 "Artificial Intelligence Applications." After 12 years of continuous presence, it has become a well-known and recognized mature event, offering AI scientists from all over the globe the chance to present their research achievements and to cope with the AI research explosion that is taking place at a meteoric speed. The 12th AIAI was held in Thessaloniki, Greece, during September 16–18, 2016.

Following a long-standing tradition, this Springer volume belongs to the IFIP AICT series and it contains the accepted papers that were presented orally at the AIAI 2016 main conference and in the workshops that were held as parallel events. Three workshops were organized, by invitation to prominent and distinguished colleagues, namely:

- The Third MT4BD2016 (Workshop on New Methods and Tools for Big Data),
- The Fifth MHDW 2016 (Mining Humanistic Data Workshop), and
- The First 5G-PINE (Workshop on 5G-Putting Intelligence to the Network Edge).

It is interesting that two of these workshops have a continuous presence in the AIAI events, which means that they are well established in the AI community.

All papers went through a peer-review process by at least two independent academic referees. Where needed, a third and a fourth referee were consulted to resolve any potential conflicts. For the 12th AIAI conference, 65 papers were submitted. Out of these submissions, 30 papers (46.15 %) were accepted for oral presentation as full ones, whereas another eight papers (12.3 %) were accepted as short ones. The authors of the accepted papers of the main event come from 12 different countries around the globe, namely: Brazil, Canada, China, Cyprus, Denmark, UK, Greece, India, Italy, Norway, Portugal, and USA.

As the title of the conference denotes, there are two core orientations of interest, basic research AI approaches and also applications in real-world cases. The diverse nature of the papers presented demonstrates the vitality of AI computing methods and proves the wide range of AI applications as well. The accepted papers of the 12th AIAI conference are related to the following thematic topics:

- Artificial Neural Networks
- Classification
- Clustering
- Control Systems – Robotics
- Data Mining
- Engineering Applications of AI
- Environmental Applications of AI
- Feature Reduction
- Filtering
- Financial-Economics Modeling
- Fuzzy Logic

- Genetic Algorithms
- Hybrid Systems
- Image and Video Processing
- Medical AI Applications
- Multi-Agent Systems
- Ontology
- Optimization
- Pattern Recognition
- Support Vector Machines
- Text Mining
- Web-Social Media Data AI Modeling

Three distinguished keynote speakers were invited to deliver lectures at the 12th AIAI conference.

Professor Barbara Hammer (Bielefeld University, Germany) gave a talk entitled "Discriminative Dimensionality Reduction for Data Inspection and Classifier Visualization." Barbara Hammer received her PhD in computer science in 1995 and her venia legendi in computer science in 2003, both from the University of Osnabrück, Germany. From 2000 to 2004, she was chair of the junior research group Learning with Neural Methods on Structured Data at University of Osnabrück before accepting the position of professor of theoretical computer science at Clausthal University of Technology, Germany, in 2004. Since 2010, she has held a professorship for theoretical computer science for cognitive systems at the CITEC cluster of excellence at Bielefeld University, Germany. Several research stays have taken her to Italy, UK, India, France, The Netherlands, and the USA. Her areas of expertise include hybrid systems, self-organizing maps, clustering, and recurrent networks as well as applications in bioinformatics, industrial process monitoring, or cognitive science. She chaired the IEEE CIS Technical Committee on Data Mining in 2013 and 2014, and she is chair of the Fachgruppe Neural Networks of the GI and vice-chair of the GNNS. She has been elected as IEEE CIS AdCom member for 2016–2018. Barbara has published more than 200 contributions to international conferences/journals, and she is coauthor/editor of four books.

Professor Aristidis Likas (University of Ioannina, Greece) delivered a talk entitled "Number of Clusters Estimation, Multi-view Clustering and Their Use for Video Summarization." Aristidis Likas is a professor in the Department of Computer Science and Engineering of the University of Ioannina, Greece. He received his diploma in electrical engineering from the National Technical University of Athens, Greece, in 1990 and his PhD degree in electrical and computer engineering from the same university in 1994. Since 1996, he has been with the Department of Computer Science and Engineering, University of Ioannina, Greece. He is interested in developing methods for machine learning/data mining problems (mainly classification, clustering, statistical and Bayesian learning) and in the application of these methods to video analysis, computer vision, medical diagnosis, bioinformatics, and text mining. His recent research focuses on techniques for estimating the number of clusters, kernel-based clustering, and multi-view clustering. He has published more than 80 journal papers and more than 80 conference papers attracting over 5,000 citations. Recently, he received a Best Paper Award at the ICPR 2014 conference. He has participated in several national and European research and development projects. He is a senior member of the IEEE. He served as an associate editor of the *IEEE Transactions on Neural Networks* journal and as general co-chair of the ECML PKDD 2011 and the SETN 2014 conferences.

Professor Jan Peters (Max Planck Institute for Intelligent Systems, TU Darmstadt, Germany) gave a talk on "Machine Learning of Motor Skills for Robots: From Simple Skills to Table Tennis and Manipulation." Jan Peters is a full professor (W3) for intelligent autonomous systems at the Computer Science Department of the Technische Universität Darmstadt and at the same time a senior research scientist and group leader at the Max Planck Institute for Intelligent Systems, where he heads the interdepartmental Robot Learning Group. Jan Peters has received the Dick Volz Best 2007 US PhD Thesis Runner-Up Award, the Robotics: Science & Systems – Early Career Spotlight, the INNS Young Investigator Award, and the IEEE Robotics & Automation Society's Early Career Award. Jan Peters has being honored for the development of new approaches to robot learning, robot architecture, and robotic methods and their applications for humanoid robots. In 2015, he was awarded an ERC Starting Grant. Jan Peters has studied computer science, electrical, mechanical, and control engineering at TU Munich and FernUni Hagen in Germany, and at the National University of Singapore (NUS) and the University of Southern California (USC). He has received four master's degrees in these disciplines as well as a PhD in computer science from USC.

We are grateful to Professors Spyros Sioutas, Katia Lida Kermanidis (Ionian University, Greece), Christos Makris (University of Patras Greece), and Phivos Mylonas (Ionian University, Greece). Thanks to their invaluable contribution and hard work, the 5th MHDW workshop was held successfully once more and it has already become a well-accepted event running in parallel with AIAI.

It was a great pleasure to host the Third MT4BD in the framework of the AIAI conference. We wish to sincerely thank its organizers for their great efforts and for their invaluable contribution. More specifically we wish to thank Spiros Likothanassis (University of Patras, Greece) and Dimitris Tzovaras (CERTH/ITI, Thessaloniki, Greece) for this well-established event.

The First 5G-PINE (Putting Intelligence to the Network Edge) workshop was an important part of the AIAI 2016 conference and it was driven by the hard work of Ioannis P. Chochliouros (Hellenic Telecommunications Organization - OTE, Greece), Leonardo Goratti (CREATE-NET, Italy), Oriol Sallent (UPC Spain), Haris Mouratidis

(University of Brighton, UK), Ioannis Neokosmidis (INCITES, Luxembourg), and Athanasios Dardamanis (SmartNet, Greece).

All workshops had a high attendance from scientists from all parts of Europe and some from Asia (e.g., UK, Greece, India, Italy, Spain, and Turkey) and we would like to thank all participants for this. The workshops received 33 submissions of which 17 were accepted as full papers, while seven were selected to be presented as short ones.

The 12th organization of AIAI is a real proof of the brand name that the conference has gained among the circles of the international scientific community. After so many years of hard effort, it is recorded as a mature event with loyal followers and it has plenty of new and qualitative research results to offer to the international scientific community. We hope that the readers of these proceedings will be highly motivated and stimulated for further research in the domain of AI in general.

September 2016   Lazaros Iliadis
                                                                      Ilias Maglogiannis

Organization

General Chair

John MacIntyre   University of Sunderland, UK

Honorary Chair

Tharam Dillon   Latrobe University, Melbourne, Australia

Program Co-chairs

Lazaros Iliadis   Democritus University of Thrace, Greece
Ilias Maglogiannis   University of Piraeus, Greece

Workshop Chairs

Christos Makris   University of Patras, Greece
Spyros Sioutas   Ionian University, Greece
Harris Papadopoulos   Frederick University, Cyprus

Tutorial Chairs

Chrisina Jayne   Robert Gordon University, Aberdeen, Scotland, UK
Richard Chbeir   University of Pau and Adour Countries, France

Special Session Chair

Giacomo Boracchi   Politecnico di Milano, Italy

Advisory Chairs

Plamen Angelov   Lancaster University, UK
Nikola Kasabov   KEDRI Auckland University of Technology,
                          New Zealand

Organizing Chair

Yannis Manolopoulos   Aristotle University of Thessaloniki, Greece

Website and Advertising Chair

Ioannis Karydis   Ionian University, Greece

Program Committee

Michel Aldanondo	Toulouse-University-CGI, France
Athanasios Alexiou	Ionian University, Greece
George Anastassopoulos	Democritus University of Thrace, Greece
Ioannis Andreadis	Democritus University of Thrace, Greece
Andreas Andreou	University of Cyprus, Cyprus
Costin Badica	University of Craiova, Romania
Zbigniew Banaszak	Warsaw University of Technology, Poland
Ramazan Bayindir	Gazi University, Turkey
Bartlomiej Beliczynski	Warsaw University of Technology, Poland
Nik Bessis	Edge Hill University, UK
Farah Bouakrif	University of Jijel, Algeria
Antonio Padua Braga	Universidade Federal de Minas Gerais, Brazil
Peter Brida	University of Zilina, Slovakia
Frantisek Capkovic	Slovak Academy of Sciences, Slovakia
George Caridakis	National Technical University of Athens, Greece
Ioannis Chamodrakas	National and Kapodistrian University of Athens, Greece
Aristotelis Chatziioannou	National Hellenic Research Foundation, Greece
Jefferson Rodrigo De Souza	FACOM/UFU, Brazil
Ruggero Donida Labati	Università degli Studi di Milano, Italy
Georgios Evangelidis	University of Macedonia, Thessaloniki, Greece,
Javier Fernandez De Canete	University of Malaga, Spain
Maurizio Fiasché	Politecnico di Milano, Italy
Mauro Gaggero	National Research Council of Italy, Italy
Alexander Gammerman	Royal Holloway University, UK
Christos Georgiadis	University of Macedonia, Thessaloniki, Greece
Giorgio Gnecco	IMT – Institute for Advanced Studies, Lucca, Italy
Hakan Haberdar	University of Houston, USA
Petr Hajek	University of Pardubice, Czech Republic
Ioannis Hatzilygeroudis	University of Patras, Greece
Martin Holena	Institute of Computer Science, Czech Republic
Chrisina Jayne	Robert Gordon University, UK
Raúl Jiménez-Naharro	Universidad de Huelva, Spain
Jacek Kabziński	Technical University of Lodz, Poland
Antonios Kalampakas	American University of the Middle East, Kuwait
Achilles Kameas	Hellenic Open University, Greece
Ryotaro Kamimura	Tokai University, Japan
Stelios Kapetanakis	University of Brighton, UK

Konstantinos Karpouzis	ICCS, Greece
Petros Kefalas	The University of Sheffield International Faculty, CITY College, Greece
Katia Lida Kermanidis	Ionian University, Greece
Muhammad Khurram Khan	King Saud University, Saudi Arabia
Yiannis Kokkinos	University of Macedonia, Thessaloniki, Greece
Mikko Kolehmainen	University of Eastern Finland, Finland
Petia Koprinkova-Hristova	Bulgarian Academy of Sciences, Bulgaria
Dimitrios Kosmopoulos	ATEI Crete, Greece
Konstantinos Koutroumbas	National Observatory of Athens, Greece
Paul Krause	University of Surrey, UK
Ondrej Krejcar	University of Hradec Kralove, Czech Republic
Stelios Krinidis	CERTH/ITI, Thessaloniki, Greece
Adam Krzyzak	Concordia University, Canada
Pekka Kumpulainen	TUT/ASE, Finland
Vera Kurkova	Academy of Sciences of the Czech Republic, Czech Republic
Efthyvoulos Kyriacou	Frederick University, Cyprus
Florin Leon	University "Gheorghe Asachi" of Iasi, Romania
Helmut Leopold	AIT Austrian Institute of Technology, Austria
José María Luna	University of Córdoba, Spain
Ilias Maglogiannis	University of Piraeus, Greece
George Magoulas	Birkbeck College, UK
Mario Natalino Malcangi	Università degli Studi di Milano, Italy
Manolis Maragoudakis	University of the Aegean, Greece
Francesco Marcelloni	University of Pisa, Italy
Konstantinos Margaritis	University of Macedonia, Thessaloniki, Greece
Nikolaos Mitianoudis	Democritus University of Thrace, Greece
Haralambos Mouratidis	University of Brighton, UK
Phivos Mylonas	Ionian University, Greece
Eva Onaindia	Universitat Politecnica de Valencia, Spain
Mihaela Oprea	University Petroleum-Gas of Ploiesti, Romania
Stefanos Ougiaroglou	University of Macedonia, Thessaloniki, Greece
Harris Papadopoulos	Frederick University, Cyprus
Elpiniki Papageorgiou	ATEI of Lamia, Greece
Antonios Papaleonidas	Democritus University of Thrace, Greece
Efi Papatheocharous	SICS Swedish ICT, Sweden
Nicos G. Pavlidis	Lancaster University, UK
Miltos Petridis	Brighton University, UK
Vassilis Plagianakos	University of Thessaly, Greece
Bernardete Ribeiro	University of Coimbra, Portugal
Alexander Ryjov	Moscow State University, Russia
Ilias Sakellariou	University of Macedonia, Thessaloniki, Greece
Marcello Sanguineti	Università di Genova, Italy
Alexander Sideridis	Agricultural University of Athens, Greece
Dragan Simic	University of Novi Sad, Serbia

Invited Talks

Discriminative Dimensionality Reduction for Data Inspection and Classifier Visualization

Barbara Hammer

Bielefeld University, Germany

Abstract. The amount of electronic data available today increases rapidly; hence humans rely on automated tools which allow them to intuitively scan data volumes for valuable information. Dimensionality reducing data visualization, which displays high dimensional data in two or three dimensions, constitutes a popular tool to directly visualize data sets on the computer screen. Dimensionality reduction, however, is an inherently ill-posed problem, and the results vary depending on the chosen technology, the parameters, and even random aspects of the algorithms - there is a high risk to display noise instead of valuable information.

In the presentation, we discuss discriminative dimensionality reduction techniques, i.e. methods which enhance a dimensionality reduction method by auxiliary information such as class labels. This allows the practitioner to easily focus on those aspects he is interested in rather than noise. We will discuss two different approaches in this realm, which rely on a parametric resp. non-parametric metric-learning scheme, and display their effect in several benchmarks. We discuss how these methods can be extended to non-vectorial and big data, and how they open the door to a visualization of not only the given data but any given classifier.

Multimodality in Data Clustering: Application to Video Summarization

Aristidis Likas

University of Ioannina, Greece

Abstract. Clustering constitutes an essential problem in machine learning and data mining with important applications in science, technology and business. In this talk, multimodality is related to clustering in two different ways.

The first part of this talk focuses on the clustering of data that are multimodal in the sense that multiple representations (views) are available for each data instance, coming from different sources and/or feature spaces. Typical multi-view clustering approaches treat all available views as being equally important. We will present approaches that assign a weight to each view. Such weights are automatically tuned to reflect the quality of the views and lead to improved clustering solutions.

In the second part of this talk the term 'multimodality' is used to express dataset inhomogeneity given some similarity or distance measure. We will present criteria for estimating the homogeneity of a group of data instances based on statistical tests of unimodality. Then we will describe the use of such criteria for developing incremental and agglomerative clustering methods that automatically estimate the number of clusters. We will also discuss methods for sequence segmentation that use the above criteria for deciding on segment boundaries.

Finally we will present results from the application of the above clustering and segmentation methods for video summarization, and, more specifically, for video sequence segmentation and extraction of representative key-frames.

Machine Learning of Motor Skills for Robots: From Simple Skills to Table Tennis and Manipulation

Jan Peters

Max Planck Institute for Intelligent Systems, Germany
TU Darmstadt, Germany

Abstract. Autonomous robots that can assist humans in situations of daily life have been a long standing vision of robotics, artificial intelligence, and cognitive sciences. A first step towards this goal is to create robots that can learn tasks triggered by environmental context or higher level instruction. However, learning techniques have yet to live up to this promise as only few methods manage to scale to high-dimensional manipulator or humanoid robots. In this talk, we investigate a general framework suitable for learning motor skills in robotics which is based on the principles behind many analytical robotics approaches. It involves generating a representation of motor skills by parameterized motor primitive policies acting as building blocks of movement generation, and a learned task execution module that transforms these movements into motor commands. We discuss learning on three different levels of abstraction, i.e., learning for accurate control is needed to execute, learning of motor primitives is needed to acquire simple movements, and learning of the task-dependent "hyperparameters" of these motor primitives allows learning complex tasks. We discuss task-appropriate learning approaches for imitation learning, model learning and reinforcement learning for robots with many degrees of freedom. Empirical evaluations on a several robot systems illustrate the effectiveness and applicability to learning control on an anthropomorphic robot arm. These robot motor skills range from toy examples (e.g., paddling a ball, ball-in-a-cup) to playing robot table tennis against a human being and manipulation of various objects.

Machine Learning Based Bioinformatics as a Tool for Big-Bata Analytics on Molecular Biology Datasets

Seferina Mavroudi

Technological Institute of Western Greece, Greece

Abstract. Deciphering the underlying biological mechanisms that lead to disease could pave the way for personalized medicine hopefully leading to early prevention of disease and drugs with minimal side-effects. Fulfilling this premise however is very demanding since Biology is complex, with thousands of key players interacting with each other in systems at various scales. In the light of the curse of dimensionality it is obvious that only the advent of big data in modern molecular biology provides the ground for building meaningful models that could formulate novel hypothesis. Moreover, extracting valuable biological knowledge in such environments is usually not feasible with simple statistical methods and sophisticated machine learning paradigms have to be encountered.

In the present talk we will briefly introduce the systems biology perspective according to which all essential biological molecules from genes, proteins, metabolites to cells and organs form "a network of networks". We will mention the genomic, proteomic and other heterogeneous medical data sources of big data production and we will ultimately elaborate on the analysis of these kinds of data with modern machine learning techniques. The challenges, pitfalls and perspectives of the analysis will be discussed.

Specific case studies concerning proteomic and transcriptomic data analysis aiming at biomarker discovery will be presented. The first case study is related to big data proteomics analysis and specifically to the case of analyzing TMT based Mass Spectrometry datasets which is not only a big data problem but is also related to complex analysis steps. Due to the huge amount of the processing data, standard approaches and serial implementations fail to deliver high quality biomarkers while being extremely time consuming. For this task machine learning and more specifically meta-heuristic methods were deployed combined with high performance parallel computing techniques to provide biomarkers of increased predictive accuracy with feasible and realistic time requirements.

The second case study which will be presented includes big data analytics on transcriptomics data related to the diagnosis of early stage Parkinson disease. Specifically, a unique network medicine pipeline has been used to combine multiple gene expression datasets created from both microarrays and RNA-sequencing experiments. The proposed methodology not only uncovered significantly fewer biomarkers than the standard approach but also came out with a set of biomarkers which present higher predictive performance and are highly relevant to the underlying mechanisms of Parkinson disease. Cloud computing technology has been used to ease the application of the proposed pipeline in multiple datasets.

Contents

Ontology-Web and Social Media AI Modeling (OWESOM)

Environmental AI Modeling (ENAIM)

Agents-Robotics-Control (AROC)

Artificial Neural Network Modeling (ANNMO)

Mining Humanistic Data Workshop (MHDW)

New Methods and Tools for Big Data Wokshop (MT4BD)

5G – Putting Intelligence to the Network Edge (5G-PINE)

Medical Artificial Intelligence Modeling (MAIM)

A Cumulative Training Approach
to Schistosomiasis Vector Density Prediction

Terence Fusco$^{(\boxtimes)}$ and Yaxin Bi

School of Computing and Mathematics, Ulster University, Newtownabbey, UK
Fusco-T@email.ulster.ac.uk, bi.y@ulster.ac.uk

Abstract. The purpose of this paper is to propose a framework of
building classification models to deal with the problem in predicting
Schistosomiasis vector density. We aim to resolve this problem using
remotely sensed satellite image extraction of environment feature values,
in conjunction with data mining and machine learning approaches. In
this paper we assert that there exists an intrinsic link between the density and distribution of the Schistosomiasis disease vector and the rate
of infection of the disease in any given community; it is this link that
the paper is focused to investigate. Using machine learning techniques,
we want to accumulate the most significant amount of data possible to
help with training the machine to classify snail density (SD) levels. We
propose to use a novel cumulative training approach (CTA) as a way of
increasing the accuracy when building our classification and prediction
model.

1 Introduction

The resurgence of epidemic disease breakouts in regions of Asia and South Amer-
ica in the past decade has given local governments and health organisations cause
for much concern. The devastating impact that these diseases can have on many
aspects of human, cattle and crop life incurs huge financial and social cost. This
rationale makes research into the prevention and preparation for future outbreaks, a problem that requires immediate attention and one that is crucial to
supporting the locally affected municipalities [1]. The epidemic disease Schistosomiasis is detrimental to many sections of society in China. Schistosomiasis is the
second most widely affected disease in the world as stated by the World Health
Organisation [2]. It is a disease, which is transmitted through water infected by
parasites known as Schistosomes. The intermediate host of the disease is the
Oncomelania Hupensis snail. Humans are affected mainly through freshwater
used for washing clothes and household items as well as through infected crops
and cattle. The affect it can have on many areas of human, cattle, crop life both
in terms of health and financially is a valid cause for concern [1]. To combat Schistosomiasis can be very difficult due to the fact that there is no vaccine available
against the disease and therefore it can only be treated once the patient has
been infected. Currently, the most effective way of dealing with the disease is by

© IFIP International Federation for Information Processing 2016
Published by Springer International Publishing Switzerland 2016. All Rights Reserved
L. Iliadis and I. Maglogiannis (Eds.): AIAI 2016, IFIP AICT 475, pp. 3–13, 2016.
DOI: 10.1007/978-3-319-44944-9_1

trying to establish areas that are of high risk of the disease and putting in place preventative measures such as chemical treatment to specific freshwater areas [3] in order that the disease is addressed before the vectors multiply or increase in density and distribution. An alternative solution is to plant poplar trees, which would disturb the natural vegetation and moisture factors that encourage snail life and breeding habitat [4]. Whichever method is applied to at-risk areas will have a time and financial cost incurred therefore the concerned municipalities require the most informed data available before acting and addressing the area in question. The local governments will also need to prepare those areas for any panic or influx of patients that may occur.

The environment features present in Schistosomiasis areas of interest can be shown to be intrinsically linked to the disease infection rates [5]. By using data mining methods we can assess the corollary relations between the environment feature values and the SD and distribution values. We aim to identify the environment conditions which make the Oncomelania Hupensis snail most suitable for transmission of the Schistosomiasis disease. We know that for the Oncomelania Hupensis reproduction and for life to flourish, it requires specific environment conditions. We also know the snails will not survive in strong currents and that during early years in their lives the Oncomelania Hupensis snail will live only in water. Once they are adults they then must move from the water usually to moist soil above the water line as the snail activity increases with soil moisture and that the optimum temperature for breeding is around 20 °C [6]. The Oncomelania Hupensis snail flourishes and breeds particularly well in areas with high levels specific environment features such as soil moisture (NDMI) and vegetation (NDVI) therefore we can deduce that areas which meet these specific environment conditions have a greater likelihood of high snail vector density.

By analyzing and assessing this information we can achieve greater success from our classification accuracy. With the implementation of this research approach we can make the most informed prediction on which to base information to provide to those concerned. We believe that the most promising approach to detect high-risk areas of disease outbreak is to use vector density classification techniques based on environmental features that exist in each area of interest.

Using our proposed Cumulative Training Approach (CTA), we can enhance the training potential of our limited dataset. This will help to provide a larger pool of relevant training data which we hope will increase the classification accuracy during the testing process. The process involved uses the data from a combination of collective years' data as a training set to train the machine for classifying SD based on the environment information given. Particularly the CTA also involves the pre-processing of segmenting the SD into the three or five point categories, handling of missing values, environment feature selection and correlation analysis between environment features and attributes.

This paper provides the description, rationale and results of preliminary experiments that examine the correlation and influence levels between environmental features and SD present in the Dongting Lake area of China. This lake represents a very relevant study area with which to examine the moisture and vegetation levels required for snail life to flourish. The datasets used in this paper were derived from remotely sensed image extraction information together with manually collected field survey data provided by our Chinese project partners at Academy of Opto-Electronics the and the European Space Agency (ESA). The datasets have been analysed quantitatively and results are illustrated with this paper. The aim of these studies is to discover if there exists strong correlation between individual or component environment features and the Schistosomiasis disease vector (Oncomelania Hupensis snail) density and distribution. If we can identify this, then we can make future SD classifications using our prediction models based on previously collected datasets. The resulting prediction models will be capable of making informed assumptions on future SD levels and therefore provide likelihood of outbreaks of the disease occurring based on environmental feature values on a larger scale and with greater efficiency than is currently available.

The CTA proposed in this paper is a framework we use to enhance the training potential of real-world sparse datasets. This approach is conducted using a range of pre-processing methods together with attribute ranking and data analysis, the results of which we take into consideration when building our classification and prediction models to determine the density and distribution of epidemic disease vectors. We aim to enrich our training set by investigating the various methods discussed in this paper to deduce if they can have a positive effect when applied for classification of SD in terms of accuracy performance. This includes using combined years of data instances for training against alternative testing sets. We aim to take into account and apply the optimal test conditions as a training paradigm to discover whether those criteria perform better than standard datasets for classification purposes.

2 Experiment Data

The datasets used in this report are derivative of remotely sensed satellite images ranging from between 2003–2009 in the Dongting Lake area of China. The images were processed and feature extraction was carried out by our Chinese partners to provide values for the environmental features present for each year. While we can access vast amounts of data from satellite imaging, the primary field survey data of which we can be sure is a much more time consuming process so this is why we have such a limited dataset in terms of instances. The number of common features from each year was seven with the collective number of instances being 180. While the dataset is relatively small in data mining terms, it provides a basis on which to form initial opinions and observations as to which attributes or combination of attributes have the strongest influence on SD levels. When we deduce which feature subsets are most influential on the SD levels, we can make assertions on future SD classification and therefore provide important information to those concerned for preventative measures to be put into place.

During initial assessment of the dataset, we looked at how we would categorize the SD values in terms of whether the raw value provided would constitute the label of high SD. To this end, the data was preprocessed by normalizing the values in order to gain arbitrary values into a predefined range which could then be labelled in terms of density level. We subsequently assessed whether we could achieve better classification success by using a 3 or 5 point scale of SD as in Low, Medium or High as opposed to Very Low, Low, Medium, High and Very High. We must categorize the density level in this way for classification purposes otherwise we will be restricted to using either statistical or regression models. The data was initially normalised to achieve a range from 0 to 1 then discretised into the 3 or 5 point scale with even distribution. To discretise the data, data binning was used. This method takes a set of continuous values and turns them into a set of bins which are nominal values. The number of bins was set to five for 'very low', 'low', 'medium', 'high' and 'very high'; this resulted in five data intervals which the data was then split into. It was also used for three bins, representing 'low', 'medium' and 'high'.

In addition, we used two regressive methods on our data to make initial assessment on how well the data fits and therefore how well each year fits for classification purposes. While using the linear regression and support vector regression on our unprocessed SD data, we can assess the accuracy of each year of data when predicting new instances of SD. With linear regression we assume environment features to be independent in a dataset; in this case the environment factors are in relation to the dependent variable which is the SD value.

These regressive methods do not provide specific classification percentage accuracy results as the SD value has not been pre-processed or classified to a selected scale that can be used for prediction. We can instead use the coefficient of determination calculation to give the R value which tells us how good of a fit the data is that we are experimenting with. We can assess the results of the coefficient of determination with results ranging from 0 to 1. The closer to 1 the result is, the better the fit therefore the higher likelihood of predicting a new instance of SD.

Equation 1 involves taking the average of the entire SD actual values for each year, then subtracting the average value from each individual actual SD value to the power of two and the same with each predicted value from that year. We then take the sum of results from each instance of predicted and actual values and divide the total value from the predicted SD calculation by the total actual value calculation with the value ranging between 0 and 1 with 0 being the least well-fitting data and 1 being the best fitting data.

We can see from Table 1b that the best fitting data is from 2008 training and testing data using linear regression. It scored 0.8 which is the closest result to 1 making it the most promising data combination for classifying future instances of SD. We can also see that in 2007, both the linear regression and support vector regression classifiers performed well with similar performance which can also be an indicator of potentially generating good classification models.

Once we had made initial assessment of the datasets, they were preprocessed by normalizing and then discretizing the SD information from each year. This enabled us to have more options for using different algorithms for classification

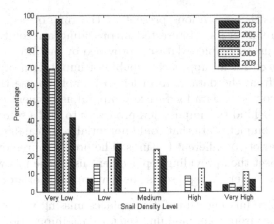

Fig. 1. Classification of SD over Time

Table 1. Statistics for yearly SD Categories and R^2 Values

Year	AVG SD	Normalised	Category
2003	0.727	0.107	V. Low
2005	0.879	0.129	V. Low
2007	2.633	0.388	Low
2008	1.396	0.206	Low
2009	1.056	0.156	V. Low
Collective	1.014	0.149	V. Low

(a) Average SD Values

	2003	2005	2007	2008	2009
LR	0.325	0.590	0.734	0.808	0.699
SVR	0.052	0.221	0.732	0.691	0.506

(b) R^2 Values for Linear Reg. and Support Vector Reg.

purposes. The SD data was separated at the beginning into 5 categories and the results are recorded in Fig. 1 and Table 1a. They show on average that density and distribution of the Oncomelania Hupensis snail during the time period from 2003–2009 were predominantly very low and low. This is what we would expect to see once the data has been preprocessed but it does not provide the entire picture so we will have to explore the dataset further and assess the relative SD levels and in conjunction with environmental features.

3   Methods

In earth observation research, weather conditions directly affect the quality of satellite imagery which causes some values of environment variables to be missing in the set and discontinuity in terms of fully recorded data relationships. These partially complete datasets are caused by anomalies in the remotely sensed image extraction process and by issues such as weather clarity from satellite imagery.

One of the major issues faced when using the data provided was that specific data particularly from 2007 was only partially complete. This problem correlates directly to the weather conditions present at the time of acquiring data from the satellites therefore we are interested in providing a resolution that can be applied to any future incomplete datasets provided by satellite images. This issue highlighted the need for an approach capable of imputation of the values that were incomplete from the dataset in order to be able to use the 2007 dataset and any other incomplete data for future temporal assessment of SD levels.

The rationale behind this imputation process is to find a solution for replacement of partially complete data that could potentially be scalable for much larger datasets with a variety of different features. The process of removing known values from our dataset then providing replacement using the following methods is documented below, where V represents the feature value of an instance.

- The Weka replace missing value filter replaces missing values with the mean and modal values from the remaining set for data imputation.
- The Single Pre-Succession Method uses the previous and following values to replace the missing value.

$$v_i = \frac{v_{i-1} + v_{i+1}}{2} \tag{1}$$

- The Mean Single Pre-Succession Method uses the previous and following values to replace the missing value together with the mean of the entire set.

$$v_i = \frac{v_{i-1} + v_{i+1} + \hat{v}}{3} \tag{2}$$

- The Double Pre-Succession Method uses the two previous and following values to replace the missing value.

$$v_i = \frac{v_{i-2} + v_{i-1} + v_{i+1} + v_{i+2}}{4} \tag{3}$$

- The Mean Double Pre-Succession Method uses the two previous and following values to replace the missing value together with the mean of the entire set.

$$v_i = \frac{v_{i-2} + v_{i-1} + v_{i+1} + v_{i+2} + \hat{v}}{5} \tag{4}$$

We can see from Table 2 that the most accurate performing method is the Mean Double PreSuccession method with an average percentage difference of 32.58 % while the lowest performance is of the PreSuccession method which has an average percentage difference of 333.36 % from the original value that was replaced. These results can now be analysed and used for future incomplete datasets to verify the accuracy of value replacement over more extensive datasets.

3.1 Feature Assessment

- We want to evaluate the dataset to discover the relevance of each attribute to SD levels individually and as subsets of features.
- To assess and rank the features of the data yearly to gain a deeper understanding of the value of the environment features to the data as a whole.

- Selection of an efficient, well performing method to handle replacement of missing values in the data as this is an ongoing issue with RS images that will be required for application in any future data that may be accessed.
- To distinguish the most effective category of SD to move forward with for future classification purposes.

3.2 Information Gain

To assess the attribute values in relation to SD, Information Gain attribute ranker was applied to the data and documented in a table for each years' data. Information Gain is a feature ranking approach that uses entropy to identify which feature in the dataset gains the most information relative to the class. This is beneficial when carrying out analysis of a dataset to extract the most influential features in relation to their corresponding SD values. It can be of significant value in order to identify any corollary inferences with regards environment features to SD levels. Once we identify which attributes have most significant influence on the SD value, then it can be established for future experiments that these specific attributes are closely connected to high levels of SD.

The results in Table 3 show relative consistency with each year having similar positions for each of the attributes. We can see certain attributes consistently trending such as the Normalised Difference Water Index (NDWI) and the Tasseled Cap Greenness (TC_G) which indicate that these attributes are of significant value in relation to SD of each of the particular years in the dataset. These results will now form an element of consideration for our CTA model.

The information Gain calculation used is shown in Eq. 5 where entropy (H) is given of the class (C) given the attribute (A) [7]. Entropy and information gain are intrinsically linked as the decrease in the entropy of the class is a direct

Table 2. Data imputation from 2007

Original	Weka	PreSucc. Method	Mean PreSucc.	Double PreSucc.	Mean Dbl PreSucc.
0.0348	0.228	0.251	0.169	0.012	0.061
0.128	0.201	0.100	0.254	0.123	0.069
−0.084	−0.084	0.236	0.052	0.401	0.064
0.024	0.201	0.097	0.106	0.027	0.050
−0.660	−0.471	0.026	−0.156	0.080	−0.083
0.242	0.228	0.386	0.214	0.103	0.072
−0.521	−0.622	0.387	−0.091	0.346	−0.062
0.410	0.657	−0.006	0.232	−0.141	0.112
0.400	0.344	0.025	0.123	−0.025	0.064
0.35545	0.657	0.007	0.236	−0.119	0.117
Avg.% Diff.	145.13 %	**333.36%**	226.81 %	132.04 %	**32.58%**

Table 3. Information Gain feature ranking

	2003	2005	2007	2008	2009	Collective
1	NDWI	NDWI	NDWI	TC_G	MNDWI	TC_G
2	TC_W	TC_W	TC_W	NDWI	NDVI	NDWI
3	TC_G	TC_G	TC_G	NDVI	NDWI	TC_W
4	NDMI	NDMI	NDMI	MNDWI	TC_G	NDMI
5	MNDWI	MNDWI	MNDWI	NDMI	NDMI	MNDWI
6	NDVI	NDVI	NDVI	TC_W	TC_W	NDVI
7	TC_B	TC_B	TC_B	TC_B	TC_B	TC_B

reflection of the added information about the class provided by the attribute and this is referred to as the information gain and therefore entropy is a pre-requisite for information gain to be calculated [8].

$$H(C|A) = - \sum_{a \in A} p(a) - \sum_{c \in C} p(c|a) \log_2 p(c|a) \tag{5}$$

Correlation analysis was applied to the combination of each of the attributes with the SD temporally. In terms of relationships, we used Pearson's r approach, which uses the covariates X and Y, this is then divided by the standard deviation of X and of Y to give a correlation value of each individual attribute and SD value. The results are shown in Fig. 2b and they indicate that data from 2008 is not in correlation with the alternate years as the trend lines show us. The combination of the SD and environmental attributes (X, Y) does not show correlate with the dataset from each year. The corollary relationship results between SD levels and environment features is an integral component of our CTA framework below for future classification of SD levels based on environment factors. As the environment feature values increase towards 1.0 in Table 4 it shows the impact factor that is present when compared with SD levels with TCG and NDVI from 2005 showing good correlation as opposed to NDWI in the same year.

$$P(x, y) = \frac{cov \ (x, y)}{\sigma x \sigma y} \tag{6}$$

4   Cumulative Training Approach (CTA)

Given the limited amount of data and the pre-processing results, we consider how to construct prediction/classification models. A caveat to address with the SD classification is the fact that for years 2003, 2005 and 2007 we have 18/19 attributes partially complete labelled whereas with years 2008 and 2009 we have eight/nine attributes given to experiment with. The most beneficial approach to dealing with this issue is to use those attributes which are common to each

Table 4. Correlation analysis between SD vs. Features

	TCB	TCG	TCW	NDMI	NDVI	MNDWI	NDWI
2003	−0.069	0.352	0.352	0.101	0.359	−0.289	−0.339
2005	0.308	0.517	−0.301	0.332	0.519	−0.4999	−0.521
2007	0.287	0.192	−0.17	0.194	0.227	−0.07	−0.195
2008	−0.32	−0.132	0.163	0.194	−0.179	0.397	0.289
2009	0.208	0.333	−0.193	0.139	0.418	−0.462	−0.387

year in order to make a comparable dataset for training and testing purposes. It was decided to use the initial year's collected research information to build a training model, which is then used as a benchmark against future data for testing purposes. This approach will enable us to enrich the dataset with variable subsets of the data being used to discover temporal relationships within the dataset. The method was used tested with five classification methods to assess accuracy. We can see from Fig. 2a that year 2003 training data with 2005 testing yields highly accurate results as the accuracy during training and prediction accuracy are in close proximity to each other, this indicates good classification performance.

This method of training will make up another component of our proposed idea referred to throughout this paper as the CTA. By applying this proposed CTA, we are combining the most promising experiments and test results from a variety of relevant areas of our data pertaining to the classification of the Schistosomiasis disease vector. Using this data we can then build a model for application with any future spatio-temporal epidemic disease environment data, which is a different approach from the standard application of classification methods.

In addition carried out testing on three ensemble learning methods of Bagging, Boosting and Stacking as the ensemble methods have been shown to provide better classification accuracy than single classifiers [9]. Using these three ensemble methods we can get a varied range of results based on training model performance (Adaboost), equal sized training set sampling (Bagging) and combined classifier prediction (Stacking). Results were recorded in Table 5.

Table 5. CTA ensemble results

Training	Testing	Boosting		Bagging		Stacking	
		Train	Predict	Train	Predict	Train	Predict
2003/05	2009	0.740	0.483	0.753	0.5	0.712	0.467
2003/05/07	2009	0.530	0.467	0.504	0.467	0.556	0.467
2003/05/07/08	2009	0.558	0.483	0.509	0.533	0.491	0.467
2003/05	2008	0.740	0.478	0.753	0.348	0.712	0.326
2003/05/07	2008	0.530	0.326	0.504	0.348	0.556	0.326
2003/05	2007	0.740	0.295	0.753	0.318	0.712	0.295

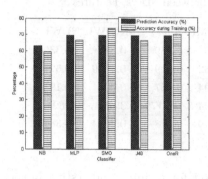 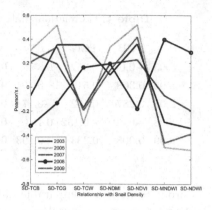

(a) 2003Train - 2005Test CTA Data   (b) Pearson's Correlation Co-Efficient

Fig. 2. CTA and Correlation figures

5   Conclusion

From the correlation analysis graph, we can see that each of the years data with the exception of 2008, follow together in a trend which shows that the correlation values of each combination of attributes together with SD, can be predictable which is of high value to this particular research area looking at future distribution and density predictions.

By handling and assessing missing value replacement in the data, we can identify the success of replacing these values based on the mean and mode of the existing data. These results can be applied to future RS data that will be accessed for research and experimentation. By testing effectiveness of replacement methods, we can identify confidence in future replacement of data.

All experiments and collective research to date have become part of the CTA for Schistosomiasis vector density and distribution prediction. This approach has provided us with a better understanding of our datasets and the classification results which it provides. In combining each aspect of the training process we have a greater understanding of the research area and we can apply this knowledge to future data obtained for classification and prediction purposes.

From the results to date, we can deduce that specific environmental attributes such as TC_G and NDWI have more influence on the SD and distribution than others. This information will be further analysed and implemented into a cluster ensemble algorithm for optimum accuracy classification for future work [10].

References

1. Ross, A.G.P., Sleigh, A.C., Li, Y., Davis, G.M., Williams, G.M., Jiang, Z., Feng, Z., Manus, D.: Schistosomiasis in the people's republic of china: prospects and challenges for the 21st century. Clin. Microbiol. Rev. **14**(2), 270–295 (2001)

2. WHO: Schistosomiasis (2015)
3. Ma, C., Dai, Q., Li, X., Liu, S.: The analysis of east dongting lake water change based on time series of remote sensing data. In: 2014 12th International Conference on Signal Processing (ICSP), Institute of Electrical & Electronics Engineers (IEEE), October 2014
4. Sun, Q., Zhang, J., Zhou, J., Wu, L., Shan, Q.: Effect of poplar forest on snail control in dongting lake area. In: 2009 3rd International Conference on Bioinformatics and Biomedical Engineering, Institute of Electrical & Electronics Engineers (IEEE), June 2009
5. Wu, J.Y., Zhou, Y.B., Li, L.H., Zheng, S.B., Liang, S., Coatsworth, A., Ren, G.H., Song, X.X., He, Z., Cai, B., You, J.B., Jiang, Q.W.: Identification of optimum scopes of environmental factors for snails using spatial analysis techniques in dongting lake region, china. Parasites Vectors $7(1)$, 216 (2014)
6. Seto, E., Xu, B., Liang, S., Gong, P., Wu, W., Davis, G., Qiu, D., Gu, X., Spear, R.: The use of remote sensing for predictive modeling of schistosomiasis in china. Photogram. Eng. Remote Sens. $68(2)$, 167–174 (2002)
7. Blake, C.L., Merz, C.J.: UCI Repository of machine learning databases (1998)
8. Quinlan, J.R.: C4.5: Programs for Machine Learning, vol. 1 (1993)
9. Pan, M., Wood, E.F.: Impact of accuracy, spatial availability, and revisit time of satellite-derived surface soil moisture in a multiscale ensemble data assimilation system. IEEE J. Sel. Top. Appl. Earth Obs. Remote Sens. $3(1)$, 49–56 (2010)
10. Elshazly, H.I., Elkorany, A.M., Hassanien, A.E., Azar, A.T.: Ensemble classifiers for biomedical data: performance evaluation. In: 2013 8th International Conference on Computer Engineering & Systems (ICCES), Institute of Electrical & Electronics Engineers (IEEE), November 2013

A Mobile and Evolving Tool to Predict Colorectal Cancer Survivability

Ana Silva[1], Tiago Oliveira[1(✉)], Vicente Julian[2], José Neves[1], and Paulo Novais[1]

[1] Algoritmi Centre/Department of Informatics, University of Minho, Braga, Portugal
a55865@alunos.uminho.pt, {toliveira,jneves,pjon}@di.uminho.pt
[2] Departamento de Sistemas Informáticos y Computación,
Universidad Politécnica de Valencia, Valencia, Spain
vinglada@dsic.upv.es

Abstract. In this work, a tool for the survivability prediction of patients with colon or rectal cancer, up to five years after diagnosis and treatment, is presented. Indeed, an accurate survivability prediction is a difficult task for health care professionals and of high concern to patients, so that they can make the most of the rest of their lives. The distinguishing features of the tool include a balance between the number of necessary inputs and prediction performance, being mobile-friendly, and featuring an online learning component that enables the automatic evolution of the prediction models upon the addition of new cases.

1 Introduction

The colorectal cancer, is a subtype of cancer, which affects the lower portion of the gastrointestinal tract and develops in the cells lining the colon and rectum [24]. It can be further divided according to the site where the pathology develops. Colon and rectum cancers are, in fact, different pathologies, with different associated genetic causes and different progressions according to distinct molecular pathways [29]. Statistics show that colorectal cancer is the most common form of cancer in the digestive system, the third most common and the fourth deadliest cancer overall [15].

The use of machine learning (ML) techniques has been growing in cancer research [17]. The accurate prediction of survivability in patients with cancer remains a challenge namely due to the heterogeneity and complexity of the disease. However, accurate survivability prediction is important for patients with cancer so that they can make the most of the rest of their lives. It is also important to help clinicians to make the best decisions, when palliative care is an essential component of the process. Given that colon and rectal are the most common cancers of the digestive system, one would expect the existence of numerous tools for ascertaining the likelihood of a patient surviving this disease. Although there are some tools for this task, few provide predictions for both colon and rectal cancer, and none of them apply ML techniques in order to

L. Iliadis and I. Maglogiannis (Eds.): AIAI 2016, IFIP AICT 475, pp. 14–26, 2016.
DOI: 10.1007/978-3-319-44944-9_2

build evolving predictive models. Furthermore, their digital support may hinder their consultation at care delivery.

The objective of this work is to present an easy to use tool that provides survivability predictions of colon and rectal cancer patients for 1, 2, 3, 4 and 5 years after diagnosis and treatment. Due to the ubiquitous presence of mobile devices in everyday life and the ease with which one is able to consult these devices and use their applications, we chose to develop this tool as a mobile application. The underlying model for survivability prediction was obtained through ML techniques applied to the data from the Surveillance, Epidemiology, and End Results (SEER) program [18], a large cancer registry in the United States, and arguably the most complete cancer database in the world. The dataset includes records of patients diagnosed with different types of cancer from 1973 to 2012. The focus of this paper will be placed on the mobile solution developed for survivability prediction, but a part of the paper will be dedicated to briefly describing its underlying ML model so as to provide a better comprehension of the work as a whole.

The paper is structured as follows. Section 2 presents related work featuring survivability prediction tools for colon and rectal cancer, with an analysis of their main strengths and limitations. Section 3 describes the selected requirements for the tool, its underlying ML-based predictive model, architecture and a comprehensive use case. Section 4 provides a reflection about the strengths and limitations of our approach. Finally, Sect. 5 presents the conclusions drawn so far and future work considerations.

2   Related Work

Existing tools for colon or rectal cancer survivability prediction are mostly available as web applications. Table 1 shows a summary of their main features, namely: (1) whether the application is used for colon or rectal cancer; (2) the number of features necessary to get a prediction; (3) the data set that its underlying model is based on; (4) the technique used to construct the predictive model; (5) the type of target prediction it produces; and (6) a measure of performance in the form of a concordance index (C-index). The C-index corresponds to the probability of giving a correct response in a binary prediction problem. It is considered to be numerically equivalent to the area under the ROC curve (AUC) [16].

There is a disparity in the number of features used in each tool. However, twelve [22] or even nine [10,25] features may be too much information for a physician to input on-the-fly. Furthermore, there are cases in which the increased number of features does not necessarily translate into a better performance, as can be seen in the direct comparison between the works in [28] and in [22].

All the underlying models are based on statistical modelling, most notably on Cox regression analysis [8]. This is the dominant multivariate approach used in survivability prediction and corresponds to a multiple linear regression of the hazard on a set of variables. This indicates that the use of soft computing techniques, namely ML, in survivability prediction, especially in colon and rectal

Table 1. Characteristics of applications for colon and rectal cancer survivability prediction.

Characteristics	Bush and Michaelson [10]	Chang et al. [11]	Weiser et al. [28]	Renfro et al. [22]	Wang et al. [27]	Valentini et al. [25]	Bowles et al. [7]
(1) Cancer type	Colon	Colon	Colon	Colon	Rectal	Rectal	Rectal
(2) Number of Features	9	6[a]	2/3/7	12	5[a]	9	7[a]
(3) Data set	SEER	SEER	SEER	Adjuvant Colon Cancer End Points (ACCENT)	SEER	Five European randomized trials	SEER
(4) Model	Regression based	Regression based	Regression based	Regression based	Regression based	Regression based	Regression based
(5) Target	0 – 15 years	1 – 10 years (disease specific survivability) 0 – 5 years (conditional survivability)	5 years	5 years	0 – 5 years	1– 10 years	1 – 10 years (disease specific survivability) 0 – 5 years (conditional survivability)
(6) Performance C-index	–	0.816	0.61/0.63/0.68	0.66	0.75	0.70	–

[a]Including months which the patient has already survived (for conditional survivability calculation).

cancer, has yet to be fully explored. Since one of the advantages of ML is having more discriminative power in identifying patterns in data and finding nuances that may escape statistical modelling, its usage for survivability prediction may result in models with better performances [17]. As such, ML was chosen as the modelling approach for this work.

Most of the target predictions, either for colon or for rectal cancer, cover a 5-year span [22,27,28]. Even though there are models that cover a wider time span [7,10,11,25], the five year barrier is an important goal for a colorectal cancer patient to overcome, and is used throughout clinical practice guidelines [4,5] as a turning point for follow-up procedures, in which the vigilance over the patient is lightened, and for the assessment of the recurrence risk. For this reason, the present work will also have a target prediction of five years. Another noteworthy observation is that only two of the tools feature conditional survivability predictions.

To determine if the tools are suitable for mobile devices, the applications were analysed using the mobile-friendly test tool from *Google*[1]. The results showed that, except for the tools reported in [22,28], all the others are unsuitable for mobile access. The test revealed that the text was too small to read, the mobile viewport was not set, links were too close to each other and usually the content

---

[1] Mobile-friendly test tool of Google is available at https://www.google.com/webmasters/tools/mobilefriendly/.

was wider than the screen. Therefore, few of these applications had a mobile-friendly design. Another goal is to address this by developing a cross-platform tool that is available to users in a practical and intuitive way, through a smart-phone or tablet.

3 CRCPredictor: An Application for Survivability Prediction

Throughout the last decade, mobile phones have gone from being simple phones to being handheld pocket-sized computers. Their capabilities, namely the processing and on-board computing capacity incite the development of applications [6]. According to data from the International Data Corporation (IDC) Worldwide Quarterly Mobile Phone Tracker, the Android of Google and iOS of Apple are the two most popular smartphone operating systems [12].

For the health care industry, mobile applications yielded new boundaries in providing better care and services to patients. Moreover, it is making a revolution in the way information is managed and made available [23]. The portability of mobile applications can increase the productivity of health care professionals. It grants a rapid access to information and multimedia resources, allowing health care professionals to make decisions more quickly with a lower error rate, increasing the quality of patient documentation and improved workflow patterns [26]. This work discloses an assistive tool to help physicians to improve their practice. The problem it addresses is predicting the survivability of colorectal cancer patients in an individualized manner.

3.1 Requirements for the Survivability Prediction Tool

Several functionalities were delineated to achieve a solution that covers the limitations mentioned in Sect. 2 and, at the same time, is able to help physicians to improve their practice. These functionalities are summarized in the following functional requirements for the prediction tool: allow the user to select the cancer type (either colon or rectal) for which he seeks a prediction; allow the user to provide inputs for a set of selected features, based on which the underlying models generate survivability predictions; allow the user to choose the value of an input for a feature from a set of pre-determined values; provide a survivability prediction, according to the inputs, for 1, 2, 3, 4 and 5 years after the diagnosis and treatment; provide a likelihood value for the prediction of each year; to allow the visualization of the predictions and likelihood values in a chart; and allow the insertion of new patient registries into the case database, thus increasing the number of cases for the periodic recalculation of the prediction models.

Additional requirements for the tool are that it should be made available in the two main mobile platforms (iOS and Android) and it should be able to recalculate the prediction models upon the addition of a significant number of new patient registries. This confers a dynamism to the prediction models and should ensure their evolution over time.

3.2   Colon and Rectal Cancer Survivability Prediction Models

Survivability prediction was approached as a binary classification problem. The goal was to produce predictions for 1, 2, 3, 4 and 5 years after treatment of colon or rectal cancer. Each classification label (there were five representing years 1, 2, 3, 4 and 5) could only have two values: *survived* or *not survived*. As such, it was necessary to build five survivability prediction models (one per year) for each type of cancer. The created models were based on the SEER dataset. The criteria for selecting patient registries was the same for both colon and rectal cancer. Only patients with age greater than or equal to 18 years old were selected. Patients who were alive at the end of the data collection whose survival time had not yet reached 60 months (five years) and those who passed away of causes other than colon or rectal cancer were sampled out. After preprocessing, 38,592 cases were isolated for colon cancer and 12,818 cases were considered for rectal cancer. From the isolated cases for each pathology, 10 % were selected for testing sets. After filtering cases with "unknown" values, the colon cancer testing set had 2,221 cases and the training set had 20,061 cases. The testing set for rectal cancer had 551 cases and the training set had 4,962 cases. In total, the training set had 61 attributes representing possible classification features.

All the phases, from preprocessing to evaluation, were executed using Rapid-Miner[2], an open source data mining software chosen for its workflow-based interface and an intuitive application programming interface (API).

Using the Optimize Selection [21] operator for feature selection with the classification labels as target, a total of 6 features were obtained from a feature selection phase for each cancer type. Their name and description are shown in Tables 2 and 3. The training sets for colon and rectal cancer with their respective selected features were used in the learning of multiple prediction models using different ML ensemble methods such as bagging, adaboost, bayesian boosting, stacking, and voting. The accuracy, the AUC and the F-measure were used as performance measures in order to evaluate the models developed for colon and rectal cancer. The accuracy is the percentage of correct responses among the examined cases [9]. The F-measure is a combination of precision (a form of accuracy, also known as positive predictive value) and recall (also known as sensitivity) [20]. The AUC can be interpreted as the percentage of randomly drawn data pairs of individuals that have been accurately classified in the two populations [16]. These measures were calculated using the training data set and 10-fold cross validation. By applying the testing sets to the models, we calculated the percentage of incorrectly classified cases. The stacking[3] [14], using k-NN, decision tree, and random forest classifiers as base learners and a naive bayes classifier as a stacking model learner, was the best performing model for both colon and rectal cancer. Upon prediction, the model is capable of providing a

---

[2] Software available at https://rapidminer.com.

[3] Stacking combines base classifiers of different types. Each base classifier generates a model using the training set, then a meta-learner integrates the independently learned base classifier models into a high level classifier by re-learning a meta-level training set.

Table 2. Features obtained by feature selection and used for colon cancer models.

Attribute	Description
Age at diagnosis	The age (in years) of the patient at time of diagnosis
Carcinoembryonic antigen	The interpretation of the highest Carcinoembryonic Antigen test results
Clinical assessment of regional lymph nodes	The clinically evident regional lymph nodes
AJCC stage	The grouping of the TNM information combined
Primary site	Identification of the site in which the primary tumor originated
Regional nodes examined	The total number of regional lymph nodes that were removed and examined by the pathologist

Table 3. Features obtained by feature selection and used for rectal cancer models.

Attribute	Description
Age at diagnosis	*
Extension of the Tumor	Information on extension of the tumor
Tumor size	Information on tumor size (in mm)
AJCC stage	*
Surgery of primary site	Describes a surgical procedure that removes and/or destroys tissue of the primary site performed as part of the initial work-up or first course of therapy.
Gender	The sex/gender of the patient at diagnosis

*Described in Table 2.

confidence value that represents the likelihood of the prediction. Table 4 shows the performance values of the best model developed, for both cancer types.

As the intent with this paper is to present the features of the developed tool and describe its inner workings, it was considered that an exhaustive description of the ML process was out of scope.

3.3 Architecture

The CRCPredictor is a hybrid mobile application targeting smartphones and tablets. The back-end of this tool includes two web services: one to give the survivability prediction responses for colon or rectal cancer to the user and another to recalculate the survivability prediction models. Figure 1 shows the architecture of the CRCPredictor system.

The *Survival Prediction App* was developed using a hybrid approach, between a web and a native methodology. This allows an abstraction from the native language of the target operating system while retaining the core features of a native app. A hybrid application is developed by applying web technologies

Table 4. Performance measures for 10-fold cross validation and the incorrectly classified cases from the test data set of the stacking model.

Cancer type	Measure	1 Year	2 Year	3 Year	4 Year	5 Year	Average
Colon	AUC	0.982	0.985	0.987	0.989	0.987	0.986
	Accuracy	95.84 %	96.62 %	96.79 %	97.09 %	97.03 %	96.67 %
	F-Measure	90.41 %	94.43 %	95.55 %	97.61 %	97.51 %	95.10 %
	Incorrectly classified (%)	3.24 %	3.02 %	2.97 %	2.84 %	3.11 %	3.03 %
Rectal	AUC	0.961	0.973	0.973	0.975	0.969	0.970
	Accuracy	95.61 %	95.81 %	93.81 %	94.82 %	94.32 %	94,87 %
	F-Measure	80.56 %	87.32 %	85.85 %	96.63 %	96.20 %	89.31 %
	Incorrectly classified (%)	3.81 %	3.99 %	6.17 %	4.36 %	4.36 %	4.54 %

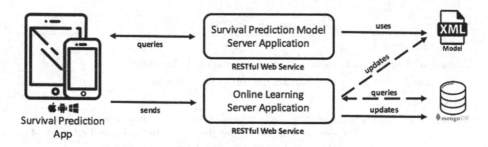

Fig. 1. Architecture of the CRCPredictor system.

(mainly, HTML5, CSS and JavaScript) and is executed inside a native container on the mobile device. It is suitable for multiple platforms and is distributable through an application store, just like native applications. This type of approach can have an inferior performance compared with native applications. However, nowadays mobile devices have powerful capabilities and the performance gap is hardly noted. The application was developed using AngularJS, Ionic Framework, and Cordova. Cordova wraps the HTML/JavaScript app into a native container which can access the device functions of several platforms [1]. These functions are exposed via a unified JavaScript API, for an easy access to the full native functionalities.

The *Survival Prediction Model Server Application* was developed to cover the need of an individualized system, able to respond according to a particular set of patient characteristics. It exposes a set of RESTful web services. This service architecture was chosen for being light-weight, easy to access and scalable [30]. The web services were developed in Java with the Java API for RESTful Web Services (JAX-RS) [2]. The data is sent over the HTML POST method when the health care professional submits the values for the prediction features on the *Survival Prediction App*. The RESTful web service, using the RapidMiner API, receives the values and feeds them to the corresponding models, encoded in XML files. The response with the survivability predictions for the five years is returned in a JSON format.

The *Online Learning Server Application* also follows a REST architecture. It handles newly submitted patient data. The outcomes are added to a database for a posterior recalculation of all the models, which keeps them up-to-date. The data is inserted into a NoSQL database and, for each 1000 new registries, the models for the five years are recalculated, generating five new XML files for the type of cancer that just got the thousandth new case. The 1000 mark was arbitrarily defined and can be subject to adjustment.

3.4   Use Case

Figure 2a shows the first screen that appears when the *Survival Prediction App* of the CRCPredictor is initiated. By clicking on the menu (Fig. 2b), all options available in this application become visible.

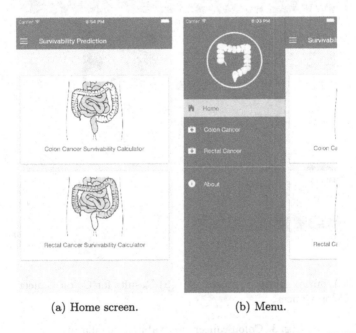

(a) Home screen.   (b) Menu.

Fig. 2. Home screen and menu of the Survival Prediction App of the CRCPredictor.

A typical use case is getting a prediction for colon cancer survivability. Supposing a physician is treating a patient diagnosed with colon cancer, once the type of cancer in the home screen is set (as shown in Fig. 2b), the health care professional inserts the values for the selected features (Fig. 3a). All features, except for the age of the patient, are filled in by choosing the value from a list of available options. By submitting a case of a patient with *55* years old, having a *positive/elevated* carcinoembryonic antigen value, with clinical assessment of regional lymph nodes of *not clinically evident*, with the primary site of the cancer being in the *sigmoid colon*, with stage *0* and with *5* as the number of regional nodes examined, the values are sent to the *Survival Prediction Model Server Application* and the outcome is calculated. The prediction is always provided in the form of confidence values for a positive prediction, i.e., the confidence that the patient will survive. This is displayed in a new screen in the form of a bar chart (Fig. 3b). For the stage of the patient, the physician can choose between the TNM system or the grouped stage, known as American Joint Committee on Cancer (AJCC) stage. The results show that, while the model was able to predict with 100 % confidence that the patient will survive the first three years, the confidence of his surviving the fourth and fifth years is 0 %. To predict the survivability of a patient diagnosed with rectal cancer, the procedure is similar to the one used for colon cancer.

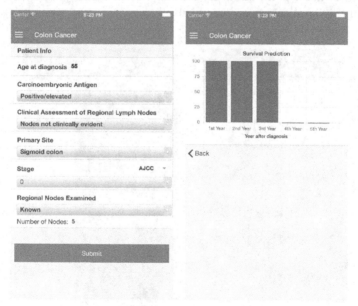

(a) Survivability Features of Colon Cancer.   (b) Results for Colon Cancer.

Fig. 3. Colon cancer survivability calculator.

4   Analysis and Discussion

In terms of inputs the constructed prediction models, for both colon and rectal cancer, require only the input of six selected features. Comparing with the related tools in Sect. 2, this number is inferior to the number of features used in the underlying models of the tools described in [7,10,22,25], and is closer to the number of inputs of the remaining prediction tools. The number of input features may be crucial to the adoption or the rejection of a tool, as it may become difficult to use it on-the-fly if too much information is needed. Another aspect to note is that, apart from the *age at diagnosis* and the *AJCC Stage*, the feature selection produced two very different sets for colon and for rectal cancer, which is in line with the notion that the two, although having aspects in common, are different pathologies. Regarding the colon cancer features, the *age at diagnosis* and the *AJCC stage* are present in most colon cancer prediction tools [11,22,28]. The other selected features are not usually present, but they are closely related to the ones that are. For instance, the *clinical assessment of regional lymph nodes* is a product of a medical evaluation of a feature widely used in the existing tools [22,28] that is the number of lymph nodes found to have cancer out of the lymph nodes isolated during surgery. The same can be said about the selected features for rectal cancer, i.e., they are, at the very least, closely related to the ones used in other prediction tools.

In [3], the use of ML ensemble models to develop survival prediction models for colon cancer is described. The modelling component of our work is similar to that approach; therefore, it is possible to compare the performance of our selected model with theirs. The classification accuracies reported in [3] for years 1, 2, and 5 were 90.38 %, 88.01 %, and 85.13 %. The reported AUCs were 0.96, 0.95, and 0.92, respectively. As shown in Table 4, we were able to improve the classification results with our models. This direct comparison is not possible for the rectal cancer model as it was not possible to find such a closely related work in the literature. However, it is possible to verify that the rectal cancer model performs worse than the colon model in every metric, possibly due to the smaller size of the training set used in the learning process. At the same time, both the colon and rectal cancer models showed low classification errors on the randomly selected test data sets. Additionally, when comparing the AUCs of the generated models in Table 4 with the C-indexes in Table 1, it is possible to conclude that the generated models show a better discriminative power than the currently available models.

Regarding the CRCPredictor system, it fulfils the requirements defined at the beginning of the work. The distinguishing features of the system's architecture are its flexibility and scalability, which make the addition of new features (services) simple and easy. The *Survival Prediction App* was developed as a mobile-friendly application, enabling the easy access of health care professionals to its functionalities on their mobile devices. Another component that distinguishes this system from established tools is the *Online Learning Server Application* which ensures the continuous evolution of the prediction models. However, the system does not provide conditional survivability predictions, which makes it less appealing when compared with the works in [7,11,25], as this is a type of information that health care professionals generally like to know.

5   Conclusions and Future Work

The main contribution of this work is a survivability prediction tool for colon and rectal cancer. Its distinguishing features are a balance between the number of necessary inputs and prediction performance, being mobile-friendly, and featuring an online learning component that enables the automatic recalculation and evolution of the prediction models upon the addition of new cases. The goal with this tool is to facilitate the access of health care professionals to instruments capable of enriching their practice and improving their results. Future work on the tool includes the development of conditional survivability models that allow the user to get a prediction knowing that the patient has already survived a number of years after diagnosis and treatment. Additionally, we intend to conduct experiments to assess how well the tool fulfils the needs of health care professionals and identify aspects to improve. Additionally, the models presented herein will be considered for inclusion in a guideline-based decision support system, described in [19], as a dynamic knowledge complement to the static recommendations of clinical practice guidelines. Since colon and rectal cancer affect mostly the elderly, this survivbility prediction apllication can be used within a technological environment, such as the one disclosed in [13], to provide better support to this population group.

Acknowledgements. This work has been supported by FCT – Fundação para a Ciência e Tecnologia within the Project Scope UID/CEC/00319/2013. The work of Tiago Oliveira is supported by a FCT grant with the reference SFRH/BD/85291/2012.

References

1. Apache cordova. http://cordova.apache.org/. Accessed 31 Mar 2016
2. Project kenai. https://jax-rs-spec.java.net/. Accessed 04 Apr 2016
3. Al-Bahrani, R., Agrawal, A., Choudhary, A.: Colon cancer survival prediction using ensemble data mining on seer data. In: 2013 IEEE International Conference on Big Data, pp. 9–16 (2013)
4. Benson, A., Bekaii-Saab, T., Chan, E., Chen, Y.J., Choti, M., Cooper, H., Engstrom, P.: NCCN clinical practice guideline in oncology colon cancer. Technical report, National Comprehensive Cancer Network (2013)
5. Benson, A., Bekaii-Saab, T., Chan, E., Chen, Y.J., Choti, M., Cooper, H., Engstrom, P.: NCCN clinical practice guideline in oncology rectal cancer. Technical report, National Comprehensive Cancer Network (2013)
6. Boulos, M.N., Wheeler, S., Tavares, C., Jones, R.: How smartphones are changing the face of mobile and participatory healthcare: an overview, with example from ecaalyx. Biomed. Eng. Online **10**(1), 24 (2011)
7. Bowles, T.L., Hu, C.Y., You, N.Y., Skibber, J.M., Rodriguez-Bigas, M.A., Chang, G.J.: An individualized conditional survival calculator for patients with rectal cancer. Dis. Colon Rectum **56**(5), 551–559 (2013)
8. Bradburn, M.J., Clark, T.G., Love, S.B., Altman, D.G.: Survival analysis part II: multivariate data analysis - an introduction to concepts and methods. Br. J. Cancer **89**(3), 431–436 (2003)

9. Bradley, A.P.: The use of the area under the ROC curve in the evaluation of machine learning algorithms. Pattern Recogn. **30**(7), 1145–1159 (1997)
10. Bush, D.M., Michaelson, J.S.: Derivation: Nodes + PrognosticFactors Equation for Colon Cancer accuracy of the Nodes + PrognosticFactors equation. Technical report (2009)
11. Chang, G.J., Hu, C.Y., Eng, C., et al.: Practical application of a calculator for conditional survival in colon cancer. J. Clin. Oncol. **27**(35), 5938–5943 (2009)
12. Corporation, I.D.: Idc: Smartphone os market share 2015, 2014, 2013, and 2012. http://www.idc.com/prodserv/smartphone-os-market-share.jsp. Accessed 30 Mar 2016
13. Costa, R., Novais, P., Machado, J., Alberto, C., Neves, J.: Inter-organization cooperation for care of the elderly. In: Wang, W., Li, Y., Duan, Z., Yan, L., Li, H., Yang, X. (eds.) Integration and Innovation Orient to E-Society Volume 2. IFIP, vol. 252, pp. 200–208. Springer, Boston (2007)
14. Džeroski, S., Ženko, B.: Is combining classifiers with stacking better than selecting the best one? Mach. Learn. **54**(3), 255–273 (2004)
15. Ferlay, J., Soerjomataram, I., Ervik, M., et al.: Globocan 2012: Estimated cancer incidence, mortality and prevalence worldwide in 2012 (2012). http://globocan. iarc.fr. Accessed 27 Dec 2015
16. Hanley, J.A., McNeil, B.J.: The meaning and use of the area under a receiver operating characteristic (ROC) curve. Radiology **143**(1), 29–36 (1982)
17. Kourou, K., Exarchos, T.P., Exarchos, K.P., Karamouzis, M.V., Fotiadis, D.I.: Machine learning applications in cancer prognosis and prediction. Comput. Struct. Biotechnol. J. **13**, 8–17 (2015)
18. National Cancer Institute: Surveillance, epidemiology and end results program (2015). http://seer.cancer.gov/data/. Accessed 10 Jan 2016
19. Novais, P., Oliveira, T., Neves, J.: Moving towards a new paradigm of creation, dissemination, and application of computer-interpretable medical knowledge. Prog. Artif. Intell. **5**(2), 77–83 (2016)
20. Powers, D.M.W.: What the F-measure doesn't measure . . . . Technical report, Beijing University of Technology, China & Flinders University, Australia
21. RapidMiner: Rapidminer documentation: Optimize selection (2016). http:// docs.rapidminer.com/studio/operators/data_transformation/attribute_space_ transformation/selection/optimization/optimize_selection.html. Accessed 03 Jan 2016
22. Renfro, L.A., Grothey, A., Xue, Y., Saltz, L.B., André, T., Twelves, C., Labianca, R., Allegra, C.J., Alberts, S.R., Loprinzi, C.L., et al.: Accent-based web calculators to predict recurrence and overall survival in stage III colon cancer. J. Natl. Cancer Inst. **106**(12), dju333 (2014)
23. Siau, K., Shen, Z.: Mobile healthcare informatics. Med. Inform. Internet Med. **31**(2), 89–99 (2006)
24. Vachani, C., Prechtel-Dunphy, E.: All about rectal cancer (2015). http://www. oncolink.org/types/article.cfm?aid=108&id=9457&c=703. Accessed 27 Dec 2015
25. Valentini, V., van Stiphout, R.G., Lammering, G., Gambacorta, M.A., Barba, M.C., Bebenek, M., Bonnetain, F., Bosset, J.F., Bujko, K., Cionini, L., Gerard, J.P., Rödel, C., Sainato, A., Sauer, R., Minsky, B.D., Collette, L., Lambin, P.: Nomograms for predicting local recurrence, distant metastases, and overall survival for patients with locally advanced rectal cancer on the basis of European randomized clinical trials. J. Clin. Oncol. **29**(23), 3163–3172 (2011)
26. Ventola, C.L.: Mobile devices and apps for health care professionals: uses and benefits. Pharm. Ther. **39**(5), 356–364 (2014)

27. Wang, S.J., Wissel, A.R., Luh, J.Y., Fuller, C.D., Kalpathy-Cramer, J., Thomas, C.R.: An interactive tool for individualized estimation of conditional survival in rectal cancer. Ann. Surg. Oncol. **18**(6), 1547–1552 (2011)
28. Weiser, M.R., Gönen, M., Chou, J.F., Kattan, M.W., Schrag, D.: Predicting survival after curative colectomy for cancer: individualizing colon cancer staging. J. Clin. Oncol. **29**(36), 4796–4802 (2011)
29. Yamauchi, M., Lochhead, P., Morikawa, T., et al.: Colorectal cancer: a tale of two sides or a continuum? Gut **61**(6), 794–797 (2012)
30. Zhao, H., Doshi, P.: Towards automated restful web service composition. In: IEEE International Conference on Web Services, ICWS 2009, pp. 189–196. IEEE (2009)

An Implementation of a Decision-Making Algorithm Based on a Novel Health Status Transition Model of Epilepsy

Mandani Ntekouli[1], Maria Marouli[2], Georgia Konstantopoulou[3],
George Anastassopoulos[4(✉)], and Dimitrios Lymperopoulos[1]

[1] Wire Communications Lab, Department of Electrical and Computer Engineering,
University of Patras, Patras, Greece
[2] Department of Medicine, University of Patras, Patras, Greece
[3] Special Office for Health Consulting Services, University of Patras, Patras, Greece
[4] Medical Informatics Lab, Department of Medicine,
Democritus University of Thrace, Alexandroupolis, Greece
anasta@med.duth.gr

Abstract. Epilepsy is one of the most common and dangerous neurological disorders, affecting millions of people around the world every year. Its symptoms are quite subtle and the transition from one phase of the disorder to another can go undetected and end to a life threatening situation, if the patient is not carefully monitored. In this paper we propose a novel health status transition model in epilepsy, as well as an implementation scheme suitable to be used in health telemonitoring systems. This model is able to monitor the patient and detect abnormalities providing a time margin for him/her to take actions and for his/her caregivers to be prepared to help and act. Based on whole model's transitions information we created a health-caring ontology. Finally, we used Java in order to develop an appropriate decision-making telemonitoring algorithm based on the proposed model.

Keywords: Epilepsy · Status transition model · Health-caring ontology · Protégé · Decision-making algorithm · Telemonitoring algorithm

1 Introduction

Epilepsy is one of the most risky chronical medical conditions [1,2]. Nowadays, 65 millions of people worldwide suffer from epilepsy and most of them live a normal and healthy life. However, we should be aware that epilepsy can be deadly. Epilepsy is a neurological malfunction which affects the nervous system and causes various seizures. Very long seizures or seizures successively occurring, while patients don't recover between them, are dangerous incidents. So, it is important that these seizures are fast detected, identified and treated.

© IFIP International Federation for Information Processing 2016
Published by Springer International Publishing Switzerland 2016. All Rights Reserved
L. Iliadis and I. Maglogiannis (Eds.): AIAI 2016, IFIP AICT 475, pp. 27–38, 2016.
DOI: 10.1007/978-3-319-44944-9_3

A modern approach of that is by building a telemonitoring system providing personalized services to patients, who have been previously diagnosed with epilepsy. These services allow patients to detect the anticipated seizures and take the necessary precautions. Actually, preventing patients from seizures is impossible while well treating of them is the target. Patients receive early warnings about the anticipated seizure gaining time for treating their condition. In parallel, other entities, like their caregivers, their doctor, or even a hospital are informed about the patient's critical condition and act accordingly.

In this paper, we organized the different phases that an epileptic patient can enter in a model. We gathered and classified various trigger factors and symptoms that are likely to lead to a seizure and according to them we identify its kind and risk level. Subsequently, and based on the proposed model we developed an event-driven epilepsy telemonitoring algorithm using ontologies in Java [3]. Finally, we evaluated the performance of this algorithm using the data of a specific medical case.

The paper is structured as follows. In Sect. 2, the proposed Transition Model of Epilepsy is presented. In Sect. 3, the implementation of the proposed model's decision making algorithm is presented. Sections 4 and 5 include the evaluation of the implemented algorithm and the discussion, and conclusions, respectively.

2   Transition Model of Epilepsy

In this study, physicians and engineers worked together in order to map the evolution of the health status of subjects (patients) suffering from epilepsy [5–9]. The aim of this study was to create an appropriate health caring algorithm suitable to be adopted by advanced epilepsy telemonitoring systems. For this purpose, we considered the following as model's primitives:

- a subject suffering from epilepsy can be characterized by fourteen (14) different health statuses, named as: Normal, Predisposition to Epilepsy, High Predisposition to Epilepsy, Generalized Absence Seizure, Generalized Myoclonic Seizure, Generalized Clonic Seizure, Generalized Atonic Seizure, Generalized Tonic Seizure, Generalized Tonic-Clonic Seizure, Partial Seizure, Complex Partial Seizure, Simple Partial Seizure, Status Epilepticus, Post Seizure
- in each health status the subject is being cared by a specific group of caregivers (nurse, volunteers, relatives)
- the subject moves from one health status to another (Transition Phase), whenever a specific symptom presents (Fig. 1).
- within each Transition Phase the model poses specific actions to be performed by both the telemonitoring algorithm and the caregivers. These actions should be strongly coupled with the health record and profile of the subject, in cases of personalized telemonitoring services.

In this section we analyze every possible Transition Phase while moving from one health status to another. Each one of these phases is activated by specific Trigger Factors or symptoms, as it is subsequently analyzed.

Transition Phase 1: This phase declares the beginning of the health caring algorithm, posing the subject to the "Normal status". In this situation no action is required.

Transition Phase 2: Transition from "Normal status" to the "Predisposition to Epilepsy status". It occurs whenever one or more of the following eleven (11) Trigger Factors $\{TF_i, i = (1, .., 11)\}$ are presented:

1. Interruption of antiepileptic drugs
2. Alcohol abuse (drunkenness)
3. Lack of alcohol or benzodiazepines
4. Lack of sleep
5. Physical exhaustion
6. Mental stress (stress)
7. Interruption of drugs acting on the central nervous system (e.g., barbiturates, benzodiazepines)
8. Febrile illness
9. Menstruation
10. Metabolic disorders (hypoglycemia, hyponatremia)
11. Photosensitivity-discontinuous and rhythmically repetitive light stimulation

Actions: In this transition, increased attention is required by the subject for a possible future occurrence of one or more of the general symptoms of "High Predisposition to Epilepsy status". The algorithm undertakes the responsibility to keep aware about this risk, both the subject and the authorized caregivers by sending regularly to them electronic notification messages. The type and the frequency of these messages are strongly tied with the subjects Electronic Health Record (EHR) and epileptic profile.

Transition Phase 3A: Transition from "Normal status" to "High Predisposition to Epilepsy status". It occurs whenever one or more of the following eleven (11) General Symptoms $\{GS_j, j = (1, .., 11)\}$ are presented:

1. Aura - usually aesthetic (visual disturbances or dysesthesias)
2. Stomach discomfort
3. Feeling of fear or panic
4. Nausea or headache
5. Breathing problem
6. Tachycardia
7. High Blood Pressure
8. Dizziness
9. High body temperature
10. Strange smell
11. Pleasant or unpleasant sensation

Actions: The algorithm takes over to send often to the subject as well as to another authorized caregiver electronic notification messages. The type and the frequency of these messages are specified in accordance with the epileptic medical history and profile of the subject, as well as his/her active context (home, work, road, driving, etc.). In addition, the algorithm takes over to call a caregiver in the current subjects place. The caregiver is in charge of recording all symptoms that are likely to happen to the subject during the forthcoming seizure, given the fact that the subject will not be able to perceive and include these symptoms in the algorithm by himself.

Transition Phase 3B: Transition from "Predisposition to epilepsy status" to "High Predisposition to epilepsy status". It occurs whenever one or more of the above eleven (11) General symptoms $\{GS_j, j = (1, .., 11)\}$, are presented. The actions remain the same as in Phase 3A.

Transition Phase 4A: Transition from "High Predisposition status" to "Generalized Absence Seizure status" occurs whenever unconsciousness with mild motor activity (patient's gaze freezes, does not answer, does not react), called in our model Absence symptom $\{AS_1\}$, is presented.

Transition Phase 4B: Transition from "High Predisposition to epilepsy status" to "Generalized Myoclonic Seizure status" occurs whenever sudden involuntary muscle contractions, called in our model Myoclonic symptom $\{MS_1\}$, are presented.

Transition Phase 4C: Transition from "High Predisposition status" to "Generalized Atonic Seizure status" occurs whenever Drop attack (suddenly the muscle tone is lost throughout the body and the subject collapses like a sack on the ground), called in our model Atonic symptom $\{ATS_1\}$, is presented.

Transition Phase 4D: Transition from "High Predisposition status" to "Generalized Clonic Seizure status" occurs whenever clonic convulsions, called in our model Clonic symptom $\{CS_1\}$, are presented.

Transition Phase 4E: Transition from "High Predisposition status" to "Generalized Tonic Seizure status" occurs whenever tonic convulsions, called in our model Tonic Symptom $\{TonS_1\}$, are presented.

If the subject is in the "Generalized Tonic Seizure status" and the $\{CS_1\}$ is presented, we suppose that the subject moves to "Generalized Tonic-Clonic Seizure status".

Transition Phase 4F: Transition from "High Predisposition status" to "Generalized Partial Seizure status". It occurs whenever one or more of the following ten (10) Partial Symptoms $\{GPS_u, u = (1, .., 10)\}$ are presented:

1. Tonic involuntary movements of one (upper end or lower) end or of the shank
2. Face deformation
3. Sensory disorders - numbness and tingling in any part of the body
4. Visual disturbances e.g. flashes, zig zag
5. Auditory disorders, such as simple sounds or even music
6. Difficulties in speaking, tangle, halting speech, salivation, chewing sounds and movements, teeth grinding
7. Olfactory disorders e.g. feeling bad smell
8. Nausea
9. Affective disorders - sudden change of sentiment
10. Other rare forms, like deja vu

In this Phase, if the subject has his/her consciousness lost, he/she moves to "Complex Partial Seizure status". Correspondingly, if the subject has his/her consciousness he/she moves to "Simple Partial Seizure status".

Common Actions for Phases 4A to 4F: Based on the epileptic history/profile of the subject and the symptoms that led to the finding of the seizure, the caregiver communicates with the doctor who is treating the subject and describes the whole course of the seizure. As far as the algorithm is concerned, it will enter in a standby mode (Delay), whose duration is equal to three (3) minutes. This delay is necessary in order to determine whether the subject suffers from successive attacks without acquiring consciousness.

Transition Phase 4G: However, if the subject does not go into any of the above six (6) seizure statuses while he is in the "High Predisposition to Epilepsy status", then we believe that the algorithm will enter in a waiting state (Delay), whose duration depends on the epileptic medical history of each subject, and then make a new display control of one (or more) new $\{RF_i\}$.

Transition Phase 5: Transition from one of the seizure statuses to "Status Epilepticus" occurs whenever the subject is in the same seizure status for duration of at least three (3) minutes.

Actions: In this Phase, the caregiver takes over to notify an ambulance of a nearby hospital because the situation is extremely serious and dangerous since the patient can even die. As far as the algorithm is concerned, it will enter in a standby mode (Delay), whose duration depends on the epileptic medical history of each patient and the particular seizure status. Then, the algorithm will take control of the restoration of patient's full consciousness.

Transition Phase 6A: Transition from "Status Epilepticus" to "Post Seizure status". It occurs whenever the subject's consciousness has been fully restored and one or more of the following eleven (11) Post Seizure symptoms $\{PS_v, v = (1, .., 11)\}$ are presented:

1. Sleep
2. Memory deficit the person does not remember what had happened
3. Dizziness
4. Drowsiness
5. Headache
6. Fear
7. Confusion
8. Feeling of Shame
9. Difficulty in speaking
10. Weakness
11. Thirst

Transition Phase 6B: Transition from one of the eight (8) seizure statuses to "Post Seizure status" occurs whenever one or more of the above eleven (11) $\{PS_v\}$ are presented.

Actions: In this Phase, we suppose that the algorithm will enter in a Standby mode (Delay), whose duration depends on the epileptic medical history of each subject and the $\{PS_v\}$ that were presented in each case. Then, the algorithm will carry out a display control of one (or more) new $\{PS_v\}$.

Transition Phase 7A: Transition from "Status Epilepticus" to "Normal status" occurs whenever the subject's consciousness has been fully restored and none of the above eleven (11) $\{PS_v\}$ are appeared.

Transition Phase 7B: Transition from "Post Seizure status" to "Normal status" occurs whenever none of the above eleven (11) $\{PS_v\}$ are presented.

3   The Implementation of the Model's Decision-Making Algorithm

3.1   The Applied Implementation Method

In order to organize the medical terms, events and actions of previously described mapping, we developed a health-caring ontology [10]. An ontology is defined as an explicit, formal specification of shared conceptualization. So, it is an abstract model of concepts and relationships that exist in a certain knowledge domain and it can be represented in an unambiguous, computer-readable and understandable way. Especially for medical expert systems, ontology is the de facto engineering artifact to develop, process and exchange such models.

An ontology uses a common language to formalize its knowledge domain. The most recognizable are the `Resource Description Framework` (RDF) and the `Web Ontology Language` (OWL), for which World Wide Web Consortium (W3C) [11] developed a standard.

To develop our domain ontology we chose to use Protégé [12], in its latest stable version 4.3. It is a free, open source editor which helps us build knowledge-based systems. In Protégé, we can model the concepts and relationships of our world and by using reasoners, such as Pellet, Hermit, etc. we can provide semantic classification of the medical terms, combine the defined ones and infer new information of the world. Reasoners, also, check for ontology inconsistencies. Moreover, it is important to mention that Protégé is based on `Java`. So, we can easily import our `.owl` file in `Java` via a `Java Application Programming Interface` (API).

Finally, in order to provide services to a specific patient we implemented a personalized telemonitoring algorithm based on our ontology using `Java`. We chose that programming language mostly due to the ease of Protégé's integration to it. For the development we used as `Java` Integrated Development Environment (IDE) Eclipse and the `API` of our choice - which will import the ontology - was `Jena` [13]. One of the main reasons for the choice of Jena was that it has plenty of libraries that can be used to create and manipulate RDF graphs. Ontology' s OWL language can be saved in a `RDF/XML` format in Protégé, in order to be understandable by `Jena`.

3.2   Ontology Engineering for Epilepsy

In this work, we developed a knowledge-based ontology for people diagnosed with epilepsy. We acquired the knowledge, necessary for the creation of the ontology,

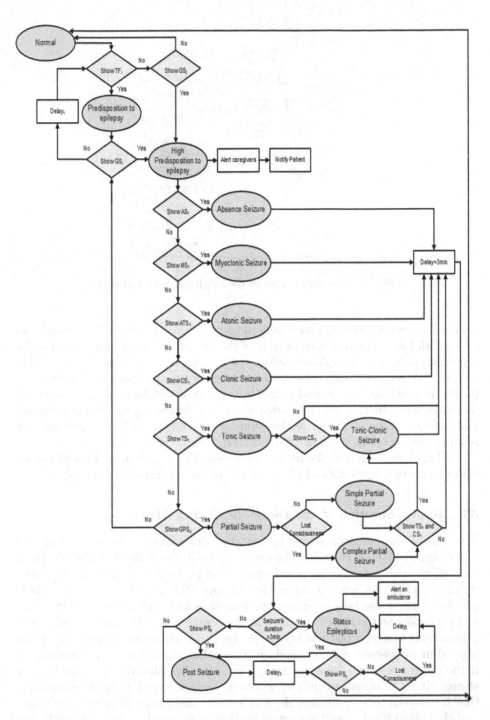

Fig. 1. Epileptic patient's status transition diagram in the proposed telemonitoring algorithm

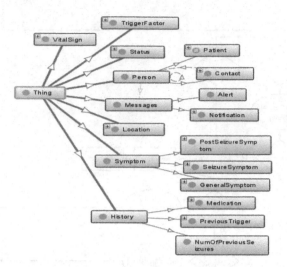

Fig. 2. The main classes in the implemented ontology

from field experts and based on it we build a model using Protégé. First of all, we defined all the required classes as well as the object and data properties for the representation of all the phases before, after, and in the course of an epileptic seizure. We set some exact cardinality restrictions on classes and axioms on properties, such as domain and range, in order to make the reasoner's inference more accurate. In the end, we added some individuals on the defined classes. For example, some of them are the symptoms of all the different categories of seizures, which are individuals of the corresponding class.

In Fig. 2, we represent the main classes used in our ontology. This visualisation was created by OntoGraf plugin that can be integrated in Protégé.

3.3   Implementation of the Decision-Making Algorithm

The aim of the ontology is not to contain information that can be updated and changed without changing the ontology itself. Basically it is the blueprint that an algorithm has to be based upon in order to provide personalized and decision-making services to a specific patient. That algorithm in our case is developed using the Java programming language with which we coded the rules and conditions so the algorithm can make the correct, in each case, decisions and take actions based on the new inputs. As far as the decisions are concerned, the algorithm will decide in which medical status the patient has been transited to. If the patient does not have any symptoms, the algorithm decides that his/her status will be "Normal status", otherwise it decides according to the symptoms that he/she exhibits. Regarding the actions, the algorithm will proceed to notify and alert the subject as well as contact his/her caregivers based on his/her health status.

After we imported the ontology via the Jena API, we added the new individuals such as our main subjects along with their properties, for instance the symptoms that he/she exhibits each moment. Finally, Java has to interpret the new inputs as part of the ontology, change the necessary parameters - like the values of the properties which our patient has - make the necessary decisions, and save it as an ontology file that can be opened again and visualized by Protégé.

3.4   Input Data to the Decision-Making Algorithm

The data that need to be inserted in the algorithm must be obtained via many ways. There are data that come from smart sensors. Wearable sensors are placed non-invasively on the patient's body and continuously collect and transmit vital signs measurements. Nowadays, these sensors can be tiny, flexible and embedded in everyday accessories like a watch. In our model, the vital signs that need to be measured are ECG (heart rate), body temperature, respiratory rate and blood pressure, etc. In addition, other data need to be input manually by the patient or the caregiver, but that is implemented in a non distracting and disturbing (for the patient) way. In case of epilepsy, most trigger factors and symptoms are by their nature quite fuzzy and cannot be monitored via an autonomous way. So, the algorithm request people to keep it updated regarding the patient's condition.

4   Evaluation of the Proposed Model

The evaluation of the proposed model is made by means of a specific medical case study [4]. In this case study, we personalize the ontology based on a female patient, named Kate.

4.1   Kate's Epileptic Medical History and Profile

Kate had a normal birth, without perinatal complications and a normal psychosocial growth. There are no significant health issues in her life. Her father seemed to have had Absence seizures as a child but he was never treated for that. In the age of 20, Kate had her first seizure. She first felt disturbed by the lights of a club she was out with her friends and then felt weak and her vision blurred. She lost consciousness, presented urinary incontinence and when she woke up her tongue was bleeding. Her friends mentioned that her limbs and body were stroking for less than a minute. Laboratory tests and Magnetic Resonance Imaging (MRI) exams produced no results. Electroencephalogram (EEG) exam was impaired only under strong pulsative light but no seizure was inducted. She was not diagnosed as epileptic and she didn't take any medication. Three months later, she had another seizure while watching television, with the same characteristics. Kate then mentioned she was tired because of exams and then was diagnosed with epilepsy and was prescribed medication. Since then, there was no other epileptic incidence.

Now, Kate is 27 years old, right-handed, with a history of symptomatic epilepsy with Tonic-Clonic seizures and under a 500 mg dosage of Depakine 1/day therapy since she was 20 years old.

4.2 Kate's Telemonitoring Through the Proposed Algorithm

One year ago, our research team selected Kate's case as the appropriate case in order to evaluate the proposed model. In our implemented algorithm, we added three different individuals, Kate, her husband (caregiver) and her doctor. The algorithm took as a fact that her current status is "Normal". Moreover, we asked Kate and her husband to notify her current active context in order the algorithm to be aware, whenever specific Trigger Factors or symptoms were detected. Some months later, Kate visited her neurologist with the request of stopping her medication in order to get pregnant. Her doctor decided to reduce the dose of medication, a fact that algorithm recognized as a Trigger Factor, then it activated the Transition Phase 2 and finally it decided that Kate transited from "Normal status" to "Predisposition to Epilepsy status". In parallel, notifications were sent regularly to her husband and to herself about her status.

Three days later, she experienced menstruation. So now she entered as an input in the algorithm the second Trigger Factor and she was notified twice a day, one in the morning and one in the afternoon to avoid stress and lights. Stress and lights were chosen by the algorithm according to her epileptic medical history (Previous Trigger Factors). Her status remained as it is, "Predisposition to Epilepsy".

Four days later, Kate visited some friends at a restaurant. The algorithm, being informed about the strong lights existing in that place, sent a warning message to her. Fifteen minutes later, when the algorithm asked her again, she felt uncomfortable, tired and her vision impaired. So, she entered her symptoms, but now she had additionally one General Symptom, Aura. Our patient transited from "Predisposition to Epilepsy status" to the "High predisposition to Epilepsy status". Notifications were sent to her and to her husband in order to check up on her.

While heading to the toilet but before reaching her destination, she lost consciousness and tonic along with clonic convulsions occurred. When her husband reached her he removed the sharp items, to prevent her from hurting herself, and he entered the new symptoms in order for the doctor to be notified and for him to receive updates as to how to proceed. Kate, at this point, transited from "High predisposition to Epilepsy status" to "Generalized Tonic-clonic Seizure status". Automatically her doctor was notified about Kate's condition.

Three minutes later she was still having seizures without regaining her consciousness. As three minutes passed Kate transited from "Generalized Tonic-Clonic Seizure status" to "Status Epilepticus" and the nearest hospital was notified and they sent an ambulance.[1]

---

[1] In this situation, in order to visualize the relationships that have been created up to this moment, the data were saved and could be again opened in a new ontology file. The results are shown in Fig. 3, created with OntoGraf in Protégé.

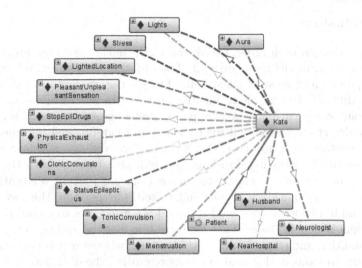

Fig. 3. Kate's relationships, in the specific case study, when she is in "Status Epilepticus" status based on the proposed algorithm

Fifteen minutes later, the ambulance arrived, the paramedics inserted her with 10 mg diazepam IV and she was transferred to the hospital. Ten minutes later the seizures stopped, and she regained consciousness. She was disoriented, she couldn't speak properly and she was sleepy. In that stage Kate showed Post Seizure symptoms, and after her husband entered them her status was updated to "Post Seizure".

She woke up four hours later and she was mentally normal. She was feeling good and she wanted to be checked out of the hospital but she had to be kept under supervision for one day. She was prescribed again Depakine of 500 mg and after entering this information in the algorithm she transited from "Post Seizure status" to "Predisposition to Epilepsy status" until all the symptoms of this category are gone. To be more exact, she would stay in the above status until menstruation had stopped. Finally, if no other symptoms appeared, she would transit to "Normal status".

4.3   Discussion

The above real world case study offered us a chance to evaluate the algorithm's performance. Despite the fact that we used non full automated data (vital signs, context conditions, etc.) acquisition procedures, the algorithm yielded the correct results in any patients living and health condition. The indicated by the algorithm transitions were the appropriate and the predicted health status after each transition was verified by the disorder's evolution. The doctor and the caregiver were, anytime, well aware about the anticipated seizures. Adopting this algorithm in a telemonitoring system for epilepsy offers advanced capabilities to the patient and the caregivers in order to deal with these seizures.

5   Conclusions

Epilepsy is a dangerous disorder, mainly due to the nature of its symptoms. As we have shown in our model there are a lot of factors that take place and contribute to its progress - and most of them being fuzzy - the prediction of a seizure is extremely difficult. Being able to identify on time the status of the disorder in which the subject is currently in, is of paramount importance, but it is also very hard to be detected, especially if the patient is not careful, well-informed or not in regular contact with a doctor. Our model offers a viable non-intrusive way to make informed life-saving decisions that will give time to both the patient and his/her doctor (or caregivers) to take action and prevent a situation that could escalate very quickly if it goes undetected. Furthermore the way that we designed and implemented our model, provides a great way to record the status of the patient in regular basis. That log gives him/her the ability to telemonitor his/her condition and also examine in that record, whenever it is necessary, past changes and previous health related events regarding the disorder.

References

1. Karl, M., Neumann, M.: Neurology. In: Nikolaos, B.D. (ed.), pp. 10–15, 69–70, 112, 135–139, 199, 387, 544–564. Rotonta, Thessaloniki (1989) (in greek)
2. Misulis, K.E., Head, T.C.: Netter's concise neurology. In: Papathanasopoulos, P. (ed.), pp. 2–5, 37–39, 57, 86–94, 187–213. Gkotsis Kon/nos & SIA E.E., Patras (2007) (in greek)
3. Hebeler, J., Fisher, M., Blace, R., Perez-Lopez, A.: Semantic Web Programming, pp. 35–299. Wiley Publishing, Inc., Indianapolis (2009)
4. Bickley, L.S.: Bates' Guide to Physical Examination and History Taking. In: Georgios, B., Charalampos, G., Antreas, K., Eustratios, M., Dimitrios, M., Theofanis, O., Gearasimos, P., Athanasios, P., Charalampos, R., Panagiotis, S., Christodoulos, S., Petros, S., Athanasios, T., Epameinondas, T., Georgios, C. (eds.) pp. 12, 606–607. Iatrikes Ekdoseis P.X. Pasxalidis, Athina (in greek) (1974)
5. Elzawahry, H., Do, C.S., Lin, K., Benbadis, S.R.: The diagnostic utility of the ictal cry. Epilepsy Behav. 18(3), 306–307 (2010). (Elsevier)
6. Kaplan, P.W.: The EEG of status epilepticus. J. Clin. Neurophysiol. **23**, 221–229 (2006). (Wolters Kluwer)
7. Drislane, F.W., Blum, A.S., Schomer, D.L.: Focal status epilepticus: clinical features and significance of different EEG patterns. Epilepsia **40**(9), 1254–1260 (1999). (Wiley Online Library)
8. Brenner, R.P.: EEG in convulsive and nonconvulsive status epilepticus. J. Clin. Neurophysiol. **21**(5), 319–331 (2004). (Wolters Kluwer)
9. Luders, H., Acharya, J., Baumgartner, C., et al.: Semiological seizure classification. Epilepsia **39**(9), 1006–1013 (1998). (Wiley Online Library)
10. Riano, D., Real, F., Lopez-Vallverdu, J.A., Campana, F., Ercolani, S., Mecocci, P., Annicchiarico, R., Caltagirone, C.: An ontology-based personalization of healthcare knowledge to support clinical decisions for chronically ill patients. J. Biomed. Inform. **45**(3), 429–446 (2012). (Elsevier Inc.)
11. World Wide Web Consortium (W3C). www.w3.org/
12. Protégé Ontology Editor. protege.stanford.edu/
13. Jena Apache Semantic Web Framework. jena.apache.org/

Integrative Bioinformatic Analysis of a Greek Epidemiological Cohort Provides Insight into the Pathogenesis of Primary Cutaneous Melanoma

Georgia Kontogianni[1,2], Olga Papadodima[1], Ilias Maglogiannis[2],
Konstantina Frangia-Tsivou[3], and Aristotelis Chatziioannou[1(✉)]

[1] Metabolic Engineering and Bioinformatics Group, Institute of Biology,
Medicinal Chemistry and Biotechnology,
National Hellenic Research Foundation, Athens, Greece
{gkontogianni,opapadod,achatzi}@eie.gr
[2] Department of Digital Systems, School of Information
and Communication Technologies,
University of Piraeus, Piraeus, Greece
imaglo@unipi.gr
[3] HistoBio Diagnosis, Athens, Greece
dfrangia@otenet.gr

Abstract. Melanoma is the most lethal type of skin cancer. In this study for the first time we analyze a Greek cohort of primary cutaneous melanoma biopsies, subjected to whole exome sequencing, in order to derive their mutational profile landscape. Moreover, in the context of big data analytical methodologies, we integrated the results of the exome sequencing analysis with transcriptomic data of cutaneous melanoma from GEO, in an attempt to perform a multi-layered analysis and infer a tentative disease network for primary melanoma pathogenesis. The purpose of this research is to incorporate different levels of molecular data, so as to expand our understanding of cutaneous melanoma and the broader molecular network implicated with this type of cancer. Overall, we showed that the results of the integrative analysis offer deeper insight in the underlying mechanisms affected by melanoma and could potentially contribute to the valuable effective epidemiological characterization of this disease.

Keywords: Data integration · Next generation sequencing · Functional analysis · Skin cancer

1 Introduction

Melanoma is the most dangerous form of skin cancer [1]. Cutaneous melanoma (or melanoma of the skin), the most common type of melanoma, is a complex multi-factorial disease as both environmental and genetic factors are involved in its manifestation [2]. It is often a fatal neoplasm, derived from melanocytes, that accounts for most skin cancer deaths. In the advanced stages of this cancer, therapeutic intervention usually fails to improve survival despite recent advances in immunotherapy.

L. Iliadis and I. Maglogiannis (Eds.): AIAI 2016, IFIP AICT 475, pp. 39–52, 2016.
DOI: 10.1007/978-3-319-44944-9_4

According to the World Health Organization, 132,000 melanoma skin cancers occur globally each year and the global incidence of melanoma continues to increase, with a main predisposal factor; sun exposure.

The complexities of cellular metabolism and regulatory pathways involved have, until recently, obstructed the formulation of a unified description for melanoma [3]. Thus, despite the descent of gene signatures for various cancers, e.g. breast or colon cancer, a similar progress remains elusive for malignant melanoma. This could be attributed to the intricate nature of the molecular basis of cutaneous melanoma, which needs neatly stratified epidemiological cohorts to effectively address the issue of the high heterogeneity of this disease. In any case, genomic studies are limited by the shortage of similar melanoma cohorts, collecting and maintaining frozen tumor tissue, therefore rendering gene expression profiling studies of melanoma relatively scarce [4]. Still, efforts have been made to overcome any issues and shed some light on the underlying mechanisms involved with melanoma pathogenesis and metastases [4, 5]. A number of important emerging biological pathways and gene targets recently identified in melanoma are reported in [6]. Key biological pathways, where several significant genes (e.g. CDKN2A, CDK4, RB1) are involved, include proliferation, transcriptional control, extracellular matrix remodeling, glutamate signaling, and apoptosis.

In this study, we have focused on integrating different levels of molecular data through functional analysis to improve our understanding of the underlying mechanisms involved with melanoma. We incorporated established microarray datasets with next generation sequencing mutational data creating a potential disease network for melanoma.

Section 2 describes the techniques and methodology used in this study for the analysis of next generation sequencing and transcriptomic data. Then, in Sect. 3 we present the results derived from the two datasets, in separate subsections, followed by another subsection for data integration. Finally, we conclude this study in Sect. 4.

2   Materials and Methods

2.1   Analysis of Next Generation Exome Sequencing Data

The data analyzed in this section derived from Whole Exome Sequencing (WES) data of paired tumor and adjacent normal tissue from 9 patients with cutaneous melanoma (manuscript under preparation). The framework for the analysis of Next Generation Sequencing (NGS) data includes various state-of-the-art tools and has been previously presented by our team [7].

We first align the reads to the reference genome (hg19, version b37), using BWA (Burrows-Wheeler Aligner) [8] for DNA reads, adjusted for paired-end sequencing and run in consecutive steps for finding the correct coordinates and generate the final alignment in proper format. Then, we preprocess reads using Picard [9], for marking duplicate reads and sorting sequences according to the reference, to allow further processing with GATK (Genome Analysis Toolkit) [10], so as to ensure the quality of reads (all reads are given quality scores and can be dismissed if needed) and perform realignments and recalibrations based on the scores and references, to optimize the output reads and permit the following variance and somatic mutation investigation. Inspection for variance is performed with MuTect [11], which exploits statistical methodologies (Bayesian classifiers) and identifies sites of somatic mutations in paired datasets (tumor vs. normal). To annotate these sites, we use Oncotator [12], which utilizes several databases to link the sites to specific genes. Finally, we perform functional analysis to identify the molecular pathways affected by the specific mutations, and gene prioritization, so as to highlight genes with central role, implicated in diverse and major mechanisms in the Gene Ontology tree. These are performed using BioInfoMiner [13], which combines the StRAnGER2 [14] and GOrevenge [15] algorithms. Figure 1 presents the workflow used here.

The complexity of NGS data is high, due to the high amount of information contained in each separate sample (compressed ~ 10 Gb per sample/20 Gb per patient/ ~ 150 Gb for all) and the fact that several distinct parameters need to be adjusted at each step, so as to optimize the performance and the quality of the results

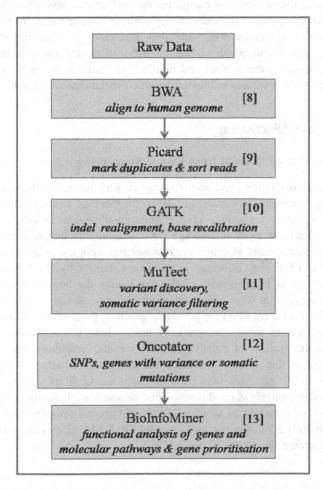

Fig. 1. Workflow of analysis for the identification of variance and somatic mutations

(i.e., BWA needs to be adjusted for paired-end sequencing and run in consecutive steps for finding the correct coordinates and generate the final alignment in proper format).

2.2 Analysis of Transcriptomic Data

Transcriptomic analyses among different groups allow the exploration and identification of alterations in gene expression profiles between them. The data used in this section were previously analyzed in [16]. Briefly, the microarray dataset was taken from the Gene Expression Omnibus (GEO) [17, 18], with accession number GDS1375. RNA was isolated from 45 primary melanoma, 18 benign skin nevi, and 7 normal skin tissue specimens and was analyzed for gene expression analysis, using the Affymetrix Hu133A microarray chip containing 22,000 probe sets. Following global normalization, gene expression values across all categories were log transformed, and the mean values of all genes in the normal skin were calculated. Afterwards, the mean gene vector concerning the normal skin categories was subtracted from all replicate vectors of the other two categories. The differentially expressed gene values of the melanoma versus skin, and nevi versus skin, were then analyzed. A false discovery rate for multiple testing adjustment, p-value 0.001 and a 2-fold change threshold for significant differential expression were applied and finally, 1425 unique genes were statistically selected, as being differentially expressed between melanoma and the normal state.

3 Results and Discussion

3.1 Mutational Data Derived from Exome Sequencing

WES data derived from tumor and normal samples were aligned to the human genome, with an average sequence coverage of >100x (number of reads aligning to known reference bases), ideal for achieving the mutational profile required. Overall, the individual samples have depth of coverage >90, with only 3 samples achieving a lower score. Still this lower score is found only in normal samples, which does not affect further analysis, since high coverage is necessary mainly by the tumor samples to overcome endogenous heterogeneity. Table 1 contains the alignment rates and coverage for all samples that were examined. After the processing of individual samples for analysis of variance based on the reference genome, pairs of data from each patient (tumor vs. normal) are jointly analyzed, so as to identify somatic point mutations. Table 2 shows the number of putative sites of somatic mutations, after the MuTect analysis, as well as the count of missense and nonsense mutations for each patient. These mutations affect gene products, by amino acid substitutions or protein truncation, and require further analysis as candidate genetic biomarkers. It is worth noting that the complete workflow for a pair of samples (tumor and normal samples from one patient) needs approximately 35 h running time on a 64 Gb RAM/12 processor cluster server (finally, summarizing the results in ~ 10 Mb).

Table 1. Alignment rates and coverage

patient	Normal		Tumor	
	Alignment rate	Coverage	Alignment rate	Coverage
1	96	129.8	95.4	118.9
2	84.2	70.5	89.5	91
3	66.7	35.9	93.6	102.2
4	93.8	103.9	92.6	111.5
5	96.8	101.5	96.7	111.4
6	96.8	123.3	96.2	104.8
7	97.4	117.4	96.8	111.9
8	87.9	88.7	88.8	92.3
9	95.6	128.4	92	120.7

Table 2. Number of somatic mutations, missense/nonsense mutations, and unique genes affected per patient

Patient	Sites of somatic mutations	Missense/nonsense mutations	Unique genes affected
1	855	224	214
2	1134	309	295
3	826	281	265
4	73	10	10
5	944	275	265
6	5985	1811	1474
7	812	226	200
8	922	224	219
9	1111	224	214

In order to discover the molecular pathways affected by the specific mutations, after annotating the mutations to specific genes, we performed functional analysis of the union of affected genes from all the patients (2685 unique genes), which revealed 40 statistically significant biological processes (p-value < 0.05), shown in Table 3.

3.2 Transcriptomic Data

The transcriptomic analysis from [16] revealed 1425 unique differentially expressed genes. Enrichment analysis showed 36 statistically significant biological processes (p-value < 0.05), which are presented in Table 4.

Table 3. Table of the significant biological processes influenced by the mutated genes. Enrichment represents the ratio of the number of genes in the input list annotated with a GO term to the total number of genes annotated to this specific term

Term id	Term definition	Enrichment	Hypergeometric p-value	Corrected p-value
GO:0007156	Homophilic cell adhesion via plasma membrane adhesion molecules	69/150	4.33E−20	0.0014
GO:0007155	Cell adhesion	148/531	2.17E−15	0.0027
GO:0050911	Detection of chemical stimulus involved in sensory perception of smell	105/389	1.68E−10	0.0037
GO:0030198	Extracellular matrix organization	84/313	1.38E−08	0.0048
GO:0086010	Membrane depolarization during action potential	17/30	1.10E−07	0.0063
GO:0007411	Axon guidance	95/375	3.64E−08	0.0068
GO:0006811	Ion transport	82/319	1.53E−07	0.0101
GO:0022617	Extracellular matrix disassembly	39/117	2.98E−07	0.0108
GO:0006814	Sodium ion transport	37/106	1.59E−07	0.0115
GO:0055085	Transmembrane transport	162/767	7.19E−07	0.012
GO:0007608	Sensory perception of smell	61/224	7.06E−07	0.0125
GO:0019228	Neuronal action potential	16/31	1.42E−06	0.0144
GO:0035725	Sodium ion transmembrane transport	30/89	5.11E−06	0.0145
GO:0007268	Synaptic transmission	97/428	5.75E−06	0.0178
GO:0042391	Regulation of membrane potential	36/117	6.79E−06	0.0195
GO:0007186	G-protein coupled receptor signaling pathway	192/976	9.61E−06	0.0198
GO:0030574	Collagen catabolic process	26/74	9.12E−06	0.0203
GO:0007605	Sensory perception of sound	39/133	1.03E−05	0.0223
GO:0034765	Regulation of ion transmembrane transport	35/118	2.18E−05	0.0257
GO:0060080	Inhibitory postsynaptic potential	8/11	2.26E−05	0.0257

(Continued)

Table 3. (*Continued*)

Term id	Term definition	Enrichment	Hypergeometric p-value	Corrected p-value
GO:0070588	Calcium ion transmembrane transport	38/129	1.19E−05	0.0258
GO:0018108	Peptidyl-tyrosine phosphorylation	37/130	3.53E−05	0.0287
GO:0016339	Calcium-dependent cell-cell adhesion via plasma membrane cell adhesion molecules	13/27	3.64E-05	0.0306
GO:0070509	Calcium ion import	13/28	5.89E−05	0.0323
GO:0007018	Microtubule-based movement	24/74	8.72E−05	0.0331
GO:0001539	Cilium or flagellum-dependent cell motility	6/7	5.96E−05	0.034
GO:0032228	Regulation of synaptic transmission, GABAergic	7/10	0.0001	0.0353
GO:0007399	Nervous system development	72/322	0.0001	0.0376
GO:0007169	Transmembrane receptor protein tyrosine kinase signaling pathway	33/119	0.0002	0.0382
GO:0034220	Ion transmembrane transport	65/286	0.0002	0.0395
GO:0001964	Startle response	10/20	0.0002	0.0399
GO:0050907	Detection of chemical stimulus involved in sensory perception	28/96	0.0002	0.0405
GO:0007416	Synapse assembly	17/47	0.0002	0.0445
GO:0071625	Vocalization behavior	7/12	0.0006	0.0447
GO:2000821	Regulation of grooming behavior	4/4	0.0005	0.0455
GO:0016337	Single organismal cell-cell adhesion	30/109	0.0003	0.0465
GO:0030534	Adult behavior	12/29	0.0004	0.0468
GO:0034332	Adherens junction organization	14/38	0.0006	0.0476
GO:0034329	Cell junction assembly	22/76	0.001	0.0492
GO:0015721	Bile acid and bile salt transport	11/27	0.0009	0.0493

Table 4. Table of the significant biological processes influenced by the differentially expressed genes. Enrichment represents the ratio of the number of genes in the input list annotated with a GO term to the total number of genes annotated to this specific term

Term id	Term definition	Enrichment	Hypergeometric p-value	Corrected p-value
GO:0030198	Extracellular matrix organization	66/313	0.00000676	0.0014
GO:0008544	Epidermis development	31/109	0.00000027	0.0033
GO:0030216	Keratinocyte differentiation	19/56	0.000003067	0.0043
GO:0006094	Gluconeogenesis	16/48	0.00002341	0.0053
GO:0048013	Ephrin receptor signaling pathway	21/91	0.0005	0.0078
GO:0060512	prostate gland morphogenesis	4/4	0.0001	0.0079
GO:0033599	Regulation of mammary gland epithelial cell proliferation	4/5	0.0006	0.0094
GO:0045861	Negative regulation of proteolysis	9/26	0.0011	0.0114
GO:0061436	Establishment of skin barrier	7/17	0.0012	0.0116
GO:0060326	Cell chemotaxis	15/57	0.0008	0.0132
GO:0071230	Cellular response to amino acid stimulus	13/48	0.0013	0.0155
GO:0051591	Response to cAMP	14/54	0.0013	0.0157
GO:0048538	Thymus development	12/45	0.0022	0.0182
GO:0045669	Positive regulation of osteoblast differentiation	14/57	0.0023	0.0199
GO:0001954	Positive regulation of cell-matrix adhesion	8/23	0.0019	0.021
GO:0042060	Wound healing	20/95	0.0024	0.022
GO:0007155	Cell adhesion	78/531	0.0028	0.0235
GO:0061036	Positive regulation of cartilage development	6/15	0.0032	0.0236
GO:0022617	Extracellular matrix disassembly	23/117	0.003	0.025
GO:0045765	Regulation of angiogenesis	9/30	0.0033	0.027
GO:0071526	Semaphorin-plexin signaling pathway	7/20	0.0036	0.0292
GO:0048661	Positive regulation of smooth muscle cell proliferation	13/54	0.004	0.0298

(*Continued*)

Table 4. (*Continued*)

Term id	Term definition	Enrichment	Hypergeometric p-value	Corrected p-value
GO:0050773	Regulation of dendrite development	5/11	0.0038	0.0313
GO:0048678	Response to axon injury	9/32	0.0053	0.0337
GO:0010951	Negative regulation of endopeptidase activity	26/144	0.0056	0.0343
GO:0061621	Canonical glycolysis	8/27	0.0059	0.0346
GO:0070373	Negative regulation of ERK1 and ERK2 cascade	12/50	0.0057	0.0374
GO:0055086	Nucleobase-containing small molecule metabolic process	16/78	0.008	0.0402
GO:0007160	Cell-matrix adhesion	18/92	0.0084	0.0402
GO:0060441	Epithelial tube branching involved in lung morphogenesis	6/17	0.0066	0.0405
GO:0030032	Lamellipodium assembly	9/33	0.0066	0.0407
GO:0030324	Lung development	20/106	0.0086	0.0435
GO:0002009	Morphogenesis of an epithelium	7/23	0.0084	0.045
GO:0043153	Entrainment of circadian clock by photoperiod	6/18	0.009	0.0454
GO:0007266	Rho protein signal transduction	13/59	0.0087	0.0465
GO:0030855	Epithelial cell differentiation	16/79	0.009	0.0483

3.3 Data Integration

To facilitate a deeper examination of our datasets, we compared the gene lists from the mutational and transcriptomic analyses. Figure 2 illustrates the total unique and common genes, from the two types of datasets. Only 5 % of the total genes were common between the two sets. Nevertheless, among the highly ranked processes, presented in Tables 3 and 4, cell adhesion, extracellular matrix organization and extracellular matrix disassembly, containing a large number of genes, are found as significantly affected in both cases.

In order to create a feasible disease network for melanoma, we merged the previous results, and carried out an additional functional analysis. This enrichment analysis revealed 45 statistically significant biological processes (p-value < 0.05), presented in Fig. 3, ranked according to their corrected p-values.

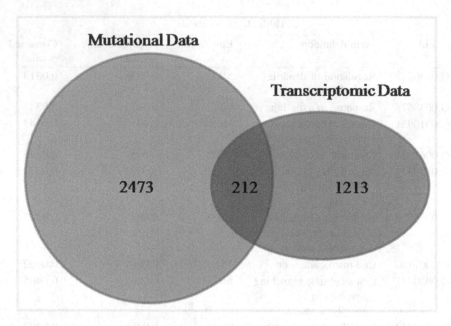

Fig. 2. Venn diagram for the significant gene lists from the two analyses

This potential Disease Network revealed several mechanisms with known signifi-cance, consistent with melanoma. Enrichment of GO terms, such as epithelial tube branching involved in lung morphogenesis, morphogenesis of an epithelium, epithelial cell differentiation, and regulation of mammary gland epithelial cell proliferation reflects the topological origin of cutaneous melanoma [19, 20]. Furthermore, cell-matrix procedures (organization, adhesion) have been previously reported as sig-nificantly altered in tumors [21, 22], as well as lamellipodium assembly, an essential structure for cell migration, which plays an important part in cell invasion and metastasis of cancer [23, 24]. In relation to the ephrin receptor and Rho protein sig-naling pathways, the Eph receptor tyrosine kinases and their ephrin ligands have specific expression patterns in cancer cells [25], while Rho-like GTPase have been identified as key regulators of epithelial architecture and cell migration, both correlated to cancer development [26, 27].

As expected, the previously discussed significant pathways from Tables 3 and 4 are complemented by the additional data, incorporating an increased number of genes, with considerable implication in melanoma manifestation and progression. Among the significant processes are several previously highlighted by the distinct datasets, but also a number of newly generated, after data integration. Figure 4 indicates the unique and common pathways in each case.

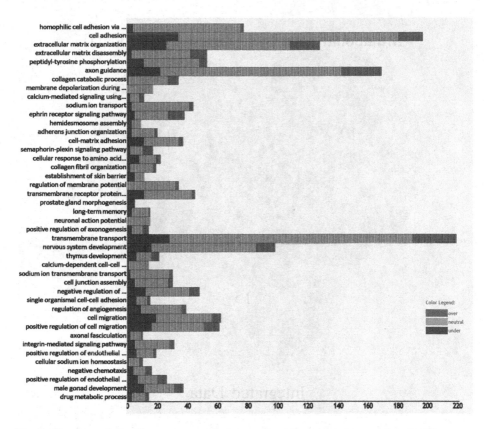

Fig. 3. Bar plot of significant terms with the number of associated genes (x-axis). Terms are ranked using the corrected p-value. The colors of the genes specify their expression fold change, green -on the left- for under-expressed genes and red -on the right- for over-expressed genes, with neutral indicating somatic mutation

4 Conclusions and Future Work

In this study, we sought to export the broader molecular network implicated with cutaneous melanoma. We integrated molecular data of different levels in order to identify the important mechanisms that are involved in this type of cancer. This integration advanced our understanding about the mechanisms implicated with melanoma, by observing the correlation between different sets and levels of data. More importantly, it allowed the manifestation of additional mechanisms previously concealed by the statistical cut-offs, thus enhancing the disease network and our general understanding of the phenomenon.

Our future aim is to expand our current dataset, including data from more patients. Apart from that, we want explore additional methodologies for data integration. Furthermore, our goal is to integrate the molecular data with imaging data from dermoscopy, to improve feature selection and classification techniques, concerning melanoma.

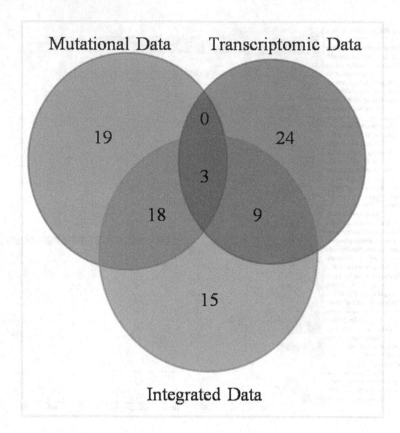

Fig. 4. Venn diagram for the significant pathway lists from the two distinct analyses, as well as their integration

Acknowledgements. This work has been supported by the 12CHN 204 Bilateral Greece-China Research Program of the Hellenic General Secretariat of Research and Technology and the Chinese Ministry of Research and Technology entitled "Personalization of melanoma therapeutic management through the fusion of systems biology and intelligent data mining methodologies-PROMISE," sponsored by the Program "Competitiveness and Entrepreneurship," Priority Health of the Peripheral Entrepreneurial Program of Attiki.

References

1. The Skin Cancer Foundation. http://www.skincancer.org/
2. Rossi, C.R., Foletto, M., Vecchiato, A., Alessio, S., Menin, N., Lise, M.: Management of cutaneous melanoma M0: state of the art and trends. Eur. J. Cancer **33**(14), 2302–2312 (1997)
3. Dummer, R., Hoek, K.: Human Melanoma: From Transcriptome to Tumor Biology, Forschungsdatenbank der Universität Zürich (2004–2008)

4. Winnepenninckx, V., Lazar, V., Michiels, S., Dessen, P., Stas, M., Alonso, S.R., Eggermont., A.M.: Gene expression profiling of primary cutaneous melanoma and clinical outcome. J. Natl. Cancer Inst. **98**(7), 472–482 (2006)
5. Raskin, L., Fullen, D.R., Giordano, T.J., Thomas, D.G., Frohm, M.L., Cha, K.B., Gruber, S.B.: Transcriptome profiling identifies HMGA2 as a biomarker of melanoma progression and prognosis. J. Invest. Dermatol. **133**(11), 2585–2592 (2013)
6. Dutton-Regester, K., Hayward, N.K.: Reviewing the somatic genetics of melanoma: from current to future analytical approaches. Pigm. Cell Melanoma Res. **25**(2), 144–154 (2012)
7. Maglogiannis, I., Goudas, T., Billiris, A., Karanikas, H., Valavanis, I., Papadodima, O., Kontogianni, G., Chatziioannou, A.: Redesigning EHRs and clinical decision support systems for the precision medicine era. In: Proceedings of the 16th International Conference on Engineering Applications of Neural Networks (INNS), p. 14. ACM (2015)
8. Li, H., Durbin, R.: Fast and accurate long-read alignment with Burrows-Wheeler transform. Bioinformatics, Epub (2010). [PMID: 20080505]
9. Picard tools. http://picard.sourceforge.net
10. McKenna, A.: The Genome analysis toolkit: a mapreduce framework for analyzing next-generation DNA sequencing data. Genome Res. **20**(9), 1297–1303 (2010)
11. Cibulskis, K., et al.: Sensitive detection of somatic point mutations in impure and heterogeneous cancer samples. Nat. Biotechnol. **31**(3), 213–219 (2013)
12. Ramos, A.H., Lichtenstein, L., Gupta, M., Lawrence, M.S., Pugh, T.J., Saksena, G., Getz, G.: Oncotator: cancer variant annotation tool. Hum. Mutat. **36**(4), E2423–E2429 (2015)
13. e-Nios BioInfoMiner tool. https://bioinfominer.com/#/welcome
14. Pilalis, E.D., Chatziioannou, A.A.: Prioritized functional analysis of biological experiments using resampling and noise control methodologies. In: IEEE 13th International Conference on Bioinformatics and Bioengineering (BIBE), pp. 1–3 (2013). doi:10.1109/BIBE.2013.6701558
15. Moutselos, K.: GOrevenge: a novel generic reverse engineering method for the identification of critical molecular players, through the use of ontologies. IEEE Trans. Biomed. Eng. **58**(12), 3522–3527 (2011)
16. Moutselos, K., Maglogiannis, I., Chatziioannou, A.: Integration of high-volume molecular and imaging data for composite biomarker discovery in the study of Melanoma. BioMed Res. Int. (2014)
17. Barrett, T., Troup, D.B., Wilhite, S.E., et al.: NCBI GEO: archive for functional genomics data sets - 10 years on. Nucleic Acids Res. **39**(Database issue), D1005–1010 (2011)
18. Talantov, D., Mazumder, A., Jack, X.Y., Briggs, T., Jiang, Y., Backus, J., Wang, Y.: Novel genes associated with malignant melanoma but not benign melanocytic lesions. Clin. Cancer Res. **11**(20), 7234–7242 (2005)
19. Martin-Belmonte, F., Perez-Moreno, M.: Epithelial cell polarity, stem cells and cancer. Nat. Rev. Cancer **12**, 23–38 (2012)
20. Jögi, A., Vaapil, M., Johansson, M., Påhlman, S.: Cancer cell differentiation heterogeneity and aggressive behavior in solid tumors. Upsala J. Med. Sci. **117**(2), 217–224 (2012)
21. Saladi, S.V., et al.: Modulation of extracellular matrix/adhesion molecule expression by BRG1 is associated with increased melanoma invasiveness. Mol. Cancer **22**(9), 280 (2010)
22. Hart, I.R., Birch, M., Marshall, J.F.: Cell adhesion receptor expression during melanoma progression and metastasis. Cancer Metastasis Rev. **10**(2), 115–128 (1991)
23. Machesky, L.M.: Lamellipodia and filopodia in metastasis and invasion. FEBS Lett. **582**(14), 2102–2111 (2008)
24. Kato, T., Kawai, K., Egami, Y., Kakehi, Y., Araki, N.: Rac1-dependent lamellipodial motility in prostate cancer PC-3 cells revealed by optogenetic control of Rac1 activity. PloS one **9**(5), e97749 (2014)

25. Pasquale, E.B.: Eph receptors and ephrins in cancer: bidirectional signalling and beyond. Nat. Rev. Cancer **10**(3), 165–180 (2010)
26. Ridley, A.J.: Rho proteins and cancer. Breast Cancer Res. Treat. **84**(1), 13–19 (2004)
27. Sander, E.E., Collard, J.G.: Rho-like GTPases: their role in epithelial cell–cell adhesion and invasion. Eur. J. Cancer **35**(9), 1302–1308 (1999)

Machine Learning Preprocessing Method for Suicide Prediction

Theodoros Iliou[1], Georgia Konstantopoulou[2], Mandani Ntekouli[3],
Dimitrios Lymberopoulos[3], Konstantinos Assimakopoulos[4],
Dimitrios Galiatsatos[1], and George Anastassopoulos[1(✉)]

[1] Medical Informatics Lab, Medical School,
Democritus University of Thrace, Komotini, Greece
anasta@med.duth.gr
[2] Special Office for Health Consulting Services,
University of Patras, Patras, Greece
[3] Wire Communications Lab, Department of Electrical Engineer,
University of Patras, Patras, Greece
[4] Department of Psychiatry, University of Patras, Patras, Greece

Abstract. The main objective of this study was to find a preprocessing method to enhance the effectiveness of the machine learning methods in datasets of mental patients. Specifically, the machine learning methods must have almost excellent classification results in patients with depression who have thoughts of suicide, in order to achieve the sooner the possible the appropriate treatment. In this paper, we establish a novel data preprocessing method for improving the prognosis' possibilities of a patient suffering from depression to be leaded to the suicide. For this reason, the effectiveness of many machine learning classification algorithms is measured, with and without the use of our suggested preprocessing method. The experimental results reveal that our novel proposed data preprocessing method markedly improved the overall performance on initial dataset comparing with PCA and Evolutionary search feature selection methods. So this preprocessing method can be used for significantly boost classification algorithms performance in similar datasets and can be used for suicide tendency prediction.

Keywords: Data preprocessing · Principal component analysis · Classification · Feature selection · Suicidal ideation · Depression · Mental illness

1 Introduction

Suicidal ideation is generally associated with depression and other mood disorders. However, it seems to have associations with many other psychiatric disorders, life events, and family events, all of which may increase the risk of suicidal ideation. For example, many people with borderline personality disorder exhibit recurrent suicidal behavior and suicidal ideation. One study found that 73 % of patients with borderline personality disorder have attempted suicide, with the average patient having 3 or 4 attempts.

© IFIP International Federation for Information Processing 2016
Published by Springer International Publishing Switzerland 2016. All Rights Reserved
L. Iliadis and I. Maglogiannis (Eds.): AIAI 2016, IFIP AICT 475, pp. 53–60, 2016.
DOI: 10.1007/978-3-319-44944-9_5

Early detection and treatment are the best ways to prevent suicidal ideation and suicide attempts. If signs, symptoms, or risk factors are detected early then the person will hopefully seek for treatment and help before attempting to take his/her own life. In a study of people who did commit suicide, 91 % of them likely suffered from one or more mental illnesses. Nevertheless, only 35 % of those people were treated or being treated for a mental illness. This emphasizes the importance of early detection; if a mental illness is detected, it can be treated and controlled to help prevent suicide attempts. Another study investigated strictly suicidal ideation in adolescents. This study found that depression symptoms in adolescents as early as of ninth (9) grade (14–15 years old) is a predictor of suicidal ideation.

2   Suicide - Suicidal Ideation

Suicide is a prevalent problem that concerns all countries in the world. However, it is rarely discussed both in the media and in everyday conversations. Many times when people make thoughts regarding one self-destructive behavior, they are attributed to the term "suicidal ideation". The suicidal ideation in some people may persist for years, and in others it may be occasional and caused by difficult events happened in their life. Suicidal thoughts that a person makes may neither be clear nor defined nor involve a very well organized suicide plan. The more persistent and intense these thoughts are, the more serious is the suicidal ideation. People who have attempted suicide even once in their life, are much more likely to try again, especially within the first year of the attempt.

The majority of people who attempt suicide show some samples of their purposes before proceeding to act. Symptoms of suicidal ideation are immediately visible, especially from those of their close environment. The sense of despair–which can be expressed through phrases like "nothing is going to change and get better"–, the feeling of helplessness, the belief that their suicide constitutes no obstacle to family life and friends, alcohol and substance abuse, the preparation of a note for their imminent suicide and tendency to accidents, such as the intentional carelessness in dangerous situations are evidence that if is perceived early can prevent these people from a possible suicide attempt.

One in two people who commit suicide had a history of depression. The rates of suicide in depressed patients are higher than in patients with other diagnosed disorders [1–3] and even higher in patients with severe depression. Undeniable is the fact that people with the disorder of depression consider themselves, the world and the future in a negative way. This indicates the relationship between suicide and the feeling of despair that they have. They believe that there are few or no alternatives for that in his life. Thus, an evaluation of depressed people should include control of suicidal behavior. The purpose of the clinical therapists is to estimate the possibility of suicidal episode, so that can be properly avoided.

3 What is Depression?

Depression is a disorder that affects mood, thoughts and is usually accompanied by physical discomfort. It affects the eating habits of the patient, his/hers sleep, the way he/she sees himself/herself and how he/she thinks and perceives the world. When someone is diagnosed with depression, he/she often describes himself/herself as a sad, desperate, discouraged and disappointed person.

However, every day we use the term depression meaning a state of unhappiness and misery, that is most of the time transient, has less intensity and probably is caused by something relatively insignificant. This "everyday" meaning of depression differs from the depression as a disorder which is characterized by symptoms that last more than two weeks and are severe enough to interfere with daily life of a person and leads him/her to functional impairment in many aspects of it. In psychiatry, the term depression can also be referred as a mental illness, even when their symptoms do not have reached a high level of severity to obtain such a diagnosis.

For example, people that experience this kind of pessimistic, and intensively sad feelings, do not even have the strength to get up in the morning and do the basic things for surviving, like eating or sleeping. So some people with depression sleep too many hours some cannot sleep at all, while others do very irregular sleep, or they wake frequently during the night or difficulty falling asleep. The most common sleep disorder is the morning awakening, in which the person wakes up very early in the morning and cannot go back to sleep. To be more specific, someone might experience when he/she has depression has depressed mood lasting most of the day and nearly every day, for a period of two weeks, loss of pleasure and reduced interest in activities that were previously the person wanted, and he liked to do.

Helplessness, pessimism, lack of hope and concern about the future are symptoms to be depressed. The person sees everything black and believes that this will remain. Difficulty in concentrating, thinking, memory and making decisions. To have feelings and thoughts of guilt, worthlessness and low self-esteem.

Sometimes the person with depression feels so desperate that commits suicide. The suicide attempt is the most serious and dangerous complication of depression. In people with severe depression, suicide risk is particularly high.

4 Data Collection

In this paper we establish a mechanism for detecting the possibilities of a patient suffering from depression to be leaded to the suicide. For this reason we measure, using real world statistical data, the effectiveness of all the above symptoms in each case. This cohort is the same one used in previous study [4] and concerns 91 patients who had come to the Special Office for Health Consulting Services University of Patras were diagnosed with different types of depression [5]. Patients were falling in one of the below categories: Major Depressive Disorder, Persistent depressive disorder (Dysthymia), Bipolar Disorders (I & II), Cyclothymic disorder, and Depressive disorder not otherwise specified (DD-NOS). Our study group included both sexes, age 18–30 and their files contained history of the last 5 years.

A key element for the validity of the disorder decision method is the confirmation of the existence of each symptom, based on interviews that were done. We examined the "symptoms" and not the "points" that had the patient. Symptoms are determined by himself/herself, while the points are independent observations people make the environment and the specialist. For example, the crying may be a point and insomnia a symptom.

With the method of interviews, we recorded the symptoms and the time period that these symptoms occur (e.g., depressed mood over two weeks, or sleep disturbances over two years) in ninety-one (91) patients who were diagnosed with a mood disorder. Then, depending on the symptoms and the time they were repeated we characterized the type of disorder (e.g. Persistent Depressive Disorder–Dysthymia).

In order to achieve our goal, we analyzed all incidents concerning emotional symptomatology and more specifically, concerning about the symptoms that are associated with mood changes.

5 Description of Machine Learning Methods

5.1 Data Pre-processing Methods

Data pre-processing is an important step in the data mining process since analysis of data that has not been carefully examined can produce misleading results. To this end, the representation and quality of data should first be ensured prior the execution of the experiments. Preprocessing tasks include data cleaning (e.g. identification or outliers' removal), data integration, data transformation (i.e. new feature generation) and data reduction. The product of a data pre-processing task is a new training set that would eventually improve the classification performance and reduce the classification time. This is due to the fact that the dimensionality of the data is reduced, which allows learning algorithms to operate faster and more effectively. In some cases, accuracy on future classification can be improved; in others, the result is a more compact, easily interpreted representation of the target concept [6].

In this paper, we used Principal Component Analysis (PCA) and a novel machine learning data preprocessing method that we have proposed in [7] in order to compare our suggested method performance with PCA.

5.1.1 Feature Selection

In order to identify if feature (attribute) selection provides better results in our problem and optimize the classification time and performance, a feature selection evaluator, the CfsSubsetEval attribute evaluator was used. Feature selection is the process of selecting a subset of relevant features for use in model construction. The central assumption when using a feature selection technique is that the data contains many redundant or irrelevant features. For the selection of the method, the WEKA 3.8 data mining software was used [8]. WEKA offers many feature selection and feature ranking methods, where each method is a combination of feature search and evaluator of currently selected features. Several combinations have been tested in order to assess the feature selection combination that gives the optimum performance for our problem.

The feature evaluator and search method (offered in WEKA) that presented the best performance in the data set were (i) Correlation-based Feature Selection Sub Set Evaluator and (ii) Evolutionary Search method.

The Correlation-based Feature Selection Sub Set Evaluator (CfsSUbsetEval) evaluates the worth of a subset of attributes by considering the individual predictive ability of each feature along with the degree of redundancy between them. Subsets of features that are highly correlated with the class while having low intercorrelation are preferred. On the other hand, Evolutionary Search explores the attribute space using an Evolutionary Algorithm (EA). The EA is a (mu, lambda) one with the following operators: uniform random initialization, binary tournament selection, single point crossover, bit flip mutation and generational replacement with elitism (i.e., the best individual is always kept). The combination of the above mentioned methods proposed four from all of the features that formed originally the feature set. These features are: (i) difficulties in functioning, (ii) unworthiness/guilt, (iii) Major Depressive Disorder and (iv) Depressive disorder not otherwise specified (DD-NOS).

5.2   Short Description of Suggested Data Pre-processing Method

The proposed method can substantially improve successful classification when applying machine learning techniques to data mining problems. It transforms the input data into a new form of data, which is more suitable and effective for the learning scheme chosen. Below follows the detailed description of the method.

Step 1. Let's assume that a dataset of a machine learning problem named dataset1 is chosen, with n instances (rows), k variables (columns) and m classes.

The differences between adjacent elements of every instance of dataset1 are calculated, and the new $k-1$ variables are added in dataset1, creating a new dataset named dataset2 with $k + (k-1)$ variables.

Step 2. Assuming that the set of attributes for every instance is a vector whose elements are the coefficients of a polynomial in descending power, step 2 estimates the derivative of the vector. The result is a new vector (one element shorter than initial one), with the coefficients of the derivative in descending power. Then, this new vector is added in dataset2 forming a new dataset named dataset3.

Step 3. In the third step of the proposed method, a new set (called from now-on Basic-Set) is created randomly selecting 10 % of data from dataset3, consisting of d instances and m classes. The remaining 90 % of dataset3 is called Rest-Set. Then, matrix right division (or slash division) of every Basic Set instance (row) with the remaining rows of the Basic Set is computed (Slash or matrix right division B/A is roughly the same as B*inv(A), more precisely, $B/A = (A'\backslash B')'$). Then, follows the calculation of mean and median values of the division result for every instance of each class with the rest instances of its class (variables $Mean\_class_m\_row_x$ and $Median\_class_m\_row_x$ respectively), producing totally $m + m = 2m$ new variables ($Total\_Mean_1$, $Total\_Mean_2$,..., $Total\_Mean_m$ and $Total\_Median_1$, $Total\_Median_2$, ..., $Total\_Median_m$ for every row of the Basic Set. Hence, we have d values for $Total\_Mean_1$, d values for $Total\_Mean_2$, ..., d values for $Total\_Mean_m$ and d values for $Total\_Median_1$, d values for $Total\_Median_2$, ..., d values for $Total\_Median_m$.

Apart from the above, the Total_Mean and Total_Median values are calculated as shown in Eqs. (1) and (2) respectively (m is the name of the class and d is the sum of Basic Set rows). Finally, m total_Mean and m total_MEDIAN values result, one for every class of the Basic set.

$$\text{Total\_Mean}_m = \frac{(\text{Mean\_class}_m\_\text{row}_1 + \text{Mean\_class}_m\_\text{row}_2 + \ldots + \text{Mean\_class}_m\_\text{row}_d)}{d} \quad (1)$$

$$\text{Total\_Median}_m = \frac{(\text{Median\_class}_m\_\text{row}_1 + \text{Median\_class}_m\_\text{row}_2 + \ldots + \text{Median\_class}_m\_\text{row}_d)}{d} \quad (2)$$

Step 4. Assuming that Rest-Set from step 3 has r instances (rows) and m classes, a similar to step 3 approach follows. Specifically, matrix right division of every single Rest-Set row with every single row of the Basic Set is performed. Then, the mean and median values of the division result of every row for each class are calculated ($\text{RS\_Mean\_class}_m\_\text{row}_j$ and $\text{RS\_Median\_class}_m\_\text{row}_j$ respectively), producing new $m + m = 2m$ variables for every row of the Rest Set. As a result, we have r values for $\text{RS\_Mean\_class}_m\_\text{row}_j$, and r values for $\text{RS\_Median\_class}_m\_\text{row}_j$.

Similarly to step 3, we compute mean and medial values ($\text{RS\_Mean\_class}_m\_\text{row}_x$ and RS_Median_class_m respectively) for every class.

Apart from the above, the $\text{Final\_Mean}_m\_\text{row}_j$ and $\text{Final\_Median}_m\_\text{row}_j$ values are also calculated as shown in Eqs. (3) and (4) respectively (m is the name of the class and j (from 1 to r) is the row of the Rest set.

$$\text{Final\_Mean}_m\_\text{row}_j = \text{total\_Mean}_{m\ (\text{step3})} - \text{RS\_Mean\_Class}_m\_\text{row}_j \quad (3)$$

$$\text{Final\_Median}_m\_\text{row}_j = \text{total\_Median}_{m\ (\text{step3})} - \text{RS\_Median\_Class}_m\_\text{row}_j \quad (4)$$

Finally, m $\text{Final\_Mean}_m\_\text{row}_j$ and $\text{Final\_Median}_m\_\text{row}_j$ values result, one for every class m and every row j of the Rest set.

Step 5. The rows (variables) RS_Mean_classm_rowj, RS_Median_classm_rowj, Final_Meanm_rowj and Final_Medianm_rowj for every class are selected from previous step and then are placed in a new table [7].

The method ends with the transposition of the Table we described in previous step and the final dataset is now ready to be forwarded in any classification schema. Concluding the description of the proposed method, it is evident that the final dataset consist of 4 variables, namely $\text{RS\_Mean\_class}_m\_\text{row}_j$, $\text{RS\_Median\_class}_m\_\text{row}_j$, $\text{Final\_Mean}_m\_\text{row}_j$ and $\text{Final\_Median}_m\_\text{row}_j$ for every class of the initial dataset. Thus, if the original dataset has m classes, the final dataset will have $4 * m$ variables.

6   Experimental Results

For our experiments we used the dataset described in Sect. 4. In order to categorize subjects into two classes (suicide tendency, no-suicide tendency), several machine learning classification algorithms were tested in this paper, selected based on their popularity and frequency in biomedical engineering problems. Each classifier was tested with initial dataset, with dataset after feature selection with Evolutionary search, with transformed dataset using PCA method and finally with the transformed dataset using our suggested data preprocessing method (Table 1). In order to better investigate the generalization of the prediction models produced by the machine learning algorithms, the repeated 10-fold cross validation method was used. We used the WEKA default parameters for the classifiers that we have used. For the classifiers with the best performance (MLP) the parameters are: Hidden layers = 8, learning rate = 0.3, momentum = 0.2, training time = 500 epochs.

Table 1. Classification results

Classifiers	Initial data set (%)	Evolutionary search (%)	PCA (%)	Suggested method (%)
MLP (Multilayer Perceptron)	64.83	75.82	64.83	92.18
MultilayerPerceptronCS	64.83	75.92	64.83	92.18
Radial Basis Function Classifier	75.82	74.72	78.02	85.93
RBFNetwork	70.32	75.82	76.92	84.37
FURIA (Fuzzy Logic algorithm)	71.42	75.82	76.92	82.81
SMO (Support Vector Machines)	64.83	76.92	75.62	76.56
HMM (Hidden Markov Models)	76.92	76.92	76.53	76.56
J48-Graft	64.83	73.62	64.83	82.81
Random Forest	73.62	74.72	76.84	85.93
IB1	54.94	76.92	67.03	93.75

In Table 1, we can observe that HMM classification results have not increased with any of the preprocessing or attribute selection methods we have used. Using Evolutionary search, classification results was increased almost in all classification algorithms except RBF classifier and HMM. Using PCA method, classification results were increased in most of the classification algorithms as well. Our suggested data preprocessing method significantly increased the classification performance (93.75 % with IB1 algorithm and 92.18 % with MLP) and achieved the best classification results comparing with all the other methods we have used.

7   Conclusions

Data pre-processing is an important step in the data mining process. If there is much irrelevant and redundant information present or noisy and unreliable data, then knowledge discovery during the training phase is more difficult. Data preparation and filtering steps can take considerable amount of processing time. The experimental results reveal that our novel proposed data preprocessing method markedly improved the overall performance in initial dataset comparing with PCA and Evolutionary search feature selection method. In our point of view, our suggested method can be used to significantly boost classification algorithms performance in similar datasets and can be used for suicide tendency prediction.

In future work, it would be preferable to make the same experiments in similar datasets consisting of more records, using different classifiers and different feature selection and data preprocessing methods. In addition, our proposed data preprocessing method could be modified so as to achieve better classification performance.

References

1. Miles, C.P.: Conditions predisposing to suicide: a review. J. Nerv. Ment. Dis. **164**, 231–246 (1977)
2. Angst, J., Angst, F., Stassen, H.H.: Suicide risk in patients with major depressive disorder. J. Clin. Psychiatry **60**, 57–116 (1999)
3. American Psychiatric Association (APA): Practice Guidelines for the treatment of Patients with Major Depressive Disorder. (3rd edn.) (2010)
4. Galiatsatos, D., Konstantopoulou, G., Anastassopoulos, G., Nerantzaki, M., Assimakopoulos, K., Lymberopoulos, D.: Classification of the most significant psychological symptoms in mental patients with depression using bayesian network. In: Proceeding of the 16th International Conference on Engineering Applications of Neural Networks (EANN 2015), 25–28 September 2015
5. Diagnostic and Statistical Manual of Mental Disorders DSM-V, pp. 123–154. American Psychiatric Publishing, Washington DC, 5th edn. (2013)
6. Kotsiantis, S.B., Kanellopoulos, D., Pintelas, P.E.: Data preprocessing for supervised leaning. World Acad. Sci. Eng. Technol. **1**, 856–861 (2007)
7. Iliou, T., Anagnostopoulos, C.N., Nerantzaki, M., Anastassopoulos, G.: A novel machine learning data preprocessing method for enhancing classification algorithms performance. In: Proceeding of the 16th International Conference on Engineering Applications of Neural Networks (EANN 2015), 25-28 September 2015
8. Waikato Environment for Knowledge Analysis, Data Mining Software in Java. http://www.cs.waikato.ac.nz/ml/index.html. Accessed 1 May 2016

Classification – Pattern Recognition (CLASPR)

Using Frequent Fixed or Variable-Length POS Ngrams or Skip-Grams for Blog Authorship Attribution

Yao Jean Marc Pokou[1], Philippe Fournier-Viger[1,2], and Chadia Moghrabi[1(✉)]

[1] Department of Computer Science, Université de Moncton, Moncton, NB, Canada
{eyp3705,philfv,chadia.moghrabi}@umoncton.ca
[2] School of Natural Sciences and Humanities,
Harbin Institute of Technology Shenzhen Graduate School,
Shenzhen, Guangdong, China

Abstract. Authorship attribution is the process of identifying the author of an unknown text from a finite set of known candidates. In recent years, it has become increasingly relevant in social networks, blogs, emails and forums where anonymous posts, bullying, and even threats are sometimes perpetrated. State-of-the-art systems for authorship attribution often combine a wide range of features to achieve high accuracy. Although many features have been proposed, it remains an important challenge to find new features and methods that can characterize each author and that can be used on non formal or short writings like blog content or emails. In this paper, we present a novel method for authorship attribution using frequent fixed or variable-length part-of-speech patterns (ngrams or skip-grams) as features to represent each author's style. This method allows the system to automatically choose its most appropriate features as those sequences being used most frequently. An experimental evaluation on a collection of blog posts shows that the proposed approach is effective at discriminating between blog authors.

Keywords: Authorship attribution · Part-of-speech patterns · Top-k POS sequences · Frequent POS patterns · Skip-grams · Ngrams · Blogs

1 Introduction

Authorship analysis is the process of examining the characteristics of a piece of work to identify its authors [1]. The forerunners of authorship attribution (AA) are Medenhall who studied the plays of Shakespeare in 1887 [2], and Mosteller and Wallace who studied the disputed authorship of the Federalist Papers in 1964 [3]. Beside identifying the authors of works published anonymously, popular applications of AA techniques include authenticating documents, detecting plagiarism, and assisting forensic investigations.

© IFIP International Federation for Information Processing 2016
Published by Springer International Publishing Switzerland 2016. All Rights Reserved
L. Iliadis and I. Maglogiannis (Eds.): AIAI 2016, IFIP AICT 475, pp. 63–74, 2016.
DOI: 10.1007/978-3-319-44944-9_6

In recent years, an emerging application of AA has been to analyze online texts, due to the increasing popularity of Internet-based communications, and the need to find solutions to problems such as online bullying, and anonymous threats. However, this application raises novel challenges, as online texts are written in an informal style, and are often short, thus providing less information about their authors. The accuracy achieved with well-established features for AA like function words, syntactic and lexical markers suffers when applied to short-form messages [4], as many of the earliest studies have been tested on long formal text such as books. It is thus a challenge to find new features and methods that are applicable to informal or short texts.

Studies on AA focused on various types of features that are either syntactic or semantic [5]. Patterns based on syntactic features of text are considered reliable because they are unconsciously used by authors. Syntactic features include frequencies of ngrams, character ngrams, and function words [1]. Baayen, van Halteren, and Tweedie rewrote the frequencies rules for AA based on two syntactically annotated samples taken from the Nijmegen corpus [6]. Semantic features take advantage of words' meanings and their likeness, as exemplified by Clark and Hannon whose classifier quantifies how often an author uses a specific word instead of its synonyms [7]. Moreover, state-of-the-art systems for AA often combine a wide range of features to achieve higher accuracy. For example, the JStylo system offers more than 50 configurable features [8].

Syntactic markers, however, have been less studied because of their language-dependent aspect. This paper studies complex linguistic information carried by part-of-speech (POS) *patterns* as a novel approach for authorship attribution of informal texts. The hypothesis is that POS patterns more frequently appearing in texts, could accurately characterize authors' styles.

The contributions of this paper are threefold. First, we define a novel feature for identifying authors based on fixed or variable length POS patterns (ngrams or skip-grams). A signature is defined for each author as the intersection of the top k most frequent POS patterns found in his or her texts, that are less frequent in texts by other authors. In this method, the system automatically chooses its most appropriate features rather than have them predefined in advance. Second, a process is proposed to use these signatures to perform AA. Third, an experimental evaluation using blog posts of 10 authors randomly chosen from the *Blog Authorship Corpus* [9] shows that the proposed approach is effective at inferring authors of informal texts. Moreover, the experimental study provides answers to the questions of how many patterns are needed, whether POS skip-grams or ngrams perform better, and whether fixed length or variable-length patterns should be used.

The rest of this paper is organized as follows. Section 2 reviews related work. Section 3 introduces the proposed approach. Section 4 presents the experimental evaluation. Finally, Sect. 5 draws the conclusions.

2   Related Work

The use of authorship attribution techniques goes back to the 19th century with the studies of Shakespeare's work. This section concentrates on those using POS tagging or informal short texts.

Stamatatos et al. exclusively used natural language processing (NLP) tools for their three-level text analysis. Along with the token and analysis levels, the phrase level concentrated on the frequencies of single POS tags. They analyzed a corpus of 300 texts by 10 authors, written in Greek. Their method achieved around 80 % accuracy by excluding some less significant markers [10].

Gamon combined syntactic and lexical features to accurately identify authors [11]. He used features such as the frequencies of ngrams, function words, and POS trigrams from a set of eight POS tags obtained using the NLPWin system. Over 95 % accuracy was obtained. However, its limitation was its evaluation using only three texts, written by three authors [11].

More recently, Sidorov et al. introduced syntactic ngrams (sngrams) as a feature for AA. Syntactic ngrams are obtained by considering the order of elements in syntactic trees generated from a text, rather than by finding n contiguous elements appearing in a text. They compared the use of sngrams with ngrams of words, POS, and characters. The 39 documents by three authors from Project Gutenberg were classified using SVM, J48 and Naïve Bayes implementations provided by the WEKA library. The best results were obtained by SVM with sngrams [12]. The limitations of this work are an evaluation with only three authors, a predetermined length of ngrams, and that all 11,000 ngrams/sngrams were used.

Variations of ngrams were also considered. For example, skip-grams were used by García-Hernández et al. [13]. Skip-grams are ngrams where some words are ignored in sentences with respect to a threshold named the skip step. Ngrams are the specific case of skip-grams where the skip step is equal to 0. The number of skip-grams is very large and their discovery needs complex algorithms [12]. To reduce their number, a cut-off frequency threshold is used [13].

POS ngrams have also been used for problems related to AA. Litvinova et al. used the frequencies of 227 POS bigrams for predicting personality [14].

Unlike previous work using POS patterns (ngrams or skip-grams), the approach presented here considers not only fixed length POS patterns but also variable length POS patterns. Another distinctive characteristic is that it finds only the k most frequent POS patterns in each text (where k is set by the user), rather than using a large set of patterns or using a predetermined cut-off frequency threshold. This allows the proposed approach to use a very small number of patterns to create a signature for each author, unlike many previous works that compute the frequencies of hundreds or thousands of ngrams or skip-grams.

3   The Proposed Approach

The proposed approach takes as input a training corpus C_m of texts written by m authors. Let $A = \{a_1, a_2, ...a_m\}$ denote the set of authors. Each author a_i

$(1 \leq i \leq m)$ has a set of z texts $T_i = \{t_1, t_2, \ldots t_z\}$ in the corpus. The proposed approach is applied in three steps.

Preprocessing. The first step is to prepare blogs from the corpus so that they can be used for generating author signatures. All information that does not carry an author's style is removed, such as images, tables, videos, and links. In addition, each text is stripped of punctuation and is split into sentences using the Rita NLP library (http://www.rednoise.org/rita/). Then, every text is tagged using the Standford NLP Tagger (http://nlp.stanford.edu/software/) as it produced 97.24 % accuracy on the Penn Treebank Wall Street Journal corpus [15]. Since the main focus is analyzing how sentences are *constructed* by authors rather than the choice of words, words in texts are discarded and only the *patterns* about parts of speech are maintained. Thus, each text becomes a set of sequences of POS tags. For example, Table 1 shows the transformation of six sentences from an author in the corpus.

Table 1. An example of blog text transformation.

#	Original sentence	Transformed sentence into POS sequences
1	Live From Ohio Its Youth Night	JJ IN NNP PRP$ NNP NNP
2	So the Youth Rally was tonight	IN **DT** NNP NNP VBD **NN**
3	I think it was a success	PRP VBP PRP VBD **DT NN**
4	Maybe not	RB RB
5	The activity looked complicated and confusing but luckily Tim and I didn't have to do it instead we got to (...) right	**DT NN** VBD JJ CC JJ CC RB <u>NNP</u> CC PRP <u>VBP</u> <u>VBP</u> TO VB <u>PRP</u> <u>RB</u> PRP VBD TO (...) JJ
6	And finally the skit	CC RB **DT NN**

Signature Extraction. The second step of the proposed approach is to extract a signature for each author, defined as a set of part-of-speech patterns (POSP) annotated with their respective frequency. Signature extraction has four parameters: the number of POS patterns (ngrams or skip-grams) to be found k, the minimum pattern length n, the maximum length x, and the maximum gap *maxgap* allowed between POS tags. Frequent patterns of POS tags are extracted from each text. The hypothesis is that each text may contain patterns of POS tags unconsciously left by its author, representing his/her writing style, and could be used to identify that author accurately. For each text t, the k most frequent POS patterns are extracted using a general-purpose sequential pattern mining algorithm [16]. Let *POS* denote the set of POS tags. Consider a sentence $w_1, w_2, \ldots w_y$ consisting of y part-of-speech tags, and a parameter *maxgap* (a positive integer). A n-skip-gram is an ordered list of tags $w_{i_1}, w_{i_2}, \ldots w_{i_n}$ where $i_1, i_2, \ldots i_n$ are integers such that $0 < i_j - i_{j-1} \leq maxgap + 1 (1 < j \leq n)$.

Table 2. Part-of-speech patterns and their relative frequencies.

Pattern	Relative frequency (%)	Part-of-speech description
NN	66.6	Noun, singular or mass
DT-NN	66.6	Determiner - Noun
NNP	50.0	Proper noun, singular
VBP-VBP-PRP	16.6	Verb, non-3rd person singular-verb, non-3rd person singular - Personal pronoun
NNP-VBP-VBP-PRP-RB	16.6	Proper noun, singular - Verb, non-3rd person singular-verb - Verb, non-3rd person singular-verb -Personal pronoun- Adverb

A n-skip-gram respecting the constraint of $maxgap = 0$ (i.e. no gaps are allowed) is said to be a ngram. For a given text, the *frequency* of a sequence *seq* is the number of sequences (sentences) from the text containing *seq*. Similarly, the *relative frequency* of a sequence is its frequency divided by the number of sequences in the text. For example, the frequency of the 5-skip-gram $\langle NNP, VBP, VBP, PRP, RB \rangle$ is 1 in the transformed sentences of Table 1, while the frequency of the 2-skip-gram $\langle DT, NN \rangle$ is 4 (this pattern appears in the second, third, fifth, and sixth transformed sentences).

In the following, the term *part-of-speech pattern* of a text t, abbreviated as $(POSPt)_{n,x}^{k}$, or *patterns*, is used to refer to the k most frequent POS patterns extracted from a text, annotated with their relative frequency, respecting the *maxgap* constraint, and having a length no less than n and not greater than x.

Then, the *POS patterns of an author* a_i in all his/her texts are found. They are denoted as $(POSPa_i)_{n,x}^{k}$ and defined formally as the union of the POS patterns found in all of his/her texts: $(POSPa_i)_{n,x}^{k} = \bigcup_{t \in T_i} (POSPt)_{n,x}^{k}$.

Then, the signature of each author a_i is extracted. The signature s_{a_i} of author a_i is the intersection[1] of the POS patterns of his/her texts T_i. The signature is formally defined as: $(s_{a_i})_{n,x}^{k} = \bigcap_{t \in T_i} (POSPt)_{n,x}^{k}$.

For instance, the part-of-speech patterns $(POSP_{1029959})_{1,5}^{4}$ of the blogger 1029959 from our corpus for $maxgap = 3$ are shown in Table 2. In this table, POS patterns are ordered by decreasing frequency. It can be seen that the patterns Noun (NN), and Determiner - Noun (DT-NN) appear in four of the six sentences shown in Table 1. Note that the relative frequency of each pattern is calculated as the relative frequency over all texts containing the pattern.

Moreover, the POS patterns of an author a_i may contain patterns having unusual frequencies that truly characterize the author's style, but also patterns representing common sentence structures of the English language. To tell apart these two cases, a set of reference patterns and their frequencies is extracted

---

[1] A less strict intersection could also be used, requiring occurrences in some or the majority of texts rather than all of them.

to be used with each signature for authorship attribution. Extracting this set of reference patterns is done with respect to each author a_i by computing the union of all parts of speech of the other authors[2]. This set is formally defined as follows. The *Common POS patterns of all authors excluding an author* a_i is the union of all the other $POSPa$, that is: $(CPOSa_i)_{n,x}^k = \bigcup_{a \in A \wedge a \neq a_i}(POSPa)_{n,x}^k$.

For example, the common POS patterns computed using authors 206953 and 2369365, and excluding author 1029959, are the patterns DT, IN and PRP. The relative frequencies (%) of these patterns for author 206953 are 67.9 %, 69.6 % and 79.1 %, while the relatives frequencies of these patterns for author 2369365 are 64.2 %, 63.0 % and 58.7 %. Note that the relative frequency of each pattern in CPOS is calculated as the relative frequency over all texts containing the pattern.

The revised signature of an author a_i after removing the common POS patterns of all authors excluding a_i is defined as: $(s'_{ai})_{n,x}^k = (s_{ai})_{n,x}^k \setminus (CPOSa_i)_{n,x}^k$. When the revised signature of each author $a_1, a_2, ...a_m$ has been extracted, the collection of revised signatures $s_{n,x}^{\prime,k} = \{(s'_{a1})_{n,x}^k, (s'_{a2})_{n,x}^k, \ldots, (s'_{am})_{n,x}^k\}$ is saved.

Authorship Attribution. The third step of the proposed approach is to use the generated signatures to perform authorship attribution, that is to identify the author a_u of an anonymous text t_u that was not used for training. The algorithm takes as input an anonymous text t_u, the sets of signatures $s_{n,x}^{\prime,k}$ and the parameters n, x and k. The algorithm first extracts the part-of-speech patterns in the unknown text t_u with their relative frequencies. Then, it compares the patterns found in t_u and their frequencies with the patterns in the signature of each author using a similarity function. Each author and the corresponding similarity are stored as a tuple in a list. Finally, the algorithm returns this list sorted by decreasing order of similarity. This list represents a ranking of the most likely authors of the anonymous text t_u. Various metrics may be used to define similarity functions. In this work, the Pearson correlation was chosen as it provided better results in initial experiments.

4   Experimental Evaluation

Experiments were conducted to assess the effectiveness of the proposed approach for authorship attribution using either fixed or variable-length ngrams or skip-grams of POS patterns. Our Corpus consists of 609 posts from 10 bloggers, obtained from the *Blog Authorship Corpus* [9]. Blog posts are written in English and were originally collected from the *Blogger* website (https://www.blogger.com/). The ten authors were chosen randomly. Note that non-verbal expressions (emoticons, smileys or interjections used in web-blogs or chatrooms such as *lol, hihi, and hahaaa* were not removed because consistent part-of-speech tags were returned by the tagger for the different blogs. The resulting corpus has a total of $265, 263$ words and $19, 938$ sentences. Details are presented in Table 3.

---

[2] A subset of all other authors can also be used if the set of other authors is large.

Table 3. Corpus statistics.

Author id	Post count	Word count	Sentence count
1029959	83	29,799	2,314
2069530	125	40,254	2,409
2369365	119	24,148	1,692
3182387	87	30,375	1,752
3298664	34	26,052	2,417
3420481	63	19,063	1,900
3454871	37	16,322	1,722
3520038	45	21,312	1,698
3535101	1	24,401	1,865
3701154	15	33,537	2,169
Total	609	265,263	19,938

Then, each text was preprocessed (as explained in Sect. 3). Eighty percent (80 %) of each text was used to extract each author's signature (training), and the remaining 20 % was used to perform the unknown authorship attribution by comparing each text t_u with each author signature (testing). This produced a ranking of the most likely author to the least likely author, for each text.

4.1 Influence of Parameters n, x, k, and $maxgap$ on Overall Results

Recall that our proposed approach takes four parameters as input, i.e. the minimum and maximum length of part-of-speech patterns n and x, the maximum gap $maxgap$, and k the number of patterns to be extracted in each text. The influence of these parameters on authorship attribution success was first evaluated. For our experiment, parameter k was set to 50, 100, and 250. For each value of k, the length of the part-of-speech patterns was varied from $n = 1$ to $x = 4$. Moreover, the $maxgap$ parameter was set to 0 (ngrams), and from 1 to 3 (skip-grams). For each combination of parameters, we measured the *success rate*, defined as the number of correct predictions divided by the number of predictions. Tables 4, 5, 6, and 7 respectively show the results obtained for $maxgap = 0, 1, 2, 3$, for various values of k, n and x. Furthermore, in these tables, results are also presented by ranks. The row R_z indicates the number of texts where the author was predicted as one of the z most likely authors, divided by the total number of texts (success rate). For example, R_3 indicates the percentage of texts where the author is among the three most likely authors as predicted by the proposed approach. Since there are 10 authors in the corpus, results are shown for R_z varied from 1 to 10.

The first observation is that the best overall results are obtained by setting $n = 2, x = 2$, $k = 250$ and $maxgap = 0$. For these parameters, the author of an anonymous text is correctly identified 73.3 % of the time, and 86.6 % as one of the two most likely authors (R_2).

The second observation is that excellent results can be achieved using few patterns to build signatures ($k = 250$). Note that our approach was also tested with other values of k such as 50, but it did not provide better results than $k = 250$ (results are not shown due to space limitation). This is interesting as it means that signatures can be extracted using a very small number of the k most frequent POS patterns (as low as $k = 100$) and still characterize well the writing style of authors. This is in contrast with previous works that generally used a large number of patterns to define an author's signature. For example, Argamon et al. have computed the frequencies of 685 trigrams [17] and Sidorov et al. computed the frequencies of 400/11,000 ngrams/sngrams [12]. By Occam's Razor, it can be argued that models with less patterns (simpler) may be less prone to overfitting.

Third, it can be observed that good results can also be obtained using POS skip-grams. This is interesting since skip-grams of words or POS have received considerably less attention than ngrams in previous studies. Moreover, to our knowledge no studies had previously compared the results obtained with ngrams and skip-grams for authorship attribution of informal and short texts. Fourth, it is found that using skip-grams of fixed length (bigrams) is better than using patterns of variable length. This provides an answer to the important question of whether fixed length POS patterns or variable-length POS patterns should be used. This question was not studied in previous works.

4.2 Influence of Parameters n, x and k on Authorship Attribution for Each Author

The previous subsection studied the influence of parameters n, x and k on the ranking of authors for all anonymous texts. This subsection analyzes the results

Table 4. Overall classification results using ngrams (maxgap = 0)

(a) Top-k, for k=100.

n, x	1,2	1,3	1,4	2,2	3,3
R_1	56.7	53.3	56.7	**63.3**	40.0
R_2	73.4	60.0	63.4	**76.6**	70.0
R_3	80.1	73.3	73.4	76.6	73.3
R_4	83.4	83.3	83.4	79.9	80.0
R_5	86.7	90.0	90.1	79.9	86.7
R_6	90.0	90.0	90.1	86.6	90.0
R_7	90.0	90.0	90.1	86.6	90.0
R_8	96.7	96.7	96.8	89.9	96.7
R_9	96.7	96.7	96.8	96.6	100.0
R_{10}	100.0	100.0	100.0	100.0	100.0

Success rate in %

(b) Top-k, for k=250.

n, x	1,2	1,3	1,4	2,2	3,3
R_1	60.0	60.0	56.7	**73.3**	46.7
R_2	73.3	76.7	73.4	**86.6**	70.0
R_3	83.3	86.7	86.7	86.6	83.3
R_4	86.6	86.7	90.0	86.6	86.6
R_5	86.6	93.4	90.0	89.9	89.9
R_6	93.3	93.4	93.3	89.9	93.2
R_7	93.3	93.4	96.6	89.9	96.5
R_8	96.6	96.7	96.6	89.9	99.8
R_9	96.6	96.7	96.6	96.6	99.8
R_{10}	100.0	100.0	100.0	100.0	100.0

Success rate in %

Table 5. Overall classification results using skip-grams with maxgap = 1

(a) Top-k, for k=100.

	Success rate in %				
n, x	1, 2	1, 3	1, 4	2, 2	3, 3
R_1	56.7	50.0	50.0	**66.7**	50.0
R_2	70.0	60.0	60.0	**80.0**	73.3
R_3	80.0	83.3	80.0	83.3	76.6
R_4	86.7	83.3	80.0	86.6	86.6
R_5	86.7	86.6	83.3	86.6	86.6
R_6	90.0	89.9	90.0	89.9	93.3
R_7	90.0	89.9	90.0	89.9	93.3
R_8	93.3	93.2	93.3	93.2	96.6
R_9	93.3	93.2	93.3	96.5	96.6
R_{10}	100.0	100.0	100.0	100.0	100.0

(b) Top-k, for k=250

	Success rate in %				
n, x	1, 2	1, 3	1, 4	2, 2	3, 3
R_1	63.3	56.7	60.0	**70.0**	56.7
R_2	80.0	76.7	76.7	**83.3**	70.0
R_3	86.7	83.4	83.4	83.3	76.7
R_4	90.0	83.4	90.1	83.3	86.7
R_5	90.0	90.1	93.4	83.3	90.0
R_6	90.0	90.1	93.4	86.6	93.3
R_7	93.3	93.4	93.4	89.9	96.6
R_8	96.6	96.7	96.7	93.2	96.6
R_9	96.6	96.7	96.7	96.5	99.9
R_{10}	100.0	100.0	100.0	100.0	100.0

Table 6. Overall classification results using skip-grams with maxgap = 2

(a) Top-k, for k=100.

	Success rate in %				
n, x	1, 2	1, 3	1, 4	2, 2	3, 3
R_1	53.3	46.7	50.0	**63.3**	56.7
R_2	73.3	70.0	63.3	**76.6**	70.0
R_3	80.0	73.3	73.3	79.9	73.3
R_4	83.3	80.0	80.0	79.9	80.0
R_5	86.6	83.3	80.0	83.2	86.7
R_6	86.6	86.6	90.0	83.2	93.4
R_7	89.9	89.9	90.0	83.2	93.4
R_8	93.2	89.9	90.0	89.9	93.4
R_9	93.2	93.2	93.3	96.6	96.7
R_{10}	100.0	100.0	100.0	100.0	100.0

(b) Top-k, for k=250.

	Success rate in %				
n, x	1, 2	1, 3	1, 4	2, 2	3, 3
R_1	60.0	56.7	56.7	63.3	**66.7**
R_2	70.0	73.4	73.4	83.3	**76.7**
R_3	83.3	80.1	76.7	83.3	80.0
R_4	83.3	83.4	83.4	86.6	80.0
R_5	86.6	86.7	86.7	86.6	90.0
R_6	89.9	90.0	90.0	89.9	93.3
R_7	93.2	93.3	93.3	89.9	93.3
R_8	96.5	96.6	96.6	89.9	100.0
R_9	96.5	96.6	96.6	96.6	100.0
R_{10}	100.0	100.0	100.0	100.0	100.0

for each author separately. Tables 8a and b respectively show the success rates attributed to each author (rank R_1) for $maxgap = 0$, $k = 100$ and $k = 250$, when the other parameters are varied.

It can be observed that for most authors, at least 66.7 % of texts are correctly attributed. For example, for $n = 2$, $x = 2$ and $k = 250$, four authors have all texts correctly identified, four have 66.7 % of their texts correctly classified, and two have 33.3 % of texts correctly classified. Overall, it can be thus found that the proposed approach performs very well.

It can also be found that some authors were harder to classify (author 3420481 and 3454871). After investigation, we found that the reason for the incorrect classification is that both authors have posted a same very long blog post, which represents about 50 % of the length of their respective corpus. This content has

Table 7. Overall classification results using skip-grams with maxgap $= 3$

(a) Top-k, for k=100

	Success rate in %				
n,x	$1,2$	$1,3$	$1,4$	$2,2$	$3,3$
R_1	46.7	46.7	46.7	**66.7**	50.0
R_2	73.4	70.0	66.7	**73.4**	66.7
R_3	73.4	76.7	76.7	80.1	70.0
R_4	83.4	80.0	76.7	80.1	80.0
R_5	83.4	83.3	80.0	86.8	80.0
R_6	83.4	83.3	86.7	86.8	83.3
R_7	90.1	86.6	86.7	86.8	83.3
R_8	90.1	89.9	90.0	90.1	90.0
R_9	93.4	93.2	93.3	96.8	96.7
R_{10}	100.0	100.0	100.0	100.0	100.0

(b) Top-k, for k=250.

	Success rate in %				
n,x	$1,2$	$1,3$	$1,4$	$2,2$	$3,3$
R_1	63.3	60.0	56.7	**66.7**	66.7
R_2	76.6	73.3	70.0	**83.4**	73.4
R_3	83.3	80.0	73.3	86.7	73.4
R_4	86.6	83.3	76.6	86.7	76.7
R_5	86.6	83.3	83.3	86.7	80.0
R_6	86.6	86.6	86.6	90.0	83.3
R_7	89.9	93.3	89.9	90.0	93.3
R_8	96.6	93.3	93.2	96.7	100.0
R_9	96.6	93.3	93.2	96.7	100.0
R_{10}	100.0	100.0	100.0	100.0	100.0

Table 8. Success rate per author using skip-grams with maxgap $= 0$ (ngrams)

(a) Top-k, for k=100.

	Success rate per author in %				
Authors	$1,2$	$1,3$	$1,4$	$2,2$	$3,3$
1029959	33.3	33.3	33.3	66.7	33.3
2069530	100.0	100.0	100.0	66.7	33.3
2369365	33.3	33.3	33.3	66.7	33.3
3182387	100.0	100.0	100.0	100.0	33.3
3298664	100.0	100.0	100.0	100.0	66.7
3420481	0.0	0.0	0.0	0.0	0.0
3454871	0.0	0.0	0.0	33.3	0.0
3520038	66.7	66.7	100.0	66.7	66.7
3535101	66.7	33.3	33.3	66.7	66.7
3701154	66.7	66.7	66.7	66.7	66.7

(b) Top-k, for k=250.

	Success rate per author in %				
Authors	$1,2$	$1,3$	$1,4$	$2,2$	$3,3$
1029959	33.3	33.3	33.3	**66.7**	0.0
2069530	100.0	100.0	100.0	**100.0**	100.0
2369365	33.3	0.0	33.3	**66.7**	0.0
3182387	100.0	100.0	100.0	**100.0**	100.0
3298664	100.0	100.0	100.0	**100.0**	100.0
3420481	0.0	33.3	0.0	**33.3**	0.0
3454871	0.0	0.0	33.3	**33.3**	33.3
3520038	100.0	66.7	66.7	**100.0**	33.3
3535101	66.7	66.7	33.3	**66.7**	66.7
3701154	66.7	100.0	66.7	**66.7**	33.3

Table 9. Accuracy per author using ngrams ($maxgap = 0$)

(a) Top-k, for k=100.

Authors	n,x				
	$1,2$	$1,3$	$1,4$	$2,2$	$3,3$
1029959	0.833	0.867	0.867	0.933	0.933
2069530	0.933	0.967	0.967	0.967	0.900
2369365	0.867	0.867	0.867	0.800	0.933
3182387	1.000	0.967	0.967	1.000	0.833
3298664	0.933	0.867	0.900	1.000	0.967
3420481	0.833	0.867	0.867	0.867	0.700
3454871	0.867	0.900	0.900	0.900	0.900
3520038	0.967	0.967	1.000	0.900	0.867
3535101	0.933	0.900	0.900	0.967	0.900
3701154	0.967	0.900	0.900	0.933	0.867

(b) Top-k, for k=250.

Authors	n,x				
	$1,2$	$1,3$	$1,4$	$2,2$	$3,3$
1029959	0.933	0.867	0.867	0.933	0.833
2069530	0.933	0.900	0.967	1.000	1.000
2369365	0.900	0.900	0.867	0.833	0.800
3182387	1.000	1.000	1.000	1.000	1.000
3298664	0.833	0.867	0.833	0.967	1.000
3420481	0.833	0.867	0.867	0.900	0.767
3454871	0.867	0.900	0.900	0.933	0.900
3520038	1.000	0.967	0.967	0.967	0.767
3535101	0.933	0.933	0.900	0.967	0.933
3701154	0.967	1.000	0.967	0.967	0.933

Table 10. Summary of best success rate results (Best R_1 Rank)

Markers	Ngrams		Skip-grams		Ngrams vs Skip-grams	
Parameters	k	50	k	250	k	250
	n, x	1, 2	n, x	3, 3	n, x	2, 2
			$maxgap$	1	$maxgap$	0
Results in %	R_1	73.3	R_1	70.0	R_1	73.3
	R_2	83.3	R_2	76.6	R_2	86.6
	R_3	90.0	R_3	86.6	R_3	86.6
No. of words	2,615,856 (30 books, 10 authors)		idem (30 books, 10 authors)		265,263 (609 posts, 10 authors)	

a distinct writing style, which suggests that it was not written by any of these authors. Thus, it had a great influence on their signatures, and led to the poor classification of these authors.

In contrast, some authors were very easily identified with high success rate and high accuracy (cf. Table 9 for accuracies). For example, all texts by authors 3182387 and 3298664 were almost always correctly classified for the tested parameter values in these tables. The reason is that these authors have distinctive writing styles in terms of part-of-speech patterns.

5   Conclusions

A novel approach using fixed or variable-length patterns of part-of-speech skip-grams or n-grams as features was presented for blog authorship attribution. An experimental evaluation using blog posts from 10 blog authors has shown that authors are accurately classified with more than 73.3 % success rate, and an average accuracy of 94.7 %, using a small number of POS patterns (e.g. $k = 250$), and that it is unnecessary to create signatures with a large number of patterns. Moreover, it was found that using fixed length POS bigrams provided better results than using POS ngrams or using a larger gap between POS tags, and that using fixed length patterns is preferable to using variable-length patterns. To give the reader a feel of the relative efficacy of this approach on traditional long texts, we present a summary table of results on a corpus of 30 books by ten 19th century authors (c.f. Table 10 for comparative results) [18,19]. In future work, we plan to develop other features and also evaluate the proposed features on other types of texts.

Acknowledgements. This work is financed by a National Science and Engineering Research Council (NSERC) of Canada research grant, and the Faculty of Research and Graduate Studies of the Université de Moncton.

References

1. Koppel, M., Schler, J., Argamon, S.: Authorship attribution: what's easy and what's hard? (2013). SSRN 2274891
2. Mendenhall, T.C.: The characteristic curves of composition. Science **9**(214), 237–246 (1887)

3. Mosteller, F., Wallace, D.: Inference and Disputed Authorship: The Federalist. Addison-Wesley, Reading (1964)
4. Grant, T.: Text messaging forensics: Txt 4n6: idiolect free authorship analysis? (2010)
5. Stamatatos, E., Fakotakis, N., Kokkinakis, G.: Automatic text categorization in terms of genre and author. Comput. Linguist. **26**(4), 471–495 (2000)
6. Baayen, H., van Halteren, H., Tweedie, F.: Outside the cave of shadows: using syntactic annotation to enhance authorship attribution. Literary Linguist. Comput. **11**(3), 121–132 (1996)
7. Clark, J.H., Hannon, C.J.: A classifier system for author recognition using synonym-based features. In: Gelbukh, A., Kuri Morales, Á.F. (eds.) MICAI 2007. LNCS (LNAI), vol. 4827, pp. 839–849. Springer, Heidelberg (2007)
8. McDonald, A.W.E., Afroz, S., Caliskan, A., Stolerman, A., Greenstadt, R.: Use fewer instances of the letter "i": toward writing style anonymization. In: Fischer-Hübner, S., Wright, M. (eds.) PETS 2012. LNCS, vol. 7384, pp. 299–318. Springer, Heidelberg (2012)
9. Schler, J., Koppel, M., Argamon, S., Pennebaker, J.W.: Effects of age and gender on blogging. In: AAAI Spring Symposium: Computational Approaches to Analyzing Weblogs, vol. 6, pp. 199–205 (2006)
10. Stamatatos, E., Fakotakis, N., Kokkinakis, G.: Computer-based authorship attribution without lexical measures. Comput. Humanit. **35**(2), 193–214 (2001)
11. Gamon, M.: Linguistic correlates of style: authorship classification with deep linguistic analysis features. In: Proceedings of the 20th International Conference on Computational Linguistics, Association for Computational Linguistics, p. 611 (2004)
12. Sidorov, G., Velasquez, F., Stamatatos, E., Gelbukh, A., Chanona-Hernández, L.: Syntactic n-grams as machine learning features for natural language processing. Expert Syst. Appl. **41**(3), 853–860 (2014)
13. García-Hernández, R.A., Martínez-Trinidad, J.F., Carrasco-Ochoa, J.A.: Finding maximal sequential patterns in text document collections and single documents. Informatica **34**(1), 93–101 (2010)
14. Litvinova, T., Seredin, P., Litvinova, O.: Using part-of-speech sequences frequencies in a text to predict author personality: a corpus study. Indian J. Sci. Technol. **8**(S9), 93–97 (2015)
15. Toutanova, K., Klein, D., Manning, C.D., Singer, Y.: Feature-rich part-of-speech tagging with a cyclic dependency network. In: Proceedings of 2003 Conference on North American Chapter of the ACL - Human Language Technologies, pp. 173–180 (2003)
16. Fournier-Viger, P., Gomariz, A., Gueniche, T., Mwamikazi, E., Thomas, R.: TKS: efficient mining of top-K sequential patterns. In: Motoda, H., Wu, Z., Cao, L., Zaiane, O., Yao, M., Wang, W. (eds.) ADMA 2013, Part I. LNCS, vol. 8346, pp. 109–120. Springer, Heidelberg (2013)
17. Argamon-Engelson, S., Koppel, M., Avneri, G.: Style-based text categorization: what newspaper am i reading. In: Proceedings of AAAI Workshop on Text Categorization, pp. 1–4 (1998)
18. Pokou, J.M., Fournier-Viger, P., Moghrabi, C.: Authorship attribution using small sets of frequent part-of-speech skip-grams. In: Proceedings of the 29th International Florida Artificial Intelligence Research Society Conference, pp. 86–91 (2016)
19. Pokou, J.M., Fournier-Viger, P., Moghrabi, C.: Authorship attribution using variable-length part-of-speech patterns. In: Proceedings of the 7th International Conference on Agents and Artificial Intelligence, pp. 354–361 (2016)

Increasing Diversity in Random Forests Using Naive Bayes

Christos K. Aridas$^{(\boxtimes)}$, Sotiris B. Kotsiantis, and Michael N. Vrahatis

Computational Intelligence Laboratory (CILab), Department of Mathematics,
University of Patras, 26110 Patras, Greece
char@upatras.gr, {sotos,vrahatis}@math.upatras.gr

Abstract. In this work a novel ensemble technique for generating random decision forests is presented. The proposed technique incorporates a Naive Bayes classification model to increase the diversity of the trees in the forest in order to improve the performance in terms of classification accuracy. Experimental results on several benchmark data sets show that the proposed method archives outstanding predictive performance compared to other state-of-the-art ensemble methods.

Keywords: Ensemble methods · Decision forests · Pattern classification

1 Introduction

In machine learning and data mining, ensemble methods make use of single or multiple learning algorithms in order to generate a diverse set of classifiers aiming to improve performance/robustness over a single underlying classifier [16]. Experimental studies and machine learning applications prove that a certain supervised learning algorithm outperforms any other algorithm for a particular problem or for a particular subset of the input dataset, but it is unusual to discover a single classifier that will reach the best performance on the overall problem domain [17].

Ensembles of classifiers can be generated via several methods [2]. Common procedures that are used to create an ensemble of classifiers include, among others, (i) Using different splits of a training data set with a single learning algorithm, (ii) Using different training parameters with a single learning algorithm, (iii) Using multi-learning methods.

Diversity [21] between the base classification models is considered to be a key aspect when constructing a classifier ensemble. In this work, we propose a variation of the Random Forests [6] algorithm that incorporates new features using Naive Bayes [8] before the construction of the forest. The new generated features aim to increase the diversity among the trees in the forest. Our empirical evaluation concludes that the new features increase the diversity and that they leed to a better final classifier.

Published by Springer International Publishing Switzerland 2016. All Rights Reserved
L. Iliadis and I. Maglogiannis (Eds.): AIAI 2016, IFIP AICT 475, pp. 75–86, 2016.
DOI: 10.1007/978-3-319-44944-9_7

The rest of the paper is organized as follows. In Sect. 2 some of the most well-known techniques for generating ensembles, that are based on a single learning algorithm, are discussed. In Sect. 3 the proposed method is presented. Furthermore, the results of the experiments on several real and laboratory data sets, after being compared with state-of-the-art ensemble methods, are portrayed and discussed. Finally, Sect. 4 concludes the paper and suggests further directions in current research.

2   Background Material

This section presents a brief survey of techniques for generating ensembles using a sole learning algorithm. These techniques rely on modifying the training data set. Methods of modifying the training data set include, among others, sampling the training patterns, sampling the feature space, a combination of the two and modifying the weight of the training patterns.

Bagging is a method for creating an ensemble of classifiers that was proposed by Breiman [5]. Bagging generates the classifiers in the ensemble by taking random subsets of the training data set with replacement and building one classifier on each bootstrap sample. The final classification prediction for an unseen pattern is constructed by taking the majority vote over the class labels produced by the base classification models.

While Bagging relies on random and independent changes in the training data implemented by bootstrap sampling, Boosting [11] encourages guided changes of the training data to direct further classifiers toward more "difficult cases". It assigns weights to the training patterns, which are then modified according to how well the coupled case is learned by the classifier. The weights for misclassified patterns are increased. Thus, re-sampling happens based on how well the training patterns are classified by the previous base classifier. Given that the training set for one classification model depends on the previous one, boosting requires sequential runs and therefore is not easily adapted to a parallel process. After several iterations, the prediction is made by taking a weighted vote of the predictions of each classifier, with the weights being relative to each classifiers accuracy on its training set. AdaBoost is a practical version of the boosting approach [11].

Ho [15] constructed a forest of decision trees named Random Subspace Method that preserves highest accuracy on the training patterns and improves on generalization accuracy as it grows in complexity. Random Subspace Method consists of global multiple decision trees created systematically by pseudorandomly selecting half of the available features, i.e. trees built in randomly chosen subspaces. The final classification prediction for an unseen pattern is constructed by averaging the estimates of posterior probabilities at the leaves of all the trees in the forest.

Random Forests [6] is an alternate method for building ensembles. It is a combination of Bagging and Random Subspace Method. In Random Forests, every tree in the forest is constructed from a bootstrapped sample from the training set. Additionally, the split that is selected in tree construction is not the best split between all the available features. Instead, it is the best split among a random selection of the features [22].

Despite the fact that Random Forests yields one of the most successful [9] classification models, the improvement of its classification accuracy remains an open problem for the machine learning research field. Several authors [3,19,24] have studied and proposed techniques that could improve the performance of Random Forests. In [19] it is illustrated that the most important improvement in the performance of Random Forests is achieved by changing the mechanism of voting in the prediction. In [24] the authors implemented a similar approach, where class votes by trees in the forest are weighed according to their performance; Therefore, heavier weights are assigned to better performing trees. The authors of [3] experimentally show that the classification accuracy is enhanced when a random forest is composed of good and uncorrelated trees with high accuracies, while correlated and bad trees with low classification accuracies are ignored.

3   The Proposed Method

In this work, a modified version of Random Forests is proposed that is constructed not only by pseudorandomly selecting attribute subsets but also by encapsulating Naive Bayes estimation in the training phase, as well as in the classification phase.

Given a class variable y and a dependent feature vector x_1 through x_n, Bayes theorem states the following relationship:

$$P(y \mid x_1, \ldots, x_n) = \frac{P(y)P(x_1, \ldots x_n \mid y)}{P(x_1, \ldots, x_n)}. \tag{1}$$

Under the assumption that features are conditionally independent

$$P(x_i \mid y, x_1, \ldots, x_{i-1}, x_{i+1}, \ldots, x_n) = P(x_i \mid y), \tag{2}$$

for all i Eq. (2) is simplified to

$$P(y \mid x_1, \ldots, x_n) = \frac{P(y) \prod_{i=1}^{n} P(x_i \mid y)}{P(x_1, \ldots, x_n)}. \tag{3}$$

Since $P(x_1, \ldots, x_n)$ is constant given the input the formula used by the Naive Bayes classifier is

$$P(y \mid x_1, \ldots, x_n) \propto P(y) \prod_{i=1}^{n} P(x_i \mid y) \Rightarrow \hat{y} = \arg\max_{y} P(y) \prod_{i=1}^{n} P(x_i \mid y). \tag{4}$$

The assumption of independence is almost always wrong. Besides this, an extensive comparison of a simple Bayesian classifier with state-of-the-art algorithms showed that the former sometimes is superior to other supervised learning algorithms even on datasets with important feature dependencies [8]. In [12] Friedman explans why the simple Bayes method remains competitive.

Our intention is to generate a forest of decision trees that will be as diverse as possible for producing better results [4]. For this reason, in the training phase, we trained a classifier using the Naive Bayes algorithm. Afterwards, we used the same training set to generate predictions and class membership probabilities using the Naive Bayes classifier. The predictions and the class membership probabilities will increase the original feature space by concatenating them as new features. As far as predictions are concerned, a new feature vector will be generated. This vector will contain only the class label for each instance that is predicted by the Naive Bayes model. In the case of class membership probabilities, new feature vectors will be generated, as many as the number of classes. Assuming that we have a sample of a dataset with four features as presented in Table 1 and three classes. An example of the above process is illustrated in Tables 2 and 3. After the concatenation, a Random Forests classifier will be trained using the new n-dimensional feature vector. The same procedure, the generation of new feature vectors, will be used in the classification phase. The predicted class of an unseen instance will be the vote by the trees in the forest, weighted by their probability estimates. The proposed method is presented in Algorithm 1.

3.1 Numerical Experiments

In order to verify the performance of the proposed method, a number of experiments on some classification tasks were conducted and the results are reported

Table 1. Original feature space

x_0	x_1	x_2	x_3	y
4.7	3.2	1.3	0.2	0
4.6	3.1	1.5	0.2	0
6.4	3.2	4.5	1.5	1
6.9	3.1	4.9	1.5	1
5.8	2.7	5.1	1.9	2
7.1	3.0	5.9	2.1	2

Table 2. Feature space augmented with Naive Bayes model's predictions

x_0	x_1	x_2	x_3	f_0	y
4.7	3.2	1.3	0.2	0	0
4.6	3.1	1.5	0.2	0	0
6.4	3.2	4.5	1.5	1	1
6.9	3.1	4.9	1.5	2	1
5.8	2.7	5.1	1.9	2	2
7.1	3.0	5.9	2.1	2	2

Table 3. Feature space augmented with Naive Bayes model's class membership probabilities

x_0	x_1	x_2	x_3	f_1	f_2	f_3	y
4.7	3.2	1.3	0.2	1.000	0.000	0.000	0
4.6	3.1	1.5	0.2	1.000	0.000	0.000	0
6.4	3.2	4.5	1.5	0.000	0.945	0.055	1
6.9	3.1	4.9	1.5	0.000	0.456	0.544	1
5.8	2.7	5.1	1.9	0.000	0.025	0.975	2
7.1	3.0	5.9	2.1	0.000	0.000	1.000	2

in this section. From the KEEL data set repository [1] fourteen data sets were chosen and used as is without any further preprocessing. In Table 4 the name, the number of patterns, the attributes, as well as the number of different classes for each data set are shown.

All data sets were partitioned using a ten-fold cross-validation procedure. This method divides the instances in ten equal folds. Each tested method was trained using nine folds and the fold left out was used for evaluation, using the metric of classification accuracy. This was repeated ten times. Then the average accuracy across all trials was computed.

Algorithm 1. NB Forest

---

    procedure Training(X, y, *generateNBProbas*, *generateNBPredictions*)
        Build a Naives Bayes (NB) model using X and y
        $X_{gen} \leftarrow$ *GenerateFeatures*(X,*generateNBProbas*,*generateNBPredictions*)
        Build a Random Forest model using X_{gen} and y
    end procedure
    procedure Classification(X, *generateNBProbas*, *generateNBPredictions*)
        $X_{gen} \leftarrow$ *GenerateFeatures*(X,*generateProbabilities*,*generatePredictions*)
        Use the X_{gen} to classify a test instance
    end procedure
    function GenerateFeatures(X, *generateNBProbas*, *generateNBPredictions*)
        $X_{gen} \leftarrow X$
        if *generateNBProbas* **then**
            Use NB to generate class membership probabilities as X_{probas}
            $X_{gen} \leftarrow X_{gen} \cup X_{probas}$
        end if
        if *generateNBPredictions* **then**
            Use NB to generate class predictions as $X_{predictions}$
            $X_{gen} \leftarrow X_{gen} \cup X_{predictions}$
        end if
        return X_{gen}
    end function

Table 4. Benchmark data sets used in the experiments

Data set	#attributes	#patterns	#classes
appendicitis	7	106	2
banana	2	5300	2
cleveland	13	297	5
ecoli	7	336	8
glass	9	214	7
led7digit	7	500	10
libras	90	360	15
phoneme	5	5404	2
ring	20	7400	2
segment	19	2310	7
spambase	57	4597	2
texture	40	5500	11
twonorm	20	7400	2
yeast	8	1484	10

Firstly, we ran experiments using different settings of the proposed method and compared them to the original Random Forests (RF) algorithm. The experiments were performed in python using the scikit-learn [18] library. All classifiers were built using the default settings in scikit-learn, which means that all ensembles were generated using ten base classifiers.

In Table 5 the results obtained using variants of the proposed method is presented. The variant that is better than the original Random Forests is reported in bold.

NBFB denotes a Random Forests classifier that was trained using the original space along with class membership probabilities generated by the Naive Bayes model. NBFD denotes a Random Forests classifier that was trained using the original space along with predictions generated by Naive Bayes model. NBFBD was trained using both class membership probabilities and predictions that were generated by the Naive Bayes model, concatenated with the original feature space. With the intention to discard the parameters of the proposed method, i.e. which new features should be generated by the Naive Bayes model, and get the most out of the generated features, we ran an experiment by selecting the best of RF, NBFB, NBFD and NBFBD using 5-fold cross-validation in the training set. The model selection was performed by selecting the variant that gave the minimum average error rate across all folds. The last variant is denoted as NBFCV. In the last row of the Table 5 counted the W(ins)/T(ies)/L(osses) of the algorithm in the column against the original Random Forests algorithm obtained by the Wilcoxon Singed Ranks Tests [23]. Apart from the combined

Table 5. Average accuracy of Random Forests variants

Data set	RF	NBFB	NBFD	NBFBD	NBFCV
appendicitis	0.85000	**0.87000**	**0.87909**	0.85000	**0.86909**
banana	0.89000	0.88830	**0.88962**	**0.89094**	**0.89170**
cleveland	0.54555	**0.54624**	**0.60362**	**0.57051**	**0.57336**
ecoli	0.81569	0.79795	**0.83957**	0.80704	**0.82487**
glass	0.74536	0.71355	0.74304	0.73080	**0.74575**
led7digit	0.70600	**0.71400**	0.70800	0.70200	**0.70800**
libras	0.78056	0.77500	**0.78611**	0.76111	**0.79722**
phoneme	0.89415	**0.89433**	**0.89896**	0.89026	**0.89637**
ring	0.92946	**0.97824**	**0.97892**	**0.97865**	**0.97905**
segment	0.97229	0.97229	**0.97662**	0.96667	**0.97576**
spambase	0.94736	0.94649	**0.94758**	0.94562	**0.94758**
texture	0.96709	0.96709	**0.97073**	0.96473	**0.96964**
twonorm	0.94054	**0.97689**	**0.97865**	**0.97878**	**0.97865**
yeast	0.57277	0.56940	**0.58159**	**0.58828**	**0.58426**
Statistic		−0.235	−2.919	−0.035	−3.296
p-value		0.814	0.004	0.972	0.001
W/T/L		6/2/6	12/0/2	5/1/8	14/0/0

methodology, it is clear that almost every variation of the proposed method performs better that the original Random Forests method in most cases.

We chose NBFCV and RF and we measured the diversity using two artificial data sets. The first data set was generated by taking a multi-dimensional standard normal distribution and defining classes separated by nested concentric multi-dimensional spheres, so that, roughly, equal numbers of samples are in each class (quantiles of the χ^2 distribution). The second data set is a binary classification problem used in [14]. The data set has ten features that are sampled from standard independent Gaussian and the target class y is defined by:

$$Y = \begin{cases} 1, & \text{if } \sum_{i=1}^{10} X^2 > 9.34 \\ -1, & \text{otherwise.} \end{cases}$$

In Table 6 the name, the number of patterns, the attributes as well as the number of different classes for each data set are shown.

We used the five-fold cross-validation procedure and measured the diversity in each test fold using the Kohavi-Wolpert Variance [21]. In Table 7 the mean of the Kohavi-Wolpert Variance across all testing folds is presented. The diversity increases as the variance decreases, so the lower the better. In parentheses the mean accuracy score across all folds is reported. The results of Table 7 indicate

Table 6. Artificial data sets for measuring diversity

Data set	#attributes	#patterns	#classes
Gaussian quantiles	10	1000	2
Hastie	10	12000	2

Table 7. Average measurement of diversity using Kohavi-Wolpert Variance

Data set	RF	NBFCV
Gaussian quantiles	0.159 (0.775)	0.040 (0.945)
Hastie	0.074 (0.855)	0.004 (0.990)

Table 8. Average accuracy of compared algorithms

Data set	NBFCV	RF	NB	BGT	BGN	BST	BSN	RT	RN	VTN
appendicitis	0.869	0.850	0.850	0.860	0.851	0.804	0.743	0.879	**0.887**	0.860
banana	**0.892**	0.890	0.613	0.888	0.614	0.747	0.600	0.585	0.605	0.889
cleveland	**0.573**	0.546	0.553	0.549	0.530	0.452	0.493	0.566	0.559	0.543
ecoli	**0.825**	0.816	0.589	0.795	0.426	0.780	0.601	0.738	0.631	0.688
glass	**0.746**	0.745	0.448	0.732	0.442	0.663	0.581	0.735	0.444	0.615
led7digit	0.708	0.706	0.672	**0.712**	0.590	0.698	0.662	0.690	0.666	0.686
libras	**0.797**	0.781	0.631	0.758	0.639	0.722	0.675	0.747	0.633	0.650
phoneme	0.896	0.894	0.761	**0.905**	0.762	0.872	0.561	0.813	0.760	0.845
ring	0.979	0.929	0.980	0.927	**0.980**	0.874	**0.980**	0.936	0.970	**0.980**
segment	**0.976**	0.972	0.798	0.975	0.798	0.964	0.851	0.951	0.750	0.894
spambase	**0.948**	0.947	0.821	0.938	0.822	0.944	0.721	0.945	0.795	0.893
texture	**0.970**	0.967	0.774	0.961	0.773	0.931	0.855	0.975	0.772	0.839
twonorm	**0.979**	0.941	**0.979**	0.937	**0.979**	0.843	0.976	0.937	0.970	0.977
yeast	**0.584**	0.573	0.141	0.570	0.183	0.503	0.350	0.513	0.406	0.228

that involving Naive Bayes for feature generation makes the ensemble more diverse and this results in an increment of classification accuracy.

Afterwards, we trained models using different ensemble methods that were presented in Sect. 2. We trained a Naive Bayes (NB) classifier, a Bagging classifier that used Decision Tree (BGT) as a base learning algorithm and a Bagging classifier that used Naive Bayes (BGN) as a base learning algorithm. Also, we trained Boosting (AdaBoost) and Random Space Method ensembles using Decision Tree (BST, RT) and Naive Bayes (BSN, RN) as base learners. Finally, we trained a voting classifier, denoted as VTN, using Random Forests and Naive Bayes that predicts the average of class membership probabilities. All ensemble methods were built by using ten base classifiers apart from Boosting methods that were built by using fifty base classifiers.

The results obtained are presented in Table 8. In the comparisons, we include the NBFCV variant because it performed better against the original Random Forests. According to Table 8 the proposed method seems to perform better than the well-known ensemble methods.

Demšar [7] recommends that the non-parametric tests should be preferred over the parametric in the context of machine learning, since they do not assume normal distributions or homogeneity of variance. Hence, in order to validate the significance of the results, the Friedman test [13], which is a rank-based non-parametric test for comparing several machine learning algorithms on multiple data sets, was used. The null hypothesis of the test states that all the methods perform equivalently and thus their ranks should be equivalent. The average rankings, according to the Friedman test, are presented in Table 9.

Table 9. Average rankings of the algorithms according to the Friedman test

Algorithm	Ranking
NBFCV	1.7143
RF	3.7500
BGT	4.0714
RT	4.6786
VTN	5.5714
BST	6.0000
NB	7.1071
RN	7.2143
BGN	7.3571
BSN	7.5357
Statistic	49.5896
p-value	$<10^{-6}$

Friedman's test ranks our algorithm in the first place. Besides this, assuming a significance level of 0.05 in Table 9, the p-value of the Friedman test implies that the null hypothesis has to be rejected. So, there is at least one method that performs statistically different from the proposed method. In order to investigate the aforesaid, Finner's [10] hoc procedure was used.

In Table 10 the p-values obtained by applying post hoc procedure, over the results of the Friedman statistical test, are presented. Finner's procedure rejects the hypotheses that have unadjusted p-values ≤ 0.05. That said, the adjusted p-values obtained through the application of the Finner's post hoc procedure are presented in Table 11.

The results obtained by Tables 9 and 11 indicate that the proposed method performs better than any other method. Nonetheless, when involving other algorithms in the comparisons the proposed variant does not seem to perform statistically better than the original Random Forest method.

Table 10. Post hoc comparison for the Friedman test

i	Algorithm	$z = (R_0 - R_i)/SE$	p	Finner
9	BSN	5.087130	$<10^{-5}$	0.005683
8	BGN	4.931083	$<10^{-5}$	0.011334
7	RN	4.806246	$<10^{-5}$	0.016952
6	NB	4.712618	$<10^{-5}$	0.022539
5	BST	3.745127	0.000180	0.028094
4	VTN	3.370614	0.000705	0.033617
3	RT	2.590379	0.009587	0.039109
2	BGT	2.059820	0.039416	0.044570
1	RF	1.778935	0.075250	0.050000

Table 11. Post hoc comparison for the Friedman test with adjusted p-values

i	Algorithm	$p_{Unadjusted}$	p_{Finner}
9	BSN	$<10^{-5}$	$<10^{-5}$
8	BGN	$<10^{-5}$	$<10^{-5}$
7	RN	$<10^{-5}$	$<10^{-5}$
6	NB	$<10^{-5}$	$<10^{-5}$
5	BST	0.000180	0.000325
4	VTN	0.000750	0.001125
3	RT	0.009587	0.012309
2	BGT	0.039416	0.044232
1	RF	0.075250	0.075250

4　Conclusions and Future Work

An ensemble of classifiers is a collection of classification models whose individual predictions are blended, typically by weighted or unweighted voting, to assign a class label to each new pattern. The creation of effective ensemble methods is an active research field in machine learning. Ensembles of classifiers are usually significantly more accurate than the individual underlying classifiers. The main explanation is that many learning algorithms apply local optimization techniques, which may get stuck in local optima.

It was demonstrated after a number of comparisons with Random Forests and other well-known ensembles, that increasing the feature space of a small random forest with predictions and class membership probabilities of a Naive Bayes model can increase the performance in terms of classification accuracy, in most cases.

In a following work, the proposed method will be investigated as far as regression problems and the problem of choosing the right number of trees in the forest are concerned. Also, the implementation and the evaluation of the proposed method in on-line learning [20] problems will be addressed.

References

1. Alcala-Fdez, J., Fernandez, A., Luengo, J., Derrac, J., Garcia, S.: KEEL data-mining software tool: data set repository, integration of algorithms and experimental analysis framework. Multiple-Valued Logic Soft Comput. **17**(2–3), 255–287 (2011)
2. Bernardini, F., Monard, M., Prati, R.: Constructing ensembles of symbolic classifiers, p. 6. IEEE (2005)
3. Bharathidason, S., Jothi Venkataeswaran, C.: Improving classification accuracy based on random forest model with uncorrelated high performing trees. Int. J. Comput. Appl. **101**(13), 26–30 (2014)
4. Biau, G.: Analysis of a random forests model. J. Mach. Learn. Res. **13**(1), 1063–1095 (2012)
5. Breiman, L.: Bagging predictors. Mach. Learn. **24**(2), 123–140 (1996)
6. Breiman, L.: Random forests. Mach. Learn. **45**(1), 5–32 (2001)
7. Demar, J.: Statistical comparisons of classifiers over multiple data sets. J. Mach. Learn. Res. **7**, 1–30 (2006)
8. Domingos, P., Pazzani, M.: On the optimality of the simple Bayesian classifier under zero-one loss. Mach. Learn. **29**(2), 103–130 (1997)
9. Fernndez-Delgado, M., Cernadas, E., Barro, S., Amorim, D.: Do we need hundreds of classifiers to solve real world classification problems? J. Mach. Learn. Res. **15**, 3133–3181 (2014)
10. Finner, H.: On a monotonicity problem in step-down multiple test procedures. J. Am. Stat. Assoc. **88**(423), 920–923 (1993)
11. Freund, Y., Schapire, R.E.: Others: experiments with a new boosting algorithm. In: ICML, vol. 96, pp. 148–156 (1996)
12. Friedman, J.H.: On bias, variance, 0/1–loss, and the curse-of-dimensionality. Data Min. Knowl. Disc. **1**(1), 55–77 (1997)
13. Friedman, M.: The use of ranks to avoid the assumption of normality implicit in the analysis of variance. J. Am. Stat. Assoc. **32**(200), 675 (1937)
14. Hastie, T., Tibshirani, R., Friedman, J.: The Elements of Statistical Learning. Springer Series in Statistics. Springer, New York (2009)
15. Ho, T.K.: The random subspace method for constructing decision forests. IEEE Trans. Pattern Anal. Mach. Intell. **20**(8), 832–844 (1998)
16. Kotsiantis, S.: Combining bagging, boosting, rotation forest and random subspace methods. Artif. Intell. Rev. **35**(3), 223–240 (2011)
17. Opitz, D., Maclin, R.: Popular ensemble methods: an empirical study. J. Artif. Intell. Res. **11**, 169–198 (1999)
18. Pedregosa, F., Varoquaux, G., Gramfort, A., Michel, V., Thirion, B., Grisel, O., Blondel, M., Prettenhofer, P., Weiss, R., Dubourg, V., Vanderplas, J., Passos, A., Cournapeau, D., Brucher, M., Perrot, M., Duchesnay, D.: Scikit-learn: machine learning in Python. J. Mach. Learn. Res. **12**, 2825–2830 (2011)
19. Robnik-Šikonja, M.: Improving random forests. In: Boulicaut, J.-F., Esposito, F., Giannotti, F., Pedreschi, D. (eds.) ECML 2004. LNCS (LNAI), vol. 3201, pp. 359–370. Springer, Heidelberg (2004)

20. Saffari, A., Leistner, C., Santner, J., Godec, M., Bischof, H.: On-line random forests, pp. 1393–1400. IEEE (2009)
21. Tang, E.K., Suganthan, P.N., Yao, X.: An analysis of diversity measures. Mach. Learn. **65**(1), 247–271 (2006)
22. Verikas, A., Gelzinis, A., Bacauskiene, M.: Mining data with random forests: a survey and results of new tests. Pattern Recogn. **44**(2), 330–349 (2011)
23. Wilcoxon, F.: Individual comparisons by ranking methods. Biometrics Bull. **1**(6), 80 (1945)
24. Winham, S.J., Freimuth, R.R., Biernacka, J.M.: A weighted random forests approach to improve predictive performance. Stat. Anal. Data Min. **6**(6), 496–505 (2013)

Identifying Asperity Patterns Via Machine Learning Algorithms

Kostantinos Arvanitakis[✉] and Markos Avlonitis

Department of Informatics, Ionian University, 49100 Corfu, Greece
{c14arva,avlon}@ionio.gr

Abstract. An asperity's location is very crucial in the spatiotemporal analysis of an area's seismicity. In literature, b-value and seismic density have been proven as useful indicators for asperity location. In this paper, machine learning techniques are used to locate areas with high probability of asperity existence using as feature vector information extracted solely by earthquake catalogs. Many machine learning algorithms are tested to identify those with the best results. This method is tested for data from the wider region of Hokkaido, Japan where in an earlier study asperities have been detected.

Keywords: Asperity · Density · b-value · Seismicity · Machine learning

1 Introduction

Asperities are considered to be large and strong patches on a seismic fault. They have dimensions ranging from less than a kilometer to tens of kilometers. These are locked inside the faults under high pressure and release most of their energy during the eventual earthquake [1]. Asperities can accumulate large portions of tectonic stress and by their rupture an earthquake of great magnitude is generated.

In literature it has been shown that the b-value, i.e. the slope of the Frequency - Magnitude distribution, is significantly lower in asperities, in comparison with other fault zones which have higher b-value [1].

In a recent study made in the region of Hokkaido (Japan), [2] the authors proposed a method for locating asperities by means of the earthquakes' density. Therein, asperities were selected as sections of a region with small number of events, at least one event with high magnitude and surrounded by sections with large number of events.

There are two popular methods of how to locate an asperity. As by definition, an asperity is a high stress area surrounded by low stress areas, the former method calculates the stress levels of a region and points the areas with higher levels [3–5].

The later method uses the surface's slip. By using GPS data of the slip distribution on the earth's surface, the asperities are located in regions where big deformations are detected [6–12].

The main problem with these methods is that they are based on reverse engineering and as a result produce non-unique probabilistic conclusions. The creation of an accurate method to locate asperities is of highly importance. As previously stated asperities are

© IFIP International Federation for Information Processing 2016
Published by Springer International Publishing Switzerland 2016. All Rights Reserved
L. Iliadis and I. Maglogiannis (Eds.): AIAI 2016, IFIP AICT 475, pp. 87–93, 2016.
DOI: 10.1007/978-3-319-44944-9_8

responsible for generating large earthquakes. Thus, an asperity's location is information of high value. Due to the high probability of generating large earthquakes, this information can help the state apparatus in decision making and strategic planning of the seismic regulations according to the building construction and also the expansive policy of cities and civil engineering projects, in order to increase the safety of human life.

Many examples of machine learning, data mining, and feature extraction methods can be found in literature, that have been used in seismicity analysis as tools for the earthquake hazards prediction and prevention. A recent study [13] used co-occurrence cluster mining to identify earthquake swarms and seismic patterns in different regions but with similar properties that probably will be correlated. Also data mining methods have been used [14] for forecasting the month or the year an earthquake will occur. Neural networks also have been used for earthquake prediction. In detail, Panakkat et al. [15] proposed a recurrent neural network, with training and testing data from the Southern California and the San Francisco Bay, to predict the time and location of seismic events. In another study [16], relating with neural networks, the authors proposed and tested in the wider region of Chile, a neural network that could predict the probability that an earthquake of magnitude larger than a threshold value will be generated and also the probability of an earthquake's magnitude with a limited magnitude interval.

The purpose of this study is to introduce a machine learning approach in locating asperities in space. Our hypothesis of whether machine learning can identify asperities will be tested on data from the region of Hokkaido based on the result of Takahashi and Kasahara [2].

2   Materials and Methods

2.1   Seismic Data

The hypocenter data used in the experiments were determined by Hokkaido University, Sapporo, Japan. The catalog contains data from July 1st, 1976 until December 31st, 2002. Every earthquake in the catalog is a record with information about the time the earthquake occurred (year, month, day, hour, minute), the earthquakes epicenter (latitude, longitude, depth), and the earthquake's magnitude. The tested area (Hokkaido region) is located between $41° - 43°$ Latitude and $142.5° - 145.5°$ Longitude.

2.2   Data Representation

Many experiments were conducted using WEKA [17], a platform that allows experimenting with state-of-the-art techniques in machine learning. Due to the complexity of the earthquake phenomenon, there are not many features that can describe an area's seismicity thorough. For the presented task, the b-value and the seismic density features were selected, which are acceptable characteristics, among seismology researchers, of an area's seismicity. Also the longitude and latitude attributes were selected to describe the data's location.

- Latitude of the corresponding area (Numeric)
- Longitude of the corresponding area (Numeric)
- Density of earthquake instances in the corresponding area (Numeric)
- b-value of the Gutenberg-Richter frequency-magnitude distribution (Numeric)
- Asperity indicator (Binary)

The attribute "Asperity indicator", is a binary variable (Yes or No) indicating if an area consist an asperity or not, and it was used as the classification class of the vector in the experiments conducted.

2.3 Feature Vector Extraction

For the purposes of this paper the wider area of Hokkaido region was separated in a grid by 0.1 latitude and longitude degrees. In order to ensure the robustness of the estimated b-values, the radius of every cell that had at least 30 events was increased in order to contain 50 events, using the data of the surrounding cells. The process of creating the grid was automated. Software was created in C language that composes a separate catalog for each cell of the grid and also measures the corresponding density. Every cell was labeled with the latitude and longitude of its centroid.

For all the sections where the number of events (density) was greater than 50, the b-value was calculated. In the formula, that describes the Gutenberg-Richter frequency-magnitude distribution (G-R FMD) (1), N is the accumulated number of events, M is the events magnitude, a-value indicates the total seismicity rate of the region, and the b-value constitutes the slope of the distribution and describes the ratio of small and big earthquakes in an earthquake catalog [18]. The most often used procedures to calculate the b-value of a G-R FMD is the Maximum Likelihood Estimate of b-value [19] method created by Utsu and the least square technique [20]. For the b-value estimation the Maximum likelihood method was chosen.

$$Log(N) = a - b * M \tag{1}$$

The calculations were made by the software ZMAP [21]. The purpose of this application is to determine the quality of seismic data, which are included in earthquake catalogs, and also to calculate and extract useful features. The application combines many basic and useful tools for seismological research.

In our feature vector, every section with density lower than 50 was marked with b-value '?' corresponding to the WEKA's missing value symbol.

3   Machine Learning Algorithms

In total 39 classification algorithms were used to test our hypothesis. The five most effective in means of precision and recall are described below.

Random Forest is a tree classification algorithm developed by Leo Breiman [22]. This algorithm creates a forest of random trees. Random vectors are created from the training set and based on them the growth of each tree is made. The algorithm ensures that every random vector will be unique. Finally each tree vote for the class that every testing instance will be registered and the most popular class is selected from the forest.

Ridor is an implementation of the Ripple-Down Rule learner [23] classification algorithm in WEKA. The Ripple Down Rule creates a binary decision tree different from the ordinary tree classifiers, were a decision can be reached in an interior node in contrast with standard trees were a decision can be made only in the root of the tree. All the rules-nodes are connected with a two way relation. If the premises for a node are all true then the testing subject will be asserted by this node. All these nodes are connected with if-true and if-false sub-nodes. The parental node can make a decision only when the correspondingly sub-nodes are fully in line.

Simple CART (Classification and Regression Tree) is an algorithm also developed by Leo Breiman [24]. The Cart algorithm consist a greedy algorithm where in each stage of the tree building process chooses the most discriminatory feature. To do so each time the attribute to be split is chosen be means of entropy. Finally the algorithm creates a binary decision tree.

BFTree (Best First Tree) [25] is a decision tree algorithm similar to the depth first tree algorithm. In a best-first tree an attribute is placed in the root of the tree and branches are created based on criteria. The training instances are split in subsets, one for every branch. The process is repeated for the branch with the "best" subset using only the attributes that reaches the branch. The construction process stops when all nodes are pure.

4 Experimental Process

Due to high imbalance of examples between the two classification classes (539 No and 61 Yes) the SMOTE [26] preprocess algorithm was used. This algorithm creates synthetic examples of the minority class. To do so, it uses the k nearest neighbors of every example of the minority class. The minority class was thus oversampled by 783 % in order to even the examples in both classes resulting with 539 "No" and 538 "Yes" examples.

With the two classes evenly matched experiments were conducted using all the available classifying algorithms of WEKA.

Every time the 10-folds cross validation technique was used. The available examples are randomly portioned in 90 % for training and 10 % for testing. This process is repeated 10 times. In every tested algorithm the results are derived from combining the output of both 10 experiments conduct with randomly created training and test samples.

The following parameters where used:

The RandomForest algorithm was set to generate 100 trees and every time to use all the vector's features.

In the Simple cart algorithm, the internal cross-validation was made with 5 folds and the minimal number of objects at the terminal nodes was 2.

For the rule based classification algorithm Ridor 1 fold of the data was used for pruning and 2 folds for growing the rules.

In the BFTree algorithm the internal cross-validation was made with 5 folds and all the available data were used for training set.

In Table 1 the building model time is presented for every algorithm.

Table 1. Building time for the algorithm's model in seconds.

Classifier	Time
RandomForest	0.37
SimpleCart	0.14
Ridor	0.9
BFTree	0.13
NBTree	0.17

5   Evaluation

The evaluation of the algorithms results is made by means of precision and recall. The top five algorithms are Random Forest, Simple CART, Ridor, BFtree, and NBTree. All five algorithms exhibit precision and recall higher than 0.9, Fig. 1 below displays the classification results.

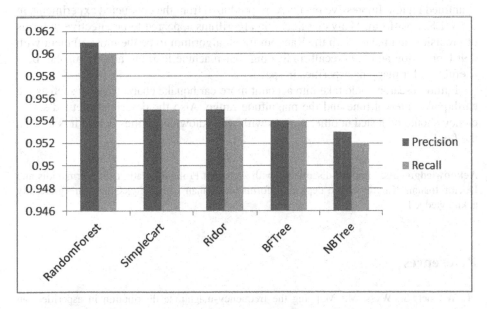

Fig. 1. Precision and recall comparison for five classification algorithms

The Random Forest algorithm really stands among the other four with precision 0.961 and recall 0.96. Simple CART, Ridor, and BFTree are really close with precision and recall ranking above 0.954. Table 2 contains precision, recall and the confusion matrix of the top five algorithms.

Table 2. Results of the top five algorithms.

Classifier	Precision	Recall	a	b	
RandomForest	0.961	0.96	509	30	a = No
			13	525	b = Yes
SimpleCart	0.955	0.955	504	35	a = No
			14	524	b = Yes
Ridor	0.955	0.954	497	42	a = No
			8	530	b = Yes
BFTree	0.954	0.954	506	33	a = No
			17	521	b = Yes
NBTree	0.953	0.952	500	39	a = No
			13	525	b = Yes

6   Conclusion

In the wider region of Hokkaido in north Japan, supervised machine learning algorithms were used to identify areas with asperity properties. The proposed feature vector consisted of geographic location information, the seismic density and b-value of the examined region. Impressive results were produced from the conducted experiments in the WEKA platform. Many classification algorithms appeared to be effective in terms of precision and recall, with the Random forest algorithm to be the most efficient with 0.961 precision and 0.96 recall, indicating that machine learning algorithms can be a useful tool for mapping asperities in space.

Future research could take into account more earthquake characteristics such as the earthquake interval time and the magnitude range. Also the described method's efficiency should be tested in other regions with well-known asperities before it is used in the field.

Acknowledgments. Fruitful discussions with Assistant Professor Katia Lida Kermanidis and Doctor Ioannis Karydis of the dept. of Informatics, Ionian University, Greece are gratefully acknowledged.

References

1. Wiemer, S., Wyss, M.: Mapping the frequency-magnitude distribution in asperities: an improved technique to calculate recurrence times? J. Geophys. Res. **102**, 115–128 (1997)
2. Takahashi, H., Kasahara, M.: Spatial relationship between interseismic seismicity, coseismic asperities and aftershock activity in the southwestern Kuril islands. In: Volcanism and Subduction: The Kamchatka Region (2013)
3. Hatzfeld, D., et al.: The Galaxidi earth-quake of 18 November 1992 a possible asperity within the normal fault system of the Gulf of Corinth (Greece). Bull. Seismol. Soc. Am. **86**, 1987–1991 (1996)

4. Park, S.C., Mori, J.: Are asperity patterns persistent Implication from large earthquakes in Papua New Guinea. J. Geophys. Res. Solid Earth. **112**, 3 (2007)
5. Dalguer, L.A., Irikura, K., Riera, J.D.: Simulation of tensile crack generation by three-dimensional dynamic shear rupture propagation during an earthquake. J. Geophys. Res. Solid Earth, **108** (2003)
6. Pulido, N.: Broadband frequency asperity parameters of crustal earthquakes from inversion of near-fault ground motion. In: 13th World Conference on Earthquake Engineering (2004)
7. Irikura, K., Miyake, H., Iwata, T., Kamae, K., Kawabe, H., Dalguer, A.: Recipe for predicting strong ground motions from future large earthquakes. In: 13th World Conference on Earthquake Engineering (2004)
8. Ozacar, A., Beck, S.L.: The 2002 Denali fault and 2001 Kunlun fault earthquakes: complex rupture processes of two large strike-slip events. Bull. Seismol. Soc. Am. **94**, 278–292 (2005)
9. Kagawa, T., Irikura, K., Somerville, P.G.: Differences in ground motion and fault rupture process between the surface and buried rupture earthquakes. Earth Planets Space **56**, 3–14 (2004)
10. Murotani, S., Satake, K., Fujii, Y.: Scaling relations of seismic moment, rupture area, average slip, and asperity size for M ~9 subduction-zone earthquakes. Geophys. Res. Lett. **40** (2013)
11. Spence, W., Mendoza, C., Engdahl, E.R., Choy, G.L., Norabuena, E.: Seismic subduction of the Nazca ridge as shown by the 1996–97 Peru earthquakes. Pure. appl. Geophys. **154**, 753–776 (1999)
12. Pulido, N., Aoi, S., Fujiwara, H.: Rupture process of the 2007 Notohanto earthquake by using an isochrones back-projection method and K-NET/KiK-net data. Earth Planets Space **60**, 1035–1040 (2008)
13. Ken-ichi, F., Daiki, I., Masayuki, N.: Discovering seismic interactions after the 2011 Tohoku earthquake by co-occurring cluster mining. Trans. Jpn. Soc. Artif. Intell. **29**(6), 493–502 (2014)
14. Otari, G.V., Kulkarni, R.V.: A review of application of data mining in earthquake prediction. Int. J. Comput. Sci. Inf. Technol. **3**, 3570–3574 (2012)
15. Panakkat, A., Adeli, H.: Neural network models for earthquake magnitude prediction using multiple seismicity indicators. Int. J. Neural Syst. **17**, 13–33 (2007)
16. Reyesa, J., Morales-Estebanb, A., Martínez-Álvarezc, F.: Neural networks to predict earthquakes in Chile. Appl. Soft Comput. **13**, 1314–1328 (2013)
17. Hall, M., Frank, E., Holmes, G., Pfahringer, P., Witten, I.H.: The WEKA data mining software: an update. ACM SIGKDD Explor. Newslett. **11**, 10–18 (2008)
18. Kulhanek, O.: Seminar on b-value. In: Department of Geophysics, Charles University, Prague (2005)
19. Aki, K.: Maximum likelihood estimate of b in the formula log N = a − bM and its confidence limits. Bull. Earthq. Res. Inst. **43**, 237–239 (1965)
20. Gutenberg, B., Richter, C.F.: Frequency of earthquakes in California. Bull. Seismol. Soc. Am. **34**, 185–188 (1944)
21. Wiemer, S.: A software package to analyze seismicity: ZMAP. Seismol. Res. Lett. **72**, 373–382 (2001)
22. Breiman, L.: Random forests. Mach. Learn. **45**, 5–32 (2001)
23. Gaines, B.R., Compton, P.: Induction of ripple-down rules applied to modeling large databases. J. Intell. Inf. Syst. **5**, 211–228 (1995)
24. Kalmegh, S.: Analysis of WEKA data mining algorithm REPTree, simple cart and randomtree for classification of indian news. Int. J. Innov. Sci. Eng. Technol. **2**, 438–446 (2015)
25. Shi, H.: Best-first decision tree learning. Thesis in Department of Computer Science, University of Waikato, Hamilton, New Zealand (2006)
26. Chawla, N.V., Bowyer, K.W., Hall, L.H., Kegelmeyer, W.P.: SMOTE: synthetic minority over-sampling technique. J. Artif. Intell. Res. **16**, 321–357 (2002)

Combining Prototype Selection
with Local Boosting

Christos K. Aridas$^{(\boxtimes)}$, Sotiris B. Kotsiantis, and Michael N. Vrahatis

Computational Intelligence Laboratory (CILab), Department of Mathematics,
University of Patras, 26110 Patras, Greece
char@upatras.gr, {sotos,vrahatis}@math.upatras.gr

Abstract. Real life classification problems require an investigation of relationships between features in heterogeneous data sets, where different predictive models can be more proper for different regions of the data set. A solution to this problem is the application of the local boosting of weak classifiers ensemble method. A main drawback of this approach is the time that is required at the prediction of an unseen instance as well as the decrease of the classification accuracy in the presence of noise in the local regions. In this research work, an improved version of the local boosting of weak classifiers, which incorporates prototype selection, is presented. Experimental results on several benchmark real-world data sets show that the proposed method significantly outperforms the local boosting of weak classifiers in terms of predictive accuracy and the time that is needed to build a local model and classify a test instance.

Keywords: Local boosting · Weak learning · Prototype selection · Pattern classification

1 Introduction

In machine learning, instance-based (or memory-based) learners classify an unseen object by comparing it to a database of pre-classified objects. The fundamental assumption is that similar instances will share similar class labels.

Machine learning models' assumptions would not necessarily hold globally. Local learning [1] methods come to solve this problem. The latter allow to extend learning algorithms, that are designed for simple models, to the case of complex data, for which the models' assumptions are valid only locally. The most common case is the assumption of linear separability, which is usually not fulfilled globally in classification problems. Despite this, any supervised learning algorithm that is able to find only a linear separation, can be used inside a local learning process, producing a model that is able to model complex non-linear class boundaries.

A technique of boosting local weak classifiers, that is based on a reduced training set after the usage of prototype selection [11], is proposed. It is common that boosting algorithms are well-known to be susceptible to noise [2]. In the case

© IFIP International Federation for Information Processing 2016
Published by Springer International Publishing Switzerland 2016. All Rights Reserved
L. Iliadis and I. Maglogiannis (Eds.): AIAI 2016, IFIP AICT 475, pp. 94–105, 2016.
DOI: 10.1007/978-3-319-44944-9_9

of local boosting, the algorithm should manage reasonable noise and be at least as good as boosting, if not better. For the experiments, we used two variants of Decision Trees [21] as weak learning models: one-level Decision Trees, which are known as Decision Stumps [12] and two-level Decision Trees. An extensive comparison over several data sets was performed and the results show that the proposed method outperforms simple and local boosting in terms of classification accuracy.

In the next Section, specifically in Subsect. 2.1, the localized experts are discussed, while boosting approaches are described in Subsect. 2.2. In Sect. 3 the proposed method is presented. Furthermore, in Sect. 4 the results of the experiments on several UCI data sets, after being compared with standard boosting and local boosting, are portrayed and discussed. Finally, Sect. 5 concludes the paper and suggests further directions in current research.

2   Background Material

For completeness purposes, local weighted learning, prototype selection methods as well as boosting classifier techniques are briefly described in the following subsections.

2.1   Local Weighted Learning and Prototype Selection

Supervised learning algorithms are considered global if they use all available training sets, in order to build a single predictive model, that will be applied in any unseen test instance. On the other hand, a method is considered local if only the nearest training instances around the testing instance contribute to the class probabilities.

When the size of the training data set is small in contrast to the complexity of the classifier, the predictive model frequently overfits the noise in the training data. Therefore, the successful control of the complexity of a classifier has a high impact in accomplishing good generalization. Several theoretical and experimental results [23] indicate that a local learning algorithm provides a reasonable solution to this problem.

In local learning [1], each local model is built completely independent of all other models in a way that the total number of local models in the learning method indirectly influences how complex a function can be estimated - complexity can only be controlled by the level of adaptability of each local model. This feature prevents overfitting if a strong learning pattern exists for training each local model.

Prototype selection is a technique that aims to decrease the training size without surfacing the prediction performance of a memory based learner [18]. Besides this, by reducing the training set size it might decrease the computational cost that will be applied in the prediction phase.

Prototype selection techniques can be grouped in three categories: preservation techniques, which aim to find a consistent subset from the training data set, ignoring the presence of noise, noise removal techniques, which aim to remove noise, and hybrid techniques, which perform both objectives concurrently [22].

2.2 Boosting Classifiers

Experimental research works have proven that ensemble methods usually perform better, in terms of classification accuracy, than the individual base classifier [2], and lately, several theoretical explanations have been advised to explain the success of some commonly used ensemble methods [13]. In this work, a local boosting technique that is based on a reduced training set, after the usage of prototype selection [11], is proposed and for this reason this section introduces the boosting approach.

Boosting constructs the ensemble of classifiers by subsequently tweaking the distribution of the training set based on the accuracy of the previously created classifiers. There are several boosting variants. These methods assign a weight to each training instance. Firstly, all instances are equally weighted. In each iteration a new classification model, named base classifier, is generated using the base learning algorithm. The creation of the base classifier has to consider the weight distribution. Then, the weight of each instance is adjusted, depending on the accuracy of the prediction of the base classifier for that instance. Thus, Boosting attempts to construct new classification models that are able to better classify the "hard" instances for the previous ensemble members. The final classification is obtained from a weighted vote of the base classifiers. AdaBoost [8] is the most well-known boosting method and the one that is used over the experimental analysis that is presented in Sect. 3.

Adaboost is able to use weights in two ways to generate a new training data set to provide to the base classifier. In boosting by sampling, the training instances are sampled with replacement with probability relative to their weights. In [26] the authors showed empirically that a local boosting-by-resampling technique is more robust to noise than the standard AdaBoost. The authors of [17] proposed a Boosted k-NN algorithm that creates an ensemble of models with locally modified distance weighting that has increased generalization accuracy and never performs worse than standard k-NN. In [10] the authors presented a novel method for instance selection based on boosting instance selection algorithms in the same way boosting is applied to classification.

3   The Proposed Algorithm

Two main disadvantages of simple local boosting are: (i) When the amount of noise is large, simple local boosting does not have the same performance [26] as Bagging [3] and Random Forest [4]. (ii) Saving the data for each pattern increases storage complexity. This might restrict the usage of this method to limited training sets [21]. The proposed algorithm incorporates prototype selection to handle, among others, the two previous problems. In the learning phase, a prototype selection [11] method based on the Edited Nearest Neighbor (ENN)

[24] technique reduces the training set by removing the training instances that do not agree with the majority of the k nearest neighbors. In the application phase, it constructs a model for each test instance to be estimated, considering only a subset of the training instances. This subset is selected according to the distance between the testing sample and the available training samples. For each testing instance, a boosting ensemble of a weak learner is built using only the training instances that are lying close to the current testing instance. The prototype selection aims to improve the classification accuracy as well as the time that is needed to build a model for each test instance at the prediction.

The proposed ensemble method has some free parameters, such as the number of neighbors (k_1) to be considered when the prototype selection is executed, the number of neighbors (k_2) to be selected in order to build the local model, the distance metric and the weak learner. In the experiments, the most well - known Euclidean similarity function was used as a distance metric.

In general, the distance between points $\mathbf{x}$ and $\mathbf{y}$ in a Euclidean space R^n is given by (1).

$$d(x, y) = \|x - y\|_2 = \sqrt{\sum_{i=1}^{n} |x_i - y_i|^2}. \tag{1}$$

The most common value for the nearest neighbor rule is 5. Thus, the k_1 was set to 5 and $k_2 = 50$. since at about this size of instances, it is appropriate for a simple algorithm to build a precise model [14]. The proposed method is presented in Algorithm 1.

Algorithm 1. PSLB(k_1, k_2, *distanceMetric, weakLearner*)

---

procedure TRAINING(k_1, *distanceMetric*)
    for each training instance **do**
        Find the k_1 nearest neighbors using the selected *distanceMetric*
        if instance does not agree with the majority of the k_1 **then**
            Remove this instance from the training set
        end if
    end for
end procedure
procedure CLASSIFICATION(k_2, *distanceMetric, weakLearner*)
    for each testing instance **do**
        Find the k_2 nearest neighbors using the selected *distanceMetric*
        Apply boosting to the base *weakLearner* using the k_2 nearest neighbors
        The answer of the boosting ensemble is the prediction for the testing instance
    end for
end procedure

---

4   Numerical Experiments

In order to evaluate the performance of the proposed method, an initial version was implemented[1] and a number of experiments were conducted using several data sets from different domains. From the UCI repository [16] several data sets were chosen. Discrete features transformed to numeric by using a simple quantization. Each feature is scaled to have zero mean and standard deviation one. Also all missing values were treated as zero. In Table 1 the name, the number of patterns, the attributes, as well as the number of different classes for each data set are shown.

Table 1. Benchmark data sets used in the experiments

Data set	#patterns	#attribues	#classes
cardiotocography	2126	21	10
cylinder-bands	512	25	2
dermatology	366	24	6
ecoli	336	7	8
energy-y1	768	8	3
glass	214	9	6
low-res-spect	531	100	9
magic	19020	10	2
musk-1	476	166	2
ozone	2536	72	2
page-blocks	5473	10	5
pima	768	8	2
synthetic-control	600	60	6
tic-tac-toe	958	9	2

All experiments were run on an Intel Core i3–3217U machine at 1.8 GHz, with 8 GB of RAM, running Linux Mint 17.3 64bit using Python and the scikit-learn [19] library.

For the experiments, we used two variants of Decision Trees [25] as weak learners. One-level Decision Trees [12], also known as Decision Stumps, and two-level Decision Trees [20]. We used the Gini Impurity [5] as criterion to measure the quality of the splits in both algorithms. The boosting process for all classifiers performed using the AdaBoost algorithm with 25 iterations in each model. In order to calculate the classifiers accuracy, the whole data set was divided into five mutually exclusive folds and for each fold the classifier was trained on the union of all of the other folds. Then, cross-validation was run five times for each algorithm and the mean value of the five folds was calculated.

---

[1] https://bitbucket.org/chkoar/pslb.

4.1   Prototype Selection

The prototype selection process is independent of the base classifier and it takes place once in the training phase of the proposed algorithm. It depends only on the k_1 parameter. The number of neighbors to be considered when the prototype selection is executed. In Table 2 the average of training patterns, the average of the removed patterns as well as the average reduction of each data set is presented. The average refers to the average of all training folds during the 5-fold cross-validation.

Table 2. Average reduction

Data set	#avg training patterns	#avg removed patterns	%avg reduction
cardiotocography	1701	268	15,73
cylinder-bands	410	60	14,60
dermatology	293	7	02,53
ecoli	269	29	10,86
energy-y1	614	17	02,77
glass	171	40	23,13
lowresspect	425	47	11,11
magic	15216	1798	11,81
musk-1	381	24	06,25
ozone	2029	50	02,48
page-blocks	4378	111	02,53
pima	614	109	17,77
synthetic-control	480	8	01,67
tic-tac-toe	766	1	00,13

4.2   Using Decision Stump as Base Classifier

In the first part of the experiments, Decision Stumps [12] were used as weak learning classifiers. Decision Stumps (DS) are one-level Decision Trees that classify instances based on the value of just a single input attribute. Each node in a decision stump represents a feature in an instance to be classified and each branch represents a value that the node can take. Instances are classified starting at the root node and are sorted based on their attribute values. In the worst case, a Decision Stump will behave as a base line classifier and will possibly perform better, if the selected attribute is particularly informative.

The proposed method, denoted as PSLBDS, is compared with the Boosting Decision Stumps, denoted as BDS and the Local Boosting of Decision Stumps, denoted as LBDS. Since the proposed method uses fifty neighbors, a 50-Nearest

Neighbors (50NN) classifier has included in the comparisons. In Table 3 the average accuracy of the compared methods is presented. Table 3 indicates that the hypotheses generated by PSLBDS are apparently better since the PSLBDS algorithm has the best mean accuracy score in nearly all cases.

Table 3. Average accuracy of the compared algorithms using an one-level decision tree as base classifier

Data set	PSLBDS	BDS	LBDS	50NN
cardiotocography	**0.682** ± 0.028	0.548 ± 0.088	0.659 ± 0.015	0.607 ± 0.029
cylinder-bands	**0.613** ± 0.030	0.560 ± 0.037	0.584 ± 0.014	0.582 ± 0.017
dermatology	**0.942** ± 0.027	0.641 ± 0.142	0.940 ± 0.022	0.902 ± 0.020
ecoli	**0.821** ± 0.029	0.622 ± 0.129	0.794 ± 0.026	0.780 ± 0.050
energy-y1	**0.844** ± 0.090	0.706 ± 0.050	0.836 ± 0.092	0.822 ± 0.091
glass	**0.582** ± 0.085	0.285 ± 0.094	0.568 ± 0.065	0.446 ± 0.169
low-res-spect	0.850 ± 0.025	0.584 ± 0.069	0.846 ± 0.012	**0.851** ± 0.023
magic	**0.849** ± 0.005	0.828 ± 0.005	0.834 ± 0.005	0.828 ± 0.004
musk-1	**0.727** ± 0.096	**0.727** ± 0.052	0.718 ± 0.085	0.618 ± 0.096
ozone	0.966 ± 0.008	0.960 ± 0.010	0.887 ± 0.133	**0.971** ± 0.001
page-blocks	**0.954** ± 0.012	0.853 ± 0.163	0.950 ± 0.013	0.942 ± 0.007
pima	**0.757** ± 0.028	0.755 ± 0.024	0.685 ± 0.024	0.749 ± 0.017
synthetic-control	**0.947** ± 0.011	0.472 ± 0.074	0.943 ± 0.020	0.887 ± 0.030
tic-tac-toe	**0.884** ± 0.084	0.733 ± 0.034	0.882 ± 0.083	0.747 ± 0.101

Demšar [6] suggests that the non-parametric tests should be preferred over the parametric in the context of machine learning problems, since they do not assume normal distributions or homogeneity of variance. Therefore, in the direction of validating the significance of the results, the Friedman test [9], which is a rank-based non-parametric test for comparing several machine learning algorithms on multiple data sets, was used, having as a control method the PSLBDS algorithm. The null hypothesis of the test states that all the methods perform equivalently and thus their ranks should be equivalent. The average rankings, according to the Friedman test, are presented in Table 4.

Assuming a significance level of 0.05 in Table 4, the p-value of the Friedman test indicates that the null hypothesis has to be rejected. So, there is at least one method that performs statistically different from the proposed method. With the intention of investigating the aforementioned, Finner's [7] and Li's [15] post hoc procedures were used.

In Table 5 the p-values obtained by applying post hoc procedures over the results of the Friedman statistical test are presented. Finner's and Li's procedure rejects those hypotheses that have a p-value ≤ 0.05. That said, the adjusted p-values obtained through the application of the post hoc procedures are presented in Table 6. Hence, both post hoc procedures agree that the PSLBDS algorithm performs significantly better than the BDS, the LBDS as well as the 50NN rule.

4.3 Using Two-Level Decision Tree as a Base Classifier

Afterwards, two-level Decision Trees were used as weak learning base classifiers. A two-level Decision Tree is a tree with max depth=2. The proposed method, denoted as PSLBDT, is compared to the Boosting Decision Tree, denoted as BDT and the Local Boosting of Decision Trees, denoted as LBDT. Since the proposed method uses fifty neighbors a 50-Nearest Neighbors (50NN) classifier has included in the comparisons. In Table 7 the average accuracy of the compared methods is presented. Table 7 indicates that the hypotheses generated by PSLBDT are apparently better, since the PSLBDT algorithm has the best mean accuracy score in most cases.

The average rankings, according to the Friedman test, are presented in Table 8. The proposed algorithm was ranked in the first place again. Assuming significance level of 0.05 in Table 8, the p-value of the Friedman test indicates that the null hypothesis has to be rejected. So, there is at least one method that performs statistically different from the proposed method. Aiming to investigate the aforesaid, Finner's and Li's post hoc procedures were used again.

In Table 9 the p-values obtained by applying post hoc procedures over the results of Friedman's statistical test are presented. Finner's and Li's procedure rejects those hypotheses that have a p-value $\leq$ 0.05. That said, the adjusted p-values obtained through the application of the post hoc procedures are presented in Table 10. Both post hoc procedures agree that the PSLBDT algorithm performs significantly better than the BDT and the 50NN rule but not significantly better than the LBDT as far as the tested data sets are concerned.

4.4 Time Analysis

One of the two contributions of this study was to improve the classification time over the local boosting approach. In order to prove this, the total time that is required to predict all instances in the test folds was recorded. Specifically,

Table 4. Average rankings of Friedman test (DS)

Algorithm	Ranking
PSLBDS	1.1429
LBDS	2.4286
50NN	2.8571
BDS	3.5714
Statistic	26.228571
p-value	0.000009

Table 5. Post hoc comparison for the Friedmans test (DS)

i	Algorithm	$z = (R_0 - R_i)/SE$	p	Finner	Li
3	BDS	4.97709	0.000001	0.016952	0.052189
2	50NN	3.51324	0.000443	0.033617	0.052189
1	LBDS	2.63493	0.008415	0.05	0.05

Table 6. Adjusted p-values (DS)

i	Algorithm	$p_{Unadjusted}$	p_{Finner}	p_{Li}
3	BDS	0.000001	0.000002	0.000001
2	50NN	0.000443	0.000664	0.000446
1	LBDS	0.008415	0.008415	0.008415

Table 7. Average accuracy of the compared algorithms using a two-level decision tree as base classifier

Data set	PSLBDT	BDT	LBDT	50NN
cardiotocography	0.683 ± 0.020	0.584 ± 0.072	$\mathbf{0.686 \pm 0.017}$	0.607 ± 0.029
cylinder-bands	$\mathbf{0.609 \pm 0.049}$	0.608 ± 0.034	0.564 ± 0.025	0.582 ± 0.017
dermatology	$\mathbf{0.958 \pm 0.040}$	0.800 ± 0.041	0.951 ± 0.020	0.902 ± 0.020
ecoli	$\mathbf{0.813 \pm 0.032}$	0.753 ± 0.036	0.800 ± 0.030	0.780 ± 0.050
energy-y1	$\mathbf{0.845 \pm 0.060}$	0.830 ± 0.064	0.844 ± 0.071	0.822 ± 0.091
glass	0.608 ± 0.048	0.569 ± 0.112	$\mathbf{0.652 \pm 0.057}$	0.446 ± 0.169
low-res-spect	$\mathbf{0.877 \pm 0.042}$	0.573 ± 0.136	0.872 ± 0.023	0.851 ± 0.023
magic	0.849 ± 0.006	$\mathbf{0.856 \pm 0.007}$	0.841 ± 0.006	0.828 ± 0.004
musk-1	0.738 ± 0.072	$\mathbf{0.752 \pm 0.024}$	0.746 ± 0.074	0.618 ± 0.096
ozone	0.967 ± 0.008	0.925 ± 0.064	0.888 ± 0.132	$\mathbf{0.971 \pm 0.001}$
page-blocks	$\mathbf{0.960 \pm 0.010}$	0.924 ± 0.023	0.956 ± 0.010	0.942 ± 0.007
pima	$\mathbf{0.763 \pm 0.023}$	0.742 ± 0.014	0.730 ± 0.018	0.749 ± 0.017
synthetic-control	0.950 ± 0.011	0.830 ± 0.036	$\mathbf{0.953 \pm 0.016}$	0.887 ± 0.030
tic-tac-toe	$\mathbf{0.893 \pm 0.078}$	0.665 ± 0.126	0.889 ± 0.081	0.747 ± 0.101

the prediction of each test fold was executed three times and the minimum time was recorded for each fold. Then, the average of all folds was calculated. In Table 11 the average prediction time in seconds of LBDS, PSLBDS, LBDT and PSLBDTS is presented. In the case of one-level decision trees (LBDS, PSLBDS) the proposed method reduced the expected prediction time in more than 15 % in 6 of 14 cases, while in the case of two-level decision trees (LBDT, PSLBDT) the proposed method reduced the expected prediction time in more than 15 % in 7 of 14 cases. In Fig. 1 the absolute percentage changes are presented.

Table 8. Average rankings of Friedman test (two-level tree)

Algorithm	Ranking
PSLBDT	1.5
LBDT	2.2857
50NN	3.0714
BDT	3.1429
Statistic	15
p-value	0.001817

Table 9. Post hoc comparison for the Friedmans test (two-level tree)

i	Algorithm	$z = (R_0 - R_i)/SE$	p	Finner	Li
3	BDT	3.366855	0.00076	0.016952	0.046982
2	50NN	3.22047	0.00128	0.033617	0.046982
1	LBDT	1.610235	0.107347	0.05	0.05

Table 10. Adjusted p-values (two-level tree)

i	Algorithm	$p_{Unadjusted}$	p_{Finner}	p_{Li}
3	BDT	0.00076	0.002279	0.000851
2	50NN	0.00128	0.002279	0.001432
1	LBDT	0.107347	0.107347	0.107347

Table 11. Average prediction times, in seconds

Data set	LBDS	PSLBDS	LBDT	PSLBDT
cardiotocography	33.89	33.26	32.43	29.20
cylinder-bands	8.16	8.07	8.45	7.86
dermatology	3.56	3.52	3.28	3.20
ecoli	5.00	3.61	4.66	2.92
energy-y1	8.58	7.19	7.59	6.25
glass	3.39	3.37	3.46	3.16
low-res-spect	6.74	6.38	5.77	3.77
magic	257.14	160.31	213.59	107.98
musk-1	9.53	9.50	8.80	7.99
ozone	14.84	4.89	7.24	1.69
page-blocks	17.27	9.34	12.28	4.27
pima	11.72	8.90	11.07	7.56
synthetic-control	6.32	6.12	3.89	3.76
tic-tac-toe	13.56	13.56	12.18	12.00

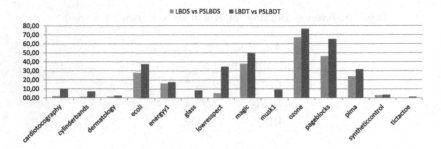

Fig. 1. Percentage change of prediction time between Local Boosting and the proposed method

5   Synopsis and Future Work

Local memory-based techniques delay the processing of the training set until they receive a request for an action like classification or local modelling. A data set of observed training examples is always retained and the estimate for a new test instance is obtained from an interpolation based on a neighborhood of the query instance.

In this research work at hand, a local boosting after prototype selection method is presented. Experiments on several data sets show that the proposed method significantly outperforms the boosting and local boosting method, in terms of classification accuracy and the time that is required to build a local model and classify a test instance. Typically, boosting algorithms are well known to be subtle to noise [2]. In the case of local boosting, the algorithm should handle sufficient noise and be at least as good as boosting, if not better. By means of the promising results obtained from performed experiments, one can assume that the proposed method can be successfully applied to the classification task in the real world case with more accuracy than the compared machine learning approaches.

In a following work the proposed method will be investigated as far as regression problems are concerned as well as the problem of reducing the size of the stored set of instances, by also applying feature selection instead of simple prototype selection.

References

1. Atkeson, C.G., Schaal, S., Moore, A.W.: Locally weighted learning. Artif. Intell. Rev. **11**(1), 11–73 (1997)
2. Bauer, E., Kohavi, R.: An empirical comparison of voting classification algorithms: bagging, boosting, and variants. Mach. Learn. **36**(1), 105–139 (1999)
3. Breiman, L.: Bagging predictors. Mach. Learn. **24**(2), 123–140 (1996)
4. Breiman, L.: Random forests. Mach. Learn. **45**(1), 5–32 (2001)
5. Breiman, L., Friedman, J., Stone, C., Olshen, R.: Classification and Regression Trees. Chapman & Hall, New York (1993)

6. Demar, J.: Statistical comparisons of classifiers over multiple data sets. J. Mach. Learn. Res. **7**, 1–30 (2006)
7. Finner, H.: On a monotonicity problem in step-down multiple test procedures. J. Am. Stat. Assoc. **88**(423), 920–923 (1993)
8. Freund, Y., Schapire, R.E.: Others: experiments with a new boosting algorithm. ICML. **96**, 148–156 (1996)
9. Friedman, M.: The use of ranks to avoid the assumption of normality implicit in the analysis of variance. J. Am. Stat. Assoc. **32**(200), 675 (1937)
10. Garca-Pedrajas, N., de Haro-Garca, A.: Boosting instance selection algorithms. Knowl. Based Syst. **67**, 342–360 (2014)
11. Garcia, S., Derrac, J., Cano, J.R., Herrera, F.: Prototype selection for nearest neighbor classification: taxonomy and empirical study. IEEE Trans. Pattern Anal. Mach. Intell. **34**(3), 417–435 (2012)
12. Iba, W., Langley, P.: Induction of one-level decision trees. In: Proceedings of the Ninth International Workshop on Machine Learning, pp. 233–240, ML 1992. Morgan Kaufmann Publishers Inc., San Francisco (1992)
13. Kleinberg, E.M.: A mathematically rigorous foundation for supervised learning. In: Kittler, J., Roli, F. (eds.) MCS 2000. LNCS, vol. 1857, pp. 67–76. Springer, Heidelberg (2000)
14. Kotsiantis, S.B., Kanellopoulos, D., Pintelas, P.E.: Local boosting of decision stumps for regression and classification problems. J. Comput. **1**(4), 30–37 (2006)
15. Li, J.: A two-step rejection procedure for testing multiple hypotheses. J. Stat. Plann. Infer. **138**(6), 1521–1527 (2008)
16. Lichman, M.: UCI Machine Learning Repository (2013)
17. Neo, T.K.C., Ventura, D.: A direct boosting algorithm for the k-nearest neighbor classifier via local warping of the distance metric. Pattern Recogn. Lett. **33**(1), 92–102 (2012)
18. Olvera-Lpez, J.A., Carrasco-Ochoa, J.A., Martnez-Trinidad, J.F., Kittler, J.: A review of instance selection methods. Artif. Intell. Rev. **34**(2), 133–144 (2010)
19. Pedregosa, F., Varoquaux, G., Gramfort, A., Michel, V., Thirion, B., Grisel, O., Blondel, M., Prettenhofer, P., Weiss, R., Dubourg, V., Vanderplas, J., Passos, A., Cournapeau, D., Brucher, M., Perrot, M., Duchesnay, D.: Scikit-learn: machine learning in Python. J. Mach. Learn. Res. **12**, 2825–2830 (2011)
20. Quinlan, J.R.: Induction of decision trees. Mach. Learn. **1**(1), 81–106 (1986)
21. Rokach, L.: Ensemble-based classifiers. Artif. Intell. Rev. **33**(1–2), 1–39 (2010)
22. Segata, N., Blanzieri, E., Delany, S.J., Cunningham, P.: Noise reduction for instance-based learning with a local maximal margin approach. J. Intell. Inf. Syst. **35**(2), 301–331 (2010)
23. Vapnik, V.N.: Statistical Learning Theory: Adaptive and Learning Systems for Signal Processing, Communications, and Control. Wiley, New York (1998)
24. Wilson, D.L.: Asymptotic properties of nearest neighbor rules using edited data. IEEE Trans. Sys. Man Cybern. **2**(3), 408–421 (1972)
25. Witten, I.H., Frank, E., Hall, M.A.: Data Mining: Practical Machine Learning Tools and Techniques. Morgan Kaufmann Series in Data Management Systems, 3rd edn. Morgan Kaufmann, Burlington (2011)
26. Zhang, C.X., Zhang, J.S.: A local boosting algorithm for solving classification problems. Comput. Stat. Data Anal. **52**(4), 1928–1941 (2008)

Convolutional Neural Networks
for Pose Recognition in Binary
Omni-directional Images

S.V. Georgakopoulos[1], K. Kottari[1], K. Delibasis[1], V.P. Plagianakos[1],
and I. Maglogiannis[2(✉)]

[1] Department of Computer Science and Biomedical Informatics,
University of Thessaly, Lamia, Greece
{spirosgeorg, vpp}@dib.uth.gr,
kottarikonstantina@gmail.com, kdelibasis@gmail.com
[2] Department of Digital Systems, University of Piraeus, Piraeus, Greece
imaglo@unipi.gr

Abstract. In this work, we present a methodology for pose classification of silhouettes using convolutional neural networks. The training set consists exclusively from the synthetic images that are generated from three-dimensional (3D) human models, using the calibration of an omni-directional camera (fish-eye). Thus, we are able to generate a large volume of training set that is usually required for Convolutional Neural Networks (CNNs). Testing is performed using synthetically generated silhouettes, as well as real silhouettes. This work is in the same realm with previous work utilizing Zernike image descriptors designed specifically for a calibrated fish-eye camera. Results show that the proposed method improves pose classification accuracy for synthetic images, but it is outperformed by our previously proposed Zernike descriptors in real silhouettes. The computational complexity of the proposed methodology is also examined and the corresponding results are provided.

Keywords: Computer vision · Convolutional neural networks (CNNs) · Omnidirectional image · Fish-eye camera calibration · Pose classification · Synthetic silhouette

1 Introduction

Several computer vision and Artificial Intelligence applications require classification of segmented objects in digital images and videos. The use of object descriptors is a conventional approach for object recognition though a variety of classifiers. Recently, many reports have been published supporting the ability of automatic feature extraction by Convolutional Neural Networks (CNNs) that achieve high classification accuracy in many generic object recognition tasks, without the need of user-defined features. This approach is often referred to as deep learning.

More specifically, CNNs are state of the art classification methods in several problems of computer vision. They have been suggested for pattern recognition [2], object localization [3], object classification in large-scale database of real world images

© IFIP International Federation for Information Processing 2016
Published by Springer International Publishing Switzerland 2016. All Rights Reserved
L. Iliadis and I. Maglogiannis (Eds.): AIAI 2016, IFIP AICT 475, pp. 106–116, 2016.
DOI: 10.1007/978-3-319-44944-9_10

[4], and malignancy detection on medical images [5–7] Several reports exist in literature for the problem of human pose estimation and can be categorized in two approaches. The first approach relies on leveraging images local descriptors (HoG [8], SHIFT [9], Zernike [1, 10, 11, 12] to extract features and subsequently constructing a model for classification. The second approach is based on model fitting processes [13, 14].

CNNs are trainable multistage architectures that belong to the first approach of classification methods [15]. Basically each stage comprises of three types of layers, the convolution, the pooling and the classic neural network layer, which is commonly refereed as fully-connected layer. Each stage consists of one of the previous layer type or an arbitrary combination of them. The trainable components of a convolutional layer are mapped as a batch of kernels and perform the convolution operator on the previous layer's output. The pooling layer performs a subsampling to its input with most commonly used pooling function to be the max-pooling, taking the maximum value of the local neighborhoods. Finally, the fully-connected layer can be treated as a special case of kernel with size 1×1. To train this network, the Stochastic Gradient Descent is usually utilized with the usage of mini-batches [16]. However, a drawback of the CNNs is the extensive training time required because of the amount of trainable parameters. Due to the inherent parallelism in their structure, the usage of the graphics processing units (GPUs) has been established to perform the training phase [4]. To achieve high quality results, the CNNs require training dataset of large size.

Very recently, methods that use deep neural networks in order to tackle the problem of human pose estimation have started to appear in the literature. In [17] the use the CNN as a regressor for rough joint locator hand-annotations of the Frame Labeled In Cinema (FLIC) dataset is reported. While the results are very promising, the existence of hand-annotations on training set is needed.

In this work, we test the suitability of a well-established CNN methodology for pose recognition in synthetic binary silhouette images. To the best of our knowledge CNNs have not been utilized to deal with binary images problems. These types of problems are distinct, because information is limited by the lack of RGB data. Furthermore, the silhouettes of this work are rendered through an omni-directional camera – fisheye. Fisheye cameras are dioptric omni-directional cameras, increasingly used in computer vision applications [18, 19], due to their 180 degree field of view (FoV). In [20–22] the calibration of fish-eye camera is reported to emulate the strong deformation introduced by the fish-eye lens. In [23] a methodology for correcting the distortions induced by the fish-eye lens is presented. The high volume of artificial silhouettes required for training the CNN, is produced by using 3D human models rendered by a calibrated fish-eye camera. Comparisons are provided for synthetic and real data with our recent method proposed in [1].

2 Methodology

2.1 Overview of the Method

The main goal of this work is the assessment of the popular CNN technique ability to recognize different poses of binary human silhouettes from indoor images acquired by a roof-based omnidirectional camera. An extensive dataset of binary silhouettes is created using 3D models [24, 25] of a number of subjects in 5 different standing positions. The 3D models are placed in different positions and at different rotations round the Z-axis in the real world room. Then they are rendered through the calibration of the fish-eye camera, generating binary silhouettes. The dataset of binary silhouettes is separated into training and testing subsets. Classification results are calculated using the testing subset, as well as real segmented silhouettes of approximately the same poses. The aforementioned steps of the proposed methodology are shown in the block diagram of Fig. 1.

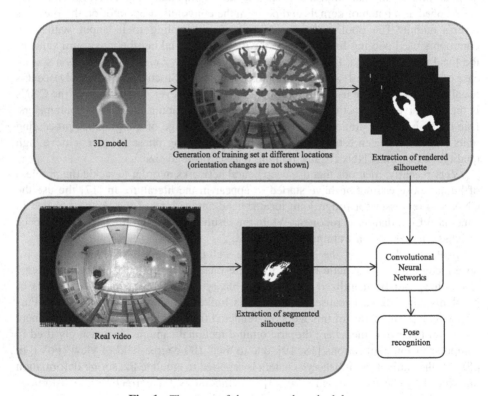

Fig. 1. The steps of the proposed methodology

2.2 Calibration of the Fish-Eye Camera

The fish-eye camera is calibrated using a set of manually provided points, as described in detail in [26]. The achieved calibration compares favourably to other state of the art methodologies [27] in terms of accuracy. In abstract level, the calibration process defines a function F that maps a real world point $\mathbf{x_{real}} = (x_{real}, y_{real}, z_{real})$ to coordinates (i,j) of an image frame:

$$F(\mathbf{x_{real}}) = (i,j) \tag{1}$$

The resulting calibration is visualized in Fig. 2 for a grid of points virtually placed on the floor and on the walls of the room.

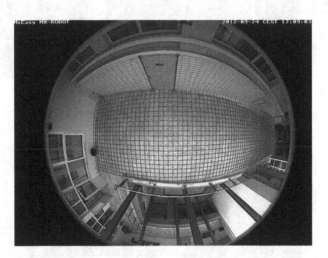

Fig. 2. Visualization of the resulting fish-eye model calibration, on the FoV of the indoor environment in which experiments have taken place. The landmark points defined by the user are shown as circles and their rendered position on the frame marked by stars.

Let B be the binary frame. If we apply the above equation for each $\mathbf{x_{real}}$ point of a 3D model and set $B(i,j) = 1$, then $\mathbf{B}$ will contain the silhouette as imaged by the fish-eye camera.

2.3   Synthetically Generated Silhouettes

In this work, a number of 3D models (Fig. 3), obtained from [24, 25], are utilized. These models have known real-world coordinates stored in the form of triangulated surfaces. However, only the coordinates of the vertices are used for rendering the binary silhouette frames. Rendering of the models and generation of a silhouette in a binary frame B was achieved by using the parameterized camera calibration as following:

$$B\bigl(F\bigl(\mathbf{x_{real,k}}\bigr)\bigr) = 1 \tag{2}$$

where $\{\mathbf{x_{real,k}}, k = 1, 2, \ldots, N\}$ are the points of a 3D model. Every model was placed in the viewed room at different locations, having different orientations (rotation round the Z-axis).

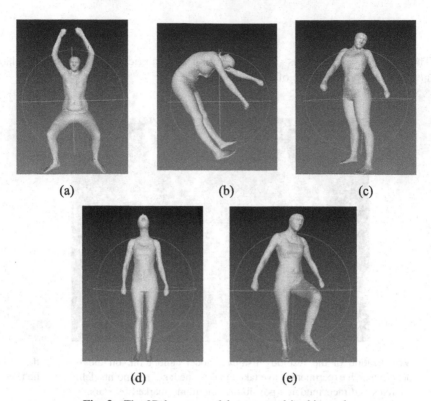

(a)   (b)   (c)

(d)   (e)

Fig. 3. The 3D human model poses used in this work.

2.4   Convolutional Neural Networks

The CNN was implemented with four convolutional layers; each one of the first three is followed by a max-pooling layer, while the fourth convolutional layer is followed by a fully-connected feed-forward Neural Network with two hidden layers (see Fig. 4). There exist four convolutional layers, each one consists of a $n \times n$, filter, for $n = 5, 4, 3$ and 2 respectively. Pooling filters of a 2×2 dimension exists between each two successive convolutional layers. The convolutional layers consist of 16, 16, 32, and 32 filters, respectively, while the max-pooling layers utilize 16, 16, and 32 filters. Finally, the Feed Forward Neural Network consists of one layer with 64 neurons, followed by 20 hidden neurons and one output layer of the generalization of logistic regression function, the softmax functions, where maps the CNN's input into the class probabilities and is commonly used in CNNs. In order to exploit the power of the CNNs that relies on the depth of their layers and at the same time considering the limitations of the GPU memory, we feed the network with binary images sized to 113×113 pixels that contain the synthetic silhouettes. The structure of the CNN used in this work is constructed as following: The inputs of the network are binary image silhouettes. The network consists of four convolutional layers, two fully connected layers and the output layer, which are succeeded by a max-pooling layer. The aforementioned construction is illustrated in Fig. 4.

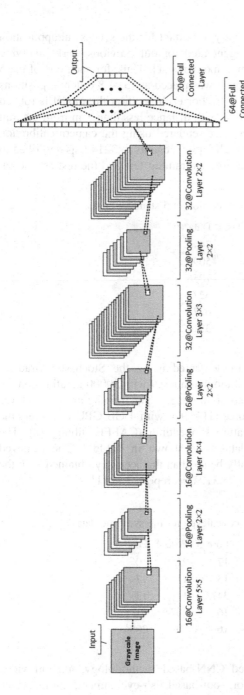

Fig. 4. An illustration of the employed CNN architecture.

3   Results

The generation of an extensive dataset is very important for the successful application of CNNs. We used the dataset that we generated for our previous work involving evaluation of geodetically corrected Zernike moments [1]. Thus, for each one of the 5 poses, the 3D model of each available subject was placed at 13×8 different positions defined on a grid with constant step of 0.5 m. For each position the model is rotated round the Z-axis every $\pi/5$ radians. Positions with distance less than 1 m from the trail of the camera were excluded. Silhouettes were rendered using the camera calibration (Subsect. 2.2).The total number of data for all 5 poses equals to 32142, as described in Table 1. The 50 % of the silhouettes was used as training test and the rest as test set.

Table 1.   Dataset for the classification of 5 standing poses.

Pose (Class)	Data
1	7446
2	5096
3	7056
4	6272
5	6272
Sum	32142

The CNN learning algorithm was implemented using the Stochastic Gradient Descent (SGD) with learning rate 0.01 and momentum 0.9 with 30000 iterations on the 50 % of the whole dataset and a mini-batch of 50 images. The training of the CNN was performed using the GPU NVIDIA GeForce GTX 970 with 4 GB GPU-RAM and the Convolutional Architecture for Fast Feature Embedding (CAFFE) library [2]. The confusion matrix for the 5 synthetic datasets is shown in Table 2. The achieved accuracy was 98.08 %, which is marginally better than the accuracy obtained with the geodetically corrected Zernike moments (95.31 % as reported in [1]).

Table 2.   The confusion matrix achieved for the synthetic data.

	Pose 1	Pose 2	Pose 3	Pose 4	Pose 5
Pose 1	3663	12	17	21	10
Pose 2	17	2490	13	11	17
Pose 3	13	12	3473	17	13
Pose 4	18	17	16	3074	11
Pose 5	13	11	16	10	3086

In order to further test the proposed CNN-based methodology, 4 short video sequences were acquired using the indoor, roof-based fish-eye camera, during which the subjects assumed two generic poses: "standing" and "fallen". Each frame of the real video was manually labelled. The training set that was used for training the CNN and the other classifiers in the case of ZMI [1] for the two generic poses (standing and fallen) was obtained as following: The dataset for the generic "standing" pose consists of the union of the 5 standing poses (of Fig. 3). The dataset for the generic "fallen" pose consists of prone/back and side falling, generated from the generic standing pose by interchanging Z-axis with Y-axis coordinate and Z-axis with X-axis, respectively.

This dataset (which does not contain any real silhouettes) was used in order to train the CNN and the other classifiers to evaluate the 2 generic poses and recognise them in the real silhouettes. Comparative results with our previous approach (geodetically corrected Zernike moments [1]) are shown in Table 3. It can be seen that in real data, our previously proposed geodesically-corrected ZMI (GZMI) have better discriminating power compared to the CNN. Based on our results so far, it appears that the GZMI are more immune to imperfect segmentation than the CNN. Possible explanations on this observation are suggested in the next section.

Table 3. The classification results for the standing/fallen generic classes.

	Accuracy	Sensitivity	Specificity
real video, CNN	0.6715	0.5281	0.6948
real video, GZMI	0.8173	0.8854	0.7864

The achieved accuracy of the CNN compares favorably with the accuracy achieved using the recently proposed GZMI [1] for different sizes of the training subset. Figure 5 shows, comparatively, the accuracies achieved by the two methods for training set equal to 10 % up to 50 % of the total dataset of the synthetically generated poses.

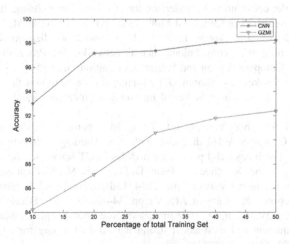

Fig. 5. The accuracy of the test set as a function of the size of the training set comparatively for the two methods. The sizes of the indicated training sets are equal to 10 % up to 50 % (with a step of 10 %) of the total dataset. It can be observed that the accuracy of the proposed CNN (upper blue line) compares favorably to the accuracy of the GZMI (lower magenta line) of [1].

4   Conclusions and Further Work

The purpose of this work was to assess the ability of convolutional neural networks – CNNs – to correctly classify synthetically generated, as well as real silhouettes, using an extensive synthetic training set. The training set consists exclusively from the synthetic images that are generated from three-dimensional (3D) human models, using a calibrated omni-directional camera (fish-eye). Our results show that the proposed CNN approach is marginally better than the geodesic Zernike moment invariants (GZMI) proposed in our recent work [1], but appears to be outperformed in the problem of real silhouette classification. The GZMI features were adapted from their classic definition, using geodesic distances and angles defined by the camera calibration. On the other hand, the CNN generates features that minimize classification error during the training phase, but do not correspond directly to physical aspects of omni-directional image formation. The results that are reported in this work indicate that the proposed GZMI features appear to be more robust to noise induced by imperfect segmentation, than the features generated by CNN. However, more experimentation is required to draw more definite conclusions. Thus, it is in our future steps to further investigate the structure and the parameters of the utilized CNN in order to improve the performance.

References

1. Delibasis, K.K., Georgakopoulos, S.V., Kottari, K., Plagianakos, V.P., Maglogiannis, I.: Geodesically-corrected Zernike descriptors for pose recognition in omni-directional images. Integr. Comput. Aided Eng. **23**(1), 1–15 (2016)
2. Jia, Y., Shelhamer, E., Donahue, J., Karayev, S., Long, J., Girshick, R., Guadarrama, S., Darrell, T.: Caffe: convolutional architecture for fast feature embedding. In: Proceedings of the ACM International Conference on Multimedia, pp. 675–678. ACM, November 2014
3. Oquab, M., Bottou, L., Laptev, I., Sivic, J.: Is object localization for free?-weakly-supervised learning with convolutional neural networks. In: Proceedings of the IEEE Conference on Computer Vision and Pattern Recognition, pp. 685–694 (2015)
4. Krizhevsky, A., Sutskever, I., Hinton, G. E.: Imagenet classification with deep convolutional neural networks. In: Advances in Neural Information Processing Systems, pp. 1097–1105 (2012)
5. Cheng, J. Z., Ni, D., Chou, Y. H., Qin, J., Tiu, C. M., Chang, Y. C., Huang, C. S., Shen, D., Chen, C. M.: Computer-Aided diagnosis with deep learning architecture: applications to breast lesions in us images and pulmonary nodules in CT scans. Sci. Rep., 6, 1–13 (2016)
6. Li, Q., Cai, W., Wang, X., Zhou, Y., Feng, D. D., Chen, M.: Medical image classification with convolutional neural network. In: 2014 13th International Conference on Control Automation Robotics & Vision (ICARCV), pp. 844–848). IEEE, December 2014
7. Suk, H.I., Lee, S.W., Shen, D.: Alzheimer's disease neuroimaging initiative: hierarchical feature representation and multimodal fusion with deep learning for AD/MCI diagnosis. NeuroImage **101**, 569–582 (2014)

8. Junior, O. L., Delgado, D., Gonçalves, V., Nunes, U.: Trainable classifier-fusion schemes: an application to pedestrian detection. In: Intelligent Transportation Systems, vol. 2, October 2009

9. Tamimi, H., Andreasson, H., Treptow, A., Duckett, T., Zell, A.: Localization of mobile robots with omnidirectional vision using particle filter and iterative sift. Robot. Auton. Syst **54**(9), 758–765 (2006)

10. Hwang, S. K., Billinghurst, M., Kim, W. Y.: Local descriptor by zernike moments for real-time keypoint matching. In: 2008 Congress on Image and Signal Processing, CISP 2008, vol. 2, pp. 781–785. IEEE, May 2008

11. Zhu, H., Shu, H., Xia, T., Luo, L., Coatrieux, J.L.: Translation and scale invariants of Tchebichef moments. Pattern Recogn. **40**(9), 2530–2542 (2007)

12. Shutler, J. D., Nixon, M. S.: Zernike velocity moments for description and recognition of moving shapes. In: BMVC, pp. 1–10, September 2001

13. Yang, Y., Ramanan, D.: Articulated pose estimation with flexible mixtures-of-parts. In: 2011 IEEE Conference on Computer Vision and Pattern Recognition (CVPR), pp. 1385–1392. IEEE, June 2011

14. Kottari, K., Delibasis, K., Plagianakos, V., Maglogiannis, I.: Fish-eye camera video processing and trajectory estimation using 3d human models. In: Iliadis, L., Maglogiannis, I., Papadopoulos, H. (eds.) Artificial Intelligence Applications and Innovations. IFIP Advances in Information and Communication Technology, vol. 436, pp. 385–394. Springer, Heidelberg (2014)

15. LeCun, Y., Bottou, L., Bengio, Y., Haffner, P.: Gradient-based learning applied to document recognition. Proc. IEEE **86**(11), 2278–2324 (1998)

16. Bottou, L.: Stochastic gradient descent tricks. In: Montavon, G., Orr, G.B., Müller, K.-R. (eds.) Neural Networks: Tricks of the Trade, 2nd edn. LNCS, vol. 7700, pp. 421–436. Springer, Heidelberg (2012)

17. Toshev, A., Szegedy, C.: Deeppose: human pose estimation via deep neural networks. In: Proceedings of the IEEE Conference on Computer Vision and Pattern Recognition, pp. 1653–1660 (2014)

18. Kemmotsu, K., Tomonaka, T., Shiotani, S., Koketsu, Y., Iehara, M.: Recognizing human behaviors with vision sensors in a network robot system. In: Proceedings of IEEE International Conference on Robotics and Automation, pp. 1274–1279 (2006)

19. Zhou, Z., Chen, X., Chung, Y., He, Z., Han, T. X. Keller, M.: Activity analysis, summarization and visualization for indoor human activity monitoring. IEEE Trans. on Circuit and Systems for Video Technology, 18(11): 1489–1498 (2008)

20. Mei, C., Rives, P.: Single view point omnidirectional camera calibration from planar grids. In: IEEE International Conference on Robotics and Automation, (ICRA), pp. 3945–3950. IEEE, Rome (2007)

21. Li, H., Hartley, R.I.: Plane-based calibration and auto-calibration of a fish-eye camera. In: Narayanan, P.J., Nayar, S.K., Shum, H.-Y. (eds.) ACCV 2006. LNCS, vol. 3851, pp. 21–30. Springer, Heidelberg (2006)

22. Shah, S., Aggarwal, J.: Intrinsic parameter calibration procedure for a high distortion fish-eye lens camera with distortion model and accuracy estimation. Pattern Recogn. **29**(11), 1775–1788 (1996)

23. Wei, J., Li, C.F., Hu, S.M., Martin, R.R., Tai, C.L.: Fisheye video correction. IEEE Trans. Vis. Comput. Graph. **18**(10), 1771–1783 (2012)

24. Hasler, N., Ackermann, H., Rosenhahn, B., Thormahlen, T., Seidel, H.P.: Multilinear pose and body shape estimation of dressed subjects from image sets. In: IEEE Conference on Computer Vision and Pattern Recognition (CVPR 2010), pp. 1823–1830 (2010)

25. http://resources.mpi-inf.mpg.de/scandb/
26. Delibasis, K., Plagianakos, V., Maglogiannis, I.: Refinement of human silhouette segmentation in omni-directional indoor videos. Comput. Vision Image Underst. **128**, 65–83 (2014)
27. Rufli, M., Scaramuzza, D., Siegwart, R.: Automatic detection of checkerboards on blurred and distorted images. In: Proceedings of the IEEE/RSJ International Conference on Intelligent Robots and Systems (IROS 2008), pp. 3121–3126. Nice, France (2008)

Ontology-Web and Social Media AI Modeling (OWESOM)

The eLOD Ontology:
Modeling Economic Open Data

Michalis Vafopoulos[1], Gerasimos Razis[2],
Ioannis Anagnostopoulos[2(✉)], Georgios Vafeiadis[1],
Dimitrios Negkas[3], Eleftherios Galanos[3], Aggelos Tzani[3],
Ilias Skaros[3], and Konstantinos Glykos[3]

[1] Software and Knowledge Engineering Laboratory,
IIT, NCSR-"Demokritos", Athens, Greece
vaf@aegean.gr, vafeiadis.giorgos@gmail.com
[2] Department of Computer Science and Biomedical Informatics,
School of Sciences, University of Thessaly, Lamia, Greece
{razis, janag}@dib.uth.gr
[3] University of Piraeus Research Centre, Piraeus, Greece
linkedeconomy@gmail.com

Abstract. For decades, valuable economic data was out of reach for most of the people. Gradually, public budgets and tenders are becoming openly available and global initiatives promote financial transparency and data innovation. But, yet the poor quality of open data undermines their potential to answer interesting questions (e.g. efficiency of public funds and market processes). The eLOD ontology has been initiated as a top-level conceptualization that interlinks the publicly available economic open data by modelling the flows incorporated in public procurement together with the market process to address complex policy issues. This paper presents the basic aspects of eLOD ontology in interlinking and querying diverse open data ranging from budget execution to prices. Already, eLOD ontology is used by two EU projects in order to to develop new systems, to enable information exchange between systems, to integrate data from heterogeneous sources and to publish open data related to economic activities.

Keywords: Linked data · Semantic web · Knowledge representation · Economy

1 Introduction – Related Works

In this paper, we introduce the eLOD Ontological schema as the "heart" and basic data modelling and handling infrastructure of LinkedEconomy[1] portal. LinkedEconomy aims at providing a universal access to Greek and international economy data, as well as at promoting the benefits of linking heterogeneous sources under the concept of open and reusable data in this critical domain [1]. In addition, it targets to offer a unique platform of rich linked economy data for enhancing the citizens' awareness in respect to economic issues in Greece and worldwide, as well as to provide curated and

---

[1] http://linkedeconomy.org/en/.

© IFIP International Federation for Information Processing 2016
Published by Springer International Publishing Switzerland 2016. All Rights Reserved
L. Iliadis and I. Maglogiannis (Eds.): AIAI 2016, IFIP AICT 475, pp. 119–130, 2016.
DOI: 10.1007/978-3-319-44944-9_11

semantified data to the Linked Open Data research community. It is the ancestor of the Public Spending (PSNET) initiative[2], which was the first attempt to harvest, align, interlink, analyze and distribute as Linked Open Data massive amounts of public spending data from Greece and from six other (local and national) governments [2]. Related worldwide initiatives are driven from similar Open Government Data projects. Below, we classify them in four major fields according to their applications in the economy domain.

In the field of open budgets, the International Budget Partnership advocates for public access to accountable budget systems. One of its projects, the Open Budget Survey Tracker[3] allows citizens to monitor whether central governments are releasing the requisite information on how the government is managing public finances. Starting from 2015, EU funds the Open Budgets project[4] to provide a scalable platform for public administrations to publish open budget data that is easy-to-use, flexible, and attractive for all. In addition, the Open Spending Initiative[5] offers an easy system to upload, explore and share public finance data (e.g. budgets or expenditure databases). In 2013, more than 16 million transactions from 363 different datasets around the world have been uploaded. A similar project from the Sunlight Foundation[6] analyzes the spending data uploaded in the official portal of US government. OpenTED[7] is an initiative that provides data dumps of tenders from the joint European procurement system more easily accessible to journalists and researchers.

In the field of data standards, the Fiscal Data Package[8] is developed as a simple, open technical specification for government budget and spending data. Public Contracts Ontology has been introduced by the LOD2 project [3] to provide an ontological basis for representing key concepts in tenders and expressing structured data about public contracts. The Core vocabularies have been developed by European Commission's Interoperability Solutions for European Public Administrations (ISA) programme as simplified, generic, re-usable and extensible data standards that model the characteristics of basic entities in a context-neutral and syntax-neutral fashion. Most of them have been incorporated in our modelling, as it will be described below. Open Contracting Partnership[9] is a consultation process to create a set of global principles that can serve as a guide to advance open contracting around the world. In this context, the Open Contracting Data Standard (OCDS) sets out key documents and data that should be published at each stage of a contracting process.

In the field of company data, Open Corporates[10] aggregates company information from different countries and jurisdictions. The specific team is working on creating

---

[2] http://publicspending.net.
[3] http://obstracker.org.
[4] http://openbudgets.eu.
[5] http://openspending.org.
[6] http://clearspending.org.
[7] http://ted.openspending.org/.
[8] http://fiscal.dataprotocols.org.
[9] http://www.open-contracting.org/.
[10] http://opencorporates.com.

Linked Data representations out of their databases, by mapping company metadata to certified ontologies such as the Core Business Vocabulary and linking them to other data hubs, such as DBpedia.org and Geonames. Financial statements and accounting reports that are published by companies contain important data.

In the area of product data, Open Product Data - POD[11] hosted by OKF as a community project, stands as a public database of product data connected to barcodes, in order to empower consumers with useful and machine-readable product information (e.g. prices).

Other related research approaches have a two-fold orientation. In the first fold, we have efforts that analyse the benefits of using open data in economy, trying in parallel to provide a unified framework and modelling strategy [4–6], while, in the second fold, we have the contributions that integrate the benefits of openness, thus creating a flourishing linked ecosystem of economy data [7–10].

2 eLOD Ontology: Modelling Economic Data Under Semantics

In this section, we describe some well-known and established structured vocabularies, which are used in the economy domain, as well as the model that incorporates interconnections between public finances and market processes in Greece and consists of the main classes and properties of our ontology.

2.1 Description of Sources and Vocabularies Used

Table 1 depicts the data sources used as well as some information regarding them (API support, URI). During the ontology design phase, our basic aim was to reuse other well-known and established vocabularies and ontologies that cover our needs. When existing models were inadequate we introduced and defined new concepts. The well-established ontologies and vocabularies we used are FOAF[12], GoodRelations[13], Public Contracts[14], Organization Ontology[15], Registered Organization Vocabulary[16], Dublin Core[17], SKOS[18] and vCard[19], while we also use some properties from the DBpedia Ontology[20].

---

[11] http://product.okfn.org.

[12] http://www.foaf-project.org.

[13] http://purl.org/goodrelations.

[14] https://code.google.com/p/public-contracts-ontology.

[15] http://www.w3.org/TR/vocab-org.

[16] http://www.w3.org/TR/vocab-regorg.

[17] http://dublincore.org.

[18] http://www.w3.org/2004/02/skos.

[19] http://www.w3.org/TR/vcard-rdf.

[20] http://wiki.dbpedia.org/services-resources/ontology.

Table 1. Related information for data sources used

Data sources used

Name	Provision through API	URI
Transparency	Yes	diavgeia.gov.gr
Central electronic registry of public procurement	No	www.eprocurement.gov.gr
National strategic reference framework	No	www.espa.gr/en
Central market of Thessaloniki	No	www.kath.gr
Prices observatory	Yes	www.e-prices.gr
Fuel prices observatory	Yes	www.fuelprices.gr
Municipality of Athens	No	www.cityofathens.gr/khe/ proypologismos
Municipality of Thessaloniki	No	http://www.thessaloniki.gr

FOAF, which is an acronym of "Friend of a Friend", is a vocabulary for describing people, their activities and relationships with other people and objects. This concept can be generalized in order to describe all types of entities, named "agents", who are responsible for specific actions. FOAF aims to describe the world by using simple ideas inspired by the Web. In our economic context, this vocabulary is used in order to define and describe the agents who are responsible for specific actions. Two categories of agents are used, namely "Persons" and "Organizations".

GoodRelations is an ontology, which aims at defining a data structure for products related to electronic commerce, the prices, the stores and the data of the companies. Its use allows the expression of the commercial and operational details of scenarios for e-commerce. The main entities in this domain are the involved agents, in terms of persons or companies, the objects of the commercial activities, the items for sale, lease or repair, as well as the locations where such offers are available. In our economic context, this ontology is used in order to identify and describe the "Business Entities" which are involved in a commercial activity (their legal names and their Vat Ids), the type of their services, and the financial details of the contract or of the payment (i.e. the wholesale or retail price, whether tax is included in the price and the expressed currency of the price). Moreover, it is also used to express a point or area of interest from which a particular product or service is available.

Public Contracts is an ontology that aims at describing the contracts in the public sector. It is based on the "GoodRelations" ontology for the modeling of business entities and price specifications.

In our economic context, this ontology is used in order to define and describe the following:

- the public contracts during all stages of their existence (tender, contract, payment),
- the procedures specifying how the details of a contract are published and how a supplier is chosen,
- the basic focus of the contract (i.e. works, supplies or services),

- the price of the contract, depending on its phase (before or after the offer),
- the award criteria that define the conditions under which the best offer will be selected and awarded along with their weights, and
- the main and supplementary products or services purchased by the contract (determined by the CPV codes).

The *Organization* ontology aims at describing the organizational structures in order to support the disclosure interconnected data of organizational information in various sectors. It is designed as to allow extensions in specific areas for classifying the organizations and roles as well as for extensions in order to support relevant information, such as organizational activities. Its design allows the publication of information on organizations and their organizational structures, including governmental ones. In our economic context, this ontology is used in order to define and describe the organizations and their organizational units. Their structure is also represented using the properties of this ontology. Additional properties are also provided in order to illustrate the members and their structures within an organization, as well as the roles, positions and relationships between people and organizations.

The *Registered Organization* vocabulary aims at describing the entities, which have obtained the status of a legal entity through a formal registration process, usually at the national or regional registry. It includes a minimum number of classes and properties that are designed to depict the typical information recorded by the business' registries, thus facilitating the exchange of information between them, despite having considerable variation between the recorded and the published data. In our economic context, this vocabulary is used in order to identify and describe the business entities and their properties, including their type, status and activity.

Dublin Core and *Dublin Core Terms* is a small vocabulary for the description of general metadata of the Web and of natural resources. In our economic context, this vocabulary is used in order to define the entities that are responsible for publishing a contract, the subject of the contract, and the date of its formal issue.

SKOS, which is the acronym of "Simple Knowledge Organization System", is a vocabulary, which is designed in order to represent thesaurus, classification schemes, classifications, lists or any other type of structured controlled vocabulary. Its main aim is to allow the easy publication and use of these vocabularies as interconnected data. In our economic context, this dictionary is used in order to define numerous controlled vocabularies and to represent code lists (e.g. Currencies).

The specification of *vCard* is generally used for describing people and organizations. Usually, vCard objects are encoded based on their own syntax or in their XML format. The vCard ontology can be used for the semantic representation of any vCard data and is also focused on describing people and organizations, including location information and groups of such entities, as "FOAF" and "Organization" do. In our economic context this ontology is used in order to represent geographical information regarding agents and products, such as delivery addresses.

DBpedia is an ontology, which covers different domains and is created manually based on the most frequently used information from infoboxes of Wikipedia. For the purposes of our case, properties of this ontology are used in order to define different type of information about businesses (e.g. subsidiaries of a company).

2.2   The ELOD Ontological Schema

The ontology of each data source along with its description and example queries can be found on Github[21]. In this repository, we describe in an analytical way the ontological schema created for the modeling of each data source described in Sect. 2. Apart from the ontology description, SPARQL queries are provided along with a sample of their responses. The model of our ontological schema is depicted in Fig. 1. As it shows, government forms and publishes budgets, parts of which include projects and works that are assigned through calls for tenders. After contracts have being signed and projects are fulfilled, funds are transferred. Spending data are often used to assess the completion of public budgets. However, another type of added-value fiscal information that has been started to provided publicly is subsidies and aid data. Subsidies include government payments to firms and households based on a development plan (e.g. in Greece we have the Greek National Strategic Reference Framework (anaptyxi.gov.gr/), while in EU the farmsubsidy.openspending.org). Procurements, subsidies and aid awarding processes are involving the exchange of information among authorities (e.g. tax offices, business registries and various public agencies) and the official publication of relevant information (e.g. call for tenders, payments).

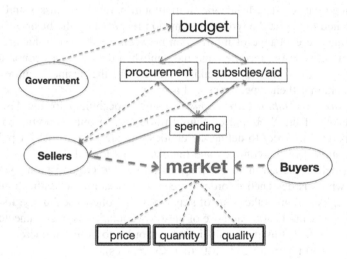

Fig. 1. Ontological model that incorporates interconnections between public finances and market processes

In addition to the above, we have also modeled the Market domain. Information about the market process (e.g. price, value of sales and quality) is an important source of business value and is mostly closed into corporate environments. Market process information is partially shared with government authorities (e.g. tax office), suppliers and consumers. Although, there is a small but crucial part of business information that

---

[21] https://github.com/LinkedEcon/LinkedEconomyOntology-ELOD.

should shared publicly as open data to ensure that quality and competition are best served. This set of publicly available information should include the vector of prices (at least for basic goods and services), aggregated quantities sold in wholesale and retail markets and all the relevant input to assess quality of provisioned products and services. In Greece, a representative sampling of retail prices for thousands basic consumer goods and fuels can be retrieved in regular basis. In some cases, (e.g. Central Market of Thessaloniki) wholesale prices and quantities are provided for fruit, vegetables and meat. In order to make clearer to the reader the reuse and incorporation of the aforementioned ontologies and vocabularies to the eLOD Ontology, two informative tables are provided. Table 2 presents the percentage of the classes and properties belonging to each vocabulary.

Table 2. Classes/properties percentage use of established vocabularies in eLOD ontology

Vocabulary	Distribution of classes in eLOD ontology (%)	Distribution of properties in eLOD ontology (%)
{Own semantics}	50 %	67.7 %
Good relations	15.9 %	8.7 %
Public contracts	9.1 %	7.7 %
Organizations	9.1 %	3.1 %
SKOS	4.5 %	3.1 %
Registered organization	4.5 %	2.6 %
FOAF	4.5 %	1.0 %
vCard	2.3 %	2.6 %
Dublin core	–	3.1 %
DBpedia	–	0.5 %

2.3   Approach and Reuse

The eLOD ontology is based more on a theory-driven than in data- or statistics-driven approach. It has been designed to better balance the trade-off of being as generic as to be scalable to future open data categories and as specific as to be compatible with existing initiatives. It could be an opportunity for many of the diverse communities, which are working on transparency, global standards and economic data, to join forces in addressing useful and not-yet-answered questions. For example, we can't yet answer, if public spending is expensive and comparable across countries? or can we compare financial ratios in public budgets? Can we comparatively analyze wholesale and retail prices? Due to its theory-driven approach eLOD is used by two major European projects related to open economic data. Your Data Stories (YDS, your-datastories.eu) is an EU funded project (Grant Agreement No. 645886) that aims to convert publically available economic open data into re-usable and interoperable building blocks that can be used to construct applications. YDS will allow any actor to design and implement personalised public services. Based on the eLOD model feeds from trusted sources will be interconnected with the new and re-purposed data feeds

Table 3. Query 1 results [Hellenic telecommunications organization S.A.]

Diavgeia	Number of payments	11,288
	Total amount (in Euros)	44,550,925.34
e-Procurement	Number of payments	289
	Total amount (in Euros)	30,570,036.85
NSRF	Number of subsidies	5
	Total amount (in Euros)	208,311

provided by users of the social web in order to form a meaningful, searchable, customizable, re-usable and open hyper–market of data, feeds, and services. The second project is Big Data Europe (http://www.big-data-europe.eu/) which is funded by EU (grant agreement n. 644564) for enabling European companies to build innovative multilingual products and services based on semantically interoperable, large-scale, multi-lingual data assets and knowledge, available under a variety of licenses and business models. LOE model has been selected to support the implementation of one of the seven societal challenges that Big Data Europe focuses. In particular, the social sciences challenge refers to statistical and research data linking and integration and will focus on citizens budgeting and control (Table 4).

Table 4. Federated Query 2 results [Hellenic telecommunications organization S.A.]

Expense approval counter	7,738
Expense approval amount	35,746,746.58
Founding year	1949
Net income	676,400,000
Number of employees	29,330

3 Asking the Data: A Case Study

The aim of LinkedEconomy.org is not only to transform and semantically enrich the input data into RDF graphs, but also to apply a unified ontological model in order to treat and query distinct datasets as one.

As already mentioned, a public endpoint[22] allows the search of the collected semantic data and their combination and enrichment with other sources. SPARQL Query 1 displays the advantages of having a unified ontological model, as it combines such data from three different sources, namely Diavgeia, e-Procurement and NSRF. The query returns the number of decisions and subsidies referring to payments and their total amount in Euros for the case of the Hellenic Telecommunications Organization S.A

---

[22] http://elod7.linkedeconomy.org:8890/sparql.

(member of the Deutsche Telekom Group). For these decisions this organization appears as "seller", meaning that it receives payments for offering a service or from selling a product.

One of the main characteristics of the semantic data is that other publicly available datasets from the LOD cloud can at the same time be queried and combined with relevant information. Such an example is the federated SPARQL Query 2, which combines the economical data of Diavgeia, as stored in our graph, along with DBpedia information. Table 3 depicts the results we receive when we "ask" our data combined with DBpedia for the case of the Hellenic Telecommunications Organization S.A. Apart from the information from Diavgeia, the user receives information in respect to the organization's founding year, net income and number of employees.

4 Discussion - Future Work

This paper presents (i) the economic open data sources used, (ii) the ontological model that orchestrate linked data flows, (iii) the common controlled vocabularies that are incorporated in eLOD ontology in order to place it the global economic Linked Data cloud. All consist of the "data engine" of LinkedEconomy, a project that lead to publishing data of high added value not only for exploration purposes, but also for exploiting the benefits of transparency and openness to the citizens, the research community, and the government itself. Through this research initiative, our team tackled the challenge of building a common terminology for the basic financial and economy activities, which will -in turn- facilitate the research over new linked data and sources. All described components form a system capable of linking economy-related data at large scale, creating in parallel a framework for collect, validate, clean and publish linked data streams. Our efforts produced a CKAN repository, which publishes datasets from sources that are being updated regularly and contain valuable information in respect to social and economic research[23]. Citizens and economy stakeholders (government, local authorities, etc.) can exploit 27 datasets from 14 classified data sources in machine-readable format (xlsx, csv, rdf). Examples include economy data, such as public procurements, budgets, prices, expenditures, data from the insurance domain and employment, as well as financial and macro-economic data. As we envisage the use of Web 3.0 technologies and we acknowledge the benefits of openness, we offer a publicly available SPARQL endpoint that consists of more than 210M triplets in total, and we share all ontological schemas and related specifications in Github.

Finally, as publicly available open data are growing rapidly worldwide, we currently work on modeling many foreign economic datasets according the already semantic knowledge we developed. We expect to support really soon the provision of more than 15 different economic datasets from Europe, UK, Australia, USA and Canada, while we plan to further extend this economy-linked data cloud in the near future.

---

[23] http://ckan.linkedeconomy.org.

Acknowledgements. This work has been supported by a national co-funded project (LinkedEconomy.org) under the 4th National Strategic Reference Framework (NSRF) period between 2007 and 2013. It was implemented by the Ministry of Education and Religious Affairs and the University of Piraeus Research Centre. The project aimed at providing a universal access to Greek and International economy data, as well as at promoting the benefits of linking heterogeneous sources under the concept of open and reusable data in this critical domain.

Appendix

Query 1: Retrieving information from three different Greek data sources (Diavgeia – e-Procurement, and NSRF)

```
PREFIX elod: <http://linkedeconomy.org/ontology#>
PREFIX dcterms: <http://purl.org/dc/terms/>
PREFIX gr: <http://purl.org/goodrelations/v1#>

SELECT (COUNT(?spendingD) AS ?paymentDiavgeiaCounter)
(SUM(xsd:decimal(?amountD)) AS ?paymentDiavgeiaAmount)
(COUNT(?spendingK) AS ?paymentKhmdhsCounter) (SUM(xsd:decimal(?amountK))
AS ?paymentKhmdhsAmount) (COUNT(?subsidy) AS ?subsidyCounter)
(SUM(xsd:decimal(?siAmount)) AS ?subsidyAmount)
FROM <http://linkedeconomy.org/DiavgeiaII/2015>
FROM <http://linkedeconomy.org/EprocurementProper>
FROM <http://linkedeconomy.org/Subsidies>
WHERE {
#Transparency
{
?spendingD elod:hasExpenditureLine ?expLine ;
      dcterms:issued ?date ;
      rdf:type elod:SpendingItem .
?expLine elod:amount ?ups ;
      elod:seller <http://linkedeconomy.org/resource/Organization/094019245> .
?ups gr:hasCurrencyValue ?amountD .
FILTER (?date >= "2015-01-01T00:00:00Z"^^xsd:dateTime) .
FILTER (?date < "2016-01-01T00:00:00Z"^^xsd:dateTime) .
FILTER NOT EXISTS {?spendingD elod:hasCorrectedDecision ?correctedDecision}
.
FILTER NOT EXISTS {?ups elod:riskError "1"^^xsd:boolean} .
}
UNION
#KHMDHS
{
?spendingK elod:hasRelatedContract ?contract ;
      elod:hasExpenditureLine ?expLine .
?expLine elod:amount ?ups ;
      elod:seller <http://linkedeconomy.org/resource/Organization/094019245> .
```

```
?ups gr:hasCurrencyValue ?amountK .
?contract elod:signatureDate ?date .
FILTER (?date >= "2014-01-01T00:00:00Z"^^xsd:dateTime) .
FILTER (?date < "2015-01-01T00:00:00Z"^^xsd:dateTime) .
}
UNION
#NSRF
{
?subsidy elod:hasRelatedSpendingItem ?spendingItem ;
     elod:beneficiary
<http://linkedeconomy.org/resource/Organization/094019245> .
?spendingItem elod:amount ?siUps .
?siUps gr:hasCurrencyValue ?siAmount.
}
}
```

Query 2: Combining eLod and Linked Open Data cloud information (from DBpedia)

```
PREFIX elod: <http://linkedeconomy.org/ontology#>
PREFIX gr: <http://purl.org/goodrelations/v1#>
PREFIX dbo: <http://dbpedia.org/ontology/>
PREFIX dcterms: <http://purl.org/dc/terms/>

SELECT (COUNT(?expenseApproval) AS ?expenseApprovalCounter)
(SUM(xsd:decimal(?amount)) AS ?expenseApprovalAmount) ?foundingYear
?netIncome ?numOfEmployees
WHERE
{
#Transparency
GRAPH <http://linkedeconomy.org/DiavgeiaII/2015> {
?expenseApproval elod:hasExpenditureLine ?expLine ;
          dcterms:issued ?date ;
          rdf:type elod:ExpenseApprovalItem .
?expLine elod:amount ?ups ;
     elod:seller <http://linkedeconomy.org/resource/Organization/094019245> .
?ups gr:hasCurrencyValue ?amount .
FILTER (?date >= "2015-01-01T00:00:00Z"^^xsd:dateTime) .
FILTER (?date < "2016-01-01T00:00:00Z"^^xsd:dateTime) .
FILTER NOT EXISTS {?expenseApproval elod:hasCorrectedDecision
?correctedDecision} .
}
#DBpedia
SERVICE <http://dbpedia.org/sparql> {
<http://dbpedia.org/resource/OTE> dbo:foundingYear ?foundingYear ;
dbo:netIncome ?netIncome ; dbo:numberOfEmployees ?numOfEmployees .
}
}
```

References

1. Vafopoulos, M.: The web economy: goods, users, models and policies. Found. Trends Web Sci. **3**(1–2), 1–136 (2011)
2. Vafopoulos, M., Meimaris, M., Anagnostopoulos, I., et al.: Public spending as LOD: the case of Greece. Semant. Web **6**(2), 155–164 (2015)
3. Klimek, J., Knap, T., Mynarz, J., Necaský, M., Svátek, V.: LOD2 - creating knowledge out of interlinked data framework for creating linked data in the domain of public sector contracts (2012)
4. O'Riain, S., Curry, E., Harth, A.: XBRL and open data for global financial ecosystems: a linked data approach. Int. J. Account. Inf. Syst. **13**(2), 141–162 (2012)
5. Hodess, R., Budgets, O.: The political economy of transparency, participation and accountability. J. Econ. Lit. **52**(2), 545–548 (2014)
6. Tygel, A.F., Attard, J., Orlandi, F., Campos, M.L., Auer, S.: How much? Is not enough-an analysis of open budget initiatives (2015). arXiv preprint arXiv:1504.01563
7. Petrou, I., Meimaris, M., Papastefanatos, G.: Towards a methodology for publishing linked open statistical data. JeDEM – eJ. eDemocracy Open Gov. **6**(1), 97–105 (2014)
8. Höffner, K., Martin, M., Lehmann, J.: LinkedSpending: openspending becomes linked open data. Semant. Web **7**(1), 95–104 (2015)
9. Alvarez-Rodríguez, J.M., Vafopoulos, M., Llorens, J.: Enabling policy making processes by unifying and reconciling corporate names in public procurement data. The CORFU technique. Comput. Stan. Interfaces **41**, 28–38 (2015)
10. Vafopoulos, M.N., Vafeiadis, G., Razis, G, Anagnostopoulos, I., Negkas, D., Galanos, L.: Linked Open economy: take full advantage of economic data. SSRN 2732218 (2016)

Web Image Indexing Using WICE and a Learning-Free Language Model

Nicolas Tsapatsoulis[✉]

Department of Communication and Internet Studies,
Cyprus University of Technology, 30, Arch. Kyprianos Str., 3036 Limassol, Cyprus
nicolas.tsapatsoulis@cut.ac.cy

Abstract. With the advent of Web 2.0 and the rapidly increasing popularity of online social networks that make extended use of visual information, like Facebook and Instagram, web image indexing regained great attention among the researchers in the areas of image indexing and information retrieval. Web image indexing is traditionally approached, by commercial search engines, using text-based information such as image file names, anchor text, web-page keywords and, of course, surrounding text. In the latter case, for effective indexing, two requirements should be met: Correct identification of the related text, known as image context, and extraction of the right terms from this text. Usually, researchers working in the field of web image indexing consider that once the image context is identified extraction of indexing terms is trivial. However, we have shown in our previous work that this is not the rule of thumb.

In this paper we get advantage of Web Image Context Extraction (WICE) using visual web-page parsing and specific distance metrics and following this we locate key terms within this text to index the image using language models. In this way, the proposed method is totally learning free, i.e., no corpus need to be collected to train the keyword extraction component, while the identified indexing terms are more descriptive for the image since they are extracted from a portion of web-page's text. This deviates from the traditional web image indexing approach in which keywords are extracted from all text in the web-page. The evaluation, performed on a dataset of 978 manually annotated web images taken from 243 web pages, shows the effectiveness of the proposed approach both in image context extraction and indexing.

Keywords: Image retrieval · Web image indexing · Web page parsing · Language models

1   Introduction

Since the beginning of the World Wide Web (WWW) and the development of cheap digital recording and storage devices the amount of available on-line digital images, continuously increases. The increasing popularity of online social

© IFIP International Federation for Information Processing 2016
Published by Springer International Publishing Switzerland 2016. All Rights Reserved
L. Iliadis and I. Maglogiannis (Eds.): AIAI 2016, IFIP AICT 475, pp. 131–140, 2016.
DOI: 10.1007/978-3-319-44944-9_12

networks, like Instagram, that are based on visual information push further this tendency. As a result, effective and efficient web image indexing and retrieval schemes are of high importance and a lot of research has been devoted towards this end.

In general, image retrieval research efforts are falling into two broad categories: content-based and text-based. Content-based methods retrieve images by analyzing and comparing the content of a given image example as a starting point. Text-based methods are similar to document retrieval and retrieve images using keywords. The latter is the approach of preference both for ordinary users and search engine engineers. Besides the fact that the majority of users are familiar with text-based queries, content-based image retrieval lacks semantic meaning. Furthermore, image examples that have to be given as a query are rarely available. From the search engine perspective, text-based image retrieval methods get advantage of the well established techniques for document indexing and are integrated into a unified document retrieval framework. However, for text-based image retrieval to be feasible, images must be somehow related with specific keywords or textual description. In contemporary search engines this kind of textual description is, usually, obtained from the web page, or the document, containing the corresponding images and includes HTML alternative text, the file names of the images, captions, metadata tags and surrounding text [1, 2]. Text metadata are not always available, and in most cases are not accurate (i.e., they do not fully describe the visual content of the image). In addition, disambiguation of different visual aspects of the same term is very difficult using text metadata without taking into consideration the context.

Surrounding text, is the text that surrounds a Web image inside an HTML document. This text is, indeed, a very important source of semantic information for the image. However, automatic localization of surrounding text is by no means easy mainly due to the modern web-page layout formatting techniques which are based on external files (stylesheets). As a result, visual segmentation (parsing) of the rendered web-page is required in order to identify the surrounding text of an image. The need to automatically extract the semantically related, to an image, textual blocks and assign them to this image led to what we call Web Image Context Extraction (WICE). In that terms, WICE is the process of automatically assigning the textual blocks of a web document to the images of the same document they refer to [3].

In content-based image retrieval features such as color, shape or texture are used for indexing and searching web images. The user provides a target image and the system retrieves the best ranked images based on their similarity from the user's query. Although it has been a long time since the scientists, working on this area, defined the semantic gap [4], i.e., the inability of a system to interpret images based on automatically extracted low-level features, a solution still does not exist. WICE methods may be used as a means of bridging this gap. For instance, in [5] the authors propose an auto-annotation system which combines a content-based search stage to image annotation along with a text-based stage in order to leverage the image dataset in learning from similar annotations.

Despite the fact that image context identification and text-based image indexing is very important per se, the huge amount of images which do no appear in web-pages or they do not have a clearly related context, either as surrounding text or as specific keywords, puts another challenge. Recently, the idea of visual concept modeling [6,7] was proposed as possible solution to this problem. In these approaches keywords are modeled through via low-level features and non-annotated images are passed through these models in order to identify possible matches with the visual representation of the models and assigned the corresponding keywords. The approach is promising but the keyword models require proper training, usually approached as a learning by example procedure, and, thus, they depend heavily on the selected corpus and keyword identification. So far the training set was created using manually annotated data and, for quality assurance on the keyword selection, crowdsourcing methods were adopted [8,9]. Recently, Giannoulakis and Tsapatsoulis [10] investigated the case of Instagram as a source for annotated images concluding that an overall 26% of hashtags accompanying photos in Instagram are related with photos' visual content. Thus, filtering approaches are still required for a proper training set to be created.

In this paper we investigate the automatic extraction of keywords, from a web-image's context, that can be used either for image indexing or for the automatic creation for training datasets for the visual modeling of keywords mentioned above. Our method deviates from existing approaches in a very important aspect: It does not require any sort of training since it is based on a priori fixed English language model to identify the importance of a keyword in a text fragment. The latter is identified through a computationally efficient visual-based html parsing algorithm. The fact that image context is in most cases concise leads traditional approaches, like probabilistic, *tf-idf* based and clustering based ones, to failure. Thus, image context needs to extended, but, in this case the correlation of the selected text with a specific image in the web page decreases and all images in the web-page tend to share the same indexes (which are also similar to the indexes of the web-page itself). Finally, probabilistic, *tf-idf* and clustering based approaches require training. As a result the problem is recycled: In order to automatically index non-explicitly annotated images you need training examples but in order to automatically create the training examples you need train the indexing extraction algorithms!

2   Related Work

2.1   WICE Methods

In text-based web image retrieval, image file names, anchor texts, surrounding paragraphs or even the whole text of the hosting web page are traditionally used. The user provides keywords or key phrases and text retrieval techniques are used for the retrieval of the best ranked image. Early examples of these methods include [11–13]. However, it turned out very soon that the relevant, to each image, text fragment of the hosting web page must be extracted for better accuracy of retrieval. This is the well-known WICE problem [3], already mentioned in introduction. The high diversity of designing patterns in web pages, the noisy environment (advertisements, graphics, navigational objects etc.), end the existence of too much textual and visual information in single documents are prohibiting factors a WICE system must overcome.

Several WICE methods have been proposed in the literature. A first category of approaches as [14,15], make use of the DOM tree structure of the hosting web page. In general these methods are not adaptive and they are designed for specific design patterns.

Web page visual segmentation is a second category of approaches to the WICE problem. This kind of approaches was initially proposed in [16], where the authors use Vision based Page Segmentation (VIPS) [17] in order to extract blocks, which contain both image and text, and construct an image graph using link structures. Web page segmentation is indeed a more adequate solution to the problem of text extraction since it is adaptable to different web page styles and depends on the visual cues that form each web page. Most of the proposed algorithms falling within this approach they are computationally heavy [18] and they are not designed specifically for the problem of image indexing [19]; there-fore, they often deliver poor results [20]. In addition the creators of VIPS stopped its maintenance as early as in 2011.

In the proposed approach, the WICE problem is tackled through HTML code parsing of the rendered web page. This approach is computationally light and easily applicable and in modern web pages that include several CSS files, for styling purposes, as well as dynamic elements (such as PHP code), is quite effective. It executes html parsing by combining the ideas and the tool presented in [21] with the open source code of Mozilla web browser[1].

2.2   Web Image Indexing from Concise Text Fragments

The text fragments identified by WICE methods are usually very concise; as a result traditional keyword extraction (*tf-idf* like methods) does not apply. Web image retrieval based on clickthrough data is more relevant and effective. Clickthrough data are usually collected from search logs and include text queries and data from relevance feedback [22]. These methods [23–26] are quite effective but they are based on machine learning; thus, they suffer from the scalability problem and they are inappropriate for large scale web image retrieval.

The proposed method uses a learning-free language model for web image indexing using text fragments located by a new WICE method explained next.

3   The Proposed Method

The overall architecture of the proposed method is shown in Fig. 1 along with an illustrative example. It consists of three main steps: (a) html parsing, obtained with the aid of the lxml parser[2], (b) the WICE algorithm, and (c) an English language model accompanied by an keyword extraction algorithm.

---

[1] http://www.mozilla.org/en-US/.
[2] http://lxml.de/parsing.html.

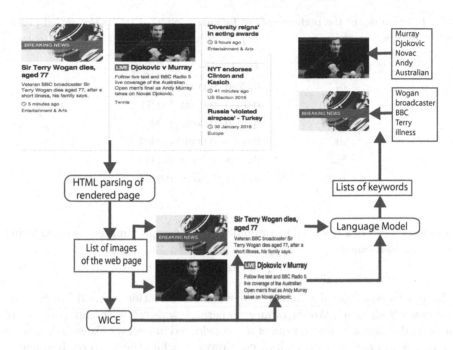

Fig. 1. The architecture of the proposed method through an example

3.1   The WICE Algorithm

The aim of the WICE algorithm is to identify the context (text fragment) of an image given its URL. First the position of the image and its (rendered) dimensions are computed with the aid of the Mozilla open source code. Small images and graphic types (i.e., gif) corresponding to logos and banners are discarded in this stage. Next the nearby sentences (text belonging to the same level as the image in the DOM tree and being within a radius equal to 0.3 of web page's rendered height from image's center) of the image are selected along with the caption, alternative text (if exists) and the hyperlink text. All these text data are merged together to form the text fragment (referred to as image context in the following) related to the given image.

3.2   An English Language Model for Image Retrieval

The relative frequency f_w of appearance of a word 'w' in natural (human) languages follows the well known Zipf law [27]; that is the product of f_w with the ranking R_w of word 'w' is approximately constant:

$$f_w \cdot R_w = c \qquad (1)$$

Given that the number of words in web pages is very large and continuously increasing, due to new documents, misspelling, slang, etc., the probability of

Table 1. Examples of the performance of language model. $P(w)$ is the actual probability of word w

Word (w)	R_w	$P(w)$	$\frac{c}{R_w}$	$\frac{c}{R_w + k^{R_w}}$
of	2	0.0280	0.0600	0.0374
to	3	0.0260	0.0400	0.0277
with	14	0.0060	0.0086	0.0067
at	15	0.0056	0.0080	0.0063
his	24	0.0038	0.0050	0.0035
but	25	0.0038	0.0048	0.0034

appearance $P(w)$ of a word 'w' in a web page can be approximated by considering $P(w) = f_w$ as follows:

$$P(w) = \frac{c}{R_w} \tag{2}$$

In text-based retrieval a common keyword identification method involves the well-known *tf-idf* score. Words, or more general, tokens [27] with high *tf-idf* score in a given document or text fragment are considered important (keywords) for its description and can be used as indices. However, while the *tf* (term frequency) value depends only in the specific document the *idf* (inverse document frequency) value is computed based on a relative large number of relevant documents. Thus, in order to compute *tf-idf* we need a training corpus.

In this paper, we argue that web images appear in every type of document in the web, and, as a result it is not necessary to collect a specific domain corpus to compute *idf* and find the keywords of a text fragment. Therefore, we approximate the *idf* value with $\frac{1}{P(w)}$ and we arrive in a learning free indexing method for text fragments related with web images. Equation 2 gives a rough approximation of $P(w)$. After a little experimentation (see Table 1 for some examples) we found that a more accurate language model can be obtained by:

$$P(w) = \frac{c}{R_w + k^{R_w}} \tag{3}$$

where $c = 0.12$ and $k = 1.1$.

In any case, in order for a language model given above to be useful the ranking R_w[3] of every word should be available. In this work we get advantage of the 'Wordcount' project[4] for this purpose.

Keyword extraction is facilitated by the language model and involves a series of steps: (a) text fragment segmentation into sentences, (b) stopwords removal, (c) part of speech (POS) tagging, (e) noun and proper noun selection as candidate indices, (f) ranking of selected terms based on the adopted language model, and (g) final selection of the indexing terms (keywords) based on the $S = \frac{tf}{P(w)}$

---

[3] http://www.wordcount.org/main.php.
[4] http://www.wordcount.org/about.html.

Table 2. Evaluation results for context extraction. GS denotes gold standard

	# tokens in GS	# tokens found	TP	FP	FN	R	P	F-measure
Context localization	61683	65918	52267	13651	9416	0.847	0.793	0.819

score (terms whose score exceeds an empirically derived threshold T are kept as keywords).

4   Experimental Evaluation

In order to evaluate the proposed method 978 web images taken from 243 web pages were used. Three annotators (students of the Cyprus University of Technology) were asked, independently, to: (a) for each image identify and copy its context (text fragment), and (b) select the keywords from context that best describe the image. The contexts and keywords from the three annotators were merged and used as the gold standard for the evaluation of the proposed method. Indicative examples of images, their context and the keywords chosen by the annotators and found by the algorithm are online available[5].

Table 2 summarizes the results of context extraction. By TP we denote the 'True Positive' rate, that is, the tokens that were in the gold standard and found by the algorithm. Similarly FP denotes 'False Positive' rate, i.e., tokens found by the algorithm but not in the gold standard and FN denotes 'False Negative' rate, that is, tokens in the gold standard that were not found. Recall (R), Precision (P) and F-score (F) values are computed as usual by:

$$R = \frac{TP}{TP + FN} \tag{4}$$

$$P = \frac{TP}{TP + FP} \tag{5}$$

$$F = \frac{2 \cdot P \cdot R}{P + R} \tag{6}$$

We can see in Table 2 that the algorithm tends to include in image context more tokens than those identified by the annotators. This is mainly caused by our decision to include in image context not only the nearby sentences but the tokens in image caption and image's hyperlink. The latter was typically never selected by the annotators although from several studies we know that the information in hyperlinks is of utmost importance in information retrieval. Overall, the results are satisfactory given the simplicity of the proposed method. For comparison see the results reported by Alcic and Conrad [15].

---

[5] http://cis.cut.ac.cy/~nicolas.tsapatsoulis/ckasapi/showImages.php.

Table 3. Evaluation results for keyword identification

	# keywords in GS	# keywords found	TP	FP	FN	R	P	F-measure
Keyword identification	5966	7237	2836	3130	4401	0.475	0.392	0.430

Table 3 shows the results of keyword identification. We observe that the recall rate is close to 50 % which is very promising compared to similar methods (see for instance [26]). Human annotators tend to use more 'emotional' words to describe images even in cases where the visual content does not correspond clearly to these terms. On the hand the proposed algorithm promotes nouns as keywords (a choice made during the design of the algorithm) and especially named entities (as a result of the use of the proposed language model). Similarly to the context extraction case the algorithm identifies, in general, more keywords than the annotators causing the recall to become higher than precision. Nevertheless, in information retrieval the tendency is to pursuit higher recall than precision (irrelevant results are better than no results).

5   Conclusion and Further Work

In this paper we have presented a method for web page image indexing which is based on language models. The method can be applied to identify keywords for any web image without training on a particular corpus. It is based on raw html parsing of web pages to identify the nearest (to the image) text block and then a metric which combines the frequency of terms in the block with their frequency ranking, in the corresponding language, is used. Preliminary results, on an especially designed and annotated web image database, are promising and show the effectiveness of the proposed method. However, there are also some limitations that need to be surpassed for the method to be widely applied while some improvements are planned for (near) future work. The effectiveness of the proposed system on 'carousel' type web images needs to be tested. Furthermore, the algorithm is currently applied only on English web pages since we get advantage of the ranking of English words to create our language model. However, once such a study for any other natural language exists the extension of the proposed method is straightforward.

An improvement of the proposed method is to consider the structure of the surrounding text to further weight the terms. Thus, terms that appear in headers, subheaders, weblinks, etc., will be given higher importance than the terms in the plain text. Finally, the method will be used in the context of visual concept modeling [8] for automatic creation of image-keywords pairs that are required for training purposes. In this context, the other basic limitation of this work, that is, the inability of applying it to non-web images, will be overcome. For more information on this, please see [6,7].

References

1. Souza Coelho, T.A., Calado, P.P., Souza, L.V., Ribeiro-Neto, B., Munt, R.: Image retrieval using multiple evidence ranking. IEEE Trans. Knowl. Data Eng. **16**(4), 408–417 (2004)
2. Datta, R., Joshi, D., Li, J., Wang, J.Z.: Image retrieval: ideas, influences, and trends of the new age. ACM Comput. Surv. **40**(2), 5:1–5:60 (2008)
3. Tryfou, G., Theodosiou, Z., Tsapatsoulis, N.: Web image context extraction based on semantic representation of web page visual segments. In: Proceedings of International Workshop on Semantic and Social Media Adaptation and Personalization, pp. 63–67. IEEE (2012)
4. Del Bimbo, A.: Visual Information Retrieval. Morgan Kaufmann Publishers Inc., San Francisco (1999)
5. Wang, X.J., Zang, L., Jing, F., Ma, W.Y.: Annosearch: image auto-annotation by search. In: IEEE Computer Society Conference on Computer Vision and Pattern Recognition, vol. 2, pp. 1483–1490. IEEE (2006)
6. Theodosiou, Z., Tsapatsoulis, N.: Image retrieval using keywords: the machine learning perspective. In: Semantic Multimedia Analysis and Processing, pp. 3–30. CRC Press (2014)
7. Xu, G.Q., Mu, Z.C.: Automatic image annotation using modified keywords transfer mechanism base on image-keyword graph. Int. J. Comput. Sci. Issues **10**(2), 267–272 (2013)
8. Theodosiou, Z., Tsapatsoulis, N.: Modelling crowdsourcing originated keywords within the athletics domain. In: Iliadis, L., Maglogiannis, I., Papadopoulos, H. (eds.) Artificial Intelligence Applications and Innovations. IFIP AICT, vol. 381, pp. 404–413. Springer, Heidelberg (2012)
9. Theodosiou, Z., Tsapatsoulis, N.: Crowdsourcing annotation: modelling keywords using low level features. In: Proceedings of the 5th International Conference on Internet Multimedia Systems Architecture and Application, pp.1–4. IEEE (2011)
10. Giannoulakis, S., Tsapatsoulis, N.: Instagram hashtags as image annotation metadata. In: Chbeir, R., Manolopoulos, Y., Alhajj, R. (eds.) AIAI 2015. IFIP AICT, vol. 458, pp. 206–220. Springer, Heidelberg (2015). doi:10.1007/978-3-319-23868-5_15
11. Alexandre, L., Pereira, M., Madeira, S., Cordeiro, J., Dias, G.: Web image indexing: combining image analysis with text processing. In: Proceedings of the 5th International Workshop on Image Analysis for Multimedia Interactive Services, WIAMIS 2004 (2004)
12. Smith, J.R., Chang, S.F.: An image and video search engine for the world-wide web. In: Proceedings of SPIE Storage and Retrieval for Image and Video Databases, pp. 84–95. SPIE (1997)
13. Ortega-Binderberger, M., Mexico, A.: Webmars: a multimedia search engine for the world wide web. University of Illinois at Urbana-Champaign (1999)
14. Fauzi, F., Hong, J.L., Belkhatir, M.: Webpage segmentation for extracting images and their surrounding contextual information. In: Proceedings of the 17th ACM International Conference on Multimedia, pp. 649–652. ACM (2009)
15. Alcic, S., Conrad, S.: A clustering-based approach to web image context extraction. In: Proceedings of the 3rd International Conferences on Advances in Multimedia, pp. 74–79. IARIA (2011)
16. He, X., Cai, D., Wen, J.R., Ma, W.Y., Zhang, H.J.: Clustering and searching www images using link and page layout analysis. ACM Trans. Multimedia Comput. Commun. Appl. **3**(2) (2007)

17. Cai, D., Yu, S., Wen, J.R., Ma, W.Y.: Vips: a vision based page segmentation algorithm. Technical report, Microsoft Research (2003)
18. Tryfou, G., Tsapatsoulis, N.: Using visual cues for the extraction of web image semantic information. In: Zaphiris, P., Buchanan, G., Rasmussen, E., Loizides, F. (eds.) TPDL 2012. LNCS, vol. 7489, pp. 396–401. Springer, Heidelberg (2012)
19. Tryfou, G., Tsapatsoulis, N.: Extraction of web image information: semantic or visual cues? In: Iliadis, L., Maglogiannis, I., Papadopoulos, H. (eds.) Artificial Intelligence Applications and Innovations. IFIP AICT, vol. 381, pp. 368–373. Springer, Heidelberg (2012)
20. Alcic, S., Conrad, S.: Measuring performance of web image context extraction. In: Proceedings of the 10th International Workshop on Multimedia Data Mining, pp. 1–8. ACM (2010)
21. Pappas, N., Katsimpras, G., Stamatatos, E.: Extracting informative textual parts from web pages containing user-generated content. In: Proceedings of 12th International Conference on Knowledge Management and Knowledge Technologies, vol. 4, pp. 1–8. ACM (2012)
22. Park, J.Y., O'Hare, N., Schifanella, R., Jaimes, A., Chung, C.W.: A large-scale study of user image search behavior on the web. In: Proceedings of the 33rd Annual ACM Conference on Human Factors in Computing Systems, pp.985–994. ACM (2015)
23. Tsikrika, T., Diou, C., de Vries, A.P., Delopoulos, A.: Image retrieval using multiple evidence ranking. Multimedia Tools Appl. **55**(1), 27–52 (2011)
24. Fang, Q., Xu, H., Wang, R., Qian, S., Wang, T., Sang, J., Xu, C.: Towards msr-bing challenge: ensemble of diverse models for image retrieval. In: Proceedings of the 2013 MSR-Bing Image Retrieval Challenge, Microsoft (2013)
25. Hua, X.S., Yang, L., Wang, J., Wang, J., Ye, M., Wang, K., Rui, Y., Li, J.: Clickage: towards bridging semantic and intent gaps via mining click logs of search engines. In: Proceedings of the 21st ACM International Conference on Multimedia, pp. 243–252. ACM (2013)
26. Sarafis, I., Diou, C., Tsikrika, T., Delopoulos, A.: Weighted svm from clickthrough data for image retrieval. In: 2014 IEEE International Conference on Image Processing (ICIP), pp. 3013–3017. IEEE (2014)
27. Manning, C.D., Schütze, H.: Foundations of Statistical Natural Language Processing. MIT Press, Cambridge (1999)

An Intelligent Internet Search Assistant Based on the Random Neural Network

Will Serrano[✉] and Erol Gelenbe

Intelligent Systems and Networks Group, Electrical and Electronic Engineering,
Imperial College London, London, UK
{g.serranoll, e.gelenbe}@imperial.ac.uk

Abstract. Even web services that are free of charge, typically offer access to online information based on some form of economic interest of the web service itself. Thus advertisers who put the information on the web will make a payment to the search services based on the clicks that their advertisements receive. Thus end users cannot know that the results they obtain from Web search engines are exhaustive, or that they actually respond to their needs. To fill the gap between user needs and the information that is presented to them on the web, Intelligent Search Assistants have been proposed to act at the interface between users and search engines to present data to users in a manner that reflects their actual needs or their observed or stated preferences. This paper presents an Intelligent Internet Search Assistant based on the Random Neural Network that tracks the user's preferences and makes a selection on the output of one or more search engines using the preferences that it has learned. We also introduce a "relevance metric" to compare the performance of our Intelligent Internet Search Assistant against a few search engines, showing that it provides better performance.

Keywords: Intelligent Internet Search Assistant · World Wide Web · Random Neural Network · Web search · Search engines

1 Introduction

The need to search for specific information in the ever expanding Internet has led the development of Web search engines. Whereas their benefit is the provision of a direct connection between users and the information or products sought, any search outcome will be influenced by a commercial interest as well as by the users' own ambiguity in formulating their requests or queries. An example of this situation is travel services. The Internet has made accessible real time travel industry's information and services; customers can purchase flight tickets, hotels and holiday packs online. Distribution costs have been reduced due a shorter value chain; however businesses not shown on the top positions within the search results may lose potential customers. A similar scenario also occurs within academic search; the Internet has allowed the democratization of academic publications. Authors can upload their work onto their personal Webpages bypassing the traditional model of the journal peer review. There is the biased interest from authors to get their publications in top search positions in order to reach a bigger audience so they will be cited more. In both examples ranking

© IFIP International Federation for Information Processing 2016
Published by Springer International Publishing Switzerland 2016. All Rights Reserved
L. Iliadis and I. Maglogiannis (Eds.): AIAI 2016, IFIP AICT 475, pp. 141–153, 2016.
DOI: 10.1007/978-3-319-44944-9_13

algorithms are essential as they decide the relevance; they make information visible or hidden to customers or users. Under this model, Web search engines or recommender systems can be tempted to artificially rank results from some specific businesses for a fee whereas also authors or business can be tempted to manipulate ranking algorithms by "optimizing" the presentation of their work or products. The main consequence is that irrelevant results may be shown on top positions and relevant ones "hidden" at the very bottom of the search list.

In order to address the presented search issues; this paper proposes an Intelligent Internet Search Assistant (ISA) that acts as an interface between an individual user's query and the different search engines. Our ISA acquires a query from the user and retrieves results from one or various search engines assigning one neuron per each Web result dimension. The result relevance is calculated by applying our innovative cost function based on the division of a query into a multidimensional vector weighting its dimension terms with different relevance parameters. Our ISA adapts and learns the perceived user's interest and reorders the retrieved snippets based in our dimension relevant centre point. Our ISA learns result relevance on an iterative process where the user evaluates directly the listed results. We evaluate and compare its performance against other search engines with a new proposed quality definition, which combines both relevance and rank. We have also included two learning algorithms; Gradient Descent learns the centre of relevant dimensions and Reinforcement Learning updates the network weights based on rewarding relevant dimensions and punishing irrelevant ones. We have validated our ISA against other Web search engines and metasearch engines using travel services and open user queries. We have also analysed the Gradient Descent and Reinforcement Learning algorithms based on result relevance and learning speed.

We describe the application of neural networks in Web search in Sect. 2. We define our Intelligent Internet Search Assistant mathematical model in Sect. 3 and we have validated it against other Web search engines in Sect. 4. Finally, we present our conclusions in Sect. 5.

2 Related Work

The ability of neural networks to learn iteratively from different inputs to acquire the desired outputs as a mechanism of adaptation to users' interest in order to provide relevant answers have already been applied in the World Wide Web and recommender systems.

F. Scarselli et al. [1] and M. Chau et al. [2] use a neural network by assigning a neuron to each Web page; they create a graph where the neural links are the equivalent of the hyperlinks. S. Bermejo et al. [3] use a similar approach to our proposal, the allocation of one neuron per Web search result, however the main difference is that the network is trained to cluster results by meaning. C. Burgues et al. [4] define RankNet which uses neural networks to evaluate Web sites by training the neural network based on query-document pairs. Shu, B. et al. [5] retrieve results from different Web search engines and train the network following the assumption that a result in a top position would be relevant. J. Boyan et al. [6] use reinforcement learning to rank Web pages using their HTML properties and hyperlink connections between them. X. Wang et al. [7] use a back propagation neural network with its input nodes corresponding to an specific quantified user profile and one output node which it is the a probability the user would consider the Web page relevant.

3 The Intelligent Internet Search Assistant Model

The search assistant we design is based on the Random Neural Network (RNN) [8–10]. This is a biologically inspired spiking recurrent stochastic model for neural networks. Its main analytical properties are the "product form" and the existence of the unique network steady state solution. The RNN represents more closely how signals are transmitted in many biological neural networks where they actual travel as spikes or impulses, rather than as analogue signal levels. It has been used in different applications including network routing with cognitive packet networks, using reinforcement learning, which requires the search for paths that meet certain pre-specified quality of service requirements [11, 17], search for exit routes for evacuees in emergency situations [12, 13], pattern based search for specific objects [14], video compression [15], and image texture learning and generation [16].

3.1 Search Model

In the case of our own application of the RNN, the search for information or for some meaning needs requires us to specify some elements: an M-dimensional universe of X entities or ideas to be searched, a high level query that specifies the N-properties or concepts requested by a user and a method that searches and selects Y entities from the universe showing the first Z results to user according to an algorithm or rule. Each entity or concept in the universe is distinct from the others in some recognizable way; for instance two entities may be different just in the date or time-stamp that characterizes the time when they were last stored or in the ownership or origin of the entities. On the other hand, we consider concepts to be distinct if they contain any different meaning, even though if they are identical with respect to a user's query.

We consider that the universe which we are searching within as a relation U that consists of a set of X M-tuples, $U = \{v_1, v_2 \ldots v_X\}$, where $v_i = (l_{i1}, l_{i2} \ldots l_{iM})$ and li are the M different attributes for $i = 1, 2 \ldots X$. The relation U is a very large relation consisting on $M >> N$ attributes. The important concept in the development of this paper is a query can be defined as $R_t(n(t)) = (R_t(1), R_t(2), \ldots, R_t(n(t)))$ where n(t) is a variable N-dimension attribute vector with $1 < N < M$ and t is the search iteration being $t > 0$; n(t) is variable so that attributes can be added or removed based on their relevance as the search progresses, i.e. as t increases. Each $R_t(n(t))$ takes its values from the attributes within the domain D(n(t)), where D is the corresponding domain that forms the universe U. Thus D(n(t)) is a set of properties or meanings based in words or integers, but also words in another language, or a set of icons, images or sounds.

The answer A to the query $R_t(n(t))$ is a set of Y M-tuples $A = \{v_1, v_2 \ldots v_Y\}$ where $v_o = (l_{o1}, l_{o2} \ldots l_{oM})$ and lo are the M different attributes for $o = 1, 2 \ldots Y$.

Our Intelligent Internet Search Assistant only shows to the user the first set of Z tuples that have the highest neuron potentials among the set of Y tuples. The neuron potential that represents the relevance of each M-tuple v_o is calculated at each t iteration. The user or the high level query itself is limited mainly by two main factors: the user's lack of information about all the attributes that form the universe U of entities and ideas, or the user's lack of precise knowledge about what he is looking for.

3.2   Result Cost Function

We consider the universe U is formed of the entire results that can be searched. We assign each result provided by a search engine to an M-tuple v_o of the answer set A. We calculate the result relevance based on a cost function described within this section. The query $R_t(n(t))$ is a variable N-dimension vector that specifies the attributes the user consider relevant. The number of dimensions of the attribute vector n(t) varies as the iteration t increases. Our Intelligent Internet Search Assistant associates an M-tuple v_o to each result provided by the Search Engine creating an answer set A of Y M-tuples. Search Engines select their results from the universe U. We apply our cost function to each result or M-tuple v_o from the answer set A of Y M-tuples. We consider each v_o as a M-dimensional vector. The cost function is firstly calculated based on the relevant N attributes the user introduced on the High Level Query, $R_1(n(1))$ within the domain D(n (1)) however, as the search progresses, $R_t(n(t))$, attributes may be added or removed based on the perceived relevance within the domain D'(n(t)). We calculate the overall Result Score, RS, by measuring the relationship between the values of its different attributes:

$$RS = RV * HW \tag{1}$$

where RV is the Result Value which measures the result relevance and HW the Homogeneity Weight. The Homogeneity Weight (HW) rewards results that have relevance or scores dispersed along their attributes. This parameter is also based on the idea that the first dimensions or attributes of the user query $R_t(n(t))$ are more important than the last ones:

$$HW = \frac{\sum_{n=1}^{N} HF[n]}{N} \tag{2}$$

where HF[n], homogeneity factor, is a N-dimension vector associated to the result and n is the attribute index from the query $R_t(n(t))$:

$$HF[n] = \begin{vmatrix} \frac{N-n}{N} & \text{if } SD[n] > 0 \\ 0 & \text{if } SD[n] = 0 \end{vmatrix} \tag{3}$$

We define Score Dimension SD[n] as a N-dimension vector that represents the attribute values of each result or M-tuple v_0 in relation with the query $R_t(n(t))$. The Result Value (RV) is the sum of each dimension individual score:

$$RV = \sum_{n=1}^{N} SD[n] \tag{4}$$

where n is the attribute index from the query $R_t(n(t))$. Each dimension of the Score Dimension vector SD[n] is calculated independently for each n-attribute value that forms the query $R_t(n(t))$:

$$SD[n] = S * PPW * RPW * DPW \tag{5}$$

We consider only three different types of domains of interest: words, numbers (as for dates and times) and prices. S is the score calculated depending if the domain of the attribute is a word (WS), number (NS) or price (PS). If the domain D(n) is a word, our ISA calculates the score Word Score (WS) following the formula:

$$S = \frac{WR}{NW} \tag{6}$$

where the value of WR is 1 if the word of the n-attribute of the query $R_t(n(t))$ is contained in the search result or 0 otherwise. NW is the number of words in the search result. If the domain D(n) is a number, our ISA selects the best Number Score (NS) from the numbers they are contained within the search result that maximizes the cost function:

$$S = \frac{\left(1 - \left(\frac{|DV-RV|}{|DV|+|RV|}\right)\right)}{NN} \tag{7}$$

where DV is the value of the n-attribute of the query $R_t(n(t))$, RV is the value of a number in the result and NN is the total number of numbers in the result. If the domain D(n) is a price, our ISA chooses the best Price Score (PS) from the prices in the result that maximizes the cost function:

$$S = \frac{\left(\frac{DV}{RV}\right)}{NP} \tag{8}$$

where DV is value of the n-attribute of the query $R_t(n(t))$, RV is the value of a price in the result and NP is the total number of prices in the result. We penalize if the search result provides unnecessary information by dividing the score by the total amount of elements in the Web result. The dimension Score Dimension vector, SD[n] is weighted according to different relevance factors:

$$SD[n] = S * PPW * RPW * DPW \qquad (9)$$

The Position Parameter Weight (PPW) is based on the idea that an attribute value shown within the first positions of the search result is more relevant than if it is shown at the final:

$$PPW = \frac{NC - DVP}{NC} \qquad (10)$$

where NC is the number of characters in the result and DVP is the position within the result where the value of the dimension is shown. The Relevance Parameter Weight (RPW) incorporates the user's perception of relevance by rewarding the first attributes of the query $R_t(n(t))$ as highly desirable and penalising the last ones:

$$RPW = 1 - \frac{PD}{N} \qquad (11)$$

where PD is the position of the n-attribute of the query $R_t(n(t))$ and N is the total number of dimensions of the query vector $R_t(n(t))$. The Dimension Parameter Weight (DPW) incorporates the observation of user relevance with the value of domains $D(n(t))$ by providing a better score on the domain values the user has more filled on the query:

$$DPW = \frac{NDT}{N} \qquad (12)$$

where NDT is the number of dimensions with the same domain (word, number or price) on the query $R_t(n(t))$ and N is the total number of dimensions of the query vector $R_t(n(t))$. We assign this final Result Score value (RS) to each M-tuple v_o of the answer set A. This value is used by our ISA to reorder the answer set A of Y M-tuples, showing to the user the first set of Z results which have the higher potential value.

3.3   User Iteration

The user, based on the answer set A can now act as an intelligent critic and select a subset of P relevant results, C_P, of A. C_P is a set that consists of P M-tuples $C_P = \{v_1, v_2 \ldots v_P\}$. We consider v_P as a vector of M dimensions; $v_p = (l_{p1}, l_{p2} \ldots l_{pM})$ where l_p are the M different attributes for p = 1, 2 … P. Similarly, the user can also select a subset of Q irrelevant results, C_Q of A, $C_Q = \{v_1, v_2 \ldots v_Q\}$. We consider v_q as a vector of M dimensions; $v_q = (l_{q1}, l_{q2} \ldots l_{qM})$ where lq are the M different attributes for q = 1, 2 … Q. Based on the user iteration, our Intelligent Internet Search Assistant provides to the user with a different answer set A of Z M-tuples reordered to MD, the minimum distance to the Relevant Centre for the results selected, following the formula:

$$RCP[n] = \frac{\sum\limits_{p=1}^{P} SD_p[n]}{P} = \frac{\sum\limits_{p=1}^{P} l_{pn}}{P} \qquad (13)$$

where P is the number of relevant results selected, n the attribute index from the query $R_t(n(t))$ and $SD_p[n]$ the associated Score Dimension vector to the result or M-tuple v_P formed of l_{pn} attributes. An equivalent equation applies to the calculation of the Irrelevant Centre Point. Our Intelligent Internet Search Assistant reorders the retrieved Y set of M-tuples showing only to the user the first Z set of M-tuples based on the lowest distance (MD) between the difference of their distances to both Relevant Centre Point (RD) and the Irrelevant Centre Point (ID) respectively:

$$MD = RD - ID \qquad (14)$$

where MD is the result distance, RD is the Relevant Distance and ID is the Irrelevant Distance. The Relevant Distance (RD) of each result or M-tuple v_q is formulated as below:

$$RD = \sqrt{\sum_{n=1}^{N} (SD[n] - RCP[n])^2} \qquad (15)$$

where SD[n] is the Score Dimension vector of the result or M-tuple v_q and RCP[n] is the coordinate of the Relevant Centre Point. Equivalent equation applies to the calculation of the Irrelevant Distance. Therefore we are presenting an iterative search progress that learns and adapts to the perceived user relevance based on the dimensions or attributes the user has introduced on the initial query.

3.4   Dimension Learning

The answer set A to the query $R_1(n(1))$ is based on the N dimension query introduced by the user however results are formed of M dimensions therefore the subset of results the user has considered as relevant may have other relevant concepts hidden the user did not considered on the original query. We consider the domain D(m) or the M attributes from which our universe U is formed as the different independent words that form the set of Y results retrieved from the search engines. Our cost function is expanded from the N attributes defined in the query $R_1(n(1))$ to the M attributes that form the searched results. Our Score Dimension vector, SD[m], is now based on M-dimensions. An analogue attribute expansion is applied to the Relevance Centre Calculation, RCP[m]. The query $R_1(n(1))$ is based on the N-Dimension vector introduced by the user however the answer set A consist of Y M-tuples. The user, based on the presented set A, selects a subset of P relevant results, C_P and a subset of Q irrelevant results, C_Q.

Let us consider C_P as a set that consists of P M-tuples $C_P = \{v_1, v_2 \dots v_P\}$ where v_P is a vector of M dimensions; $v_P = (l_{p1}, l_{p2} \dots l_{pM})$ and l_p are the M different attributes for p = 1, 2 ... P. The M-dimension vector Dimension Average, DA[m], is the average value of the m-attributes for the selected relevant P results:

$$DA[m] = \frac{\sum\limits_{p=1}^{P} SD_p[m]}{P} = \frac{\sum\limits_{p=1}^{P} l_{pm}}{P} \tag{16}$$

where P is the number of relevant results selected, m the attribute index of the relation U and $SD_p[m]$ the associated Score Dimension vector to the result or M-tuple v_P formed of l_{pm} attributes. We define ADV as the Average Dimension Value of the M-dimension vector DA[m]:

$$ADV = \frac{\sum\limits_{m=1}^{M} DA[m]}{M} \tag{17}$$

where M is the total number of attributes that form the relation U. The correlation vector $\sigma[m]$ is the difference between the dimension values of each result with the average vector:

$$\sigma[m] = \frac{\sum\limits_{p=1}^{P} (SD_p[m] - DA[m])}{P} = \frac{\sum\limits_{p=1}^{P} (l_{Pm} - DA[m])}{P} \tag{18}$$

where P is the number of relevant results selected, m the attribute index of the relation U and $SD_p[m]$ the associated Score Dimension vector to the result or M-tuple v_P formed of l_{pm} attributes. We define C as the average correlation value of the M-dimensions of the vector $\sigma[m]$:

$$C = \frac{\sum\limits_{m=1}^{M} \sigma[m]}{M} \tag{19}$$

where M is the total number of attributes that form the relation U. We consider an m-attribute relevant if its associated Dimension Average value DA[m] is larger than the average dimension ADV and its correlation value $\sigma[m]$ is lesser than the average correlation C. We have therefore changed the relevant attributes of the searched entities or ideas by correlating the error value of its concepts or properties represented as attributes or dimensions. On the next iteration, the query $R_2(n(2))$ is formed by the attributes our ISA has considered relevant. The answer to the query $R_2(n(2))$ is a different set A of Y M-tuples. This process iterates until there are not new relevant results to be shown to the user.

3.5   Gradient Descent Learning

Gradient Descent learning is based on the adaptation to the perceived user interests or understanding of meaning by correlating the attribute values of each result to extract similar meanings and cancel superfluous ones. The ISA Gradient Descent learning algorithm is based on a recurrent model. The inputs $i = \{i_1, ..., i_P\}$ are the M-tuples v_P corresponding to the selected relevant result subset C_P and the desired outputs $y = \{y_1, ..., y_P\}$ are the same values as the input. Our ISA then obtains the learned random neural network weights, calculates the relevant dimensions and finally reorders the results according to the minimum distance to the new Relevant Centre Point focused on the relevant dimensions.

3.6   Reinforcement Learning

The external interaction with the environment is provided when the user selects the relevant result set C_P. Reinforcement Learning adapts to the perceived user relevance by incrementing the value of relevant dimensions and reducing it for the irrelevant ones. Reinforcement Learning modifies the values of the m attributes of the results, accentuating hidden relevant meanings and lowering irrelevant properties. We associate the Random Neural Network weights to the answer set A; $W = A$. Our ISA updates the network weights W by rewarding the result relevant attributes by:

$$w(p, m) = l_{pm}^{s-1} + l_{pm}^{s-1} * \left(\frac{l_{pm}^{s-1}}{\sum_{m=1}^{M} l_{pm}^{s-1}} \right) \tag{20}$$

where p is the result or M-tuple v_P formed of l_{pm} attributes, m the result attribute index, M the total number of attributes and s the iteration number. ISA also updates the network weights by punishing the result irrelevant attributes by:

$$w(p, m) = l_{pm}^{s-1} - l_{pm}^{s-1} * \left(\frac{l_{pm}^{s-1}}{\sum_{m=1}^{M} l_{pm}^{s-1}} \right) \tag{21}$$

where p is the result or M-tuple v_P formed of l_{pm} attributes, m the result attribute index, M the total number of attributes and s the iteration number. Our ISA then recalculates the potential of each of the result based on the updated network weights and reorders them, showing to the user the results which have a higher potential or score.

4   Validation

The Intelligent Internet Search Assistant we have proposed emulates how Web search engines work by using a very similar interface to introduce and display information. We validate our ISA algorithm with a set of three different experiments. Users in the experiments can both choose between the different Web search engines and the N number of results they would to retrieve from each one. We propose the following formula to measure Web search quality; it is based on the concept that a better search engine provides with a list of more relevant results on top positions. In an list of N results, we score N to the first result and 1 to the last result, the value of the quality proposed is then the summation of the position score based of each of the selected results. Our definition of Quality, Q, can be defined as:

$$Q = \sum_{i=1}^{Y} RSE_i \tag{22}$$

where RSE_i is the rank of the result i in a particular search engine with a value of N if the result is in the first position and 1 if the result is the last one. Y is the total number of results selected by the user. The best Web search engine would have the largest Quality value. We define normalized quality, $\overline{Q}$, as the division of the quality, Q, by the optimum figure which it is when the user consider relevant all the results provided by the Web search engine. On this situation Y and N have the same value:

$$\overline{Q} = \frac{Q}{\frac{N(N+1)}{2}} \tag{23}$$

We define I as the quality improvement between a Web search engine and a reference:

$$I = \frac{QW - QR}{QR} \tag{24}$$

where I is the Improvement, QW is the quality of the Web search engine and QR is the quality reference; we use the Quality of Google as QR in our validation exercise.

In our first experiment we have asked to our validators to search for different queries using only Google; ISA provides with a set of reordered results from which the user needs to select the relevant results. We show the average values for the 20 different queries, the average number of results retrieved by Google and the average number of results selected by the user. We represent the normalized quality of Google and ISA with the improvement of our algorithm against Google. In our second experiment, ISA provides with a reordered list from where the user needs to select which results are relevant. Our ISA reorders the results using the dimension relevant centre point providing to the user with another reordered result list from where the user needs to select the relevant ones. We show the average values for the 16 different queries, the average number of results selected by the user and the average number of results selected. We also represent the normalized quality of Google, ISA and the ISA with the relevant circle iteration including the improvement against Google in both scenarios. In our third experiment, validators can select from which Web search engine they would their results to be retrieved from; as in our first experiment, the users need to select the relevant results. Our ISA combines the results retrieved from the different Web search engines selected. We present the average values for the 18 different queries. We show the normalized quality of each Web search engine selected including our ISA; because users can choose any Web search engine; we are not introducing the improvement value as we do not have a unique reference Web search engine (Table 1).

Table 1. Web search engine validation

Results retrieved	Results selected	Google Q	ISA Q	ISA I	ISA Circle Q	ISA Circle I
Experiment 1–20 queries						
19.35	8.05	0.4626	0.4878	15.39 %	–	–
Experiment 2–16 queries						
21.75	8.75	0.4451	0.4595	18 %	0.4953	26 %
Web	Google	Yahoo	Ask	Lycos	Bing	ISA
Experiment 3–18 queries						
Q	0.2691	0.2587	0.3454	0.3533	0.3429	0.4448

4.1   ISA Learning

Users in the experiments can choose between Google and Bing with either Gradient Descent or Reinforcement Learning type. Our ISA then collects the first 50 results from the Web search engine selected, reorders them according to its cost function and finally show to the user the first 20 results. We consider 50 results is a good approximation of search depth as more results can add clutter and irrelevance; 20 results is the average number of results read by a user before he launches another search if he does not find any relevant one. ISA reorders results while learning on the two step iterative process showing only the best 20 results to the user. We present the average Quality values of the Web search engine and ISA for the 29 different queries searched by different users, the learning type and the Web search engine used. The first I represents the improvement from ISA against the Web search; the second I is between ISA iterations 2 and 1 and finally the third I is between the ISA iterations 3 and 2 (Table 2).

Table 2. ISA learning validation

Gradient descent learning: 17 queries								
Web	ISA	I	Web	ISA	I	Web	ISA	I
0.41	0.58	43 %	0.45	0.61	14 %	0.46	0.62	8 %
Reinforcement learning: 12 queries								
Web	ISA	I	Web	ISA	I	Web	ISA	I
0.42	0.57	34 %	0.47	0.67	36 %	0.49	0.68	0.0 %

5 Conclusions

We have proposed a novel approach to Web search where the user iteratively trains the neural network while looking for relevant results. We have also defined a different process; the application of the Random Neural Network as a biological inspired algorithm to measure both user relevance and result ranking based on a predetermined cost function. Our Intelligent Internet Search Assistant performs generally slightly better than Google and other Web search engines however, this evaluation may be biased because users tend to concentrate on the first results provided which were the ones we showed in our algorithm. Our ISA adapts and learns from user previous relevance measurements increasing significantly its quality and improvement within the first iteration. Reinforcement Learning algorithm performs better than Gradient Descent. Although Gradient Descent provides a better quality on the first iteration; Reinforcement Learning outperforms on the second one due its higher learning rate. Both of them have a residual learning on their third iteration. Gradient Descent would have been the preferred learning algorithm if only one iteration is required; however Reinforcement Learning would have been a better option in the case of two iterations. It is not recommended three iterations because learning is only residual.

References

1. Scarselli, F., Liang, S., Hagenbuchner, M., Chung, A.: Adaptive page ranking with neural networks. In: Proceeding of WWW 2005 Special Interest Tracks and Posters of the 14th International Conference on World Wide Web, pp. 936– 937 (2005)
2. Chau, M., Chen, H.: Incorporating web analysis into neural networks: an example in Hopfield net searching. IEEE Trans. Syst Cybern. C Appl. Rev. **37**(3), 352–358 (2007)
3. Bermejo, S., Dalmau, J.: Web metasearch using unsupervised neural networks. In: IWANN 2003 Proceedings of the 7th International Work-Conference on Artificial and Natural Neural Networks: Part II: Artificial Neural Nets Problem Solving Methods, pp. 711–718 (2003)
4. Burgues, C., Shaked, T., Renshaw, E., Lazier, L., Deeds, M., Hamilton, N., Hullender, G.: Learning to rank using gradient descent. In: ICML 2005 Proceedings of the 22nd International Conference on Machine Learning, pp. 89–96 (2005)
5. Shu, B., Kak, S.: A neural network-based intelligent metasearch engine. Inf. Sci. Inform. Comput. Sci. **120**, 1–11 (2009)
6. Boyan, J., Freitag, D., Joachims, T.: A machine learning architecture for optimizing web search engines. In: Proceedings of the AAAI Workshop on Internet-based Information Systems (1996)
7. Wang, X., Zhang, L.: Search engine optimization based on algorithm of BP neural networks. In: Proceedings of the Seventh International Conference on Computational Intelligence and Security, pp. 390–394 (2011)
8. Gelenbe, E.: Random Neural Network with negative and positive signals and product form solution. Neural Comput. **1**, 502–510 (1989)
9. Gelenbe, E.: Learning in the recurrent Random Neural Network. Neural Comput. **5**, 154–164 (1993)
10. Gelenbe, E., Timotheou, S.: Random Neural Networks with synchronized interactions. Neural Comput. **20**(9), 2308–2324 (2008)

11. Gelenbe, E., Lent, R., Xu, Z.: Towards networks with cognitive packets. In: Goto, K., Hasegawa, T., Takagi, H., Takahashi, Y. (eds.) Performance and QoS of next generation networking, pp 3–17, Springer, London (2011)
12. Gelenbe, E., Wu, F.J.: Large scale simulation for human evacuation and rescue. Comput. Math Appl. **64**(12), 3869–3880 (2012)
13. Filippoupolitis, A., Hey, L., Loukas, G., Gelenbe, E., Timotheou, S.: Emergency response simulation using wireless sensor networks. In: Proceedings of the 1st International Conference on Ambient Media and Systems, no. 21 (2008)
14. Gelenbe, E., Koçak, T.: Area-based results for mine detection. IEEE Trans. Geosci. Remote Sens. **38**(1), 12–24 (2000)
15. Cramer, C., Gelenbe, E., Bakircloglu, H.: Low bit-rate video compression with neural networks and temporal subsampling. Proc. IEEE **84**(10), 1529–1543 (1996)
16. Atalay, V., Gelenbe, E., Yalabik, N.: The Random Neural Network model for texture generation. Int. J. Pattern Recogn. Artif. Intell. **6**(1), 131–141 (1992)
17. Gelenbe, E.: Steps towards self-aware networks. Commun. ACM **52**(7), 66–75 (2009)
18. Gelenbe, E.: Analysis of single and networked auctions. ACM Trans. Internet Technol. **9**(2), 1–24 (2009)

Deep Neural Networks for Web Page Information Extraction

Tomas Gogar[✉], Ondrej Hubacek, and Jan Sedivy

Department of Cybernetics, Czech Technical University in Prague,
Karlovo namesti 13, Prague, Czech Republic
{gogartom,hubacon2,sedivja2}@fel.cvut.cz

Abstract. Web wrappers are systems for extracting structured information from web pages. Currently, wrappers need to be adapted to a particular website template before they can start the extraction process. In this work we present a new method, which uses convolutional neural networks to learn a wrapper that can extract information from previously unseen templates. Therefore, this wrapper does not need any site-specific initialization and is able to extract information from a single web page. We also propose a method for spatial text encoding, which allows us to encode visual and textual content of a web page into a single neural net. The first experiments with product information extraction showed very promising results and suggest that this approach can lead to a general site-independent web wrapper.

Keywords: Information extraction · Web wrappers · Convolutional neural networks

1 Introduction

The Internet is the biggest and the fastest growing source of data in today's world. Many information systems that gather structured data need to acquire information from web pages. However, HTML files are designed to be processed by web browsers and do not contain information in a structured form[1]. Therefore, systems that can extract structured information from web pages receive special attention in the research community. Such tools are usually referred to as *web wrappers*.

Although people can easily extract information from different web pages, the task of creating an automatic wrapper that can extract information from multiple websites is considered as a very complex problem. It is mainly because the semantics of elements depends on many properties such as textual content,

---

[1] There are efforts to include structured data in HTML, such as schema.org project, but it is still not widely used by web developers.

© IFIP International Federation for Information Processing 2016
Published by Springer International Publishing Switzerland 2016. All Rights Reserved
L. Iliadis and I. Maglogiannis (Eds.): AIAI 2016, IFIP AICT 475, pp. 154–163, 2016.
DOI: 10.1007/978-3-319-44944-9_14

visual appearance and relative positioning. Therefore, the research community is mainly focused on wrappers that need to be adapted to a particular website and then they can extract information from its web pages [3,5,8,10,16]. However, such approach brings many disadvantages, such as difficult scalability and maintenance. In this work, we show that a combination of visual and textual data in a single model can help us to create general (multi-site) wrapper. The three main contributions of this work are: (1) We propose a method of encoding data from a web rendering engine into a deep neural net - i.e. a method for spatial encoding of text. (2) On the task of product information extraction, we show that the neural net could be trained to extract information in non-trivial cases. (3) We make our dataset, source codes and final model public, in order to provide a benchmark for future work[2]. This paper is organized as follows: Sect. 2 briefly summarizes related work. In Sect. 3 we give an overview of our system, which is then described in detail in Sects. 4 and 5. In Sect. 6 we summarize our experiments and discuss achieved results. Finally, Sect. 7 summarizes this work and suggests future research.

2   Related Work

As we have mentioned in Sect. 1, current wrappers need to be adapted to a particular website before the extraction process starts. These wrappers make use of the fact that the web pages are generated from templates and thus have similar structure. The first group of wrappers uses manually labeled examples for their initialization [3,8,10]. Although these approaches achieve high accuracy, their manual nature makes them unusable in large-scale extractions. Mainly because initialization and maintenance of such wrapper for thousands of websites is not feasible. More recent works address these issues, some propose automatic wrapper maintenance [6,11], some propose methods for automatic initialization. These methods use tree-matching algorithms in order to find repeated patterns either across multiple pages [4,12,14] or within a single page [5,16] (more extensive survey of wrappers can be found in [6]). The disadvantage of these methods is their dependence on repeated patterns, which makes them unable to automatically extract information from unique document (such as invoice or product description). To the best of our knowledge, this is the first published work that addresses this problem by creating domain-specific wrapper that generalizes across previously unseen templates and does not need any site-specific initialization. Our model is inspired by deep convolutional neural networks (DCNNs) used in computer vision. Since 2012, when Krizhevsky et al. presented DCNN, which established new state-of-the-art result in image classification task [9], many other methods using convolutional nets have appeared and many of them have achieved very good results in other computer vision tasks - such as object detection [7,13] or visual segmentation [1]. These works have motivated our research, where we try to apply similar principles to a different area - Information Extraction.

---

[2] https://github.com/gogartom/TextMaps.

3   Architecture Overview

The semantics of elements in a semi-structured document depends on the textual content as well as on many other visual properties. Therefore, we convert data from web page DOM tree to a 2D plane and we use object detection techniques from computer vision to find DOM elements that contain the requested information. Although visual data and textual data were combined in other works on information extraction [5,16], we believe that this is the first work in this field that encodes visual and textual data into a single model. The architecture of the whole classification process is depicted in Fig. 1.

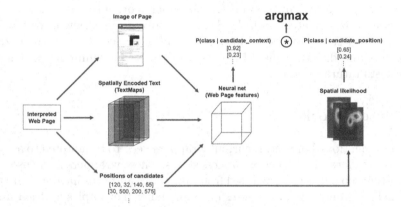

Fig. 1. System architecture overview

The following paragraphs briefly describe basic workflow of our method. In the first phase, we use the web rendering engine to fully interpret the web page and we save its screenshot and a DOM tree. From the DOM tree we are able to extract all the nodes together with their positions. Text nodes are then used to create textual input for the net (see Sect. 4.1) and leaf nodes are used as candidates for classification (candidate elements are represented by the rectangles they occupy in rendered web page). Then, we use convolutional neural net, which processes visual and textual context of candidate elements and predicts their probabilities $P(c|element\_context)$ of being in one of the predefined classes c. The target classes c are task-specific, for example the product information extraction system would classify DOM elements into four classes: *Product name*, *Main product image*, *Current final product price* and *Others*.

However, not only the close context of elements plays a role, but also absolute positioning in the web page is important. Unfortunately, convolutional networks with large inputs do not capture absolute positioning of their features and so we use training data to model spatial probability distributions $P(c|element\_position)$ (see Sect. 5), which are used to classify elements based on their absolute position. The resulting probability $P(c|element\_context)$ is assumed to be independent of $P(c|element\_position)$ and thus we get the final prediction $P(c|element\_position, element\_context)$ by multiplication. The element with the highest resulting probability is selected as our final prediction for a given class.

4 Neural Network

Our neural network is implemented in Caffe framework and is based on a model for object detection described in [7]. In this object detection network visual features are extracted by multiple convolutional layers and a list of object proposals (represented as rectangles in input space) is then classified into particular classes. In our work, we use the same principle for classification of proposed rectangles, however, the overall architecture of our network is very different. The main difference is that our network does not process only visual but also textual data.

4.1 Spatial Text Encoding

In the field of natural language processing, texts are usually encoded in a form of vectors [2]. However, these models are not very suitable for processing of semi-structured documents, because they are designed to capture content of paragraphs and do not encode exact positions of words. In the following paragraphs, we describe our approach to text encoding, which we call *Spatial bag of words* or simply *Text Maps*.

DOM tree stores texts in *text nodes*, which can include text of various length and we can compute their bounding boxes, i.e. rectangles where all the text from the nodes is displayed (see Fig. 2). In order to encode texts effectively, we divide a page with a grid (granularity is an adjustable parameter, see Fig. 2). Using this grid we can store texts into a sparse tensor with dimensions $(N, H/g, W/g)$, where N is the size of vocabulary, H and W are page dimensions (in pixels) and g is the size of the grid cell (in pixels).

The encoding process treats each text node of a web page individually. It splits the text to individual words and for each word adds 1 to the resulting web page tensor at positions $(i, \{X\}, \{Y\})$, where $\{X\}$ and $\{Y\}$ represents indices of grid cells that are covered by text node, i is a feature index for a particular word. The feature index i is not looked up in a dictionary (as in ordinary bag-of-words settings), but as in [15], it is computed automatically by hashing a word into values between 0 and $(N - 1)$ (see example in Fig. 2). In our work we have used Murmurhash3 (32-bit signed), $N = 128$ and $g = 8$.

Note that Text Maps compress textual information a lot, we lose information about word ordering, exact positioning (because of grid size) and even exact word identity (because of index collisions). However, we get a compact representation of text, which is the same as representation of images (tensors of size $[3, H, W]$). This property is very important because it gives our model a chance to easily combine both types of information.

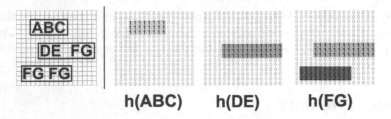

Fig. 2. Example of TextMap creation: (Left) Text nodes with words "ABC", "DE", "FG", (Right) Resulting text maps for N = 3, h() is a hashing function.

4.2 Network Architecture

This section describes architecture of our neural net which is depicted in Fig. 3. Our network has three types of inputs: *Screenshot* (we use upper crop of the page with dimensions 1280×1280, however this net can work with arbitrarily sized pages), *TextMaps* (tensor with dimensions $128 \times 160 \times 160$) and *Candidate boxes* (list of box coordinates of arbitrary length).

A screenshot is processed by three convolutional layers (the first two layers are initialized with pretrained weights from BVLC AlexNet). TextMaps are processed with one convolutional layer with kernel size 1×1 and thus its features capture various combinations of words. These two layers are then concatenated and processed by final convolutional layer. The parameters of these filters are summarized in Table 1.

The last convolutional layer has 96 channels and represents the final features of the web page (see Fig. 3). We then use ROI (Region of Interest) Pooling method [7] which allows us to take a list of bounding boxes, project them to the last layer and extract features which spatially correspond to the particular area on the page. Since DOM elements are sometimes very small, the ROI Pooling layer is set to max pool only one value from each feature map. This results in one feature vector (with 96 elements) per each candidate. These candidate vectors are then individually classified with linear classifier and softmax into final classes (4 classes in our test case). For more information on ROI Pooling, please see [7].

Table 1. Parameters of convolutional layers. C - convolutional layer, R - ReLU nonlinearity, MP - max-pooling layer

Stage	Image				Text	Both
Type	C+R	MP	C+R	C+R	C+R	C+R
# Channels	96	96	256	384	48	96
Filter size	11	3	5	3	1	5
Stride	4	2	1	1	1	1

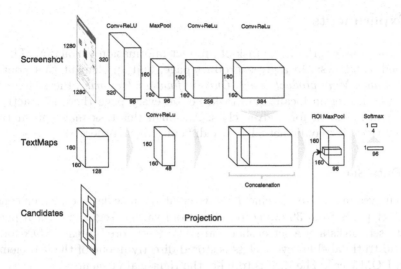

Fig. 3. Architecture of our CNN: upper part represents visual feature extraction, which is then combined with textual features (middle). Coordinates of candidate boxes are projected to final feature tensor and resulting vector is extracted using ROI MaxPooling. Finally, linear model with softmax classifies vector into predefined classes.

4.3 Training

The net was trained using stochastic gradient descent with cross-entropy loss, with learning rate $5e^{-4}$ and momentum 0.9. Mini-batch contains two images each with 100 candidate elements (3 ground truths + 97 randomly chosen others). In order to prevent overfitting, we augmented the dataset by randomly changing hue of screenshots and also by inverting page colors (with 15 % probability).

5   Spatial Probability Distribution

As mentioned in Sect. 3, we need a second model that approximates spatial probability distribution of class c. This distribution is modeled as:

$$P(c|x,y) = (f_c * g)(x,y) = \sum_{m=-M}^{M} \sum_{n=-M}^{M} f_c(x-m, y-n) \cdot g(m,n) \qquad (1)$$

where $*$ is convolution, g is 2D discrete gaussian kernel (variance was chosen experimentally). f_c is frequency matrix for class c defined as $f_c(x,y) = \frac{n_c(x,y)}{N}$, where $n_c(x,y)$ is number of samples where element of class c occupied pixel at (x,y) and N is total number of training samples. When candidate element is received (with bounding box [l,r,b,t]), its class probability given its position is computed as an average over pixels it covers:

$$P(c|elem\_position) = P(c|(l,r,b,t)) = \frac{1}{(r-l) \cdot (b-t)} \sum_{l \leq x \leq r} \sum_{t \leq y \leq b} p(c|x,y) \qquad (2)$$

6 Experiments

We test our framework on the task of product information extraction. The task is defined as follows: On a given product page, find an element that contains: *Product name, Main product image, Current final product price.* Please note, that we test the model on localization task, i.e. on every page there is exactly one ground truth element for a given class. Although this task may appear trivial on some pages, it may be actually very difficult on the others.

6.1 Data Set

Since our system requires specific data, we created a new data set, which consists of product pages from 39 online retailers from various segments. Each page in the data set consists of a screenshot and DOM tree stored in the JSON format. A ground truth label for every class is stored directly in one of the leaf elements in each DOM tree[3]. The task is to label the right leaf elements.

6.2 Baseline Models

Unfortunately, we are not aware of any model that extracts information from a single web page and that would be suitable for comparison with our work. However, we would like to explore whether our neural net was able to learn some non-trivial dependencies between elements. Therefore we compare its results with two baseline algorithms:

Spatial Distribution Baseline: The first baseline is trivial, given candidate leaf elements it selects those with highest spatial probability $P(C|candidate\_position)$.

Heuristic+Spatial Distribution Baseline: The second baseline is more complex. First, it filters candidate elements using simple textual heuristics[4]. Then, from these prefiltered elements, it selects the one with the highest spatial probability $P(C|candidate\_position)$.

6.3 Results

Using 10-fold crossvalidation we test how our model works on previously unseen sites. In Table 2 accuracy achieved on test sites after 10 thousand training iterations is presented. We can see the best results are achieved by Neural Net in combination with Spatial model. These results are comparable with web wrappers with automatic site initialization.

---

[3] Sources and data set: https://github.com/gogartom/TextMaps.

[4] *Price* candidates contain a dollar sign ($) and arbitrary numbers, *Image* candidates does not contain any text, *Name* candidates contain at least two unique words.

Table 2. Comparison of algorithms: mean and standard deviation of accuracy across 10 splits (in percents).

Algorithm	Image accuracy	Price accuracy	Name accuracy
NeuralNet+Spatial	**98.7 ± 1.6**	**95.3 ± 6.6**	**87.1 ± 15.0**
NeuralNet	95.9 ± 2.9	86.2 ± 9.3	78.4 ± 19.0
Baseline: Heuristic+Spatial	63.7 ± 20.1	73.6 ± 18.8	34.4 ± 20.5
Baseline: Spatial	46.5 ± 18.7	9.7 ± 14.4	12.2 ± 12.0

Table 3. Neural Net with different input data: mean and standard deviation of accuracy across 10 splits (in percents).

Neural net inputs	Image accuracy	Price accuracy	Name accuracy
Screenshot+TextMap	95.9 ± 2.9	86.2 ± 9.3	78.4 ± 19.0
Screenshot	93.5 ± 7.4	73.3 ± 19.4	73.4 ± 16.0
TextMap	41.4 ± 18.6	77.0 ± 17.9	49.4 ± 18.0

Interesting observation is the gap between Heuristic baseline and our framework in price accuracy (73.6 % vs. 95.4 %). This result suggests that Neural Net can capture price elements in non-trivial situations. Manual inspection of results confirms this hypothesis and some examples are shown in Fig. 4. We can see that our system is able to recognize *current price* elements of different sizes and distinguish them from other price tags.

Another very important observation is that localizing product name appears to be more complicated task. When examining results, we have observed that some sites divide name into two parts (manufacturer+product name) in DOM tree. Unfortunately, our net does not have capacity to distinguish between the two and we leave this issue to future work.

The last group of experiments address the influence of different input data. We have tried to train our neural net ignoring either visual or textual data and compare these results with the original net that combines both. The results are summarized in Table 3. We can see that textual data itself are not sufficient for the task, while visual data perform better. The best results are achieved by the combination of both inputs, however the difference is not significant. The biggest improvement was achieved in *current price* detection. We experimentally

Fig. 4. Examples of *current price* detection.

verified that neural net that combines both inputs can detect spatially smaller price tags, which it is not able to recognize while using the visual features only.

7   Conclusions

In this work we have proposed a novel method for learning information extraction system that is able to generalize across previously unseen pages and therefore does not need any site-specific initialization. Since the ability of processing both - textual and visual data is crucial for this task, we have proposed a method for spatial text encoding (spatial bags-of-words). This approach allowed us to combine both types of information in one convolutional neural net. We have shown on a task of product information extraction that our model is able to generalize across web sites and can extract information in non-trivial situations (with overall accuracy 93.7 %). Achieved results are very promising and allow for immediate practical applications. However, our approach might still be improved in several ways. (i) We plan to replace the hashing function in *Text Maps* with learned representations of paragraphs. (ii) The detection algorithm may be improved in order to extract information that is stored in multiple leaf elements. (iii) And finally, simple Spatial Probability model can be replaced with more robust attention-based neural model.

Acknowledgments. This work was supported by the Grant Agency of the CTU in Prague, No. SGS16/086/OHK3/1T/13. Comp. resources provided by the CESNET LM2015042 and the CERIT Scientific Cloud LM2015085.

References

1. Badrinarayanan, V., Kendall, A., Cipolla, R.: Segnet: a deep convolutional encoder-decoder architecture for image segmentation. arXiv preprint (2015)
2. Baudiš, P., Šedivý, J.: Sentence pair scoring: towards unified framework for text comprehension. ArXiv preprints, March 2016
3. Califf, M.E., Mooney, R.J.: Bottom-up relational learning of pattern matching rules for information extraction. J. Mach. Learn. Res. **4**, 177–210 (2003)
4. Dalvi, N., Kumar, R., Soliman, M.: Automatic wrappers for large scale web extraction. Proc. VLDB Endow. **4**(4), 219–230 (2011)
5. Fan, S., Wang, X., Dong, Y.: Web data extraction based on visual information and partial tree alignment. In: 2014 11th Web Information System and Application Conference (WISA), September 2014, pp. 18–23 (2014)
6. Ferrara, E., Meo, P.D., Fiumara, G., Baumgartner, R.: Web data extraction, applications and techniques: a survey. Knowl. Based Syst. **70**, 301–323 (2014)
7. Girshick, R.: Fast R-CNN. In: Proceedings of the IEEE International Conference on Computer Vision, pp. 1440–1448 (2015)
8. Hsu, C.N., Dung, M.T.: Generating finite-state transducers for semi-structured data extraction from the web. Inf. Syst. **23**, 521–538 (1998)
9. Krizhevsky, A., Sutskever, I., Hinton, G.E.: Imagenet classification with deep convolutional neural networks. In: Advances in Neural Information Processing Systems (2012)

10. Kushmerick, N.: Wrapper induction: efficiency and expressiveness. Artif. Intell. **118**(1–2), 15–68 (2000)
11. Ortona, S., Orsi, G., Buoncristiano, M., Furche, T.: Wadar: joint wrapper and data repair. Proc. VLDB Endow. **8**(12), 1996–1999 (2015)
12. Qiu, D., Barbosa, L., Dong, X.L., Shen, Y., Srivastava, D.: Dexter: large-scale discovery and extraction of product specifications on the web. VLDB Endow. (2015)
13. Sermanet, P., Eigen, D., Zhang, X., Mathieu, M., Fergus, R., LeCun, Y.: Overfeat: integrated recognition, localization and detection using convolutional networks. ArXiv preprint (2013)
14. Brambilla, M., Tokuda, T., Tolksdorf, R. (eds.): ICWE 2012. LNCS, vol. 7387. Springer, Heidelberg (2012)
15. Weinberger, K., Dasgupta, A., Langford, J., Smola, A., Attenberg, J.: Feature hashing for large scale multitask learning. In: Proceedings of the 26th Annual International Conference on Machine Learning, ICML 2009 (2009)
16. Zhai, Y., Liu, B.: Automatic wrapper generation using tree matching and partial tree alignment. In: Proceedings of the National Conference on Artificial Intelligence (2006)

Environmental AI Modeling (ENAIM)

Modeling Beach Rotation Using a Novel Legendre Polynomial Feedforward Neural Network Trained by Nonlinear Constrained Optimization

Anastasios Rigos[1], George E. Tsekouras[1(✉)], Antonios Chatzipavlis[2], and Adonis F. Velegrakis[2]

[1] Department of Cultural Technology and Communication, University of the Aegean, Mitilini, Greece
a.rigos@aegean.gr, gtsek@ct.aegean.gr
[2] Department of Marine Sciences, University of the Aegean, Mitilini, Greece
a.chatzipavlis@marine.aegean.gr, beachtour@aegean.gr

Abstract. A Legendre polynomial feedforward neural network is proposed to model/predict beach rotation. The study area is the reef-fronted Ammoudara beach, located at the northern coastline of Crete Island (Greece). Specialized experimental devices were deployed to generate a set of input-output data concerning the inshore bathymetry, the wave conditions and the shoreline position. The presence of the fronting beachrock reef (parallel to the shoreline) increases complexity and imposes high non-linear effects. The use of Legendre polynomials enables the network to capture data non-linearities. However, in order to maintain specific functional requirements, the connection weights must be confined within a pre-determined domain of values; it turns out that the network's training process constitutes a constrained nonlinear programming problem, solved by the barrier method. The performance of the network is compared to other two neural-based approaches. Simulations show that the proposed network achieves a superior performance, which could be improved if an additional wave parameter (wave direction) was to be included in the input variables.

Keywords: Beach rotation · Feedforward neural network · Legendre polynomials · Perched beach · Nonlinear constrained optimization

1 Introduction

Beach rotation refers to the realignment of the beach shoreline due mainly to lateral (alongshore) sediment movement caused by shifts in incident wave energy [1]. The phenomenon is controlled by the wave-coastal morphology interaction that can result in large localized changes in shoreline position (retreat or advance) and, thus, in changes of the beach planform which, however, may not lead to long term sediment loss or gain; beaches often return to their initial platform with the changes being often seasonal [2–4]. Although beach rotation has been considered/modeled as an alongshore sediment transport process, recent research suggests a more complex beach response to wave

© IFIP International Federation for Information Processing 2016
Published by Springer International Publishing Switzerland 2016. All Rights Reserved
L. Iliadis and I. Maglogiannis (Eds.): AIAI 2016, IFIP AICT 475, pp. 167–179, 2016.
DOI: 10.1007/978-3-319-44944-9_15

energy, whereby alongshore variability in cross-shore sediment fluxes may also be significant [2, 5]. Beach rotation processes are expected to be more complicated in the case of perched beaches i.e. beaches that are fronted by natural or artificial reefs [6], as the sediment dynamics and morhodynamics of these beaches are controlled also by the reef's depth and morphology. Wave transformation and breaking over the reef can induce high non-linear effects [7, 8]. As a result, the standard modeling methodologies require complex mathematical structures with extremely high computational costs [3–5, 9].

On the other hand, polynomial functions are in the position to effectively model data nonlinearities [10]. Polynomial neural networks utilize polynomials to represent the nodes' activation functions and, thus, increase modeling capabilities. Ma and Khorasani [11] incorporated into the network's structure Hermite polynomials, where the corresponding parameters were optimized in terms of an adaptive learning scheme. Lee and Jeng [12] used tensor products to develop a Chebysev polynomial type network, whereas Patra et al. [13] performed nonlinear channel equalization for wireless communication systems in terms of a Legendre polynomial-based neural network. Although the above approaches show good testing performances, they use high number of nodes and thus, can hardly be applied to high dimensional nonlinear problems. In comparison, Chebyshev polynomial radial basis function and neural-fuzzy networks were developed to perform efficient shoreline extraction from coastal imagery [10] and predict coastal erosion [14].

In this paper, we propose a feed-forward neural network that employs Legendre polynomials as activation functions to model the shoreline beach rotation of a perched beach (Ammoudara, Crete Fig. 1). Linear combinations of the input variables are generated and appropriately scaled by constraining the corresponding weighting parameters. The scaled functions maintain the linearity of the input variables (which is an important issue when regression analysis is to be applied) and the output is expanded into truncated Legendre series. Since the network's connection weights are constrained, the training process becomes a constrained nonlinear optimization problem solved by the barrier method.

The paper is organized as follows. Section 2 describes the experimental setup and the data acquisition process. Section 3 provides a detailed analysis of the proposed network and the training process used. Section 4 illustrates the simulation experiments, and the paper concludes in Sect. 5.

2 Experimental Setup and Raw Data Extraction

The study area is the eastern sector of Ammoudara beach, a 6.1 km long microtidal, urban perched beach, located at the west of the port of Heraklion, Crete, Greece (Fig. 1). The beach, is fronted by a submerged beachrock reef, oriented almost parallel to the shoreline, the width of which and its distance from the shoreline vary between 15–50 m and 40–70 m, respectively. Before detailing the experimental setup and the data acquisition process, it is convenient to discuss some concepts involved in the analysis as well as the physical meaning of the input-output variables used in this paper.

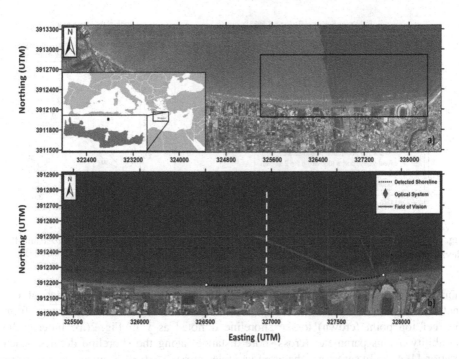

Fig. 1. (a) Ammoudara beach, Heraklion, Crete; the position of the offshore POSEIDON E1-M3A wave buoy is illustrated as a black dot in the inset. (b) Optical system location (diamond point), and field of vision of the 3 deployed cameras (confined within the red lines); the shoreline section examined by video imagery (i.e. detected shoreline) is shown by the dashed black line along the shoreline, with the white vertical dashed-line corresponding to one cross-shore section (out of the 52 sections studied); the offshore dark grey zone parallel to the shoreline delineates the beachrock reef. (Satellite image source: Bing Maps, Microsoft) (Color figure online)

In view of Fig. 1(b) the length of the shoreline studied is defined by the black dashed line lying on the shoreline (denoted as "Detected Shoreline" in the figure). The vertical white dashed line corresponds to one out of 52 cross-shore sections used in our experiments. Each cross-shore section is associated with a bathymetry profile, which is shown in Fig. 2(a). It is widely accepted that specific morphological characteristics of a reef can affect beach rotation; the reef acts in a similar manner to a submerged breakwater, absorbing the incoming wave energy [8, 9, 15]. The reef morphological characteristics used as inputs are enumerated as follows (Fig. 2): the reef depth (in meters) from the sea surface is denoted as d (Fig. 2(a)), the reef inshore and offshore slopes as ω_1 and ω_2 (Fig. 2(b)) and the reef width (in meters) at 1.2 m water depth as w (Fig. 2(b)). In the latter case, the depth of 1.2 m was decided after a specialized data processing that showed that this reef width at this water depth had the most substantial effects.

Apart from the above parameters that quantify the bathymetry characteristics, we use two more parameters that describe the wave conditions namely, the significant wave height denoted as H_S (in meters), and the peak wave period symbolized T_P (in seconds). Note that the last two are important as they impose a direct control on beach

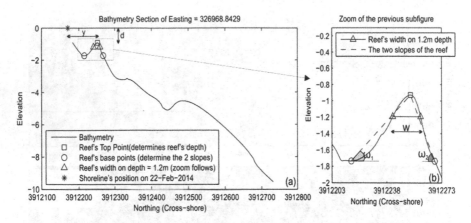

Fig. 2. (a) Cross-shore bathymetric profile (see Fig. 1(b)), showing the beachrock reef and the elevation parameters. (b) Zoom on the reef showing its structural parameters.

morphodynamics [1, 3, 7]. All above parameters form the input variables of our analysis. The output variable quantifying beach rotation is the distance (in meters) from the reef top point (crown) to the shoreline denoted as y in Fig. 2(a). Indeed, the variability of this parameter (cross-shore distance) along the shoreline defines beach rotation [1–4, 7]. In summary, the input variables are: $x_1 = d$, $x_2 = \tan \omega_1$, $x_3 = \tan \omega_2$, $x_4 = w$, $x_5 = H_S$, and $x_6 = T_P$, and the output variable is y (input variables $p = 6$ and one output).

The experimental methodology consists of high detailed nearshore bathymetric data, and a long-term time series (10-month period from January 2014 to November 2014) of shoreline position and wave conditions. More specifically, bathymetric data were obtained through a single beam digital Hi-Target HD 370 echo-sounder and a Differential GPS (Topcon Hipper RTK-DGPS) deployed from a very shallow draft inflatable boat. Using interpolation, from these bathymetric data 52 cross-shore sections were derived (a sample is given in Figs. 1 and 2). Information on the shoreline position for the 10-month period was obtained from coastal video imagery provided by a system consisting of 3 PointGrey FLEA-2 video cameras, installed on the study area, monitoring a beach stretch of 1400 m long (the fields of vision of these 3 cameras are shown in Fig. 1(b)). A detailed description of the system and the automated procedure developed to extract the shoreline from the video images is provided in Velegrakis et al. [8]. The above experiments provided the raw data for the input variables $x_1 - x_4$, and for the output variable y. Data concerning the variables $x_5 = H_S$, and $x_6 = T_P$ were obtained by an offshore wave buoy (POSEIDON E1-M3A buoy) located about 35 km to the north of the beach (35.66^0 N and 24.99^0 E) at 1440 m water depth (see Fig. 1 (a)), installed/operated by the Greek National Centre for Marine Research (GNCMR).

In total, the experimental setup generated $N = 4148$ input-output data of the form $\{x_k; y_k\}|_{k=1}^{N}$ with $x_k = \begin{bmatrix} x_{k1} & x_{k2} & x_{k3} & x_{k4} & x_{k5} & x_{k6} \end{bmatrix}^T$. These data are going to be elaborated by the proposed neural network in order to model/predict beach rotation.

3   The Proposed Legendre Polynomial Feedforward Network

In this section, we introduce a feed-forward neural network (FFNN), the nodes of which utilize Legendre polynomials as activation functions. The Legendre polynomials possess powerful function approximation capabilities, and they are defined by the subsequent formula [16],

$$P_n(x) = \frac{1}{2^n n!} \frac{d^n}{dx^n} \left[(x^2 - 1)^n \right] \tag{1}$$

where $n = 0, 1, \ldots$ is the polynomial order. The Legendre polynomials are orthogonal for $x \in [-1, 1]$ satisfying the following inner product condition [13, 16],

$$\int_{-1}^{1} P_m(x) P_n(x) \, dx = \begin{cases} \dfrac{2}{2n+1}, & m = n \\ 0, & m \neq n \end{cases} \tag{2}$$

In addition, they can be generated by the next recurrent relations [16],

$$P_0(x) = 1; \ P_1(x) = x; \ P_n(x) = \frac{1}{n}[(2n-1)xP_{n-1}(x) - (n-1)P_{n-2}(x)] \ \text{for} \ n \geq 2 \tag{3}$$

Let us assume that the available input-output dataset is denoted as,

$$S = \left\{ (x_k, y_k) : x_k = [x_{k1}, x_{k2}, \ldots, x_{kp}]^T, k = 1, 2, \ldots, N \right\} \tag{4}$$

where p is the dimension of the input space, and N is the number of the training data (note that in the application discussed in this paper: $p = 6$ and $N = 4148$).

The proposed neural network is illustrated in Fig. 3. There are four layers involved. Given that the desired order of the Legendre polynomials is n, the Layer 1 comprises n nodes, each of which generates a linear combination of the input variables,

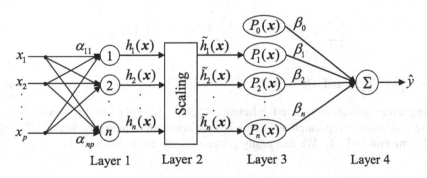

Fig. 3. The Legendre polynomial feedforward neural network.

$$h_\ell(x) = \sum_{j=1}^{p} a_{\ell j} x_j \tag{5}$$

where $1 \leq \ell \leq n$, $1 \leq j \leq p$, and $a_{\ell j}$ are the weight parameters.

The Layer 2 applies a scaling procedure, which maps the values of $h_\ell(x)$ $(1 \leq \ell \leq n)$ in the interval $[-1, 1]$.

As mentioned above, the reason for this scaling procedure is that the Legendre polynomials are orthogonal in the interval $[-1, 1]$ and therefore, they are able to operate only in this interval. To accomplish this task, we introduce a specialized methodology that maintains the linearity with respect to the original inputs. We denote the domain of values associated with j-th input variable as $D_j = \left[x_j^L, x_j^U\right]$, meaning that $x_j^L \leq x_j \leq x_j^U$. Note that the lower and upper bounds x_j^L and x_j^U are fixed and depend on the system data, only. Relationally, we can define an interval $A = [a_L, a_U]$ to confine the weight parameters so that $a_L \leq a_{\ell j} \leq a_U$ (i.e. $a_{\ell j} \in A$) for every ℓ and j. The values for lower bound a_L and the upper bound a_U are pre-selected in terms of a trial-and-error approach as to obtain the best possible results, and are kept fixed throughout the whole learning process.

The question is to find a transformation to map the functions $h_\ell(x)$ in the interval $[-1, 1]$. Based on the interval arithmetic [17] the multiplication of the intervals $D_j = \left[x_j^L, x_j^U\right]$ and $A = [a_L, a_U]$ gives the interval $[L_j, U_j]$ with,

$$L_j = \min\left\{a_L x_j^L, \ a_L x_j^U, \ a_U x_j^L, \ a_U x_j^U\right\} \tag{6}$$

$$U_j = \max\left\{a_L x_j^L, \ a_L x_j^U, \ a_U x_j^L, \ a_U x_j^U\right\} \tag{7}$$

Therefore,

$$L_j \leq a_{\ell j} x_j \leq U_j \qquad (1 \leq j \leq p) \tag{8}$$

For all input variables we add the above inequalities, and using Eq. (5),

$$\sum_{j=1}^{p} L_j \leq \sum_{j=1}^{p} a_{\ell j} x_j \leq \sum_{j=1}^{p} U_j \Rightarrow Q_L \leq h_\ell(x) \leq Q_U \tag{9}$$

where $Q_L = \sum_{j=1}^{p} L_j$ and $Q_U = \sum_{j=1}^{p} U_j$. Thus, $h_\ell(x) \in [Q_L, Q_U] \ \forall \ell$. The question of finding a transformation to map the functions $h_\ell(x)$ $(1 \leq \ell \leq n)$ in the interval $[-1, 1]$ is now equivalently rephrased as finding a transformation to map the interval $[Q_L, Q_U]$ on the interval $[-1, 1]$. We can easily prove that this transformation is,

$$\tilde{h}_\ell(x) = \frac{2}{Q_U - Q_L} h_\ell(x) - \frac{Q_U + Q_L}{Q_U - Q_L} \tag{10}$$

where $\tilde{h}_\ell(x) \in [-1, 1]$.

By setting $R = 2/(Q_U - Q_L)$, $\Omega = (Q_U + Q_L)/(Q_U - Q_L)$ and taking into account the Eq. (5), the Eq. (10) yields,

$$\tilde{h}_\ell(x) = R \sum_{j=1}^{p} a_{\ell j} x_j - \Omega = \sum_{j=1}^{p} R a_{\ell j} x_j - \Omega \tag{11}$$

The above equation directly indicates that the scaled functions $\tilde{h}_\ell(x)$ $(1 \le \ell \le n)$ are linear combinations of the input variables, something very important for the regression analysis that follows.

The Layer 3 includes n nodes with activation functions the Legendre polynomials of the linear combinations reported in Eq. (11),

$$P_\ell(x) = P_\ell(\tilde{h}_\ell(x)) = P_\ell\left(\sum_{j=1}^{p} R a_{\ell j} x_j - \Omega \right) \tag{12}$$

Note that, based on (3), the zero order polynomial $P_0(x)$ is always equal to one and has no effect on the input variables. Therefore, it is used as the network's bias.

Finally, the Layer 4 produces the network's estimated output by intertwining the outputs of the Layer 3, in order to expand the linear combinations of Eq. (11) into the subsequent truncated Legendre series,

$$\hat{y} = \beta_0 + \sum_{\ell=1}^{n} \beta_\ell P_\ell\left(\sum_{j=1}^{p} R a_{\ell j} x_j - \Omega \right) \tag{13}$$

The network's learning process carries out the estimation of weights $a_{\ell j}$ and β_ℓ through the minimization of the network's square error: $J_{SE} = \sum_{k=1}^{N} |y_k - \hat{y}_k|^2$, which based on the form of Eq. (13), constitutes a regression analysis. However, to maintain $\tilde{h}_\ell(x) \in [-1, 1]$, the following relation must hold,

$$a_{\ell j} \in A \Rightarrow a_L \le a_{\ell j} \le a_U \qquad \forall \ell, j \tag{14}$$

Thus, while the estimation of β_ℓ $(1 \le \ell \le n)$ is unconstrained, the estimation of $a_{\ell j}$ $(1 \le \ell \le n;\ 1 \le j \le p)$ is constrained. By splitting the double inequality in (14), the constrained optimization problem is now given as follows:

Minimize

$$J_{SE} = \sum_{k=1}^{N} |y_k - \hat{y}_k|^2 = \sum_{k=1}^{N} \left| y_k - \left(\beta_0 + \sum_{\ell=1}^{n} \beta_\ell P_\ell \left(\sum_{j=1}^{p} R \, a_{\ell j} x_j - \Omega \right) \right) \right|^2 \quad (15)$$

Subject to

$$\phi(a_{\ell j}) = -a_{\ell j} + a_L \le 0 \quad (\ell = 1, 2, \ldots, n; \ j = 1, 2, \ldots, p) \quad (16)$$

$$\psi(a_{\ell j}) = a_{\ell j} - a_U \le 0 \quad (\ell = 1, 2, \ldots, n; \ j = 1, 2, \ldots, p) \quad (17)$$

It can be easily verified that the feasible area of the problem is convex, because it forms the intersection of plane surfaces (i.e. convex sets). To perform the optimization we could use the well-known *penalty method*, which approaches to the feasible area from outside [18]. However, we are required not to leave the feasible area; otherwise the Legendre polynomials would be forced to operate outside of the orthogonality region. Thus, we choose to use the *barrier method* [18], which searches for a solution only in the feasible area without leaving it. According to the *barrier method*, the above constrained problem can be exactly resolved by the unconstrained minimization of the following function,

$$F = \sum_{k=1}^{N} \left| y_k - \left(\beta_0 + \sum_{\ell=1}^{n} \beta_\ell P_\ell \left(\sum_{j=1}^{p} R \, a_{\ell j} x_j - \Omega \right) \right) \right|^2 - \frac{1}{\gamma} \sum_{\ell=1}^{n} \sum_{j=1}^{p} \left(\frac{1}{\phi(a_{\ell j})} + \frac{1}{\psi(a_{\ell j})} \right)$$

$$(18)$$

where $\sum_{\ell=1}^{n} \sum_{j=1}^{p} \left(\frac{1}{\phi(a_{\ell j})} + \frac{1}{\psi(a_{\ell j})} \right)$ is the barrier function and γ is the barrier constant, which is required to take a sufficiently large positive value.

To perform the unconstrained minimization of F we use the steepest-descent method based on Armijo's rule [19]. To do so we define the vector,

$$z = \left[z_1, z_2, \ldots, z_{np}, z_{np+1}, z_{np+2}, \ldots, z_{n(p+1)} \right]^T = \left[a_{11}, a_{12}, \ldots, a_{np}, \beta_1, \beta_2, \ldots, \beta_n \right]^T$$

$$(19)$$

For the $t + 1$ iteration the learning rule is,

$$z(t+1) = z(t) - \eta(t) \, \nabla F(z(t)) \quad (20)$$

where $\eta(t) = \lambda^\tau$ with $\lambda \in (0, 1)$. The parameter τ is the smallest positive integer such that,

$$F(z(t) - \eta(t) \, \nabla F(z(t))) - F(z(t)) < -\varepsilon \eta(t) \, \|\nabla F(z(t))\|^2 \quad (21)$$

with $\varepsilon \in (0, 1)$. Finally, the partial derivatives in Eqs. (20) and (21) can easily be derived.

4 Simulation Study

Based on the analysis described in Sect. 2, the data set includes $N = 4148$ input-output data pairs (corresponding to 52 beach cross-sections) of the form $\{x_k; y_k\}|_{k=1}^{N}$ with $x_k = [x_{k1} \quad x_{k2} \quad \ldots \quad x_{k6}]^T$ and $y_k \in \Re$. The data set was divided into a training set consisting of the 60 % of the original data, and a testing set consisting of the remainder 40 %. Table 1 depicts the parameter setting for the Legendre polynomial neural network.

Table 1. Parameter setting for the proposed network

Parameter	Value	Parameter	Value
a_L	-10	λ	0.05
a_U	10	ε	0.10
γ	10^8		

For comparison, two more neural networks were designed. The first one was a radial basis function (RBF) network. The parameters of the basis functions were estimated in terms of the conditional fuzzy clustering, which was developed in [20], while the connection weights were calculated by the least-squares method.

The second one was a feedforward neural network (FFNN), the activation functions of which read as follows,

$$f(x) = \tanh \frac{x}{2} \tag{22}$$

To train the FFNN we applied the steepest-descent based on the Armijo's rule (see previous Section) in order to minimize the network's square error. All networks were implemented using the Matlab software.

The performance index to conduct the simulations was the root mean square error,

$$RMSE = \sqrt{\frac{1}{N}\sum_{k=1}^{N}|y_k - \hat{y}_k|^2} \tag{23}$$

For the three networks we considered various numbers of nodes, while for each number of nodes we run 20 different initializations.

The results are shown in Table 2. The Legendre polynomial neural network appears to have superior performance compared with the other two networks. The best result for both the training and testing data sets is obtained by the proposed network when $n = 4$.

The results reported in Table 2 are visualized also in Fig. 4. There are some interesting remarks: (a) the difference between the Legendre polynomial and the remainder of the networks tested appears to be significant particularly in the testing data; (b) the RBF outperforms the FFNN in both cases; (c) although the best result for

Table 2. Comparative results in terms of the *RMSE* mean values and the corresponding standard deviations obtained by the three networks for various numbers of nodes in the hidden layer

		$n=3$	$n=4$	$n=5$
Proposed network	Training Data	9.838 ± 0.196	9.569 ± 0.208	9.594 ± 0.152
	Testing Data	10.053 ± 0.183	9.438 ± 0.186	9.525 ± 0.249
RBF	Training Data	10.637 ± 0.113	10.319 ± 0.081	10.298 ± 0.092
	Testing Data	10.627 ± 0.168	10.346 ± 0.142	10.299 ± 0.135
FFNN	Training Data	11.106 ± 0.838	11.056 ± 0.799	10.711 ± 0.740
	Testing Data	11.107 ± 0.926	10.856 ± 0.748	10.938 ± 0.846

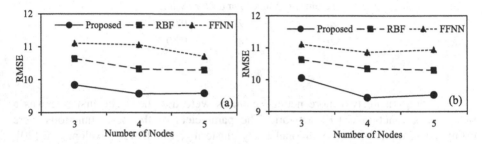

Fig. 4. Mean values of the RMSE as a function of the number of nodes for: (a) the training data, and (b) the testing data.

the proposed network was obtained for $n = 4$, the general tendency is to obtain smaller RMSEs as the number of polynomial nodes increases.

It is very interesting to see how the above results are translated into meaningful observations as far as the beach rotation is concerned. Figure 5 concerns a specific testing data and shows the predictions obtained by the three methods. Based on this figure, the proposed network clearly achieves the best beach rotation prediction.

Although the overall prediction performance of the proposed Legendre polynomial neural network may not appear to be very satisfactory on the basis of the RMSE (9.5 m), the following should be noted. First, Ammoudara shoreline position is characterized by high spatiotemporal variability, with the difference between the most inshore and most offshore recorded shoreline position during the 10-month monitoring period being between 3 and 8 m [8]; the proposed network's predictions are of the same order of magnitude and thus may be considered as satisfactory in this high non-linear coastal system, particularly as in many cross-shore sections, the network's predictions were much closer to the observed shoreline position (see Fig. 5). Secondly, adjacent sections of the shoreline showed large differences in terms of beach erosion/accretion patterns, suggesting significant control by small differences in reef morphology and the direction of wave approach. Hydrodynamic modeling has shown that small differences in the angle of wave approach result in quite different inshore hydrodynamic regimes (waves and wave-induced currents), even in the case of offshore waves with the same significant wave heights (H_S) and periods (T_P) [8]. As the offshore

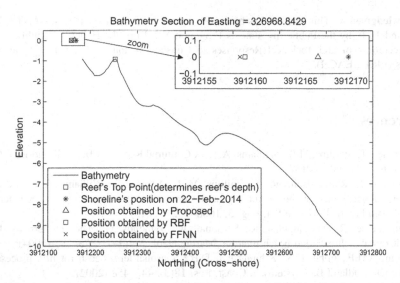

Fig. 5. A sample of the cross-shore shoreline postion predicted by the three networks.

wave data set did not include details on wave direction, the offshore waves used were grouped collectively as northerly waves (those waves affecting the beach); thus, the observed discrepancies could be mainly due to the absence of an additional wave parameter (angle of wave incidence) in the network input variables. As a future research, it would be interesting to test how the above results would be in the case that wave direction was included as an input variable.

5 Summary and Conclusions

In this paper we present a systematic methodology that includes a sophisticated experimental setup and a novel feedforward neural network to model beach rotation in a reef fronted (perched) beach (Ammoudara, Crete). A set of significant morphological and wave variables were identified that can directly affect beach rotation, which together with records of shoreline position from a coastal video imagery system were used to generate the network's input-output training data. The proposed network consists of four layers. The main task of the first and the second layers has been to obtain linear combinations of the input variables and then, to appropriately scale them before entering the third layer that comprises the Legendre polynomial activation functions. This scaling process was deemed necessary due to limitations imposed by the orthogonality of the Legendre polynomial. As a result, the weights of the linear combinations must be confined in a predetermined domain of values. Therefore, the training process of the network becomes a constrained nonlinear optimization problem, resolved by the barrier method. The comparative simulation experiments carried out showed that the proposed network can effectively model beach rotation, particularly if detailed wave direction data are available to be included as an additional input variable.

Acknowledgments. This research has been co-financed in 85 % by the EEA GRANTS, 2009–2014, and 15 % by the Public Investments Programme (PIP) of the Hellenic Republic. Project title: Recording of and Technical Responses to Coastal Erosion of Touristic Aegean island beaches (ERA BEACH).

References

1. Thomas, T., Phillips, M.R., Williams, A.T.: A Centurial Record of Beach Rotation. J. Coast. Res. **65**, 594–599 (2013)
2. Thomas, T., Rangel-Buitrago, N., Phillips, M.R., Anfuso, G., Williams, A.T.: Mesoscale morphological change, beach rotation and storm climate influences along a macrotidal embayed beach. J. Marine Sci. Eng. **3**, 1006–1026 (2015)
3. Ranasinghe, R., McLoughlan, R., Seasonal, A., Symonds, G.: The southern oscillation index, wave climate and beach rotation. Marine Geol. **204**(3–4), 273–287 (2004)
4. Klein, A.H.F., Filho, L.B., Schumacher, D.H.: Seasonal-term beach rotation processes in distinct Headland Bay systems. J. Coast. Res. **18**(3), 442–458 (2002)
5. Harley, M.D., Turner, I.L., Short, A.D.: New insights into embayed beach rotation: The importance of wave exposure and cross-shore processes. J. Geophys. Res. **120**(8), 16 (2015)
6. Gallop, S.L., Bosserelle, C., Eliot, I., Pattiaratchi, C.B.: The influence of lime-stone reefs on storm erosion and recovery of a perched beach. Cont. Shelf Res. **47**, 16–27 (2012)
7. Gallop, S.L., Bosserelle, C., Eliot, I., Pattiaratchi, C.B.: The influence of coastal reefs on spatial variability in seasonal sand fluxes. Marine Geol. **344**, 132–143 (2013)
8. Velegrakis, A.F., Trygonis, V., Chatzipavlis, A.E., Karambas, Th., Vousdoukas, M.I., Ghionis, G., Monioudi, I.N., Hasiotis, Th., Andreadis, O., Psarros, F.: Shoreline variability of an urban beach fronted by a beachrock reef from video imagery. Natural Hazards (2016). doi:10.1007/s11069-016-2415-9
9. Lowe, R.J., Hart, C., Pattiaratchi, C.B.: Morphological constraints to wave-driven circulation in coastal reef-lagoon systems: a numerical study. J. Geophys. Res. **115**, C09021 (2010)
10. Rigos, A., Tsekouras, G.E., Vousdoukas, M.I., Chatzipavlis, A., Velegrakis, A.F.: A Chebyshev polynomial radial basis function neural network for automated shoreline extraction from coastal imagery. Integr. Comput. Aided Eng. **23**, 141–160 (2016)
11. Ma, L., Khorasani, K.: Constructive feedforward neural networks using Hermite polynomial activation functions. IEEE Trans. Neural Netw. **16**(4), 821–833 (2005)
12. Lee, T.T., Jeng, J.T.: The Chebyshev-polynomials-based unified model neural networks for function approximation. IEEE Trans. Syst. Man Cybern. Part B Cybern. **28**(6), 925–935 (1998)
13. Patra, J.C., Meher, P.K., Chakraborty, G.: Nonlinear channel equalization for wireless communication systems using legendre neural networks. Sig. Process. **89**(11), 2251–2262 (2009)
14. Tsekouras, G.E., Rigos, A., Chatzipavlis, A., Velegrakis, A.: A neural-fuzzy network based on Hermite polynomials to predict the coastal erosion. Commun. Comput. Inf. Sci. **517**, 195–205 (2015)
15. Alexandrakis, G., Ghionis, G., Poulos, S.E.: The Effect of beach rock formation on the morphological evolution of a beach. The case study of an Eastern Mediterranean beach: Ammoudara, Greece. J. Coast. Res. **69**(SI), 47–59 (2013)
16. Bell, W.W.: Special Functions for Scientists and Engineers. D. Van Nostrand Company Ltd., London (1968)

17. Moore, R.E.: Interval Analysis. Prentice-Hall, Englewood Cliff (1966)
18. Luenberger, D.G., Ye, Y.: Linear and Nonlinear Programming, 3rd edn. Springer, New York (2008)
19. Armijo, L.: Minimization of functions having Lipschitz continuous first partial derivatives. Pacific J. Math. **16**(1), 1–3 (1966)
20. Pedrycz, W.: Conditional fuzzy clustering in the design of radial basis function neural networks. IEEE Trans. Neural Netw. **9**(4), 601–612 (1998)

Environmental Impact on Predicting Olive Fruit Fly Population Using Trap Measurements

Romanos Kalamatianos$^{(\boxtimes)}$, Katia Kermanidis, Markos Avlonitis, and Ioannis Karydis

Department of Informatics, Ionian University, 49100 Corfu, Greece
{cl4kala, kerman, avlon, karydis}@ionio.gr

Abstract. Olive fruit fly trap measurements are used as one of the indicators for olive grove infestation, and therefore, as a consultation tool on spraying parameters. In this paper, machine learning techniques are used to predict the next olive fruit fly trap measurement, given as input environmental parameters and knowledge of previous trap measurements. Various classification algorithms are employed and applied to different environmental settings, in extensive comparative experiments, in order to detect the impact of the latter on olive fruit fly population prediction.

Keywords: Olive fruit fly · Machine learning · Population prediction · Classification · Naive bayes · Nearest neighbors · Decision trees · Random forests · Support vector machines

1 Introduction

The olive fruit fly is a pest that has been recorded to infest solely the olive fruits since at least the third century BC [1]. Such infestations cause great damage to the production of both olive oil or table olives [2] in many olive oil producing countries, including Greece. The olive fruit fly is active during the summer and reaches its population peak during autumn, while during the winter and in the first months of spring it hibernates, until environmental conditions are favorable for it to reemerge [1].

The population growth of the olive fruit fly and, by extension, the level of infestation of an olive grove are affected by various environmental factors. However, the two primary factors that affect the activity of the olive fruit fly are temperature [3] and relative humidity [4, 5].

Population control of the olive fruit fly can be achieved through spraying of the olive trees, either with bait or universal [1, 6]. However, in order for the spraying to have any effect, it has to be applied when conditions are appropriate. Two factors indicate when spraying should commence [6]: (a) the ripeness level of the olive fruit, as the fruit needs to be ripe in order for it to be susceptible to the olive fruit fly and (b) the population of the fly, i.e. when a certain population threshold (recorded via sampling) is exceeded. Sampling is achieved through McPhail traps or yellow sticky traps [1, 6]. The threshold is set to seven olive fruit flies per trap per week during the summer and is decreased to five olive fruit flies per trap per week during autumn [6].

© IFIP International Federation for Information Processing 2016
Published by Springer International Publishing Switzerland 2016. All Rights Reserved
L. Iliadis and I. Maglogiannis (Eds.): AIAI 2016, IFIP AICT 475, pp. 180–190, 2016.
DOI: 10.1007/978-3-319-44944-9_16

Machine learning techniques have been used to detect oil spills on the surface of the sea by scanning radar images [7], to automatically identify species by sound [8] and to monitor flood protection systems [9]. Machine learning techniques have also been applied in numerous agriculture processes such as the prediction of when a cow should be culled in a dairy herd [10], the estimation of soil moisture [11], the estimation of a cow's oestrus [12] and the prediction of olive fruit fly infestation using information about olive tree health as well as trap measurements [13].

The aim of this paper is to predict future olive fruit fly trap measurements, and by extension olive fruit fly infestations/outbreaks, using machine learning algorithms. Our approach differentiates from previous work [13] by constructing a feature vector that consists of environmental factors e.g. temperature, instead of the olive tree health, as well as trap measurements.

2 Data Collection

2.1 Environmental Data

The data used in the experiments were collected from environmental sensors and olive fruit fly traps that were installed at 16 locations on the north-western side of the island of Corfu, Greece. Readings on the olive fruit fly traps show the total number of olive fruit flies caught by the trap. Each reading was conducted every five days for the period from 10th June 2015 to 29th September 2015, at all locations. All sensors at all locations logged temperature values at a 15 min interval, while a few of these also logged relative humidity values.

2.2 Feature Selection

In order to perform classification experiments, the aforementioned environmental data were transformed into the following set of attributes (in order to represent readings as feature-value learning vectors):

- Mean temperature of the last five days before next trap reading
- Mean maximum temperature of the last five days before next trap reading
- Mean minimum temperature of the last five days before next trap reading
- Day 1 Mean Temperature
- Day 1 Maximum Temperature
- Day 1 Minimum Temperature
- Day 2 Mean Temperature
- Day 2 Maximum Temperature
- Day 2 Minimum Temperature
- Day 3 Mean Temperature
- Day 3 Maximum Temperature
- Day 3 Minimum Temperature
- Day 4 Mean Temperature
- Day 4 Maximum Temperature

- Day 4 Minimum Temperature
- Day 5 Mean Temperature
- Day 5 Maximum Temperature
- Day 5 Minimum Temperature

Apart from the environmental attributes, one more attribute, namely the trap measurement of the last reading (number of flies caught), was used as input. All aforementioned attributes are numeric. Finally another attribute, denoting the next trap reading was used as the classification class.

2.3   Feature Vector Extraction

The process of extracting the attributes for the feature vectors from the sensor data was automated by use of a script. The script was written in Python that automatically computes the mean, mean maximum and mean minimum temperature for the five day period before the next trap reading, as well as the mean, maximum and minimum temperature for each day in the aforementioned five day period. The script exports all vectors in a CSV (Comma Separated Values) file. Finally trap readings are added manually at each corresponding vector instance.

The temperature-related attributes, initially numeric, were discretized into the following three bins:

- <15, temperature is lower than 15 °C,
- 15 to 32, temperature is between 15 °C and 32 °C,
- >32, temperature is greater than 32 °C.

The discretization of the temperature values was based on the temperature range (between 15 °C and 32 °C [14]), in which the olive fruit fly is active. If the temperature of the environment is below the lower or exceeds the upper threshold, then the olive fruit fly is motionless due to extreme cold or heat. Accordingly, herein we assume that outside the optimal temperature range of the olive fruit fly, the traps will not capture any olive fruit flies.

Trap reading related attributes have also been discretized into the following bins:

- 0 to 4, none or up to 4 olive fruit flies inside the trap,
- 5 to 6, five or six olive fruit flies inside the trap,
- >=7, greater than or equal to seven olive fruit flies inside the trap.

The use of a ternary quantisation of the number of olive fruit flies is based on trap measurements analysis, i.e. the infestation threshold depends on the season the measurements are made. Specifically, in the summer months the infestation threshold is set to seven olive fruit flies per trap per week. On the other hand, from September onwards the infestation threshold is decreased to five olive fruit flies per trap per week, due to cooler weather [6]. Therefore, although the last bin value would always indicate infestation, the second bin value would be depended on the season.

3   Machine Learning Algorithms

The WEKA machine learning workbench[1] was used for running the classification experiments. In the sequel, a number of classification algorithms that were selected for experimentation are shortly presented.

J48 [15] is a decision tree induction algorithm and it is a version of C4.5, an earlier algorithm developed by J. Ross Quinlan [16]. C4.5 generates a decision tree based on information gain of the attributes in the available training data. More specifically, the attribute whose values discriminate most clearly the training examples according to their class label is identified in each iteration. The algorithm stops when there are no further attributes to explore or when all the training examples are separated according to their class label. Additionally, J48 incorporates two tree pruning methodologies: The first one is known as subtree replacement and it replaces a node in a decision tree with the corresponding leaf, if the given subtree does not help classification accuracy. This pruning process starts from the leaves of the fully formed tree, and moves bottom up toward the root. The second methodology is known as subtree raising in which a node may replace other nodes while it is moved towards the root. This type of pruning most of the times has insignificant effect on decision tree models (Table 1).

Table 1.   J48 parameter values

Binary splits	No
Confidence factor	0.25
Minimum instances per leaf	2
Reduced error pruning	No
Subtree raising	Yes
Pruned	Yes
Laplace smoothing	No

Sequential Minimal Optimization or SMO [17] is an ameliorated algorithm for training support vector machines. SMO cuts in pieces a large quadratic programming optimization problem converting it into smaller problems (sub-problems of quadratic programming). The sub-problems are solved quickly because they are solved analytically which means that SMO avoids to use extra time for arithmetical quadratic programming optimization as an inner loop. So SMO manages to reduce computation time significantly (Table 2).

Naïve Bayes [15] is a probabilistic classifier based on the assumption of conditional independence [18], which assumes that the appearance of a specific feature given the class value is unrelated to the appearance of any other feature in the dataset. Though not valid in reality, this assumption has been proven to cope well with several classification problems. Additionally, this algorithm needs a small amount of training data to determine the parameters necessary for classification. Due to the hypothesis of

---

[1] http://www.cs.waikato.ac.nz/ml/weka/.

Table 2. SMO parameter values

Complexity parameter	1.0
Round-off error	1.0E-12
Filter type	Normalize training data
Kernel	PolyKernel
Random seed for cross validation	1
Tolerance parameter	0.001

Table 3. Naive Bayes parameter values

Use kernel estimator	No
Use supervised discretization	No

independent variables; there is no need to estimate the entire covariance matrix but only the differentiations of the variables for each class (Table 3).

The RandomForest [19] is a meta-learning classification algorithm that runs iteratively. In each iteration a decision tree is induced from a randomly selected subset of the features. The number of iterations is pre-defined. The final classification error is the mean error over all iterations (Table 4).

Table 4. RandomForest parameter values

Maximum depth	Unlimited
Number of attributes	0
Number of trees to be generated	100
Seed	1

AdaBoost [20] is another meta-learning algorithm, that iteratively changes instance weights, based on whether they were classified correctly (or not) in a previous iteration. Thereby, the learner is forced to focus on instances that are hard to classify. The final classification is derived from the weighting of the models induced after every iteration (Table 5).

Table 5. AdaBoost parameter values

Classifier	SMO
Number of iterations	10
Seed	1
Use resampling	No
Weight threshold	100

The IBk algorithm [15] is an alternate version of the k-nearest neighbor algorithm. Using the Euclidean distance as a distance metric, it identifies the k training examples that are closest to a given test instance, and, via majority voting selects its class label.

The value of k can be explicitly pre-defined, or estimated optimally using cross validation. In our experiments, the number of nearest neighbors ranged from 1 to 33 neighbors, where only odd values were selected (Table 6).

Table 6. IBk parameter values

Cross validate	No
Distance weighting	No
Use of mean squared error	No
Nearest neighbor search algorithm	LinearNNSearch
Window size	0

The Multilayer Perceptron is an artificial neural network [21], where each of the multiple layers of nodes is fully connected to the following one. Multilayer perceptrons are able to distinguish data that are not linearly separable. Experiments were conducted for one hidden layer with the number of nodes ranging from 1 to 10 (Table 7).

Table 7. Multilayer Perceptron parameter values

Decrease learning rate	No
Hidden layers	1
Learning rate	0.3
Momentum	0.2
Nominal to binary filter	Yes
Normalize attributes	Yes
Normalize numeric class	Yes
Reset	Yes
Seed	0
Training time	500
Validation set size	0
Validation threshold	20

4   Experimental Process

184 training instances were supplied. Due to the small size of the training data, no test set could be supplied for the validation of the results. Therefore the 10-fold cross-validation method was used. The original sample is randomly partitioned into ten subsamples. One out of ten subsamples is kept as validation data for testing the model, and the remaining nine subsamples are used as training data. The cross-validation process is then repeated ten times, with each of the ten subsamples being used only once as validation data. Results are averaged across the ten experiments.

5 Evaluation

Figure 1 displays the classification results of the five aforementioned machine learning algorithms. The SMO algorithm produces the best results in both precision and recall, with the J48 algorithm being close by. AdaBoostM1 produces the same results with SMO using as a base learner the SMO algorithm. On the other hand, NaiveBayes produces the worst results in comparison with the other algorithms, with a significant decrease in precision and a quite noteworthy low recall.

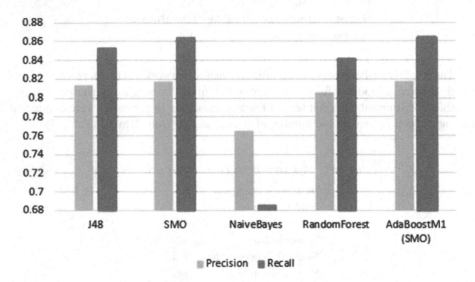

Fig. 1. Precision and recall comparison for five classification algorithms

Figure 2 presents the pruned tree constructed by the J48 algorithm. It is clear that the algorithm considers the "previous reading" as the most important attribute for classification. With the minimum temperature of day 3 being used when the previous reading has a value of "5 to 6".

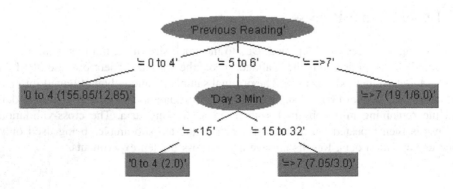

Fig. 2. Tree view constructed by J48

The next experiment is using the IBk algorithm, as shown in Fig. 3, with maximum recall for one and for five nearest neighbors. The best results in both recall and precision are achieved when using one nearest neighbor. From three nearest neighbors and onwards the precision of the algorithm decreases, with a significant drop after nine nearest neighbors. After eleven nearest neighbors, precision and recall stay stable.

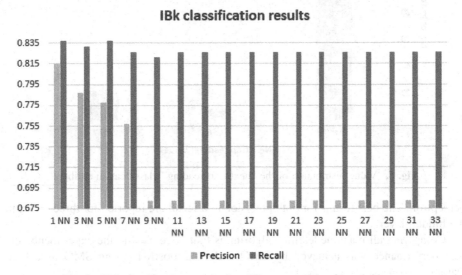

Fig. 3. IBk classification results for different number of k neighbours

Finally, the Multilayer Perceptron algorithm, shown in Fig. 4, exhibits a great difference between recall and precision values for any number of nodes in the

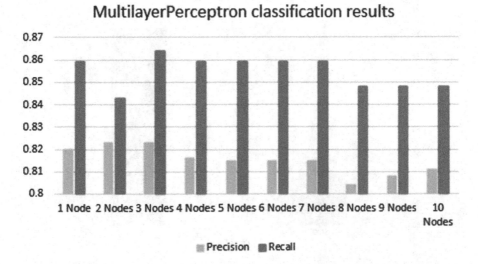

Fig. 4. Multilayer Perceptron classification results for different number of nodes

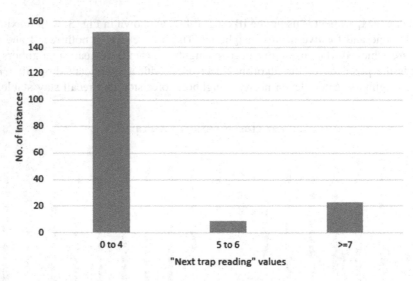

Fig. 5. Value distribution of the "next trap reading" classification attribute

perceptron. The best results for both precision and recall are reported for three nodes in the hidden layer.

Comparing all machine learning algorithms that were used in the experiments, the best performance was achieved by SMO and AdaBoostM1, using SMO as a base learner.

It is important to note that the values of the classification attribute are not balanced. Figure 5 depicts the distribution of the three values among the 184 instances. Over 80 % of the instances have a value of "0 to 4", while the value "5 to 6" is the least encountered

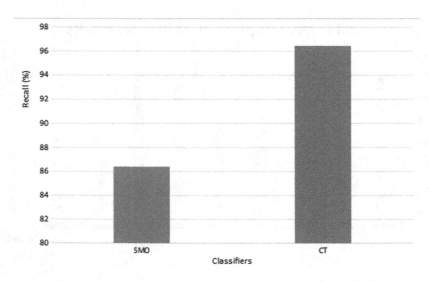

Fig. 6. Results comparison with previous work [13]

value, with nine instances. This fact explains why the J48 algorithm, as shown by the constructed tree in Fig. 2, can't classify any instance to the value "5 to 6".

Figure 6 shows the comparison of the best classifier presented herein against the best classifier of [13] in terms of recall. It is obvious that the Classification Trees (CT) have a far better recall compared to the SMO algorithm by almost 10 %. This difference is attributed to the significant imbalance in the distribution of the class value in the training data. Additionally, the results produced by the CT were achieved using only six attributes and the classification class had only two values, whether to treat or not. Finally, a number of the used attributes contained information about the pheno-logical state of the olive tree.

6 Conclusion

In this work, supervised machine learning is used to predict future olive fruit fly trap's measurements. The proposed feature vector consists of environmental parameters, specifically temperature, and information about previous trap measurements. Results produced by the conducted experiments were promising with the support vector machine algorithm providing the best classification results, although the values of the classification attribute were unbalanced.

Our approach differentiates from previous work by using environmental parameters instead of information about the health of the olive trees. When results between the best classifiers of the proposed and existing work were compared, the proposed approach produced results with 10 % ameliorated recall.

Future research will take into account more environmental parameters such as relative humidity and the amount of light the olive fruit flies are exposed to. Fur-thermore, the experiments described are planed to be conducted again in more training instances as measurement data accumulate.

Acknowledgments. Financial support of the European Union and of National Funds of Greece and Albania under the IPA Cross-Border PROGRAMME "Greece - Albania 2007–2013", project title "Enhancing Olive Oil Production with the use of Innovative ICT" with the acronym "e-Olive", is gratefully acknowledged.

References

1. Vossen, P., Varel, L., Devarenne A.: Olive fruit fly. University of California Cooperative Extension, Sonoma County (2006)
2. Rice, R.: Bionomics of the olive fruit fly bactrocera (Dacus) olea. University California Plant Protection Quarterly **10**, 1–5 (2000)
3. Fletcher, B.S.: Temperature development rate relationships of the immature stages and adults of tephritid fruit flies. In: Robinson, A.S., Hooper, G. (eds.) Fruit Flies: Their Biology, Natural Enemies and Control, vol. 3A, pp. 273–289. Elsevier, Amsterdam (1989)

4. Yokoyama, V.Y., Rendon P., Sivinski, J.: Biological control of olive fruit fly (Diptera: Tephritidae) by reseases of Psyttalia cf. concolor (Hymenoptera: Braconidae) in California, parasitoid longevity in presence of the host, and host status of walnut husk fly. In: 7th International Symposium on Fruit Flies of Economic Importance, pp. 157–164. Salvador, Brazil (2006)
5. Broufas, G.D., Pappas, M.L., Koveos, D.S.: Effect of relative humidity on longevity, ovarian maturation, and egg production in the olive fruit fly (Diptera: Tephritidae). Ann. Entomol. Soc. Am. **102**(1), 70–75 (2009)
6. Patsias, A.: EE Katapolmeesee tou THkou tee Elee [The Fighting of Olive fruit fly]. Publicity Department of Agricultural Sector Applications and Publicity. Nicosia, Cyprus (2005)
7. Kubat, M., Holte, R.C., Matwin, S.: Machine learning for the detection of oil spills in satellite radar images. Mach. Learn. **30**, 195–215 (1998)
8. Acevedo, M.A., Corrada-Bravo, C.J., Corrada-Bravo, H., Villanueva-Rivera, L.J., Aide, T.M.: Automated classification of bird and amphibian calls using machine learning: a comparison of methods. Ecol. Inf. **4**, 206–214 (2009)
9. Pyayt, A.L., Mokhov, I.I., Lang, B., Krzhizhanovskaya, V.V., Meijer, R.J.: Machine learning methods for environmental monitoring and flood protection. World Acad. Sci. Eng. Technol. **78**, 118–124 (2011)
10. McQueen, R.J., Garner, S.R., Nevill-Manning, C.G., Witten, I.H.: Applying machine learning to agricultural data. Comput. Electron. Agric. **12**(4), 275–293 (1995)
11. Ahmad, S., Kalra, A., Stephen, H.: Estimating soil moisture using remote sensing data: a machine learning approach. Adv. Water Res. **33**(1), 69–80 (2010)
12. Mitchell, R.S., Sherlock, R.A., Smith, L.A.: An investigation into the use of machine learning for determining oestrus in cows. Comput. Electron. Agric. **15**(3), 195–213 (1996)
13. del Sagrado, José, del Águila, I.M.: Olive fly infestation prediction using machine learning techniques. In: Borrajo, D., Castillo, L., Corchado, J.M. (eds.) CAEPIA 2007. LNCS (LNAI), vol. 4788, pp. 229–238. Springer, Heidelberg (2007)
14. Vossen, P.: Monitoring and Control of Olive Fruit Fly (OLF) for Olive Production in California. University of California Cooperative Extension (2014)
15. Witten, I.H., Frank, E.: Data Mining: Practical Machine Learning Tools and Techniques, 2nd edn. Morgan Kaufmann, San Francisco (2005)
16. Quinlan, J. R.: Bagging, boosting and C4.5. In: 13th National Conference on Artificial Intelligence, pp. 725–730. AAAI Press, Portland (1996)
17. Platt, J.C., Sequential minimal optimization: a fast algorithm for training support vector machines. Technical report MSR-TR-98–14, Microsoft Research (1998)
18. Russell, S., Norvig, P.: Artificial Intelligence: A Modern Approach, 3rd edn. Prentice Hall, Upper Saddle River (2009)
19. Breiman, L.: Random Forests. Mach. Learn. **45**, 5–32 (2001)
20. Freund, Y., Schapire, R.E.: Experiments with a new boosting algorithm. In: Proceedings of the Thirteenth International Conference on Machine Learning, pp. 148–156 (1996)
21. Rosenblatt, F.: Principles of Neurodynamics: Perceptrons and the Theory of Brain Mechanisms. Spartan Books, Washington, D.C. (1961)

A Hybrid Soft Computing Approach Producing Robust Forest Fire Risk Indices

Vardis-Dimitris Anezakis[1], Konstantinos Demertzis[1(✉)],
Lazaros Iliadis[1], and Stefanos Spartalis[2]

[1] Lab of Forest-Environmental Informatics and Computational Intelligence,
Democritus University of Thrace, 193 Pandazidou st., 68200 Orestiada, Greece
{danezaki,kdemertz,liliadis}@fmenr.duth.gr
[2] Laboratory of Computational Mathematics, School of Engineering,
Department of Production and Management Engineering,
Democritus University of Thrace,
V.Sofias 12, Prokat, Building A1, 67100 Xanthi, Greece
sspart@pme.duth.gr

Abstract. Forest fires are one of the major natural disaster problems of the Mediterranean countries. Their prevention - effective fighting and especially the local prediction of the forest fire risk, requires the rational determination of the related factors and the development of a flexible system incorporating an intelligent inference mechanism. This is an enduring goal of the scientific community. This paper proposes an Intelligent Soft Computing Multivariable Analysis system (ISOCOMA) to determine effective wild fire risk indices. More specifically it involves a Takagi-Sugeno-Kang rule based fuzzy inference approach, that produces partial risk indices (PRI) per factor and per subject category. These PRI are unified by employing fuzzy conjunction T-Norms in order to develop pairs of risk indices (PARI). Through Chi Squared hypothesis testing, plus classification of the PARI and forest fire burned areas (in three classes) it was determined which PARI are closely related to the actual burned areas. Actually we have managed to determine which pairs of risk indices are able to determine the actual burned area for each case under study. Wild fire data related to specific features of each area in Greece were considered. The Soft computing approach proposed herein, was applied for the cases of Chania, and Ilia areas in Southern Greece and for Kefalonia island in the Ionian Sea, for the temporal period 1984–2004.

Keywords: Takagi-Sugeno-Kang · Fuzzy inference system · T-Norms · Chi-Square test · Individual indices · Unified index · Forest fires

1 Introduction

Greece has a very important forest capital, as 50 % of the territory is covered by woodland. About 25 % of it is characterized by high vegetation coniferous and broadleaf high biodiversity, the remaining of low trees and shrubs near inhabited areas. Also there are approximately 2 million acres of rangelands. During the last 20 years, the average annual burned areas in the country are higher than 45,000 acres as a result

© IFIP International Federation for Information Processing 2016
Published by Springer International Publishing Switzerland 2016. All Rights Reserved
L. Iliadis and I. Maglogiannis (Eds.): AIAI 2016, IFIP AICT 475, pp. 191–203, 2016.
DOI: 10.1007/978-3-319-44944-9_17

of 1500 forest fires. The determination of the factors that favor ignition and contribute to the spread of wild fires (WF) requires a detailed spatiotemporal analysis of the historical data for each area under study. Moreover, the specification of the correlations between these parameters is absolutely necessary. This research paper proposes an innovative hybrid forest fire modeling system operating on a local basis. The reasoning of the ISOCOMA employs Computational Intelligence approaches in order to produce an overall fire risk index.

1.1 Literature Review

Iliadis and Betsidou [9] have implemented an intelligent rule based fuzzy inference system (FIS) evaluating wild fire risk for the forest departments of Greece. The estimation of the risk indices was done by using fuzzy triangular membership functions and Einstein fuzzy conjunction T-Norms. Iliadis and Zigkrika [11] have also developed a FIS that performs and evaluates scenarios (by assigning weights to the involved features) towards the estimation of a characteristic overall forest fire risk index in Greece. Papakonstantinou et al. [17] have proposed a fuzzy rule based system to produce the drought risk indices vectors for the forest regions of Cyprus under study. Iliadis et al. [10] have developed a fuzzy inference system under the MATLAB platform. The system uses three distinct Gaussian distribution fuzzy membership functions in order to estimate the partial and the overall risk indices due to wild fires in the southern part of Greece. Özbayoğlu and Bozer [16] estimated the potential burned areas using geographical and meteorological data. Several computational intelligence approaches were used namely: Multilayer Perceptron (MLP), Radial Basis Function Networks (RBFN), Support Vector Machines (SVM) and fuzzy logic. Shidik and Mustofa [18] used a Back-Propagation Neural Network which was trained based on meteorological and forest weather indices, so as to classify the burned area in three categories. Aldrich et al. [1] investigated the effect of variations in land use and climate in the occurrence of forest fires. Catry et al. [6] used logistic regression models to predict the relative probability of ignition occurrence, as a function of the resulting fire size.

1.2 Innovations of the Proposed Methodology

The main innovation of the ISICOMA is the development of four partial risk indices (PRI), which are derived from the respective analysis of separate parameters, creating and analyzing meaningful relationships and rules of correlations between them. This raises the problem of wild fires (WF) on an absolutely realistic basis. In addition, it is for the first time that an intelligent system combines the use of an adaptive fuzzy inference Takagi-Sugeno-Kang (AFITS) system with the wide use of fuzzy conjunction T-Norms in order to obtain higher fitting rates between PRI and TPRI with the actual burned areas.

1.3   Data

The first step towards the development of an overall wild fire risk index model was the determination of all factors that affect the behavior of a forest fire. The data collected from the forest inspections and from the Hellenic national meteorological service. According to [12] the following factors have been identified as playing a key role (Table 1).

Table 1. Factors affecting fire behavior

Flammability of vegetation	Monthly rainfall
Canopy density	Previous month rainfall
Vegetation density	Altitude
Air temperature	Slope
Relative humidity	Ground orientation
Wind speed, daily rainfall	Exposure

Utilizing and analyzing in-depth studies in the raw meteorological, topographical and vegetative data of the areas concerned [3, 4] the following categories were obtained (Table 2).

Table 2. Classification of the fire parameters

	Parameters	Class1	Class2	Class3	Class4	Class5	Class6
Meteorological	Wind	0–1 bf	1.1–4 bf	4.1–7 bf	7.1–9 bf	>9.1 bf	
	Air temperature/relative humidity	Low risk	Medium risk	High risk			
Topographic	Slope	0–20 %	21–40 %	41–60 %	61–80 %	81–100 %	>100 %
	Ground Orientation Exposure	Unspecified	North	South	East	West	
	Altitude	Low	Medium	High			
Vegetation	Canopy Density	Absent	Rare	Full			
	Vegetation Density	Absent	Rare canopy < 0.4	Dense canopy > 0.4			
	Flammability of vegetation	Low risk	Medium risk	High risk			
Drought	Rainfall (Daily, Monthly, Previous Month)	Low	Medium	High			

1.4   Areas of Study

Kefalonia (island in the Ionian Sea) Ilia (prefecture in Peloponnese) and Chania (prefecture in Crete island) have been chosen as the areas of interest. They have rich vegetation, they have protected areas (under Natura network) and their climate is dry and hot with low rain height. Also both Chania and Kefalonia are characterized by high touristic development and growth with high land value. On the other hand ancient Olympia is located in Ilia prefecture. Thus, it is an area of high cultural and touristic value. During the period 1984–2014, totally 1397 wild fires occurred in Ilia, 857 in Chania and 1298 in Kefalonia.

The Fire Ignition Indicator (FIGI) which emerges by combining the effect of temperature and humidity and the Spread Index which considers the effect of wind and slope (SPRI) have been used to produce significant evidence of forest fire risk. In a previous research effort of our team [2] have found that the SPRI is "High" in the 30–50 % of the cases, whereas the FIGI has shown smaller high and medium hazard rates.

2   Theoretical Framework and Methodology

2.1   Fuzzy Inference Systems

The Sugeno Fuzzy implication is the basic modeling approach used by the ISOCOMA. Introduced in 1985 [19], it is similar to the Mamdani method. While Mamdani FIS uses the technique of defuzzification of a fuzzy output, Sugeno FIS uses weighted average to compute the crisp output. The fuzzy membership functions (FMF) of the output are either linear (first order polynomials or constant crisp values). A typical rule in a Sugeno fuzzy model if the outputs are first-order linear has the form:

$$\text{If Input 1 } = \text{x and Input 2 } = \text{y then Output is z } = ax + by + c \qquad (1)$$

For a zero-order Sugeno model, the output level z is a constant crisp value c (a = b=0).

The output level z_i of each rule is weighted by the firing strength w_i of the rule. For an AND rule with Input 1 = x and Input 2 = y, the firing strength is

$$w_i = \text{AndMethod} (F_1(x), F_2(y)) \qquad (2)$$

where $F_{1,2}$ are the membership functions for Inputs 1 and 2. The final output of the system is the weighted average of all rule outputs, computed as in (3).

$$Final\, output = \frac{\sum_{i=1}^{N} w_i z_i}{\sum_{i=1}^{N} w_i} \qquad (3)$$

where N is the number of rules.

2.2   T-Norms

This paper attempts to calculate the Unique Overall Risk Index (UORI), resulting from the cumulative effect of all the related factors, after performing integration operations on all individual fuzzy sets. This task is carried out, by the use of specific fuzzy conjunction "AND" operators (CONO) known as T-Norms in the literature. The Min, the Algebraic, the Drastic, the Einstein and the Hamacher Products act as T-Norms [5, 7, 13–15]. The T-Norms are the unifiers of partial risk indices and they are quite optimistic as they are assigning the minimum risk value to the overall index [8].

2.3   Chi-Square Test

The Chi-Squared hypothesis-testing is a non-parametric statistical test in which the sampling distribution of the test statistic is a chi-square distribution when the null hypothesis is true. The null hypothesis H_0 usually refers to a general statement or default position that there is no relationship between two measured phenomena, or no difference among groups. The H_0 is assumed to be true until evidence suggest otherwise [20]. The statistical control index used for this assessment is the test statistic X^2.

$$X^2 = \sum \frac{(f_{o-}f_e)^2}{f_e} \qquad (4)$$

Where f_e is the expected frequency and fo the observed one. The degrees of freedom are estimated as follows (based on the rXc table of labeled categories):

$$df = (r-1)(c-1) \qquad (5)$$

For the H_0 the critical values for the test statistic X^2 are estimated by the X^2 distribution after considering the degrees of freedom. If the result of the test statistic is less than the value of the Chi-Square distribution, then we accept H_0 otherwise we reject it.

3   Description of the Proposed Methodology

The core of the modeling approach proposed herein was based on the grouping of twelve initial fire risk indices in four classes of partial risk indices and subsequently their integration into a unique overall one, the Conceptual Risk Index (CRI), for each fire incident and for each area of study. Then based on the burned area, the Actually Burned Surfaces Index (ACBUS) was determined. Moreover, we performed fuzzy conjunction (with T-Norms) of the four CRIi indices (i = 1 … 4) selecting all combinations of pairs in order to find those that belong to the same Linguistics with the ones of ACBUS. For example, we tried to estimate the forest fire incidents that were assigned "High Risk" pairs of indices and at the same time "High Risk" ACBUS ones.

3.1   The Algorithm

The proposed algorithmic process involves 9 distinct steps, which are discussed below:

1. Evaluating the twelve initial features influencing the phenomenon of forest fires and subsequently grouping them conceptually, in four thematic areas (feature categories). A distinct Conceptual Risk Index (CRI) has been developed for each feature category, totally four of them CRIi (i = 1 … 4). More specifically, the Weather Risk Index (WRI) was constructed from the contribution of temperature, humidity and wind speed. Correspondingly the drought index (DRI) comprises of the daily plus the monthly precipitation and of the precipitation in the previous month.

The topographic Risk index (TRI) is related to the slope to the altitude and to the exposure. The vegetation Risk index (VRI) is defined by the flammability of forest species, the canopy density and the vegetation density. The same methodology has been followed for all three areas under study.

2. The Fuzzy Inference Engine Takagi-Sugeno-Kang (FIETS) has been used. According to it, each feature is fuzzified based on properly designed fuzzy Trapezoidal membership functions. In this way each parameter of each incident has been assigned fuzzy risk linguistics.

3. This process determines the corresponding fuzzy set (linguistic) for each wild fire incident.

4. Proper fuzzy weighted rule sets have been designed and implemented. The number of the trapezoidal fuzzy membership functions used for the determination of each partial risk index related to each one of the 12 parameters, was determined by the number of its corresponding classes in Table 2. The number of fuzzy sets created for each feature, were exactly as many as its corresponding classes in Table 2. For the topographic parameter "Ground Orientation Exposure" the first class was removed because it was declared officially as unspecified (Tables 3 and 4).

Table 3. Fuzzy Sets and the corresponding Linguistics of each feature

	Kefalonia	Chania	Ilia
Air temperature	3 Low(L), Medium (M), High(H)	3 L, M, H	3 L, M, H
Relative humidity,	3 L, M, H	3 L, M, H	3 L, M, H
Daily rainfall	3 L, M, H	3 L, M, H	3 L, M, H
Monthly rainfall	2 M, H	3 L, M, H	3 L, M, H
Previous month rainfall	3 L, M, H	3 L, M, H	3 L, M, H
Altitude	3 L, M, H	3 L, M, H	3 L, M, H
Flammability of vegetation	3 L, M, H	3 L, M, H	3 L, M, H
Canopy density, vegetation density	3 Very Low(VL), L, H	3 VL, L, H	3 VL, L, H
Wind speed	4 L, M, H, Very High(VH)	5 L, M, H, VH, Ultra High(UH)	5 L, M, H, VH, UH
Slope	5 VL, L, M, H, VH,	6 VL, L, M, H, VH, UH	6 VL, L, M, H, VH, UH
Ground orientation exposure	4 VL, L, M, H	4 VL, L, M, H	4 VL, L, M, H

It should be clarified that in the following Table 5, the fuzzy values of the involved features are connected with the Min T-Norm (Table 6).

5. Each of the four CRIs has been calculated. The output of the Sugeno-type FIS takes values in the closed interval [0, 1] (Table 7).

Table 4. Conceptual Fuzzy Sets and the Linguistics of the (CRI)

	WRI	DRI	TRI	VRI	UORI
Kefalonia	5L, M, H, VH, UH	3L, M, H	4L, M, H, VH	4 L, M, H, VH	5 L, M, H, VH, UH
Chania	5L, M, H, VH, UH	3L, M, H	4L, M, H, VH	4 L, M, H, VH	5 L, M, H, VH, UH
Ilia	4L, M, H, VH	3L, M, H	4L, M, H, VH	4 L, M, H, VH	5 L, M, H, VH, UH

Table 5. Fuzzy rule set (T-Norms Fuzzy-AND) for the determination of the WRI

Humidity	Air temperature	Wind speed	WRI
H	L	L	L
H	L	M	L
H	M	L	L
H	M	M	M
H	M	H	M
H	M	VH	M
H	H	H	H
H	H	VH	UH
M	L	L	L
M	L	M	L
M	L	H	L
M	M	L	L
M	M	M	M
M	M	H	L
M	M	VH	H
M	M	UH	H
M	H	L	L
M	H	M	M
M	H	H	H
M	H	VH	VH
M	H	UH	VH
L	L	L	L
L	L	M	L
L	L	H	L
L	M	L	L
L	M	M	L
L	M	H	L
L	M	VH	L
L	M	UH	M
L	H	M	L
L	H	H	L

Table 6. Number of rules in the fuzzy rule sets use for the determination of each CRI and for the UORI for each area.

	WRI	DRI	TRI	VRI	UORI
Kefalonia	23	8	44	20	46
Chania	23	11	50	18	61
Ilia	33	12	47	21	68

Table 7. Four CRIs and UORI membership functions

Linguistics of CRIs and UORI	Constant values
Low	0
Medium	0.5
High	0.9
Very High	0.95
Ultra High	1

6. The UORI has been produced by adjustment of the weights of the fuzzy rules (Table 7).
7. The actual burned area for each area under study has been fuzzified by obtaining four fuzzy Linguistics representing the Actually Burned Surfaces (ACBUSij) $i = 1 \ldots 4$ and $j = 1 \ldots N$ (N is the number of examined cases) namely: Low Burned (LBUR), Average Burned (ABUR), High Burned (HBUR) and Extremely Burned (ExBUR). Subsequently, a comparative analysis has been performed between the four CRIs and the UORI with the ACBUS for each case (Table 8).

Table 8. ACBUS is a two dimensional matrix as it is shown below. N is the number of the areas under study

$\mu_{11} = LBUR_{11}$	$\mu_{12} = ABUR_{12}$	$\mu_{13} = HBUR_{13}$	$\mu_{14} = ExBUR_{14}$
$\mu_{21} = LBUR_{21}$	$\mu_{22} = ABUR_{22}$	$\mu_{23} = HBUR_{23}$	$\mu_{24} = ExBUR_{24}$
....	.....	.....	.....
$\mu_{N1} = LBUR_{N1}$	$\mu_{N2} = ABUR_{N2}$	$\mu_{N3} = HBUR_{N3}$	$\mu_{N4} = ExBUR_{N4}$

8. The fuzzy T/Norms (Algebraic, Drastic, Einstein, Hamacher - products, Min relation) have been applied through the combination of all four (4) ACBUS membership values (in pairs) in order to perform three distinct risk scenarios.

Totally six Pair Risk Indices (PARI) have been developed by considering the partial indices in pairs.

   a. Conjunction between the:
   (i) Meteorological Indices (MI) AND the Drought indices (DRI) which produces the (MIDRI)

(ii) The Topographic indices (TI) AND the Vegetation indices (VEGI) that produces the (TIVEGI). Thus, two Pair Risk Indices (PARI) have been developed namely: the MIDRI and the TIVEGI.

b. Conjunction between the:

(i) MI AND the TI which produces the (MITI)

(ii) The DRI AND the VEGI which produces the (DRIVEGI). Also two Pair Risk Indices have been developed the MITI and the DRIVEGI.

c. Conjunction between the:

(i) MI AND the VEGI which produces the (MIVEGI)

(ii) DRI AND the TI that produces the (DRITI). Moreover two Pair Risk Indices have been designed MIVEGI and DRITI.

Totally for each research area 5 (the number of T-Norms) * 6 (the number of PARI) = 30 cases were obtained.

9. Use of the Chi-Square Statistical Test at significance level of a = 0.05 in order to obtain the PARI where the three Linguistics (low, medium, high risk) are depended on the three corresponding linguistics (low, medium, high Burned) of the ACBUS. The tables used were 3 * 3 and the degrees of freedom $df_4 = 9,488$ (Fig. 1).

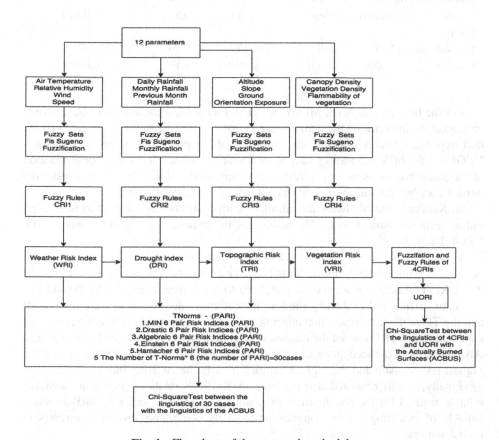

Fig. 1. Flowchart of the proposed methodology

4 Results and Discussion

After extensive testing on multiple scenarios and different methods for the examination of all possible coupling combinations of forest fire severity for Sites Kefallonia Chania and Ilia, important conclusions were drawn on what indicators determine the risk of burned areas. The best Linguistic fitting with the ACBUS indices was derived from the UORI and from the combination of MI-TI PARI for all areas. The DRI and other combinations like the DRIVEGI did not show high convergence with the actual burned areas (Table 9).

Table 9. Test statistic and P-Value between four CRIs and UORI with ACBUS

T-Norms	MI- ACBUS	DRI- ACBUS	TI- ACBUS	VEGI- ACBUS	UORI- ACBUS
Ileia					
Test statistic	27.58	0.82	31.37	15.45	180.18
P-Value	0.00001	0.9344	0.00001	0.0038	0.00001
Kefalonia					
Test statistic	105.22	0.316	1.4	3.34	716.04
P-Value	0.00001	0.988	0.843	0.5	0.00001
Chania					
Test statistic	42.45	0.68	67.67	7.76	258.07
P-Value	0.00001	0.9531	0.00001	0.1	0.00001

For the Ilia prefecture, the MI and the TI and all of their combinations have shown an important influence on the actual burned areas. According to the X squared test the null hypothesis was rejected which has proved the dependency. Moreover, the DRI-VEGI and the MIVEGI indices have shown a rather minor effect on the determination of the actual burned areas (small Test Statistic-higher P-Value). On the other hand, the burned area has proven to be independent from the MIDRI PARI.

In Kefalonia the MI index and all of its combinations have proved to have great influence in the burned areas. The result was the opposite for the other indices (TI, VEGI, DRI).

In Chania, the TI and MI indices and all of their combinations have shown a high correlation with the total burned areas (high Test Statistic values and Low P-Value). The DRIVEGI does not seem to be correlated to the burned areas (Tables 10 and 11).

Summarizing and evaluating the four indicators studied it was proved that the MI and the TI are the key factors that affect the severity of forest fires whereas beyond any expectation the DRI was not the catalyst or retarding factor in the spread of forest fires. Also from the examined pairs of indices, the combination of MI-TI has shown the highest test Statistic and the highest correlation with the ACBUS index.

Finally, it was observed that the fuzzy T-NORMS relations, constitute a highly reliable method for the development of a unified overall risk index (UORI) which is capable of modeling a very complex problem by combining several parameters or partial indices.

Table 10. P-Values between the linguistics of the PARI and the linguistics of the ACBUS

T-Norms	MIDRI	TIVEGI	MITI	DRIVEGI	MIVEGI	DRITI
Ileia						
MIN	0.33	0.00001	0.00001	0.004	0.01	0.00001
Algebraic	0.29	0.00001	0.00001	0.004	0.0029	0.00001
Drastic	0.51	0.00001	0.00001	0.02	0.01	0.00001
Einstein	0.289	0.00001	0.00001	0.0018	0.012	0.00001
Hamacher	0.33	0.00001	0.00001	0.004	0.011	0.00001
Kefalonia						
MIN	0.00001	0.63	0.00001	0.50	0.005	0.84
Algebraic	0.00001	0.52	0.00001	0.49	0.00001	0.84
Drastic	0.00001	0.83	0.00001	0.55	0.00001	0.78
Einstein	0.00001	0.93	0.00001	0.47	0.00001	0.84
Hamacher	0.00001	0.61	0.00001	0.50	0.00001	0.84
Chania						
MIN	0.00001	0.00001	0.00001	0.10	0.00001	0.00001
Algebraic	0.00001	0.00001	0.00001	0.09	0.47	0.37
Drastic	0.00001	0.00001	0.00001	0.09	0.00001	0.00001
Einstein	0.00001	0.00001	0.00001	0.09	0.00001	0.00001
Hamacher	0.00001	0.00001	0.00001	0.10	0.00001	0.00001

Table 11. Test statistic between the linguistics of the PARI and the linguistics of the ACBUS.

T-Norms	MIDRI	TIVEGI	MITI	DRIVEGI	MIVEGI	DRITI
Ileia						
MIN	4.6	59.06	291.5	15.01	12.55	35.6
Algebraic	4.91	265.12	484.55	15.05	16.05	40.67
Drastic	3.28	67.51	475.69	11.57	13.18	33.78
Einstein	4.97	35.82	468.27	11.88	12.8	40.97
Hamacher	4.6	60.78	329.19	15.01	12.97	41.58
Kefalonia						
MIN	71.89	2.57	646.13	3.33	14.55	1.4
Algebraic	71.93	3.19	646.62	3.39	215.76	1.4
Drastic	59.04	1.45	646.53	3.02	323.87	1.72
Einstein	78.19	0.79	649.64	3.53	215.81	1.4
Hamacher	78.15	2.65	646.28	3.3	217.39	1.4
Chania						
MIN	45.62	49.63	153.57	7.68	38.55	66.85
Algebraic	42.46	54.67	124.43	7.87	3.5	4.25
Drastic	42.98	39.86	113.23	7.87	34.88	70.65
Einstein	43.5	55.01	113.38	7.87	39.48	67.88
Hamacher	42.46	33.25	118.89	7.7	41.35	66.98

5 Conclusions and Future Work

This research proposes the use of an innovative method for the analysis and study of the main parameters related to forest fires. The proposed approach was developed towards the estimation of the UORI (forest fire severity index) which is based on advanced soft computing techniques. More specifically, a fuzzy inference system was developed. After the performance of extensive testing, the actual interrelationships between the involved parameters were discovered and hidden knowledge was revealed. This modeling research effort has yielded high rates of accurate classifications as a result of a comparative analysis between the obtained indices and the ACBUS. The function of the model was tested in consideration with various scenarios and presented important outcome regarding those indices or their pairs which directly determine the forest fires risk.

It is important to mention that it is the first time that so many parameters are used and combined to estimate the potential severity of wildfires, creating numerous individual indicators in order to construct a unified index that highly reflects the ACBUS.

As future directions that could improve the proposed model we suggest the potential use of more parameters directly related to forest fires in order to create more combinations of fuzzy rules and sub-indicators. In this way it will be possible to derive an even stronger final adaptive unified index compatible with the ACBUS.

Finally, we propose the future use of other machine learning methods (unsupervised - competitive learning) or hybrid soft computing approaches (fuzzy-neural networks) and optimization algorithms aimed at even higher rates of correct classification.

References

1. Aldrich, S.R., Lafon, C.W., Grissino-Mayer, H.D., DeWeese, G.G.: Fire history and its relations with land use and climate over three centuries in the central Appalachian Mountains, USA. J. Biogeogr. **41**, 2093–2104 (2014)
2. Anezakis, B., Iliadis, L.: Estimation of fire ignition and fire spread risk indices with fuzzy conjunction models MAX-MIN and MAX-PROD. In: Environmental Policy: Theory and Practice. Published by the Democritus University of Thrace, pp. 20–32 (2015). (in Greek) http://utopia.duth.gr/~emanolas/files/Dervitsiotis.pdf
3. Bougoudis, I., Dermetzis, K., Iliadis, L.: HISYCOL a hybrid computational intelligence system for combined machine learning: the case of air pollution modeling in Athens. J. Neural Comput. Appl. 1–16 (2015). Springer
4. Bougoudis, I., Dermetzis, K., Iliadis, L.: Fast and low cost prediction of extreme air pollution values with hybrid unsupervised learning. J. Integr. Comput. Aided Eng. 1–13 (2015)
5. Calvo, T., Mayor, G., Mesira, R.: Aggregation Operators: New Trends and Applications. Studies in Fuzziness and Soft Computing. Physica-Verlag, Heidelberg (2002)
6. Catry, F.X., Rego, F.C., Moreira, F., Bacao, F.: Characterizing and modelling the spatial patterns of wildfire ignitions in Portugal: fire initiation and resulting burned area modelling. Monit. Manag. For. Fires I WIT Trans. Ecol. Environ. **119**, 213–221 (2008)

7. Cox, E.: Fuzzy Modeling and Genetic Algorithms for Data Mining and Exploration. Elsevier Science, USA (2005)
8. Huang, C.E., Ruan, D., Kerre, E.: Fuzzy risks and updating a fuzzy risk with new observations. J. Risk Anal. Int. J. (2007, in press)
9. Iliadis, L., Betsidou, T.: Soft computing modeling of wild fire risk indices: the risk profile of Peloponnesus region in Greece. In: E-Agriculture and Rural Development: Global Innovations and Future Prospects. IGI Global Publishers of Science and Technology Pennsylvania, New York, USA Publisher Professor Maumbe, B.M., Davis College of Agriculture, Natural Resources and Design, W. Virginia University, USA, pp. 220–234 (2012)
10. Iliadis, L., Skopianos, S., Tachos, S., Spartalis, S.: A fuzzy inference system using Gaussian distribution curves for forest fire risk estimation. In: Papadopoulos, H., Andreou, A.S., Bramer, M. (eds.) AIAI 2010. IFIP AICT, vol. 339, pp. 376–386. Springer, Heidelberg (2010)
11. Iliadis, L., Zigkrika, N.: Evaluating fuzzy multi-feature scenarios for forest fire risk estimation. J. Inf. Technol. Agric. 4 (2011)
12. Kailidis, D.: Forest Fires, 3rd edn, p. 510. Giahoudi-Giapouli editions, Thessaloniki (1990). (in Greek)
13. Kecman, V.: Learning and Soft Computing. MIT Press, London England (2001)
14. Leondes, C.: Fuzzy Logic and Expert Systems Applications. Academic Press, San Diego (1998)
15. Nguyen, H., Walker, E.: A First Course in Fuzzy Logic. Chapman and Hall, Boca Raton (2000)
16. Özbayoğlu, A.M., Bozer, R.: Estimation of the burned area in forest fires using computational intelligence techniques. Complex Adapt. Syst. Procedia Comput. Sci. 12, 282–287 (2012)
17. Papakonstantinou, X., Iliadis, L.S., Pimenidis, E., Maris, F.: Fuzzy modeling of the climate change effect to drought and to wild fires in cyprus. In: Iliadis, L., Jayne, C. (eds.) EANN/AIAI 2011, Part I. IFIP AICT, vol. 363, pp. 516–528. Springer, Heidelberg (2011)
18. Shidik, G.F., Mustofa, K.: Predicting size of forest fire using hybrid model. In: Linawati, Mahendra, M.S., Neuhold, E.J., Tjoa, A.M., You, I. (eds.) ICT-EurAsia 2014. LNCS, vol. 8407, pp. 316–327. Springer, Heidelberg (2014)
19. Sugeno, M.: Industrial Applications of Fuzzy Control. Elsevier Science Publishing Company, Amsterdam (1985)
20. http://www.actuar.aegean.gr/notes/22-Katsanos-Avouris.pdf (in Greek)

Applying Artificial Neural Networks to Short-Term $PM_{2.5}$ Forecasting Modeling

Mihaela Oprea$^{(\boxtimes)}$, Sanda Florentina Mihalache, and Marian Popescu

Automatic Control, Computers and Electronics Department,
Petroleum-Gas University of Ploiesti, Ploiesti, Romania
{mihaela,sfrancu,mpopescu}@upg-ploiesti.ro

Abstract. Air pollution with suspended particles from $PM_{2.5}$ fraction represents an important factor to increasing atmospheric pollution degree in urban areas, with a significant potential effect on the health of vulnerable people such as children and elderly. $PM_{2.5}$ air pollutant concentration continuous monitoring represents an efficient solution for the environment management if it is implemented as a real time forecasting system which can detect the $PM_{2.5}$ air pollution trends and provide early warning or alerting to persons whose health might be affected by $PM_{2.5}$ air pollution episodes. The forecasting methods for PM concentration use mainly statistical and artificial intelligence-based models. This paper presents a model based protocol, *MBP – $PM_{2.5}$ forecasting* protocol, for the selection of the best ANN model and a case study with two artificial neural network (ANN) models for real time short-term $PM_{2.5}$ forecasting.

Keywords: Artificial neural networks · Forecasting modeling · Air pollution · $PM_{2.5}$ air pollutant short-term forecasting · Model based forecasting protocol

1 Introduction

Climate change is a modern topic nowadays. Air pollution is one of the most important environmental problems on the globe, and causes many types of allergies, respiratory illnesses, cardiovascular diseases, acute bronchitis diseases, etc. [1, 2]. Particulate matter (PM) is an air pollutant with high impact on humans because short-term and long-term exposure to high concentrations may produce severe health effects and premature mortality [3, 4].

Short-term forecasting of $PM_{2.5}$ air pollution trends can use different methods: deterministic, statistical, neural, hybrid (e.g. neuro-fuzzy) etc. The statistical models include linear regression, ARIMA, principal components analysis, etc., and have been used for their forecasting skills [5, 6]. The forecasted results generated using these linear statistical models are in general not satisfactory. An alternative is the use of computational intelligence approaches, such as artificial intelligence-based models [5, 7]. Artificial neural networks [8] and adaptive neuro-fuzzy inference systems (ANFIS) have been successfully applied in air pollution forecasting domain [9–11]. The chosen of an efficient forecasting method is done by experiment, depending on the available time series databases with measurements of $PM_{2.5}$ concentration, meteorological parameters, other air pollutants concentration that influence $PM_{2.5}$. Depending on the

© IFIP International Federation for Information Processing 2016
Published by Springer International Publishing Switzerland 2016. All Rights Reserved
L. Iliadis and I. Maglogiannis (Eds.): AIAI 2016, IFIP AICT 475, pp. 204–211, 2016.
DOI: 10.1007/978-3-319-44944-9_18

correlation degree with PM$_{2.5}$, a part of these parameters can be considered as inputs in the PM$_{2.5}$ forecasting model. We are applying such a model under the ROKIDAIR research project (http://www.rokidair.ro) whose goal is to provide an intelligent tool (ROKIDAIR DSS) for early warning/alerting of PM$_{2.5}$ air pollution episodes in urban areas (in two pilot cities from Romania, Ploiesti and Targoviste), in order to reduce the potential negative effects of air pollution on children health. Within this project we are developing a model based on artificial intelligence, named ROKIDAIR IA which has two main components: a short-term PM$_{2.5}$ forecasting component and an intelligent decision support component, based on knowledge. In this paper we focus on short-term PM$_{2.5}$ forecasting modeling based on ANN.

2 The Artificial Neural Network Approach for Short-Term PM$_{2.5}$ Forecasting

Artificial neural networks are universal approximators that can learn complex mapping between the input and the output data [12]. An ANN is composed by a set of artificial neurons which are connected according to a topology. Each connection between two neurons has a weight (a numerical value in the interval [0, 1]) showing the degree of that connection which is derived during the ANN training stage. The number of input neurons is given by the input parameters of the forecasting problem, the output neurons are the PM$_{2.5}$ forecasted values in the time window t + k (named also, forecast horizon), while the number of hidden neurons is derived by experiment during training. Some of the ANNs types most used to solve forecasting problems are feed forward artificial neural networks [13], recurrent ANNs [14] and radial basis ANNs [12]. Some recent research results reported in the literature confirmed the good performance of the neural predictors used to detect the air pollution evolution [15–17].

Figure 1 shows an example of a feed forward ANN for PM$_{2.5}$ forecasting. The model uses past measurements of PM$_{2.5}$ concentration and other atmospheric parameters. The ANN has an input layer, an output layer and one or more hidden layers. Usually, one hidden layer is enough to capture the evolution of the forecasted parameter according to the data sets available for ANN training. Feed forward ANNs are trained with a backpropagation algorithm which can be improved by choosing the right learning parameters, adjusted during training. The generation of an ANN model must follow three steps: (1) ANN training with a training algorithm on a training data set; (2) ANN validation on a training data set; (3) ANN testing on a testing data set.

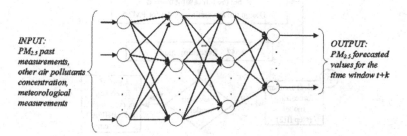

INPUT:
PM$_{2.5}$ past measurements, other air pollutants concentration, meteorological measurements

OUTPUT:
PM$_{2.5}$ forecasted values for the time window t+k

Fig. 1. Example of a feed forward ANN for PM$_{2.5}$ forecasting

The PM$_{2.5}$ ANN forecasting model is derived by training the ANN on a training set selected from the data sets that are available for the urban area that is studied. After training the ANN model is validated and tested on specific data sets. A recent comparison between some ANN models applied to PM$_{2.5}$ prediction is described in [18]. The main advantage of an ANN forecasting model is given by its capability to capture with good accuracy the forecasting function when enough large data sets are used. Our proposed approach for PM$_{2.5}$ short-term forecasting is based on the *MBP - PM$_{2.5}$ forecasting* protocol, developed under the ROKIDAIR project.

3   The PM$_{2.5}$ Forecasting Model Development Protocol

We have developed a protocol, *MBP - PM$_{2.5}$ forecasting*, for building the PM$_{2.5}$ forecasting model under the ROKIDAIR project. The main purpose of the protocol is to facilitate the systematic construction of the short-term PM$_{2.5}$ forecasting model that will be used by the ROKIDAIR Decision Support System in order to provide decisions under the form of warning/alerting messages regarding the potential negative effects on children health of the PM$_{2.5}$ air pollution episodes. The *MBP - PM$_{2.5}$ forecasting* protocol defines the steps of PM$_{2.5}$ forecasting model design. The air pollution forecasting module determines the evolution for short term PM$_{2.5}$ concentration.

Figure 2 presents the logic diagram of the *MBP - PM$_{2.5}$ forecasting* protocol (with 4 main steps) for the short-term PM$_{2.5}$ forecasting module of the ROKIDAIR Decision Support System.

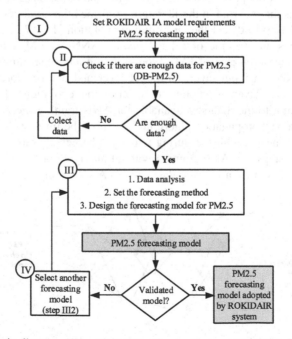

Fig. 2. Logic diagram of the *MBP-PM$_{2.5}$ forecasting* protocol (steps I, II, III, IV)

In the first step are set the PM$_{2.5}$ forecasting model requirements (as e.g. past measurements window time, forecasting horizon, input/output parameters, forecasting accuracy). In step II it is checked if the database with PM$_{2.5}$ concentration measurements and other PM$_{2.5}$ related atmospheric parameters measurements (DB-PM$_{2.5}$) has enough data. If there are not enough data, it is started a process of data collection (usually, for the analyzed urban area or a similar PM$_{2.5}$ air polluted urban area). When enough data are stored in the database, step III is performed with a data analysis sub-step (III.1), followed by the setting of the forecasting method (III.2) and the design of the PM$_{2.5}$ forecasting model (III.3) according to the methodology of selecting the best solution. After step III, the short-term PM$_{2.5}$ forecasting model is generated. If the model is not validated, another forecasting method is chosen in step III.2. If the model is validated than it is adopted by the ROKIDAIR system. The model validation is performed according to the desired forecasting performance which is measured with some indicators: mean absolute error (MAE), index of agreement (IA), root mean square error (RMSE), and coefficient of determination (R^2).

As we are focusing on the artificial neural network based forecasting method, we present the main steps of the methodology proposed for feed forward and radial basis ANN model selection which were integrated in the ROKIDAIR *MBP – PM$_{2.5}$ forecasting* protocol.

MBP – PM$_{2.5}$ forecasting protocol – ANN Selection Methodology

Step 1. Time series data processing (i.e. DB-PM$_{2.5}$) – in order to be used by the PM$_{2.5}$ ANN forecasting method;

Step 2. Select the most relevant atmospheric input parameters to short-term PM$_{2.5}$ forecasting (e.g. by using principle component analysis);

Step 3. Select the training, validation and testing data sets for the ANN model;

Step 4. Set the ANN architecture (e.g. input nodes, output nodes, hidden nodes, radial function, cluster seed, number of clusters etc.);

Step 5. Adjust the training parameters according to the training algorithm;

Step 6. ANN training, validation, testing using the training data set chosen in step 3;

Step 7. Analyze the performances of the designed ANN model (i.e. RMSE, IA, R^2);

Step 8. Select the best ANN model for real time short-term PM$_{2.5}$ forecasting.

A good PM$_{2.5}$ forecasting model should have a smaller error (RMSE, MAE), a coefficient of determination and an index of agreement close to 1. In order to keep the PM$_{2.5}$ short-term forecasting model as simple as possible for an efficient real time PM$_{2.5}$ forecasting, a minimum number of the atmospheric parameters (e.g. temperature and relative humidity) most relevant to PM$_{2.5}$ concentration evolution are chosen.

4   Experimental Results

The data sets used in this study come from an air quality monitoring station from an urban area of Ploiesti, Romania, and each data set contains approximately 4200 samples for PM$_{2.5}$ concentrations and temperature. From all meteorological parameters the temperature is correlated with PM$_{2.5}$ evolution. The data from Ploiesti monitoring station referring to PM$_{2.5}$ concentrations has the maximum of 36.45 µg/m^3, and a minimum of 0.19 µg/m³. In the same time, the temperature data set has the maximum of 37.24 °C, and the minimum of −0.2 °C.

The proposed forecasting models use normalized data for both $PM_{2.5}$ concentrations and temperature. The data were randomly divided with the following percentages: 70 % for training, 15 % for validation and 15 % for testing. We propose two types of forecasting models in this study, based on ANNs. One model has as inputs the four previous $PM_{2.5}$ hourly concentrations (Fig. 3a) and the other has one more input than the first one, namely the current hourly temperature (Fig. 3b). The output of the models is the same in both cases - short term forecasted value for the next hour $PM_{2.5}$ concentration.

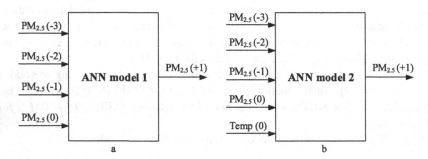

Fig. 3. ANN models

The structure of the proposed neural network contains four neurons in the input layer, one hidden layer and one neuron in the output layer. In the study there were used two types of neural networks, namely feed forward backpropagation (FFwd) and layer recurrent (LRec). As training algorithm the preferred method is Levenberg-Marquardt, and for the adaptive learning functions there are studied the gradient descent with momentum weight and bias (learngdm) and gradient descent weight and bias (learngd). The simulations were performed modifying also the number of neurons in the hidden layer.

The training and validation errors have values around 0.001 and 0.0007 respectively. The accuracy of the models can be evaluated based on the comparison between the actual value and forecasted value of $PM_{2.5}$ concentration, with mean error and standard deviation criteria. The performances of the designed ANN models are compared using statistical indices such as RMSE, IA, R^2, and R.

The two models are compared using statistical criteria and a selection of the results are presented in Tables 1 and 2, the best configuration for each ANN model being highlighted.

For the first model using only PM concentrations as inputs the best results are obtained in the case of layer recurrent structure with 5 neurons in the hidden layer and the *learngdm* adaptation learning function. In this case the root mean squared error have the smallest value, and IA, R^2 and R indices have the biggest values.

Table 1. Statistical indices for ANN model 1

ANN model 1		RMSE [μg/m^3]	IA	R^2	R
4 × 3 × 1/	FFwd	1.1106	0.9902	0.9620	0.9809
Learngdm	LRec	1.1268	0.9899	0.9609	0.9803
4 × 3 × 1/	FFwd	1.1123	0.9902	0.9619	0.9808
Learngd	LRec	1.1132	0.9902	0.9619	0.9808
4 × 4 × 1/	FFwd	1.1188	0.9901	0.9615	0.9806
Learngdm	LRec	1.1257	0.9899	0.9610	0.9803
4 × 4 × 1/	FFwd	1.1188	0.9901	0.9615	0.9806
Learngd	LRec	1.1182	0.9900	0.9615	0.9806
4 × 5 × 1/	**FFwd**	1.1128	0.9902	0.9619	0.9808
Learngdm	**LRec**	**1.0908**	**0.9905**	**0.9634**	**0.9815**
4 × 5 × 1/	FFwd	1.1132	0.9901	0.9618	0.9808
Learngd	LRec	1.1193	0.9901	0.9614	0.9806
4 × 6 × 1/	FFwd	1.1057	0.9903	0.9624	0.9810
Learngdm	LRec	1.1110	0.9902	0.9620	0.9809
4 × 6 × 1/	FFwd	1.1064	0.9903	0.9623	0.9810
Learngd	LRec	1.0933	0.9905	0.9632	0.9814

Table 2. Statistical indices for ANN model 2

ANN model 2		RMSE [μg/m^3]	IA	R^2	R
4 × 3 × 1/	FFwd	1.1106	0.9902	0.9620	0.9809
Learngdm	LRec	1.1110	0.9902	0.9620	0.9808
4 × 3 × 1/	FFwd	1.1123	0.9902	0.9619	0.9808
Learngd	LRec	1.1198	0.9900	0.9614	0.9806
4 × 4 × 1/	FFwd	1.0985	0.9904	0.9629	0.9813
Learngdm	LRec	1.1206	0.9900	0.9613	0.9805
4 × 4 × 1/	FFwd	1.1188	0.9901	0.9615	0.9806
Learngd	LRec	1.1046	0.9903	0.9624	0.9811
4 × 5 × 1/	FFwd	1.0966	0.9905	0.9630	0.9813
Learngdm	LRec	1.1134	0.9901	0.9618	0.9808
4 × 5 × 1/	FFwd	1.1221	0.9900	0.9612	0.9805
Learngd	LRec	1.0998	0.9904	0.9628	0.9812
4 × 6 × 1/	FFwd	1.1074	0.9902	0.9622	0.9810
Learngdm	LRec	1.1256	0.9899	0.9610	0.9803
4 × 6 × 1/	**FFwd**	**1.0951**	**0.9905**	**0.9631**	**0.9814**
Learngd	LRec	1.0966	0.9904	0.9630	0.9814

The second model with temperature as additional input has the best results (comparing the same statistical indices) in the case of feed forward structure with 6 neurons in the hidden layer and the *learngd* adaptation learning function.

The best results from the two models showed that no significant enhancement has been produced when current hourly temperature is included as additional input variable to the second ANN model. The best structure between the two is the one from the first model with PM concentrations as inputs ($4 \times 5 \times 1$ – Learngdm – Layer Recurrent) with: RMSE = 1.0908 $\mu g/m^3$, IA = 0.9905, R^2 = 0.9634 and R = 0.9815.

Figure 4 presents a partial view of the comparison between testing and forecasted data for the best ANN structure.

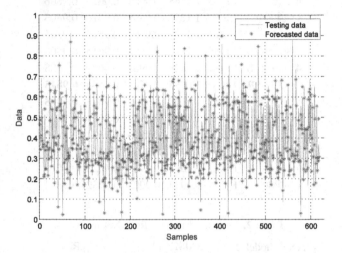

Fig. 4. Comparison between testing and forecasted data

5   Conclusions

The paper presented two ANN models proposed for real time $PM_{2.5}$ short-term forecasting in the case of a polluted town in Romania. In order to select the best ANN forecasting model, we have designed a model based $PM_{2.5}$ forecasting protocol, named *MBP – $PM_{2.5}$ forecasting* protocol, which is integrated in the ROKIDAIR MBP protocol for the development of the ROKIDAIR DSS. The first proposed model uses as inputs only hourly PM concentrations and the second one uses an additional input the current hourly temperature. The conclusions are that the accuracy of both ANN models are almost the same, so both models can be considered appropriate approaches to real time short term forecast. As future work we propose to include other meteorological variables into the model, use additional hybrid modelling techniques such as FIR with genetic algorithm, or expand the forecasting window to next day.

Acknowledgements. The research leading to these results has received funding from EEA Financial Mechanism 2009-2014 under the project ROKIDAIR *"Towards a better protection of children against air pollution threats in the urban areas of Romania"* contract no. 20SEE/ 30.06.2014.

References

1. Kampa, M., Castanas, E.: Human health effects of air pollution. Environ. Pol. **151**, 362–367 (2008)
2. Qin, G., Meng, Z.: Effects of sulfur dioxide derivatives on expression of oncogenes and tumor suppressor genes in human bronchial epithelial cells. Food Chem. Toxicol. **47**, 734–744 (2009)
3. Baker, K.R., Foley, K.M.: A nonlinear regression model estimating single source concentrations of primary and secondarily formed PM2.5. Atmos. Environ. **45**, 3758–3767 (2011)
4. Nebot, A., Mugica, F.: Small-particle pollution modeling using fuzzy approaches. In: Obaidat, M.S., Filipe, J., Kacprzyk, J., Pina, N. (eds.) Simulation and Modeling Methodologies. AISC, vol. 256, pp. 239–252. Springer, Heidelberg (2014)
5. Oprea, M., Dragomir, E.G., Mihalache, S.F., Popescu, M.: Prediction methods and techniques for PM2.5 concentration in urban environment (in Romanian). In: Iordache, S., Dunea, D. (eds.) Methods to Assess the Effects of Air Pollution with Particulate Matter on Children's Health (in Romanian), pp. 387–428. MatrixRom, Bucharest (2014)
6. Kumar, N., Chu, A., Foster, A.: An empirical relationship between PM2.5 and aerosol optical depth in Delhi metropolitan. Atmos. Environ. **41**(21), 4492–4503 (2007)
7. Akkoyunlu, A., Yetilmezsoy, K., Erturk, F., Oztemel, E.: A neural network-based approach for the prediction of urban SO2 concentrations in the Istanbul metropolitan area. Inter. J. Environ. Pol. **40**, 301–321 (2010)
8. Yilmaz, I., Kaynar, O.: Multiple regression, ANN (RBF, MLP) and ANFIS models for prediction of swell potential of clayey soils. Expert Syst. Appl. **38**, 5958–5966 (2011)
9. Morabito, F.C., Versaci, M.: Fuzzy neural identification and forecasting techniques to process experimental urban air pollution data. Neural Netw. **16**, 493–506 (2003)
10. Yildirim, Y., Bayramoglu, M.: Adaptive neuro-fuzzy based modelling for prediction of air pollution daily levels in city of Zonguldak. Chemosphere **63**, 1575–1582 (2006)
11. Ashish, M., Rashmi, B.: Prediction of daily air pollution using wavelet decomposition and adaptive-network-based fuzzy inference system. Int. J. Environ. Sci. **2**(1), 185–196 (2011)
12. Haykin, S.: Neural networks. A comprehensive foundation. Pearson Education Inc., New Delhi (1999)
13. Hornik, K.: Approximation capabilities of multilayer feed-forward networks. Neural Netw. **4**, 251–257 (1991)
14. Mandic, D., Chambers, J.: Recurrent Neural Networks for Prediction Learning Algorithms, Architectures and Stability. Wiley, New York (2001)
15. Kurt, A., Oktay, A.B.: Forecasting air pollutant indicator levels with geographic models 3 days in advance using neural networks. Expert Syst. Appl. **37**, 7986–7992 (2010)
16. Fernando, H.J., Mammarella, M.C., Grandoni, G., Fedele, P., Di Marco, R., Dimitrova, R., Hyde, P.: Forecasting PM10 in metropolitan areas. Efficacy of neural networks. Environ. Pollut. **163**, 62–67 (2012)
17. Feng, X., Li, Q., Zhu, Y., Hou, J., Jin, L., Wang, J.: Artificial neural networks forecasting of PM2.5 pollution using air mass trajectory based geographic model and wavelet transformation. Atmos. Environ. **107**, 118–128 (2015)
18. Oprea, M., Mihalache, S.F., Popescu M.: A comparative study of computational intelligence techniques applied to PM$_{2.5}$ air pollution forecasting. In: Proceedings of 2016 6th International Conference on Computers Communications and Control (ICCCC), pp. 103–108. Baile Felix, Oradea, Romania (2016)

AIRuleBased Modeling (AIRUMO)

Modeling Mental Workload Via Rule-Based Expert System: A Comparison with NASA-TLX and Workload Profile

Lucas Rizzo, Pierpaolo Dondio, Sarah Jane Delany, and Luca Longo$^{(\boxtimes)}$

School of Computing, Dublin Institute of Technology, Dublin, Ireland
`luca.longo@dit.ie`

Abstract. In the last few decades several fields have made use of the construct of human mental workload (MWL) for system and task design as well as for assessing human performance. Despite this interest, MWL remains a nebulous concept with multiple definitions and measurement techniques. State-of-the-art models of MWL are usually ad-hoc, considering different pools of pieces of evidence aggregated with different inference strategies. In this paper the aim is to deploy a rule-based expert system as a more structured approach to model and infer MWL. This expert system is built upon a knowledge-base of an expert and translates into computable rules. Different heuristics for aggregating these rules are proposed and they are elicited using inputs gathered in an user study involving humans performing web-based tasks. The inferential capacity of the expert system, using the proposed heuristics, is compared against the one of two ad-hoc models, commonly used in psychology: the NASA-Task Load Index and the Workload Profile assessment technique. In detail, the inferential capacity is assessed by a quantification of two properties commonly used in psychological measurement: sensitivity and validity. Results show how some of the designed heuristics can over perform the baseline instruments suggesting that MWL modelling using expert system is a promising avenue worthy of further investigation.

Keywords: Rule-based expert system · Mental workload · Heuristics

1   Introduction

Mental workload (MWL) is a multi-faceted phenomenon with no clear and widely accepted definition. Intuitively, it can be described as the amount of cognitive work expended to a certain task during a given period of time. However, this is a simplistic definition and other factors such as stress, time pressure and mental effort can all influence MWL [11]. The principal reason for measuring MWL is to quantify the mental cost of performing a task in order to predict operator and system performance [1]. It is an important construct, mainly used in the fields of

© IFIP International Federation for Information Processing 2016
Published by Springer International Publishing Switzerland 2016. All Rights Reserved
L. Iliadis and I. Maglogiannis (Eds.): AIAI 2016, IFIP AICT 475, pp. 215–229, 2016.
DOI: 10.1007/978-3-319-44944-9_19

psychology and ergonomics, mainly with application in aviation and automobile industries [5,20] and in interface and web design [15,16,23]. According to Young and Stanton, underload and overload can weaken performance [28]. However, optimal workload has a positive impact on user satisfaction, system success, productivity and safety [12]. Often the information necessary for modelling the construct of MWL is uncertain, vague and contradictory [13]. State-of-the-art measurement techniques do not take into consideration the inconsistency of data used in the modelling phase, which might lead to contradictions and loss of information. For example, if the time spent on a certain task is low it can be derived that the overall MWL is also low, however, if the effort invested in the task is extremely high, then the contrary can be inferred. The aim of this study is to investigate the use of rule-based expert systems for the modelling and inference of MWL. An expert system is a computable program designed to model the problem-solving ability of a human expert [3]. This human expert has to provide a knowledge base, then in turn is translated into computable rules. These rules are used by an inference engine aimed at inferring a numerical index of MWL. Since there is no ground truth indicating if such index is fully correct, the inferential capacity of the defined expert system needs to be investigated in order to gauge its quality. To solve this, the proposal is to adopt some of the most commonly used criteria used in psychometrics such as validity and sensitivity [4,22,24]. In simple terms, these criteria are aimed at assessing whether a technique is measuring the construct under investigation and whether it is capable of differentiating variations in workload. From this, the following research question can be defined: *can implementations of rule-based expert systems, compared to state-of-the-art MWL inference techniques, enhance the modelling of mental workload according to sensitivity and validity?*

The remainder of this paper is organised as follows: Sect. 2 describes related works on MWL, its assessment techniques and provides a general view on rule-based expert systems. Section 3 presents the design of an experiment, the methodology adopted. Findings are discussed in Sect. 4 while Sect. 5 concludes our contribution and introduces future work.

2   Related Work

2.1   Mental Workload Assessment Techniques

As stated by several authors, there is no simple and agreed definition of mental workload [6,20,27]. It is thought to be multidimensional and multifaceted, resulting from the aggregation of many different factors thus difficult to be uniquely defined [1]. The basic intuition is that mental workload is the necessary amount of cognitive work for a person to accomplish a task over a period of time. Nevertheless, a large number of measures have been developed [7,29] and practitioners have found measuring MWL to be useful [25]. Most empirical classification assessment procedures can be divided in three major categories [19]:

- *Subjective measures*: operators are required to evaluate their own MWL according to different rating scales or a set of questionnaires.

- *Performance-based measures*: these infer an index of MWL from objective notions of performance on the primary task, such as number of errors, completion time or reaction time to respond to secondary tasks.
- *Physiological measures*: these infer a value of MWL according to some physiological response from the operator such as pupillary reflex or muscle activity.

Further details for each category can be found in [5,17]. This study makes use of two of the subjective measures of MWL that have been largely employed for the last four decades [7,21,24]. These are used as base-lines and are: NASA-Task Load Index (TLX) [7] and Workload Profile (WP) [24].

The NASA-TLX is a multidimensional scale, initially developed for the use in the aviation industry. Its application has been spread across several different areas, such as automobile drivers, medical profession, users of computers and military cockpits. Also, it has achieved great importance and is considered a reference point for the development of new measures and models [6]. NASA-TLX consists of six sub scales: mental demand, physical demand, temporal demand, frustration, effort and performance (Table 4, in the Appendix, questions 1–5 plus physical demand). The computation of an overall MWL index is made through a weighted average of these six dimensions d_i quantified using a questionnaire. The weights w_i are provided by the operator according to a comparison of each possible pair of the six dimensions, for example "which contributed more for the MWL: mental demand or effort?", "which contributed more for the MWL: performance or frustration?", giving a total of 15 preferences. The number of times each dimension is chosen defines its weight (Eq. 1).

The Workload Profile is another MWL assessment technique based on the Multiple Resource Theory (MRT) [26]. In contrast to the NASA-TLX, it is built upon 8 dimensions: perceptual/central processing, response processing, spatial processing, verbal processing, visual processing, auditory processing, manual responses and speech responses (Table 4, question 6–13). The operator is asked to rate the proportion of attentional resources, in the range 0 to 1, for each dimension, then summed. For comparison purpose, this sum is averaged (Eq. 2).

$$\text{TLX}_{\text{MWL}} = \left(\sum_{i=1}^{6} d_i \times w_i \right) \frac{1}{15} \quad (1) \qquad \text{WP}_{\text{MWL}} = \sum_{i=1}^{8} d_i \quad (2)$$

According to [22] WP is preferred to NASA-TLX if the goal is to compare the MWL of two or more tasks with different levels of difficulty, while NASA-TLX is preferred if the goal is to predict the performance of a particular individual in a single task. Several criteria have been proposed for the selection and development of measurement techniques [19]. In this study the focus is on two of them:

- *validity*: to determine whether the MWL measurement instrument is actually measuring MWL. Two variations of validity are usually employed in psychology: concurrent and convergent. The former aims at determining to what extent a technique can explain objective performance measures, such as task execution time. The second indicates whether different MWL techniques correlate to each other [24]. In literature, concurrent and convergent validity are calculated adopting statistical correlation coefficients [12,22].

- *sensitivity*: the capability of a technique to discriminate significant variations in MWL and changes in resource demand or task difficulty [19]. Formally, sensitivity has been assessed in two different ways: multiple regression [24] and ANOVA [12,22]. The aim was to identify statistically significant differences of the MWL indexes associated to each task under examination.

2.2   Mental Workload and Rule-Based Expert System

An expert system is a computer program created in order to emulate an expert in a given field [3]. The goal is to imitate the experts capability of solving different tasks in its area. Unlike usual procedural algorithms, an expert system normally has two modules: a *knowledge base* and an *inference engine*. The knowledge base is provided by the expert and translated into a set of rules, which will be utilised by an inference engine. A typical rule is of the form "*IF ... THEN ...*" and the engine will elicit and aggregate all the rules in order to infer a conclusion. In [9], a literature review of many areas in which expert systems have been applied is provided, while [8,18] are examples of works in the more general field of knowledge representation. To the best of our knowledge, the only study that attempted to model MWL employing inference rules by Longo [10]. Here, modelling MWL has been proposed as a defeasible reasoning process, which is a kind of reasoning built upon inference rules that are defeasible. Defeasible reasoning does not produce a final representation of MWL, but rather a dynamic representation that might change in the light of new evidence and rules. Following this approach, rule-based expert systems might be suitable complements because of their capacity to imitate the problem-solving ability of an expert and facilitate the justification of the inferred conclusion.

3   Design and Methodology

In order to answer the research question an experiment is designed as it follows:

1. acquisition of a knowledge base (KB) related to MWL from an expert;
2. KB translation into different types of rule (forecast, undercutting, rebutting)
3. construction of models ($e_1 - e_4$, $fr_1 - fr_4$) based on two variations of KB, each employing different types of rules and heuristics ($H_1, ..., H_4$);
4. comparison of the inferential capacity of each model against selected baseline instruments (NASA-TLX and WP) according to validity and sensitivity:
   - validity is measured to investigate if the implemented rule-based expert system is capable of inferring MWL as well as the baseline instruments.
   - sensitivity is measured to determine the quality of the inference made by the implemented expert system.

Table 1. Experiments set up: types of rules employed by two variations of the same knowledge base (left) and name of each model, variation used, heuristic adopted (right).

Types of rules Knowledge base variations	Model	KB variation 1	KB variation 2	Heuristics h_1 h_2 h_3 h_4
Forecast — ① Undercutting — ② Rebutting	e_1	✓		h_1 ✓
	e_2	✓		h_2 ✓
	e_3	✓		h_3 ✓
	e_4	✓		h_4 ✓
	fr_1		✓	h_1 ✓
	fr_2		✓	h_2 ✓
	fr_3		✓	h_3 ✓
	fr_4		✓	h_4 ✓

3.1 Knowledge Base (KB)

Research studies performed by Longo et al. have developed a knowledge base for the inference of MWL in the field of human computer interaction [11,12,16]. The goal was to investigate the impact of structural changes of web interfaces on the imposed mental workload on end-users after interacting with them. The knowledge base developed comprises by 21 attributes (Table 4), containing a set of features believed to be useful for modelling MWL, each of them quantified, through a subjective question, in the range $[0, 100] \in \mathbb{R}$. The MWL has four possible levels, as per Definition 1.

Definition 1 *(Mental workload level). Four MWL levels are defined: underload (U), fitting$^-$ (F$^-$), fitting$^+$ (F$^+$) and overload (O).*

The set of rules built from the knowledge-base of the expert [11] can be seen in the Appendix and a formal definition follows.

Definition 2 *(Rules). Three types of rules are defined.*

– *Forecast rule (FR): takes a value α of an attribute X and infers a MWL level β if α is in a predefined range $[x_1, x_2]$ with $x_1, x_2 \in \mathbb{N}$ and $x_2 > x_1$.*

$$FR: \ IF \ \alpha \in [x_1, x_2] \ THEN \ \beta$$

– *Undercutting rule (UR): takes one or more attributes values, $\alpha_1, \cdots, \alpha_n$, and undercuts what is inferred by a forecast rule Y if $\alpha_1 \in [x_1^1, x_2^1], \cdots, \alpha_n \in [x_1^n, x_2^n]$. In this case it is said that rule Y is discarded, $d(Y)$, and will not be considered for future inferences of MWL.*

$$UR: \ IF \ \alpha_1 \in [x_1^1, x_2^1] \ and \ \cdots \ and \ \alpha_n \in [x_1^n, x_2^n] \ THEN \ d(Y)$$

– *Rebutting rule (RR): is a relationship between two forecast rules, Y_1 and Y_2, that can not coexist.*

$$RR: \;\; \textbf{IF } Y_1 \textbf{ and } Y_2 \textbf{ THEN } d(Y_1) \textbf{ and } d(Y_2).$$

Example 1. *An example of possible rules are:*

– *Forecast rules*
  EF1: [$\textbf{IF}$ effort $\in [0, 32]$ $\textbf{THEN}$ U] EF4: [$\textbf{IF}$ effort $\in [67, 100]$ $\textbf{THEN}$ O]
  MD1: [$\textbf{IF}$ mental demand $\in [0, 32]$ $\textbf{THEN}$ U]
  PK1: [$\textbf{IF}$ past knowledge $\in [0, 32]$ $\textbf{THEN}$ O]
– *Undercutting rule*
  DS1: [$\textbf{IF}$ task difficulty $\in [67, 100]$ $\textbf{and}$ skills $\in [67, 100]$ $\textbf{THEN}$ d(EF4)]
– *Rebutting rule - r5: [$\textbf{IF}$ PK1 $\textbf{and}$ EF1 $\textbf{THEN}$ d(PK1) $\textbf{and}$ d(EF1)].*

3.2 Inference Engine

Having defined the set of rules, the next step for inferring MWL is to implement an inference engine. Our inference engine starts with the activation of rules in the set of FR. These will be called *activated rules*. This activation is based on the inputs provided by the user. Afterwards, rules from the set of UR and RR might discard activated rules, solving some part of the contradictory information. This step is not compulsory. The implementation of rule-based expert systems without UR and RR is also provided. Activated rules that are not discarded are called *surviving rules*. After defining the set of surviving rules, there still might be some inconsistent inferences. Surviving rules will likely be inferring different MWL levels, even with the application of UR and RR. The expert system, therefore, must be able to aggregate the surviving rules and produce a final inference of MWL. Next an example follows:

Example 2. *Following rules from Example 1 and given a numerical input it is possible to define the set of activated rules and the set of surviving rules.*

– ***Inputs:*** *[effort = 80, past knowledge = 15, task difficulty = 90,*
  mental demand = 20, skills = 70, temporal demand = 10]
– ***Rules:*** *Activated: [EF4, PK1, MD1, TD1, DS1] Discarded: [EF4]*
  Surviving: [PK1, MD1, TD1].

Example 2 illustrates a set of surviving rules inferring underload MWL (MD1, TD1) and overload MWL (PK1) at the same time. At this stage, a typical set of conflict resolutions strategies for expert systems include: deciding a priority for each rule, firing all possible lines of reasoning or choosing the first rule addressed. However, none of these strategies is applicable in our experiment, since there is no preference among rules, order of evaluation or possibility to compute more than one output. The knowledge base does not provide sufficient information for performing this computation and because of that four heuristics are defined to accomplish the aggregation of the surviving rules. The strategies are developed in order to extract different pieces of information from the surviving rules, which are aggregated or not in different fashions. The final MWL will be a value in the range $[0, 100] \in \mathbb{R}$. Before presenting such heuristics it is necessary to define the value of a surviving rule (Definition 3).

Definition 3 *(Surviving rule value). The value of a surviving rule $r \in FR$, with input $0 \leq \alpha \leq 100$ related to attribute X, is given by the function*

$$f(r) = \begin{cases} \alpha, & if\ X \propto MWL \\ 100 - \alpha, & if\ X \propto \frac{1}{MWL} \end{cases}$$

with $X \propto MWL$ a direct relationship, $X \propto \frac{1}{MWL}$ an inverse relationship[1].

Given Definition 3 the following heuristics are designed:

- h_1: the average of the surviving rules of the MWL level with the largest cardinality of surviving rules. In case of two or more levels with equal cardinality, it computes the mean of the averages. The idea is to give importance to the largest point of view (largest set of surviving rules) to infer MWL.
- h_2: the highest average value of the surviving rules for each MWL level. This is a pessimistic point of view, and infers the highest MWL according to the different sets of surviving rules of each MWL level.
- h_3: average value of all surviving rules. This is to give equal importance to all surviving rules, regardless of which level of MWL they were supporting.
- h_4: average of average of surviving rules of each MWL level. This is to give equal importance to all sets of MWL levels.

Example 3. *Following Example 2, the value of the surviving rules is given by $f(PK1) = 85$, $f(MD1) = 20$ and $f(TD1) = 10$. Finally, the overall MWL computed by each heuristic is: $h_1: \frac{20+10}{2} = 15$, $h_2: max(85, \frac{20+10}{2}) = 85$, $h_3: \frac{20+10+85}{3} = 38.3$ and $h_4: \frac{\frac{20+10}{2}+85}{2} = 50$.*

4   Data Collection, Elicitation of Models and Evaluation

Nine information seeking web-based tasks of varying difficulty and demand (Table 3), were performed by participants over three websites: Google, Wikipedia and Youtube. Two alterations of the interface of each web-site were proposed, having overall ($9 \times 2 = 18$) configurations. 40 volunteers performed 9 tasks (on a random alteration) and after each, they answered each question of Table 4 using a paper-based scale in the range $[0..100] \in \aleph$, partitioned in 3 regions delimited at 33 and 66. Due to loss of data or partial completion of questionnaires, 406

---

[1] Only the attributes past knowledge, skills and performance of Table 4 have an inverse relationship with MWL (the higher the answer the lower the MWL level) while the others have a direct relationship.

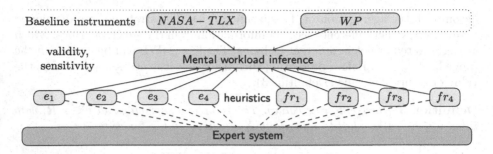

Fig. 1. Evaluation strategy schema

instances were valid. Collected answers, for each instance, were used to elicit the rules of each model (Sect. 3), aggregated with their heuristic, that in turn, produced an index of MWL, in the scale $[0..100] \in \Re$. The outputs formed a distribution of MWL indexes, one for each model, and these were compared against the ones of the baseline models according to validity and sensitivity (Fig. 1).

4.1 Validity

In line with other studies [12,22], validity was assessed using correlation coefficients. In order to select the most suitable statistic, a test of the normality of the distributions of the MWL indexes, produced by each model, was performed using the Shapiro-Wilk test. This test did not achieve a significance greater than 0.05 for most of the models, underlying the non normality of data. As a consequence, the Spearman's rank-order correlation was selected.

Convergent validity: aimed at determining to what extent a model correlate with other model of MWL. As it can be seen from Fig. 2, the baseline instruments (NASA-TLX and WP) achieved a correlation of .538 (dashed reference line) with each other. When correlated with NASA-TLX, e_3 and fr_3 obtained a higher correlation than this. These two models both apply the heuristic h_3, which is the average of all surviving rules, a similar computational method used by the baseline instruments. Just in two other cases (e_1, fr_1) a good correlation (close to the reference line) with WP was obtained. These 2 models implement heuristic h_1, which is the average of the surviving rules of the MWL level (set of rules) with the largest cardinality. The above 4 cases demonstrate how models can be built using rule-based expert system showing similar validity than other baseline MWL assessment instruments believed to shape the construct of MWL.

Concurrent validity: aimed at determining the extent to which a model correlate with task completion time (objective performance measure)[2]. From Fig. 3, it is possible to note that even the baseline instruments do not have a high correlation with task completion time. The first dashed line represents the correlation of 0.178 between NASA-TLX and Time while the second represents

---

[2] Due to measurement errors, only 281 instances have an associated time.

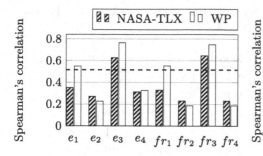 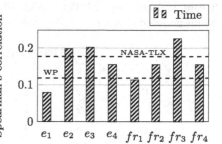

Fig. 2. Convergent validity: $p < 0.05$.   **Fig. 3.** Concurrent validity: $p < 0.05$.

the correlation of 0.119 between WP and Time. Similarly to convergent validity, the models applying heuristic h_3 (e_3, and fr_3) plus the model e_2 were the ones that better correlated with task completion time, Fig. 3, over performing the NASA-TLX. Almost all the models over performed also the WP baseline. These findings suggest that computational models of MWL can be built as rule-based expert systems, and these are capable of enhancing the concurrent validity of the assessments when compared with state-of-the-art models.

4.2 Sensitivity

In line with other studies [12, 22], sensitivity was assessed by analysis of variance. In particular, the non-parametric Kruskal-Wallis H test was performed over the MWL distributions generated by each model, and this was selected because some of the assumptions behind the equivalent of one-way ANOVA were not met. Only model e_4 was not capable of rejecting the null hypothesis of same distribution of MWL indexes across tasks ($p < 0.01$). This means that, for the other models, statistical significant differences exist. The Kruskal-Wallis H test, however, does not tell exactly which pairs of tasks are different from each other. As a consequence, post hoc analysis was performed and the Games-Howell test was chosen because of unequal variances of the distributions under analysis. Table 2 depicts how many pairs of tasks each model was capable of differentiating at different significance levels ($p < 0.05$ and $p < 0.01$). As is can be observed, models applying heuristic h_3 (fr_3 and e_3) outperformed the WP but underperformed the NASA-TLX. This result is a confirmation that sensitive mental workload rule-based expert systems can be successfully built and compete with existing benchmarks in the field.

4.3 Summary of Findings

Quantifications of the validity and the sensitivity of developed models suggest that rule-based expert systems can be successfully built for mental workload modelling and assessment because their inferential capacity lies between the

Table 2. Sensitivity of MWL models with Games-Howell post hoc analysis. The maximum pairwise comparisons of 9 tasks is $\binom{9}{2} = 36$).

Model	$p < 0.05$	$p < 0.01$	Model	$p < 0.05$	$p < 0.01$
NASA-TLX	18	12	NASA-TLX	18	12
WP	9	4	WP	9	4
e_1	2	1	fr_1	2	0
e_2	5	3	fr_2	4	1
e_3	13	10	fr_3	17	10
e_4	0	0	fr_4	4	1

inferential capacity of two state-of-the-art assessment instruments, namely the Nasa Task Load Index and the Workload profile. However, here it is argued that these systems are more appealing and dynamic than selected state-of-the-art approaches. Firstly, they use rules built with terms that are closer to the way humans reason and that imitate experts problem-solving ability. Secondly, they embed heuristics for aggregating rules in a more dynamic way, with a better capacity of handling uncertainty and conflicting pieces of information compared to fixed formulas of state-of-the-art models. Thirdly, they allow the comparison of knowledge-bases and beliefs of different MWL designers thus increasing the understanding of the construct of Mental Workload itself.

5   Conclusion and Future Work

This research presents a new way of modelling and assessing the construct of Mental Workload (MWL) by means of rule-based expert systems. A knowledge base of a MWL designer was elicited and translated into computational rules of various typology. Different heuristics for aggregating these rules were designed aimed at inferring MWL as a numerical index. Inferred indexes were systematically compared with those generated by two state-of-the-art MWL assessment techniques: the NASA Task Load Index and the Workload Profile. This comparison included the quantification of two properties of each distribution of MWL indexes, namely sensitivity and validity, commonly employed in the literature. Findings suggest that rule-based expert systems are promising not only because they can approximate the inferential capacity of selected state-of-the-art MWL assessment techniques. They also offer a flexible approach for translating different knowledge-bases and beliefs of MWL designers into computational rules supporting the creation of models that can be replicated, extended and falsified, thus enhancing the understanding of the construct of mental workload itself. Future works will be focused on the replication of the approach adopted in this study using other knowledge bases elicited from other MWL experts. Additionally, this approach will be extended incorporating fuzzy representation of rules and acceptability semantics, borrowed from argumentation theory [2,14], with the aim of improving conflict resolution of rules and building models expected to have an even higher sensitivity and validity.

Acknowledgments. Lucas Middeldorf Rizzo would like to thank CNPq (Conselho Nacional de Desenvolvimento Científico e Tecnológico) for his Science Without Borders scholarship, proc n. 232822/2014-0.

Appendix

Knowledge Base

For the attribute mental demand the forecast rules are:

MD1: [**IF** mental demand $\in [0, 32]$ **THEN** U]   MD3: [**IF** mental demand $\in [50, 66]$ **THEN** F^+]
MD2: [**IF** mental demand $\in [33, 49]$ **THEN** F^-]   MD4: [**IF** mental demand $\in [67, 100]$ **THEN** O]

The same principle applies to the attributes temporal demand, physical demand, solving and deciding, selection of response, task and space, verbal material, visual resources, auditory resources, manual response, speech response, effort, parallelism, and context bias, forming 52 other rules. For psychological stress, motivation, past knowledge, skills and performance the forecast rules are:

PS1: [**IF** psychol. stress $\in [0, 32]$ **THEN** U]   SK2: [**IF** skills $\in [67, 100]$ **THEN** U
PS2: [**IF** psychol stress $\in [67, 100]$ **THEN** O]   PF1: [**IF** performance $\in [0, 32]$ **THEN** O]
MV1: [**IF** motivation $\in [0, 32]$ **THEN** U]   PF2: [**IF** performance $\in [33, 49]$ **THEN** F^+]
PK1: [**IF** past knowledge $\in [0, 32]$ **THEN** O]
PK2: [**IF** past knowledge $\in [67, 100]$ **THEN** U]   PF3: [**IF** performance $\in [50, 66]$ **THEN** F^-]
SK1: [**IF** skills $\in [0, 32]$ **THEN** O]   PF4: [**IF** performance $\in [67, 100]$ **THEN** U]

The undercutting rules and rebutting rules are:

AD1a: [**IF** arousal $\in [0, 32]$ **and** task difficulty $\in [0, 32]$ **THEN** d(PF4)]
AD1b: [**IF** arousal $\in [0, 32]$ **and** task difficulty $\in [0, 32]$ **THEN** d(PF3)]
AD1c: [**IF** arousal $\in [0, 32]$ **and** task difficulty $\in [0, 32]$ **THEN** d(PF2)]
AD2a: [**IF** arousal $\in [0, 32]$ **and** task difficulty $\in [67, 100]$ **THEN** d(PF4)]
AD2b: [**IF** arousal $\in [0, 32]$ **and** task difficulty $\in [67, 100]$ **THEN** d(PF3)]
AD2c: [**IF** arousal $\in [0, 32]$ **and** task difficulty $\in [67, 100]$ **THEN** d(PF2)]
AD3a: [**IF** arousal $\in [33, 49]$ **and** task difficulty $\in [0, 32]$ **THEN** d(PF1)]
AD3b: [**IF** arousal $\in [33, 49]$ **and** task difficulty $\in [0, 32]$ **THEN** d(PF4)]
AD4a: [**IF** arousal $\in [33, 49]$ **and** task difficulty $\in [67, 100]$ **THEN** d(PF1)]
AD4b: [**IF** arousal $\in [33, 49]$ **and** task difficulty $\in [67, 100]$ **THEN** d(PF3)]
AD4c: [**IF** arousal $\in [33, 49]$ **and** task difficulty $\in [67, 100]$ **THEN** d(PF4)]
AD4d: [**IF** arousal $\in [50, 66]$ **and** task difficulty $\in [67, 100]$ **THEN** d(PF1)]
AD4e: [**IF** arousal $\in [50, 66]$ **and** task difficulty $\in [67, 100]$ **THEN** d(PF3)]
AD4f: [**IF** arousal $\in [50, 66]$ **and** task difficulty $\in [67, 100]$ **THEN** d(PF4)]
AD5a: [**IF** arousal $\in [50, 66]$ **and** task difficulty $\in [0, 32]$ **THEN** d(PF1)]
AD5b: [**IF** arousal $\in [50, 66]$ **and** task difficulty $\in [0, 32]$ **THEN** d(PF2)]
AD5c: [**IF** arousal $\in [50, 66]$ **and** task difficulty $\in [0, 32]$ **THEN** d(PF3)]
AD5d: [**IF** arousal $\in [67, 100]$ **and** task difficulty $\in [0, 32]$ **THEN** d(PF1)]
AD5e: [**IF** arousal $\in [67, 100]$ **and** task difficulty $\in [0, 32]$ **THEN** d(PF2)]
AD5f: [**IF** arousal $\in [67, 100]$ **and** task difficulty $\in [0, 32]$ **THEN** d(PF3)]
AD6a: [**IF** arousal $\in [67, 100]$ **and** task difficulty $\in [67, 100]$ **THEN** d(PF2)]
AD6b: [**IF** arousal $\in [67, 100]$ **and** task difficulty $\in [67, 100]$ **THEN** d(PF3)]
AD6c: [**IF** arousal $\in [67, 100]$ **and** task difficulty $\in [67, 100]$ **THEN** d(PF4)]
MV2: [**IF** motivation $\in [0, 32]$ **THEN** d(EF3)] - MV3: [**IF** motivation $\in [0, 32]$ **THEN** d(EF4)]
MV4: [**IF** motivation $\in [67, 100]$ **THEN** d(EF1)] - MV5: [**IF** motivation $\in [67, 100]$ **THEN** d(EF2)]
DS1: [**IF** task difficulty $\in [67, 100]$ **and** skills $\in [67, 100]$ **THEN** d(EF4)]
DS2: [**IF** task difficulty $\in [67, 100]$ **and** skills $\in [67, 100]$ **and** effort $\in [0, 32]$ **THEN** d(PF1)]

DS3: [**IF** task difficulty ∈ [67, 100] **and** skills ∈ [67, 100] **and** effort ∈ [33, 49] **THEN** d(PF1)]
DS4: [**IF** task difficulty ∈ [67, 100] **and** skills ∈ [67, 100] **and** effort ∈ [50, 66] **THEN** d(PF1)]
r1: [**IF** MD1 **and** SD4 **THEN** d(MD1), d(SD4)] - r2: [**IF** MD4 **and** SD1 **THEN** d(MD4), d(SD1)]
r3: [**IF** PK1 **and** SK4 **THEN** d(PK1), d(SK4)] - r4: [**IF** PK4 **and** SK1 **THEN** d(PK4), d(SK1)]
r5: [**IF** PK1 **and** EF4 **THEN** d(PK1, d(EF1)] - r6: [**IF** PK2 **and** EF4 **THEN** d(PK2), d(EF4)]
r7: [**IF** SK1 **and** EF1 **THEN** d(SK1), d(EF1)] - r8: [**IF** SK4 **and** EF4 **THEN** d(SK4), d(EF4)]
r9: [**IF** CB4 **and** PS1 **THEN** d(CB4), d(PS1)]

Tasks and Questionnaire

Table 3. List of experimental tasks

Task	Description	Task condition	Web-site
T_1	Find out how many people live in Sidney	Simple search	Wikipedia
T_2	Read http://simple.wikipedia.org/wiki/Grammar	No goals, no time pressure	Wikipedia
T_3	Find out the difference (in years) between the year of the foundation of the Apple Computer Inc. and the year of the 14th FIFA world cup	Dual-task and mental arithmetical calculations	Google
T_4	Find out the difference (in years) between the foundation of the Microsoft Corp. & the year of the 23rd Olympic games	Dual-task and mental arithmetical calculations	Google
T_5	Find out the year of birth of the 1st wife of the founder of playboy	Single task + time pressure (2-min limit). Each 30 secs user is warned of time left	Google
T_6	Find out the name of the man (interpreted by Johnny Deep) in the video www.youtube.com/watch?v=FfTPS-TFQ_c	Constant demand on visual and auditory modalities. Participant can replay the video if required	Youtube
T_7	(a) Play the song www.youtube.com/watch?v=Rb5G1eRIj6c. While listening to it, (b) find out the result of the polynomial equation $p(x)$, with $x = 7$ contained in the wikipedia article http://it.wikipedia.org/wiki/Polinomi	Demand on visual modality and inference on auditory modality. The song is extremely irritating	Wikipedia
T_8	Find out how many times Stewie jumps in the video www.youtube.com/watch?v=TSe9gbdkQ8s	Demand on visual resource + external interference: user is distracted twice & can replay video	Youtube
T_9	Find out the age of the blue fish in the video www.youtube.com/watch?v=H4BNbHBcnDI	Demand on visual and auditory modality, plus time-pressure: 150-sec limit. User can replay the video. There is no answer.	Youtube

Table 4. Experimental study questionnaire [11]

Dimension	Question
Mental demand	How much mental and perceptual activity was required (e.g., thinking, deciding, calculating, remembering, looking, searching, etc.)? Was the task easy (low mental demand) or complex (high mental demand)?
Temporal demand	How much time pressure did you feel due to the rate or pace at which the tasks or task elements occurred? Was the pace slow and leisurely (low temporal demand) or rapid and frantic (high temporal demand)?
Effort	How much conscious mental effort or concentration was required? Was the task almost automatic (low effort) or it required total attention (high effort)?
Performance	How successful do you think you were in accomplishing the goal of the task? How satisfied were you with your performance in accomplishing the goal?
Frustration	How secure, gratified, content, relaxed and complacent (low psychological stress) versus insecure, discouraged, irritated, stressed and annoyed (high psychological stress) did you feel during the task?
Selection of response	How much attention was required for selecting the proper response channel and its execution? (manual - keyboard/mouse, or speech - voice)
Task and space	How much attention was required for spatial processing (spatially pay attention around you)?
Verbal material	How much attention was required for verbal material (eg. reading or processing linguistic material or listening to verbal conversations)?
Visual resources	How much attention was required for executing the task based on the information visually received (through eyes)?
Auditory resources	How much attention was required for executing the task based on the information auditorily received (ears)?
Manual Response	How much attention was required for manually respond to the task (eg. keyboard/mouse usage)?
Speech response	How much attention was required for producing the speech response(eg. engaging in a conversation or talk or answering questions)?
Context bias	How often interruptions on the task occurred? Were distractions (mobile, questions, noise, etc.) not important (low context bias) or did they influence your task (high context bias)?
Past knowledge	How much experience do you have in performing the task or similar tasks on the same website?
Skill	Did your skills have no influence (low) or did they help to execute the task (high)?
Solving and deciding	How much attention was required for activities like remembering, problem-solving, decision-making and perceiving (eg. detecting, recognizing and identifying objects)?
Motivation	Were you motivated to complete the task?
Parallelism	Did you perform just this task (low parallelism) or were you doing other parallel tasks (high parallelism) (eg. multiple tabs/windows/programs)?
Arousal	Were you aroused during the task? Were you sleepy, tired (low arousal) or fully awake and activated (high arousal)?
Task difficult	Task difficult was given by the formula: $\text{Task}_{difficult} = \frac{1}{8}((\text{solving/deciding}) + (\text{auditory resources}) + (\text{manual response}) + (\text{speech response}) + (\text{response}) + (\text{task/space}) + (\text{verbal material}) + (\text{visual resources}))$
Physical demand	The physical demand was considered 0 for all instances

References

1. Cain, B.: A review of the mental workload literature. Technical report, Defence Research and Development Canada Toronto, Human System Integration Section (2007)
2. Dung, P.M.: On the acceptability of arguments and its fundamental role in non-monotonic reasoning, logic programming and n-person games. Artif. Intell. **77**(2), 321–357 (1995)
3. Durkin, J., Durkin, J.: Expert Systems: Design and Development. Prentice Hall PTR, Upper Saddle River (1998)
4. Eggemeier, F.T.: Properties of workload assessment techniques. Adv. Psychol. **52**, 41–62 (1988)
5. Gartner, W.B., Murphy, M.R.: Pilot workload and fatigue: a critical survey of concepts and assessment techniques. National Aeronautics Space Performance (1976)
6. Hart, S.G.: NASA-task load index (NASA-TLX); 20 years later. In: Proceedings of the Human Factors and Ergonomics Society Annual Meeting, vol. 50, pp. 904–908. Sage Publications (2006)
7. Hart, S.G., Staveland, L.E.: Development of NASA-TLX (task load index): results of empirical and theoretical research. Adv. Psychol. **52**, 139–183 (1988)
8. Hatzilygeroudis, I., Prentzas, J.: Integrating (rules, neural networks) and cases for knowledge representation and reasoning in expert systems. Expert Syst. Appl. **27**(1), 63–75 (2004)
9. Liao, S.H.: Expert system methodologies and applications a decade review from 1995 to 2004. Expert Syst. Appl. **28**(1), 93–103 (2005)
10. Longo, Luca: Formalising human mental workload as non-monotonic concept for adaptive and personalised web-design. In: Masthoff, J., Mobasher, B., Desmarais, M.C., Nkambou, R. (eds.) UMAP 2012. LNCS, vol. 7379, pp. 369–373. Springer, Heidelberg (2012)
11. Longo, L.: Formalising Human Mental Workload as a Defeasible Computational Concept. Ph.D. thesis, Trinity College Dublin (2014)
12. Longo, L.: A defeasible reasoning framework for human mental workload representation and assessment. Behav. Inf. Technol. **34**(8), 758–786 (2015)
13. Longo, L., Barrett, S.: A computational analysis of cognitive effort. In: Nguyen, N.T., Le, M.T., Świątek, J. (eds.) Intelligent Information and Database Systems. LNCS, vol. 5991, pp. 65–74. Springer, Heidelberg (2010)
14. Longo, L., Dondio, P.: Defeasible reasoning and argument-based medical systems: an informal overview. In: 27th International Symposium on Computer-Based Medical Systems, pp. 376–381, New York, USA. IEEE (2014)
15. Longo, L., Dondio, P.: On the relationship between perception of usability and subjective mental workload of web interfaces. In: IEEE/WIC/ACM International Conference on Web Intelligence and Intelligent Agent Technology, WI-IAT 2015, Singapore, December 6–9, vol. 1, pp. 345–352 (2015)
16. Longo, L., Rusconi, F., Noce, L., Barrett, S.: The importance of human mental workload in web-design. In: 8th International Conference on Web Information Systems and Technologies, pp. 403–409, April 2012
17. Meshkati, N., Hancock, P.A., Rahimi, M., Dawes, S.M.: Techniques in mental workload assessment. In: Wilson, J.R., Corlett, E.N. (eds.) Evaluation of Human Work: A Practical Ergonomics Methodology, pp. 749–782. Taylor & Francis (1995)
18. Mitra, R.S., Basu, A.: Knowledge representation in mickey: an expert system for designing microprocessor-based systems. IEEE Trans. Syst. Man Cybern. Part A Syst. Hum. **27**(4), 467–479 (1997)

19. O'Donnell, R.D., Eggemeier, F.T.: Workload assessment methodology. In: Boff, K.R., Kaufman, L., Thomas, J.P. (eds.) Handbook of Perception and Human Performance, vol. 2, chap. 42, pp. 1–49. Wiley, New York (1986)
20. Paxion, J., Galy, E., Berthelon, C.: Mental workload and driving. Front. Psychol. **5**, 1344 (2014)
21. Reid, G.B., Nygren, T.E.: The subjective workload assessment technique: a scaling procedure for measuring mental workload. Adv. Psychol. **52**, 185–218 (1988)
22. Rubio, S., Díaz, E., Martín, J., Puente, J.M.: Evaluation of subjective mental workload: a comparison of SWAT, NASA-TLX, and workload profile methods. Appl. Psychol. **53**(1), 61–86 (2004)
23. Tracy, J.P., Albers, M.J.: Measuring cognitive load to test the usability of web sites. Ann. Conf. Soc. Tech. Commun. **53**, 256–260 (2006)
24. Tsang, P.S., Velazquez, V.L.: Diagnosticity and multidimensional subjective workload ratings. Ergonomics **39**(3), 358–381 (1996)
25. Tsang, P.S., Wilson, G.F.: Mental workload measurement and analysis. In: Salvendy, G. (ed.) Handbook of Human Factors and Ergonomics, 2nd edn, pp. 417–449. Wiley, New York (1997)
26. Wickens, C.D.: Processing resources and attention. Multiple-task performance, pp. 3–34 (1991)
27. Young, M.S., Brookhuis, K.A., Wickens, C.D., Hancock, P.A.: State of science: mental workload in ergonomics. Ergonomics **58**(1), 1–17 (2015)
28. Young, M.S., Stanton, N.A.: Attention and automation: new perspectives on mental underload and performance. Theor. Issues Ergonomics Sci. **3**(2), 178–194 (2002)
29. Radu, Vasile: Stochastic Modeling of Thermal Fatigue Crack Growth. ACM, vol. 1. Springer, Switzerland (2015)

Convolutive Audio Source Separation Using Robust ICA and Reduced Likelihood Ratio Jump

Dimitrios Mallis, Thomas Sgouros, and Nikolaos Mitianoudis[✉]

Department of Electrical and Computer Engineering,
Democritus University of Thrace, 67100 Xanthi, Greece
malldimi1@gmail.com, {tsgouros,nmitiano}@ee.duth.gr

Abstract. Audio source separation is the task of isolating sound sources that are active simultaneously in a room captured by a set of microphones. Convolutive audio source separation of equal number of sources and microphones has a number of shortcomings including the complexity of frequency-domain ICA, the permutation ambiguity and the problem's scalabity with increasing number of sensors. In this paper, the authors propose a multiple-microphone audio source separation algorithm based on a previous work of Mitianoudis and Davies [1]. Complex FastICA is substituted by Robust ICA increasing robustness and performance. Permutation ambiguity is solved using the Likelihood Ration Jump solution, which is now modified to decrease computational complexity in the case of multiple microphones.

1 Introduction

The problem of Blind Audio Source Separation (BASS) implies the extraction of independent audio sources from an audio mixture that has been observed by a number of microphones, without any prior knowledge regarding the involved sources or the mixing system. In recent years, many methods have been proposed for resolving this problem with relative success. BASS becomes more complicated when we are dealing with real-room audio recordings. In reverberant rooms, each source is recorded multiple times by each microphone under different time delays and amplifications, due to sound waves' reflections on the room surfaces. The mixing system can thus be modelled using a room impulse response of finite length (FIR filter). In the general case of M microphones that capture a mixture of N sources, a common representation of the aforementioned mixing is $\boldsymbol{x}(t) = \boldsymbol{A} * \boldsymbol{s}(t)$, where $*$ denotes linear convolution, $\boldsymbol{s}(n) = [s_1(n), s_2(n), ..., s_N(n)]^T$ are the source signals, $\boldsymbol{x}(n) = [x_1(n), x_2(n), ..., x_M(n)]^T$ are the observation signals, n is the time index and $\boldsymbol{A}$ is a matrix, whose elements $\boldsymbol{a}_{ij}$ are FIR filters, describing the room impulse responses between the j-th source and the i-th microphone.

A classic decomposition method for performing BASS is Independent Component Analysis (ICA). ICA extracts Independent Components (ICs) from a linear,

© IFIP International Federation for Information Processing 2016
Published by Springer International Publishing Switzerland 2016. All Rights Reserved
L. Iliadis and I. Maglogiannis (Eds.): AIAI 2016, IFIP AICT 475, pp. 230–241, 2016.
DOI: 10.1007/978-3-319-44944-9_20

instantaneous mixture, assuming independence between the original sources. In real rooms, where we deal with convolutive mixtures, ICA can also be applied by moving the separation to the frequency domain, where the convolution between the sources and the room transfer function is reduced to multiplication [2] for a number of discrete frequency bins L i.e. $\boldsymbol{x}(f, n) = \boldsymbol{A}_f \boldsymbol{s}(f, n)$, where $f = 1, \ldots, L$. In other words, we transform a difficult convolution problem to a number of easier instantaneous problems, that can be solved using ICA.

By solving the separation problem in the frequency domain, ICA introduces 2 ambiguities: scale and permutation. The first results into random scaling of the extracted ICs, which can cause spectral deformations and reduce separation quality. The latter results into arbitrary source permutations along the discrete frequency bins and as a result inability to achieve separation. The scale ambiguity can be resolved easily as a post processing step, by mapping the estimated sources back to the microphones' domain and recover the signals as they have been originally observed by the microphones [1].

The permutation ambiguity, on the other hand, is a difficult problem, and various techniques have been proposed without featuring robust performance in all cases. In [3], Mazur and Mertins align permutations by using generalised Gaussian Distribution in order to find differences between neighbouring frequency bins. Sawada et al. [4] exploit the correlation coefficients of amplitude envelopes, which if maximised show the correct source alignment, while in [5] Saito et al. utilise, for the same purpose, the correlation between interfrequency power ratios. A different approach was followed by Sarmiento et al. [6] who find the spectral similarities between the separated components in the frequency domain by employing a contrast function. A region-growing approach, to minimise the spreading of possible misalignments, in order to improve permutation alignment, was introduced by Wang et al. [7].

Most of the available methods for tackling the convolutive source separation problem are focused on the two-source two-microphone (2×2) case. However, low-cost commercially available hardware, such as the Microsoft Kinect interface, has been developed to offer low-latency four-microphone recordings and can be used to process 4×4 cases. In this paper, we focus on the problem of audio source separation for determined cases (equal number of microphones and sources) that involves more than two sources. The presented methodology offers a computationally efficient solution for both the separation task as well as the permutation ambiguity. Based on the Kinect interface, we created a set of recordings containing mixtures of multiple sources as well as the original sources for evaluation purposes. This dataset is publicly available for further evaluation of audio separation methods[1]. In this dataset, we will apply a novel framework that is optimized for multiple sources. This is then compared with the previous work of Mitianoudis and Davies [1] to observe its efficiency for multiple sources. The proposed framework includes a robust complex ICA separation algorithm, called RobustICA [8], that has not been used before for convolutive audio source separation. In addition, we present a new technique to tackle the permutation

---

[1] Dataset available at http://utopia.duth.gr/nmitiano/download.html.

ambiguity problem, especially for large number of sources, based on the Likelihood Ratio Jump solution [1]. We show that this new technique can reduce the computational cost of addressing the permutation problem in comparison to the original Likehood Ratio Jump and can produce the same, if not better separation quality.

2 Instantaneous Complex Source Separation

In the instantaneous case, we consider the following mixing process: $\boldsymbol{x}(n) = \boldsymbol{A} \cdot \boldsymbol{s}(n)$. In order to separate the sources, we have to estimate an unmixing matrix $\boldsymbol{W}$, such that $\boldsymbol{u}(n) = \boldsymbol{W} \cdot \boldsymbol{x}(n) \approx \boldsymbol{s}(n)$.

2.1 The FastICA Algorithm

In the determined case, where the number of sources is equal to the number of observations ($N = M$), the most popular method of estimating the unmixing matrix $\boldsymbol{W} = [\boldsymbol{w_1}, \boldsymbol{w_2}, ..., \boldsymbol{w_N}]^T$ is the FastICA algorithm. There are many implementations of the FastICA algorithm, that are based on optimization of a contrast function emphasizing nonGaussianity using a fixed-point iteration algorithm. One common fixed point algorithm is the following [9],

$$\Delta \boldsymbol{W}_f = \boldsymbol{D}[\text{diag}(-\alpha_i) + \mathcal{E}\{\phi(\boldsymbol{u})\boldsymbol{u}^H\}]\boldsymbol{W}_f \tag{1}$$

where $\phi(u) = u/|u|$ is an activation function for superGaussian sources, α_i an adaptive parameter and $\mathcal{E}\{\cdot\}$ denotes the expectation operator. This iterative update of the unmixing matrix $\Delta \boldsymbol{W}_f$ is calculated using a maximum likelihood estimator. The method also needs the data of every frequency bin to be prewhitenned. Even though this method has been initially introduced for real-data mixture, it has shown to work well with complex data in [1].

2.2 The RobustICA Algorithm

In this section, we examine a source separation algorithm, named RobustICA [8]. RobustICA optimizes the following generalized form of kurtosis.

$$\mathcal{K}(\boldsymbol{w}) = \frac{\mathcal{E}\{|y|^4\} - 2\mathcal{E}^2\{|y|^2\} - |\mathcal{E}\{y^2\}|^2}{\mathcal{E}^2\{|y|^2\}} \tag{2}$$

The above definition of kurtosis can be applied to both real and complex data. In addition, prewhitening is not necessary for RobustICA. RobustICA uses exact line search optimization of the absolute kurtosis contrast function, instead of fixed-point optimization, used by FastICA [10].

$$\mu_{opt} = \arg \max_{\mu}(\mathcal{K}(\boldsymbol{w} + \mu\boldsymbol{g})) \tag{3}$$

The search direction can be given by the gradient of the kurtosis $\boldsymbol{g} = \boldsymbol{\nabla}_{\boldsymbol{w}}\mathcal{K}(\boldsymbol{w})$. Exact line search is often a computationally expensive optimization technique that requires additional numerical analysis algorithms. In the case of kurtosis, the optimal step size μ_{opt} is calculated algebraically with a minimum computational cost. It is shown in [8] that μ_{opt} can be calculated from the root of a low-degree polynomial that maximizes the absolute value of the contrast function along the search direction. RobustICA has a number of advantages [8] compared to the original FastICA:

- RobustICA does not make any assumption regarding the sources' statistical profile, and can deal with real and complex sources alike.
- Prewhitening is not mandatory before RobustICA. Multiple ICs in that case can be extracted with the method of linear regression in contrast to symmetric orthogonalization that is used by FastICA.
- The method can target sub-Gaussian or super-Gaussian sources in a specific order. This feature is useful in the audio separation case, where we know in advance that data in the frequency domain can be mostly modelled as super-Gaussian [1].
- The method is robust to the presence of saddle points and spurious local extrema of the contrast function [8].
- RobustICA can achieve great separation performance with relatively small additional computational cost, compared to other ICA implementations. This feature is demonstrated in [8] and will be verified by the experimental results in this paper.

Despite the fact that prewhitening is not mandatory for RobustICA, it will be used as a preprocessing step in our proposed framework. This is due to the observation that it leads to a more computationally efficient implementation in the case of multiple sources. Since the prewhitened components lay on an orthogonal structure, every ICA iteration that attracts one IC towards an original source, forces the rest of the ICs to converge faster to other sources. This can be achieved with the use of symmetric orthogonalization, as in (4).

$$W_f^+ \leftarrow W_f(W_f^H W_f)^{-0.5} \qquad (4)$$

On the contrary, in linear regression, after the extraction of an IC, we have to separate a reduced mixture from a random position, which can be rather slow. As a result, we use prewhitening to improve the convergence speed of our method, in expense of the separation performance limitations that prewhitening can introduce.

3   Frequency-Domain Source Separation

Frequency-domain source separation methods apply the Short-Time Fourier Transform (STFT) to the mixture recordings $x(t)$. Consequently, the convolutive mixture is transformed to L instantaneous mixture via the STFT, i.e. $x(t) = A * s(t) \Rightarrow X(f,t) = A_f S(f,t)$. The separation problem can be solved independently using any complex ICA algorithm, such as the RobustICA.

ICA's inherent scale and permutation ambiguities impose severe problems in this framework and must be resolved. Scale ambiguity is tackled using a mapping to the microphone domain [1]. There exist many methods to tackle the permutation ambiguity of frequency-domain BASS methods.

3.1   Likelihood Ratio Jump

Mitianoudis and Davies introduced in [1] the Likelihood Ratio Jump method for the alignment of frequency bins to the correct source. This method can be used either after each iteration of the ICA algorithm, or even better as a post-processing mechanism. The method works iteratively and in each iteration forms a likelihood ratio test to decide, which permutation is the most probable for each frequency bin. It uses a set of rescaling parameters γ_{ij} that model the probability that the ith source has moved to the jth position. For each frequency bin, it calculates the probabilities for all the possible permutations. For example, in a mixing of 3 sources a possible permutation of the extracted ICs: $IC3 \rightarrow IC1$, $IC1 \rightarrow IC2$, $IC2 \rightarrow IC3$, forms the probability:

$$L = -\log(\gamma_{31}\gamma_{12}\gamma_{23}) \tag{5}$$

The correct permutation is the one that produces the maximum probability as in (5). For the case of three sources, there are $3! = 6$ possible permutations for the extracted ICs, which have to assessed probabilistically, to conclude which permutation is the most likely to be correct. The parameters γ_{ij} are produced through a maximum likelihood estimator and can be calculated as follows:

$$\gamma_{ij} = \frac{1}{T}\sum_t \frac{|u_i(f,t)|}{\beta_j(t)} \tag{6}$$

where $u_i(f,t)$ is the value of IC i for the discrete frequency bin f and time index t and β_j is a non-stationary time-varying scale parameter that is calculated for the source j. Finally, T is the number of observations.

The parameter $\beta_j(t)$ incorporates information related to the signal's spectral envelope over time, thus it can be interpreted as a volume measurement. Literally, it measures the overall signal amplitude along the frequency axis, emphasizing the fact that one source is "louder" than others at a certain time slot. This "temporal energy burst" can force alignment of the permutations along the frequency axis. A possible estimation for the β_j parameter can be the following:

$$\beta_j = \frac{1}{L}\sum_f |u_j(f,t)| \tag{7}$$

where L is the number of frequency bins. The Likelihood Ratio Jump (LRJ) method has demonstrated very stable performance in solving the permutation ambiguity for a large number of cases [1]. However, this was mainly demonstrated for 2×2 cases.

3.2 Reduced Likelihood Ratio Jump

One major disadvantage of this method is its computational cost that increases rapidly with the number of sources, as for each iteration of the algorithm we need to make $N!$ comparisons. For example, we can consider a case of 5 sources, where the FFT has 4096 frequency bins, and the post-processing permutation method needs to spend 15 iterations for the system to converge to the correct permutation for most of the bins. In total, we will need $5! \times 2048 \times 15 = 125 \times 4096 \times 15$ calculations of the expressions in (5) and as a result the whole task is computationally inefficient, if not prohibitive.

In this section, we propose a new "suboptimal" method, named Reduced Likelihood Ratio Jump. This technique selects to perform a few major comparisons in contrast to full set of $N!$ comparisons in the original method, thus the term "suboptimal". Nonetheless, we witnessed that it can produce the same, if not better separation quality with a considerable reduction of the computational cost.

The Reduced Likelihood Ratio Jump is based on the iterative nature of the original method. The Likelihood Ratio Jump needs, in most of the examined examples, some dozen iterations for every frequency bin to converge to the correct permutation. This is mainly due to the parameter $\beta(t)$. As previously mentioned, this parameter incorporates information about the time envelope of the signal. As more permutations are sorted in each iteration, the time envelope of each signal becomes more distinct and as a result, the parameter $\beta(t)$ has a stronger impact in the calculation, that helps resolving the permutation for frequency bins, where this task is more difficult.

During extensive experimentation, we witnessed that there are many frequency bins that feature correct permutation from the first or second iteration of the method. For the remaining frequency bins, the algorithm after some iterations needs to swap only one IC pair to restore the correct permutation, since the others have already been sorted to the correct sources. This situation is common for cases with many sources, as for most of the ICs the correct permutation is clear after a small number of iterations, and only one or two pairs may need an improved calculation of the parameter $\beta(t)$ to be permuted correctly.

Based on the above observation, we propose to reduce the number of examined permutations that our method considers in every iteration of the algorithm. As the method works iteratively, we propose to calculate the most probable permutation from a set permutation that only includes the swapping of one pair of ICs at a time. In the case of N sources, in one iteration we can calculate the most probable from $N-1$ swaps between the ICs, as happens in Likelihood Ratio Jump, but with the progression of the method only one swapping will be needed to ensure the correct permutation of the sources. Even if more than one pairs of ICs are permuted incorrectly, as the method progress the correct permutation will be restored, one pair at a time. As a result, we employ the iterations needed to make a more accurate estimation of the parameter $\beta(t)$, to reduce the set of examined permutations and consequently the required computational time. By examining only one pair of permutations at a time, we reduce the complexity of the method from $N!$ to $\frac{1}{2}N(N-1)+1$, which for 5 sources means a reduction from 120 to 11 comparisons per iteration. As we will show in the experimental section this suboptimal method does not undermine the quality of the separation but may also enhance it instead.

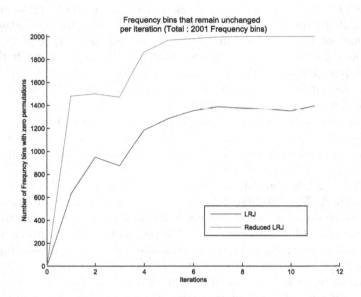

Fig. 1. Comparison between Reduced LRJ and original LRJ. More frequency bins remained unchanged using the Reduced LRJ for the same number of iterations.

In Fig. 1, we can see that the Reduced Likelihood Ratio Jump converges for more frequency bins, compared to the original method for the same number of iterations. In an example of a total of 2001 frequency bins the Reduced LRJ has concluded the permutation sorting of all the frequency bins from the 8th iteration, in contrast to the original method that always needs to sort about 600 more frequency bins. For some bins, the original LRJ changes permutation in every iteration. This phenomenon is due to the very small differences between the likelihood values that force the original method to toggle between 2 permutations for specific frequency bins.

4   Experiments

4.1   Evaluation Process

To evaluate the performance of our proposed framework, we created an evaluation dataset of 7 audio recordings. This evaluation dataset contains 5 mixtures of 3 sources and 2 mixtures of 4 sources, featuring both speech and music. The recordings were made using the Microsoft Kinect interface. The sources-microphones were placed in different positions in a real reverberant room. It also includes separate, recordings of the corresponding sources under the same recording conditions (position and loudness) in order to be used as ground truth for the evaluation of the separation quality achieved by the separation frame-work[2].

We used the RobustICA MATLAB implementation, as provided freely by the authors[3], and we made the necessary modifications to work in a convolutive source separation framework with original and the Reduced LRJ. For the experiments presented in this section, we assume a room impulse response of 90.7 ms length, which we reckon is a valid model for the recording positions in the reverberant room. To measure the separation quality, we used the metrics SDR, SIR and SAR that are designed specifically for Performance Measurement in Blind Audio Source Separation [11]. These metrics are implemented as a publicly-available MATLAB Toolbox and were designed to allow a time-invariant filter distortion of 512 samples length. We manually changed this value to 2000 samples, to cater for the wrong synchronization between the original and the estimated sources that are extracted from the mixture [11].

Metrics SDR and SAR measure the distortion and artifacts that are created by the separation method. Both examined frameworks (the proposed framework and Mitianoudis-Davies) produce similar values as they employ similar contrast functions with different optimization methods. To avoid the repetition of similar experimental results, we will therefore present SIR measurements only, and the separation quality in terms of interference elimination between the extracted sources.

4.2   Performance Comparison

In this section, we present several experiments to demonstrate the efficiency of the 2 examined frameworks using FastICA, RobustICA, the LRJ and the reduced LRJ. Firstly, in Table 1 we compare the separation quality of the two ICA implementations in terms of SIR. To tackle the permutation ambiguity, we use, in this experiment, 5 iterations of the original LRJ of Mitianoudis and Davies [1] for the two frameworks. We can see that:

- As the 2 methods perform optimization of different criteria, they do not perform the same for the examined cases. RobustICA performs better for cases (2,3,5), FastICA for (4,6) and for the rest of the recordings we observe similar separation qualities. In general, we can say that for the examined cases RobustICA with prewhitenning, can reach and outperform slightly the original method of FastICA in separation quality.
- RobustICA presents very fast convergence. In all examined cases, it produces very good separation quality in only 3 iterations. This feature of RobustICA can be a great advantage, compared to previous ICA implementations, as also mentioned in [8]. In contrast, FastICA needs more iterations to produce stable results. We can see, in Table 1, the major differences in separation quality

---

[2] Dataset available from http://utopia.duth.gr/nmitiano/download.html.
[3] http://www.i3s.unice.fr/~zarzoso/robustica.html.

Table 1. Efficiency comparison between the Frequency Domain RobustICA and FastICA implementations in terms of Signal-Interference-Ratio (dB). Here, we compare the performance of 3 and 15 iterations of RobustICA and FastICA. RobustICA reaches a better separation result faster than FastICA.

Recording	Source	3 iterations		15 iterations	
		RobustICA	FastICA	RobustICA	FastICA
1	1	−0.464	−0.09	−0.09	0.88
	2	1.698	0.99	0.60	1.51
	3	−7.30	−7.98	−8.84	−7.99
2	1	−6.36	−4.25	−6.79	−3.37
	2	5.53	0.23	3.46	1.07
	3	1.12	−2.00	0.67	−0.38
3	1	5.40	3.20	6.44	4.20
	2	0.18	−1.17	−0.63	−2.38
	3	−10.11	−8.98	−10.31	−9.78
4	1	−2.40	−2.16	−3.34	−0.10
	2	2.26	1.75	2.82	3.53
	3	−6.78	−6.22	−6.24	−10.09
5	1	3.81	−2.47	4.12	1.12
	2	−5.78	−6.05	−6.61	−8.02
	3	4.01	0.99	4.37	3.05
6	1	−6.80	−1.33	−6.79	−4.02
	2	−3.78	−6.82	−4.57	−3.72
	3	1.18	−0.05	1.73	4.27
	4	−0.63	−2.17	−0.95	−1.50
7	1	2.44	1.19	0.50	0.64
	2	−1.68	0.30	−0.25	−0.77
	3	−6.99	−7.68	−6.72	−7.87
	4	−7.83	−3.38	−5.17	−5.02

Table 2. Comparison between the Likelihood Ratio Jump and the Reduced Likelihood Ratio Jump for 8 available iterations in terms of SIR (dB).

Method	Source	Rec 1	Rec 2	Rec 3	Rec 4	Rec 5
LRJ	1	−0.24	−6.46	6.18	−1.55	3.82
	2	1.67	5.42	0.29	1.13	−5.78
	3	−6.98	1.03	−10.71	−6.49	4.01
Reduced LRJ	1	2.72	−2.82	6.44	−0.85	3.18
	2	1.58	3.82	1.91	1.46	−7.64
	3	−7.87	−2.57	−5.78	−6.49	2.41

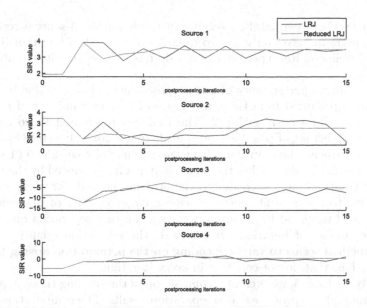

Fig. 2. Separation Quality in terms of SIR for the LRJ and the Reduced LRJ methods.

from 3 to 15 iterations of FastICA, in comparison to RobustICA that reaches very good separation quality in only 3 iterations, which is then only slightly improved as the iterations increase. Despite the fact that a RobustICA iteration is more costly than the FastICA equivalent, its fast convergence improves the total computational efficiency.

In the next experiment, we compare the separation quality with 8 iterations for both the new Reduced and the original Likelihood Ratio Jump method of Mitianoudis and Davies [1]. For the separation of frequency bins in this experiment, we have used the RobustICA with prewhitenning as it has shown to be the most efficient method. In Table 2, we can see the separation performance

Table 3. Running time comparison between the two frameworks (seconds)

Recording	Framework 1			Framework 2		
	Separation	Perm	Total	Separation	Perm	Total
1 (3 sources)	5.58	5.89	13.04	14.38	7.16	23.18
2 (3 sources)	4.89	5.21	11.43	11.30	6.00	18.61
3 (3 sources)	4.45	2.71	7.70	5.27	2.56	8.39
4 (3 sources)	4.63	2.96	8.20	5.92	2.92	9.48
5 (3 sources)	5.00	5.24	11.60	11.61	6.16	19.17
6 (4 sources)	6.73	8.93	17.36	22.25	17.86	41.82
7 (4 sources)	6.99	10.26	19.27	22.68	20.58	45.29

produced by the 2 permutation solving methods for the 3-sources recordings. The LRJ perform better only in recording 2. For the rest of the recordings, the Reduced Likelihood Ratio performs better despite the fact that it is a suboptimal method.

This improved performance of our proposed method can be due to the stability that is produced from the convergence of a greater number of frequency bins, as shown previously in Fig. 2. The Reduced Likelihood Ratio Jump, by allowing a smaller set of possible swaps, leads to a more stable state for a larger number of frequency bins. In the separation example of recording 6 (4 sources), shown in Fig. 2, we observe that the separation quality, produced by the Reduced Likelihood Ratio Jump, is greater to the original method, for every extracted source. Due to the accurate convergence for larger number of frequency bins, the Reduced Likelihood Ratio Jump reaches a constant separation quality from a smaller number of iterations. In contrast, the separation quality using the original method seems to vary, depending on the permutation arising from the frequency bins that do not converge in every iteration.

Finally, in Table 3, we present a comparison of the running times required by the 2 frameworks to produce stable separation results. The computational times of Table 3 refer to MATLAB R2013 implementations of the examined frameworks on a Pentium i7 3.4 GHz PC with 8 GB RAM. As explained previously, RobustICA requires sufficiently less iterations than FastICA and in the results of Table 1 we use 3 iterations for framework 1 (RobustICA) and 15 iterations for framework 2 (FastICA). For the permutation ambiguity, both the Reduced Likelihood Ratio Jump and the Likelihood Ratio Jump required 9 iterations. We used 9 iterations that seemed to be a good choice in Fig. 2 since after 9 iterations on average, all sources seems to present a relevant stability in the calculated SIR values. We can see that the improvement in computational time for the examined recordings is important, with similar separation performance as shown in previous experiments. RobustICA with much less iterations requires about 1/3 of the FastICA computational time, while the Reduced Likelihood Ratio Jump can solve the permutation ambiguity in about half the time of the original LRJ. As an example, for the seventh recording we need 19 sec with the proposed framework and 45 with the original one, which demonstrates the efficiency of the proposed approach.

5   Conclusion

In this paper, we presented an extension of the previous work of Mitianoudis and Davies for convolutive audio source separation. First of all, the FastICA separation algorithm was replaced with the RobustICA algorithm, which improves the performance and stability of the framework. The next improvement was the proposal of a Reduced LRJ solution, in order to reduce the increased computational cost in the case of more than 2 sensors and sources. The Reduced LRJ, although a suboptimal solution, seems to achieve better separation, since it doesn't allow more source flippings than necessary. The new framework has been tested on a newly recorded dataset using the cost-efficient Microsoft Kinect platform with success. For future work, the authors would like to extend this framework for underdetermined recordings, i.e. dealing with real-room recordings containing more sources than sensors.

References

1. Mitianoudis, N., Davies, M.: Audio source separation of convolutive mixtures. IEEE Trans. Speech Audio Process. **11**(5), 489–497 (2003)
2. Smaragdis, P.: Blind separation of convolved mixtures in the frequency domain. Neurocomputing **22**(1), 21–34 (1998)
3. Mazur, R., Mertins, A.: An approach for solving the permutation problem of convolutive blind source separation based on statistical signal models. IEEE Trans. Audio Speech Lang. Process. **17**(1), 117–126 (2009)
4. Sawada, H., Araki, S., Makino, S.: Underdetermined convolutive blind source separation via frequency bin-wise clustering and permutation alignment. IEEE Trans. Audio Speech Lang. Process. **19**(3), 516–527 (2011)
5. Saito, S., Oishi, K., Furukawa, T.: Convolutive blind source separation using an iterative least-squares algorithm for non-orthogonal approximate joint diagonalization. IEEE Trans. Audio Speech Lang. Process. **23**(12), 2434–2448 (2015)
6. Sarmiento, A., Duran-Diaz, I., Cichocki, A., Cruces, S.: A contrast function based on generalised divergences for solving the permutation problem in convolved speech mixtures. IEEE Trans. Audio Speech Lang. Process. **23**(11), 1713–1726 (2015)
7. Wang, L., Ding, H., Yin, F.: A region-growing permutation alignment approach in frequency-domain blind source separation of speech mixtures. IEEE Trans. Audio Speech Lang. Process. **19**(3), 2434–2448 (2011)
8. Zarzoso, V., Comon, P.: Robust independent component analysis by iterative maximization of the Kurtosis contrast with algebraic optimal step size. IEEE Trans. Neural Netw. **21**(2), 248–261 (2010)
9. Hyvärinen, A.: The fixed-point algorithm and maximum likelihood estimation for independent component analysis. Neural Process. Lett. **10**(1), 1–5 (1999)
10. Hyvärinen, A.: Fast and robust fixed-point algorithms for independent component analysis. IEEE Trans. Neural Netw. **10**(3), 626–634 (1999)
11. Févotte, C., Gribonval, R., Vincent, E.: BSS EVAL toolbox user guide. Technical report, IRISA Technical Report 1706, Rennes, France, April 2005. http://www.irisa.fr/metiss/bss_eval/

Association Rules Mining by Improving the Imperialism Competitive Algorithm (ARMICA)

S. Mohssen Ghafari and Christos Tjortjis[⊠]

School of Science and Technology, International Hellenic University,
14th km Thessaloniki – Moudania, 57001 Thermi, Greece
{m.ghafari, c.tjortjis}@ihu.edu.gr

Abstract. Many algorithms have been proposed for *Association Rules Mining* (ARM), like *Apriori*. However, such algorithms often have a downside for real word use: they rely on users to set two parameters manually, namely minimum *Support* and *Confidence*. In this paper, we propose *Association Rules Mining by improving the Imperialism Competitive Algorithm* (ARMICA), a novel ARM method, based on the heuristic *Imperialism Competitive Algorithm* (ICA), for finding frequent itemsets and extracting rules from datasets, whilst setting support automatically. Its structure allows for producing only the strongest and most frequent rules, in contrast to many ARM algorithms, thus alleviating the need to define minimum support and confidence. Experimental results indicate that ARMICA generates accurate rules faster than Apriori.

Keywords: Association rules mining · Data mining · Knowledge engineering

1 Introduction

With the dramatic increase in the amount of available data, searching for information in databases can be done manually with queries, but takes a long time and is not always efficient [1]. Alternatively, Data mining and particularly *Association Rules Mining* (ARM) have been proposed to analyze databases and extract frequent patterns and rules. A well-known method for ARM algorithms employs a two-step approach: find the *Frequent List* and extract *Rules* from that list [2].

However, there are still challenges faced by many ARM algorithms, as they require the user to define two parameters named *Minimum Support* and *Minimum Confidence* [3–5]. Another challenge for *Apriori* style ARM algorithms is that they use a time and resource consuming process, named *Pruning* which checks the feasibility of a rule and removes weak rules [5]. ARM algorithms could improve by approaches that automatically determine the value for parameters like minimum support and confidence [6–8], as well as by replacing pruning with more efficient methods. Although there have been researches addressing this topic, more needs to be done.

In this paper, we propose *Association Rules Mining by improving the Imperialism Competitive Algorithm* (ARMICA), a novel dynamic method based on the heuristic *Imperialism Competitive Algorithm* (ICA) [9]. ARMICA customizes itself according to

L. Iliadis and I. Maglogiannis (Eds.): AIAI 2016, IFIP AICT 475, pp. 242–254, 2016.
DOI: 10.1007/978-3-319-44944-9_21

the database that it operates on, and eliminates the dependability on initial parameters like minimum support and confidence. It automatically selects the strongest rules from the database. Moreover, in contrast with algorithms like *Apriori*, it does not involve time-consuming *pruning*. Because it eliminates all the weak items, it results in eliminating the need for pruning. ARMICA improves ICA in that it selects *imperialists* among the most *powerful countries*, rather than randomly, thus alleviating the need for a *revolution* part. In addition, having found the minimum support, ARMICA disqualifies and removes countries that have less *cost* than this minimum support. Finally, ARMICA has a different fitness function and method for creating initial countries. It was evaluated on publicly available datasets against *Apriori*. Experimental results show that it produces accurate results, up to 3 times faster than *Apriori*.

The rest of the paper is organized as follows: Sect. 2 discusses previous related work. Section 3 introduces and details the *Imperialism Competitive Algorithm* (ICA). Section 4 presents our proposed method, ARMICA. Section 5 presents the evaluation process and Sect. 6 discusses this work, its contribution and how it complements the state of the art. The paper concludes with directions for future work in Sect. 7.

2   Related Work

Association Rules Mining (ARM) is a well-known data mining method [2]. It aims at extracting frequent patterns and structure from data. Given T, a set of transactions in a transactional database: $\{t_1, t_2, \ldots, t_n\}$ and I, a set of items in this database: $\{i_1, i_2, \ldots, i_n\}$, association rules mining goal is to find all $\{X, Y\}$ where X and Y are both sets of some items (*itemsets*). In other words, ARM finds all the rules of the form: $X \rightarrow Y$, where X and Y are both subsets of I and $X \cap Y = \Phi$. This rule indicates that whenever X occurs then, Y may occur with certain probability. As a result, extracting such rules could help enrich our knowledge about a database, which supports solving real world problems.

Most of ARM approaches consist of two stages: finding the *Frequent Itemset List* and generating *rules*. One of the best-known ARM algorithms is Apriori [5]. It starts with finding the *candidate Itemset list*. Then, with a technique, named *pruning* and *joining*, it iteratively creates a candidate list and eliminates weak itemsets, employing the concept of Minimum Support. *Support* is the percentage of itemset occurrences in the transactions. Therefore, Apriori uses a user-defined threshold to check itemsets in the candidate list. If there is an itemset that has a support value less than the minimum, it removes it. Moreover, at each level there is a procedure, named pruning, which checks that every itemset in the candidate list satisfies this condition: "all the non-empty subsets of each itemset should be in the previous candidate list". If there is an itemset which does not meet this condition, Apriori removes it from the list. Eventually the candidate list becomes the frequent list. At that point, every non-empty subset of each item in the frequent list can be extracted. The algorithm removes the weaker rules based on a parameter, named Minimum Confidence. The *confidence* of a rule is calculated by:

$$\text{Support}\,(X \cup Y)\,/\text{Support}\,(X) \qquad (1)$$

Therefore, Apriori uses a user-defined threshold for min. confidence removing rules with lower confidence. Apriori is known for its efficiency in extracting frequent itemsets [5, 10]. Its main concept is the *Apriori principle*: any superset of an infrequent itemset must be an infrequent itemset. There were many algorithms that were proposed based on Apriori, which can be classified as follows [11]:

- Candidate itemsets reduction:

A smaller number of candidate itemsets incurs reduced computational cost. The hash-based DHP is such example [12]. Its main advantages are: efficient generation of frequent itemsets, effective declining in transaction database size and reduction in number of database scans. Before the next database scan, it uses a hash table to reduce the size of the candidate k-itemsets (that is C_k, for $k > 1$). The number of candidate k-itemsets may be reduced substantially, especially when the length of itemsets is equal to 2. However, DHP has a similar to Apriori behavior, as it needs to scan the database as many times as the length of the longest frequent itemset in it.

- Database scans reduction:

A large number of database scans increases the cost of time-consuming disk I/O. In order to tackle this problem, Savasere et al. proposed the Partition algorithm [13], which produces all frequent itemsets with two database scans. The Partition algorithm divides the database into several chunks in such a way that each portion can fit in memory and can be processed by Apriori. However, one of its drawbacks is that the number of candidate itemsets examined is much bigger than that of Apriori. The Dynamic Itemset Counting (DIC) algorithm [14], also divides the database into several parts. Once some frequent itemsets are produced, DIC can generate new candidate itemsets and immediately starts counting them. It should also be noted that the data distribution of a database has great impact on the performance of DIC [14].

- Combination of bottom-up and top-down search:

This algorithm also employs the Apriori principle to identify frequent itemsets in the bottom-up search, while in each round the infrequent itemsets found in the same direction are used to separate the maximal frequent candidate itemsets in the top-down direction. The advantage is that all subsets of the maximal frequent itemsets are ready as long as the maximal frequent itemsets are identified. As a result, there is no need to extract subsets of the maximal frequent itemsets again from the bottom-up direction. The Pincer-Search algorithm [15] and the MaxMiner algorithm [12] are two examples of this concept, but they are not efficient in finding the maximal frequent candidate itemsets. Meanwhile, the advance made by this approach is not remarkable when the length of the longest frequent itemset is relatively short [11].

- Proposing Dynamic ARM Algorithms:

Researchers proposed dynamic algorithms to deal with the drawbacks of setting the minimum support and confidence manually. For instance, Djenouri et al. proposed

HBSO-TS, a method that uses two heuristic algorithms: Artificial Bee Colony (ABC) for the main processing and Tabu Search (TS) for searching the neighbors of each food source for each bee [6]. The authors concede that setting many parameters required by ABC and TS is hard. In addition, they do not mention anything about possible advantages of their method concerning pruning.

Kuo et al. [7] propose an algorithm that tries to determine minimum support and confidence automatically. It uses a heuristic algorithm, named Particle Swarm Optimization (PSO). First, they apply particle swarm encoding, which they claim to be similar to chromosome encoding of genetic algorithms. Next, they calculate the fitness value, and, based on that, they generate a population of particle swarms. Finally, PSO searching is applied until it satisfies the stop condition that is finding the best particle. The support and confidence of the best particle can represent the minimal support and minimal confidence. It should be noted that setting the best solution as minimum support may not be efficient. Besides, they use minimum confidence to evaluate their final rules, like most ARM algorithms.

Nandhini et al. [8] propose an algorithm also based on PSO and a domain ontology. Like PSO, it determines min. support automatically. It also considers support and confidence of the best particle as minimum support and confidence. However, it also needs to set the minimum confidence in order to evaluate rules. Next, we detail the Imperialism Competitive Algorithm (ICA).

3 Imperialism Competitive Algorithm (ICA)

ICA is a powerful heuristic algorithm based on the notion of the relation between imperialistic countries and their competition for acquiring colonies [9]. ICA involves key concepts that we present and discuss here:

- *Country*: the variables that we want to optimize
- *Imperialist*: A powerful country that can control other countries in the form of colonies.
- *Colony*: less powerful country that imperialists try to control.
- *Empire*: a group of countries headed by an imperialist.
- *Cost*: a parameter depending on the nature of the problem at hand. In this paper, we consider it the number of occupancies of item in all the transactions.
- *Normalized cost*: the cost of each country calculated using formula (3) below.
- *Power*: a factor calculated based on the cost of each country. A country could be considered as more powerful than another, depending on their costs.
- *Total empire Power*: The sum of the powers of the imperialist and its colonies in one empire.
- $N_{Country}$: the number of all countries (items)
- N_{imp}: the number of Imperialists, which is a parameter [16].
- N_{col}: the number of colonies given by: (Number of all countries – number of Imperialists)

ICA firstly considers all the variables that we want to optimize, as a concept named country. Then it randomly selects some countries as Imperialists. Then it calculates the cost of imperialists and calculates their powers. The cost of each country could be different depending on the case. For instance, in this paper we consider each item as a country and the number of occurrences of each item in all the transactions as the cost of each country. Then, based on the power of imperialists, it divides other countries (colonies) among imperialists. In this step, the algorithm assumes that each imperialist and its colonies are part of an Empire.

Next, ICA calculates the power of each empire based on the power of its colonies and imperialist and establishes a competition between empires. In this step, the most powerful empires remove colonies from the weakest empires. If there is only one colony left for the weakest empire, then the imperialist and the last colony of the weakest empire become colonies of the strongest empire. This competition continues until there is only one empire left, as shown in Table 1.

Table 1. Imperialism competitve algorithm

---

1) Initialize the empires
2) Calculate the total Power of the empires.
3) Competition between empires. Select a colony from the weakest empires and assign it to the most powerful one.
4) Eliminate the weakest empires.
5) If there is only one empire left, stop the process. Otherwise, go to step2

3.1 Creating the Initial Empires

ICA considers an array of variables (*countries*) which should be optimized. We calculate the *cost* of each country by a function f at variables $(p_1, p_2, p_3, \ldots, p_{N_{var}})$ [9]:

$$cost_i = f(country_i) = f(p_1, p_2, p_3, \ldots, p_{N_{var}}) \tag{2}$$

At the beginning, ICA initializes countries of size N_{pop} and randomly selects N_{imp} the countries as initial imperialists. It also considers all the remaining countries (N_{col}) as potential colonies. They should be distributed across the imperialists, based on their power. The definition of *normalized cost* of an imperialist is defined in [16]. We apply this formula to calculate the costs of Items:

$$C_n = \max_i \{c_i\} - c_n \tag{3}$$

where c_n is the cost of the nth imperialist and C_n is its normalized cost. After having calculated the normalized cost of all imperialists, the normalized power of each imperialist can be computed as follows [16]:

$$p_n = \left| \frac{C_n}{\sum_{i=1}^{N_{imp}} C_i} \right| \tag{4}$$

3.2   Total Empire Power

We sum up the power of the imperialist of an empire and the mean power of its colonies as *the power of an empire* with formula (5) [9].

$$T.C.n = Cost\ of\ Imperialis + mean\ (Cost\ of\ Empire's\ Colonies) \tag{5}$$

4   Proposed Method

Section 3 introduced the original ICA algorithm. This section discusses our proposed method, ARMICA, which employs ICA, suitably modified for association rules mining. Some parts of original ICA are not suitable for ARM, thus we had to change them. For that reason, we modified ICA to adapt it with the characteristics of an ARM algorithm. Firstly, we select imperialists among the most powerful countries, in contrast to ICA, which selects them randomly. Next, we remove the revolution part of ICA. Revolution is about checking if a colony becomes more powerful than the imperialist of its empire is. In that situation, a revolution takes place and ICA exchanges that colony with the imperialist. However, in ARMICA, since we select the weakest colonies and imperialists are the most powerful countries, there is no chance for a colony to become more powerful than its imperialist. As a result, ARMICA does not need incorporate revolution. In addition, having found the minimum support in the first run of ICA, we are using this value to disqualify countries that have less cost than this minimum support. This helps to remove many unusable countries; there is no such mechanism in the original ICA. Finally, there are other modifications, which were necessary for ARMICA in order to improve ICA, like changing the fitness function and creating the initial countries. Table 2 illustrates pseudo code for ARMICA.

As Table 2 indicates, ARMICA uses the dataset as input and delivers association rules as output. In the first step, it creates a set of initial countries based on dataset attributes (initial countries). In other words, it defines each item as a country. As a result, there are as many items (attributes) in the dataset as there are countries for ICA. At this point, the algorithm calculates the cost for each country based on the number of occurrences of each item in the set of transactions. ICA considers countries with lower cost as more powerful.

However, in this paper, since we consider the cost of a country as the number of its occurrences in the set of transactions, it is more suitable that the countries with higher costs are items that are more desirable. As a result, in contrast to ICA, ARMICA

Table 2. Pseudo code for ARMICA

Input: Dataset   Output: Rules
1) Run ICA with the dataset and create initial countries based on items (attributes)
2) Calculate the cost for each country
3) Create the empires and calculate the power of each empire
4) Establish competition among empires based on their power
5) The Candidate List consists of the imperialist of the last empire and its colonies
6) In the first candidate list, consider the cost of each country as its support and then sort the countries based on their support. Select the median support as the primary min support for the next steps.
7) While there is at least one country that has support more than that primary min support do: Extract all the combinations of countries and store them in a list, send it again to ICA and get the next Candidate List
8) The last candidate list is the Frequent List
9) Extract all the subsets of items in the FrequentList and send them to ICA to get the best rules.

---

assumes that the most powerful countries are those with the highest cost. After calculating the cost for all the empires, the algorithm selects the most powerful countries as imperialists and considers the remaining countries as colonies. Hence, imperialists are the most frequent items in the dataset. The number of Imperialists is a free parameter in the original ICA, but for ARMICA, after experimentation, we consider it the number of countries divided by 10. In addition, if the initial number of countries is $N_{Country}$, and we select N_{imp} countries as imperialists, we have N_{Col} colonies.

In the next step, ARMICA calculates the power of each imperialist and distributes the colonies (other countries or other items) among them based on their power. As a result, by gathering imperialists and colonies together, the initial empires are built. ARMICA calculates the power of each empire based on the sum of powers of its colonies and their imperialist. For this purpose, the algorithm calculates the number of occurrences of each item in all the transactions, considers this value as the cost of that item (country), and calculates its power.

Next, ARMICA establishes a *competition* between empires, where the more powerful empires "steal" colonies from the weakest ones. Another difference between ARMICA and ICA is that in the original algorithm, each colony removed from the weakest empire is placed in the most powerful empire. However, ARMICA removes the colonies and stores them in a list, named *Reserve List*. If there is only one colony left in the weakest empire, the colony and its imperialist, become members (colonies) of the most powerful empire. This procedure continues until all the empires are eliminated and only the most powerful one remains. For this purpose, ARMICA removes the weakest colonies (items) of the weakest empire and keeps them in the *Reserve List*.

If the weakest empire has only one colony, then ARMICA removes that empire and considers its imperialist and its last colony as colonies of the most powerful empire. After several iterations, all the empires will be eliminated and only one empire remains. Now, if there is a member of the Reserve List which is more powerful, i.e. it has higher frequency than any colony in the last empire, the algorithm exchanges them. Hence, members of the final empire constitute our candidate list.

Overall, even though ARMICA uses a minimum support threshold, it sets it automatically and removes the weaker items. The procedure for setting the minimum support involves storing support values, which is the cost (frequency) of each country (item) in the dataset, for all the items of the first generated candidate list and sorting them. At that point, the algorithm considers the median support of all the countries as the universal minimum support. This value is also used to remove countries that have less cost than the minimum support in the next levels.

Then, ARMICA extracts all the combinations of items, stores them in a list, and sends them to ICA. This continues until there is at least one item with support higher than the minimum support. As a result, each time the algorithm produces an updated candidate list. It is noticeable that, here is another significant improvement compared to Apriori. In Apriori there is a step, named Pruning which eliminates itemsets with at least one of their subsets is not present in previous candidate lists. As it is clear, this process also needs to calculate all the subsets of the items and check their existence in the previous candidate list. In ARMICA, no such step is needed, as the algorithm selects the most powerful (frequent) items, thus eliminating the need for pruning.

Finally, the last candidate list becomes the Frequent List. At that point, ARMICA extracts all the non-empty subsets of all the items and sends them to ICA. If "I" is the item, then for every nonempty subset of s, we have a rule as:

s→ (I-s), in the original Apriori, we have a condition:

$$\text{Support (I)} / \text{Support (s)} >= \text{Minimum Confidence} \qquad (6)$$

If a rule satisfies this condition, it is selected as a frequent rule. However, since the members of the last empire in ICA are the most powerful rules (have more confidence compared to other rules), ARMICA only selects the most powerful (frequent) subsets and returns the most powerful rules. Again, it is worth mentioning that, ARMICA does not need to have a minimum confidence parameter. Apriori and many other ARM algorithms need a user defined fixed threshold or even a dynamic threshold for minimum confidence. However, our approach only selects the most powerful rules among all the rules and removes all other rules.

4.1   Example

In this section, we explain our algorithm using a small example. Table 3 illustrates a set of six transactions and eight items.

T_1, T_2,..., T_6 are 6 transactions and I_1, I_2, ..., I_8 are items in these transactions. First, the algorithm selects some of the most powerful countries as Imperialists. Since there are only 8 items, if we do not use just one imperialist (given by: number of

Table 3. Transaction details

	I1	I2	I3	I4	I5	I6	I7	I8
T1	t	t	t	t		t		t
T2		t	t	t		t		
T3	t		t				t	
T4	t		t	t		t		t
T5		t						t
T6	t	t	t		t		t	t

countries / 10) as there would be no competition resulting in selecting all the items, but instead assume that there are 3 imperialists. Their power is calculated and the remaining countries are distributed among the imperialists based on their power:

Imperialist 1: I_3		Imperialist 2: I_2		Imperialist 3: I_8
I_1	I_7	I_4	I_6	I_5

Initial empires are now built. ARMICA calculates the cost of countries and the power of each empire based on the sum of the powers of its colonies plus its imperialist. In this occasion, we have:

Empire 1's Power: 11(Powerful Empire), Empire 2's Power: 10, Empire 3's Power: 5 (Weakest Empire)

Then, the algorithm establishes a *competition* between empires. As a result, a powerful empire tries to "steal" colonies from the weakest empire. Finally, if that weak empire has only one colony left, the imperialist and colony of that empire become members of the powerful empire and the algorithm eliminates that empire. As a result, Empire 3 is eliminated and I_8 and I_5 now belong to Empire 1.

Imperialist 1: I_3				Imperialist 2: I_2	
I_7	I_1	I_5	I_8	I_6	I_4

The competition goes on. This time Empire 1 tries to get one of the colonies from Empire 2. I_6 is removed from Empire 2 and saved in the *Reserve List*. Since Empire 2 has only 2 members, it joins Empire 1.

Imperialist 1: I_3						Reserve List
I_7	I_1	I_2	I_8	I_4	I_5	I_6

Now the members of the Reserve List are compared with the Last Empire colonies. If there is a colony that has lower cost than the members of the reserve list do, then they are replaced. In this case, I_6 has higher cost compared to I_5, so the algorithm replaces them and removes I_5. Clearly, ARMICA did not use minimum support and automatically eliminated weak countries. In this step, the algorithm considers the last empire as the Candidate List. At that point, it sorts the items of the candidate list based on their cost and selects the median cost of them as Minimum Support. As a result, the cost of the I_7 becomes the minimum support, that is 4.

Next, the algorithm joins all the countries in a candidate list and calculates their costs. It is noticeable that the cost of a country like I_1I_2 is calculated by the number of occurrences of I_1 and I_2 in the same transaction, that is 2. The algorithm follows the same pattern that we mentioned before, repeatedly until there is no item left in the produced candidate list that has an equal or higher support (cost) than the minimum support. In this situation, the previous candidate list is the Frequent List. Clearly, ARMICA does not require pruning.

Frequent List: I_1I_3.

Finally, it is time to extract all the non-empty subsets of the frequent list items and after that send them to ICA in order to find countries that are more powerful. Like previously, the last empire consists of the most powerful countries that represent the strongest rules. It is clear that we did not need to define Minimum Confidence. The last empire, whose members are the final rules, is here:

I_1	$\rightarrow$	I_3
I_3	$\rightarrow$	I_1

5 Evaluation

In this section, we evaluate ARMICA, the proposed method. We implemented it using Java 1.7 in Netbeans IDE 7.2 and ran it on an Intel (R) Core (TM) i5 CPU at 2.40 GHz and 4 GB RAM. We also used the implementation of Apriori from Weka 3.6 [17] along with the Supermarket, Breast Cancer and Nursery datasets from the UCI Machine Learning Repository [18] to benchmark our method. Supermarket has 217 attributes and 4627 transactions, Breast Cancer has 10 attributes and 286 transactions, whilst Nursery has 8 attributes and 12960 transactions. We considered two factors for this evaluation: accuracy and execution time. An improvement on Apriori should decrease the execution time without decreasing accuracy. ARMICA would be a good approach, if it produced accurate rules faster.

The minimum support that ARMICA determined automatically was 27.5 %, 36.18 % and 33.33 % for the Supermarket, Breast Cancer and Nursery datasets, whilst minimum confidence was 71 %, 47 % and 1 %, respectively. We used the same values for Apriori. Table 4 illustrates the characteristics of the evaluation process. Apriori and ARMICA produced the same rules. In addition, we compared the two algorithms concerning their execution time. ARMICA took 6 s compared to Apriori's 17 s for Supermarket and 0,76 s compared to Apriori's 0,78 for Breast Cancer. Finally, it required 1.15 s for Nursery whereas Apriori required 1.22 s. This means that ARMICA generated the same rules as Apriori, but in a shorter period.

Table 4. Evaluation characteristics

Data set	Items	Transactions	Algorithm	Predefined min. support	Predefined min. confidence	Generated rules	Time (sec)
Supermarket	217	4627	Apriori	27.5 %	71 %	104	17
			ARMICA	—	—	104	6
Breast Cancer	10	286	Apriori	36.18 %	47 %	75	0,78
			ARMICA	—	—	75	0,76
Nursery	8	12960	Apriori	33.33 %	1	2	1.22
			ARMICA	—	—	2	1.15

6 Discussion

Many ARM algorithms need two parameters to be defined in advance by the user, who has no guidelines on selecting appropriate values, thus resulting in a trial and error exercise or guesswork. These parameters are used to eliminate infrequent itemsets and weak rules. With the increasing amounts of data available, it is not effective to use an algorithm that requires users to define these parameters. In this paper, we proposed an algorithm, named ARMICA, which applies the ICA algorithm, suitably modified, to find the most frequent rules of a dataset. In contrast to many algorithms, ARMICA sets minimum support automatically and does not need a minimum confidence parameter in order to eliminate the weakest rules.

ARMICA uses ICA, which is a heuristic algorithm to analyze a database. For this purpose, at the beginning, it produces the first candidate list. At that point, it considers the minimum support of itemsets in that list as the global *Minimum Support*. After that, it joins itemsets and produces a new candidate list. This process continues until there is no item left in the candidate list that has support that is equal or more than the Minimum Support. Next, ARMICA considers the last candidate list as Frequent List, extracts all the non-empty subsets of the items, and sends them to ICA. ARMICA return the strongest rules without requiring minimum confidence setting. It is worth mentioning that, since ARMICA sends all the extracted subsets to ICA, ICA only returns subsets that according to formula (1) has higher confidence.

Another advantage of the proposed method is that instead of having a time con-suming step, like Pruning, which involves extracting all the subsets of all the items, and trace their existence in the previous candidate list, ARMICA selects the most valuable items. Experimental results indicate that ARMICA has up to 3 times less execution time than Apriori in the Supermarket dataset and shorter execution time in Breast Cancer and Nursery datasets. However, since each improvement in execution time of Apriori may cause some decreasing in accuracy of generated rules, we also considered the accuracy of the ARMICA. The proposed method generates all the rules that Apriori generates. Hence, there is not any accuracy reduction in ARMICA.

7 Conclusion and Future Work

Data mining has attracted attention partly because of the enormous amount of data available. A well-established approach for extracting frequent rules and patterns from data is Association Rules Miming (ARM). Many algorithms have been proposed in this area [1–5, 19, 20]. However, there are still areas that need to be investigated further.

In this paper, we proposed a new approach, named ARMICA, based on the Imperialism Competition Algorithm. It defines minimum support automatically and extracts the best rules without the need to define minimum confidence, either. In addition, it does not need a pruning stage, as the one used by many Apriori-like ARM algorithms. However, it requires setting of the number of imperialists, a free parameter in the original ICA. We explored ways to calculate it, and found empirically that the number of countries, divided by 10, is a sensible solution, which does not require domain knowledge nor details about the dataset. In the future, we envisage optimizing this selection, alleviating the need for the user to pre-define this parameter.

Finally, ARMICA produces accurate rules fast. In the other words, it generates all the rules that Apriori produces, in a shorter time. Initial experiments have indicated that it is also faster than FP-growth, but proper evaluation is required. We plan to focus on other potential capabilities of ARMICA such as dealing with massive and/or dynamic data sets, benchmark it against state of the art ARM algorithms and investigate performance related areas, such as memory usage and time consumption.

Acknowledgements. The authors thank the Hellenic Artificial Intelligence Society (EETN) for covering part of their expenses to participate in AIAI 16.

References

1. Bhandari, A., Gupta, A., Das, D: Improvised apriori algorithm using frequent pattern tree for real time applications in data mining. In: Proceedings of the International Conference on Information Communication Technologies (ICICT), vol. 46, pp. 644–651 (2014)
2. Qodmanan, H.R., Nasiri, M., Minaei-Bidgoli, B.: Multi objective association rule mining with genetic algorithm without specifying minimum support and minimum confidence. Expert Syst. Appl. (Elsevier) **38**(1), 288–298 (2011)
3. Dong, L., Tjortjis, C.: Experiences of using a quantitative approach for mining association rules. In: Liu, J., Cheung, Y.-m., Yin, H. (eds.) IDEAL 2003. LNCS, vol. 2690. Springer, Heidelberg (2003)
4. Wang, C., Tjortjis, C.: PRICES: an efficient algorithm for mining association rules. In: Yang, Z.R., Yin, H., Everson, R.M. (eds.) IDEAL 2004. LNCS, vol. 3177, pp. 352–358. Springer, Heidelberg (2004)
5. Agrawal, R., Srikant, R.: Fast Algorithms for mining association rules in large databases. In: VLDB 1994 Proceedings of the 20th International Conference on Very Large Data Bases, pp. 487–499 (1994)
6. Djenouri, Y., Drias, H., Chemchem, A.: A hybrid bees swarm optimization and tabu search algorithm for association. In: IEEE World Congress on Nature and Biologically Inspired Computing (NaBIC), pp. 120–125 (2013)

7. Kuo, R., Chao, C., Chiu, Y.: Application of particle swarm optimization to association rule mining. Appl. Soft Comput. **11**, 326–336 (2011)
8. Nandhini, J.M., Janani, M., Sivanandham, S.N.: Association rule mining using swarm intelligence and domain ontology. In: IEEE International Conference on Recent Trends in Information Technology (ICRTIT), pp. 537–541 (2012)
9. Atashpaz-Gargari, E., Lucas, C.: Imperialist competitive algorithm: an algorithm for optimization inspired by imperialistic competition. In: IEEE Congress on Evolutionary Computation, pp. 4661–4667 (2007)
10. Agrawal, R., Imielinski, T., Swami, A.: Mining association rules between sets of items in large databases. In: ACM SIGMOD Conference on Management of Data, pp. 207–216 (1993)
11. Yang, D.L., Pan, C.T., Chung, Y.C.: An efficient hash-based method for discovering the maximal frequent set. In: Proceedings of the 25th Annual International Computer Software and Applications Conference, pp. 511–516. Chicago (2002)
12. Park, J.S., Chen, M.S., Yu, P.S.: An effective hash based algorithm for mining association rules. In: Proceedings of the 1995 ACM SIGMOD International Conference on Management of Data, pp. 175–186, (1995)
13. Savasere, A., Omiecinski, E., Navathe, S.: An efficient algorithm for mining association rules in large databases. In: VLDB 1995 Proceedings of the 21st International Conference on Very Large Data Bases, pp. 432–444 (1995)
14. Brin, S., Motwani, R., Ullman, J. D., Tsur, S.: Dynamic itemset counting and implication rules for market basket data. In: ACM SIGMOD Conference on Management of Data, pp. 255–264 (1997)
15. Lin, D.-I., Kedem, Z.M.: Pincer search: a new algorithm for discovering the maximum frequent set. In: Schek, H.-J., Saltor, F., Ramos, I., Alonso, G. (eds.) EDBT 1998. LNCS, vol. 1377, pp. 105–119. Springer, Heidelberg (1998)
16. Talatahari, S., Farahmand, A.B., Sheikholeslami, R., Gandomi, A.: Imperialist competitive algorithm combined with chaos for global optimization. Commun. Nonlinear Sci. Numer. **17** (3), 1312–1319 (2012)
17. Witten, I.H., Eibe, F., Hall, M.A.: Data Mining: Practical Machine Learning Tools and Techniques, 3rd edn. Morgan Kaufmann, Burlington (2011)
18. Bache, K., Lichman, M.: UCI machine learning repository (2013)
19. Han, J., Pei, J., Yin, Y.: Mining frequent patterns without candidate generation. In: ACM SIGMOD International Conference on Management of Data, pp. 1–12. USA (2000)
20. Scheffer, T.: Finding association rules that trade support optimally against confidence. In: Siebes, A., De Raedt, L. (eds.) PKDD 2001. LNCS (LNAI), vol. 2168, pp. 424–435. Springer, Heidelberg (2001)

Use of Flight Simulators
in Analyzing Pilot Behavior

Jan Boril[1(✉)], Miroslav Jirgl[2], and Rudolf Jalovecky[1]

[1] University of Defence, Brno, Czech Republic
{jan.boril, rudolf.jalovecky}@unob.cz
[2] Brno University of Technology, Brno, Czech Republic
miroslav.jirgl@phd.feec.vutbr.cz

Abstract. This paper describes simulation technologies in their current state as they are in the general aviation and at the University of Defence. The authors present a methodology of measurement, evaluation and modelling of pilots' behavior, both before and after flight training (20 h) on real airplanes. By means of transfer function equation of human behavior model and time constants resulting from them, the authors analyze their change depending on the continuing training of students – pilots of military aircraft. Based on the results of these analyses, they have found a positive trend in the development of pilots' transport delay and their abilities to adapt themselves to flight dynamics of an airplane.

Keywords: Flight simulators · Pilot training · Human behavior model · Time constants · MATLAB® · X-Plane

1 Introduction

In modern aviation flight simulations play an irreplaceable role both in pilot training and maintaining their needed habits. The level of safety in transport as well as individual aviation now considerably depends on simulation technologies that are used not only for training but also for investigation of aircraft accidents, studying airplane designs, or better understanding of ergonomic relations [1]. The importance of flight simulators is increasing proportionally with growing complexity of modern aviation systems, so training on simulators has become an inseparable part of pilot training, their professional growth and research focused on adaptability of aircraft environment to human capabilities and limits. As a result of the constant development of pilot natural working environment, flight simulators are becoming an invaluable instrument.

Indispensable part of training is currently being transferred from flight training on planes to simulators for the reasons of undisputable advantages that this type of training offers. The growing significance of flight simulations is also evidenced in the latest international regulations stipulating requirements for aviation synthetic training equipment. Training efficiency can be achieved with use of synthetic equipment of a low reliability for some types of missions. Thanks to the interconnection of flight simulations with the rapidly growing field of computer technologies, the synthetic methods offer more and more opportunities in running effective training of flying crews and maintenance personnel [2].

© IFIP International Federation for Information Processing 2016
Published by Springer International Publishing Switzerland 2016. All Rights Reserved
L. Iliadis and I. Maglogiannis (Eds.): AIAI 2016, IFIP AICT 475, pp. 255–263, 2016.
DOI: 10.1007/978-3-319-44944-9_22

As a result of progress made in the field of avionics, the complexity of civilian and military airplanes have increased too, which puts higher demands on crew training and strengthens the dependency on flight simulations. For the reasons of crew training and retaining their licenses many civilian airlines have established and are running large aviation training centers. Not only have flight simulation radically changed flight training methods in terms of reducing risks and increasing quality of training, but they have also resulted in a significantly higher flight safety, decreased traffic density, and positive effect on environment, when all of that is achieved in reducing cost of flight training. These trends will continue in the foreseeable future [2, 3].

The main goal of this paper is to verify the potentiality of using flight simulators and subsequent application of mathematical methods to evaluation of pilot training experience. Currently, the authors are in the phase of repeated measurements running under the equal conditions and with the same pilots. The aim of which is to verify the accuracy of all analyses depending on repeatability with an emphasis laid on changes resulting from continuous training.

2 Mathematical Equation of Human Behavior

The literature offers us several works dealing with modelling of human behavior, including pilots. Most of these papers are based on the description (1) first presented by D.T. McRuer in the 1970s which models human behavior primarily at the level of the feedback regulator as described above. It is a linear model (transfer function) of a proportionally-derivative regulator with a second order lag and time (reaction) delay, in which each of the constants has a certain physiological or neurological interpretation [4–6]:

$$F(s) = \frac{Y(s)}{X(s)} = K \frac{(T_3 s + 1)}{(T_1 s + 1)(T_2 s + 1)} e^{-\tau s}. \tag{1}$$

Where

K   Pilot gain representing a pilot's habits in response to a certain action. It is also connected with the input and output signal ratio.

T_1 Neuromuscular lag time constant expressing a pilot's delayed reaction resulting from his neuromuscular system. It ranges from 0.05 to 0.2 s and is not dependent on the training intensity.

T_2 Lag time constant characterizing a pilot's promptness and agility. Thus it is related to performing acquired stereotypes and routine procedures. Ranges from 0.1 s to 5 s.

T_3 Lead time constant connected with a pilot's experience. It reflects a pilot's capability to predict a situation that may occur. This capability is developed through training and experience and ranges from 0.2 s to 15 s.

τ Time constant expressing a delayed reaction in pilot's brain to a motional and visual perception. As a result of fatigue this constant can be extended and the pilot's regulating capabilities may subsequently fail, thus the regulating system becomes unstable. This constant most frequently ranges from 0.3 to 1 s.

S   Laplace operator.

The model in the form of Eq. (1) is a general model that may be used in a wide spectrum of activities connected with driving [7] or piloting. Individual constants most frequently reach the values within the ranges as shown above, and depend on pilots' abilities to adapt to the controlled dynamics (adaptability).

Validity of this model was verified on the basis of many experiments, and primarily thanks to its simplicity it turned out to be a very efficient tool for description and modelling of a pilot's reactions [6, 8].

The identification algorithm as written in MATLAB environment is used to identify model parameters. This algorithm is based on applying the library function *fminsearch* resulting from Nelder-Mead simplex algorithm [9]. The criterial function for searching for parameter is then in the form of the sum of squared deviations, see Eq. (2):

$$KRIT = \sum_{i=1}^{n} (y(i) - y_m(i))^2. \tag{2}$$

Where
y   real (measured) value of the output quantity (pilot's intervention),
y_m modelled (estimated) value of the output quantity (pilot's intervention),
n   length of data.

The Best fit [%] parameter is used for expressing the accuracy of identification because it gives an opportunity for making an easy comparison of identification results with those obtained through MATLAB - *System Identification Toolbox*. The equation for calculating the Best fit parameter is as follows [10]:

$$fit = 100 \left(1 - \frac{\|y - y_m\|}{\|y - mean(y)\|} \right). \tag{3}$$

Where
y   real (measured) value of the output quantity (pilot's intervention),
y_m modelled (estimated) value of the output quantity (pilot's intervention).

3 Description of Experimental Workplace and Measurement Procedure

Most of armed forces also use training centers equipped with flight simulators designed for basic training, transitions to another type of aircraft, and tactical training on the platform of both fixed-wing and rotary-wing aircraft. Beyond doubt simulations provide a wide range of true advantages, therefore a growing number of synthetic equipment is being introduced into modern air forces to be used in a broad spectrum of activities, mainly for training of aviation personnel [11]. In addition, flight simulators are used by aircraft and airplane equipment manufacturers, system developers, research organizations and academies whose effort is aimed at proving the validity of studies, designs, developments and calculations of flight systems.

3.1   Flight Simulator at the University of Defence

The Department of Air Electrical Systems strives to keep up with these above mention trends, therefore the flight simulator based on X-Plane-10 software was built up to raise the quality of training and create a capacity for scientific research. The mission of the flight simulator at the University of Defence does not lie in competing with or replicating modern sophisticated simulators designed for pilot basic or improvement training. Using this simulator the authors prepare their procedures and methods for data gathering and evaluation. Currently, as described in the chapter Introduction, the authors are in the phase of repeated measurements running under the equal conditions and with the same pilots. The aim of which is to verify the accuracy of all analyses depending on repeatability with an emphasis laid on changes resulting from continuous training. Based on the knowledge obtained in such a manner, they will be able to implement the methodology of measurement and data evaluation on commercial full flight simulators.

Throughout the whole world this simulator program is currently rated as a comprehensive, complex and highly efficient PC flight simulator that offers most advanced flight models and tools. This engineering tool has been designed for an easy estimate of a flight model of all categories of aircraft and their construction types. The information acquired from mathematical and physical calculations in real time are relatively precise. The basic version of X-Plane software offers dynamics of 30 airplanes, such as Bell 206 JetRanger helicopter, Cessna 172, King-Air C90 etc. From the previous experience the King-Air C90 airplane (Fig. 1) was chosen for testing a pilot's response to an unexpected flight situation. The simulator is capable to record the flight trajectory (selected parameters) and safe them into a text file in 20 Hz frequency.

3.2   Experimental Flight Task

A flight mode was defined - altitude of 2900 ft, speed 170 mph, angle of attack and pitch angle, including their change, was approximately zero (note 1 ft = 0.305 m and 1 mph = 0.447 m.s^{-1}). At a certain time the altitude was step-changed to 2600 ft and the task of the pilot was to correct the altitude back to the original flight level 2900 ft A total of 6 trainee pilots were tested in this flight mode, all of them had about 60–80flight hours of experience in the Zlin Z-142C AF aircraft. Each pilot had the altitude changed 10 times one after another, always after putting the aircraft back into the initial flight state.

The whole experiment was repeated after 6 months. The tested subjects were the same pilots. These pilots were trained within this time and the experience was increased by about 20 h.

Fig. 1. The simulator at the Department of Air Electrical Systems, University of Defence, built on the basis of X-Plane 10 simulation program.

4 Measurement and Data Analysis

For the purpose of demonstrating the results, an analysis of pilot no. 1 was made, as shown below. The evaluation of other flights was carried out in a similar manner.

Figure 2 presents data measured on the simulator by pilot's no. 1 containing individual measurements performed (a) in spring 2015, (b) in fall 2015 (after the training). On the left side are always altitude records by airplane H depending on the stick deflection (pilot's reaction) dv, as shown on the right side. The red curve indicates the arithmetic average of individual measurements, i.e. average pilot's response and flight altitude reached by the plane calculated out of 10 measurements.

In order to evaluate the level of pilot training, i.e. compare changes in their responses, it is advisable to create and define the so-called average pilot behavior model. The parameters of an average pilot behavior model as described by the Eq. (1) were identified from the average responses – red curves in Fig. 2 using the identification algorithm, as described above.

Thus, the identification results in the average pilot behavior model in the following form:

- spring 2015

$$F_{1A}(s) = 6.84.10^{-4} \frac{(1.62s+1)}{(0.13s+1)(0.46s+1)} \cdot \exp(-0.7s)$$

$$\text{Best fit} = 47.43\%$$

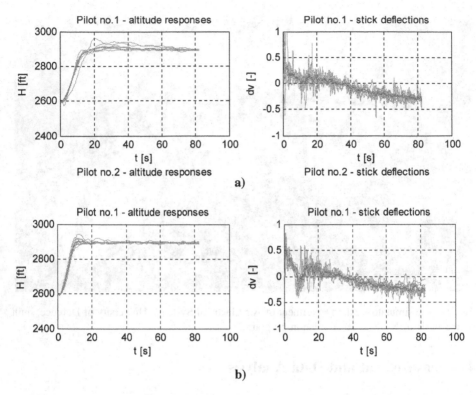

Fig. 2. Simulator measured data – pilot no. 1: (a) measured in spring 2015, (b) fall 2015. (Color figure online)

- fall 2015

$$F_{1B}(s) = 8.02.10^{-4} \frac{(2.21s + 1)}{(0.13s + 1)(0.80s + 1)} \cdot \exp(-0.62s)$$

Best fit = 45.62%

The approximation of the original responses (measured data) with use of the responses of identified models $F_{1A}(s)$ and $F_{1B}(s)$ is shown in Fig. 3.

Apparently the results indicate that practically all parameters of the model were changed. Both lead time constant T_3 and lag time constant T_2, being in relation with pilot's experience and length of flying practice, increased. On the contrary, the value of transport delay τ dropped, which means that average pilot reaction is faster by almost 0.1 s. Neuromuscular lag time constant T_1 remained unchanged. This constant represents dynamics of spreading impulses throughout human neuromuscular system. Its numeric value, i.e. approximately 0.13 s, also conforms to the theory.

The testing measurements and data analysis were carried out similarly with other pilots. The below-stated table summarizes the results achieved from the parameter

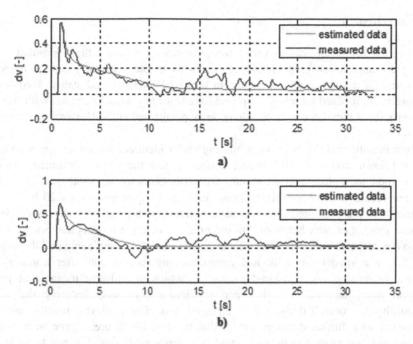

Fig. 3. The result achieved from parameter identification of pilot behavior model (pilot no. 1): (a) measured in 2015, (b) fall 2015.

identification of pilot behavior model [12, 13]. Letter A stands for the measurement taking place in spring 2015, whereas B for the measurement in fall 2015.

The results as shown in Table 1 clearly indicate that the values of neuromuscular lag time constant T_1 remain in repeated measurements practically unchanged, which confirms the premise that their value is independent of the training intensity and is characteristic for all pilots. All pilots, except for pilot no. 3, display decreased value in response time (transport delay) τ, which means that pilots are able to response to a changed situation faster. The ratio between regulating constants T_2 and T_3 determines adaptability to controlled dynamics. Changes were mostly recorded even in this case.

Table 1. Parameters of the identified transfer functions for pilots 1–6.

	A $K.10^{-4}$ [–]	B $K.10^{-4}$ [–]	A T_1 [s]	B T_1 [s]	A T_2 [s]	B T_2 [s]	A T_3 [s]	B T_3 [s]	A τ [s]	B τ [s]
Pilot no. 1	6.84	8.02	0.13	0.13	0.46	0.80	1.62	2.21	0.70	0.62
Pilot no. 2	6.33	7.39	0.13	0.14	1.15	1.51	3.16	5.22	0.78	0.73
Pilot no. 3	8.61	5.41	0.07	0.07	0.98	1.11	2.81	2.32	0.68	0.87
Pilot no. 4	5.67	6.99	0.17	0.17	1.44	1.29	2.69	2.50	0.85	0.84
Pilot no. 5	7.43	7.44	0.12	0.11	1.14	1.12	2.44	3.41	0.76	0.72
Pilot no. 6	8.06	7.39	0.08	0.07	1.25	1.49	3.41	2.96	0.82	0.72

5 Conclusion

The aim of this paper was to verify the potentiality of using flight simulators and subsequent application of mathematical methods in order to carry out an objective assessment of pilot training experience in terms of intensity and status. Two sets of measurements focused on pilots' response (reaction) to a visual stimulus with use of a stationary flight simulator were simultaneously performed at the University of Defence in Brno.

Both measurements were always running under identical conditions, i.e. with pilots who had flown around 60–80 h in real airplane before the first measurement. Another measurement took place after six months when pilots completed another training phase and their recorded number of flying hours increased approximately by 20 h.

Based on the measured data the authors made mathematical analyses needed to generate models of pilot behavior. On the grounds of these models authors were then able to create the so-called average pilot behavior model. Having compared the average pilot behavior model for individual measurements (before and after training), the authors, in most cases, registered a change, which specifically means that pilots' transport delay and the ratio between regulating constants, featuring the pilots' adaptability to controlled dynamics, changed too. The achieved results outline a potentiality of a further development in this field in the future. Hence more sets of measurement are needed to be performed in a foreseeable time horizon to verify the applied methodology.

Acknowledgments. The paper was written within project Technology Agency of the Czech Republic n.TA04031376 research/development methodology training aviation specialists L410UVP E20 and grant No. FEKT-S-14-2429 – "The research of new control methods, measurement procedures and intelligent instruments in automation", and the related financial assistance was provided from the internal science fund of Brno University of Technology.

References

1. Boril, J., Leuchter, J., Smrz, V., Blash, E.: Aviation simulation training in the Czech AirForce. In: 34th Digital Avionics Systems Conference, pp. 9A2-1–9A2-13. ALR International, Florida, Orlando (2015)
2. Allerton, D.: The impact of flight simulation in aerospace. Aeronautical J. **114**(1162), 747–756 (2010)
3. Foyle, D.C., Hooey, B.L.: Human Performance Modeling in Aviation. CRC Press, Boca Raton (2008)
4. McRuer, D.T., Krendel, E.S.: Mathematical Models of Human Pilot Behavior. System Technology, INC. AGARD AG 188, Paper No. 146, Hawthorne California (1974)
5. Lone, M.M., Cooke, A.K.: Development of a pilot model suitable for the simulation of large aircraft. In: 27th International Congress of Aeronautical Sciences, Paper ICAS 2010-6.7.2 (2010)
6. Hess, R.A., Marchesi, F.: Analytical assessment of flight simulator fidelity using pilot models. J. Guidance Control Dyn. **32**, 760–770 (2009)

7. Havlikova, M.: Diagnostic of systems with a human operator, Doctoral thesis (in Czech). Brno University of Technology (2008)
8. Glodek, M., Honold, F., Geier, T., Krell, G., Nothdurft, F., Reuter, S., et al.: Fusion paradigms in cognitive technical systems for human-computer interaction. Neurocomputing **161**, 17–37 (2015)
9. Lagarias, J.C., Reeds, J.A., Wright, H., Wright, P.E.: Convergence properties of the Nelder-Mead simplex method in low dimensions. SIAM J. Optim. **9**(1), 112–117 (1998)
10. Ljung, L.: Experiments with identification of continuous time models. In: 15th IFAC Symposium on System Identification, pp. 1175–1180 (2009)
11. Allerton, D.: Principles of Flight Simulation. Wiley, Chichester (2009)
12. Boril, J., Jalovecky, R.: Mathematical analysis of human factors using experimental parameter identification of human behaviour model. Eng. Intell. Syst. **21**(2), 1–11 (2013)
13. Boril, J., Jalovecky, R.: Experimental identification of pilot response using measured data from a flight simulator. In: Iliadis, L., Maglogiannis, I., Papadopoulos, H. (eds.) Artificial Intelligence Applications and Innovations. IFIP AICT, vol. 381, pp. 126–135. Springer, Heidelberg (2012)

Machine Learning-Learning (MALL)

Active Learning Algorithms for Multi-label Data

Everton Alvares Cherman[1]([⊠]), Grigorios Tsoumakas[2],
and Maria-Carolina Monard[1]

[1] Institute of Mathematics and Computer Sciences,
University of Sao Paulo, Sao Carlos, SP, Brazil
{echerman,mcmonard}@icmc.usp.br
[2] Department of Informatics,
Aristotle University of Thessaloniki, 54124 Thessaloniki, Greece
greg@csd.auth.gr

Abstract. Active learning is an iterative supervised learning task where
learning algorithms can actively query an oracle, i.e. a human annotator
that understands the nature of the pro blem, for labels. As the learner
is allowed to interactively choose the data from which it learns, it is
expected that the learner will perform better with less training. The
active learning approach is appropriate to machine learning applications
where training labels are costly to obtain but unlabeled data is abun-
dant. Although active learning has been widely considered for single-label
learning, this is not the case for multi-label learning, where objects can
have more than one class labels and a multi-label learner is trained to
assign multiple labels simultaneously to an object. We discuss the key
issues that need to be considered in pool-based multi-label active learning and discuss how existing solutions in the literature deal with each of
these issues. We further empirically study the performance of the existing solutions, after implementing them in a common framework, on two
multi-label datasets with different characteristics and under two different
applications settings (transductive, inductive). We find out interesting
results that we attribute to the properties of, mainly, the data sets, and,
secondarily, the application settings.

Keywords: Supervised learning · Multi-label learning · Active learning · Pool-based strategies

1 Introduction

Different approaches to enhance supervised learning have been proposed over
the years. As supervised learning algorithms build classifiers based on labeled
training examples, several of these approaches aim to reduce the amount of time
and effort needed to obtain labeled data for training. Active learning is one of
these approaches [6]. The key idea of active learning is to minimize labeling
costs by allowing the learner to query for the labels of the most informative
unlabeled data instances. These queries are posed to an oracle, e.g. a human

© IFIP International Federation for Information Processing 2016
Published by Springer International Publishing Switzerland 2016. All Rights Reserved
L. Iliadis and I. Maglogiannis (Eds.): AIAI 2016, IFIP AICT 475, pp. 267–279, 2016.
DOI: 10.1007/978-3-319-44944-9_23

annotator, which understands the nature of the problem. This way, an active learner can substantially reduce the number of labeled data required to construct the classifier.

Active learning has been developed substantially to support single-label learning, where each object (instance) in the dataset is associated with only one class label. However, this is not the case in multi-label learning, where each object is associated with a subset of labels. Due to the large number of real-world problems which fall into this category, and the interesting challenges that it poses, multi-label learning has attracted great interest in the last decade [9].

We here focus on the pool-based active learning scenario [6], where a pool of unlabeled data is available to the learning algorithm. The first contribution of this paper is the presentation of the key issues that have to be considered when applying active learning on (multi-label) data, as well as the particular decisions of existing algorithms in the literature with respect to these issues (Sect. 2). We implemented existing algorithms in a common framework within the Mulan library [8] and empirically investigated their performance on two multi-label data sets with different properties and under two different application settings (transductive, inductive). The second contribution of this paper is the presentation of these experimental results, where novel and interesting conclusions are drawn with respect to the factors that affect the performance of the different algorithms (Sect. 3).

2 Active Learning from Multi-label Data

There are a number of issues that need to be considered when attempting to apply active learning on multi-label data. In the following sections we focus on the most important ones.

2.1 Manual Annotation Approaches and Effort

Similarly to a single-label active learning system, a multi-label active learning system can request the annotation of one or more objects. If the request is for just one object, then the annotator will observe (look at, read, hear, watch) the object in an attempt to understand it and characterize it as relevant or not to each of the labels. In practice, requests are made for a batch of objects. For example, ground truth acquisition for the ImageCLEF 2011 photo annotation and concept-based retrieval tasks was achieved via crowd-sourcing in batches of 10 and 24 images [4]. In such cases, there are two ways that an annotator can accomplish the task:

1. *object-wise*, where for each object the annotator determines the relevancy to each label; and
2. *label-wise*, where for each label the annotator determines relevancy to each object[1].

---

[1] Object-wise and label-wise annotation have been called global and local labeling respectively in [2].

Consider a request for the annotation of n objects with q labels. Let c_o be the average cost of understanding an object, c_l be the average cost of understanding a label and c_{lo} be the average cost of deciding whether an object should be annotated with a particular label or not. If we set aside the cognitive and psychological aspects of the annotation process, such as our short-term memory capacity, then a rough estimation of the total cost of object-wise annotation is:

$$n[c_o + q(c_l + c_{lo})] = nc_o + nqc_l + nqc_{lo}$$

Similarly, a rough estimation of the total cost of label-wise annotation is:

$$q[c_l + n(c_o + c_{lo})] = qc_l + nqc_o + nqc_{lo}$$

Assuming that the cost of label-wise annotation is smaller than that of object-wise annotation, we have:

$$qc_l + nqc_o + nqc_{lo} < nc_o + nqc_l + nqc_{lo}$$
$$qc_l + nqc_o < nc_o + nqc_l$$
$$n(q - 1)c_o < q(n - 1)c_l$$
$$c_o < \frac{q(n - 1)}{n(q - 1)}c_l \approx \frac{qn}{nq}c_l = c_l$$

This means that choosing the annotation approach, largely depends on the object and label understanding costs. If object (label) understanding is larger, then the object (label) wise approach should be followed.

As Fig. 1 illustrates, object understanding is less costly than label understanding only for images, which humans understand in milliseconds. Documents, audio and video require far more time to understand than typical label concepts.

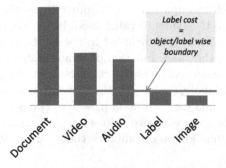

Fig. 1. The cost of understanding a label in different types of data.

2.2 Full and Partial Annotation Requests

In a classical supervised learning task, the active learning system requests the value of the target variable for one or more objects. What can the learning system request in multi-label learning?

Normally it should request the values of all binary target variables (labels) for one or more objects. Then a (batch) incremental multi-label learning algorithm can update the current model based on the new examples. A different approach is taken in [5], where the system requests the values for only a subset of the labels and subsequently infers the values of the remaining labels based on label correlations.

Sticking to the values of just a subset of the labels would require an algorithm that is incremental in terms of partial training examples. Binary relevance (BR) is perhaps the sole algorithm fulfilling this requirement, but it is a standard and often strong baseline. Therefore, the development of active learning strategies that request partial labeling of objects could be a worthwhile endeavor. However, there is an implication on annotation effort that has to be considered. If the system requests the labeling of the same object at two different annotation requests, then the cost of understanding this object would be incurred twice. As discussed in Sect. 2.1, this is inefficient for most data types.

2.3 Evaluation of Unlabelled Instances

The key aspect in a single-label active learning algorithm is the way it evaluates the informativeness of unlabelled instances. In multi-label data, the evaluation function (*query*) of active learning algorithms comprises two important parts:

1. a *scoring* function to evaluate object-label pairs; and
2. an *aggregating* function to aggregate these scores.

Algorithm 1 shows the general procedure for a batch-size $= t$, *i.e.*, t examples are annotated in each round. The evaluation function *query* calculates the evidence value of each example $E_i \subset D_u$ and returns the t most informative instances, according to the evidence value used. In each round, these t examples will be labeled by the oracle and included in the set D_l of labeled examples.

Algorithm 2 shows the *query* function of a multi-label active learning procedure. The *scoring* function considers object-label pairs (E_i, y_j) and evaluates the participation $(e_{i,j})$ of label y_j in object E_i. It returns an evidence value $e_{i,j}$ for all instances $E_i \subset D_u$ and for each label $y_j \in L = \{y_1, y_2, ..., y_q\}$. The *aggregating* function considers the q evidence values $e_{i,1}, e_{i,2}, ..., e_{i,q}$ of each instance E_i given by the *scoring* function, and combines these values into a unique evidence value e_i.

The following three families of measures have been proposed in the related work for evaluating object-label pairs (*scoring*):

1. Confidence-based score [1,2,7]. The distance of the confidence of the prediction from the *average* value is used. The nature of this value depends

input : D_l: labeled pool; D_u: unlabeled pool; E_i: multi-label example;
  L: set of labels; Y_i: subset of labels associated to E_i; t: batch size;
  R: number of rounds; F: multi-label learner; $Oracle$: the annotator;
for $r = 1, 2, .., R$ **do**
  $H \leftarrow F(D_l)$
  $\{E_i\}_{i=1}^t \leftarrow query(H, L, D_u, t)$
  $\{Y_i\}_{i=1}^t \leftarrow Oracle(\{E_i\}_{i=1}^t)$
  $D_l \leftarrow D_l \cup \{(E_i, Y_i)\}_{i=1}^t$
  $D_u \leftarrow D_u - \{E_i\}_{i=1}^t$
end

Algorithm 1. Multi-label active learning procedure for the object-wise annotation approach.

input : D_u: unlabeled pool; L: set of labels; H: multi-label classifier
output: The t instances with higher evidences

for $E_i \in D_u$ **do**
  for $y_j \in L$ **do**
    $e_{i,j} \leftarrow scoring(D_u, H, E_i, y_j)$
  end
  $e_i \leftarrow aggregating(e_{i,1}, e_{i,2}, ..., e_{i,q})$
end
$query \leftarrow best(e_1, e_2, ..., t, D_u)$

Algorithm 2. The *query* function

on the bias of learner. It could be a margin-based value (distance from the hyper-plane), a probability-based value (distance from 0.5) or other. The value returned by this approach represents how far an example is from the boundary decision threshold between positive and negatives examples. We are interested in examples that minimize this score.

2. Ranking-based score [7]. This strategy works like a normalization approach for the values obtained from the confidence-based strategy. The confidences given by the classifier are used to rank the unlabeled examples for each label. We are interested in examples that maximize this score.

3. Disagreement-based score [3,10]. Unlike the other approaches, this strategy uses two base classifiers and measures the difference between their predictions. We are interested in maximizing this score. The intuitive idea is to query the examples that most disagree in their classifications and could be most informative. Three ways to combine the confidence values output by the two base classifiers have been proposed:

    i. MMR uses a major classifier which outputs confidence values and an auxiliary classifier that outputs decisions (positive/negative). The auxiliary classifier is used to determine how conflicting the predictions are.

    ii. HLR considers a more strict disagreement using the decisions output by both classifiers to decide if there is disagreement or agreement between each label prediction of an example.

iii. SHLR tries to make a balance between MMR and HLR through a function that defines the influence of each approach in the final score.

After having obtained the object-label scores, there are two main aggregation strategies for combining the object-label scores to an overall object score:

1. AVG averages the object-label scores across all labels. Thus, given the q object-label scores $e_{i,j}$ of object E_i, the overall object-label score of object E_i is given by:

$$e_i = aggregating_{avg}(\{e_{i,j}\}_{j=1}^q) = \frac{\sum_{j=1}^q e_{i,j}}{q}$$

2. MIN/MAX, on the other hand, considers the optimal (minimum or maximum) of the object-label scores, given by:

$$e_i = aggregating_{min/max}(\{e_{i,j}\}_{j=1}^q) = min/max(\{e_{i,j}\}_{j=1}^q)$$

Note that for HLR, only the average aggregation strategy makes sense, as taking the maximum would lead to a value of 1 for almost all unlabeled instances and would not help in discriminating among them.

2.4   Experimental Protocol

Besides the multi-label active learning strategies themselves, the way that they are evaluated is another important issue to consider. Some aspects to be considered are the size of the initial labeled pool, the batch's size, the set of examples used as testing, the sampling strategy and also the evaluation approach. Next, these aspects are described with reference to related work.

Regarding the initial labeled pool, different papers built it in different ways. In [7], the examples are chosen to have at least one example positive and one negative for each label. In [10], 100 to 500 examples were selected randomly to compose the initial labeled pool. In [2], the first 100 chronologically examples were selected. In [1], the author choose randomly 10 examples to compose the initial labeled pool.

The batch size defines how many examples are queried in each round of active learning. In [1,7], only one example was queried per round. In [2] 50 examples were chosen in each round, while in [10] experiments with both 50 and 20 examples were performed.

There are basically two different ways to define the test set. The first one is to consider a totally separated test set. This was followed in [2] and though not explicitly mentioned, it seems to have also been followed in [1]. The second way is to use the remaining examples in the unlabeled pool for testing. This approach was used in [7,10].

It is worth noting that the quality of the model assessed using this second approach holds for examples in the unlabeled pool, and does not necessarily hold for new unlabeled data. Although there is a lack of discussion about this topic in the active learning literature, the decision of which evaluation approach to use depends on the application's nature. Most learning applications are interested in building a general model from a training set of examples to predict future new examples, e.g., this kind of application uses inductive inference algorithms to make its predictions. An experimental protocol using a separate test set is the correct evaluation approach for the performance assessment for the inductive inference setting. The remaining evaluation approach is biased by the active learner and hence the evaluation on these remaining examples will not be representative of the actual distribution of new unseen examples, which is the case for inductive inference.

However, there are active learning applications that want to predict labels of an *a priori* known specific set of examples. For example, in a real world personal image annotation scenario, the user would like to annotate some images of his/her collection and after few rounds of active learning, the system would annotate the remaining image in the collection [7]. For such an application, the learning assessment should use the remaining examples in the query pool.

The learning curve is the most common evaluation approach used to assess active learning techniques. A learning curve plots the evaluation measure considered as a function of the number of new instance queries that are labeled and added to D_l. Thus, given the learning curves of two active learning algorithms, the algorithm which dominates the other for more or all the points along the learning curve is better than the other. Besides the learning curve, [2,7,10] also used the value of the evaluation measure in the end of some specific number of rounds to assess the active learning techniques.

3 Experiments

The active learning algorithms described in Sect. 2.3, as well as the active learning evaluation framework, were implemented using Mulan[2] [8], a Java package for multi-label learning based on Weka[3]. Our implementation is publicly available to the community at http://www.labic.icmc.usp.br/pub/mcmonard/Implementations/Multilabel/active-learning.zip.

3.1 Setup

The experiments were performed using the datasets *Scene* and *Yeast*, two classic multi-label datasets, which can be found in the Mulan website[4]. *Scene* dataset addresses the problem of semantic image categorization. Each instance in this dataset is an image associated with some of the six available semantic classes (beach, sunset, fall foliage, field, mountain, and urban). *Yeast* is a biological dataset for gene function classification. Each instance is a yeast gene described

---

[2] http://mulan.sourceforge.net.
[3] http://www.cs.waikato.ac.nz/ml/weka.
[4] http://mulan.sourceforge.net/datasets.html.

by the concatenation of micro-array expression data and phylogenetic profile associated with one or more different functional classes.

Table 1 describes the datasets, where CL (cardinality) and DL (density) are defined as $CL(D) = \frac{1}{|D|} \sum_{i=1}^{|D|} |Y_i|$ and $DL(D) = \frac{1}{|D|} \sum_{i=1}^{|D|} \frac{|Y_i|}{q}$, respectively.

Table 1. Datasets description and label frequency statistics

Dataset	Domain	#ex	#feat	q	CL	DL	#dist	Min	1Q	Med	3Q	Max
Scene	Image	2407	294	6	1.074	0.179	15	364	404	429	432	533
Yeast	Biology	2417	103	14	4.237	0.303	198	34	324	659	953	1816

These two datasets have different properties. Although both datasets have similar number of examples, *Scene* dataset has low number of labels (6), few different multi-labels (15) and low cardinality (1.074). On the other hand, *Yeast* dataset has 14 labels, 198 different multi-labels, and a reasonably high cardinality (4.237). This means that instances in the *Yeast* dataset have more complex label space than the instances in the *Scene* dataset. Thus, learning from the *Yeast* dataset would be more difficult than learning from the *Scene* dataset.

Information related to label frequency is also important to characterize multi-label datasets. To this end, Table 1 also shows summary statistics related to labels frequency, where (Min) Minimum, (1Q) 1st Quartile, (Med) Median, (3Q) 3rd Quartile and (Max) Maximum. Recall that 1Q, Med and 3Q divide the sorted labels frequency into four equal parts, each one with 25 % of the data. Note that *Yeast* dataset in unbalanced.

Figure 2 shows a graphic distribution of the datasets label frequency using the Violin plot representation, which adds the information available from local density estimates to the basic summary statistics inherent in box plots. Note that the Violin plot may be viewed as boxplots whose boxes have been curved to reflect the estimated distribution of values over the observed data range. Moreover, observe that the boxplot is the black box in the middle, the white dot is the median and the black vertical lines are the whiskers, which indicate variability outside the upper and lower quartiles.

As mentioned in Sect. 2.3, the active learning algorithms implemented in this work are combinations of functions to evaluate object-label pairs and to aggregate these scores. The functions to evaluate the object-label pairs, *i.e.*, the *scoring* function, are: Confidence-based (CONF), Ranking-based (RANK), HLR Disagreement-based (HLR), MMR Disagreement-based (MMR), SHLR Disagreement-based (SHLR). The functions to aggregate the outputted scores, *i.e.*, the *aggregating* function, are: average (AVG) and maximum or minimum (MAX/MIN), depending on the score function.

In this work, the initial labeled pool of examples was built by randomly choosing examples until having $N_{ini} \times q$ positive single labels, *i.e.*, until $N_{ini} \times q \geq \sum_{i=1}^{|D_i|} Y_i$, where N_{ini} is user-defined. This strategy allows for fairer comparison

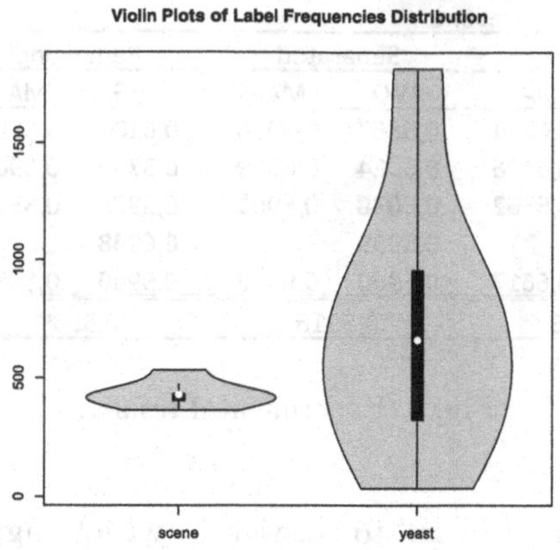

Fig. 2. Violin plots of label frequencies distribution.

across the datasets. $N_{ini} = 5, 10, 20$ was used in order to evaluate the influence of different sizes of the initial labeled pool. The general procedure — Algorithm 1 — was executed with a batch size $t = 1$, *i.e.*, one example is annotated in each run. The Binary Relevance approach was used as the multi-label classifier, using stochastic gradient descent with hinge loss as the base classifier. For the disagreement-based approaches, we used the sequential minimal optimization algorithm with a linear kernel. Both learners, are implemented in the Weka framework, and are named SGD and SMO respectively.

3.2 Results and Discussion

We report results in terms of the micro F_1 measure, and in particular its average over 1500 iterations of active selection of one example in each iteration. This is proportional to the area under the corresponding learning curve of the different algorithms. Figure 3 presents the results. Bold typeface is used to highlight the relative best performance of the different scoring functions and aggregation strategies for each particular experimental setting (dataset and protocol pair). All results were obtained using 10-folds cross-validation. The full experimental results are available online as supplementary material[5].

The first question we want to answer is **how does the size of the initial pool of training examples affect the performance of the methods?** Here we noticed the same strange general pattern across both data sets and application settings and across all algorithms: Having 5 and 20 examples per label leads to similar performance, which is slightly better compared to having 10 examples per label. In the rest of the experiments we removed this factor by considering the average results of the three different sizes of the initial pool.

---

[5] http://www.labic.icmc.usp.br/pub/mcmonard/ExperimentalResults/AIAI2016-AL LRESULTS.xls.

	Scene				Yeast			
	Remaining		Separated		Remaining		Separated	
	AVG	MAX	AVG	MAX	AVG	MAX	AVG	MAX
Confidence	0,6478	0,6274	0,5867	0,5918	0,6100	0,5556	0,5557	0,5415
Rank	0,6377	0,6376	0,6024	0,6029	0,5790	0,5907	0,5670	0,5689
MMR	**0,6979**	0,6562	**0,6038**	0,5961	0,5878	0,5586	0,5622	0,5534
HLR	0,6885		0,5999		**0,6248**		**0,5705**	
SHLR	0,6815	0,6617	0,5960	0,5970	0,5900	0,5571	0,5609	0,5583
Random	0,5928		0,5916		0,5589		0,5602	

Fig. 3. Experimental results.

The next question we want to answer is **which aggregation strategy works best for each scoring function and under what conditions?** For the confidence-based score function, in the separated protocol *min* is the best aggregation strategy in both yeast and scene, while for the remaining protocol, *avg* works best in both yeast and scene. Taking the *min* of the confidence-based score stresses more the labels for which the classifier is most uncertain (e.g. rare labels that it has not seen yet), while *avg* treats all labels equally. We hypothesize that in the remaining protocol instances with rare (difficult to be predict) labels are removed from the test set and hence stressing the performance in such labels is meaningless. In contrast, in the separate protocol, rare labels in the test set remain rare and important. Figure 4 shows the learning curves of *min* and *avg* in scene for the remaining protocol. It confirms our hypothesis, as in the initial steps, *avg* does not perform as well as *min*, but as more and more rare labels are being removed from the test set, it eventually does better. This is an important conclusion for researchers developing methods for a particular protocol, or practitioners applying methods in a particular protocol setting.

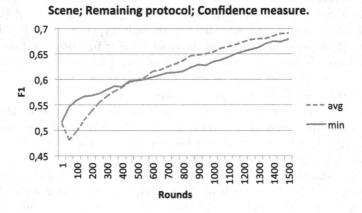

Fig. 4. Average vs minimum in scene for the confidence-based score function.

For the rank-based score function, in the scene data set, *max* and *avg* work equally well for both the remaining/separate protocols, while in the yeast database *max* gives slightly better results. The rank-based score function normalizes the absolute values of uncertainty across the labels and hence makes itself all labels equal in this sense. This alleviates the issue we discussed above. We hypothesize that the aforementioned difference between yeast and scene is due to the corresponding differences in label frequencies, as shown in Fig. 2. The *max* aggregation pays more attention to labels where the *relative* uncertainty with respect to other labels is higher, and this pays-off better, in accordance with the theory of active learning. In scene, as there are fewer labels with similar distributions, it makes no difference in focusing on all labels or only on the most uncertain one.

In terms of the disagreement-based score functions, in MMR *avg* works better than *max* for both data sets and protocols, while for SHLR, *avg* works better than *max* in the remaining setting, while they perform similarly in the separate setting. Here, we would expect similar results with confidence-based scoring and indeed we see that *avg* does better than *max* in the remaining setting. However, in contrast with confidence-based scoring, here *avg* dominates also in the separate protocol. It seems that while uncertainty is maximized in the case of rare labels, the same does not happen for the disagreement between the two classifiers. We hypothesize that this occurs because with limited training data for rare labels both classifiers' output is similarly uncertain. We also argue that the disagreement of classifiers per label, again in itself, brings all labels to the same measurement level (normalization). This also explains the good results of the average strategy.

The next question we want to answer is **which scoring function works best and under what conditions?** Comparing the different scoring functions with each other, we notice that the disagreement-based functions do best overall, with MMR giving the best results in scene and HLR the best results in yeast for both protocols. HLR is the most robust method, delivering near-top results also in scene. Recall that HLR takes into account crisp decisions instead of confidences. This shows that looking at actual confidence values can be misleading, particularly in the presence of rare labels and imbalanced distributions across the labels. In scene, where labels are similar in frequency, MMR did best, hence in these - rare in practice - cases, we expect actual confidences to offer benefits.

Further interesting results are obtained by **comparing random selection of unlabeled instances (passive learning) with the active learning approaches.** In particular, we notice that large gains are achieved in the transductive setting, while active learning methods are struggling to beat passive learning in the separated setting. This shows that in the remaining setting, the benefits of active learning are coming mostly from the removal of difficult instances from the test set rather than from the incorporation of useful instances to the training set, an interesting conclusion for active learning in general (non multi-label) that to the best of our knowledge has not been previously discussed in the literature.

4   Summary and Future Work

Although active learning in single-label learning has been investigated over several decades, this is not the case for multi-label learning. This work discussed key issues in pool-based (multi-label) active learning based on existing algorithms in the literature, which were implemented in a common framework and experimentally evaluated in two multi-label data sets with different properties and under two different application settings (transductive, inductive).

Results show that taking the average across all labels of disagreement-based scoring functions perform best, and that in particular the MMR function works better in the absence of imbalance among the labels, while HLR works better in the presence of such imbalance. Moreover, the transductive setting was found to be easier for active learning due to the removal of difficult examples.

In the future, we plan to expand our empirical study with more data sets, in order to assess the generality of our conclusions.

Acknowledgment. This research was supported by the São Paulo Research Foundation (FAPESP), grants 2010/15992-0 and 2011/21723-5, and Brazilian National Council for Scientific and Technological Development (CNPq), grant 644963.

References

1. Brinker, K.: On active learning in multi-label classification. In: Spiliopoulou, M., Kruse, R., Borgelt, C., Nurnberger, A., Gaul, W. (eds.) From Data and Information Analysis to Knowledge Engineering. Studies in Classification, Data Analysis, and Knowledge Organization, pp. 206–213. Springer, Heidelberg (2006)
2. Esuli, A., Sebastiani, F.: Active learning strategies for multi-label text classification. In: Boughanem, M., Berrut, C., Mothe, J., Soule-Dupuy, C. (eds.) ECIR 2009. LNCS, vol. 5478, pp. 102–113. Springer, Heidelberg (2009)
3. Hung, C.W., Lin, H.T.: Multi-label active learning with auxiliary learner. In: 3rd Asian Conference on Machine Learning, Taoyuan, Taiwan (2011)
4. Nowak, S., Nagel, K., Liebetrau, J.: The CLEF 2011 photo annotation and concept-based retrieval tasks. In: CLEF (Notebook Papers/Labs/Workshop), pp. 1–25 (2011)
5. Qi, G.J., Hua, X.S., Rui, Y., Tang, J., Zhang, H.J.: Two-dimensional multilabel active learning with an efficient online adaptation model for image classification. IEEE Trans. Pattern Anal. Mach. Intell. **31**(10), 1880–1897 (2009). http://dx.doi.org/10.1109/TPAMI.2008.218
6. Settles, B.: Active learning literature survey. Technical report 1648. University of Wisconsin-Madison (2010)
7. Singh, M., Brew, A., Greene, D., Cunningham, P.: Score normalization and aggregation for active learning in multi-label classification. Technical report. University College Dublin (2010)
8. Tsoumakas, G., Spyromitros-Xioufis, E., Vilcek, J., Vlahavas, I.: Mulan: a java library for multi-label learning. J. Mach. Learn. Res. **12**, 2411–2414 (2011)
9. Tsoumakas, G., Zhang, M.L., Zhou, Z.H.: Introduction to the special issue on learning from multi-label data. Mach. Learn. **88**(1–2), 1–4 (2012)

10. Yang, B., Sun, J.T., Wang, T., Chen, Z.: Effective multi-label active learning for text classification. In: Proceedings of the 15th ACM SIGKDD International Conference on Knowledge Discovery and Data Mining, KDD 2009, NY, USA, pp. 917–926 (2009) http://doi.acm.org/10.1145/1557019.1557119

Automated Determination of the Input Parameter of DBSCAN Based on Outlier Detection

Zohreh Akbari[✉] and Rainer Unland

Institute for Computer Science and Business Information Systems (ICB),
University of Duisburg-Essen, Essen, Germany
{zohreh.akbari,rainer.unland}@icb.uni-due.de

Abstract. During the last two decades, DBSCAN (Density-Based Spatial Clustering of Applications with Noise) has been one of the most common clustering algorithms, that is also highly cited in the scientific literature. However, despite its strengths, DBSCAN has a shortcoming in parameter detection, which is done in interaction with the user, presenting some graphical representation of the data. This paper introduces a simple and effective method for automatically determining the input parameter of DBSCAN. The idea is based on a statistical technique for outlier detection, namely the empirical rule. This work also suggests a more accurate method for detecting the clusters that lie close to each other. Experimental results in comparison with the old method, together with the time complexity of the algorithm, which is the same as for the old algorithm, indicate that the proposed method is able to automatically determine the input parameter of DBSCAN quite reliably and efficiently.

Keywords: Clustering · DBSCAN · Empirical rule · Machine learning · Outlier detection · Parameter determination · Unsupervised learning

1 Introduction

Machine Learning (ML) is one of the core fields of Artificial Intelligence (AI) and is concerned with the question of how to construct computer programs that automatically improve with experience [1]. Depending on the nature of the learning data available to the learning system, machine learning methods are typically classified into three main categories [2, 3]: supervised, unsupervised and reinforcement learning. In supervised learning example inputs and their desired outputs are given and the goal is to learn a general rule that maps these inputs to their desired outputs. In unsupervisaed learning, on the other hand, no labels are given to the learning algorithm, leaving it on its own to find the hidden structure of the data, e.g. to look for the similarities between the data instances (i.e. clustering [4]), or to discover the dependencies between the variables in large databases (i.e. association rule mining [5]). In reinforcement learning the desired input/output pairs are again not presented, however, the algorithm is able to estimate the optimal actions by interacting with a dynamic environment and based on the outcomes

© IFIP International Federation for Information Processing 2016
Published by Springer International Publishing Switzerland 2016. All Rights Reserved
L. Iliadis and I. Maglogiannis (Eds.): AIAI 2016, IFIP AICT 475, pp. 280–291, 2016.
DOI: 10.1007/978-3-319-44944-9_24

of the more recent actions, while ignoring experiences from the past, that were not reinforced recently.

This research focuses on the most common unsupervised learning method (i.e. cluster analysis [4, 6]), and more specifically on one of its successful algorithms the Density-Based Spatial Clustering of Applications with Noise (DBSCAN) [7]. As mentioned above, in unsupervised learning, learner processes the input data with the goal of coming up with some summary or compressed version of the data [4]. Clustering a dataset is a typical example of this type of learning. Clustering is the task of grouping a set of objects such that similar objects end up in the same group and dissimilar objects are diverted into different groups. Clearly, this description is quite imprecise and possibly ambiguous. However, quite surprisingly, it is not at all clear how to come up with a more rigorous definition [4], and since no definition of cluster is widely accepted many algorithms have been developed to suit specific domains [8], each of which using a different induction principle [9].

Due to their diversity, clustering methods are classified into different categories in the scientific literature [9–12]. However, despite the slight differences between these classifications, they all mention the DBSCAN algorithm as one of the eminent methods available. DBSCAN owes its popularity to the group of capabilities it offers [7]: (1) it does not require the specification of the number of clusters in the dataset beforehand, (2) it requires little domain knowledge to determine its input parameter, (3) it can find arbitrarily shaped clusters, (4) it has good efficiency on large datasets, (5) it has a notion of noise, and is robust to outliers, (6) it is designed in a way that it can be supported efficiently by spatial access methods such as R*-trees [13], and so on.

DBSCAN algorithm requires two input parameters, namely *Eps* and *MinPts*, which are considered to be the density parameters of the thinnest cluster acceptable, specifying the lowest density which is not considered to be noise. These parameters are hence respectively the radius and the minimum number of data objects of the least dense cluster possible. The algorithm supports the user in determining the appropriate values for these parameters offering a heuristic method, which imposes the user interaction based on some graphical representation of the data (presented in Sect. 2.2). However, since DBSCAN is sensitive to its input parameters and the parameters have significant influences on the clustering result, an automated and more precise method for the determination of the input parameters is needed.

Some notable algorithms targeting this problem are: (1) GRPDBSCAN, which combines the grid partition technique and DBSCAN algorithm [14], (2) DBSCAN-GM, that combines Gaussian-Means and DBSCAN algorithms [15], and (3) BDE-DBSCAN, which combines Differential Evolution and DBSCAN algorithms [16]. Opposed to these methods, which all intend to solve the problem using some other techniques, this paper remains with the original idea of the DBSCAN algorithm and just tries to omit the user interaction needed, allowing the algorithm to detect the appropriate value itself. This is done using some basic statistical techniques for outlier detection. Two different approaches are mentioned in this paper, which apply the concept of standard deviation to the problem of outlier detection, namely the empirical rule for normal distributions and the Chebyshev's inequality for non-normal distributions [17, 18]. This work, however, focuses mainly on the application of the empirical rule to outlier detection in normal distributed data, and addresses the Chebyshev's inequality only as a possible solution for non-normal distributions.

The rest of the paper is organized as follows. Section 2 describes the DBSCAN algorithm and its supporting technique for the determination of its input parameters. In Sect. 3, the above mentioned statistical techniques for outlier detection are presented (i.e. the empirical rule and the Chebyshev's inequality). Section 4 describes the automated technique for the determination of the parameter *Eps*. Experimental results and the time complexity of the automated technique are then discussed in Sect. 5. Section 6 concludes with a summary and some directions for the feature researches.

2   DBSCAN: Density-Based Spatial Clustering of Applications with Noise

According to [7], the key idea of DBSCAN algorithm is that for each point of the cluster the neighborhood of a given radius has to contain at least a minimum number of points, i.e. the density in the neighborhood has to exceed some threshold. The following definitions support the realization of this idea.

Definition 1 (*Eps − neighborhood* of a point): The *Eps − neighborhood* of a point p, denoted by $N_{Eps}(p)$, is defined by $N_{Eps}(p) = \{q \in D | dist(p, q) \leq Eps\}$.

Definition 2 (directly density-reachable): A point p is directly density-reachable from a point q, w.r.t. *Eps* and *MinPts*, if

1. $p \in N_{Eps}(q)$ and

2. $\left| N_{Eps}(q) \geq MinPts \right|$

The second condition is called core point condition (There are two kinds of points in a cluster, points inside of the cluster, called core points, and points on the border of the cluster, called border points).

Definition 3 (density-reachable): A point p is density-reachable from a point q, w.r.t. *Eps* and *MinPts*, if there is a chain of points $p_1, \ldots, p_n, p_1 = q, p_n = p$ such that p_{i+1} is directly density-reachable from p_i.

Definition 4 (density-connected): A point p is density-connected to a point q, w.r.t. *Eps* and *MinPts*, if there is a point o such that both, p and q are density-reachable from o, w.r.t. *Eps* and *MinPts*.

Definition 5 (cluster): Let D be a database of points. A cluster C, w.r.t. *Eps* and *MinPts*, is a non-empty subset of D satisfying the following conditions:

1. $\forall p, q$:if $p \in C$ and q is density-reachable from p, w.r.t. *Eps* and *MinPts*, then $q \in C$. (Maximality)
2. $\forall p, q \in C$: p is density-connected to q, w.r.t. *EPS* and *MinPts*. (Connectivity)

Definition 6 (noise): Let $C_1, \ldots, C_k$ be the clusters of the database D, w.r.t. parameters Eps_i and $MinPts_i, i = 1, \ldots, k$. Then the noise is defined as the set of points in the database D not belonging to any cluster C_i, i.e. $noise = \{p \in D | \forall i : p \notin C_i\}$.

The following lemmata are important for validating the correctness of the algorithm. Intuitively, they state that having the parameters Eps and $MinPts$, a cluster can be discovered in a two-step approach. First, choose an arbitrary point from the database satisfying the core point condition as a seed. Second, retrieve all points that are density-reachable from the seed, obtaining the cluster containing the seed.

Lemma 1: Let p be a point in D and $|N_{Eps}(p)| \geq MinPts$. Then the set $O = \{o | o\ D\ and\ o\ is\ density - reachable\ from\ p,\ w.r.t.\ Eps\ and\ MinPts\}$ is a cluster, w.r.t. Eps and $MinPts$.

Lemma 2: Let C be a cluster, w.r.t. Eps and $MinPts$, and let p be any point in C with $|N_{Eps}(p)| \geq MinPts$. Then C equals to the set $O = \{o | o\ is\ density - reachable\ from\ p, w.r.t.\ Eps\ and\ MinPts\}$.

2.1 The Algorithm

The DBSCAN algorithm can be described as follows (Table 1):

Table 1. Algorithm 1: Pseudo-code of the DBSCAN

DBSCAN Algorithm (Input: $D, Eps, MinPts$)
1. While (D has an unclassified[a] point)
2. Select an arbitrary unclassified point p.
3. If p does not satisfy the core point condition, mark it as a noise.
4. Else retrieve all the density-reachable points from $N_{Eps}(p)$ forming a cluster containing $N_{Eps}(p)$ and mark all the member of this cluster as classified.
5. End While

[a] Note that the term unclassified here indicates that it is not determined yet if the point is a noise or not.

2.2 Determining the Parameters *Eps* and *MinPts*

DBSCAN offers a simple but effective heuristic method to determine the parameters Eps and $MinPts$ of the thinnest cluster in the dataset. For a given k function $k - dist$ is defined from the Database D to the real numbers, mapping each point to the distance from its $k - th$ nearest neighbor. When sorting the points of the dataset in descending order of their $k - dist$ values, the graph of this function gives some hints concerning the density distribution in the dataset. This graph is called the sorted $k - dist$ graph. It is clear that the first point in the first valley of the $MinPts - dist$ graph can be the threshold point with the maximal $MinPts - dist$ value in the thinnest cluster. All points with a larger $MinPts - dist$ value are considered to be noise, and all the other points are assigned to some clusters.

DBSCAN states that according to experiments, the $k - dist$ graphs for $k > 4$ do not significantly differ from the $4 - dist$ graph and, furthermore, they need considerably more computation. Therefore, it eliminates the parameter *MinPts* by setting it to 4 for all datasets (for 2-dimensional data). The parameter determination method also explains, that since in general, it is very difficult to detect the first valley of the $k - dist$ graph automatically, but it is relatively simple for the user to see this valley in a graphical representation, it is suggested to follow an interactive approach for determining the threshold point.

3   Statistical Techniques for Outlier Detection

The term noise in DBSCAN algorithm is equivalent to an outlier in statistics, which is an observation that is far removed from the rest of the observations [19]. One of the basic statistical techniques for outlier detection is called the empirical rule. The empirical rule is an important rule of thumb, that is used to state the approximate percentage of values that lie within a given number of standard deviation from the *mean* of a set of data if the data are normally distributed. The empirical rule, also called the 68-95-99.7 rule or the three-sigma rule of thumb states that 68.27 %, 95.45 % and 99.73 % of the values in a normal distribution lie within one, two and three standard deviations of the mean [17]. One of the practical usages of the empirical rule is as a definition of outliers as the data that fall more than three standard deviations from the norm in normal distributions [20] (Fig. 1).

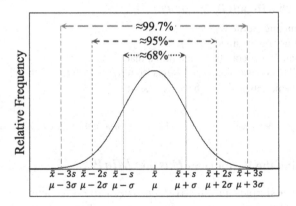

Fig. 1. The Empirical Rule [21]

If there are many points that fall more than three standard deviations from the norm, then the distribution is most likely non-normal. In this case, Chebyshev's inequality, which applies to non-normal distributions, is applicable. Chebyshev's inequality states that in any probability distribution, at least $1 - \dfrac{1}{k^2}$ of the values are within k standard deviations of the *mean* [17] (e.g. in non-normal distributions at least 99 % of the values lie within 10 standard deviations of the *mean*). Hence, using the Chebyshev's inequality, the outlier can also be defined as the data that fall outside an appropriate number of standard deviations from the mean [22][1].

4   Automated Determination of the Parameter Eps

Setting the *MinPts* to 4, determining the parameter *Eps*, the algorithm is aiming a radius that covers the majority of the 4 − *dist* values and stands well as a threshold for the specification of the noise values. As mentioned above, the term noise in DBSCAN algorithm is equivalent to an outlier in statistics, which is an observation that is far removed from the rest of the observations [19]. Thus, the idea here is to use statistical rules in order to find the threshold value between the accepted 4 − *dist* values and the values considered for the noise points.

As mentioned above, one of the practical usages of the empirical rule is as a definition of outliers as the data that fall more than three standard deviations from the norm in normal distributions [20]. Thus, considering the 4 − *dist* values, the value of parameter *Eps* can be set to their *mean* plus three standard deviations. This would cover even more than 99.73 % of the calculated 4 − *dist* values, since the 4 − *dist* values smaller than *mean* − 3 × *SD* are also covered here.

Border points and even in general, points closer to the border of the clusters usually have greater *k* − *dist* values, which lead to larger *Eps* values and thus might cause two close clusters to be detected as one cluster (Since the parameter *MinPts* or *k* is set to 4, this problem may be caused mostly by the border points). These relatively greater *k* − *dist* values, however, do not have any positive effect on the process of cluster detection, as the *k* − *dist* values of the core points are actually the ones forming the right clusters and at the same time covering the border points. Figure 2 shows a case in which the 4 − *dist* value of border point *p* is much larger than the 4 − *dist* value of the core point *q*, which can actually cover *p* in its 4 − *dist* − *neighborhood*.

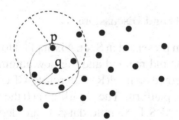

Fig. 2. 4 − *dist* values for example core (*q*) and border point (*p*)

In order to eliminate the negative effect of the *k* − *dist* values of the border points, the algorithm presented here considers any point with minimum *k* − *dist* value which

---

[1] This work focuses solely on the empirical rule and the normal distributions. However, the possibility of using the Chebyshev's inequality is given here, in order to show that the general idea of using outlier detection techniques for the reason of parameter determination in DBSCAN is not limited to the distribution of the data.

covers the border point in its $k - dist - neighborhood$ and replaces the $k - dist$ value of this border point with the $k - dist$ value of this core point. Thus for a given k, function $k - dists'$ is defined from the Dataset D to the real numbers, mapping each point to the $k - dist$ value of any core point, covering this point in its $k - dist - neighborhood$, with minimum $k - dist$ value. Actually, following this technique, points are considered in ascending order of their $4 - dist$ values, then taking each point p, if the $4 - dist'$ value for any point in its four nearest neighbors is not set so far, this value will be set to the $4 - dist$ value of point p. Using this technique for each point, the $k - dist$ value of the smallest cluster, the point can join, would be considered. At the end the *mean* and the standard deviation of these $k - dist'$ values which are saved for all points are calculated and the *Eps'* value is set to *mean* $+ 3 \times SD$. The following pseudo-code indicates this method (Table 2).

Table 2. Algorithm 2: Pseudo-code of the *EpsFinder*

EpsFinder (Input: D)
1. For each point p find the four nearest neighbors.
2. Sort the points in ascending order of theirs $4 - dist$ values.
3. Following the ascending order, take each point p and if the $4 - dist'$ value for any of its four nearest neighbors is not set so far, set this value to the $4 - dist$ value of the point p.
4. Calculate the *mean* of the $4 - dist'$ values: *mean*
5. Calculate the standard deviation of the $4 - dist'$ values: *SD*
6. Set the *Eps'* value to *mean* $+ 3 \times SD$.

5   Experimental Results and Time Complexity

In this section the experimental results and the time complexity of the automated technique proposed in Sect. 4 (*EpsFinder*) are discussed.

5.1   Experimental Results and Discussions

In this section, the algorithm presented in Sect. 4 is applied to some datasets. This makes the comparison between the old method and the new automated method possible. All the experiments were performed on Intel(R) Celeron(R) CPU 1.90 GHz with 2 GB RAM on the Microsoft Windows 8 platform. The algorithm and the datasets were implemented in Java on Eclipse IDE, MARS.1. Sample datasets are depicted in Fig. 3. The noise percentage for datasets 1 and 2 is 0 %, however, datasets 3 and 4 do have noise values.

In order to show the results of the clustering, each cluster is presented by a different shade of gray in Fig. 4. Noise points are marked using black color.

Figure 5 shows the sorted $4 - dist'$ graphs of the sample datasets. Here, *Eps* indicates the value determined by the user, according to the visual representation of the data, and *Eps'* represents the value calculated automatically by the algorithm presented in Sect. 4 (*EpsFinder*).

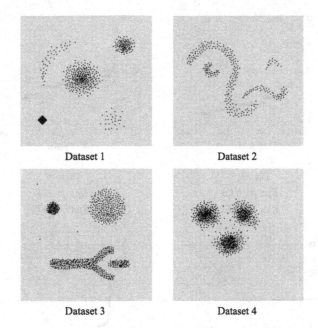

Dataset 1   Dataset 2

Dataset 3   Dataset 4

Fig. 3. Sample datasets

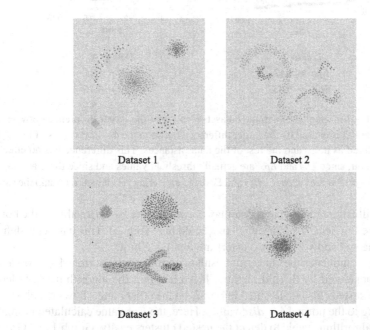

Dataset 1   Dataset 2

Dataset 3   Dataset 4

Fig. 4. Detected clusters

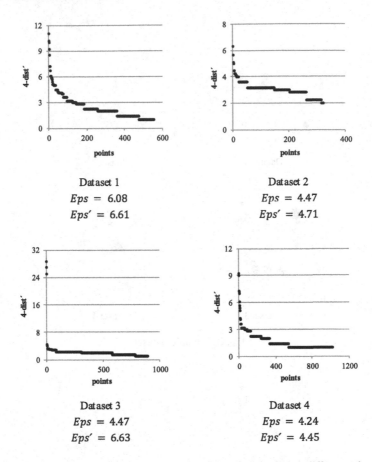

Fig. 5. Sorted 4 – *dist'* graphs for sample datasets (Note that the larger difference between *Eps* and *Eps'* for Dataset 3 is caused by the larger difference between the 4 – *dist'* values of those data instances considered as noise and the rest of the data instances. This difference has no effect on the clustering result, since *Eps* and *Eps'* are actually threshold values and since there are no data instances with 4 – *dist'* values between *Eps* and *Eps'*, the clustering result would remain the same.)

In order to illustrate the problem that may occur with the $k - dist$ value of the border points (discussed in Sect. 4), dataset 5 is presented here (Fig. 6). This dataset is defined in a way that nested and very close clusters are available in it.
Result 1 in Fig. 7, indicates the clustering result according to the normal 4 – *dist* values, which were considered by the old method. It is clear that the algorithm has failed to distinguish the nested clusters. Result 2 in Fig. 7, on the other hand, shows the clustering result according to the normal 4 – *dist'* values. Here, the *Eps* value calculated is smaller and hence the algorithm is able to detect the nested clusters easily. Graph 1 and Graph 2 in Fig. 7 show here the 4 – *dist* and 4 – *dist'* values calculated using each of the techniques, together with the corresponding *Eps* and *Eps'* values.

Dataset 5

Fig. 6. Dataset 5

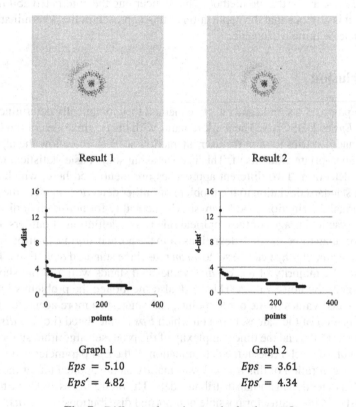

Result 1	Result 2

Graph 1	Graph 2
Eps = 5.10	*Eps* = 3.61
Eps′ = 4.82	*Eps′* = 4.34

Fig. 7. Different clustering results for dataset 5

It should be pointed out that even though the experiments presented here were all for 2-dimensional datasets, the idea can be applied to high-dimensional datasets as well. This is clearly possible, since the calculation of the distance between the points and the application of standard deviation remains the same for high-dimensional datasets. The only point that must be considered is that, the DBSCAN has suggested 4 as the *MinPts* value just for 2-dimensional datasets. However, as mentioned before, *Eps* and *MinPts* are the density parameters of the thinnest cluster; therefore it is always possible to determine the *Eps* by keeping the *MinPts* parameter small enough (or even just by setting it to one). The diversity of the density may always be described with different radii containing a predefined number of points (*MinPts*).

5.2 Time Complexity

Since the algorithm needs to find the four nearest neighbors of each point in the dataset, the time complexity of the algorithm cannot be less than $O(n^2)$. Of course, since these points should have been also retrieved in the user interaction technique, and the only difference here is the calculation of the *mean* and the standard deviation, which can be done in $O(n)$, it is clear that the time complexity of the automated technique presented here, is the same as for the old method. Thus concerning the automated abilities of this technique, it is obvious that the application of this approach in the determination of the *Eps* parameter is quite reasonable.

6 Conclusion

This paper proposes a simple and effective method to automatically determine the input parameter *Eps* of DBSCAN. The work remains with the original idea of the DBSCAN algorithm and just tries to omit the user interaction needed, and allow the algorithm to detect the appropriate value itself. This is done using some basic statistical techniques for outlier detection. Two different approaches are mentioned here, which apply the concept of standard deviation to the problem of outlier detection, namely the empirical rule for normal distributions and Chebyshev's inequality for non-normal distributions. One of the practical usages of the empirical rule is as a definition of outliers as the data that fall more than three standard deviations from the norm in normal distributions. Thus, the value of parameter *Eps* can be set to *mean* plus three standard deviations. This value would cover the majority of the $k - dist'$ values and stands well as a threshold for the specification of the noise values. This work also mentioned the problem which occurs with the $k - dist$ values of the border points, and suggests a more accurate method for the determination of the values, based on which *Eps* is calculated (i.e. $k - dist'$ values). Experimental results and the time complexity of the proposed algorithm suggest that the application of this technique in the determination of the *Eps* parameter is quite reasonable. The concentration of this research was mainly on the application of the empirical rule to outlier detection in normal distributed data. The future works will have to consider the Chebyshev's inequality for possible non-normal distributions of $k - dist'$ values.

References

1. Mitchell, T.M.: Machine Learning. McGraw Hill, New York (1997)
2. Russell, S., Norvig, P.: Artificial Intelligence: A Modern Approach, 2nd edn. Prentice Hall, Englewood Cliffs (2002)
3. Hertzmann, A., Fleet, D.: Machine Learning and Data Mining Lecture Notes, CSC 411/D11, Computer Science Department, University of Toronto (2012)

4. Shalev-Shwartz, S., Ben-David, S.: Understanding Machine Learning: From Theory to Algorithm, Cambridge University Press, New York (2014)
5. Ventura, S., Luna, J.M.: Pattern Mining with Evolutionary Algorithms. Springer, Heidelberg (2016)
6. Murphy, K.P.: Machine Learning: A Probabilistic Perspective. The MIT Press, Cambridge (2012)
7. Ester, M., Kriegel, H.–P., Sander, J., Xu, X.: A density-based algorithm for discovering clusters in large spatial databases with noise. In: Simoudis, E., Han, J., Fayyad, U.M. (eds.) Proceedings of the Second International Conference on Knowledge Discovery and Data Mining (KDD-96), pp. 226–231. AAAI Press (1996)
8. Estivill-Castro, V., Yang, J.: A Fast and robust general purpose clustering algorithm. In: Pacific Rim International Conference on Artificial Intelligence, pp. 208–218 (2000)
9. Rokach, L., Maimon, O.: Clustering methods. In: Rokach, L., Maimon, O.: The Data Mining and Knowledge Discovery Handbook, pp. 321–352. Springer Science + Business Media, Inc., Heidelberg (2005)
10. Berkhin, P.: Survey of Clustering Data Mining Techniques, Technical Report, Accrue Software, San Jose, CA (2002)
11. Han, J., Kamber, M.: Data Mining Concepts and Techniques, pp. 335–391. Morgan Kaufmann Publishers, San Francisco, CA (2001)
12. Everitt, B.S., Landau, S., Leese, M., Stahl, D.: Cluster Analysis, 5th edn. Wiley, Chichester (2011)
13. Beckmann, N., Kriegel, H.-P., Schneider, R., Seeger, B.: The R*-tree: an efficient and robust access method for points and rectangles. In: Proceedings of the ACM SIGMOD International Conference on Management of Data, Atlantic City, NJ, pp. 322–331 (1990)
14. Darong, H., Peng, W.: Grid-based DBSCAN algorithm with referential parameters. In: Proceedings of the International Conference on Applied Physics and Industrial Engineering (ICAPIE-2012), Phys. Procedia, vol. 24(B), pp. 1166–1170 (2012)
15. Smiti, A., Elouedi, Z.: DBSCAN-GM: An improved clustering method based on Gaussian means and DBSCAN techniques. In: 16th International Conference on Intelligent Engineering Systems (INES), pp. 573–578 (2012)
16. Karami, A., Johansson, R.: Choosing DBSCAN parameters automatically using differential evolution. Int. J. Comput. Appl. 91(7), 1–11 (2014)
17. Black, K.: Business Statistics: For Contemporary Decision Making (7th Edn.), Wiley, Hoboken, NJ (2011)
18. Ott, R. L., Longnecker, M.T.: An Introduction to Statistical Methods and Data Analysis (7th Edn.), Cengage Learning, Boston (2015)
19. Maddala, G.S.: Outliers. Introduction to Econometrics, 2nd edn, pp. 88–96. MacMillan, New York (1992)
20. Coolidge, F.L.: Statistics: A Gentle Introduction, p. 458. SAGE Publications, Inc., Thousand Oaks (2012)
21. Shafer, D. S., Zhang, Z.: Introductory Statistics, v. 1.0, Flatworld Knowledge, Washington, D.C. (2012)
22. Amidon, B.G., Ferryman, T.A., Cooley, S.K.: Data outlier detection using the Chebyshev theorem. In IEEE Aerospace Conference Proceedings, pp. 3814–3819 (2005)

Exemplar Selection via Leave-One-Out Kernel Averaged Gradient Descent and Subtractive Clustering

Yiannis Kokkinos and Konstantinos G. Margaritis[(⊠)]

Parallel and Distributed Processing Laboratory,
Department of Applied Informatics, University of Macedonia,
156 Egnatia str., P.O. Box 1591, 54006 Thessaloniki, Greece
kmarg@uom.gr

Abstract. Scalable data mining and machine learning require data abstractions. This work presents a scheme for automatic selection of representative real data points as exemplars. Currently few algorithms can select representative exemplars from the data. K-medoids and Affinity Propagation are such algorithms. K-medoids requires the number of exemplars to be given in advance, as well as a dissimilarity matrix in memory. Affinity propagation automatically finds exemplars as well as their k number but it requires a similarity matrix in memory. A fast algorithm, which works without the need of any matrix in memory, is Subtractive Clustering, but it requires user-defined bandwidth parameters. The essence of the proposed solution relies on a leave-one-out kernel averaged gradient descent that automatically estimates a suitable bandwidth parameter from the data in conjunction with Subtractive Clustering algorithm that further uses this bandwidth for extracting the most representative exemplars, without initial knowledge of their number. Experimental simulations and comparisons of the proposed solution with Affinity propagation exemplar selection on various benchmark datasets seem promising.

Keywords: Subtractive clustering · Kernel averaged · Leave-one-out · Gradient descent · Automatic exemplar selection

1 Introduction

A common problem in applications that collect and store their data is that the number of training examples may be large. Hence, many machine learning and data mining algorithms become slow [1, 2]. One of the solutions is to select most representative exemplars from the data. These exemplars are real data points that form an abstract view of the whole dataset, can represent the structure of the data and can also be used for recognizing patterns [2]. Finding exemplars is a hard problem [3] but is more interesting and informative than dividing data into clusters. Detecting exemplars goes beyond simple clustering, as the exemplars store compressed information [3]. Hence, exemplar selection techniques try to find additional regional information in order to extract representative k-exemplars or k-medoids or k-centers which are close to any given training point so as to minimize the maximum distance from a point to its nearest

© IFIP International Federation for Information Processing 2016
Published by Springer International Publishing Switzerland 2016. All Rights Reserved
L. Iliadis and I. Maglogiannis (Eds.): AIAI 2016, IFIP AICT 475, pp. 292–304, 2016.
DOI: 10.1007/978-3-319-44944-9_25

exemplar. The first exemplar-based algorithm was k-medoids [4] which requires the number k of exemplars to be given in advance, as well as a dissimilarity matrix in memory. Yet, finding exemplars without knowing the k number is a challenge since this k-centers problem or k median objective is NP hard [5–7].

Currently, Affinity Propagation (AP) introduced by Frey and Dueck [8] is the state-of-the-art algorithm for detecting exemplars and subsequently clustering the data around them. AP has been applied in various fields and many applications. In AP all data points are simultaneously considered as exemplars, but exchange deterministic messages until a good set of exemplars gradually emerges. AP finds an approximate solution by using this message passing optimization strategy that is based on max-sum algorithm in a factor graph [8]. Hence, AP does not require the number of exemplars, since this number gradually emerges automatically during the process. However AP does require a similarity matrix in main memory as well as a user defined parameter, the preferences, which are the diagonal values of the similarity matrix.

A fast algorithm, which works without the need of any similarity matrix in main memory, is Subtractive Clustering (SC) [9, 10]. This algorithm was also employed in RBF neural network training [11, 12]. Subtractive Clustering can determine both the exemplars and their number [10] but it requires carefully selected user-defined parameters for the bandwidth and the stopping criteria.

In this work we propose a leave-one-out kernel average gradient descent procedure that estimates a bandwidth parameter from the data, and then we use this bandwidth in a modified subtractive clustering algorithm. We demonstrate that the proposed scheme can provide an automatic estimate of most representative exemplars from the data and in the same time can recognize shapes of patterns.

The rest of the paper is organized as follows. Section 2 provides the basics for Subtractive Clustering. Section 3 introduces the proposed gradient descent of the leave-one-out kernel averaged regression function. Section 4 describes all the initializations and the parameter settings for the proposed scheme. Section 5 presents several experimental simulations and comparisons, while Sect. 6 concludes the paper.

2 Subtractive Clustering Basics

Subtractive clustering algorithm [9–12] selects a set of exemplars from the most representative real data points by using their density. Subtractive clustering can work without any priori information about the number of exemplars. In the first step it computes a density-based potential for every point and then gradually subtracts exemplars by updating all the remaining potentials. The potential $P(i)$ for each point $\mathbf{x}_i$ is defined as a sum of Gaussian kernels over all the N data points as:

$$p(i) = \sum_{j=1}^{N} \exp(-a\|\mathbf{x}_j - \mathbf{x}_i\|^2) \tag{1}$$

where $a = (2/\sigma_a)^2$ and the bandwidth σ_a represents a neighbourhood radius. A data point will have high potential $P(i)$ and high density if it has many neighbour points.

After finding all $P(i)$ the algorithm iteratively executes an updating cycle as:

(1) Find data point $\mathbf{x}^*$ (cluster center) with the highest potential value P^*
(2) Revise the potential of all other points using $P(i) = P(i) - P^*\exp(-b\|\mathbf{x}^* - \mathbf{x}_i\|^2)$

The updating cycle for the potentials $P(i)$ terminates if the current max potential P^* drops below a certain value and the algorithm stops if ($P^* < e\ P_1^*$) [10–12] where P_1^* is the first max potential and e a small percentage. In each iteration the highest potential P^* of the selected point $\mathbf{x}^*$ will substantially affect all the revised potentials of the points near by. Thus, the data points near the selected point $\mathbf{x}^*$ will have significantly reduced density. The updates of the potentials use $b = (2/\sigma_b)^2$ where bandwidth σ_b is another positive constant which also defines a neighbourhood radius. Usually σ_b is taken to be as $1.5\sigma_a$, in order to avoid the selection of closely located exemplars.

The main problem is choosing an appropriate value for the bandwidth parameter σ_a. This choice is of crucial importance and is usually done via extensive experimentation and trial-and-error. The potentials $P(i)$ represent density. So, one can subjectively try to choose a bandwidth σ_a by looking at potentials produced by a wide range of bandwidths, starting with large values of σ_a and gradually decreasing them until a reasonable density is reached. However, such an approach is impractical and too many validations are needed, since there is no way to define a-priori a suitable density value. This is what we are looking for in the first place. A more important issue is that the potentials affect the number of exemplars and their locations. If the bandwidth is very small this will result in neglecting the effect of neighbouring points and then all points will be selected as exemplars. If the bandwidth is small then many exemplars will be selected. If the bandwidth is large then the density function will be affected by accounting all the points and few exemplars will be selected. If the bandwidth is too large then even fewer exemplars will be selected. It is very easy for anyone to see these limits by using trial-and-error. Furthermore, the bandwidth is dataset dependent and the previous limits depend on the formation of a given dataset. An automatic or semi-automatic process is essential as part of a more global analysis in order to avoid many user-defined parameters. In our scheme the proposed leave-one-out gradient descent provides proper bandwidth values for Subtractive Clustering automatically.

3   Proposed Gradient Descent of Leave-One-Out Kernel Averaged

We propose gradient descent learning of the kernel averaged (or weighted average) regression function to automatically estimate a bandwidth parameter. Given a training set $\{\mathbf{x}_i y_i\}_{i=1}^N$ where $\mathbf{x}_i$ are the points and y_j are the desired labels (which we will define later in Eq. 4), the conventional kernel averaged regression function $f(\mathbf{x}_i)$ is:

$$f(\mathbf{x}_i) = \sum_k^N g_k(\mathbf{x}_i)y_k \tag{2a}$$

$$g_k(\mathbf{x}_i) = \varphi_k(\mathbf{x}_i)\Big/\Big(\sum_j^N \varphi_j(\mathbf{x}_i)\Big) \tag{2b}$$

where $\varphi_k(\mathbf{x}) = \exp(-\delta_k(\mathbf{x})/\sigma^2)$ are Gaussian kernels and $\delta_k(\mathbf{x}) = \|\mathbf{x}_k - \mathbf{x}\|^2$ is the squared Euclidean distance. The kernel averaged $f(\mathbf{x}_i)$ has a nominator $\Sigma\varphi_j(\mathbf{x}_i)y_j$, and a denominator $\Sigma\varphi_j(\mathbf{x}_i)$ defined as a sum of $\varphi_k(\mathbf{x}_i)$ Gaussian kernels over all N data points. Since in subtractive clustering the potential $P(i) = \Sigma\varphi_j(\mathbf{x}_i)$ we can see that actually this potential is the normalization factor of $f(\mathbf{x}_i)$.

3.1   Gradient of the Leave-One-Out Kernel Averaged

The proposed *leave-one-out kernel averaged* regression function $f_{loo}(\mathbf{x}_i,\gamma)$ is given by leaving out from the sum in Eq. 2a percentage γ of the self-contribution of $\mathbf{x}_i$ as:

$$f_{loo}(\mathbf{x}_i, \gamma) = \sum_k^N g_k(\mathbf{x}_i)y_k - \gamma g_i(\mathbf{x}_i)y_i \tag{3}$$

where γ is the small leave-one-out parameter which takes values in the range [0, 1].

The proposed method uses desired labels y_i for the points $\mathbf{x}_i$. We define them as:

$$y_i = (1/N)\sum_{j=1}^N \|\mathbf{x}_j - \mathbf{x}_i\|^2 \tag{4}$$

Thus, each desired label y_i is considered as the variance of the corresponding $\mathbf{x}_i$, if this $\mathbf{x}_i$ was the center of the training set. So

The gradient $\partial E(\sigma,\mathbf{x})/\partial\sigma$, with respect to bandwidth σ, is computed from the squared error $E(\sigma,\mathbf{x})$ which is a convex function defined as $E(\sigma,\mathbf{x}) = (f_{loo}(\mathbf{x},\gamma)-y)^2$ where $f_{loo}(\mathbf{x},\gamma)$ is the leave-one-out kernel averaged regression function.

Without the leave-one-out such a gradient will not work. Taking a gradient of the kernel averaged with respect to the bandwidth will not result in a suitable solution, since eventually all points will converge to tiny bandwidth values (they will be correct for predicting themselves).

The classical squared error $E^i(\sigma,\mathbf{x}_i)$ for each $\mathbf{x}_i$ is:

$$E^i(\sigma, \mathbf{x}_i) = (1/2)(f_{100}(\mathbf{x}_i, \gamma) - y_i)^2 \tag{5}$$

The gradient descent update for the σ parameter can be defined from the gradient of the squared error as:

$$\Delta\sigma = -\xi\partial E^i(\sigma, \mathbf{x}_i)/\partial\sigma \tag{6}$$

The chain rule of the gradient gives:

$$\begin{aligned}\partial E^i(\sigma, \mathbf{x}_i)/\partial\sigma &= \left(\partial E^i(\sigma, \mathbf{x}_i)/\partial f_{loo}(\mathbf{x}_i, \gamma)\right)\left(\partial f_{loo}(\mathbf{x}_i, \gamma)/\partial\sigma\right)\\ &= (f_{loo}(\mathbf{x}_i, \gamma) - y_i)(\partial f_{loo}(\mathbf{x}_i, \gamma)/\partial\sigma)\end{aligned} \tag{7}$$

where the derivate $(\partial f_{loo}(\mathbf{x}_i, \gamma)/\partial \sigma)$ is:

$$\frac{\partial}{\partial \sigma} f_{loo}(\mathbf{x}_i, \gamma) = \sum_k^N \left(\frac{\partial}{\partial \sigma} g_k(\mathbf{x}_i) y_k\right) - \gamma \frac{\partial}{\partial \sigma} g_i(\mathbf{x}_i) y_i \tag{8}$$

where we only need to find the derivate $\partial g_k(\mathbf{x}_i)/\partial \sigma$ given by:

$$\frac{\partial}{\partial \sigma} g_k(\mathbf{x}_i) = \frac{\partial}{\partial \sigma}\left[\varphi_k(\mathbf{x}_i) \cdot \left(\sum_j^N \varphi_j(\mathbf{x}_i)\right)^{-1}\right]$$
$$= \left(\frac{\partial}{\partial \sigma} \varphi_k(\mathbf{x}_i)\right) \cdot \left(\sum_j^N \varphi_j(\mathbf{x}_i)\right)^{-1} - \varphi_k(\mathbf{x}_i) \cdot \left(\sum_j^N \varphi_j(\mathbf{x}_i)\right)^{-2}\left(\sum_j^N \frac{\partial}{\partial \sigma} \varphi_j(\mathbf{x}_i)\right) \tag{9}$$

This equation by using $\frac{\partial}{\partial \sigma} \varphi_k(\mathbf{x}_i) = \varphi_k(\mathbf{x}_i) \delta_k(\mathbf{x}_i)/\sigma^3$ becomes:

$$\frac{\partial}{\partial \sigma} g_k(\mathbf{x}_i) = (\varphi_k(\mathbf{x}_i)\delta_k(\mathbf{x}_i)/\sigma^3) \cdot \left(\sum_j^N \varphi_j(\mathbf{x}_i)\right)^{-1} - \varphi_k(\mathbf{x}_i) \cdot \left(\sum_j^N \varphi_j(\mathbf{x}_i)\right)^{-2}\left(\sum_j^N (\varphi_k(\mathbf{x}_i)\delta_j(\mathbf{x}_i)/\sigma^3)\right)$$
$$= (1/\sigma^3)\varphi_k(\mathbf{x}_i)\left(\sum_j^N \varphi_j(\mathbf{x}_i)\right)^{-1}\left[\delta_k(\mathbf{x}_i) - \left(\sum_j^N \varphi_j(\mathbf{x}_i)\right)^{-1}\left(\sum_j^N (\varphi_j(\mathbf{x}_i)\delta_j(\mathbf{x}_i))\right)\right] \tag{10}$$

and by replacing the expression for $g_k(\mathbf{x}_i)$ from Eq. 2b into Eq. 10 it gives:

$$\frac{\partial}{\partial \sigma} g_k(\mathbf{x}_i) = (1/\sigma^3) g_k(\mathbf{x}_i)\left[\delta_k(\mathbf{x}_i) - \left(\sum_j^N (g_j(\mathbf{x}_i)\delta_j(\mathbf{x}_i))\right)\right] \tag{11}$$

Equation 11 is the general derivate for any function $g_k(\mathbf{x}_i)$.

The derivate $\partial g_i(\mathbf{x}_i)/\partial \sigma$ (of the contribution of $\mathbf{x}_i$ to itself) has a shorter expression produced by Eq. 11 which after simplifications (by setting $\delta_i(\mathbf{x}_i) = 0$ and $\varphi_i(\mathbf{x}_i) = 1$) is:

$$\frac{\partial}{\partial \sigma} g_i(\mathbf{x}_i) = -(1/\sigma^3)\left(\sum_j^N (\varphi_j(\mathbf{x}_i)\delta_j(\mathbf{x}_i))\right)\left(\sum_j^N \varphi_j(\mathbf{x}_i)\right)^{-2} \tag{12}$$

Finally by substituting Eq. 11 and Eq. 12 into Eq. 8 we can compute $(\partial f_{loo}(\mathbf{x}_i, \gamma)/\partial \sigma$. In a more shorthanded notation it gives:

$$\frac{\partial}{\partial \sigma} f_{loo}(\mathbf{x}_i) = (1/\sigma^3)\sum_k^N \left(\left(\delta_k - \frac{\Sigma(\varphi_j \delta_j)}{\Sigma \varphi_j}\right) \cdot \frac{\varphi_k}{\Sigma \varphi_j} y_k\right) + \gamma \frac{\Sigma(\varphi_j \delta_j)}{\sigma^3(\Sigma \varphi_j)^2} y_i \tag{13}$$

The small leave-one-out parameter $\gamma \in [0, 1]$ prevents the gradient from converging into tiny values of the bandwidth σ. There exists a trade-off between $\gamma = 1$ which gives large bandwidths and $\gamma = 0$ which gives tiny bandwidths.

Stochastic mode (or online) of gradient descent learning computes the gradient by using a single example at a time. The algorithm randomly selects an example $\mathbf{x}_i$ and its label y_i and updates the current parameter σ by using:

$$\sigma^{(t+1)} = \sigma^{(t)} - \xi \partial E(\sigma^{(t)}, \mathbf{x}_t)/\partial \sigma \quad t = 1, \ldots N \tag{14}$$

Hence, an epoch ends after all examples are introduced in a random order. Then the gradient updates of σ are averaged over all N examples as $\sigma^{epoch} = \text{avg}(\sigma^{(t)})$ with $t = 1$, ...,N. The learning rate ξ can be constant or can vary at each epoch. For one epoch step the *leave-one-out kernel averaged gradient descent* is:

for $t = 1$ to N

pick randomly a point $\mathbf{x}_t$ without replacement

update the parameter σ by using $\sigma^{(t+1)} = \sigma^{(t)} - \xi \, \partial E(\sigma^{(t)}, \mathbf{x}_t)/\partial \sigma$

end for

4 Initializations and Parameter Settings

As usual the first thing to do is to scale the data features into the range [0, 1]. Without scaling the gradient might not converge, since the learning rate ξ depends on the scale of the feature space. By scaling the data features first, we can then use a fixed value for ξ for all datasets and hence avoid searching for suitable learning rates each time we use a different dataset. Such scaling also avoids over-fitting which occurs when some features are in large numeric ranges.

In Subtractive Clustering (SC) the potential updating cycle terminates if the current max potential $P*$ become less that a threshold ($P^* < e \, P_1^*$). If e is selected to be very small, a large number of exemplars will be selected. On the contrary, a large value of e will lead to a small exemplar set. In order to avoid any other user-defined parameter we set $e = 1/P_1^*$. That is, Subtractive Clustering terminates at j-th iteration when $P_j^* < 1$. Thus, every point starts with potential $P(i) >= 1$ and finally ends up with potential $P(i) < 1$. There is a theoretical justification for this limit since $P(i) = 1$ is the self-contribution of every i-th point to itself.

For 2-dimensional datasets in Subtractive Clustering we set $\sigma_b = 1.5\sigma_a$ as recommended. High dimensional density estimates may suffer from the curse of dimensionality. For higher dimensions there is a problem since the 1.5 % influences more strongly the nearby points and we use a variable $\sigma_b = \sigma_a + 0.5 (1.0 - k_{sofar}/N) \sigma_a$, which starts from $\sigma_b = 1.5\sigma_a$ and decays. As k_{sofar} (the number of selected exemplars so far) increases from 1 to k during the $P(i)$ updating cycle of SC, the parameter σ_b gradually decreases and in the theoretical limit $k = N$ the value σ_b becomes equal to σ_a.

For the online gradient descent we set a fixed learning rate $\xi = 0.2$ and maximum epochs = 10. Usually it converges after the first epoch if the dataset size is larger than 10000. So, for larger datasets we can set maximum epochs = 2.

For the leave-one-out kernel averaged regression function we set the leave-one-out parameter $\gamma = 0.1$. The value $\gamma = 1$ removes the self-contribution completely and will give a large bandwidth and very few exemplars, while $\gamma = 0$ will give a tiny bandwidth and almost all points as exemplars. Since the goal is just to avoid this, we found after some experimentation that a value $\gamma = 0.1$ is always sufficient enough to prevent bandwidth from converging into tiny values, so as to provide a stable solution without producing large bandwidths.

Initializing the bandwidth σ in the beginning of gradient descent (epoch = 0) is an issue, since for different datasets we may need to search for different initial values of σ each time. However there is a simple automatic way that works around this. We set the initial bandwidth equal to the trace of covariance matrix **R**. Hence, given N points $\mathbf{x}_n$ each one in d dimension, with their mean $\boldsymbol{\mu} = (1/N) \sum_n^N \mathbf{x}_n$ the covariance matrix is $R = (1/N) \sum_n^N (\boldsymbol{\mu} - \mathbf{x}_n)(\boldsymbol{\mu} - \mathbf{x}_n)^T$ and the initial value of σ is $(1/d) \sum_i^d \sqrt{r_{ii}}$, where r_{ii} are the diagonal elements of **R**. Thus, the gradient descent starts with a relative large bandwidth σ which decreases immediately after the first epoch, until it converges.

It is important to note that we use the same settings for all the datasets and no user-defined parameter is needed.

5 Experimental Simulations

The first set of experimental simulations present results for visual comparisons of AP with the proposed algorithm using four 2-d datasets. The second set present performance comparisons and quality analysis on several real world benchmark datasets.

The code for Affinity Propagation (AP) was downloaded from the official site (http://www.psi.toronto.edu/affinitypropagation). AP uses as input a similarity matrix S in which the pair-wise similarities between data points are defined from their distances as $s(i,k) = -\|\mathbf{x}_i - \mathbf{x}_k\|^2$ for every $i \neq k$, as suggested in [8]. There are two more parameters: the damping factor λ and the prior preferences $s(k,k)$ which are the diagonal values of the similarity matrix. The dumping factor is usually $\lambda = 0.5$ as suggested. For the preferences, a good choice [8] is to set all the diagonal elements $s(k,k)$ equal to the median value of all the similarities between data points. We use as preference the one half of the mean value of all similarities $(1/(2N^2)) \sum_i^N \sum_k^N s(i,k)$ that results in a moderate number of exemplars which emerge automatically. This choice selects much more exemplars than the median choice while it still avoids selecting outliers.

5.1 Evaluation Criteria and Quality Indexes

The *sum of squared errors (SSE)* which quantifies the *clustering error* is the most widely used quality criterion [8] and is given by the sum of the squared distance between each point $\mathbf{x}_i$ and its corresponding exemplar $\mathbf{c}(\mathbf{x}_i)$ as:

$$SSE = \sum_{i=1}^{N} \|\mathbf{x}_i - \mathbf{c}(\mathbf{x}_i)\|^2 \tag{15}$$

The *maximum distance (maxD)* between any point $\mathbf{x}_i$ and its exemplar $\mathbf{c}(\mathbf{x}_i)$ that can quantify if all points are compactly represented (no cluster is larger than *maxD*) is:

$$maxD = \max_i^N \|\mathbf{x}_i - \mathbf{c}(\mathbf{x}_i)\|^2 \tag{16}$$

The *normalized Hubert gamma statistic* [13] is a well known cluster evaluation criterion which is invariant to the number of clusters, given by:

$$\hat{\Gamma} = \frac{\frac{1}{M}\sum_{i=1}^{N-1}\sum_{j=i+1}^{N}(P(i,j) - \mu_P) \times (Q(i,j) - \mu_Q)}{\sigma_P \times \sigma_Q} \tag{17}$$

where $M = N(N-1)/2$, and it uses two proximity matrices P and Q both of size $N{\times}N$. An element $P(i, j)$ is the distance between points $\mathbf{x}_i$ and $\mathbf{x}_j$. An element $Q(i, j)$ is the distance between the cluster representative centroids to which $\mathbf{x}_i$ and $\mathbf{x}_j$ belong. μ_P is the mean of all elements of matrix P, μ_Q is the mean of all elements of matrix Q, while σ_P and σ_Q are their standard deviations from their means. A high value of this statistic (close to 1) indicates the existence of well-separated compact clusters.

The *net similarity cost* is defined as a cost function specifically for AP [8, 14] and it is the sum of similarities $s(i,k)$ between data points and their exemplars, minus the exemplar costs $s(k,k)$, (the preferences of the exemplars). AP identifies a set of exemplars K so as to maximize this cost given by [14]:

$$\sum_{i\notin K}\max_{k\in K} s(i, k) + \sum_{k\in K} s(k, k) \tag{18}$$

5.2   Visual Comparisons of AP with the Proposed KG-SC

For the visual comparisons we use four datasets with 2 dimensions each. We compare the results of Affinity Propagation (AP) algorithm with the proposed leave-one-out kernel gradient subtractive clustering (KG-SC in short).

Table 1 illustrates the quality indexes that correspond to the exemplar selections and clustering solutions of the datasets in Figs. 1, 2, 3 and 4 for the AP and KG-SC.

From Table 1 it seems that both algorithms can provide high quality results for the 2-dimensional datasets, while a slight precedence could be given to KG-SC.

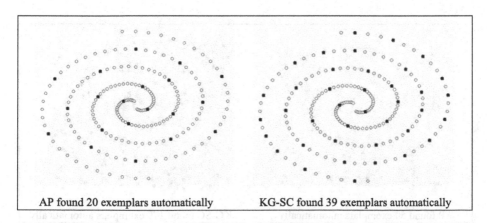

AP found 20 exemplars automatically   KG-SC found 39 exemplars automatically

Fig. 1. Dataset 1 (two spirals) has 200 points. Exemplars are marked as black squares, while the other points are marked as white circles.

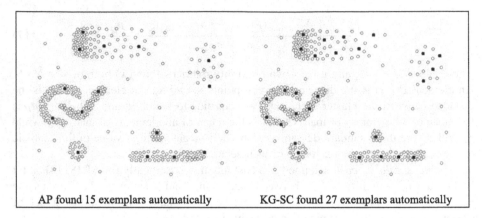

AP found 15 exemplars automatically   KG-SC found 27 exemplars automatically

Fig. 2. Dataset 2 has 322 points. Exemplars are black squares, other points are white circles.

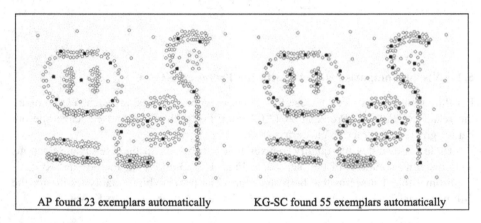

AP found 23 exemplars automatically   KG-SC found 55 exemplars automatically

Fig. 3. Dataset 3 has 523 points. Exemplars are black squares, other points are white circles.

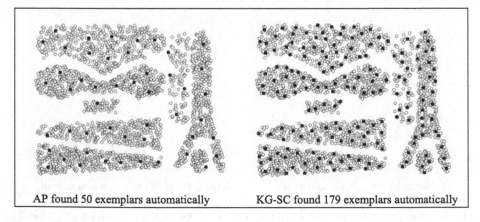

AP found 50 exemplars automatically   KG-SC found 179 exemplars automatically

Fig. 4. Dataset 4 has 2551 points. Exemplars are black squares, other points are white circles.

Table 1. Quality indexes for the exemplar selection and clustering solutions of the datasets in Figs. 1, 2, 3 and 4. For each algorithm (AP and KG-SC) we illustrate the net similarity cost, maximum Distance, clustering error, normalized Hubert statistic. Best indexes are marked in bold.

	N	Affinity propagation algorithm				Proposed leave-one-out kernel gradient subtractive clustering			
		netSim	maxD	error	Hubert	netSim	maxD	error	Hubert
Dataset 1	200	**−3.55**	0.028	1.49	0.915	−4.63	**0.016**	**0.62**	**0.979**
Dataset 2	322	**−4.01**	0.061	1.61	0.959	−5.49	**0.058**	1.03	**0.974**
Dataset 3	523	**−5.84**	0.058	2.41	0.954	−9.23	**0.042**	0.89	**0.980**
Dataset 4	2551	**−14.4**	0.015	6.61	0.976	−31.2	**0.003**	1.44	**0.995**

In addition, it is apparent in Table 1 that AP delivers exactly what it promises, that is to identify a set of exemplars K so as to maximize the net similarity cost [14]. The net similarity cost is better for AP than KG-SC. So AP remains the best algorithm for the k-centers problem.

However the clustering error that quantifies the distortion and the normalized Hubert index that quantifies the cluster compactness are better for KG-SC. Note that ideal clustering solutions usually have the normalized Hubert index close to 1 as they are in the last column of Table 1. So KG-SC delivers more well defined exemplars.

It is the k-centers problem itself which might not be able to guarantee the best exemplar selection. That is why KG-SC takes a different path; the density based, and tries to find the most important representatives from the densest ones. The better quality of the KG-SC solutions is evident from the maximum Distance, clustering error and normalized Hubert statistic in Table 1.

The computational complexity cost of the proposed KG-SC is quadratic $O(N^2)$ of the same order with the cost of SC. Actually, for large datasets and max epochs = 2 it is two times that of SC. This cost is much lower than the AP cost. The memory requirements for KG-SC is $O(Nd)$, since only the dataset is needed in main memory. On the other hand, AP does require three matrices (similarities, availabilities, responsibilities) of size $N \times N$ in main memory and this could limit the algorithm. However, one can argue that for the special case of ultra high-dimensional datasets where the data dimension is of the same order with the number of examples ($d \approx N$) the memory requirements become the same.

What will happen in a case where someone needs a fixed number of exemplars less than the KG-SC algorithm finally selects is a question that could be answered. Note that an advantage of Subtractive Clustering is that it returns exemplars in decreasing order from the most important to the least important. So, picking the first K in this list is one simple solution.

5.3 Quality Comparisons on Real World Benchmark Datasets

Quality comparisons are also performed on a number of publicly available real-world benchmark problems which are downloaded from the UCI machine learning data repository (http://archive.ics.uci.edu/ml). The specific details of these datasets (dataset name, N examples before duplicate removal, d dimensions) are illustrated in Table 2 together with the results.

Table 2. Quality indexes for the exemplar selection of various benchmark datasets with N examples and d dimensions. For each algorithm (AP and KG-SC) we illustrate the k number of selected exemplars which emerge automatically, the maximum Distance, the clustering error and the normalized Hubert gamma statistic. Best quality indexes are marked in bold.

Name	N	d	Affinity propagation algorithm k	maxD	error	Hubert	Proposed leave-one-out kernel gradient subtractive clustering k	maxD	error	Hubert
CPU	209	7	21	0.93	4.69	0.953	20	**0.60**	**4.33**	**0.960**
Yacht	308	6	33	0.14	7.73	0.958	54	**0.05**	**3.37**	**0.980**
Housing	506	13	35	0.66	39.9	0.965	45	**0.61**	**37.8**	**0.967**
Concrete	996	8	80	0.38	30.4	0.943	114	**0.18**	**25.0**	**0.960**
Airfoil	1503	6	79	0.19	20.7	0.980	138	**0.05**	**11.2**	**0.989**
Abalone	4177	7	127	0.12	**16.2**	**0.991**	45	**0.11**	21.9	**0.991**
RedWine	1599	11	116	0.65	**50.0**	**0.876**	104	**0.53**	56.2	0.857
Bodyfat	252	15	29	**0.28**	**18.7**	**0.923**	21	1.59	24.8	0.759
WhiteWine	4898	11	292	**0.11**	**77.6**	**0.917**	265	0.47	92.4	0.886
Iris	150	4	10	0.14	2.95	0.962	22	**0.09**	**1.61**	**0.978**
Haberman	306	3	23	0.20	4.02	0.907	47	**0.17**	**2.37**	**0.938**
Ecoli	336	7	27	0.21	9.29	0.931	44	**0.20**	**7.45**	**0.949**
Blood Trans	742	4	28	0.35	4.13	0.948	34	**0.11**	**3.28**	**0.963**
Banknote	1372	4	55	0.06	7.38	0.977	115	**0.05**	**3.58**	**0.989**
Phoneme	5404	5	165	0.12	32.3	0.950	374	**0.07**	**20.5**	**0.969**
Yeast	1484	8	107	**0.11**	**22.3**	**0.960**	119	0.17	23.2	**0.960**
Diabetes	768	8	83	**0.18**	**30.6**	**0.895**	65	0.62	39.1	0.834
Wine	178	13	27	**0.54**	**26.6**	**0.885**	18	0.98	37.6	0.828
Heart	297	13	43	**1.38**	**92.1**	**0.904**	25	2.13	138	0.855
Wisconsin	683	9	37	**0.97**	**83.9**	**0.931**	27	1.93	109	0.909
Dermatology	358	34	41	**2.50**	**320**	**0.909**	36*	4.34	371	0.877
Shuttle	58000	9	*	*	*	*	956	**0.002**	**1.01**	**0.999**

We found that while KG-SC as a density-based algorithm does not suffer from the existence of duplicates, AP does. Thus for a fair comparison we first remove all duplicates from the benchmark datasets. Also, since the net similarity is not a quality index but a specific cost suitable only for AP (it was always better for AP) we do not illustrate it in Table 2. Note, for future considerations that we detect several duplicates in the datasets Banknote Authentication, Blood Transfusion, Phoneme, Wisconsin Breast Cancer, Haberman, Yacht Hydrodynamics, Red Wine Quality, White Wine Quality, Concrete Compressive Strength.

Both algorithms find well defined representative exemplars and deliver high quality solutions, since the normalized Hubert gamma index is very high in both of them for all benchmark datasets. In low dimensional datasets KG-SC seems better, while in high dimensional datasets AP seems better.

There are some limitations. AP has limits in the number of examples, while the proposed KG-SC is density-based and might limited by the number of dimensions (features). The Dermatology dataset has many dimensions ($d = 34$) and the leave-one-out gradient could not converge for $\gamma = 0.1$, so we use a minimum value $\gamma = 0.01$. The Shuttle dataset has quite many examples ($N = 58000$) and the Affinity Propagation runs out of memory (it needs 39 GB). For the Shuttle dataset the proposed leave-one-out Kernel Gradient Subtractive Clustering produces 956 exemplars and a normalized Hubert gamma statistic 0.999 which indicates very well formed compact clusters.

6 Conclusions

We present a scheme that can potentially permit automatic selection of representative exemplar points from the data without the need of any used-defined parameter. By computing a gradient descent for a simple leave-one-out kernel averaged regression function that can automatically estimate a suitable bandwidth parameter for the density-based Subtractive Clustering algorithm we can extract most representative exemplars, without initial knowledge of their number. Evaluating with classical quality indexes the data clustering solutions around these exemplars reveal that the proposed KG-SC algorithm produce well separated compact and dense clusters. Experimental comparisons with the state-of-the-art Affinity Propagation exemplar selection algorithm show that both algorithms select well defined representative exemplars and can deliver high quality solutions. KG-SC is simply parallelizable, a point worthwhile studying in the future. We also plan to explore the possibility of using either mini-batch gradients, or a dual tree for speeding up KG-SC. Interesting future works could extend KG-SC in order to explore a possible automation in other density based algorithms. Currently we study the proposed KG-SC for training Neural Networks.

References

1. Dunham, M.H.: Data mining introductory and advanced topics. Prentice Hall, Upper Saddle River (2004)
2. Duda, R., Hart, P.E., Stork, D.G.: Pattern Classification. Wiley, New York (2001)
3. Mézard, M.: Where are the exemplars? Science **315**, 949–951 (2007)
4. Kaufman, L., Rousseeuw, P.: Clustering by means of medoids. In: Statistical Data Analysis Based on the L1 Norm and Related Methods, pp. 405–416 (1987)
5. Kariv, O.: Hakimi, S.L: An algorithmic approach to network location problems. The p-medians. Siam J. Appl. Math. **37**, 539–560 (1979)
6. Hochbaum, D., Shmoys, D.: A best possible heuristic for the k-center problem. Math. Oper. Res. **10**(2), 180–184 (1985)

7. Bern, M., Eppstein, D.: Approximation algorithms for NP-hard problems. PWS Publishing, Boston (1997). Chapter Approximation algorithms for geometric problems
8. Frey, B.J., Dueck, D.: Clustering by passing messages between data points. Science **315**, 972–976 (2007)
9. Chiu, S.L.: Fuzzy model identification based on cluster estimation. J. Intell. Fuzzy Syst. **2**, 267–278 (1994)
10. Kothari, R., Pittas, D.: On finding the number of clusters. In: Pattern Recognition Letters (1999)
11. Sarimveis, H., Alexandridis, A., Bafas, G.: A fast training algorithm for RBF networks based on subtractive clustering. Neurocomputing **51**, 501–505 (2003)
12. Yang, P., Zhu, Q., Zhong, X.: Subtractive clustering based RBF neural network model for outlier detection. J. Comput. **4**(8), 755–762 (2009)
13. Halkidi, M., Batistakis, Y., Vazirgiannis, M.: On clustering validation techniques. J. Intell. Inf. Syst. **17**(2/3), 107–145 (2001)
14. Dueck, D., Frey, B.J., Jojic, N., Jojic, V., Giaever, G., Emili, A., Musso, G., Hegele, R.: Constructing treatment portfolios using affinity propagation. In: Vingron, M., Wong, L. (eds.) RECOMB 2008. LNCS (LNBI), vol. 4955, pp. 360–371. Springer, Heidelberg (2008)

Design of an Advanced Smart Forum
for Tesys e-Learning Platform

Paul Ştefan Popescu[1](✉), Mihai Mocanu[1], Costel Ionaşcu[2],
and Marian Cristian Mihăescu[1]

[1] Faculty of Automation, Computers and Electronics,
Department of Computers and Information Technology,
University of Craiova, Craiova, Romania
{stefan.popescu,mmocanu,mihaescu}@software.ucv.ro
[2] Faculty of Economics and Business Administration,
University of Craiova, Craiova, Romania
icostelm@yahoo.com

Abstract. This paper presents an application of Intelligent Data Analysis techniques in the area of online educational environments and more exactly, the discussion forums within them. The research area is also referred as Educational Data Mining and have many tools and techniques already developed. This work concentrates on the improvement that can be provided by the design and implementation of a forum that has "smart" capabilities and aims to be proactive to the user's needs. The main issues addressed are the interaction design, student's academic performance and the achievement of better models by completing the already gathered data with the logged data offered by the forum. We present here three methods that can solve the above mentioned issues: recommending subjects of interest, computing trends and offering smart alerts for users that are at risk for academic failure. Every method represents a tool that will be integrated in the forum and will take benefit from the extra logged data.

Keywords: e-Learning · Machine learning · Forum

1 Introduction

Online educational environments are a constantly growing field of education. This field is important because it creates the opportunity to learn even when there is no physical presence like in regular educational environments. The physical presence in the classroom is important in terms of interaction because students are able to talk easily, change opinions and help each others. From the teachers perspective they can easily response to questions, and create some sort of ranking in terms of knowledge. There will always be some learners that have better knowledge in some educational areas; these learners can help share their knowledge to others.

In online educational environments interaction of any kind is different [1] than in regular environments, the engagement in educational activities is also different [2] and this may lead to lower student's results [3]. Going further from this lack of interaction

© IFIP International Federation for Information Processing 2016
Published by Springer International Publishing Switzerland 2016. All Rights Reserved
L. Iliadis and I. Maglogiannis (Eds.): AIAI 2016, IFIP AICT 475, pp. 305–316, 2016.
DOI: 10.1007/978-3-319-44944-9_26

e-learning platforms integrates more and more smart capabilities; from better data logging that allows data analysts to accomplish more advanced data mining and machine learning tasks to advanced smart tools that offers many facilities (i.e. subjects of interest recommendations, predicting students failure or academic performance, better data logging).

In this paper we present a forum that encapsulates smart capabilities aimed to improve the student's interaction and to create a friendlier virtual environment included in e-learning platforms. The system's design aims to ensure enough flexibility to easily adjust and install in any e-Learning platform but first we will refer Tesys e-Learning platform that runs at the University of Craiova and offers us data. Tesys is a custom e-learning platform built at the University of Craiova which fulfils the professors and students needs in terms of online educational environments necessary for distance learning. The platform is under continuous development and based on the knowledge gained from we can make further development like this Smart Forum.

The forum is designed to be strongly integrated in the e-learning platform, being available in every discipline, and well isolated in terms of implementation. We don't design it as a plug-in that uses some sort of bridge to relate between forum and e-learning platform. This sort of integration ensures more specific and subject related topics of discussion.

2   Related Work

In the area of educational data mining there has been a lot of work. This paper lies on the border of machine learning and information retrieval offering a design of a system that encapsulates feature extraction techniques that are the prerequisites for employing data mining and machine learning algorithms.

Some other forums were used to employ data mining techniques and to gain more knowledge from discussions. The purpose of [4] is to present some data mining techniques that offer a strategy for data representation. For their approach the instructor's view of the output of a thread forum is somehow limited as the can review a transcript of the written dialogue produced by participants. Because of big amount of data that consists of forum contributions the paper seeks to intersect the information an instructor wish to extract from the forum with some useful information that the system may extract from the instructor's query. This will help the instructor to improve his ability to evaluate the progress of a threaded discussion.

There are also some other paper [5] that show that students who use discussion forums have a higher chance of finishing a course. This implies MOOCs (Massive Online Open Courses) and they use some machine learning techniques to extrapolate a small set of annotation to the whole forum. These annotations help in two main ways: summarize the state of the forum and the also allow researchers to deeper understand how the forum is implied in the learning process.

Some recent research [6] predicts the students performance based on on-line discussion forums which are considered as communities of people learning from each other. They aim that forums not only inform the students about their peers' doubts or problems but it is a way to inform the professors about the learner's knowledge. In the paper they select some instances and attributes, run different supervised learning algorithms and then they measure the accuracy and comprehensibility of the prediction.

Some related work was done for the design and development of several machine learning algorithms [7]. For the supervised learning tasks we can use decision trees [8] or regression analysis [9] and we aim to evaluate the system using several machine learning metrics like area under the ROC curve [10].

3   Goals

Building and integrating *Smart Forum* aims to achieve three goals: create a virtual space for improving the users' interaction, generate more high quality logged data that can be analyzed and offering useful functionalities for improving the students' performance.

3.1   Interaction Improvement

Improving the users' interaction can be achieved by extending the means of interaction by number and quality. Forums are environments that allow message exchange, the particularity refers that the messages are most public or available for registered users and also the messages are grouped on subjects of discussion. For our approach, messages wrote on Smart Forum are available for students and instructors from a specific course. Adding a new mean of communication with specific features creates a environments which impact can be measured in terms of interaction. After the system is integrated and used for at least a semester, we evaluate the interaction and adjust some parameters like notification frequency or messages of interest alert frequency.

3.2   Data Logging

Data already logged in e-learning platforms refers users and learning resources. We threat Smart Forum as an important learning resource which can produce a important amount of data that refers both other learning resources and users. We can analyze what chapters of course were more discussed and what concepts from which chapters are referred in the most discussed topics. For user profiling we can add more data based on the actions that are perform on the forum. Logging more data can reveal several new patterns and also improve the actual models.

3.3   Improving the Student's Performance

This is a permanent topic of interest in the Educational Data Mining Area. There are so many solutions out there but there is also enough space for improvement. Collecting more relevant data and adding it to the existing models may lead the data analyst to better result. There are also many correlations that can be made between the forum activity and educational results, forum activity and the activity performed over other learning resources or the overall activity and final grades. One thing that may influence the student's performance is also the interaction between them and what they can learn from each other. By using the described platform one study can reveal if students that interact more and learn more from their colleagues will have better results or at least an ascending learning curve.

4 System Design

Smart forum creates a virtual space for interaction between all entities that perform activities in e-Learning platforms: students, professors and administrative staff. Every entity has his specific tasks that are related to their topics of interests; students discuss about the subjects related to the disciplines, professors may response to questions and may launch challenges, the administrative staff may respond to administrative questions of common interest.

Figure 1 presents how the disciplines are distributed over the Tesys e-Learning platform's infrastructures. The e-learning platform holds 3(but it can be more) years of study, every year of study have several modules (study programs) which are marked with M and every study program have a set of disciplines attached to it D. Because of the space problems we added in the figure disciplines only on some modules but there are usually a set of around five disciplines on every module. This figure is important to understand because smart forum is attached to every discipline being well distributed. The forum distribution over the disciplines ensures that the subjects discussed in the forum are strongly related to the course chapters that are used to create de discipline.

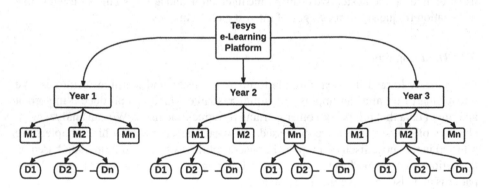

Fig. 1. Tesys disciplines distribution

Figure 2 presents the components that are encapsulated in a discipline. On the upper part of the picture we have the main actors that perform their activities on one discipline: students, professors, and secretaries. Every actor has his own interface and access to specific features of the e-Learning platform. Chapters, Homework, questions and references come at the same level with the forum and are related to the discipline covering several learning areas. The chapters are represented by documents that form the course, homework are used to assign learners tasks that take more time while questions are used for tests and exams. The forum is used to discus subjects related to a specific discipline; this makes the data collected from it to be more specific. For example, suppose a data analyst wants to analyze the subjects related to discipline "x", he needs to find the specific subjects related to the discipline "x" by using some analysis techniques which will most likely not offer 100 % accuracy or the topics will have tags that were allocated by the topic initiator which will also may not offer 100 % accuracy all the time. Distributing the forum over the discipline offers one more topic focus attribute over the existing ones. On the forum there will always be moderators which are represented by professors or/and best students, these moderators will make sure there will not be topics out of the discipline area on a specific forum.

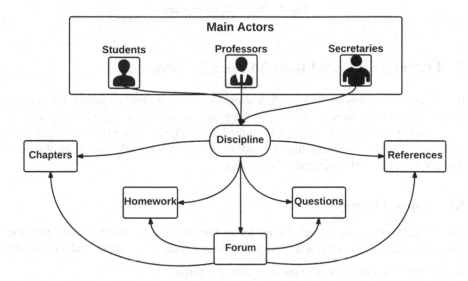

Fig. 2. Forum integration

Figure 3 presents the main architecture of the data analysis pipeline. There are two main components that contribute to data logging: Smart Forum and Tesys. Although the forum is integrated in the e-learning platform, it provides a specific amount of data that can be converted in a set of features that can be combined with the features extracted from Tesys in order to build better models. There are also some specific data analysis tasks that will be performed only on the data gathered from Tesys or Smart Forum and there will be some specific tasks that will be performed only on the combined model and these tasks will not be used for validation purposes.

Validation of the system in terms o data analysis can be performed by comparing the results obtained using the combined models with the models obtained using only the features from Tesys.

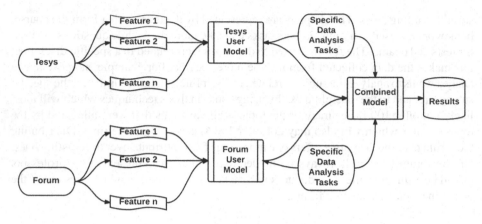

Fig. 3. Data analysis pipeline

5 Functionalities and Data Analysis Processes

In this section we present the main functionalities of the proposed system. These capabilities are specific to a forum that is integrated in an e-learning platform and are dependent on the data that is logged from the forum. There are two types of actions that are available in this forum: regular one and smart ones that includes the usage of several data mining and machine learning algorithms.

5.1 Proposed Features

Several features may be extracted using Smart Forum. These features ensure better user modelling for forum activity analysis and also may improve the overall student's profile.

- NO_POSTS – number of posts written on the forum
- NO_TOPICS – the number of topics started on the forum
- NO_QUESTIONS – the number of questions asked on forum
- NO_USEFUL_ANSWER – the number of answers written on the forum considered to be useful by other users
- NO_ACTIVE_DAYS _FORUM – the number of days when the user was active on the forum
- AVG_TIME_READ – the average time that passed between a new post/topic appeared and the moment of reading that post/topic
- NO_CONCEPTS_COURSE – the number of concepts extracted from a course that were referred by a students in his posts. This feature will give us a view of how focused on the course is a student when he posts.
- FRQ_POSTS – the frequency of posting messages on the forum
- FRQ_TOPICS – the frequency of posting new topics on the forum

Based on these features we can model the users and the activity but only after the experiments we can say which of the features will remain. Feature selection is a complex problem because we need to choose the features that are strongly related for the learner's activity but also we need to choose the ones that offer better results for the chosen algorithms.

5.2 Regular Functionalities

Posting New Topics. Posting messages and replies to a topic, in this capability we include the possibility to edit or delete a message for a specific amount of time. It is also important here to save the original message in the database for a period so the moderators can solve any complains.

Well Defined Profiles. User profiles include the number of messages posted on the forum, for the current discipline but also the number of the other posts. The profiles also refer the total number of answered questions and the number of questions that were correctly answered. Still regarding the user profiles we may have the number of questions from the forum that were answered by user and was marked as useful answer by the topic initiator or other users. This capability provides us with a vision over how trustworthy the student is when he answers at questions from the forum.

5.3 Smart Functionalities

The smart functionalities make the difference between a regular discussion forum and Smart Forum. These functionalities are enhanced by some data mining and machine learning algorithms. Extracting several features from this Smart Forum creates the possibility to implement the features that makes the forum to get the "smart" attribute. Below we present the main functionalities that can be implemented and a short description.

Subjects of Interest. Based on the educational results the forum may offer some subjects of interest for students. This capability will be accomplished by performing an analysis over the student's answered questions and text analysis over the posts. If we have a match between the concepts that can be extracted from the wrong answered questions and the concepts extracted from some specific posts or topics from the forum we can recommend those posts to the students. Going further for this capability we can also see what concepts are considered interesting by the student and we can recommend him/her some topics that weren't read by him/her.

Approach. This capability can be accomplished also by having a match between concepts extracted from the forum and the concepts extracted from the topics that were read or answered by the student and the questions that were correctly answered.

There are two directions to gain the subjects that may be interesting for a student: offering the ones that are addressed a lot or the ones that reveals lower results. In order to get the ones that are the most addressed we need to perform text analysis (i.e. concept extraction) on both the questions answered and on the messages posted on forum.

In Fig. 4 is presented an overview o the system. First we need to use clustering algorithms (i.e. SKM clustering algorithm [11]) for getting the learning that are most addressed by the students and the area from the forum that is most addressed then the concepts can be extracted using stemming algorithms from both collections. Having a bigger percent of concepts that match means a better matching and then we can recommend them to be read.

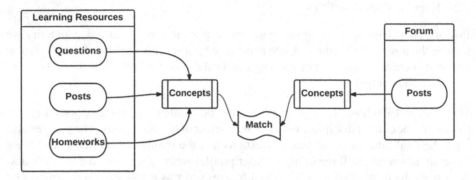

Fig. 4. Overview of the matching mechanism

In order to improve students' performances we need to get the learning resources (questions and home works) on which the learner got lower results, then using stemming getting the common concepts from them. Based on posts analysis we can get the most related messages and offer them to the learner to be read.

Computing Trends. Computing trends for students will also be a smart capability available in Smart Forum. This capability is referring by drawing some charts that will reveal how the number of posts, number of questions answered that were marked as useful, number of questions asked or number of topics read are varying over the time. This can be useful for predicting but also for measuring the student's engagement over the platform and, more specific, over the smart forum.

In Fig. 5 we present an example of trend that can be computed for a student over a full semester. On the OX axe we have the weeks from 1 to 12 and on OY axe we have the student's activity. At a deeper analysis we can see that the student starts from a activity of 4.5 and have a small progress to the week 6, then there is a decline of two weeks and then he has an ascending trend until the last week of the semester. The green line with no dots on it is the trend computed based on the student's activity. As we can see in this case is an ascending trend based on the student's activity. This activity can be approximated using several attributes that defines it.

Figure 6 presents the trend discretized for three periods, on the OX axe we have the weeks and on OY on every of the third graphs we have the level of activity. In Fig. 6 the grey line with dots represents the acitivity, the green line withought dots ascending trends, the blue line represents a stable trend and the red line represents an descendent trend computed for the student.

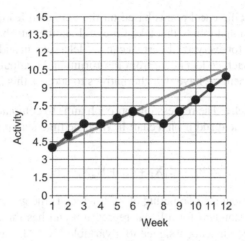

Fig. 5. Trend example

a)   b)

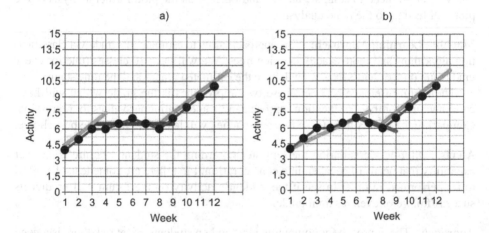

Fig. 6. Trend divided in periods

For the section (a) of the Fig. 6 the first period is represented by the first three weeks and have an ascending trend. We choose week number four to start the second period because there the trend is changing and we have stable trend untill we reach week number 8 where is another trend switch. The last period consists of the last four weeks and is an ascending trend.

Another aproach of computing trends is presented in (b) were we estimate the first 6 week at an ascending trend, then we have a descending trend between week six and 8 and then we have an ascending trend for the last weeks. This approach is easier to find if we parse the activity line during the period and brings up an allert for the user in week six as he might fail.

Approach. Computing trends based on students actions offers an overview of the student's evolution. This functionality can be accomplished using linear regression or decision trees at specific checkpoints based on the extracted features. Usually we can produce two dimensional trends that take time and another attribute on both axes but that will be one chart for every attribute that can model a student and it is hard to estimate the trend. Our approach collects data from the forum and computes the features from Sect. 5.1; based on them we have two alternatives to address this issue.

Regression Approach: The regression approach aims to compute an estimated grade based on the features and add it on a chart having the time stamp and the grade on the axes.

$$Y_i = B_0 + B_1 X_1 + \ldots + B_n X_n + E \tag{1}$$

In Eq. (1) we present the computing of "Y_i" which is the grade computed for week "i". The formula is standard for a linear regression and has three main components: "B_i" which are the coefficients assigned to a variable "X_i", "E" which is the intercept and the variables "X" which represents one of the attributes that models a instance. For every week we need to compute an "Y" and place it on the plot; further analysis of the plot will lead us to the trend analysis.

Machine Learning Approach: This approach aims to estimate the grade via classification algorithms and more exactly decision trees. We will have 10 classes (one for every grade) and on the chart's axes we will have the estimated grades and the time stamp.

The main difference between these two approaches is the grade which will have certain values for the classification approach and some continuous values for regression. Comparing the results from these two approaches will lead us to make the best choice.

Alerts. Offering alerts during the semester regarding the student's failure is another capability that needs to be taken into consideration. First there are some tests that need to be performed in order to see if the student's activity on the forum will predict its success/failure or even his final grade.

Approach. The approach for computing alerts when students are at risk there is a need for some pattern matching techniques. We plan to cover this approach with classification algorithms that can be employed using two or three classes. Based on the previous gathered data we can predict if a student is "at_risk" or "not_at_risk" but we also nned to perform some tests having three classes like "at_risk", "not_at_risk" and "avg_risk" where the last class can be used for students as a first warning level and then alert "at_risk" if the situation gets really bad.

6   Conclusions and Future Work

In this paper we presented the design and the integration approach for a forum that implements several smart functionalities that are based on underlying machine learning obtained models. Adding smart functionalities that address the goals described in the paper can improve the student's experience of the learners that perform their activities in on-line educational environments. From the many possibilities of implementing the prototype of smart forum we have chosen core algorithms that fit presented problems in terms of available input and needed task. Currently, the challenge is to choose proper machine learning algorithms and integrate with the data provided by Tesys e-Learning platform such that tasks within the smart forum may get practical and interpretable output.

The prototype version of the smart forum trains linear regression model in a predictive context for the output variable represented by the final grade. The segmented analysis provides finer grained trends for fixed of variable length timeframes and this approach is a more realistic one for the learning period of a student. Further analysis of trends in observed data is necessary as a mean to get an insight in the learning activity patterns of the students.

The second smart capability modelled as a classification problem and the linear regression model is used as a decision boundary between students "at_risk" or "not_at_risk". Further validation of derived models is needed such that confidence values in obtained predictions and classification are obtained.

As future work, improving the logging mechanism from Tesys and smart form will provide better and comprehensive real world datasets and thus the ability to extract features for better describing the observations from the training dataset.

Another improvement regards implementing more functionality in smart forum for processing the activity traces with the task of creating better student models.

References

1. Chih-Hsiung, T., McIsaac, M.: The relationship of social presence and interaction in online classes. Am. J. Distance Educ. **16**(3), 131–150 (2002). doi:10.1207/S15389286AJDE1603_2
2. Jennifer, C.: Richardson & tim newby, the role of students' cognitive engagement in online learning. Am. J. Distance Educ. **20**(1), 23–37 (2006). doi:10.1207/s15389286ajde2001_3
3. Robinsona, C.C., Hullingera, H.: New benchmarks in higher education: student engagement in online learning. J. Educ. Bus. **84**(2), 101–109 (2008). doi:10.3200/JOEB.84.2.101-109
4. Dringus, L.P., Ellis, T.: Using data mining as a strategy for assessing asynchronous discussion forums. Comput. Educ. **45**(1), 141–160 (2005). http://dx.doi.org/10.1016/j.compedu. 2004.05.003. ISSN 0360-1315
5. Liu, W., Kidziński, L., Dillenbourg, P.: Semiautomatic annotation of MOOC forum posts. Chapter State-of-the-Art and Future Directions of Smart Learning Part of the series Lecture Notes in Educational Technology, pp 399–408. doi:10.1007/978-981-287-868-7_48 Print ISBN 978-981-287-866-3 Online ISBN 978-981-287-868-7
6. Romero, C., López, M.-I., Luna, J.-M., Ventura, S.: Predicting students' final performance from participation in on-line discussion forums. Comput. Educ. **68**, 458–472 (2013). http:// dx.doi.org/10.1016/j.compedu.2013.06.009. ISSN 0360-1315
7. Huang, G.-B., Zhu, Q.-Y., Siew, C.-K.: Extreme learning machine: theory and applications. Neurocomputing **70**(1–3), 489–501 (2006). http://dx.doi.org/10.1016/j.neucom. 2005.12.126. ISSN 0925-2312
8. Purdila, V., Pentiuc, S.-G.: Fast decision tree algorithm. Adv. Electr. Comput. Eng. **14**(1), 65–68 (2014). doi:10.4316/AECE.2014.01010

9. Chao-Ying, J.P., Leea, K.L., Ingersolla, G.M.: An introduction to logistic regression analysis and reporting. J. Educ. Res. **96**(1), 3–14 (2002)
10. Bradley, A.P.: The use of the area under the ROC curve in the evaluation of machine learning algorithms. Pattern Recogn. **30**(7), 1145–1159 (1997). http://dx.doi.org/10.1016/S0031-3203(96)00142-2. ISSN 0031-3203
11. Kanungo, T., Mount, D.M., Netanyahu, N.S., Piatko, C.D., Silverman, R., Wu, A.Y.: An efficient k-means clustering algorithm: analysis and implementation. IEEE Trans. Pattern Anal. Mach. Intell. **24**(7), 881–892 (2002). doi:10.1109/TPAMI.2002.1017616

Optimization-SVM (OPSVM)

Leaf Disease Recognition in Vine Plants Based on Local Binary Patterns and One Class Support Vector Machines

Xanthoula Eirini Pantazi[1(✉)], Dimitrios Moshou[1],
Alexandra A. Tamouridou[1], and Stathis Kasderidis[2]

[1] Agricultural Engineering Laboratory, School of Agriculture, Aristotle University,
54124 Thessaloniki, Greece
renepantazi@gmail.com
[2] NOVOCAPTIS, i4G Business Incubator, Antoni Tritsi 21, P.O. Box 22461
55102 Thessaloniki, Greece

Abstract. The current application concerns a new approach for disease recognition of vine leaves based on Local Binary Patterns (LBPs). The LBP approach was applied on color digital pictures with a natural complex background that contained infected leaves. The pictures were captured with a smartphone camera from vine plants. A 32-bin histogram was calculated by the LBP characteristic features that resulted from a Hue plane. Moreover, four One Class Support Vector Machines (OCSVMs) were trained with a training set of 8 pictures from each disease including healthy, Powdery Mildew and Black Rot and Downy Mildew. The trained OCSVMs were tested with 100 infected vine leaf pictures corresponding to each disease which were capable of generalizing correctly, when presented with vine leave which was infected by the same disease. The recognition percentage reached 97 %, 95 % and 93 % for each disease respectively while healthy plants were recognized with an accuracy rate of 100 %.

Keywords: Image processing · Novelty detector · Classifier conflict · Texture descriptors

1 Introduction

Plant recognition is regarded as a puzzling computer vision problem which requires dealing with irregular shapes and textures. Interest in methods for visual classification of plants has grown recently [1] as devices equipped with cameras became ubiquitous, making intelligent field guides, education tools and automation in forestry and agriculture practical. Belhumeur et al. [1] presented how using such a system in the field enables a botanist to quickly search vast collections of plant species - a process that was time consuming and laborious can now be implemented in seconds. Recognition of leaves is usually applied on broad leaves and needles are treated separately. Several techniques have been introduced for leaf description, utilizing shape features and colour features. The leaf recognition approach which was presented by Fiel and Sablatnig [2] was based on a Bag of Words model with SIFT descriptors and reached 93.6 % accuracy on a leaf dataset of 5 Austrian tree species. Kumar et al. [3] proposed Leafsnap, which is a

L. Iliadis and I. Maglogiannis (Eds.): AIAI 2016, IFIP AICT 475, pp. 319–327, 2016.
DOI: 10.1007/978-3-319-44944-9_27

computer vision system capable of identifying plant species, They introduced a pre-filter on input images, numerous speed-ups and additional post-processing within the segmentation algorithm, the use of a simpler and more efficient recognition algorithm based on curvature instead of Inner Distance Shape Context (IDSC); an image dataset with more images, and an interactive system that can be used by non-expert users. The final product was an electronic field guide, available as a free mobile app for iOS devices. Even though the app runs on iPad and iPhone devices, the images of the leaves have to be processed on a server. This requires an internet connection to achieve recognition, so it is problematic for use in natural areas without access to internet or a limited bandwidth connection. Another severe limitation concerns the need to obtain photos of the leaves with a white background. The contribution of the current paper concerns an effective combination of pre-processing stages and One Class Classifiers applied on complex natural scenes with aim to discover early stages of infections in field grown crops enabling faster and more effective prevention of crop epidemics that threaten food supply and quality. The presented approach consists of training one classifier per disease and presenting the feature vector to a committee of one class classifiers which decide autonomously by having been trained only on target data. In the event that there is a multiple activations of one class classifiers the resultant conflict is overcome by an original conflict resolution method that utilizes the proximity of the nearest support vector to make an accurate decision which reached 97 % for testing with unforeseen examples. The mentioned technique can generalize in multiple crops for three diseases while it has been trained only with vine leaves.

2   Materials and Methods

2.1   Image Segmentation

The segmentation operation needed to extract the leaf that has the symptoms was implemented by making use of the GrabCut algorithm. The basic steps for implementing the GrabCut algorithm in the leaf segmentation [4] are presented as follows:

The initial information provided about the foreground and the backgrounds are given by the user is a rectangular selection around the object of interest. Pixels that reside outside this selection are considered as known background and other pixels inside are marked as unknown. From this data we wish to create a model which can be used to decide if the unknown pixels are foreground or background. In the GrabCut algorithm this is achieved by creating K components of multivariate Gaussian Mixture Models (GMM) corresponding to the two regions. K components are created from the known background and K components correspond to the region that could belong to the foreground, giving a total of 2*K components. The GMM components have the same dimensions as the color space and they are defined based on the color statistics from each cluster. In order to obtain a good segmentation we need to add components with low variance because this makes cluster separation easier (Figs. 1 and 2).

Fig. 1. The original image of a vine leaf that has been fed to the model of powderyHSV

Fig. 2. Segmentation of vine leaf by using the GrabCut algorithm on the original image.

2.2   The Local Binary Patterns (LBPs)

The local binary pattern is a simple but very effective texture operator which tags the pixels of an image by thresholding the region of each pixel and considers the outcome as a binary number. The LBP scheme can be seen as a fusing approach to the traditionally deviating statistical and structural representations of texture analysis. Perhaps the most significant property of the LBP operator in actual applications is its invariance beside monotonic gray level deviations caused, e.g., by illumination differences. Another like-wise important is its computational effortlessness, which makes it possible to analyze images in puzzling real-time situations. The original local binary pattern operator, presented by Li et al. [5] and Llado et al. [6] was founded on the conjecture that texture has locally two balancing aspects, a pattern and its strength. The operator works in a 3×3 locality, using the center value as a threshold. An LBP code is created my multi-plying the thresholded values with loads given by the corresponding pixels, and summing up the outcome. As the neighborhood comprises of 8 pixels, a total of $28 = 256$ dissimilar labels can be obtained dependent on the relative gray values of the epicenter and the pixels in the locality. The contrast measure (C) is acquired by subtracting the average of the gray levels under the center pixel after that of the gray levels above (or equal to) the center pixel. If the eight thresholded neighbors of the middle pixel have an identical value (0 or 1), the value of contrast is set to zero. The distributions of LBP codes are used as features in grouping or segmentation.

The local binary pattern (LBP) operator is an image operator which converts an image into a group or image of integer labels relating to small-scale appearance of the image. These labels or their statistics, and one of the most commonly used is the histo-gram, are then applied for further image analysis. The most frequently used versions of the operator are aimed for monochrome still images nevertheless it has been extended likewise for color (multi channel) images and also for videos and volumetric data. An example of an H-plane image and corresponding LBP image are shown in Figs. 3 and 4.

Fig. 3. H-plane of the vine leaf after the background of the initial image has been removed.

Fig. 4. The corresponding LBP image of the vine leaf.

2.3   The Creation of the LBP Histogram

The segmented image acquired from the GrabCut algorithm is utilized to create different monochrome descriptions as the LBP algorithms operates in single channels. Three diverse options were tried, the gray level, Hue channel and saturation channel. The most performing was the Hue channel LBP histogram as the results showed in the recognition of diseased plants against healthy plants. The image of the Hue channel is shown in Fig. 4. By applying the LBP transformation on the Hue image the textural features corresponding to contrast and hue change are exploited by labelling neighboring pictures according to the LBP algorithm explained above. The LBP image obtained from this process can be shown in Fig. 5. The histogram of the LBP image is constructed according to the frequency of the occurrence of similar values between 0 and 255, a low length histogram containing 32 bins, has been used to partition the range of 255, so as to more rough features instead of local high resolution features which might demonstrate high variability.

Fig. 5. The segmented image has been used to obtain the HSV planes. Here, the Hue plane is shown.

2.4 One Class Support Vector Machines (OCSVMs)

To support One Class SVM Classification a suitable description of the configuration of SVM as a model to designate target data was presented by Tax and Duin [7] in the form of Support Vector Data Description (SVDD). The objective of SVDD is to learn a decision function in order to predict if an example is actually a target or an outlier. SVDD is created based on the supposition that targets are bounded by a close boundary in the feature space. The OCSVM creates a model from performing adjustment by normal data according to the SVDD description. At the following stage, orders test data based on the deviance from normal training data as being either normal or outlier [9].

2.4.1 One-Class Support-Vector-Learning for Multi Class Problems

The described one-class approach can be applicable to multiclass problems in the following way. It is anticipated that data samples from k different classes are given. Let $\Omega = \{1,\ldots, k\}$ denotes the set of classes. For each distinct class $t \in \Omega$ a one-class classifier f_t is created. Specifically, k centres and corresponding radii R_t are considered. To determine the class relationship of a new undetected pattern z all decision functions are calculated in parallel leading to class memberships. In the classification stage of an undetected input pattern z three probable resulting circumstances have to be distinguished:

1. If just one classifier shows its bond for the new pattern, the resultant class label is the class membership of this classifier.
2. If more classifiers raise a claim for this pattern, a conflict situation has arrived.
3. On the other hand it is possible that no classifier matches, so in this case we can characterize this outcome as an outlier situation.

The problematic with such uncertain (2 and 3) classification assignments can be disentangled in two ways: Discarding the classification of the novel pattern or finding the matching class label through an outlier-/conflict processing approach. In the subsequent section we will present a strategy to crack such problems [9].

2.5 Nearest-Support-Vector Strategy

The class label of a novel pattern z is assessed by comparing the remoteness to the support vectors of the involved classifiers. The classifier that has the nearest support vector defines the class membership of z assuming that are the support vectors, and i_t is the amount of support vectors of classifier t.

$$SV(t) = \hat{x}_1^t, \ldots\ldots, \hat{x}_{1t}^t \tag{1}$$

$$f(z) = \operatorname{argmin}_{l \in C} \| \hat{x}_i^l - z \| \tag{2}$$

The decision function is then demarcated by where the minimum will be taken over all the support vectors of classes after the candidate set [8].

3   Results and Discussion

3.1   Application of LBP in Disease Recognition of Infected Vine Leaves

Initially, smartphone photo is obtained from infected plants. The standard requirement is that a leaf demonstrating symptoms of powdery mildew is situated in the center of the image. To obtain only the leaf area and subtract the background, a segmentation operation is performed. Then, the HSV image is obtained and the Hue plane is shown in Fig. 5. The corresponding LBP image obtained from the Hue plane of Fig. 5 is shown in Fig. 6. Then the LBP image is used in order to calculate the histogram with 32 bins. The global LBP histogram equation is given as follows:

$$H(i) = \sum_{x,y \in image} I(f(x,y) = i) \qquad (3)$$

where $i = 1,...32$, I represents the indicator function used for thresholding and $f(x,y)$ represents the LBP label. The histograms have been used as input in One Class Classifiers in the form of One Class SVMs in order to recognize specific leaf samples as a target class and treat healthy samples and other samples from different diseases as outliers. The Kernel function that has been used was RBF. The default width of the tube was $e = 0.1$. Eight samples were enough for building powdery mildew model in vines that can recognize this disease. A similar procedure was followed for healthy leaves of vines, by building an one Class SVM model that could recognize only healthy leaves. Cross testing of each One Class SVM with leaves from the other category showed perfect classification for the calibration sets and additional samples that were picked randomly from the internet and smartphone. In the current work OpenCV Computer Vision Library Version 2.4.11 has been used for image processing and classifier development.

Fig. 6. The corresponding LBP image of the vine leaf.

The One Class SVM that has been trained on vines to recognize powdery mildew is called 'powderyHSV' and it has been stored in XML format. It has been tested in various infected vine leaves successfully. Here, it is tested on vine leaves.. Initially, the powderyHSV One Class SVM recognizes the disease successfully (Fig. 7). However, in order to test the selectivity of the model, a second model has been utilized, that is called 'blackrotHSV' and recognizes black rot (Fig. 8). As it is shown below (Fig. 7), both models are activated from the observed symptoms as belonging to the sphere of confidence of one Class SVM of each model. Conflict resolution takes place by calculating the distances from the support vectors of the test image feature vector (Hue histogram) (Fig. 7). The result of the conflict resolution is that the actual model, it is identified to be powderyHSV, which is correct (Fig. 7).

Fig. 7. Conflict resolution takes place and the image is labelled correctly as infected with powdery mildew.

Fig. 8. Vine leaf image without conflict resolution is classified correctly as Black Rot infected.

One hundred samples from each class were used (actual infection from powdery mildew, black rot and downy mildew). Without applying nearest vector based conflict resolution, it was impossible to achieve correct classification in 50 % of the cases. Through conflict resolution, 100 % was reached in classifying both infected and healthy state. The numbers reached 97 % for powdery mildew identification, compared to support vector machine model for black rot and downy mildew. The recognition percentage reached 95 % and 93 % for black rot and downy mildew respectively. In all disease cases, the test samples came from vine leaves that were not included to the training set. Support Vector Machine training was accomplished, as mentioned above, with 7 samples that all came from vine cultivations' leaves.

This fact alone indicates the high generalization capability of the model, demonstrated and validated through the efficient performance on different vine leaf images that were gathered under varying conditions. The most significant advantage of this method does not only regard the application of One Class Classifiers, but also the highly effective way of discerning between infected leaves and image background. This property derives from the application of the Gaussian mixtures based segmentation method, remarkably effective for isolating image areas of sharp variation in color. The effectiveness of the presented application is a result of the sharp color variation that occurs at infected leaves, in comparison with healthy ones, thus an immediate differentiation is possible. In cases with a complex background, GrabCut algorithm can be trained to recognize a certain foreground not corresponding to the central area of the image. This ability was not required for use in this application, but can be optionally activated according to the criterion of right choice of each infected leaf in examination.

4 Conclusions

In the current paper an effective combination of pre-processing stages and One Class Classifiers applied on complex natural scenes is presented. The main aim of the presented research was to discover early stages of infections in vine leaves in order to enable faster and more effective prevention of vine epidemics that threaten food supply and quality. Four One Class Support Vector Machines (OCSVMs) were trained with a training set of 8 pictures from each disease including healthy, Powdery Mildew and Black Rot and Downy Mildew. The trained OCSVMs were tested with 100 infected vine leaf pictures corresponding to each disease. The recognition percentage reached 97 %, 95 % and 93 % for each disease respectively while healthy plants were recognized with an accuracy rate of 100 %.

Acknowledgements. The presented study has been funded by PPP-FI/FRACTALS project.

References

1. Belhumeur, P.N., et al.: Searching the world's herbaria: a system for visual identification of plant species. In: Forsyth, D., Torr, P., Zisserman, A. (eds.) ECCV 2008, Part IV. LNCS, vol. 5305, pp. 116–129. Springer, Heidelberg (2008)
2. Fiel, S., Sablatnig, R.: Automated identication of tree species from images of the bark, leaves and needles. In: Proceedings of 16th Computer Vision Winter Workshop, Mitterberg, Austria, pp. 1–6 (2011)
3. Kumar, N., Belhumeur, P.N., Biswas, A., Jacobs, D.W., Kress, W., Lopez, I.C., Soares, J.V.: Leafsnap: a computer vision system for automatic plant species identification. In: Fitzgibbon, A., Lazebnik, S., Perona, P., Sato, Y., Schmid, C. (eds.) ECCV 2012, Part II. LNCS, vol. 7573, pp. 502–516. Springer, Heidelberg (2012)
4. Rother, C., Kolmogorov, V., Blake, A.: GrabCut: interactive foreground extraction using iterated graph cuts. ACM Trans. Graph. **23**, 309–314 (2004)

5. Li, J., Wu, W., Wang, T., Zhang, Y.: One step beyond histograms: image representation using Markov stationary features. In: Proceedings of IEEE Conference on Computer Vision and Pattern Recognition, pp. 1–8 (2008)
6. Llado, X., Oliver, A., Freixenet, J., Marti, R., Marti, J.: A textural approach for mass false positive reduction in mammography. Comput. Med. Imaging Graph. **33**, 415–422 (2009)
7. Tax, D., Duin, R.: Support vector data description. Mach. Learn. **54**(1), 45–66 (2004)
8. Schölkopf, B., Platt, J.C., Shawe-Taylor, J., Smola, A.J., Williamson, R.C.: Estimating the support of a high-dimensional distribution. Neural Comput. **13**(7), 1443–1471 (2001)
9. Sachs, A., Thiel, C., Schwenker, F.: One-class support-vector machines for the classification of bioacoustic time series. ICGST Int. J. Artif. Intell. Mach. Learn. (AIML) **6**(4), 29–34 (2006)

Efficient Support Vector Machine Classification Using Prototype Selection and Generation

Stefanos Ougiaroglou[1($\boxtimes$)], Konstantinos I. Diamantaras[1],
and Georgios Evangelidis[2]

[1] Department of Information Technology,
Alexander TEI of Thessaloniki, 57400 Sindos, Greece
stoug@uom.gr, kdiamant@it.teithe.gr
[2] Department of Applied Informatics,
University of Macedonia, 54006 Thessaloniki, Greece
gevan@uom.gr

Abstract. Although Support Vector Machines (SVMs) are considered effective supervised learning methods, their training procedure is time-consuming and has high memory requirements. Therefore, SVMs are inappropriate for large datasets. Many Data Reduction Techniques have been proposed in the context of dealing with the drawbacks of k-Nearest Neighbor classification. This paper adopts the concept of data reduction in order to cope with the high computational cost and memory requirements in the training process of SVMs. Experimental results illustrate that Data Reduction Techniques can effectively improve the performance of SVMs when applied as a preprocessing step on the training data.

Keywords: Support Vector Machines · k-NN classification · Data reduction · Prototype abstraction · Prototype generation · Condensing

1 Introduction

The effectiveness, efficiency and scalability of machine learning and data mining algorithms are crucial research issues that have attracted the attention of both the industry and academia. Many proposed algorithms cannot handle high volumes of data that nowadays is easily available from several data sources. For those algorithms, data reduction[1] is an important preprocessing step.

In classification tasks data reduction processes are guided by the class labels. Many Data Reduction Techniques (DRTs) have been proposed in the context of dealing with the drawbacks of k-Nearest Neighbors (k-NN) classifier [9]. These drawbacks are: (i) the high computational cost during classification, (ii) the high memory requirements and (iii) the noise sensitivity of the classifier. However,

---

[1] Data Reduction has two points of view: (i) item reduction, and, (ii) dimensionality reduction. We consider them from the first point of view.

© IFIP International Federation for Information Processing 2016
Published by Springer International Publishing Switzerland 2016. All Rights Reserved
L. Iliadis and I. Maglogiannis (Eds.): AIAI 2016, IFIP AICT 475, pp. 328–340, 2016.
DOI: 10.1007/978-3-319-44944-9_28

that kind of data reduction has not been adopted by other classification methods which cannot manage large datasets.

A DRT can be either a Prototype Selection algorithm (PS) [10] or a Prototype Generation algorithm (PG) [20]. PS algorithms select representative instances from the training set. PG algorithms generate representatives by summarizing similar training instances. These representatives are called Prototypes. PS algorithms can be either editing or condensing. Editing aims at improving accuracy by removing noise, outliers and mislabeled instances and by smoothing the decision boundaries between classes. PG and PS-condensing algorithms try to build a small condensing set that represents the initial training data. Using a condensing dataset instead of the original dataset has the avail of low cost while accuracy remains almost as high as that achieved by using the original data. Please note that some PG and PS-condensing algorithms are called hybrid because they integrate the concept of editing.

To the best of our knowledge, DRTs have been used in the context of k-NN classification. There is no work that explores the application of data reduction on large datasets in order to render the usage of SVMs applicable on them. This is the key observation behind the motivation of the present work. Another motive is to check whether PG algorithms we proposed in the past can aid the development of fast and accurate SVM based classifiers.

This paper also contributes an experimental study on several datasets where SVM based classifiers, which are trained by the original training data and the corresponding condensing sets built by state-of-the-art DRTs, are compared to each other and against the corresponding k-NN classifiers. The paper reviews in detail the algorithms that are used in the experimental study. Our study reveals that the usage of DRTs leads to fast and accurate SVM-based classifiers.

The rest of this paper is organized as follows. Section 2 briefly reviews SVMs and the k-NN classifier and Sect. 3 presents in detail the PG and PS-condensing algorithms that we use in our experimental setup. Section 4 presents the experimental study and results. Finally, Sect. 5 concludes the paper.

2   Support Vector Machines and the k-NN Classifier

2.1   Support Vector Machines

SVMs are supervised learning models introduced in 1995 by Cortes and Vapnik [7] although the roots of the idea lie in the theory of statistical learning introduced by Vapnik almost two decades earlier [21]. They are suitable for pattern classification but can be easily extended to handle nonlinear regression problems in which case they are known as Support Vector Regressors (SVRs). The separating surface offered by an SVM classifier maximizes the *margin*, i.e., the distance of the closest patterns to it. This helps the generalization performance of the model and in fact it is related to the idea of Structural Risk Minimization [22,23] which avoids over-fitting. With the use of nonlinear kernel functions such as Gaussian (RBF) or n-th order polynomials, SVM models can produce nonlinear separating surfaces achieving very good performance in complex problems.

Due to their good generalization performance these models have become very popular with a wide range of applications, including document classification, image classification, bio-informatics, handwritten character recognition, etc. One of the major drawbacks of these models is the memory and the computational complexity requirements for large datasets. The reason is that the separating surface is obtained by solving a quadratic programming problem involving an $N \times N$ matrix, where N is the number of items in the dataset. Although there are techniques that can reduce the complexity to $O(N^2)$ [5], the problem remains hard and the size of the problem can easily become prohibitively large calling for methods for data reduction such as the ones discussed in the following sections.

2.2 k-Nearest Neighbor Classifier

The k-NN classifier [9] is an extensively used lazy (or instance-based) classification algorithm. Contrary to eager classifiers, it does not build any classification model. Some of its major properties are: (i) it is a quite simple and easy to implement algorithm, (ii) contrary to many other classifiers, it is easy to understand how a prediction has been made, (iii) it is analytically tractable and (iv) for $k = 1$ and unlimited instances the error rate is asymptotically never worse than twice the minimum possible, which is the Bayes rate [8].

The algorithm classifies a new instance by retrieving from the training set the k nearest instances to it. These instances are called neighbors. Subsequently, the algorithm assigns the new instance to the most common class among the classes of the k nearest neighbors. This class is called the major class. The process that indicates the major class is usually called nearest neighbors voting. Although any distance metric can be used, the Euclidean distance is the commonly-used distance metric. The k-NN classifier does not spend time in training any model. However, the classification step is time-consuming because in the worst case the algorithm must compute all distances between the new instance and all the training instances.

The selection of the value of k affects the accuracy of the classifier. The value of k that has the highest accuracy depends on the data. Its determination implies tuning via trial-and-error. Usually, large k values are appropriate for datasets with noise since they examine larger neighborhoods, whereas, small k values render the classifier noise-sensitive. In binary problems, an odd value for k should be used. Hence, possible ties in the nearest neighbors voting are avoided. In problems with more than two classes, ties are resolved by choosing a random "most common" class or the class voted by the nearest neighbor. The later is adopted in the experimental study of this paper.

3   Prototype Generation and Condensing Algorithms

Several PG and PS-condensing algorithms are available in the literature. Here we review only the ones used in our experimental study. For the interested reader, abstraction and selection algorithms are reviewed, categorized and compared to each other in [10,20]. Other interesting reviews are presented in [4,13,19,24].

3.1   Condensing Nearest Neighbor Rule

The Condensing Nearest Neighbor (CNN) rule [11] is the earliest condensing algorithm. Its condensing set is built by the following simple idea. Instances that are far from decision boundaries ("internal") data area of a class can be removed without loss of accuracy. Thus, CNN-rule tries to keep only the instances that lie in the close-border areas. The close-border instances are selected as follows. Initially, an instance of the training set (TS) is moved to the condensing set (CS). CNN-rule uses the 1-NN rule and classifies the instances of TS by examining the instances of CS. If an instance is wrongly classified, it is probably close to decision boundaries. Therefore, it is moved from TS to CS. This procedure is repeated and if there are no moves from TS to CS in a complete pass of TS, the algorithm terminates.

CNN-rule is misled by noise. It wrongly selects "noisy" instances with their neighborhood. Consequently, noise affects the reduction rates. CNN-rule determines the number of the prototypes automatically, without user-defined parameters. Another property is that the multiple passes over the data guarantees that the removed training instances are correctly classified by 1-NN classifier in the context of the condensing set. A disadvantage is that CNN-rule builds a different condensing set by examining the same training instances in a different order.

3.2   The IB2 Algorithm

IB2 is an one pass version of CNN-rule. IB2 is one of the Instance-Based Learning (IBL) algorithms presented in [1,2]. Each training instance $x \in TS$ is classified by the 1-NN rule on the current CS. If x is classified correctly, x is discarded. Else, x is moved to CS.

IB2 determines the size of the condensing set automatically. However, the condensing set highly depends on the order of training instances. Since it is a one-pass algorithm, it is very fast. Also, IB2 does not guarantee that the removed instances can be correctly classified by the condensing set. In addition, it builds its condensing set in an incremental manner. This means than new training instances can update an existing condensing set without considering the "old" instances that had been used for the creation of the condensing set. Hence, IB2 can be applied in streaming environments where new instances are gradually available.

3.3   The AIB2 Algorithm

In [15] we presented a PG variation of IB2. It is called Abstraction IB2 (AIB2) and inherits all the properties of IB2. AIB2 considers that the prototypes should be close to the center of the data area they represent. Contrary to IB2, AIB2 does not ignore the instances that were correctly classified. These instances contribute to the condensing set by repositioning the nearest prototype. To achieve this, each prototype has a weight value that denotes the number of instances it represents.

In an early step, a random training instance is placed in the condensing set and its weight becomes one. For each training instance x, AIB2 fetches its nearest prototype P from the current condensing set. If x has a class label different than the one of P, it is moved to the condensing set and plays the role of a prototype. Its weight becomes one. if x has the class label of P, the attributes of P are updated by taking into account the attributes of x and its weight. More formally, each attribute $attr(i)$ of P becomes $P_{attr(i)} \leftarrow \frac{P_{attr(i)} \times P_{weight} + x_{attr(i)}}{NN_{weight} + 1}$. Thus, P moves towards x. Of course, the weight of P is increased by one and x is discarded.

3.4   The Reduction by Space Partitioning Algorithms

Chen's Algorithm. The ancestor of the Reduction by Space Partitioning (RSP) algorithms is the PG algorithm proposed by Chen and Jozwik (Chen's algorithm) [6]. Chen's algorithm retrieves the instances that define the diameter of the training data, in other words the two most distant instances, a and b. Then, the algorithm splits the training data into two subsets. All the instances that are closer to a are moved to C_a. All other instances are placed in C_b. Subsequently, Chen's algorithm selects to split the non-homogeneous subset with the largest diameter. Non-homogeneous are called the subsets that have instances of more than one class. If there is no non-homogeneous subsets, the algorithm proceeds by spitting the homogeneous subsets. When the number of subsets is equal to a value specified by the user, the aforementioned procedure ends. The final step is the generation of prototypes. Each subset C is replaced by its mean instance. The class label of the mean instance is the major class in C. The mean instances constitute the condensing set.

The idea of splitting the homogeneous subset with the largest diameter is based on that this subset probably has more instances and thus, if it is split first, higher reduction will be achieved. Chen's algorithm generates the same condensing set regardless of the ordering of the instances. A drawback is that the user has to specify the number of subsets. Chen and Jozwik claim that this allows the user to define the trade-off between reduction rate and accuracy. However, the determination of this parameter implies costly trial-and-error procedures. Another weak point is that the instances that do not belong to the major class of the subset are not represented in the condensing set (they are ignored).

The RSP1 Algorithm. RSP1 [18] is similar to Chen's algorithm, but it does not ignore instances. It computes as many means as the number of distinct classes in the non-homogeneous subsets. RSP1 builds larger condensing sets than Chen's algorithm. However, it tries to improve the quality of the condensing set by taking into account all training instances.

The RSP2 Algorithm. RSP2 selects the subset that will be split first by examining the overlapping degree. The overlapping degree of a subset is the ratio of the average distance between instances belonging to different classes and the average distance between instances that belong to the same class. This splitting criterion assumes that instances that belong to a class are as close to each other as possible whereas instances that belong to different classes lie as far as possible. As stated in [18], it is better to split the subset with the highest overlapping degree than that with the largest diameter.

The RSP3 Algorithm. RSP3 [18] is the only RSP algorithm (Chen's algorithm included) that builds its condensing set without any user specified parameter. RSP3 eliminates both weaknesses of Chen's algorithm. It splits all the non-homogeneous subsets. In other words, it terminates when all subsets become homogeneous. RSP3 can use either the diameter or the overlapping degree as spiting criterion. In effect, the selection of splitting criterion is an issue of secondary importance because all non-homogeneous subsets are eventually split. Certainly, the order of the training instances is irrelevant.

RSP3 generates many prototypes for close-border areas and few prototypes for "internal" areas. The size of the condensing set depends on the level of noise in the data. The higher the level of noise, the smaller subsets constructed and the lower reduction is achieved. Please note that the discovery of the most distant instances is a time-consuming procedure since all distances between the instances of the subset should be estimated. Thus, the usage of RSP3 may be prohibitive in the case of large datasets. Since we wanted to consider only non-parametric algorithms in our experimental study, we used only RSP3.

3.5   Reduction Through Homogeneous Clusters

The RHC Algorithm. We have recently proposed the Reduction through Homogeneous Clusters (RHC) algorithm [14,16]. It belongs to PG algorithms. Like RSP3, RHC is based on the concept of homogeneity but employs k-means clustering [12,25]. Initially, the training data is considered as a non-homogeneous cluster in C. The algorithm computes a mean instance for each class in C. These mean instances are called class-means. Subsequently, RHC uses k-means clustering on C by adopting the class-means as initial means for k-means. The result is the creation of as many clusters as the number of discrete classes in C. This clustering process is applied on each non-homogeneous cluster. In the end, all clusters are homogeneous and each cluster contributes a prototype in the condensing set that is constructed by averaging the instances of the cluster.

RHC generates many prototypes for close-border areas and fewer for the "internal" areas. RHC uses the class-means as initial means for the k-means clustering in order to quickly find large homogeneous clusters. This property has the advantage of achieving a high reduction rate (the larger clusters discovered, the higher reduction rates achieved). Obviously, the instances that are noise can affect the reduction rates. Since RHC is based on k-means clustering, it is fast.

Also, its condensing set does not depend on the ordering of the training data. The experimental study presented in [14,16] shows that RHC has higher reduction rates and is faster than RSP3 and CNN-rule, whereas accuracy remains high. Please note, that dRHC [16] is a variation of RHC that handles large datasets that cannot reside in the main memory.

The ERHC Algorithm. The Editing and Reduction through Homogeneous Clusters (ERHC) [17] algorithm is a simple variation of RHC that tries to deal with noisy data. ERHC differs from RHC on the following point: Whenever a homogeneous cluster with only one instance is discovered, ERHC discards it. Thus, the final condensing set contains the means of the homogeneous clusters that have more than one instance. Obviously, ERHC integrates an editing mechanism. It simultaneously removes noise and reduces the size of the training set. Therefore, it can be characterized as hybrid PG algorithm. The experimental study in [17] proves that this simple editing mechanism can improve classification performance when data contains noise.

4   Performance Evaluation

4.1   Experimental Setup

We conducted several experiments on thirteen datasets distributed by the KEEL repository[2] [3]. Their profiles are presented in Table 1. Five datasets do not contain noise. All the other datasets have noise of various levels (see column "Noise" in Table 1). We do not use any editing algorithm for noise removal. For each dataset, we built six condensing sets. They were built by applying the algorithms presented in Sect. 3. More specifically, we used CNN-rule, IB2, RSP3, RHC, ERHC and AIB2.

We trained several SVMs on the original training set (without data reduction) and for each condensing set by using several parameter values. Finally, we kept the most accurate SVMs. In Subsect. 4.2, we report only the accuracy measurements for that SVM. The RBF kernel was used and the hyper-parameters γ, C where obtained through grid-search. Due to space restrictions, the parameter values we adopted are not reported[3].

For the five "noise-free" datasets, the k-NN classifier was run over the original training data and over the six condensing sets by setting $k = 1$. Most of the time $k = 1$ is the best choice for noise-free data. For the other eight datasets, we adopted four k values, namely, 1, 5, 9 and 13.

All measurements presented in Subsect. 4.2 are average values obtained via a five-fold cross-validation. We used the Euclidean distance as distance metric. Since CNN-rule, IB2 and AIB2 depend on the order of class labels in the training set, we randomized all the datasets. Excluding CAR, we did not perform any

---

[2] http://sci2s.ugr.es/keel/datasets.php.
[3] They can be found in http://users.uom.gr/~stoug/aiai2016.pdf.

Table 1. Dataset details

Dataset	Size (items)	Attributes	Classes	Noise
Letter Image Recognition (LIR)	20000	16	26	False
Pen-Digits (PD)	10992	16	10	False
Landsat Satellite (LS)	6435	36	6	True
Banana (BN)	5300	2	2	True
Balance (BL)	625	4	3	True
Texture (TXR)	5500	40	11	False
Yeast (YS)	1484	8	10	True
Phoneme (PH)	5404	5	2	False
MONK2 (MN2)	432	6	2	True
Twonorm (TN)	7400	20	2	True
Magic Gamma Telescope (MGT)	19020	10	2	True
Shuttle (SH)	58000	9	7	False
Car (CAR)	1728	6	4	True

other transformations. The CAR dataset has ordinal attributes. We transformed the attribute values into numerical values. Furthermore, we normalized to the interval [0-1] all attribute values of CAR.

4.2 Experimental Results

We compared the six DRTs to each other by estimating the Preprocessing Cost (PC) and the Reduction Rate (RR) that they achieved. Since the larger the training set used, the higher the cost for k-NN classifier to classify a new item and the higher the cost of the training procedure of SVMs (it is at least $O(N^2)$ see Subsect. 2.1), the RR measurements reflect the computational cost (the higher the RR, the lower the computational cost of k-NN classification and SVM training). Therefore, we do not include time measurements in our study.

Table 2 presents the RR and PC measurements. Best measurements are in bold. The last row shows the averages values. We observe that ERHC achieved the highest RR. This means that the SVM that uses the condensing set built by ERHC requites the least time for its training. AIB2 is the fastest DRT. It builds its condensing set by computing the fewest distances. On the other hand, RSP3 needs the highest computational cost in order to build its condensing set. In addition RSP3 seems to build the largest condensing sets. As expected, ERHC achieves higher RR than RHC and AIB2 is better in terms of RR and PC than IB2.

Tables 3 and 4 show the accuracy measurements achieved by the SVM and k-NN classifiers (Table 4 is the continuation of Table 3). Both tables contain seven rows for each dataset. Each row represents the different versions of the same dataset. The first one concerns the original data (i.e., without data reduction).

Table 2. Comparison of DRT algorithms in terms of Reduction Rate (RR(%)) and Preprocessing Cost (PC (millions of distance computations))

Dataset		CNN	IB2	RSP3	RHC	ERHC	AIB2
LIR	RR	83.54	85.66	61.98	88.08	**92.03**	88.12
	PC	163.03	23.37	326.52	41.85	41.85	**20.10**
PD	RR	95.36	96.23	89.22	96.52	**97.45**	97.19
	PC	11.75	1.78	86.66	2.88	2.88	**1.38**
LS	RR	80.22	84.62	73.19	89.84	**92.95**	86.72
	PC	17.99	2.22	37.70	**1.69**	**1.69**	1.92
BN	RR	77.44	83.27	75.21	79.68	**90.33**	83.40
	PC	11.49	1.58	18.76	**0.56**	**0.56**	1.53
BL	RR	65.72	69.36	64.64	78.00	**86.68**	70.36
	PC	0.21	**0.04**	0.3	0.05	0.05	**0.04**
TXR	RR	91.90	93.33	83.31	94.71	**95.94**	94.95
	PC	5.65	0.84	27.63	3.63	3.63	**0.66**
YS	RR	32.68	44.82	27.36	49.83	**79.34**	46.94
	PC	1.41	0.39	2.12	0.84	0.84	**0.37**
PH	RR	76.04	80.85	69.94	80.71	**88.05**	81.75
	PC	13.45	1.96	20.31	**0.66**	**0.66**	1.84
MN2	RR	87.23	91.68	61.33	96.47	**96.76**	92.54
	PC	0.04	0.006	0.13	0.007	0.007	**0.005**
TN	RR	82.09	88.25	84.56	96.63	**97.58**	93.44
	PC	22.13	2.07	37.13	1.64	1.64	**1.10**
MGT	RR	60.08	70.60	53.70	73.76	**84.46**	71.90
	PC	281.49	34.61	511.67	**4.08**	**4.08**	33.05
SH	RR	99.37	99.44	98.59	99.55	**99.69**	99.46
	PC	45.30	8.26	17410.18	16.83	16.83	**7.89**
CAR	RR	75.82	81.61	68.65	85.87	**90.31**	83.63
	PC	1.50	0.19	1.94	0.18	0.18	**0.17**
AVG	RR	77.50	82.29	70.13	85.36	**91.66**	83.88
	PC	44.26	5.95	89.24	5.76	5.76	**5.36**

The other six rows concern the condensing set constructed by the DRTs. Each column of the table concerns a classifier. In particular, the third column concerns the SVM classifiers while the other columns concern the k-NN classifiers. The best accuracy measurements of the different classifiers are in bold. The best accuracy among the different condensing sets is emphasized with italic style.

The results depicted by both tables are quite interesting. Almost in all cases, SVM classifiers are more accurate than the k-NN classifier. In addition, all DRTs seem to not affect accuracy achieved by SVMs. In most cases, a SVM trained

Table 3. Comparison of DRT algorithms in terms of accuracy (%) - Datasets LIR through MN2

Dataset	DRT	SVM	1-NN	5-NN	9-NN	13-NN
LIR	None	**97.58**	95.83	-	-	-
	CNN	**95.10**	92.84	-	-	-
	IB2	**94.75**	91.98	-	-	-
	RSP3	*96.51*	95.43	-	-	-
	RHC	**93.80**	93.59	-	-	-
	ERHC	92.60	**92.69**	-	-	-
	AIB2	93.71	**94.12**	-	-	-
PD	None	**99.65**	99.35	-	-	-
	CNN	**99.21**	98.68	-	-	-
	IB2	**99.02**	98.04	-	-	-
	RSP3	*99.50*	99.05	-	-	-
	RHC	**98.81**	98.30	-	-	-
	ERHC	**98.84**	98.63	-	-	-
	AIB2	**98.74**	98.33	-	-	-
LS	None	**92.40**	90.60	90.69	90.62	90.21
	CNN	**90.83**	88.21	89.99	89.39	88.00
	IB2	**88.73**	86.87	88.45	87.55	86.23
	RSP3	*91.14*	90.57	90.16	89.64	89.50
	RHC	88.95	88.95	**89.54**	88.21	86.05
	ERHC	88.37	**89.01**	88.90	86.81	84.26
	AIB2	88.19	**89.40**	87.69	86.03	84.34
BN	None	**90.57**	86.91	89.02	89.85	89.87
	CNN	*90.53*	85.62	88.15	89.09	88.77
	IB2	**90.06**	83.81	87.57	88.08	88.00
	RSP3	**90.30**	84.00	87.83	89.11	88.91
	RHC	**90.25**	83.28	87.23	88.19	88.38
	ERHC	**90.23**	88.00	89.09	88.64	88.00
	AIB2	**90.49**	82.96	87.89	88.57	89.26
BL	None	**90.96**	78.40	84.16	87.84	88.32
	CNN	*95.36*	70.88	76.32	82.72	84.16
	IB2	**95.36**	70.72	77.28	81.60	83.20
	RSP3	**94.72**	73.28	76.64	82.88	82.88
	RHC	**93.44**	68.04	75.36	82.56	84.32
	ERHC	**89.60**	76.00	83.52	83.68	83.68
	AIB2	**94.72**	68.64	77.12	83.20	81.92
TXR	None	**99.84**	99.02	-	-	-
	CNN	**99.58**	97.16	-	-	-
	IB2	**99.56**	96.35	-	-	-
	RSP3	*99.69*	98.29	-	-	-
	RHC	**99.24**	97.04	-	-	-
	ERHC	**99.10**	97.36	-	-	-
	AIB2	**99.45**	97.69	-	-	-
YS	None	**60.11**	52.02	57.01	58.15	59.43
	CNN	**59.84**	49.06	52.97	55.66	56.94
	IB2	**59.03**	46.02	52.29	55.53	55.66
	RSP3	*59.84*	50.47	54.99	57.08	57.75
	RHC	**58.89**	48.85	52.29	54.65	54.92
	ERHC	**59.09**	53.17	56.00	55.86	56.80
	AIB2	**58.22**	48.25	51.48	56.06	56.81
PH	None	89.19	**90.10**	-	-	-
	CNN	87.56	*87.82*	-	-	-
	IB2	**86.47**	85.57	-	-	-
	RSP3	**87.10**	86.94	-	-	-
	RHC	**85.88**	85.59	-	-	-
	ERHC	86.08	**86.57**	-	-	-
	AIB2	**85.83**	84.92	-	-	-
MN2	None	**100.00**	90.51	99.31	98.84	99.07
	CNN	**97.21**	95.84	90.97	84.03	84.26
	IB2	**94.66**	93.75	81.47	80.33	78.00
	RSP3	*99.08*	91.22	98.38	97.69	96.76
	RHC	91.43	**94.68**	80.09	67.12	47.12
	ERHC	90.26	**95.14**	77.53	63.65	47.12
	AIB2	**93.04**	91.43	85.65	78.23	66.70

Table 4. Comparisson of DRT algorithms in terms of Accuracy (%) - Datasets TN through CAR

Dataset	DRT	SVM	1-NN	5-NN	9-NN	13-NN
TN	None	**97.89**	94.88	96.91	97.31	97.38
	CNN	*97.84*	92.00	95.47	96.50	96.82
	IB2	**97.81**	89.15	94.95	95.87	96.45
	RSP3	**97.70**	92.68	96.30	96.88	97.31
	RHC	**97.39**	88.69	95.74	96.69	97.10
	ERHC	**97.45**	91.53	96.50	96.92	97.07
	AIB2	**97.73**	93.47	96.69	97.28	97.28
MGT	None	**83.79**	78.14	80.48	80.84	81.12
	CNN	*83.90*	74.54	78.63	79.65	80.24
	IB2	**83.38**	71.97	76.84	78.50	79.11
	RSP3	**83.71**	74.96	78.90	80.15	80.52
	RHC	**83.08**	71.97	76.67	77.83	78.94
	ERHC	**83.12**	77.01	79.64	79.86	79.86
	AIB2	**83.13**	73.36	77.40	78.36	78.89
SH	None	**99.84**	99.82	-	-	-
	CNN	99.66	**99.76**	-	-	-
	IB2	99.62	**99.73**	-	-	-
	RSP3	*99.81*	99.75	-	-	-
	RHC	**99.64**	98.01	-	-	-
	ERHC	**99.64**	98.04	-	-	-
	AIB2	99.74	**99.72**	-	-	-
CAR	None	**94.10**	85.53	89.75	90.39	89.76
	CNN	**93.28**	84.95	88.14	84.72	81.48
	IB2	**92.24**	84.43	86.11	84.20	82.41
	RSP3	*93.75*	85.59	89.70	89.12	88.14
	RHC	**87.10**	82.75	77.66	75.29	71.81
	ERHC	**86.98**	82.69	79.92	78.00	74.76
	AIB2	**91.72**	87.09	87.56	85.30	82.40

by any condensing set is as accurate as the SVM trained by the initial training set. In eight datasets, the SVMs trained by the condensing set of RSP3 are the most accurate classifier. However, RSP3 has the highest PC and the lowest RR measurements. In the cases of the rest five datasets, the most accurate classifier is the SVM built by the condensing set of CNN-rule. The accuracy achieved by IB2 is close enough to that of CNN-rule, but IB2 is faster and achieved higher RR. A final comment is that the PG and PS-condensing algorithms can effectively be used for speeding-up the training process of SVMs without sacrifying accuracy. Furthermore, we observe that, in the case of SVMs, the editing mechanism of ERHC is not as effective as it is when the k-NN classifier is used. In addition, although AIB2 achieves higher accuracy than IB2 in the case of k-NN classification, it is not true in the case of SVMs. Consequently, for SVMs, ERHC and AIB2 are not efficient extensions of RHC and IB2 respectively.

5 Conclusions

This paper demonstrated that the DRTs proposed for the k-NN classifier can also be applied for speeding-up SVMs. More specifically, the experimental measurements of our study showed that the usage of a DRT can reduce the time needed for the training process of SVMs without negatively affecting accuracy. Although the particular DRTs have been proposed for speeding up the k-NN classifier, our study illustrated that the benefits are larger when SVMs are used. The experimental results showed that in contrast to the k-NN classifier that can be affected by data reduction, the accuracy of SVMs is not affected.

References

1. Aha, D.W.: Tolerating noisy, irrelevant and novel attributes in instance-based learning algorithms. Int. J. Man Mach. Stud. **36**(2), 267–287 (1992). http://dx.doi.org/10.1016/0020-7373(92)90018-G
2. Aha, D.W., Kibler, D., Albert, M.K.: Instance-based learning algorithms. Mach. Learn. **6**(1), 37–66 (1991). http://dx.doi.org/10.1023/A:1022689900470
3. Alcalá-Fdez, J., Fernández, A., Luengo, J., Derrac, J., García, S.: Keel data-mining software tool: data set repository, integration of algorithms and experimental analysis framework. Multiple-Valued Logic Soft Comput. **17**(2–3), 255–287 (2011)
4. Brighton, H., Mellish, C.: Advances in instance selection for instance-based learning algorithms. Data Min. Knowl. Discov. **6**(2), 153–172 (2002). http://dx.doi.org/10.1023/A:1014043630878
5. Chapelle, O.: Training a support vector machine in the primal. Neural Comput. **19**, 1155–1178 (2007)
6. Chen, C.H., Jóźwik, A.: A sample set condensation algorithm for the class sensitive artificial neural network. Pattern Recogn. Lett. **17**(8), 819–823 (1996). http://dx.doi.org/10.1016/0167-8655(96)00041-4
7. Cortes, C., Vapnik, V.: Support-vector networks. Mach. Learn. **20**(3), 273–297 (1995)
8. Cover, T., Hart, P.: Nearest neighbor pattern classification. IEEE Trans. Inf. Theor. **13**(1), 21–27 (2006). http://dx.doi.org/10.1109/TIT.1967.1053964
9. Dasarathy, B.V.: Nearest Neighbor (NN) Norms: NN Pattern Classification Techniques. IEEE Computer Society Press, Los Alamitos (1991)
10. Garcia, S., Derrac, J., Cano, J., Herrera, F.: Prototype selection for nearest neighbor classification: taxonomy and empirical study. IEEE Trans. Pattern Anal. Mach. Intell. **34**(3), 417–435 (2012). http://dx.doi.org/10.1109/TPAMI.2011.142
11. Hart, P.E.: The condensed nearest neighbor rule. IEEE Trans. Inf. Theory **14**(3), 515–516 (1968)

340   S. Ougiaroglou et al.

12. McQueen, J.: Some methods for classification and analysis of multivariate observations. In: Proceeding of 5th Berkeley Symposium on Mathematics, Statistics and Probability. pp. 281–298. University of California Press, Berkeley (1967)
13. Olvera-López, J.A., Carrasco-Ochoa, J.A., Martínez-Trinidad, J.F., Kittler, J.: A review of instance selection methods. Artif. Intell. Rev. **34**(2), 133–143 (2010). http://dx.doi.org/10.1007/s10462-010-9165-y
14. Ougiaroglou, S., Evangelidis, G.: Efficient dataset size reduction by finding homogeneous clusters. In: Proceedings of the Fifth Balkan Conference in Informatics, BCI 2012, pp. 168–173. ACM, New York (2012). http://doi.acm.org/10.1145/2371316.2371349
15. Ougiaroglou, S., Evangelidis, G.: Efficient data abstraction using weighted IB2 prototypes. Comput. Sci. Inf. Syst. **11**(2), 665–678 (2014). http://dx.doi.org/10.2298/CSIS140212036O
16. Ougiaroglou, S., Evangelidis, G.: RHC: a non-parametric cluster-based data reduction for efficient k-NN classification. Pattern Anal. Appl. **19**(1), 93–109 (2019). http://dx.doi.org/10.1007/s10044-014-0393-7
17. Ougiaroglou, S., Evangelidis, G.: Efficient editing and data abstraction by finding homogeneous clusters. Ann. Math. Artif. Intell. **76**(3), 327–349 (2015). http://dx.doi.org/10.1007/s10472-015-9472-8
18. Sánchez, J.S.: High training set size reduction by space partitioning and prototype abstraction. Pattern Recogn. **37**(7), 1561–1564 (2004)
19. Toussaint, G.: Proximity graphs for nearest neighbor decision rules: recent progress. In: 34th Symposium on the INTERFACE, pp. 17–20 (2002)
20. Triguero, I., Derrac, J., Garcia, S., Herrera, F.: taxonomy and experimental study on prototype generation for nearest neighbor classification. Trans. Sys. Man Cyber. Part C **42**(1), 86–100 (2012). http://dx.doi.org/10.1109/TSMCC.2010.2103939
21. Vapnik, V.: Estimation of Dependencies Based on Empirical Data. Nauka, Moscow (1979). English translation: Springer Verlag, New York (1982)
22. Vapnik, V.: Statistical Learning Theory. Wiley, New York (1998)
23. Vapnik, V., Chervonenkis, A.: Theory of pattern recognition (1974)
24. Wilson, D.R., Martinez, T.R.: Reduction techniques for instance-based learning algorithms. Mach. Learn. **38**(3), 257–286 (2000). http://dx.doi.org/10.1023/A:1007626913721
25. Wu, J.: Advances in K-means Clustering: A Data Mining Thinking. Springer, Heidelberg (2012)

MeLiF+: Optimization of Filter Ensemble Algorithm with Parallel Computing

Ilya Isaev and Ivan Smetannikov[⊠]

Computer Science Department, ITMO University, 49 Kronverksky Pr.,
197101 St. Petersburg, Russia
isaev@rain.ifmo.ru, ismetannikov@corp.ifmo.ru

Abstract. Search of algorithms ensemble – that is, best algorithms combination is common used approach in machine learning. MeLiF algorithm uses this technique for filter feature selection. In our research we proposed parallel version of this algorithm and showed that it is not only improves algorithm performance significantly, but also improves feature selection quality.

Keywords: Feature selection · Variable selection · Attribute selection · Ensemble learning · Feature filters · Metrics aggregation · MeLiF · Parallel computing

1 Introduction

In modern world, machine learning became one of the most promising and studied science areas, mainly, because of its universal application to any data-related problem. One example of such an area is bioinformatics [3,4,6,10], which produces giant amount of data about gene expression of different organisms. This data could potentially allow to determine which DNA pieces are responsible for some visual change of indiviual, or for reactions to particular environment change. The main problem of such data is its huge number of features and relatively low amount of objects. Because of high-dimensional space, it is very hard to build a model which generalizes such data well. Furthermore, a lot of features in such datasets have nothing in common with results, so, they should be treated as noize.

$$A^* = 4$$

It seems to be logical in this case to select somehow the most relevant features and to learn a classifier on these only. This idea is implemented in such area of machine learning as feature selection. There are three main methods of feature selection: filter selection based on statistical measures of every single feature or features subsets, wrapper selection based on subspace search with classifier result as an optimization measure, and embedded selection that uses classificators inner properties [12].

© IFIP International Federation for Information Processing 2016
Published by Springer International Publishing Switzerland 2016. All Rights Reserved
L. Iliadis and I. Maglogiannis (Eds.): AIAI 2016, IFIP AICT 475, pp. 341–347, 2016.
DOI: 10.1007/978-3-319-44944-9_29

The main peculiarity of filter methods is their speed. This leads to the fact that they are frequently used for preprocessing, and resulting subsets of features further passed to other wrapper or embedded method. This is especially important for bioinformatics, where number of features in datasets is sometimes dozens and hundrends of thousands.

These days, many machine learning algorithms use ensembling [1,4,8]. MeLiF algorithm [13] tries to apply this method to feature selection. It builds a linear combination of basic filters, that selects the most relevant features. MeLiF has a structural characteristic that it can be easily modified to work in concurrent or distributed manner. At this research, we implemented parallel version of MeLiF called MeLiF+ and achieved significant speed improvement without losing in selection quality.

The remainder of the paper is organized as follows: MeLiF algorithm is described in Sect. 2, parallelization scheme is proposed in Sect. 3, experiment setup and used quality measures are outlined in Sect. 4, and finally experiment results are contained in Sect. 5.

2   MeLiF

Algorithm treats some linear combinations of basic filters as starting points. It has been observed during experiments that the best option is this following choice of starting points: $(1, 0, ..., 0), (0, 1, ..., 0), ..., (0, 0, ..., 1)$ – only one basic filter matters at the beginning, and $(1, 1, ..., 1)$ – all basic filters are equal at the beginning. Algorithm iterates over the starting points and tries to shift each coordinate value to small constants $+\delta$ and $-\delta$ – value of grid spacing for each point. If some of applied changes succeed, i.e. quality measure for a point after a shift is greater than the maximum value: the algorithm chooses that point and starts searching from its first coordinate. If, all coordinates were shifted to $+\delta$ and $-\delta$ and no quality improvement observed, algorithm stops.

Algorithm 1. MeLiF algorithm

    Input: points, delta, evaluate
1:   $q^* \leftarrow 0$
2:   $p^* \leftarrow 0$
3:   **for each** $p \in points$ **do**
4:      $q \leftarrow evaluate(p)$
5:      **if** $q > q^*$ **then**
6:          $p^* \leftarrow p$
7:          $q^* \leftarrow q$
8:   $smthChanged =$ **true**
9:   **while** smthChanged **do**
10:      **for each** $dim \in p.size$ **do**
11:          $p^+ \leftarrow p$
12:          $p^+[dim] \leftarrow p^+[dim] + delta$
13:          $q^+ \leftarrow evaluate(p^+)$

14:    **if** $q^+ > q^*$ **then**
15:        $q^* \leftarrow q^+$
16:        $p^* \leftarrow p^+$
17:        $smthChanged =$ **true**
18:        **break**
19:    $p^- \leftarrow p$
20:    $p^-[dim] \leftarrow p^-[dim] - delta$
21:    $q^- \leftarrow evaluate(p^-)$
22:    **if** $q^- > q^*$ **then**
23:        $q^* \leftarrow q^-$
24:        $p^* \leftarrow p^-$
25:        $smthChanged =$ **true**
26:        **break**
27: **return**(p^*, q^*)

Then, for each point obtained during coordinate descent, the algorithm measures value of resulting linear combination of basic filters for each feature in dataset. After that, results are sorted, and the algorithm selects N best features. Then, the algorithm runs some classifier only with that feature subset. The obtained result is saved for comparing with other points and caching. It helps to reduce working time due to visited points usage.

3   MeLiF+

We proposed the following improvements to the MeLiF method: each starting point is processed in a distinct thread with global maximum maintained through synchronization point. Moreover, *evaluate* submethod is run concurrently for $+\delta$ and $-\delta$, and selects the best point after retrieving both results. We showed that it not only improves the algorithm performance on multicore system, but also usually improves feature selection quality.

This fact has the following explanation: the original MeLiF algorithm is greedy, so it assumes that if each point it steps in is a local optimum then resulting point will be the global optimum, adding an ability to lookup for two deltas simultaneously allows algorithm to select better local optimum. Also, as starting points are processed in parallel, one thread can find a local optimum. This causes other threads to stop their work even if further descent leads to the better result. This can cause different selection result, better or worse (both cases are presented in Sect. 5), but experiments show that avarage MeLiF+ results are better.

4   Experiments

We used SVM [5] from WEKA [14] library, with polynomial kernel and soft margin parameter $C = 1$ as classifier. To improve stability, we used 5-fold cross-validation. The number of selected features was constant: $N = 100$. In order to compare our method with the old one, we used F_1 score [11] of SVM classifier. As we wanted to know how much our method differs from the original one in terms of space search strategy, we calculated z-score for each dataset.

We ran our experiments on a machine with following characteristics: 32-core CPU AMD Opteron 6272 @ 2.1 GHz, 128 GB RAM. We used $N = 50$ threads, $N = 2 \cdot p \cdot f$ threads, where p is the number of starting points, f is the number of folders used for cross-validation.

As basic filters, we used Spearman Rank Correlation (SPC), Symmetric Uncertainty (SU), Fit Criterion (FC) and Value Difference Metric (VDM) [2,9]. For each dataset, we executed MeLiF and MeLiF+ and stored their working time and points with the best classification result.

We used 50 datasets of different sizes: 33 datasets have been taken from Gene Expression Omnibus, 5 from Kent Ridge Bio-Medical Dataset, 5 from RSCTC'2010 Discovery Challenge, 4 from Broad institute Cancer Program Data Sets, 3 from Feature Selection Datasets at Arizona State University. Some datasets were multi-labeled, therefore we splitted them into several derivative binary datasets with commonly used one-versus-all technique. Then we excluded datasets that contained too few instances of one of the classes. After that, we used standard feature scaling and discretized all features to 11 different values from −5 to 5.

5   Results

Table below contains experiment results. All the datasets are sorted by their total size which is basically a multiplication of their features and objects number.

Table 1. MeLiF in comparison with swarm algorithms

Dataset	Size	F_1 score		Time		z-score
		MeLiF	MeLiF+	MeLiF	MeLiF+	
SRBCT30	191k	0.900	0.891	13	2	0.23
SRBCT31	191k	1.000	1.000	17	3	0
GDS2960	417k	0.971	0.980	33	7	-10.98
CNS	427k	0.742	0.791	33	6	-1.06
Leuk3c0	513k	0.986	0.986	34	5	0
Leuk3c1	513k	0.933	0.933	33	5	0
GDS2961	566k	0.845	0.845	49	13	0
GDS2962	566k	0.784	0.887	45	11	-11.15
DLBCK	962k	0.799	0.734	65	13	4,67
GDS2901	1337k	1.000	1.000	88	17	0
Prostate	1713k	0.925	0.903	93	34	7.27
GDS4109	1760k	0.936	0.936	142	38	0
GDS5083	2131k	0.862	0.862	195	60	0

(*continued*)

Table 1. (*continued*)

d2t0	2339k	0.765	0.765	172	71	0
d2t1	2339k	0.779	0.844	180	60	-9.73
breast	2346k	0.769	0.812	161	24	-11.71
GDS4261	2367k	1.000	1.000	130	16	0
GDS3257	2384k	0.980	1.000	131	17	-33.65
GDS4901	2515k	0.899	0.946	220	60	-7.6
GDS3553	2543k	1.000	1.000	142	18	0
GDS3116	2584k	0.826	0.826	142	23	0
GDS4336	2598k	0.865	0.865	200	66	0
GDS5047	2925k	0.989	0.989	185	41	0
GDS3995	3200k	1.000	1.000	181	24	0
GDS496840	3296k	0.890	0.890	226	40	0
GDS496841	3296k	0.953	0.936	224	40	4.04
GDS2947	3499k	1.000	1.000	217	28	0
GDS43181	3591k	0.913	0.913	275	64	0
arizona5	3738k	0.730	0.754	219	67	-7.47
Ovarian	3833k	1.000	1.000	192	23	0
GDS4103	4264k	0.918	0.918	265	71	0
GDS2771	4265k	0.760	0.760	299	81	0
GDS503730	4428k	0.738	0.826	243	140	-16.77
GDS503732	4428k	0.782	0.782	293	69	0
GDS3929	4488k	0.821	0.821	376	74	0
GDS483731	4811k	0.920	0.920	413	130	0
GDS483733	4811k	0.951	0.951	316	48	0
d5t	4860k	0.869	0.869	370	70	0
GDS3622	4961k	1.000	1.000	266	33	0
GDS2819	5412k	0.991	1.000	436	149	-4.05
d6t	5428k	0.792	0.792	381	69	0
GDS4130	5685k	1.000	1.000	315	39	0
d4t	6178k	0.719	0.719	513	124	0
GDS4129	6561k	1.000	1.000	354	43	0
GDS4222	7107k	0.965	0.971	454	84	-7.42
GDS4431	7973k	0.802	0.802	537	100	0
arizona1	8847k	0.823	0.823	558	85	0
GDS4600	9294k	0.979	0.968	472	124	23.27
GDS3244	9787k	1.000	1.000	505	65	0

In F_1 score comparison of MeLiF and MeLiF+ better results for each dataset are highlighted in grey, equal results are not highlighted. Runtime is presented in seconds. At the last column, z-score is provided.

As it can be seen from the table above, MeLiF+ is always at least 3 times faster than the MeLiF, and this difference gets up to 6 times for some datasets.

Although MeLiF and MeLiF+ have almost the same results in F_1 score, there is some difference in their work on 15 datasets as provided via z-score. But only in 5 cases MeLiF+ had worse results than original the MeLiF algorithm. But on 36 datasets, they performed equally and at 11 datasets new algorithm outperformed the original one.

6   Conclusion

The proposed parallelization scheme made algorithm in average to work 5.5 times faster without affecting selection quality. Unforunately, in this research we did not achieved linear speed improvement because of the fixed maximum of parallel processed points. In our future work, we are planning to use threads pool which is limited by the testing system and achieve linear speed growth with using exploration and exploitation [7] strategy to spread the search points in the search space. Also this should lead to high increase in optimized measure.

Acknowledgements. Authors would like to thank Julia Ugarkina and Andrey Filchenkov for useful comments and proofreading. This work was financially supported by the Government of Russian Federation, Grant 074-U01.

References

1. Abeel, T., Helleputte, T., Van de Peer, Y., Dupont, P., Saeys, Y.: Robust biomarker identification for cancer diagnosis with ensemble feature selection methods. Bioinformatics **26**(3), 392–398 (2010)
2. Auffarth, B., López, M., Cerquides, J.: Comparison of redundancy and relevance measures for feature selection in tissue classification of CT images. In: Perner, P. (ed.) ICDM 2010. LNCS, vol. 6171, pp. 248–262. Springer, Heidelberg (2010)
3. Bolón-Canedo, V., Sánchez-Maroño, N., Alonso-Betanzos, A., Benítez, J., Herrera, F.: A review of microarray datasets and applied feature selection methods. Inform. Sci. **282**, 111–135 (2014)
4. Bolón-Canedo, V., Sánchez-Maroño, N., Alonso-Betanzos, A.: An ensemble of filters and classifiers for microarray data classification. Pattern Recogn. **45**(1), 531–539 (2012)
5. Burges, C.J.: A tutorial on support vector machines for pattern recognition. Data Min. Knowl. Disc. **2**(2), 121–167 (1998)
6. Chuang, L.Y., Yang, C.H., Wu, K.C., Yang, C.H.: A hybrid feature selection method for dna microarray data. Comput. Biol. Med. **41**(4), 228–237 (2011)
7. Desautels, T., Krause, A., Burdick, J.W.: Parallelizing exploration-exploitation tradeoffs in gaussian process bandit optimization. J. Mach. Learn. Res. **15**(1), 3873–3923 (2014)
8. Dietterich, T.G.: Ensemble methods in machine learning. In: Meyers, R.A. (ed.) Multiple Classifier Systems, pp. 1–15. Springer, New York (2000)
9. Filchenkov, A., Dolganov, V., Smetannikov, I.: Pca-based algorithm for constructing ensembles of feature ranking filters. In: Proceedings of ESANN Conference, pp. 201–206 (2015)

10. Haury, A.C., Gestraud, P., Vert, J.P.: The influence of feature selection methods on accuracy, stability and interpretability of molecular signatures. PloS ONE **6**(12), e28210 (2011)
11. Huang, H., Xu, H., Wang, X., Silamu, W.: Maximum f1-score discriminative training criterion for automatic mispronunciation detection. Trans. Audio Speech Lang. Process. **23**(4), 787–797 (2015)
12. Saeys, Y., Inza, I., Larrañaga, P.: A review of feature selection techniques in bioinformatics. Bioinformatics **23**(19), 2507–2517 (2007)
13. Smetannikov, I., Filchenkov, A.: MeLiF: filter ensemble learning algorithm for gene selection. In: Advanced Science Letters. American Scientific Publisher (2016, to appear)
14. Waikato, T.U.: Weka 3: Data Mining Software in Java (2016). http://www.cs. waikato.ac.nz/ml/weka/. Accessed 7 May 2016

Genetic Search of Pickup and Delivery Problem Solutions for Self-driving Taxi Routing

Viacheslav Shalamov$^{(\boxtimes)}$, Andrey Filchenkov, and Anatoly Shalyto

Computer Technologies Lab, ITMO University,
49 Kronverksky Pr., 197101 St. Petersburg, Russia
shalamov@rain.ifmo.ru, afilchenkov@corp.ifmo.ru, shalyto@mail.ifmo.ru

Abstract. Self-driving cars belong to rapidly growing domain of cyber-physical systems with many open problems. In this paper, we study routing problem for taxis. In mathematical terms, it is well-known Pickup and Delivery problem (PDP). We use with the standard small-moves technique, which is to apply small changes to a solution for PDP in order to obtain a better one; and an approach that works with small-moves as mutations in genetic algorithms. We propose a strategy-based framework for managing set of small changes and suggest different strategies. We tested algorithms for routing on real-world dataset on taxi orders to airports in United Kingdom. The results show that algorithms using mixed strategies outperform algorithms using a single small move.

Keywords: Self-driving car · Autonomous car · Routing · Pickup and delivery · Genetic algorithms · City taxi

1 Introduction

Usage of cyber-physical systems (CPS) such as medical devices, industrial robots or smart grid, showed extensive growth in the recent years [1]. It leads to high demand on algorithms for CPSs improving their performance and interaction with environment. Control systems of this type are thought to show high autonomy and to be able to solve typical tasks met in the domain of their application.

Self-driving cars (SDC), also known as autonomous cars, are one of the CPSs with rapidly growing importance: SDCs share market with ordinary (human-driving) vehicles, and potential effect of its introduction is often referred as a revolution [9,21]. Consequences include improvements in traffic, vehicle usage and even implementation of new social strategy, so-called mobility as a service [5,25]. List of organizations racing for SDC includes not only well-known vehicle manufacturing companies, but also Google, Uber [15] and Baidu [10].

One of the possible applications of SDC is a taxi service in airports. Airports are usually a very intensive hub for taxi routing, due to many people use taxis to get to an airport or get to a city after arrival. It explains why many airports all over the world have their own taxi services or at least taxi services accredited

L. Iliadis and I. Maglogiannis (Eds.): AIAI 2016, IFIP AICT 475, pp. 348–355, 2016.
DOI: 10.1007/978-3-319-44944-9_30

to be official for this airport. This policy shows several advantages. First of all, airport-associated taxis are less affected by the traffic situation in the city. Second, tariffs for taxi are usually fixed or can be simpler adjusted to the current demand, since regular taxi services have to take into account all the demands for mobility across the city.

Many studies are devoted to SDC problem. They include:

- map generation and navigation [13,19,29];
- sensor system [3,12];
- motion planning [2,17,20,23,31];
- distributed and parallel computational infrastructure [18,24,26];
- routing in different conditions [6,11,14,32].

The last problems, namely routing, is actual for taxis. In its mathematical statement, this problem is known as Pickup and delivery problem (PDP) [16]. PDP is an optimization problem, relative to the most known travelling sales-man problem (TSP). Both of them belong to the vehicle routing problem class, determining the optimal solution for problems of this class is known to be NP-hard [22]. The TSP can be considered as a case of the PDP, since one is required not only to visit each point, but also pickup items in several points and deliver them to other points [27]. A lot of constraints can be added to the general case of PDP [28]. The taxi routing problem may be understood as last-in-first-out single vehicle pickup and delivery problem (SVPDPL): (1) we assume that we need to build a route for a single vehicle; (2) it cannot pick another item before it had delivered current one [8]. SVPDPL is known to be NP-hard problem, thus, a number of heuristics for solving it are suggested.

In this paper, we use variable neighborhood search (VNS) approach [4] for solution search that constructs a new solution by applying local changes (small moves) to an already found one. We use the framework suggested in [30] involving usage of strategies.

The remainder of the paper is organized as follows. In Sect. 2, we define SVPDPL, its solution via local changes and genetic algorithms. In Sect. 3, we describe a concept of strategy and different strategies we use in this paper. In Sect. 4, we describe experiment setups and the results of algorithm comparison. Section 5 concludes.

2   Problem Statement and Small Moves

2.1   Pickup and Delivery Problem Statement

We will follow the notation from [7] and [30]. The Pickup and delivery problem is represented with a weighted graph $G = (V, E, W)$, where V is vertex set corresponding to the points, E is edge set corresponding to the routes connecting points, and W is a weight function that is defined on V and E. It corresponds to loads of vertices and cost of travelling for edges. In this paper, we use W representing the taxi routing problem:

– $W(v)$ equals 1 if v is a vertex with a person to be picked up (*source*), and -1
if v is a vertex, to which a person should be delivered (*destination*);
– $W(\{u, v\})$ is a distance between the points corresponding to u and v.

Let P denote the source set and D denote the destination set, then V is splited
into pairs $(P, D) = ((p_1, d_1), \ldots, (p_n, d_n))$, $P, D \in V$.

We enumerate all the vertices: a source vertex p_i receives index i, and the
corresponding destination vertex d_i receives index $i+n$. An edge (i, j) can belong
to E if 1) $j = i+n$; or 2) $i > n$ and $j \le n$, $i \ne n+j$. We need to find a Hamiltonian
route, which minimizes its cost: $\sum_{j=1}^{2n-1} W(\{i_j, i_{j+1}\})$.

2.2   Small Moves

Since the search of an exact solution is NP-hard, an heuristic known as "small
moves" was suggested [4]. The main idea is to apply small changes to a found
solution in order to obtain a new, better one. Today it is state-of-the-art approach. In this work, we use the following five types of the most commonly used
small moves.

Lin-2-Opt substitutes two edges (i_j, i_{j+1}) and (i_k, i_{k+1}) with other two edges
(i_j, i_k) and (i_{j+1}, i_{k+1}), then it inverses the subpath $(i_{j+1}, \ldots, i_k)$. Our illustrated example is presented in web[1].

Double-bridge exchanges edges (i_j, i_{j+1}), (i_k, i_{k+1}), (i_l, i_{l+1}) and (i_h, i_{h+1})
with $(i_j, i_l + 1)$, (i_k, i_{h+1}), (i_l, i_{j+1}) and (i_h, i_{k+1}). Our illustrated example is
presented in web[2].

Couple-exchange selects two orders $(i, n + i)$ and $(j, n + j)$, then it replaces
i with j and $n + i$ with $n + j$. Our illustrated example is presented in web[3].

Point-exchange is analogous to the couple-exchange. The only difference is
that not the pairs, but single points are exchanged.

Relocate-block (RB) inserts a sequence $(i_t, \ldots, i_{t+n})$ into the best possible
place in the route. Our illustrated example is presented on the web [5 times][4].

2.3   Applying Genetic Algorithms

The small moves described in the previous subsection can be easily understood
as mutation operations, which can be applied to a solution in order to obtain a
new, better one. This idea was realized in [30] for applying genetic algorithms
for the solution search.

The application of genetic algorithms requires the following problem formalization: spices are solutions of the SVPDPL, and fitness function is the quality
of such a solution (the length of the route, which should be minimized). Mutation operation is the application of a small move. Each generation consists of K

---

[1] http://genome.ifmo.ru/files/papers_files/AIAI2016/fig_L2O.pdf.
[2] http://genome.ifmo.ru/files/papers_files/AIAI2016/fig_DB.pdf.
[3] http://genome.ifmo.ru/files/papers_files/AIAI2016/fig_CE.pdf.
[4] http://genome.ifmo.ru/files/papers_files/AIAI2016/fig_RB.pdf.

solutions, each solution gives a birth to n new solutions by applying a mutation operation, then some of them are chosen via the selection operation to form the next generation. Totally, there are g generations.

Let N be equal to 50, which is the number of vertices in the graph. In [30], five different genetic algorithms were tested, we will use all of them.

1. 1 + 1: each generation consists of one solution ($K = 1$), it gives a birth to one solution ($n = 1$), both the solutions are subjects of selection, the resulting solution forms the next generation ($E = 1$).
2. 1 + N: $K = 1$, $n = \sqrt{N/2}$, $E = 1$.
3. 1, N: $K = 1$, $n = \sqrt{N/2}$, only the children are subjects of selection, $E = 1$.
4. 1 + N + Big mutation (1+N+BM): $K = 1$, $n = \sqrt{N/2}$, $E = 1$. In addition, every $g/4$ generations a Big mutations fires, during which all the species are mutated $\sqrt{(N)}$ times.
5. K+KN: $K = \sqrt{N/4}$, $n = \sqrt{N/2}$, $E = K$.

3   Strategies

The question arises, which small moves can and should be used as mutation operations. The simplest way is to apply a small move of a certain type — this is a common practice. However, as it was shown in [30], applying different small moves is more beneficial. In that paper, one group of the genetic algorithms was allowed to pick mutation on each step randomly from the entire small move set (with the uniform distribution).

In this paper, we generalize such an approach with *strategy* concept. Assume that we are given a set $\mathcal{M} = \{m_1, \ldots, m_M\}$ of small moves. Then let strategy $S = \{q_1, \ldots, q_M\}$ be a vector of probabilities, with which each of small move should be applied on the next step.

If we apply only a single small move, then this strategy is *pure*. In our case, we have 5 small moves: L20, DB, CE, PE, and RB. Thus, there are five pure strategies: $S_{L20} = (1, 0, 0, 0, 0)$, $S_{DB} = (0, 1, 0, 0, 0)$, $S_{CE} = (0, 0, 1, 0, 0)$, $S_{PE} = (0, 0, 0, 1, 0)$, $S_{RB} = (0, 0, 0, 0, 1)$. However, as we have stated, mixed strategies will outperform pure strategies. The uniform, mixed strategy is: $S_{mix} = (1/5, 1/5, 1/5, 1/5, 1/5)$.

The second group of strategies is formed according to the following empirics: (1) CE and PE perform similarly, this is why we use only one of them; and (2) according to the results in [30], as well as this study, L2O seems to be the strongest algorithm. This is why it makes sense to consider strategies, which use L20 with combination of other small moves. Thus, we add five new strategies for comparison:

$$S_{mix-PE} = (1/4, 1/4, 0, 1/4, 1/4).$$
$$S_{L20+DB} = (1/2, 1/2, 0, 0, 0);$$
$$S_{L20+CE} = (1/2, 0, 0, 1/2, 0);$$
$$S_{L20+RB} = (1/2, 0, 0, 0, 1/2);$$
$$S_{L20+mix} = (1/2, 1/6, 0, 1/6, 1/6).$$

The last group of strategies we discuss in this paper is based on small-moves properties. We run experiments on 150 different subsets and collected the following statistical evaluations: n_j is the number of times, when ith small move is applied; f_i is frequency of solution improvement by applying ith small move, and c_i is frequency of solution improvement by applying ith small move ignoring cases, when it violates the problem constraints. Also let n_i denote the computational complexity of the small move application. Thus:

$$S_{\mathrm{N}} \propto \{(1/n_i\} = (1, 1, 1, 1, 1/N); S_N = (0.249, 0.249, 0.249, 0, 249, 0.005),$$

$$S_{\mathrm{F}} \propto \{f_i\}; S_F = (0.327, 0.09, 0.158, 0.16, 0.265),$$

$$S_{\mathrm{C}} \propto \{c_i\}; S_C = (0.348, 0.166, 0.160, 0.16, 0.166),$$

$$S_{\mathrm{F/N}} \propto \{f_i/n_i\}; S_{F/N} = (0.4417, 0.1215, 0.2134, 0.2162, 0.007).$$

4 Experiments

4.1 Experiment Setup and Test Data

VeeRoute company[5] kindly gave us dataset containing information on 8,000 orders for taxis to transfer people from airports or to airports in the United Kingdom in March 2015. We used airport names and addresses to obtain geographical coordinates of the corresponding points using webservice mapquest[6]. Then we used the Mercator projection implementation library in library JMapProjLib[7] to transform them to 2D coordinates on a plane. Thus, we transformed the datasets into the dataset of two sets of points. Then we generated small datasets containing 50 random addresses as source and an airport as destination.

Each algorithm was run 10 times on 5 different subsets. Each experiment was held 10 times. We estimate *optimization ratio* (OR), which represents, how much we have decreased the length of the route in comparison with the original one:

$$\mathrm{OR} = \frac{d_0 - d_1}{d_0},$$

where d_0 is the length of the initial solution, and d_1 is the achieved length. We tested algorithms on $g = 12 \cdot N^2$ generations.

4.2 Strategies Comparison

We compared all the strategies presented in the previous Section, for each of genetic scheme presented in Sect. 2. Results are presented in Table 1.

As we can see, 1+1 genetic scheme outperforms all the other genetic schemes in general. Strategies of the second type outperform corresponding strategies of

---

[5] http://veeroute.com/.
[6] www.mapquest.com/.
[7] https://github.com/OSUCartography/JMapProjLib.

Table 1. Comparison of stretagies in OR.

Strategy	1+1		1,N	1+N,BM	K+KN
S_{L20}	0.637	0.636	0.639	0.642	0.638
S_{DB}	0.410	0.406	0.437	0.416	0.391
S_{CE}	0.537	0.538	0.354	0.527	0.543
S_{PE}	0.576	0.572	0.397	0.573	0.580
S_{RB}	0.676	0.677	0.670	0.667	0.641
S_{mix}	0.729	0.726	0.695	0.715	0.700
S_{mix-CE}	0.716	0.704	0.703	0.708	0.693
S_{L20+DB}	0.647	0.639	0.655	0.633	0.606
S_{L20+PE}	0.679	0.683	0.622	0.674	0.686
S_{L20+RB}	0.702	0.698	0.701	0.697	0.674
$S_{L20+mix}$	0.722	0.717	0.726	0.704	0.702
S_N	0.736	0.705	0.631	0.709	0.679
S_F	0.726	0.716	0.712	0.717	0.701
S_C	0.733	0.717	0.696	0.731	0.704
$S_{F/N}$	0.736	0.710	0.648	0.710	0.678

the first type, with the only exception for the mixed strategy. However, statistic strategies show the best performance. Usage of small-moves complexity leads to the two best results. It can be simply explained by the fact that implying a small move (RB) consumes time N greater than any other small move, thus, applying such a normalization makes the mixed strategy more fair. However, it has no such a big improvement in other genetic schemes. When a more complex scheme is used, this difference becomes not so important. In these cases, statistic considerations have a higher impact. As we can see, for the most of other schemes, S_F, S_C are the best, due to the breed is bigger than one, and then unsuccessful application of RB will not be so crustal for reaching the optima.

5   Conclusion and Future Work

In this paper, we have suggested a novel solution for self-driving car routing, which is based on applying strategies: small-moves with assigned probabilities, which changes a found solution. We have suggested and analyzed several strategies with several genetic schemes for generating them.

We have shown that small-moves ensembling is a very profitable step for results improvement. Also we have shown that the strategies that are based on the previous experience are very powerful. Another strong improvement was to connect probabilities of applying a small move with its computational complexity (time consumed).

We see several directions for future work. First, it may be useful to apply reinforcement learning in order to choose, which small move to apply next. Second, we plan to use meta-learning to predict the initial strategy and genetic algorithm scheme. Finally, it seems to be useful to choose the size of breed dynamically on each step (decreasing it), since strategies with big breed where good in the beginning of the breed.

Acknowledgements. Authors would like to thank Daniil Chivilikhin for useful comments. This work was financially supported by the Government of the Russian Federation, Grant 074-U01.

References

1. Alur, R.: Principles of cyber-physical systems. MIT Press, Cambridge (2015)
2. Bender, P., Tas, O.S., Ziegler, J., Stiller, C.: The combinatorial aspect of motion planning: Maneuver variants in structured environments. In: Intelligent Vehicles Symposium (IV), 2015 IEEE. pp. 1386–1392. IEEE (2015)
3. Broggi, A., Bombini, L., Cattani, S., Cerri, P., Fedriga, R.: Sensing requirements for a 13,000 km intercontinental autonomous drive. In: Intelligent Vehicles Symposium (IV), 2015 IEEE, pp. 500–505. IEEE (2010)
4. Carrabs, F., Cordeau, J.F., Laporte, G.: Variable neighborhood search for the pickup and delivery traveling salesman problem with lifo loading. INFORMS J. Comput. **19**(4), 618–632 (2007)
5. Chong, Z., Qin, B., Bandyopadhyay, T., Wongpiromsarn, T., Rebsamen, B., Dai, P., Rankin, E., Ang Jr., M.H.: Autonomy for mobility on demand. In: Intelligent Autonomous Systems 12, pp. 671–682. Springer (2013)
6. Chu, K., Lee, M., Sunwoo, M.: Local path planning for off-road autonomous driving with avoidance of static obstacles. IEEE Trans. Intell. Trans. Syst. **13**(4), 1599–1616 (2012)
7. Cordeau, J.F., Laporte, G., Ropke, S.: Recent models and algorithms for one-to-one pickup and delivery problems. In: Golden, B., Raghavan, S., Wasil, E. (eds.) The Vehicle Routing Problem: Latest Advances and New Challenges. Operations Research/Computer Science Interfaces, vol. 43, pp. 327–357. Springer, Heidelberg (2008)
8. Cordeau, J., Iori, M., Laporte, G., Salazar Gonzlez, J.: A branch-and-cutalgorithm for the pickup and delivery traveling salesman problem with lifo loading. Networks **55**, 46–59 (2010)
9. Dallegro, J.A.: How google's self-driving car will change everything (2014). http://www.investopedia.com/articles/investing/052014/how-googles-selfdriving-car-will-change-everything.asp, Accessed 15 Feb 2016
10. Davies, A.: Baidu's self-driving car has hit the road (2014). http://www.wired.com/2015/12/baidus-self-driving-car-has-hit-the-road/, Accessed 15 Feb 2016
11. Dolgov, D., Thrun, S., Montemerlo, M., Diebel, J.: Path planning for autonomous vehicles in unknown semi-structured environments. Int. J. Rob. Res. **29**(5), 485–501 (2010)
12. Ercan, Z., Sezer, V., Heceoglu, H., Dikilitas, C., Gokasan, M., Mugan, A., Bogosyan, S.: Multi-sensor data fusion of dcm based orientation estimation for land vehicles. In: 2011 IEEE International Conference on Mechatronics (ICM), pp. 672–677. IEEE (2011)
13. Göhring, D., Wang, M., Schnürmacher, M., Ganjineh, T.: Radar/lidar sensor fusion for car-following on highways. In: 2011 5th International Conference on Automation, Robotics and Applications (ICARA), pp. 407–412. IEEE (2011)

14. González, D., Perez, J., Lattarulo, R., Milanés, V., Nashashibi, F.: Continuous curvature planning with obstacle avoidance capabilities in urban scenarios. In: 2014 IEEE 17th International Conference on Intelligent Transportation Systems (ITSC), pp. 1430–1435. IEEE (2014)
15. Hawkins, A.J.: Google vs. uber and the race to self-driving taxis (2015). http://www.theverge.com/2015/12/16/10309960/google-vs-uber-competition-self-driving-cars, Accessed 2016–02-15
16. Iori, M., Martello, S.: Routing problems with loading constraints. Top **18**, 4–27 (2010)
17. Haider Jafri, S.M., Kala, R.: Path planning of a mobile robot in outdoor terrain. In: Berretti, S., Thampi, S.M., Dasgupta, S. (eds.) Intelligent Systems Technologies and Applications. ALSC, pp. 187–195. Springer, Heidelberg (2016)
18. Jo, K., Kim, J., Kim, D., Jang, C., Sunwoo, M.: Development of autonomous car—part i: distributed system architecture and development process. IEEE Trans. Ind. Electron. **61**(12), 7131–7140 (2014)
19. Jo, K., Sunwoo, M.: Generation of a precise roadway map for autonomous cars. IEEE Trans. Intell. Transp. Syst. **15**(3), 925–937 (2014)
20. Kala, R., Warwick, K.: Planning autonomous vehicles in the absence of speed lanes using an elastic strip. IEEE Trans. Intell. Transp. Syst. **14**(4), 1743–1752 (2013)
21. Kim, J., Kim, H., Lakshmanan, K., Rajkumar, R.R.: Parallel scheduling for cyber-physical systems: analysis and case study on a self-driving car. In: Proceedings of the ACM/IEEE 4th International Conference on Cyber-Physical Systems, pp. 31–40. ACM (2013)
22. Laporte, G.: The vehicle routing problem: an overview of exact and approximate algorithms. Eur. J. Oper. Res. **59**(3), 345–358 (1992)
23. Li, X., Sun, Z., Kurt, A., Zhu, Q.: A sampling-based local trajectory planner for autonomous driving along a reference path. In: Intelligent Vehicles Symposium Proceedings, 2014 IEEE, pp. 376–381. IEEE (2014)
24. Martínez-Barberá, H., Herrero-Pérez, D.: Multilayer distributed intelligent control of an autonomous car. Transp. Res. Part C Emerg. Technol. **39**, 94–112 (2014)
25. Parc, C.F.: Mobility-as-a-service: Turning transportation into a software industry (2014). http://venturebeat.com/2014/12/13/mobility-as-a-service-turning-transportation-into-a-software-industry/. Accessed 15 Feb 2016
26. Pérez, J., Milanés, V., Onieva, E.: Cascade architecture for lateral control in autonomous vehicles. IEEE Trans. Intell. Transp. Syst. **12**(1), 73–82 (2011)
27. Savelsbergh, M.W., Sol, M.: The general pickup and delivery problem. Transp. Sci. **29**(1), 17–29 (1995)
28. Schneider, J., Kirkpatrick, S.: Stochastic Optimization. Springer Science & Business Media, New York (2007)
29. Schreiber, M., Hellmund, A.M., Stiller, C.: Multi-drive feature association for automated map generation using low-cost sensor data. In: Intelligent Vehicles Symposium (IV), 2015 IEEE, pp. 1140–1147. IEEE (2015)
30. Shalamov, V., Filchenkov, A., Chivilikhin, D.: Small-moves based mutation for pick-up and delivery problem. In: Proceedings of the Companion Publication of the 2016 on Genetic and Evolutionary Computation Conference. ACM (2016, in press)
31. Tanzmeister, G., Friedl, M., Wollherr, D., Buss, M.: Efficient evaluation of collisions and costs on grid maps for autonomous vehicle motion planning. IEEE Trans. Intell. Transp. Syst. **15**(5), 2249–2260 (2014)
32. Valencia, R., Morta, M., Andrade-Cetto, J., Porta, J.M.: Planning reliable paths with pose slam. IEEE Trans. Rob. **29**(4), 1050–1059 (2013)

Agents-Robotics-Control (AROC)

The Role of Emotions, Mood, Personality and Contagion in Multi-agent System Decision Making

Ilias Sakellariou[1]([✉]), Petros Kefalas[2], Suzie Savvidou[2],
Ioanna Stamatopoulou[2], and Marina Ntika[3]

[1] University of Macedonia, Thessaloniki, Greece
iliass@uom.edu.gr
[2] The University of Sheffield International Faculty,
CITY College, Thessaloniki, Greece
{kefalas,ssavidou,istamatopoulou}@city.academic.gr
[3] South-East European Research Centre, Thessaloniki, Greece
mantika@seerc.org

Abstract. Emotions have attracted much interest in the Multi Agent Systems community, mainly due to their significance in creating simulations that more accurately predict crowd behaviours. Undoubtedly, infusion of agents with artificial emotions has to be supported by current psychology theories. The present work describes a formal model of artificial emotions based on the dimensionality theory, together with simulation results of an initial experimental evaluation. The model includes interesting aspects of emotions, such as emotion changes due to perception, long term affects due to mood, and emotion contagion due to social interactions.

Keywords: Emotional agents · Formal modelling and simulation · Crowd behaviour · El Farol problem

1 Introduction

It is widely accepted that emotions can change crowd behaviour and thus, in order to simulate the latter, one has to simulate the behaviour of each individual agent by taking into account artificial emotional states. Modelling of artificial emotions in multi-agent systems (MAS) is challenging by itself, let alone the fact that emotions are affected by a number of other features, such as mood, appraisal, personality, emotion contagion, etc. All these features convert a MAS rational behaviour to an emotional one with observed differences.

The aim of this paper is to present a complete formal model of agents infused with artificial emotions based on the dimensional theory as well as to demonstrate in practice that the model can result into different crowd behaviours. The main contribution is a new formal model for emotions that includes the interaction of emotions, perception, mood, personality and emotion contagion.

© IFIP International Federation for Information Processing 2016
Published by Springer International Publishing Switzerland 2016. All Rights Reserved
L. Iliadis and I. Maglogiannis (Eds.): AIAI 2016, IFIP AICT 475, pp. 359–370, 2016.
DOI: 10.1007/978-3-319-44944-9_31

This work is an extension of that reported in [12], which did not include an emotions contagion mechanism and effects of personality traits.

The rest of the paper is organised as follows: Sect. 2 presents a theory of emotions and its relation to moods, personality and contagion as described in Psychology. In Sect. 3, we discuss how artificial emotions can be formally represented in an agent. Section 4 discusses how emotions change over time. The role of emotions is demonstrated through a simulation of a well-known problem in game theory in Sect. 5. Related work is discussed in Sect. 6. Finally, Sect. 7 concludes the paper and suggests directions for future work.

2 Emotion Theories

2.1 Emotions, Moods, Personality Traits and Contagion

Although every human interaction involves emotion, psychologists have used several different definitions for emotions [3]. Some focus on its components, while others on its expressive reactions or functions. For the purposes of the present work, we adopt the definition suggested by [8]: emotions are *"passions— as defined as event-instigated or object-instigated states of action readiness with control precedence"*. These emotional states reflect an agent's readiness to take a particular decision in order to maintain or change the way they relate to the world. Some states may appear as activation states, such as staying apathetic, with the only aim to relate or not to relate. Others involve action tendencies, and are about approaching or moving away from a person, situation or event [8].

Mood is mistaken for an emotion but it is in fact different. Although both are classified as two different categories of affect, the main feature that distinguishes mood from emotion is that mood is long-lasting and not about something or someone. The cause of the mood is not always easy to identify.

Mood affects and is affected by *personality traits*. Traditional views of personality define it in two alternative ways: (a) internal factors explaining agents' constant behaviors, which are determined by a genetic basis, and (b) interpersonal factors that agents develop to relate to the world [9]. The traditional concept of personality has now been replaced by "personality traits"—the idea that someone's character consists of several dimensions that can be better understood not as a whole, but as separate characteristics, all of which constitute what we call personality. A trait (primarily neuroticism and extraversion) can determine the way an individual will emotionally react to particular circumstances.

Emotion can be contagious. Agents tend to mimic others by watching their facial, vocal or bodily expressions and this process is influenced by a number of psychophysiological, behavioural, cognitive and emotional factors. *Emotional contagion* depends, among others, on perception of emotion in others [5] and how this perception leads to imitating their expression. As perception depends on personality, mood and emotion, emotional contagion can be defined as the result of the relationship between particular personality traits, the mood they can cause on a constant basis and the intensity of emotion that they can trigger, depending on the intensity of the stimulus.

2.2 Measuring Emotions

Emotion has only recently attracted the scientific interest it deserves due to absence of reliable measurement. The dimensional approach of emotions is of interest and is widely used for measuring emotion, due to its simplicity in depicting complex emotional situations as well as the economy goals attained by using it [22]. Its two dimensions can capture the flow of an emotional episode in real time. The dimensional approach is based on the idea that emotion can be represented and measured by two dimensions in a *circumplex* [18] (Fig. 1a):

- *valence (or positivity)*, which represents how pleasurable it is for an individual to experience this state;
- *arousal (or activation)*, which represents how likely an agent is to take some action due to its particular state.

These two dimensions are the "property of affect" representing the "core affect" [19], which is the heart of every emotional experience. It can be described as "feeling good" or "feeling bad", "elevated" or "discouraged". Core affect can be either free floating or can be triggered by some stimulus and begin an emotional episode. Core affect is closely linked to emotion: feeling happy leads to perceiving objects in a congruent way and overestimating their pleasantness as more pleasant than real. The more positive the core affect is, the more pleasant the stimuli are going to be perceived and vice versa.

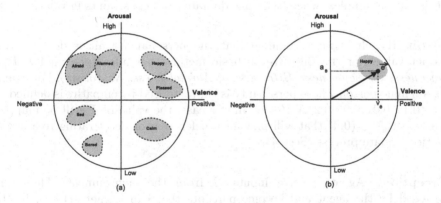

Fig. 1. (a) The circumplex of the dimensional theory of emotions with example emotions located in the two dimensions: arousal and valence. (b) Emotions as vectors or coordinates within the circumplex.

The circumplex is intended for measuring the dominant emotions and not all emotions experienced concurrently. Despite criticism, all opinions converge to the fact that this model is practical and easy to use as well as capable of capturing every day natural emotion. It is also suitable for measuring both object-less (i.e. mood or core affect) emotions as well as object-directed (primitives or basic emotions) states. Ideas about the connection between emotion, mood and particular personality traits have reached a point of convergence, relating all three components to the dimensional approach since emotion can be influenced by mood, mood can be influenced by and can also influence personality traits and personality traits are characterised by arousal and valence.

3 A Formal Representation of Dimensional Theory

Emotions. Emotions define what we will call the emotional states of an agent ^eQ. Within the circumplex, specific emotions are characterised by areas (Fig. 1a) with fuzzy limits. However, in a computational framework specific emotions are areas of points within the circumplex. A point can be formally defined with two equivalent representations (Fig. 1b):

- as a vector $\overrightarrow{e}$, where $\overrightarrow{e} = (\hat{\omega}, |\overrightarrow{e}|)$, i.e. the angle and the magnitude of the emotion vector within the circumplex ($\hat{\omega} \in [0, 360) \wedge |\overrightarrow{e}| \in [0, 1]$);
- as coordinates in the circumplex determined by the tuple (v_e, a_e), where v_e is the valence and a_e the arousal measures, with $v_e, a_e \in [0, 1] \wedge \sqrt{v_e^2 + a_e^2} \leq 1$.

Moods. Moods are considered as medium term emotions and are more long lasting than emotions. Therefore, moods, denoted by $\mathcal{M}$, can also be represented in the same way as emotions, either as $\overrightarrow{m} = (\hat{\omega}, |\overrightarrow{m}|)$ or as coordinates (v_m, a_m) within the circumplex, where the same domains and constraints in values apply.

Personality. It is most common that the personality trait is defined with distinct values for the Big Five [4] basic factors that affect personality: $P = $ *(Openness, Consciousness, Extraversion, Agreeableness, Neuroticism)*. For example, a female person whose personality is characterised by empathy is defined as $P = (0.15, 0.01, 0.09, 0.24, 0.16)$ [6]. In our case, the value of P will be mapped to a factor $f_p \in (0, 1]$ that will play an accelerating or decelerating role in the emotion update process (Sect. 4).

Perception. Agents receive inputs Σ from the environment. These are processed by the agent and become percepts that can trigger actions. In the presence of emotions, the inputs are transformed to emotional percepts $^e\Sigma$. More precisely, an input revision function ρ_σ can be defined, which given an input in Σ transforms it into an emotional percept taking into account the current emotional state, the mood and the personality, that is, $\rho_\sigma : \Sigma \times {}^eQ \times P \times \mathcal{M} \to {}^e\Sigma$.

Emotional Contagion. Emotion contagion is the impact that emotions of others have on an agent's emotion. By considering a small neighbourhood for each agent (contagion occurs when agents are in proximity of influence distance d_{inf}), the total emotional effect that the other n agents within this neighbourhood have on the former may be considered as one aggregate emotional percept (v_{con}, a_{con}). We have previously reviewed and experimented with three emotional contagion models [15]. In this work, we use the ASCRIBE model [10], where the emotional effect of one agent to another generally depends on the expressiveness of the former, the openness of the latter, and the distance between the two. This model fits naturally with our current work, since Openness and Expressiveness are two of the Big Five [4] basic factors.

4   Emotional Change

Agent behaviour is characterised by the consumption of inputs Σ that according to some state Q that the agent currently is and its internal representation of the world W (depending on the agent type, e.g. BDI, reactive, etc.), acts in a specific way in order to transform the environment with some output Γ, change its state and revise its internal representation of the world. Actions α that a rational agent does are formally defined as a function: $\alpha : \Sigma \times Q \times W \to \Gamma \times Q \times W$.

In an emotional agent, however, the above definition of actions can be revised as: $^e\alpha : \rho_\sigma(\Sigma) \times Q \times W \times {}^eQ \to \Gamma \times Q \times W \times {}^eQ$ with the triple $(\mathcal{M}, P, C)$ determining how an emotional state ^eQ changes over time. This triple defines a mood $\mathcal{M}$, a personality trait P, and a contagion model C. In the $^e\alpha$ signature, ρ_σ is an input to perception mapping (Sect. 3). A formal state-based model for emotional agents is analytically defined in [12].

The effect of change in emotions can be thought as the change of the emotion vector $\vec{e_c}$ to an updated vector $\vec{e_c}' = (v_e', a_e')$. This is depicted as a shift of the emotions vector towards a direction influenced by mood, percepts, personality and contagion (Fig. 2). The new values v_e' and a_e' are determined by functions which will be explained in the following subsections.

Personality trait f_p plays an important role in the emotion update process. A higher value indicates a more "perceptive" agent, thus it determines how quickly the emotion vector converges to an emotion percept or mood.

Change Due to Mood. As mentioned in Sect. 3, the current agent emotional state is affected by the agent's mood. Given the current emotion vector $\vec{e_c} = (v_e, a_e)$, the personality trait $f_p \in (0, 1]$ and the agent's mood $\vec{e_m} = (v_m, a_m)$, the updated emotion $\vec{e_c}' = (v_e', a_e')$ is given by the following equation:

$$(v_e', a_e') = \left(v_e + \frac{f_p^3 \cdot \Delta v_{me}}{1 + e^{2 \cdot (\Delta v_{me})}}, a_e + \frac{f_p^3 \cdot \Delta a_{me}}{1 + e^{2 \cdot (\Delta a_{me})}} \right) \quad (1)$$

where $\Delta v_{me} = v_m - v_e$ and $\Delta a_{me} = a_m - a_e$. The use of a logistic function, allows for a shift of the current emotion vector closer to the mood vector, while preserving the properties imposed by the circumplex, i.e. the magnitude of the vector being less or equal to one. In the absence of stimuli, the agent emotion vector will align with its mood vector after a number of iterations, depending on its personality trait. Since mood has a long term effect on the current emotion,

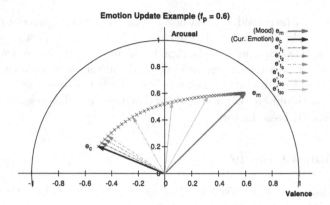

Fig. 2. Change of the emotion vector wrt time: case concerns updates due to mood.

this alignment has to occur after a significant number of steps, determined by a factor f_p^3 (recall that $f_p \in (0,1]$). Obviously, other parameters can be employed in order to calibrate the model to specific modelling requirements.

Change Due to Percepts. Emotional agents receive percepts which can be characterised as positive or negative in the sense that their emotion vector should change towards positive or negative valance, high or low arousal. Each input is associated with an emotional percept that affects the emotional state. Given an emotional percept $\overrightarrow{e_{prc}} = (v_{prc}, a_{prc})$ and the current emotion $\overrightarrow{e_c} = (v_e, a_e)$ of the agent, the updated emotion vector $\overrightarrow{e_c}' = (v_e', a_e')$ is given by:

$$(v_e', a_e') = \left(v_e + \frac{f_p^2 \cdot \Delta v}{1 + e^{-f_p \cdot (\Delta v)}}, \; a_e + \frac{f_p^2 \cdot \Delta a}{1 + e^{-f_p \cdot (\Delta a)}} \right) \tag{2}$$

where $\Delta v = v_{prc} - v_e$ and $\Delta a = a_{prc} - a_e$. If the specific input persists in the subsequent time points, then the emotion vector will eventually align with the corresponding emotion percept vector. Note that emotion change in this case depends on f_p^2 to reflect the fact that emotion change due to percepts has a greater impact than that of mood.

Change Due to Contagion. Emotional contagion is a form of perception, since agents perceive emotional stimuli that depend on the emotions of other agents in their proximity (Sect. 3). Thus, each agent is considered to have an *influence-crowd* (IC_i) that consists of all other agents in its sphere of influence, with the latter having a radius of d_{inf}, i.e. $IC_i = \{Agent_j : d(Pos_i, Pos_j) \le d_{inf}\}$.

The contagion model described below is inspired by the ASCRIBE model [10], although simpler, and is adapted to the new vector representation of emotions. *Contagion strength* w_{ij} determines the strength by which an agent j influences agent i, obviously when $j \in IC_i$. It depends on two factors: the expressiveness of agent j, $expr_j$, a measure of how much the agent manifests its emotions, and the *channel* ($channel_{ij}$) that models spatial characteristics of contagion:

$$w_{ij} = expr_j \cdot channel_{ij} \tag{3}$$

Channel strength models the fact that the closer agent j is to agent i the larger the impact it has on its emotions:

$$channel_{ij} = 1 - \frac{d(Pos_i, Pos_j)}{d_{inf}} \tag{4}$$

The overall contagion strength w_i of agent i by all agents in its influence is:

$$w_i = \sum_{j \in IC_i} (expr_j \cdot channel_{ij}) \tag{5}$$

Each emotion contagion vector coordinate is defined as the sum of the corresponding emotion vector coordinates of the influence crowd by the agents' normalised contagion strength:

$$(v_{cnt}, a_{cnt}) = \left(\sum_{j \in IC_i} (w_{ij}/w_i) \cdot v_j, \sum_{j \in IC_i} (w_{ij}/w_i) \cdot a_j \right) \tag{6}$$

The vector (v_{cnt}, a_{cnt}) forms a new kind of percept, and thus should be treated in the same manner as other emotion percepts. However, its effect on the current emotion depends on the openness (opn_i) of the agent i, and not the personality trait f_p:

$$(v'_e, a'_e) = \left(v_e + \frac{opn_i^2 \cdot \Delta v_{cnt}}{1 + e^{-opn_i \cdot (\Delta v_{cnt})}}, a_e + \frac{opn_i^2 \cdot \Delta a_{cnt}}{1 + e^{-opn_i \cdot (\Delta a_{cnt})}} \right) \tag{7}$$

where $\Delta v_{cnt} = v_{cnt} - v_e$ and $\Delta a_{cnt} = a_{cnt} - a_e$. The computation of emotion change due to contagion, concludes the mathematical realization of the emotional theory presented in Sect. 3.

5   Decision Making Based on Emotions and Contagion

In order to demonstrate how emotions can affect agent actions, we consider an example scenario, the El Farol bar problem, in which agent behaviour will be governed only by its emotions. This is an extreme case, however it serves as an indicative example of the model's behaviour. It should be noted that preliminary results of the same simulation scenario were presented in [12], however a different emotional perception model was used, and those experiments did not include emotion contagion, which is one of the main contributions of this work.

The El Farol problem is a well known problem in game theory, introduced originally by Arthur [1] as an example of how inductive reasoning, i.e. reasoning based on patterns, can be applied in "ill-defined situations". The problem concerns how each individual from a population of N people will decide on visiting the famous El Farol bar on a specific night, without knowing the intentions of the rest of the population. The decision needs a prediction of the bar attendance i.e. the number of people visiting the bar on that specific night. If bar attendance is below a threshold θ (60%) then people have a good time, i.e. the decision to visit the bar pays off, otherwise the visitors would be better off staying at home.

The approach adopted here was to base the decision only on emotions. Thus, agents decide to visit the bar when they "feel like it", i.e. their valence and arousal both have positive values ($v_e > 0 \wedge a_e > 0$). The idea is that the agents must be both in a positive emotional state (valence) and be willing to take action (arousal) in order to decide to attend. The tendency that agents have to visit the bar (not modelled explicitly in the original problem) is modelled by the agents' mood: all agents are initially assigned a mood that lies well in the positive valence, high arousal area of the circumplex:

$$\vec{m_i} = (\hat{\omega}_i, |\vec{m_i}|) \text{ with } \hat{\omega}_i \in [5, 85) \wedge |\vec{m_i}| \in (0.1, 1] \tag{8}$$

The emotion stimuli that agents perceive depends on the current bar attendance att. If the latter is below the threshold θ, agents do not receive any stimuli, since in this case their tendency to visit the bar (mood factor) is confirmed. However, in the case that the bar is overcrowded, agents perceive a "negative" emotion, i.e. an emotional percept of negative valence, low arousal, whose magnitude depends on how crowded the bar is:

$$\vec{e_{prc}} = \left(\hat{\omega_{prc}}, \frac{1 + abs(att - \theta)}{2} \right) \text{ with } \hat{\omega_{prc}} \in (180, 270) \tag{9}$$

In Eq. 9 the magnitude of the vector ranges between a value above 0.5, when the attendance is just above the threshold θ, to an extreme case of 1, when attendance is 100% and the threshold 0%. We have implemented the model in NetLogo [24], a well established platform for agent modelling and simulation. Agent behaviour was modelled using a formal X-Machine model, as in [12] and the implementation was based on the TXStates DSL [20] that allows easily encoding X-Machine agents.

5.1   Results and Discussion

In our experiments, attendance values are computed as the average of ten simulation runs with different initial conditions. Initial agent emotions are assigned to random vectors, thus in the initial steps of all simulations agent emotions converge to their mood, increasing steadily attendance, until the attendance threshold is reached. Agent mood is assigned initially (Eq. 8) and remains constant during experiments. The population was set to 100 persons and the attendance threshold was 60%. In the simulation we adopted a three-day period (simulation week), i.e. agents get a chance to visit the bar every three days, in order to allow mood and contagion effects to take place. Personality trait is drawn from a normal distribution with a mean value of 0.5 and standard deviation of 0.1.

The first set of experiments concerns simulations that do not involve contagion. The average attendance value after convergence was 61.46 with a standard deviation of 4.52, i.e. agents did manage to coordinate bar attendance successfully under the model. In Fig. 3 the line marked "No Contagion" depicts the evolution over time of the average attendance in each simulation week.

The second set of experiments concerned agent behaviour under contagion. Agents were divided into 10 clearly separable groups and had a specific influence radius d_{inf}. Clearly separable means that the groups had a distance between them such that no agent from a group could have another agent from a different group in its sphere of influence. Openness and expressiveness values were drawn from a normal distribution with a mean of 0.2 and a standard deviation 0.05. The average attendance value in this case was slightly greater, at 65.87 with a standard deviation of 4.95. Line "Contagion" in Fig. 3 shows the experimental results, as in the previous case.

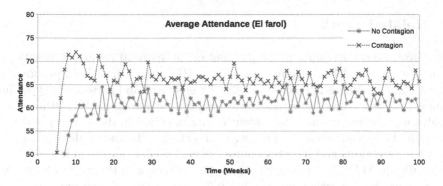

Fig. 3. The average attendance with and without contagion.

These experiments indicate that the effects of contagion increase bar average attendance and standard deviation. This is attributed to the fact that emotional states of agents tend to align to the "collective" emotion of the group, i.e. the society favours visiting the bar. This is demonstrated by the higher average line.

Probably what affects simulation results the most is the personality characteristics of the crowd. In the third set of experiments, personality trait f_p

Table 1. Average attendance wrt personality trait

f_p	Aver. Attendance No Contagion	(stdev)	Aver. Attendance Contagion	(stdev)
0.35	59.59	(3.14)	60.69	(2.82)
0.45	60.69	(4.56)	62.03	(4.08)
0.55	63.18	(5.97)	64.09	(5.56)
0.65	64.82	(6.53)	65.02	(5.61)

is constrained around different values and openness/expressiveness are drawn from a normal distribution of 0.15 with a standard deviation of 0.05, lower than that in the previous experiments. Table 1 depicts the average attendance over 10 simulation runs for different personality traits.

It is interesting to see that as the average personality trait is increasing, the average attendance and the standard deviation increases as well. Apparently, agents converge to their mood far more quickly leading to an increased average attendance, and are more perceptible to emotion percepts, thus changing their emotional state more rapidly leading to a higher standard deviation. Another interesting observation is that in low personality trait crowds, contagion results to a smaller standard deviation (albeit a constant higher average), possibly signifying that the crowd behaves in a more "coordinated" manner. However, in all experiments agent behaviour does converge to a mean value: emotions acted as a coordination mechanism for this simple society of agents.

6 Related Work

There are several attempts to model emotion in artificial agents. A number of computational models, logically formalised as BDI agents, have been proposed in [11,13,16]. Not surprisingly, the role of agent emotions and personality types has been extensively studied in evacuation simulation scenarios [23,25]. Other works involve using emotions to reduce agent communication in decision support systems [13], faster compromises in negotiation settings [21], and modelling of social aspects of emotions and its interconnection with socials norms to improve controllability in MAS [7]. Finally, emotions can be seen as a leverage to teamwork and cooperation between agents [14]. Unlike any of the related approaches, we attempt to formally model the dimensional theory of emotions including aspects like mood, emotion preception and contagion and personality trait types.

The El Farol problem has been widely studied in the literature. In the original work, attendees base their decision on past historical data of attendance and a set of k different strategies [1]. The current strategy is selected dynamically and is the one that is most accurate at the specific time point. This inductive approach leads to a mean attendance around the threshold of 60 %, with bar attendance values fluctuating around that threshold. An approach to model the problem using cognitive agents with emotions was made in [2]. The work is based on the Belief Desire Theory of Emotions (BDTE) [17] in which beliefs and desires are annotated by a degree that is updated by two operations, the Belief-Belief and the Belief-Desire comparators. The former acts as an updating mechanism of beliefs based on new percepts, while the latter updates desires according to the current newly acquired beliefs. These mechanisms act as "internal sensors" to the body, affecting perception and monitoring desire fulfilment, thus model emotions. In [2], DBTE agents model their desire to visit the bar, and update their belief, according to a strategy based on bar attendance. Strategy selection is based on their degree of belief in the adopted strategy. The main difference between our approach and the DBTE theory, is that the latter attempts to explain emotion manifestations based on belief/desire updates, whereas we consider emotions and mood as explicit parameters of the model that guide the reasoning process.

7 Conclusions and Further Work

Infusing agents with artificial emotions can be a valuable approach in order to obtain more accurate simulations of real life situations, such as evacuations, and various types of economic and social phenomena. The work described in this paper, extends our previous work [12] and presents a formal model of artificial emotions that is based on well established psychological theories (OCEAN and Dimensionality Theory) and covers important functions related to emotions, such as change to the emotional state due to percepts, the effect of mood as a long term emotional state, and the effect of social aspects such as emotion contagion. Through the El Farol example, we attempted to show the effects that the different parameters of such an emotions model have on crowd behaviour.

There is a significant number of future extensions to the current work. One of our first objectives is to experiment with different simulation scenarios that involve the economic phenomena, since those have attracted much interest in the recent years. For instance, it would be interesting to investigate the role of authoritative figures, such as government officials, news agencies, etc. in the creation or avoidance of phenomena such as bank runs. In a different direction we aim to investigate how the emotions model presented can be integrated with BDI agent architectures, leading to richer agent programming platforms.

References

1. Arthur, W.B.: Inductive reasoning and bounded rationality. Am. Econ. Rev. **84**(2), 406–411 (1994)
2. Baccan, D.D.A., Macedo, L.: Revisiting the El farol problem: a cognitive modeling approach. In: Giardini, F., Amblard, F. (eds.) MABS 2012. LNCS, vol. 7838, pp. 56–68. Springer, Heidelberg (2013)
3. Cornelius, R.: The Science of Emotion: Research and Tradition in the Psychology of Emotion. Prentice Hall, USA (1996)
4. Costa, P.T., J., McCrae, R.: Revised NEO Personality Inventory (NEO-PI-R) and NEO Five-Factor Inventory (NEO-FFI) manual (1992)
5. Doherty, W.: The emotional contagion scale. J. Nonverbal Behav. **21**(2), 131–154 (1997)
6. Durupinar, F.: From Audiences to Mobs: Crowd Simulation with Psychological Factors. Ph.D. thesis, Bilkent University, Dept. of Computer Engineering (2010)
7. Fix, J., von Scheve, C., Moldt, D.: Emotion-based norm enforcement and maintenance in multi-agent systems: foundations and petri net modeling. In: Proceedings of the 5th International Joint Conference on Autonomous Agents and Multiagent Systems, pp. 105–107. ACM (2006)
8. Fridja, N.: The psychologists point of view. In: Lewis, M., Haviland-Jones, J., Feldman-Barrett, L. (eds.) Handbook of Emotions. The Guildford Press, NY (2008)
9. Hogan, R., Hogan, J., Roberts, B.: Personality measurement and employment decisions: questions and answers. Am. Psychol. **51**(5), 469–477 (1996)

10. Hoogendoorn, M., Treur, J., van der Wal, C.N., van Wissen, A.: Modelling the interplay of emotions, beliefs and intentions within collective decision making based on insights from social neuroscience. In: Wong, K.W., Mendis, B.S.U., Bouzerdoum, A. (eds.) ICONIP 2010, Part I. LNCS, vol. 6443, pp. 196–206. Springer, Heidelberg (2010)

11. Jiang, H., Vidal, J.M., Huhns, M.N.: EBDI: an architecture for emotional agents. In: Proceedings of the 6th International Joint Conference on Autonomous Agents and Multiagent Systems, pp. 1–3. NY, USA. ACM, New York (2007)

12. Kefalas, P., Sakellariou, I., Savvidou, S., Stamatopoulou, I., Ntika, M.: The role of mood on emotional agents behaviour. In: The Proceedings of the 8th International Conference on Compuational Collective Intelligence. LNAI, vol. 9875 (accepted) (2016)

13. Marreiros, G., Santos, R., Ramos, C., Neves, J.: Context-aware emotion-based model for group decision making. Intell. Syst. IEEE **25**(2), 31–39 (2010)

14. Nair, R., Tambe, M., Marsella, S.: The Role of Emotions in Multiagent Teamwork. Who Needs Emotions?. Oxford University Press, New York (2005)

15. Ntika, M., Sakellariou, I., Kefalas, P., Stamatopoulou, I.: Experiments with emotion contagion in emergency evacuation simulation. In: Proceedings of the 4th International Conference on Web Intelligence, Mining and Semantics (WIMS 2014). pp. 49: 1–49: 11. NY, USA. ACM, New York (2014)

16. Pereira, D., Oliveira, E., Moreira, N., Sarmento, L.: Towards an architecture for emotional BDI agents. In: Proceedings of the Portuguese Conference on Artificial intelligence (EPIA2005), pp. 40–46 (2005)

17. Reisenzein, R.: Emotional experience in the computational belief-desire theory of emotion. Emot. Rev. **1**(3), 214–222 (2009)

18. Russell, J.: A circumplex model of affect. J. Pers. Soc. Psychol. **39**(6), 1161–1178 (1980)

19. Russell, J.: Core affect and the psychological construction of emotion. Psychol. Rev. **110**(1), 145–172 (2003)

20. Sakellariou, I., Dranidis, D., Ntika, M., Kefalas, P.: From formal modelling to agent simulation execution and testing. In: Proceedings of the 7th International Conference on Agents and Artificial Intelligence (ICAART-2015), pp. 87–98 (2015)

21. Santos, R., Marreiros, G., Ramos, C., Neves, J., Bulas-Cruz, J.: Personality, emotion and mood in agent-based group decision making. Int. Sys. IEEE **26**(6), 58–66 (2011)

22. Savvidou, S.: Validation of the FEELTRACE tool for recording impressions of expressed emotion. Ph.D. thesis, Queens University of Belfast, Northern Ireland, UK (2011)

23. Tsai, J., Fridman, N., Bowring, E., Brown, M., Epstein, S., Kaminka, G., Marsella, S., Ogden, A., Rika, I., Sheel, A., Taylor, M.E., Wang, X., Zilka, A., Tambe, M.: ESCAPES: Evacuation simulation with children, authorities, parents, emotions, and social comparison. In: the 10th International Conference on Autonomous Agents and Multiagent Systems - vol. 2. pp. 457–464. IFAAMAS, Richland, SC (2011)

24. Wilensky, U.: NetLogo (1999). http://ccl.northwestern.edu/netlogo/ center for Connected Learning and Computer-Based Modeling, Northwestern University. Evanston, IL

25. Zoumpoulaki, A., Avradinis, N., Vosinakis, S.: A Multi-agent simulation framework for emergency evacuations incorporating personality and emotions. In: Konstantopoulos, S., Perantonis, S., Karkaletsis, V., Spyropoulos, C.D., Vouros, G. (eds.) SETN 2010. LNCS, vol. 6040, pp. 423–428. Springer, Heidelberg (2010)

A Controller for Improving Lateral Stability in a Dynamically Stable Gait

Zhenglong Sun[(✉)] and Nico Roos

Maastricht University, Maastricht, Netherlands
z.sun@maastrichtuniversity.nl

Abstract. In the previous work we presented a new gait for humanoid robots, such as the Nao developed by Aldebaran. This new gait implemented on a Nao, reduces the energy consumption by 41 %. Then main feature of the new gait is the absence of an area of support. The foot can rotate freely around the ankle joint. This feature makes the gait suited for uneven terrains.

Stability is an important aspect of walking on uneven terrains, especially the lateral stability. During the single support phase of a step the robot balances above the stance leg. If the robot steps on a bump or in a hole, the lateral stability may be disrupted. This paper presents a controller that guarantees the lateral stability in the present of such disruptions.

1 Introduction

Bipedal walking for humanoid robots is one of the most interesting challenges in robotics. In the papers [1–3], we have investigated the possibility of creating an dynamically stable and energy efficient gait without an area of support. Here, the absence of an area of support means the ankle joint can move freely while the foot is on the ground. In the sagittal direction the robot's Center of Mass (CoM) is falling forward till the foot of the swing leg touches the ground. In the lateral direction, the robot balances above the stance foot in the single support phase, and falls towards the new stance foot in the double support phase. The falling towards the new stance foot is stopped by putting a force on the new stance leg. The resulting gait[1] was subsequently evaluated on a real Nao robot. The stability of the gait is validated on flat ground but not on uneven terrain since there is no feedback on the controller, thus robot cannot adjust the gait parameters to compensate for the uneven floor. In this paper, we improve the gait's lateral stability on uneven terrain by introducing such a controller.

The remainder of this paper is organized as follows. In the next section, we will give a brief overview of existing research about kinematics models for

---

[1] A video of the new gait at: https://project.dke.maastrichtuniversity.nl/robotlab/?attachment_id=153.

© IFIP International Federation for Information Processing 2016
Published by Springer International Publishing Switzerland 2016. All Rights Reserved
L. Iliadis and I. Maglogiannis (Eds.): AIAI 2016, IFIP AICT 475, pp. 371–383, 2016.
DOI: 10.1007/978-3-319-44944-9_32

humanoid robots, stability criteria and various approaches to obtain energy efficient bipedal walking. Section 3 briefly describes the new gait that we developed and presented in [1]. We used the Inverted Pendulum Model (IPM) to investigate the energy consumption in the sagittal plane. Subsequently, we extended the model to the lateral plane and describe a gait controller with multiple parameters for a 3D full-body humanoid robot. The controller can achieve a stable gait on a physical robot in the real world after we optimize the parameters through an Policy Gradient Reinforcement Learning (PGRL). Section 4 introduces our work on the neural network controller to enhance the lateral stability. Section 5 concludes this paper. We provide a brief summary of the results and outline the future research.

2   Related Work

2.1   Movement Models

Humanoid robots have complex bodies with irregular shape and mass distribution. Therefore, it is advantageous to obtain an elemental representation of the robot's dynamics. Ideal features of a model are simplicity, and both a conceptually and mathematically accurate representation of the dynamics of the real system. The main approaches employed to model the kinematics of humanoid robots are based on the Inverted Pendulum Model (IPM) [4] which involves a simplification compared to the body of the robot. The IPM represents the whole body of the robot as a point mass located at the center of mass (CoM) of the actual robot. The point mass is linked to the base of the robot by a telescopic massless leg. Restraining the movements of the CoM to a horizontal plane allows to simplify the motion equation of the IPM. The resulting model is known as the Linear Inverted Pendulum Model (LIPM) which [5] proposed to describe humanoid robot locomotion. The LIPM provides an efficient means to represent the kinematic behavior of the robot and it is therefore a popular tool to understand and manipulate the balance of a humanoid robot. With the LIPM and zero moment point (ZMP) stability criteria [6], institutes/companies have successfully built biped robots that can walk with various gaits adapting to different walking situations (e.g. [7–10]).

2.2   Energy Consumption

However, the movement model is not the only factor to be considered. The energy consumption of a gait is an important issue. Various approaches have been proposed to reduce the energy consumption of a gait. One of these approaches is passive-dynamic walking where the robot's dynamics are designed to enable a robot to walk down slight slopes without control input, except for the gravitational force. The paper of [11] explained this well. [12] believed that there are three primary flaws of passive-dynamic walker: they can only walk down slopes, their gaits are restricted by their dynamics, and they are sensitive to perturbations. Realizing these limitations, researchers [13] have sought to improve passive-dynamic walker by adding actuators.

A second approach to obtain energy efficient bipedal walking is the application of mechanical compliance. In the work of [13] and [14], springs were added across the hip, thigh, knee and ankle simultaneously. [15] exploited parallel knee compliance on the robot ERNIE and discussed how soft/stiff springs affect the energy efficiency at different walking speed. [16] described the implementation of series-elastic actuation on Spring Flamingo (a MIT's planar bipedal walking robot) to enable the control of the ground reaction forces during walking.

A third approach to obtaining energetically-efficient bipedal walking is the design of gaits that minimize the energetic cost of walking. The most common means of design is to use parametric optimization to the parameters that specify the gait of the robot. For example, [17] used parametric optimization to design fourth degree polynomial functions that give the joint motions over a step as functions of time. Unlike the previous example, in the work of [18] cubic splines connected at points uniformly distributed along the motion time are used to generate complete optimal steps, including a double-support phase.

Parametric optimization methods are also implemented to optimize the walking generator on humanoid Nao robots. In the work of [19], the proposed method models the omni-directional motion as the combination of a set of periodic signals. The parameters controlling the characteristics of the signals are encoded into genes and evolutionary strategies is used to learn an optimal set of parameters. Nao humanoid robots are used as the test platform. [20] augmented the 3D inverted pendulum with a spring model and use policy search to optimize the parameters of the walking engines on Nao robots. [21] introduced a two-stage learning algorithm for Central Pattern Generator (CPG) of Nao robot's bipedal walking.

3   Our New Gait

This section briefly describes our new gait presented in [1–3]. We first analyzed the gait without an area of support using an IPM with telescopic legs. Then we designed a controller which implements the gait on a real Nao robot.

3.1   Kinematics Model in Sagittal Direction

The IPM with telescopic legs allows the length of the virtual support leg to vary during a step. We proposed the leg-length policy $\delta : [-\frac{\pi}{2}, \frac{\pi}{2}] \rightarrow [0, 1]$ that determines how much the virtual support leg will be shortened as function of the angle between stance leg with vertical axis. The shortening of the stance leg is realized by bending the knee joint, see the right side of Fig. 1.

To identify the leg-length policy that minimizes the energy consumption of a robot, we make use of the fact that the robot has to bend the knee in order to shorten the leg. The knee torque is the main factor determining the energy

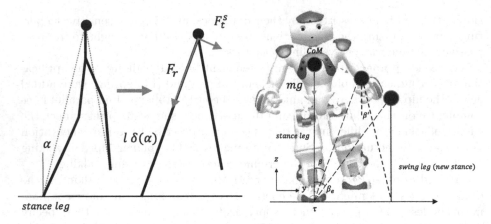

Fig. 1. The abstract Inverted Pendulum Model

Fig. 2. The CoM lateral movement during double support phase.

consumption [1]. Figure 3 shows the optimal leg-length policy $\delta(\alpha)$ as a function of the angle α from the beginning till the end of the step that we identified and Fig. 4 shows the realization using the 5-link model. The detailed information can be found in our previous publication [1].

Since we assumed the absence of an area of support and to further reduce the total energy cost, we set the stiffness on both ankle joints to almost zero. Thus, the stance leg of the robot can freely rotate around ankle joints, and the area of support reduces to a point.

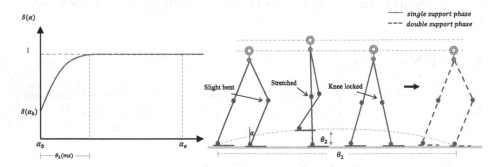

Fig. 3. The optimal leg-length policy

Fig. 4. Kinematic of sagittal motion

3.2 Kinematics Model in Lateral Direction

For a simple forward step, it is insufficient to only consider 2D dynamics in sagittal direction. To address the lateral stability, we designed a lateral controller to regulate the CoM lateral movement during double support phase which is proposed in [3]. We use the upper body tilt to initiate the lateral movement of the CoM towards the swing foot. Next, we use a force generated by the swing leg to stop the movement when CoM is balanced above the swing foot, which then becomes the new stance foot. The force generated by the swing leg is described by a force policy. In order to smooth the CoM transition trajectory, we determines the shape of the force policy by means of Quadratic Bezier curve which introduces the quadratic bezier point θ_7, one of the controller parameters in next subsection.

3.3 Controller Design

We designed a controller which implements this gait on a Nao robot. Because of the differences between the abstract model and the Nao, several parameters of the controller need to be fine-tuned. This subsection presents the parameters of a gait controller that realizes the leg-length policy described in Sect. 3.1 and the parameters that control the lateral movement of the CoM in the DPS. We identified 9 parameters that are essential in controlling a dynamic gait:

- *Step Length* (θ_1): Defines the distance which Nao moves in a singe step (sagittal).
- *Step Height* (θ_2): Defines the maximal altitude between ground and lifting feet. A high step height requires swing leg' faster move and may cause horizontal instability. A low step height increases the possibility of tripping and limits the step length.
- *Knee Bending* (θ_3): Defines the maximal bending of the swing leg at the beginning of the double support phase which determines the value of $\delta(\alpha_b)$, see Fig. 3. This parameter determines the sagittal velocity and the energy cost.
- *Step Time* (θ_4): Defines how long a single step lasts. This parameter determines the sagittal walking velocity.
- *Stretch Time* (θ_5): Defines how long it takes for the stance leg to stretch from θ_3 (angle of bent knee) to its full length at the beginning of the single support phase, see Fig. 3.
- *Torso Pitch Inclination* (θ_6): Defines the maximum angle that torso leans in sagittal direction at the beginning of the first step. If positive, it will move the center of mass (CoM) in sagittal direction. If it is set not appropriate, a fall will occur. In our experiments, the inclination lasts for 200 ms.
- *Quadratic Bezier point* (θ_7): Defines the magnitude of middle points in Quadratic Bezier Curves, which determines the force policy on swing leg (introduced in Sect. 3).
- *Torso Roll Inclination* (θ_8): Defines the maximum angle that torso leans in lateral direction. If positive, it will move the center of mass (CoM) towards the swing leg in lateral plane as discussed in Sect. 3.
- *Ratio of single support duration* (θ_9): Defines how long the single support phase lasts in one single step. The single support phase duration equals this parameter times step time θ_4.

All parameters except θ_1 (the step length) will be optimized in the experiments. We do not consider the step length for optimization because we need step length to be variable when the velocity is changing. We manually set different walking velocity v in each experiment and determined the optimal *Step Time* θ_4. The corresponding step length is given by: $\theta_1 = v\theta_4$.

Algorithm for Learning Controller Parameters. We use a policy gradient reinforcement learning method [22] to automatically search the set of possible parameters with the goal of finding the stable and low energy cost walk. In order to generate a gait that is energy efficient and stable, we considered a fitness function based on the total energy cost and the stability over a certain distance of forward walk. The energy cost determines 30 % of the fitness function value and the distance which the robot walks without falling determines 70 % of the fitness function value [1].

Learning Optimal Parameters in the Simulator. To generate the optimal gait parameters and validate the gait's performance, we uploaded the controller of our proposed gait together with an implementation of the policy gradient algorithm into the Webots simulator. We used a relatively elementary hand-tune gait as a starting policy for the policy gradient algorithm. Each new policy was evaluated by letting a robot walk at a constant distance of 0.75 m. During the walking, the energy consumption and stability were determined. The policy gradient algorithm converges to a parameters set P shown in Table 1.

Table 1. Learned parameters set P

Parameter	ϵ	Learned Value
Step length	0.1	3.9 (cm)
Step height	0.02	3.24 (cm)
Knee bending	0.1	14.2 (degree)
Step time	25	650 (ms)
Stretch time	25	78 (ms)
Torso pitch inclination	0.1	8.9 (degree)
Torso roll inclination	0.1	6.5 (degree)
Quadratic Bezier point	0.1	(0.9*DSP time, 0.2)
Ratio	0.1	0.8
Velocity		6 (cm/s)

The algorithm presented here converges to a local optimum. In order to investigate whether the results could be a global optimum, we repeated the learning experiment 500 times, each time starting from a randomly generated parameter vector x^π with the same velocity. The results of the experiments indicate that the local optimum we have in Table 1 is probably the global optimum. Therefore, the parameters set P most likely results in the most energy efficient gait.

The accompanying video material[2] shows the Nao robot walking on flat ground with our proposed gait controller at a speed of 6 cm/s. We also compare the new gait with the standard gait Aldebaran supplies with the Nao. The energy consumption of new gait is 41 % less than the Aldebaran gait.

4   Walking on Uneven Floor

The gait without support areas we proposed is validated as a dynamically stable gait on flat ground. However, the controller cannot compensate for external disturbances. This means any disturbance such as a push or stepping on uneven terrain may jeopardize its balance, because the ankle stiffness is set to almost zero. To enhance the walking stability, the gait should adapt to unknown disturbances. For example, if the robot is standing on a slope or stepping on a bump in the floor, the feet are not on the same altitude in the lateral plane, this may cause the CoM undershoot/overshoot the balance position when switching the stance leg. In order to make the problem tractable, simplifying assumptions are made. Since bipedal robot's stepping on the uneven terrain makes the altitude of robot's two foot different, the robot's walking on uneven floor can be viewed as walking on the slope in the lateral direction. The gait controller should adapt the gait parameters to compensate for the slope in the lateral direction.

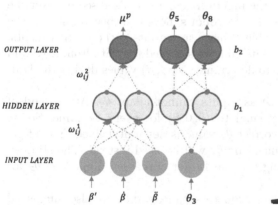
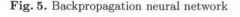

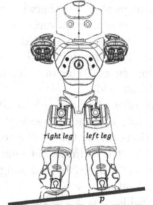

Fig. 5. Backpropagation neural network

Fig. 6. Experiment scenario: the nao robot stands on slope in angle p

---

[2] https://project.dke.maastrichtuniversity.nl/robotlab/?attachment_id=153.

4.1   Controller Design

As a first step in designing a controller that can handle disturbances influencing the lateral stability, we determined the optimal control parameters θ_2 to θ_9 when walking on a certain slope in the lateral direction. Since the left and right foot are at different height, the control parameters for the left and right leg might be different. Therefore, θ_3 to θ_9 are split into parameters for the left leg θ_i^L and the right leg θ_i^R. Next we addressed how to adapt the control parameters.

The robot does not know that it walks on a slope or about other disturbances. The only information it has available is (1) the angular speed $\dot{\beta}$ of the rotation of the CoM around the ankle of the stance leg. (2) the lateral acceleration measured by the Inertia Measurement Unit (IMU). This value approximates the angular acceleration $\ddot{\beta}$. (3) the angle β' of the CoM w.r.t the swing foot. Figure 2 shows the three parameters $(\beta', \dot{\beta}, \ddot{\beta})$. We chose to use these three parameters as inputs for the controller that can adapt the gait parameters.

We designed a series of experiment in the simulator *Webots* to obtain the optimal control parameters and input vector $(\beta', \dot{\beta}, \ddot{\beta})$ corresponding to certain slopes. In the experiments, the same policy reinforcement learning method in Subsect. 3.3 is used to find the proper control parameters that can generate the stable walking gait under different slopes. The experiments require the robot stands on various slopes where tilt angles varies from 0.00 to 0.139 (rad) in robot's lateral plane (see Fig. 6). We kept the robot walking on the slope and ran the policy search method to get the corresponding controller parameters which ensure robot's stability. Each new policy was evaluated by letting a Nao robot move for 5 seconds. The fitness function of each policy will get high score if the robot keeps stable. Otherwise, the function gets a penalized score. After the result of policy search converged while the robot's movement become stable, the control parameters are recorded. Table 2 shows the results fo those experiments. Next, from the beginning of DPS to its end, we sampled the data from IMU and joint sensors every 10 ms in order to determine $(\beta', \dot{\beta}, \ddot{\beta})$ values during the DSP for each lateral slope.

A controller that uses $(\beta', \dot{\beta}, \ddot{\beta})$ as inputs cannot adapt θ_7. At the end of the DSP, $(\beta', \dot{\beta}, \ddot{\beta})$ must become equal to $(0, 0, 0)$ for every θ_7 value. So, in the neighborhood of $(0, 0, 0)$ the correct θ_7 value is not well defined. Therefore, instead of θ_7, the stiffness determined by θ_7 will be used instead. The stiffness values are determined by sampling the the Bezier curve determined by θ_7 for different β' values, see Fig. 7.

From Table 2, we know that the parameters θ_3, θ_5 and θ_8 are also influenced by the slope angle. The parameter θ_3 solely depends on the relative elevation between stance and swing leg, and its value can be determined at end of the DSP where β' should become zero. Since θ_3 encodes the information about the slope angle, its value can be used to set the values of the other two parameters: θ_5 and θ_8.

Table 2. Learned control parameters adaptive to different slop angles

p (slop angle)	θ_3^L	θ_4^L (ms)	θ_5^L (ms)	θ_6^L	θ_8^L	θ_9^L	θ_7^L
0°	14.2°	650	78	8.9°	6.5°	0.8	(0.90,0.20)
3°	14.5°	650	75	8.9°	5.5°	0.8	(0.86 0.16)
5°	15.3°	650	76	8.9°	3.5°	0.8	(0.84 0.14)
8°	15.8°	650	74	8.9°	2.0°	0.8	(0.80 0.12)
10°	16.4°	650	75	8.9°	0.5°	0.8	(0.78 0.12)
p (slop angle)	θ_3^R	θ_4^R (ms)	θ_5^R (ms)	θ_6^R	θ_8^R	θ_9^R	θ_7^R
0°	14.2°	650	78	8.9°	6.5°	0.8	(0.90,0.20)
3°	13.5°	650	74	8.9°	6.5°	0.8	(0.90 0.22)
5°	11.7°	650	70	8.9°	7.5°	0.8	(0.91 0.26)
8°	9.6°	650	68	8.9°	8.5°	0.8	(0.92 0.28)
10°	8.2°	650	65	8.9°	9.5°	0.8	(0.94 0.29)

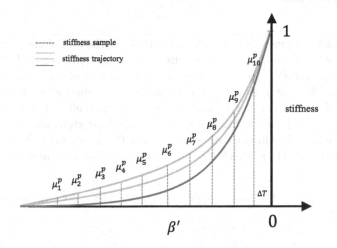

Fig. 7. Sampled stiffness over angle β'

4.2 Controller Implementation and Evaluation

We implemented two neural networks to control and improve the robot's lateral stability by adjusting the controller parameters adaptive to unknown slope. The backpropagation method has been applied for training multi-layer feedforward networks, see Fig. 5. With the trained network, we designed a simple neural network controller to maintain robot's walking stability on uneven terrain in the lateral direction. Figure 8 depicts the general architecture of the lateral stability controller that was implemented in this paper. When the robots walking on uneven floor and a new DSP begins, with the data retrieved from joint sensors (β and $\dot{\beta}$) and IMU ($\ddot{\beta}$), the first neural network takes these three variables as the input vector and outputs the new stiffness values for swing leg during the DSP.

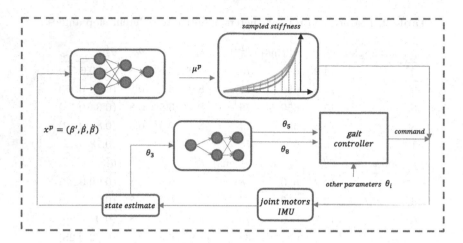

Fig. 8. Architecture for the identification process of the lateral controller parameters in double support phase

The second network uses as input the value of θ_3 at the end of the DSP and outputs the values of θ_5 and θ_8, which are used by the gait controller. Together with other fixed parameters, the gait controller generates updated joints command, to compensate for the uneven terrain. Figure 9 shows the roll angles trajectories of left/right legs when robot is walking on slope in 0.00 rad, 0.07 rad ($\approx 4°$) and 0.12 rad ($\approx 7°$). From this figure, we can see that right foot is higher than the left one when the slope exists which makes the joints on right leg rotate in less angles to let CoM approach its balance point. Moreover, under the different

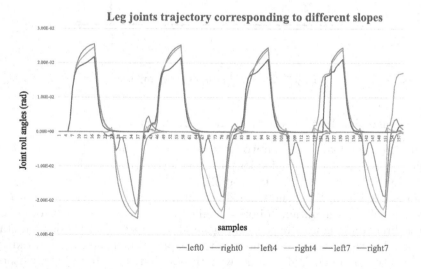

Fig. 9. The roll angle of both legs under different slope in $0°$, $4°$, $7°$.

slopes, the time for one step does not change which means our proposed gait can make a stable walk on different slopes without the loss of walking velocity. The accompanying video material[3] shows the Nao robot walking on uneven terrain with our proposed gait controller in the simulator Webots which proves our controller can handle the altitude difference of foot placement and adjust the control parameters to maintain balance.

5   Conclusion

In previous work we have presented a new gait for humanoid robots. An implementation of the gait on a Nao robot reduces the energy consumption with 41 % compared to the standard gait of the Nao. An important feature of new the gait is that it does not use an area of support. That is, the robot can rotate freely around the ankle joint while walking. This makes the new gait suited for uneven terrains because the feet can adapt to the slope of the terrain.

The absence of an area of support implies that, in principle, the robot is unstable. In the sagittal plane, the robot falls forwards in each step, and in the lateral plane, the robot balances above the stance foot in the single support phase and falls towards the swing foot in the double support phase. Nevertheless, experiment with a Nao robot on an almost flat floor consisting of wooden planks showed that the gait is stable.

In this paper, we investigate how we can improve the lateral stability of the gait when walking on an uneven terrain. The most important aspect of walking on an uneven terrain is the lateral stability. Since the robot balances above the stance leg in the single support phase while it can turn freely around the ankle joint, a bump or a hole in the walking surface may disrupt the lateral stability. Therefore, during the double support phase the robot may over- or undershoot its stable end point, namely, balancing stable above the new stance foot. The paper presents a feedback controller based on a feedforward neural network, that adapts the gait parameters in order ensure the lateral stability while walking on an uneven terrain.

The feedback controller for the lateral stability also enables the robot to handle, to some degree, slopes in the sagittal plane. In future work, we will extend the feedback controller to specifically address the effects of walking uphill or downhill.

References

1. Sun, Z., Roos, N.: An energy efficient dynamic gait for a nao robot. In: IEEE Conference on Autonomous Robot Systems and Competitions (2014)
2. Sun, Z., Roos, N.: An energy efficient gait for a nao robot. BNAIC (2013)
3. Sun, Z., Roos, N.: Dynamic lateral stability for an energy efficient gait. BNAIC (2014)

---

[3] https://www.youtube.com/watch?v=7DxHVEd8hc8.

4. Franklin, G.F., Powell, J.D., Emami-Naeini, A.: Feedback control of dynamics systems. Addison-Wesley, Reading (1994)
5. Kajita, S., Kanehiro, F., Kaneko, K., Yokoi, K., Hirukawa, H.: The 3d linear inverted pendulum mode: A simple modeling for a biped walking pattern generation. In: Proceedings of the 2001 IEEE/RSJ International Conference on Intelligent Robots and Systems, vol. 1, pp. 239–246. IEEE (2001)
6. Vukobratovic, M., Borovac, B., Surla, D., Stokic, D.: Biped Locomotion: Dynamics, Stability, Control and Application, vol. 7. Springer Science & Business Media, Heidelberg (2012)
7. Kajita, S., Kanehiro, F., Kaneko, K., Fujiwara, K., Harada, K., Yokoi, K., Hirukawa, H.: Biped walking pattern generation by using preview control of zero-moment point. In: IEEE International Conference on Robotics and Automation, 2003. Proceedings of the ICRA 2003. vol. 2, pp. 1620 1626. IEEE (2003)
8. Kajita, S., et al.: Cybernetic human HRP-4C: a humanoid robot with human-like proportions. In: Pradalier, C., Siegwart, R., Hirzinger, G. (eds.) Robotics Research. STAR, vol. 70, pp. 301–314. Springer, Heidelberg (2011)
9. Kanehiro, F., Hirukawa, H., Kajita, S.: Openhrp: Open architecture humanoid robotics platform. Int. J. Robot. Res. 23(2), 155–165 (2004)
10. Sakagami, Y., Watanabe, R., Aoyama, C., Matsunaga, S., Higaki, N., Fujimura, K.: The intelligent asimo: system overview and integration. In: IEEE/RSJ International Conference on Intelligent Robots and Systems, vol. 3, pp. 2478–2483. IEEE (2002)
11. McGeer, T.: Passive dynamic walking. Int. J. Robot. Res. 9(2), 62–82 (1990)
12. Kuo, A.D.: Choosing your steps carefully. Robot. Autom. Mag. IEEE 14(2), 18–29 (2007)
13. Collins, S.H., Ruina, A.: A bipedal walking robot with efficient and human-like gait. In: Proceedings of the 2005 IEEE International Conference on Robotics and Automation, 2005. ICRA 2005, pp. 1983–1988. IEEE (2005)
14. Farrell, K., Chevallereau, C., Westervelt, E.: Energetic effects of adding springs at the passive ankles of a walking biped robot. In: 2007 IEEE International Conference on Robotics and Automation, pp. 3591–3596. IEEE (2007)
15. Yang, T., Westervelt, E., Schmiedeler, J.P., Bockbrader, R.: Design and control of a planar bipedal robot ernie with parallel knee compliance. Auton. Robots 25(4), 317–330 (2008)
16. Pratt, J., Chew, C.M., Torres, A., Dilworth, P., Pratt, G.: Virtual model control: an intuitive approach for bipedal locomotion. Int. J. Robot. Res. 20(2), 129–143 (2001)
17. Chevallereau, C., Aoustin, Y.: Optimal reference trajectories for walking and running of a biped robot. Robotica 19(05), 557–569 (2001)
18. Saidouni, T., Bessonnet, G.: Generating globally optimised sagittal gait cycles of a biped robot. Robotica 21(02), 199–210 (2003)
19. Gökçe, B., Akln, H.L.: Parameter optimization of a signal-based omni-directional biped locomotion using evolutionary strategies. In: Ruiz-del-Solar, J. (ed.) RoboCup 2010. LNCS, vol. 6556, pp. 362–373. Springer, Heidelberg (2010)
20. Abdolmaleki, A., Shafii, N., Reis, L.P., Lau, N., Peters, J., Neumann, G.: Omnidirectional walking with a compliant inverted pendulum model. In: Bazzan, A.L.C., Pichara, K. (eds.) IBERAMIA 2014. LNCS, vol. 8864, pp. 481–493. Springer, Heidelberg (2014)

21. Shahbazi, H., Jamshidi, K., Monadjemi, A.H., Eslami, H.: Biologically inspired layered learning in humanoid robots. Knowl. Based Syst. **57**, 8–27 (2014)
22. Kohl, N., Stone, P.: Policy gradient reinforcement learning for fast quadrupedal locomotion. In: 2004 IEEE International Conference on Robotics and Automation, Proceedings ICRA 2004. vol. 3, pp. 2619–2624. IEEE (2004)

Scaled Conjugate Gradient Based Adaptive ANN Control for SVM-DTC Induction Motor Drive

Lochan Babani[(⊠)], Sadhana Jadhav, and Bhalchandra Chaudhari

College of Engineering, Pune, India
lochan212@gmail.com, {svj.elec,bnc.elec}@coep.ac.in

Abstract. In this work, an Artificial Neural Network (ANN) is developed to improve the performance of Space Vector Modulation (SVM) based Direct Torque Controlled (DTC) Induction Motor (IM) drive. The ANN control algorithm based on Scaled Conjugate Gradient (SCG) method is developed. The algorithm is tested on MATLAB Simulink platform. Results show smooth steady state operation as well as fast and dynamic transient performance. This is due to the SCG training algorithm of ANN which has the benchmarked performance against the standard Back-propagation (BP) algorithm. BP uses gradient descent optimization theory which has user selected parameters; learning rate and momentum constant. The network is trained offline and has fixed parameters. This leads to extra control effort and demands for online tuning of the parameters. SCG algorithm tunes these parameters with the use of second order approximation. Additionally, it takes less learning iterations and hence results in faster learning. Robustness to parameter variations and disturbances is the basic advantage of ANN, thus effectively controlling inherently non linear IM.

1 Introduction

The Direct Torque Control (DTC) [1] and Space Vector Modulated-Direct Torque Control (SVM-DTC) [2,3] are the latest research technologies for the VSI fed cage IMs. Both, being the special cases of vector control, have the fast and dynamic transient responses. The former has PWM-free operation and involves no coordinate transformation but 20–30 % steady state torque ripple; whereas the later has SVM based PWM, with coordinate transformation but less torque ripple upto 5–7 %. The research in SVM-DTC controlled IM drives is further advanced with respect to -

- Sensorless operation [4].
- Generation of reference stator voltage control vectors [5].
- Stator resistance compensation at low speeds [6].

© IFIP International Federation for Information Processing 2016
Published by Springer International Publishing Switzerland 2016. All Rights Reserved
L. Iliadis and I. Maglogiannis (Eds.): AIAI 2016, IFIP AICT 475, pp. 384–395, 2016.
DOI: 10.1007/978-3-319-44944-9_33

Modern control techniques like Sliding Mode Control (SMC) and intelligent control [7,8] are used to address the three issues mentioned above. Performances of various methods are compared [9] for control vector generation.

Out of the methods reported, PI has known demerits. An approach proposed in [10] gives excellent torque response and also very low torque distortions in static state, but has complex fuzzy structure. The chattering behavior of SVM can be eliminated by using ANN [11]. Apart from variety of techniques in DTC, SVM-DTC techniques based on neuro-fuzzy logic are also discussed and compared with the conventional motor control schemes [12]. Intelligent control has benefits of model-free control and hence it is robust [13]. With these advancements in intelligent control techniques, ANN is not restricted to be used as separate controller. Rather, it can be combined with other techniques to enhance the performance of the drive as in [14], where Field Oriented Control (FOC) and DTC [15] are combined to make a hybrid network and both the schemes FOC and DTC are mapped using two different ANNs. In [16], ANN is reported in control voltage vector loop. It can be seen that the ripple in torque with ANN-DTC control is very less as compared to conventional DTC at the same operating conditions [17]. In case of ANN based control, Resilient Backpropogation (RBP) and Levenberg-Marquardt (LM) [18] algorithms are very popular.

This paper presents another approach for training the ANN for the Space Vector Modulated- Direct Torque Control (SVM-DTC) based induction motor drive. The training algorithm used is Scaled Conjugate Gradient (SCG). The basic idea in this algorithm is to combine the model-trust region approach (used in the Levenberg-Marquardt algorithm), with the conjugate gradient approach [19] (Fig. 1).

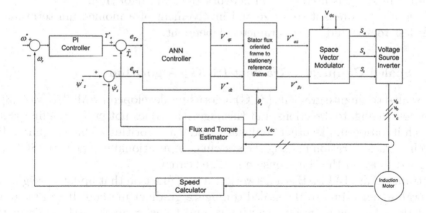

Fig. 1. ANN based SVM-DTC Scheme

2   Artificial Neural Network

Artificial Neural Networks (ANNs) have capability of recognizing the non-linear functions of their inputs. They can represent a non-linear system to the nearest possible approximation. Hence, in non-linear systems, which are difficult to control, the performance of conventional PI controller, in conditions like change in loads, disturbance and uncertainties can be improved by ANN [13].

Basically, it consist of neurons to represent inputs and outputs variables as well as intermediate layers which are interconnected via weights. The performance of ANNs depend upon the type of algorithm used, the number of neurons in the hidden layer, learning rate and the type of member function implied.

Various performance determining factors of ANNs are:

- Mean Square Error (MSE)
- Number of epochs
- Training Time
- Validation checks
- Gradient

2.1   Resilient Backpropagation (RBP) Algorithm

Resilient Backpropagation (RBP) is most suitable for pattern recognition problems. It utilizes the sign of derivative for the direction of weight update; that is, the magnitude of the derivative does not affect the weight-updating process. This eliminates the harmful effects of the magnitude of derivatives.

It generally converges much faster than other algorithms. In MATLAB, 'trainrp' function is used to train network by RBP algorithm.

The use of constant step size and involvement of a momentum term makes RBP less robust and more parameter dependent.

2.2   Scaled Conjugate Gradient (SCG) Algorithm

The scaled conjugate gradient (SCG) algorithm, developed by Moller [Moll93], is based on conjugate directions, but this algorithm does not perform a line search at each iteration unlike other conjugate gradient algorithms which require a line search at each iteration making the system computationally expensive. SCG was designed to avoid the time-consuming line search.

'trainscg' in MATLAB is a network training function that updates weight and bias values according to the scaled conjugate gradient method. It can train any network as long as its weight, net input, and transfer functions have derivative functions. In SCG algorithm, the step size is a function of quadratic approximation of the error function which makes it more robust and independent of user defined parameters.

The step size is estimating using different approach. The second order term is calculated as,

$$\bar{s}_k = \frac{E'(\bar{w}_k + \sigma_k \bar{p}_k) - E'(\bar{w}_k)}{\sigma_k} + \lambda_k \bar{p}_k \tag{1}$$

where, λ_k is a scalar and is adjusted each time according to the sign of δ_k.
The step size,

$$\alpha_k = \frac{\mu_k}{\delta_k} = \frac{-\bar{p}_j^T E'_{qw}(\bar{y}_1)}{\bar{p}_j^T E''(\bar{w}) \bar{p}_j} \tag{2}$$

where, $\bar{w}$ is weight vector in space R^n,
$E(\bar{w})$ is the global error function,
$E'(\bar{w})$ is the gradient of error,
$E'_{qw}(\bar{y}_1)$ is the quadratic approximation of error function,
$\bar{p}_1, \bar{p}_2....\bar{p}_k$ be the set of non-zero weight vectors.
λ_k is to be updated such that,

$$\bar{\lambda}_k = 2(\lambda_k - \frac{\delta_k}{|\bar{p}_k|^2}) \tag{3}$$

If $\Delta_k > 0.75$, then $\lambda_k = \lambda_k/4$
If $\Delta_k < 0.25$, then $\lambda_k = \lambda_k + \frac{\delta_k(1-\Delta_k)}{|\bar{p}_k|^2}$
where, Δ_k is comparison parameter and is given by,

$$\Delta_k = 2\delta_k[E(\bar{w}_k) - E(\bar{w}_k + \alpha_k \bar{p}_k)]/\mu_k^2 \tag{4}$$

Initially the values are set as, $0 < \sigma \leq 10^{-4}$, $0 < \lambda_l \leq 10^{-6}$ and $\bar{\lambda}_l = 0$.
Training stops when any of these conditions occurs:

- The maximum number of epochs is reached.
- The maximum amount of time is exceeded.
- Performance is minimized to the goal.
- The performance gradient falls below min-grad.
- Validation performance has increased more than max-fail times since the last time it decreased (when using validation)[20].

3   Simulation Results

A three phase induction motor with frequency = 50 Hz and power rating 3.5 kW is used. The various machine parameters are given as,

Stator resistance (rs) = 7.83 Ω
Rotor resistance (rr) = 7.55 Ω
Stator inductance (Ls) = 0.4751 H
Rotor inductance (Lr) = 0.4751 H
Mutual inductance (Lm) = 0.4535 H
No. of Poles (P) = 4
Inertia (J) = 0.013 kg-m^2

Torque of 12 Nm is applied at 0.5 s. A comparison has been done between the two algorithms Resilient Backpropogation (RBP) and Scaled Conjugate Gradient (SCG) method.

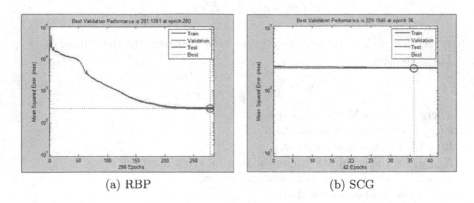

(a) RBP   (b) SCG

Fig. 2. Performance Plot for the two algorithms

SCG has been tried for various cases. Each time the conditions were varied and results are verified (Table 1).

- **Case I** The reference speed is kept to be zero.
- **Case II** A sinusoidal disturbance of amplitude 0.001 and high frequency is added in the torque and flux errors.
- **Case III** The stator resistance is increased to 150 % .
- **Case IV** The reference speed is kept to be 100 rad/s.
- **Case V** Speed is constant and torque is changed to zero at 0.7 s.
- **Case VI** Torque is constant and speed is changed from 100 rad/s to 50 rad/s at 0.8 s.

Results have been shown as a comparison between the two.

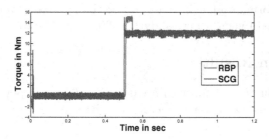

Fig. 3. Full load torque condition for Case I

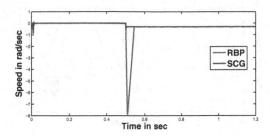

Fig. 4. Speed for the conditions referred in Case I

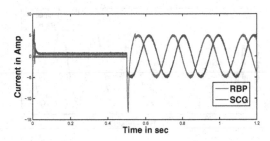

Fig. 5. Stator input current for both algorithms in Case I

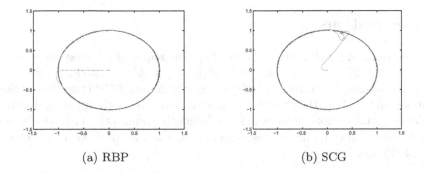

(a) RBP   (b) SCG

Fig. 6. Flux Circle for the two algorithms

Table 1. Comparison of the two algorithms in Case I

Parameter	RBP	SCG
Settling time	0.0357 s	0.0016 s
Full load steady state speed error	1–2 rpm	1–2 rpm
Full load torque error	2–3 Nm	0–1 Nm
THD	206.64 %	78.42 %
Input norm $\|Isa\|$	217.17	203.84
Mean square error (MSE)	281.13	229.19
Epochs	280	36

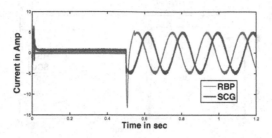

Fig. 7. Current transients for Case II

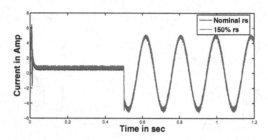

Fig. 8. Current transients for Case III

4   Observations

It is seen from Fig. 2 that the number of epochs required for least mean square error (MSE) is 280 for RBP while only 36 for SCG. Also, it is clear from Case I that total harmonic distortion (THD) is more in case of RBP than that of SCG. Figures 4 and 5 show that there exists undershoot in both speed and current graphs for RBP unlike SCG which has smooth transients. The steady state error in case of speed, current and torque is less in case of SCG as depicted by Figs. 14, 15 and 16.

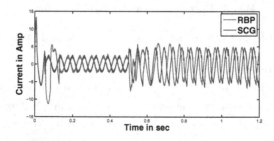

Fig. 9. Stator input current for both the algorithms in Case IV

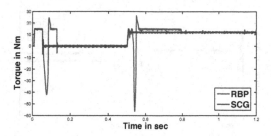

Fig. 10. Full load torque condition for Case IV

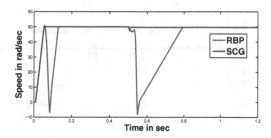

Fig. 11. Speed for the conditions referred in Case IV

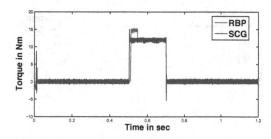

Fig. 12. Torque developed in Case V

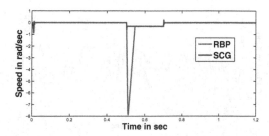

Fig. 13. Speed for the conditions referred in Case V

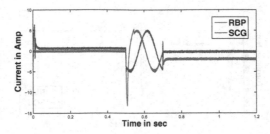

Fig. 14. Stator input current for both the algorithms in Case V

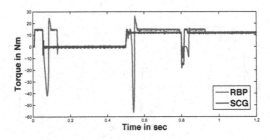

Fig. 15. Torque developed in Case VI

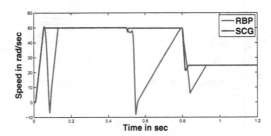

Fig. 16. Speed for the conditions referred in Case VI

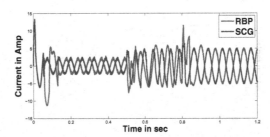

Fig. 17. Stator input current for both the algorithms in Case VI

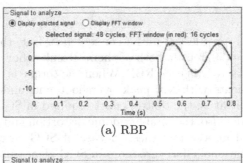

(a) RBP

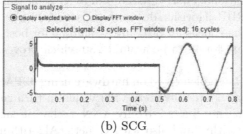

(b) SCG

Fig. 18. Selected stator current in Case I for both algorithms

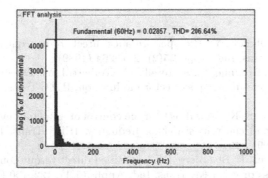

Fig. 19. Total harmonic distortion (THD) for RBP

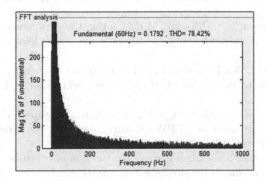

Fig. 20. Total harmonic distortion (THD) for SCG

5 Conclusion

Both the algorithms provide fast initial convergence. But, the calculations and training methods are different in both of them. Results show that the proposed controller gives better results than RBP. When the torque is applied, SCG gives more smooth transients with less peak overshoot and undershoot in case of current and speed. That is, the control effort required in SCG is comparatively less than that of RBP. Also, the total harmonic distortion and steady state errors in torque and speed appear to be less in case of SCG as compared to RBP. Though the computational efforts are more in SCG algorithm, it achieves faster learning as against RBP algorithm due to the absence of line search optimization. It is observed that the epochs for SCG are around 36 for best performance while that of RBP, they are found to be around 280 which proves the superiority of the algorithm.

This scheme can be implemented on hardware using dSPACE which produces pulses to feed Voltage Source Inverter (VSI) which in turn runs the motor. The scope of this scheme is not limited to only ANN, rather, it can be used in hybrid with some other algorithm and also with Genetic Algorithm (GA) to improve the transient response.

References

1. Takahashi, I., Ohmori, Y.: High-performance direct torque control of an induction motor. IEEE Trans. Ind. Appl. **25**(2), 257–264 (1989)
2. Habetler, T.G., Profumo, F., Pastorelli, M., Tolbert, L.M.: Direct torque control of induction machines using space vector modulation. IEEE Trans. Ind. Appl. **28**(5), 1045–1053 (1992)
3. Kang, J.-K., Sul, S.-K.: New direct torque control of induction motor for minimum torque ripple and constant switching frequency. IEEE Trans. Ind. Appl. **35**(5), 1076–1082 (1999)
4. Lascu, C., Boldea, I., Blaabjerg, F.: A modified direct torque control for induction motor sensorless drive. IEEE Trans. Ind. Appl. **36**(1), 122–130 (2000)
5. Lascu, C., Boldea, I.: Variable-structure direct torque control-a class of fast and robust controllers for induction machine drives. IEEE Trans. Ind. Appl. **51**, 785–792 (2004)
6. Husain, I., Cabrera, L.A., Elbuluk, M.E.: Tuning the stator resistance of induction motors using artificial neural network. IEEE Trans. Power Electron. **12**(5), 779–787 (1997)
7. Park, D., Kandel, A., Langholz, G.: Genetic-based new fuzzy reasoning models with application to fuzzy control. IEEE Trans. Syst. Man Cybern. **24**(1), 39–47 (1994)
8. Grabowski, P.Z., Kazmierkowski, M.P., Bose, B.K., Blaabjerg, F.: A simple direct-torque neuro-fuzzy control of PWM-inverter-fed induction motor drive. IEEE Trans. Ind. Electron. **47**(4), 863–870 (2000)
9. Jadhav, S.V., Srikanth, J., Chaudhari, B.N.: Intelligent controllers applied to SVM-DTC based induction motor drives: a comparative study. In: 2010 Joint International Conference on Power Electronics, Drives and Energy Systems (PEDES) & 2010 Power India, pp. 1–8 (2010)

10. Mir, S.A., Zinger, D.S., Elbuluk, M.E.: Fuzzy controller for inverter fed induction machines. IEEE Trans. Ind. Appl. **30**(1), 78–84 (1994)
11. Pinto, J.O.P., et al.: A neural-network-based space-vector PWM controller for voltage-fed inverter induction motor drive. IEEE Trans. Ind. Appl. **36**(6), 1628–1636 (2000)
12. Buja, G.S., Kazmierkowski, M.P.: Direct torque control of PWM inverter-fed AC motors-a survey. IEEE Trans. Ind. Electron. **51**(4), 744–757 (2004)
13. Lai, Y.-S., Chen, J.-H.: A new approach to direct torque control of induction motor drives for constant inverter switching frequency and torque ripple reduction. IEEE Trans. Energy Convers. **16**(3), 220–227 (2001)
14. Sakarung, P., Chatratana, S.: Neural network mapping of hybrid FOC-DTC induction motordrive. In: 2004 IEEE Region 10 Conference on TENCON 2004, vol. 500, pp. 467–470. IEEE (2004)
15. Kazmierkowski, M.P.: Control strategies for PWM rectifier/inverter-fed induction motors. In: Proceedings of the 2000 IEEE International Symposium on Industrial Electronics (ISIE 2000), vol. 1. IEEE (2000)
16. Jadhav, S.V., Kirankumar, J., Chaudhari, B.N.: ANN based intelligent control of induction motor drive with space vector modulated DTC. In: 2012 IEEE International Conference on Power Electronics, Drives and Energy Systems (PEDES), pp. 1–6 (2012)
17. Kumar, D., Thakur, I., Gupta, K.: Direct torque control for induction motor using intelligent artificial neural network technique. Int. J. Emerg. Trends Technol. Comput. Sci. (IJETTCS) **3**(4), 44–50 (2014)
18. Gottapu, K., Prashanth, Y., Mahesh, P., Sumith, Y., et al.: Simulation of DTC IM based on PI & artificial neural network technique. Simulation 2(7) (2013)
19. Møller, M.F.: A scaled conjugate gradient algorithm for fast supervised learning. Neural Netw. **6**(4), 525–533 (1993)
20. MATLAB

Introducing Chemotaxis to a Mobile Robot

Christina Semertzidou[1], Nikolaos I. Dourvas[1],
Michail-Antisthenis Tsompanas[1,2], Andrew Adamatzky[2],
and Georgios Ch. Sirakoulis[1,2(✉)]

[1] School of Engineering, Democritus University of Thrace, Xanthi, Greece
{chriseme,ndourvas,mtsompan,gsirak}@ee.duth.gr
[2] University of the West of England, Bristol, UK
andrew.adamatzky@uwe.ac.uk

Abstract. This paper deals with the path planning problem of a robot
in a maze based on a parallel chemotaxis bio-inspired model. The goal
is the effective search of a route, which can connect the starting position
of an autonomous robot with a final requested destination. To find this
route the robot has to take under consideration its geometry, elements of
its environment such as movements' restrictions by obstacles and other
characteristics of the topology. Chemotaxis is a term found in biology
and refers to the movement of an organism in response to a chemical
stimulus. Among numerous examples of such biological form here we
get inspiration by Physarum polycephalum, since this slime mold has
shown the ability to find the shortest path in a maze between two spots,
where chemicals exist, by following the gradient of the chemo-attractants.
Inspired by this behavior, chemotaxis will be used here to lead a robot
to its desired destination inside a labyrinth. A device transmitting signals can be considered as an equivalent chemical source and the robot's
receiver will follow the increased gradient of signal's amplitude. Moreover, an effective model, that has the ability to simulate such a problem
reducing the calculations' complexity and in the same time mimicking
the specific behavior, namely Cellular Automata (CA) is coupled with
chemotaxis. As a result, the design and implementation of a CA based
bio-inspired algorithm is proposed and an E-Puck robot uses the exact
algorithm to find the shortest path in different topologies as a proof of
concept.

Keywords: Artificial intelligence · Cellular Automata · Physarum polycephalum · Bio-inspired algorithm · Shortest path problem

1 Introduction

One of the most well known and complex problems in the field robotics is to find
the optimum path in a topology from one point to another. Many algorithms
have been successfully proposed during the previous years to control the robots
movement effectively and guide them to find the shortest path between two

© IFIP International Federation for Information Processing 2016
Published by Springer International Publishing Switzerland 2016. All Rights Reserved
L. Iliadis and I. Maglogiannis (Eds.): AIAI 2016, IFIP AICT 475, pp. 396–404, 2016.
DOI: 10.1007/978-3-319-44944-9_34

specific spots in a topology [1–3]. In many cases the nature itself provides for centuries, solutions in similar problems. In this paper, the proposed algorithm is inspired by the chemotaxis operation, meaning the guidance of an organism to a specific place using chemical stimulus. Moreover we envisage modeling of chemotaxis through Physarum, a life form which has an extremely interesting behavior during its life cycle. It has been found that during its reproductive stage called plasmodium, takes a form that has the ability to find the minimum path between two food spots in a labyrinth. Moreover, it can perform all these acts without using any central nervous control system, a common characteristic as well to all single-celled creatures. Many bio-inspired models are based on this behavior [4,5].

The algorithm that is used in this paper is inspired by the model that was proposed in [6] and modified accordingly in order to fit in the needs of the under study problem, which is the robot's guidance to an unknown environment in order to find the minimum path from one point to another. The mathematical model of chemotaxis and more specifically the diffusion equation of chemo-attractants is based on the parallel modeling tool, Cellular Automata (CA). CA can be assumed as an alternative form of microscopic reality, which supports the desired macroscopic behavior. The implementation of the proposed control algorithm has been applied on a E-Puck robot using programming environments that are created by the simulation program *Webots*. Different scenarios have been tested and presented, in order to preserve universality and effectiveness. Furthermore, a comparison is made with the results that are produced by MAT-LAB programming environment implementing the same algorithm in order to ensure their validity.

2   Cellular Automata Principles

A CA is characterized by the following ones:

- a regular lattice of cells covering a portion of a d-dimensional space;
- a set $C(\mathbf{r},t) = (C_1(\mathbf{r},t), C_2(\mathbf{r},t), ..., C_m(\mathbf{r},t))$ of variables attached to each site $\mathbf{r}$ of the lattice giving the local state of each cell at the time $t = 0, 1, 2, ...$;
- a rule $R = (R_1, R_2, ..., R_m)$ which specifies the time evolution of the states $C(\mathbf{r},t)$ in the following way: $C_j(\mathbf{r}, t+1) = R_j(C(\mathbf{r},t), C(\mathbf{r} + \boldsymbol{\delta}_1, t), C(\mathbf{r} + \boldsymbol{\delta}_2, t), ..., C(\mathbf{r} + \boldsymbol{\delta}_q, t))$, where $\mathbf{r} + \boldsymbol{\delta}_k$ designate the cells belonging to a given neighbourhood of cell $\mathbf{r}$.

The new state of the transition function during the time step $t+1$ is depended only to the previous time step t. The update of a cell is calculated using the information of its neighbors at the previous time step. However, in some cases it is needed to consider bigger memory resulting to dependence of more previous time steps $t - 1, t - 2, t - 3, ..., t - k$. The rule R is the same for all the cells and is applied simultaneously in each one of them creating a synchronous system. As a possible relaxization of CA original definition, the asynchronous updating of cell states can been also considered. Moreover, in regards to the CA rules' homogeneity, spatial or even temporal inhomogeneities can be introduced by the appliance of different rule in some specific cells in the lattice. The boundary cells are considered such an example of spatial inhomogeneity. CA are used in many fields and are capable of forming efficient models [7–9].

3   Chemotaxis Example

For readability reasons, some simple principles of chemotaxis are going to be presented through the paradigm of a living organism, namely Physarum. Physarum is a slime mould and Nakagaki *et al.* [10] were the first to observe that the plasmodium of this slime mould changes its shape as it crawls over a plain agar gel. If food is placed in two certain spots, it puts out pseudopodia that connect those food spots using the minimum path. Adamatzky in [11] placed the plasmodium in one place of the maze and, simultaneously, placed one food source (FS) in another place of the maze, before the plasmodium covers all the maze. The laboratory experiments show that the plasmodium spreads its pseudopodia trying to reach the food. Simultaneously, the food, releases the chemo-attractants to any direction in the maze. When the plasmodium finds those chemo-attractants, it follows them to the source food forming the minimum distance path between its initial site and the food site. So the plasmodium solves the maze in one pass because it is assisted by a gradient of chemo-attractants propagating from the target food. In conclusion, compared to Nakagaki *et al.* approach where thickness of the tube is proportional to flow, i.e. feedback, in Adamatzky approach Physarum acts simply as a concurrent navigator and "half" of the computation is done by diffusing chemo-attractants.

4   The Proposed CA Algorithm that the Robot Utilizes

In order to describe the chemotaxis behavior of P. polycephalum and enclose it to the robot's "brain", a bio-inspired CA algorithm has been designed when inspired by the original [6] model and modified according to the robot related short path. The study case is as follows: a robot is introduced into a specific place in the topology and a signal's transmitter is placed in another position of the maze representing the food spot. The signal is released inside the topology following the diffusion equation. When the sensors of the robot catch these "attractants", the algorithm chooses the direction with the greater attraction. After that the robot acts, turns to the specified direction and moves to the CA cell with the maximum attractants value. To simulate this experiment, the area where it takes place is divided into a matrix of squares with identical areas and each square of the surface is represented by a CA cell. The state of the (i, j) cell at time t, defined as $C_{i,j}^t$, is equal to Eq. 1:

$$C_{i,j}^t = \{ObstacleExistence_{i,j}, F_{i,j}^t, Robot_{i,j}^t\} \tag{1}$$

$ObstacleExistence_{i,j}$ is a variable that indicates the type of the area represented by the corresponding (i, j) cell. The possible values of

$ObstacleExistence_{i,j}$ are 1 for a free site of the lattice or 0 for an inaccessible site of the topology. Furthermore, $F_{i,j}^t$ represents the value of attractants (0–100) at time t in the area corresponding to the (i, j) cell. Finally, $Robot_{i,j}^t$ is a one-bit variable, which illustrates if the (i, j) cell is included in the final path and if it takes the value 1 then the robot takes the decision to move in this CA cell. The type of neighborhood that was used in this CA model is the Moore neighborhood consisting of all the eight side and diagonal neighbors of the central cell.

The diffusion equation (Eq. 2) allows us to talk about the route of randomly moving particles in n dimensions.

$$\frac{\partial}{\partial t} c(\bar{x}, t) = D \triangledown^2 c(\bar{x}, t) \tag{2}$$

But CA is a modeling tool that is based on discrete space and time. For this purpose, the discrete diffusion equation is used to describe the spread of the signal's values produced by the transmitter (Eq. 3).

$$\begin{aligned} F_{i,j}^{t+1} = \{ & F_{i,j}^t + f1 \left[(F_{i-1,j}^t - f3 \times F_{i,j}^t) \right. \\ & + (F_{i+1,j}^t - f3 \times F_{i,j}^t) \\ & + (F_{i,j-1}^t - f3 \times F_{i,j}^t) \\ & + \left. (F_{i,j+1}^t - f3 \times F_{i,j}^t) \right] \end{aligned} \tag{3}$$

where $f1, f2, f3$ are the parameters that determine the attractants' speed of expansion. After a few time steps the attractants are released inside the topology. When the robot meets them, it follows them towards the transmitter creating a route, which is the path that the robot will follow. In particular, it searches which of its neighbors has the greater value of $F_{i,j}^t$. When this value is justified, the $Robot_{i,j}^{t+1}$ value of this neighbor is changed from 0 to 1, while the robot moves to the specific cell. This procedure is repeated until the robot reaches the transmitter's position, which releases the signal inside the topology.

5   Implementation and Results

For testing reasons the *E-puck* desktop mobile robot (Fig. 1) was selected according to its properties and characteristics as introduced by EPFL [12] and previous citing works [13].

In this aspect, a variable, namely *translation* used for the implementation of the bio-inspired algorithm, is responsible for the definition of the robot's position in the CA space. It comprises three subvariables, x, y, z corresponding to the directions of the Cartesian system. The parameter *rotation* indicates the turning angle of the object to a specific direction, according to the environmental stimuli. Variable *FloorTileSize* defines the size of the CA cell, which for testing

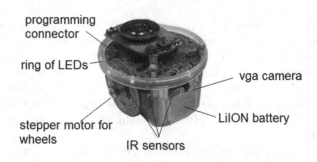

Fig. 1. The E-puck robot.

reasons is also provided in centimeters. More specifically, all simulations take into account a CA cell with side equal to 10 cm. The proposed size is chosen to fit to the exact E-Puck's dimensions which are not greater than 7.4 cm in length. Furthermore, *FloorSize* defines the size of the CA lattice and it is measured in meters. In order to reduce the calculations' complexity handled by E-puck, a 12 × 12 CA lattice is used. Another variable is *wallHeight*, which defines the height of the obstacles and is also measured in centimeters. For sake of simplicity and without loss of generality, in every simulation, the aforementioned variable takes the value 0.1. Finally, parameter *wallThickness* indicates how wide an obstacle, while in every simulation of this paper, this variable takes the value 0.074.

Three different testing topologies were used as test beds for the aforementioned CA algorithm in this paper. The first one is a topology free of obstacles (Fig. 2). The second experiment is evolved in a space with an obstacle at a specific place, depictured with the white color, (Figs. 3 and 4) and the final one is a more complex one, i.e. a maze (Fig. 5).

The robot and the food source are placed in different sites. Then the bio-inspired algorithm leads the E-puck to the transmitter using the shortest path every time. The transmitter is indicated by a cube with a *FOOD* caption on it. In order to verify the results, the same model, with the same topologies are designed in MATLAB environment.

In Fig. 2b the robot starts at coordinates (2,2) while the food source exists at (11,11). After 30 times steps the robot follows the minimum straight distance from the starting point to the final one. The robot uses the von Neumann neighborhood, consisting of solely side neighbors, and it can move only horizontally or vertically. Apparently, it will not be able to move to diagonal cells. So the 17 cells from the beginning to the ending position is the minimum distance. The same algorithm executed in MATLAB is presented in Fig. 2a. The red cells indicate the boundaries of the lattice. The black cell represents the starting position of the robot and the blue cell indicates the destination. The yellow line is the path that the robot follows. It is clear that the results are identical.

In Fig. 3b the robot starts at coordinates (2,2) and the food source exists at (11,11). However, this time, an obstacle exists inside the lattice creating a corner.

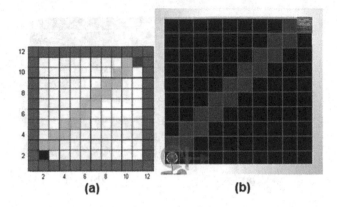

Fig. 2. Experiment in a lattice with no obstacles. (Color figure online)

The robot chooses to pass from the bottom of the obstacle towards its destination. This path consists of 17 cells. But the other minimum way from the left and upper side of the obstacle is also a distance of 17 cells and it is shown using the blue dots. As a result, these two routes are identical. The robot makes a random choice and follows the red path or the one in blue dots. After 48 times steps the robot moves on the minimum path from the starting point to the final one, avoiding the obstacle. The same algorithm executed in MATLAB is presented in Fig. 3a. Once again the results again identical.

In Fig. 4b the architecture of the topology is the same, using the same obstacle in the same position. But the initial conditions are now different. The robot starts at coordinates (2,11) and the food source exists at (11,5). After 28 times steps the robot moves on the minimum path from the starting point to the final one, avoiding the obstacle. This minimum path is the red one with 14 cells. The path passing through the other side of the obstacle is presented in blue dots. This

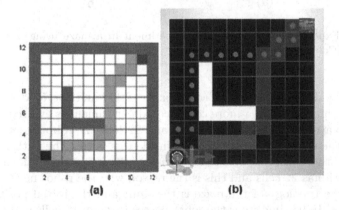

Fig. 3. Experiment in a lattice with obstacle. (Color figure online)

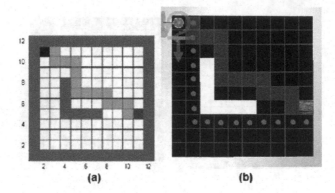

Fig. 4. Experiment in a lattice with obstacle and different starting positions. (Color figure online)

path is longer because it results to 16 cells. The robot never chooses this path. The same algorithm executed in MATLAB is presented in Fig. 4a. The results are again identical.

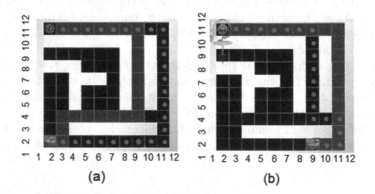

Fig. 5. (a) Experiment in a maze. (b) Experiment in a maze using different initial conditions. (Color figure online)

In Fig. 5a the robot starts at coordinates (2,11) and the food source exists at (2,2). In this situation, a more complicated experiment is evolved with many obstacles to avoid. After 140 times steps and 22 cells the robot reaches the food spot using the minimum path (red line) inside this maze. The path in blue dots present the other path that also leads from the initial position to the destination. But this path has 22 cells and this is the reason why the robot never chooses it. In Fig. 5b the topology of the maze is the same and the initial position of the food changes. In 98 time steps the robot passes through 19 cells and reaches the food using the minimun distance (red line). The line formed by blue dots is also the minimum one because it also has 19 cells. After multiple experiments, it is proven that the robot chooses randomly the one or the other route.

6 Conclusions

There are many models trying to use chemotaxis and adapt it to their solutions. It is also proven to be a very powerful tool in order to find the minimun distance between two spots in a lattice. In this paper, a proposed chemotaxis bio-inspired CA model is used into a robot, and more specific in an E-Puck. This model tries to reproduce the chemotaxis behavior, which follows the gradient of chemo-attractants towards the food source using the minimum distance. The expansion of chemo-attractants is modeled using a discrete diffusion equation. The *Webots* programming environment is used to simulate the operations of an E-Puck inside the designed topologies. After applying this model, it is shown that the robot is always capable of finding the shortest path, between itself and an attractant source using its sensors and avoiding effectively the obstacles. A MATLAB implementation verifies the Webots results. The final product proposed is a robot, which has the ability to find solutions in a complex computational problem.

References

1. Lutvica, K., Velagic, J., Kadic, N., Osmic, N., Dzampo, G., Muminovic, H.: Remote path planning and motion control of mobile robot within indoor maze environment. In: IEEE International Symposium on Intelligent Control (ISIC), pp. 1596–1601 (2014)
2. Rahnama, B., Elci, A., Metani, S.: An image processing approach to solve labyrinth discovery robotics problem. In: IEEE 36th Annual Computer Software and Applications Conference Workshops (COMPSACW), pp. 631–636 (2012)
3. Boukas, E., Kostavelis, I., Gasteratos, A., Sirakoulis, G.C.: Robot guided crowd evacuation. IEEE Trans. Autom. Sci. Eng. **12**(2), 739–751 (2015)
4. Dourvas, N., Tsompanas, M.A., Sirakoulis, G.C., Tsalides, P.: Hardware acceleration of cellular automata physarum polycephalum model. Parallel Process. Lett. **25**(01), 1540006 (2015)
5. Kalogeiton, V.S., Papadopoulos, D.P., Georgilas, I.P., Sirakoulis, G.C., Adamatzky, A.I.: Cellular automaton model of crowd evacuation inspired by slime mould. Int. J. General Syst. **44**(3), 354–391 (2015). ISO 690
6. Tsompanas, M.-A., Sirakoulis, G.C.: Modeling and hardware implementation of an amoeba-like cellular automaton. Bioinspir Biomimetics **7**(3), 036013 (19 pp.) (2012). http://iopscience.iop.org/article/10.1088/1748-3182/7/3/036013
7. Tsompanas, M.-A., Kachris, C., Sirakoulis, G.C.: Optimization of shared-memory multicore systems using game theory and genetic algorithms on cellular automata lattices. In: IEEE 27th International Parallel and Distributed Processing Symposium Workshops & PhD Forum (IPDPSW), 2013, pp. 482–490. IEEE (2013)
8. Tsompanas, M.-A., Kachris, C., Sirakoulis, G.C.: Evaluating conflicts impact over shared last-level cache using public goods game on cellular automata. In: International Conference on High Performance Computing and Simulation (HPCS), 2013, pp. 326–332. IEEE (2013)

9. Tsompanas, M.-A., Sirakoulis, G.C., Karafyllidis, I.: Modeling memory resources distribution on multicore processors using games on cellular automata lattices. In: IEEE International Symposium on Parallel & Distributed Processing, Workshops and Phd Forum (IPDPSW), 2010, pp. 1–8. IEEE (2010)
10. Nakagaki, T., Yamada, H., Tóth, Á.: Intelligence: maze solving by an amoeboid organism. Nature 407(6803), 470 (2000)
11. Adamatzky, A.: Slime mold solves maze in one pass, assisted by gradient of chemo-attractants. IEEE Trans. NanoBiosci. 11(2), 131–134 (2012)
12. Mondada, F., Bonani, M., Raemy, X., Pugh, J., Cianci, C., Klaptocz, A., Martinoli, A.: The e-puck, a robot designed for education in engineering. In: Proceedings of the 9th conference on autonomous robot systems and competitions, vol. 1, pp. 59–65. IPCB: Instituto Politcnico de Castelo Branco. ISO 690 (2009)
13. Ioannidis, K., Sirakoulis, G.C., Andreadis, I.: Cellular ants: a method to create collision free trajectories for a cooperative robot team. Rob. Auton. Syst. 59(2), 113–127 (2011)

Artificial Neural Network Modeling
(ANNMO)

Malware Detection with Confidence Guarantees on Android Devices

Nestoras Georgiou, Andreas Konstantinidis, and Harris Papadopoulos(✉)

Department of Computer Science and Engineering,
Frederick University, Nicosia, Cyprus
h.papadopoulos@frederick.ac.cy

Abstract. The evolution of ubiquitous smartphone devices has given rise to great opportunities with respect to the development of applications and services, many of which rely on sensitive user information. This explosion on the demand of smartphone applications has made them attractive to cybercriminals that develop mobile malware to gain access to sensitive data stored on smartphone devices. Traditional mobile malware detection approaches that can be roughly classified to signature-based and heuristic-based have essential drawbacks. The former rely on existing malware signatures and therefore cannot detect zero-day malware and the latter are prone to false positive detections. In this paper, we propose a heuristic-based approach that quantifies the uncertainty involved in each malware detection. In particular, our approach is based on a novel machine learning framework, called Conformal Prediction, for providing valid measures of confidence for each individual prediction, combined with a Multilayer Perceptron. Our experimental results on a real Android device demonstrate the empirical validity and both the informational and computational efficiency of our approach.

Keywords: Malware detection · Android · Security · Inductive Conformal Prediction · Confidence measures · Multilayer Perceptron

1 Introduction

The widespread deployment of smartphone devices has brought a revolution in mobile applications and services that span from simple messaging and calling applications to more sensitive financial transactions and internet banking services. As a result, a great deal of sensitive information, such as access passwords and credit card numbers, are stored on smartphone devices, which has made them a very attractive target for cybercriminals. More specifically, a significant increase of malware attacks was observed in the past few years, aiming at stealing private information and sending it to unauthorized third-parties.

Mobile malware are malicious software used to gather information and/or gain access to mobile computer devices such as smartphones or tablets. In particular, they are packaged and redistributed with third-party applications to inject

© IFIP International Federation for Information Processing 2016
Published by Springer International Publishing Switzerland 2016. All Rights Reserved
L. Iliadis and I. Maglogiannis (Eds.): AIAI 2016, IFIP AICT 475, pp. 407–418, 2016.
DOI: 10.1007/978-3-319-44944-9_35

malicious content into a smartphone and therefore expose the device's security. While the first one appeared in 2004 targeting the Nokia Symbian OS [1], in the fourth quarter of 2015 G DATA security experts reported discovering 8,240 new malware applications on average per day and a total of 2.3 million new malware samples in 2015, in just the Android OS [2]. When malware compromises a smartphone, it can illegally watch and impersonate its user, participate in dangerous botnet activities without the user's consent and capture user's personal data.

Mobile malware detection techniques can be classified into two major categories: static analysis [4,20] and dynamic analysis [7,14,18]. The former aim at detecting suspicious patterns by inspecting the source code or binaries of applications. However, malware developers bypass static analysis by employing various obfuscation techniques and therefore limiting their ability to detect polymorphic malware, which change form in each instance of the malware [13]. Dynamic analysis techniques on the other hand, involve running the application and analyzing its execution for suspicious behavior, such as system calls, network access as well as file and memory modifications. The main drawback of these techniques is that it is difficult to determine when and under what conditions the malware malicious code will be executed.

Both static and dynamic analysis techniques are typically implemented following two main approaches: signature-based approaches, which identify known malware based on unique signatures [9], and heuristic based approaches, which identify malicious malware behaviour based on rules developed by experts or by employing machine learning techniques [8,12,14,20]. Even though signature-based techniques have been successfully adopted by antivirus companies for malware detection in desktop applications, this is not a preferred solution in the case of mobile devices due to their limited available resources in terms of power and memory. Additionally, signature-based techniques cannot detect zero-day malware (not yet identified) or polymorphic malware, something that is not an issue for heuristic based techniques. On the other hand, unlike signature-based techniques, heuristic based techniques are prone to false positive detections (i.e. wrongly identifying an application as malware).

Most recent research studies focus on extending the idea of heuristic-based approaches by employing machine learning techniques. For example, Sahs and Khan [19] use a one-class Support Vector Machine to detect malicious applications based on features extracted from Android Application Packages (APKs) of benign applications only. Demertzis and Iliadis [6] propose a hybrid method that combines Extreme Learning Machines with Evolving Spiking Neural Networks using features extracted from the behaviour of applications when executed on an emulated Android environment. In [5] the same authors propose an extension to the Android Run Time Virtual Machine architecture that analyses the Java classes of applications using the Biogeography-Based Optimizer (BBO) heuristic algorithm for training a Multilayer Perceptron to classify applications as malicious or benign. Abah et. al. [11] present a detection system that uses a k-Nearest Neighbour classifier to detect malicious applications based on features extracted during execution. To the best of our knowledge, none of the machine learning based methods proposed in the literature provides any reliable indication on the likelihood of its detections being correct.

This paper proposes an approach that addresses the uncertainty arising from the use of machine learning techniques. Since there is no way of guaranteeing 100 % accuracy with any machine learning technique, this study aims at providing a reliable indication of how probable it is for a prediction to be correct. The provision of such an indication for applications identified as possible malware, would be of great value for the decision of the user on whether to remove an application or not, depending on the risk he/she is willing to take. The proposed approach utilizes a novel machine learning framework, called Conformal Prediction (CP) [21], for providing provably valid measures of confidence for each individual prediction without assuming anything more than that the data are generated independently from the same probability distribution (i.i.d.). The confidence measures produced by CP have a clear probabilistic interpretation, thus enabling informed decision making based on the likelihood of each malware detection being correct.

Specifically, we combine the inductive version of the CP framework [16,17], which is much more computationally efficient than the original version, with a Multilayer Perceptron (MLP), which is one of the most popular and well-performing machine learning techniques. We evaluated the proposed approach on a realistic malware dataset using a prototype developed for a real Android device (LG E400). Our experimental results demonstrate the empirical validity of our approach as well as its efficiency in terms of accuracy, informativeness and computational time.

The rest of the paper starts with an overview of the Conformal Prediction framework and its inductive counterpart in Sect. 2. The next section (Sect. 3) details the proposed approach. Section 4 describes the dataset used for evaluating our approach, while Sect. 5 presents our experiments and the obtained results. Finally, Sect. 6 gives our conclusions.

2   Conformal Prediction

The Conformal Prediction (CP) framework extends conventional machine learning algorithms into techniques that produce reliable confidence measures with each of their predictions. The typical classification task consists of a training set $\{(x_1, y_1), \ldots, (x_l, y_l)\}$ of instances $x_i \in \mathbb{R}^d$ together with their associated classifications $y_i \in \{Y_1, \ldots, Y_c\}$ and a new unclassified instance x_{l+1}. The aim of Conformal Prediction is not only to find the most likely classification for the unclassified instance, but to also state something about its confidence in each possible classification.

CP does this by assigning each possible classification $Y_j, j = 1, \ldots, c$ to x_{l+1} in turn and extending the training set with it, generating the set

$$\{(x_1, y_1), \ldots, (x_l, y_l), (x_{l+1}, Y_j)\}. \tag{1}$$

It then measures how strange, or non-conforming, each pair (x_i, y_i) in (1) is for the rest of the examples in the same set. This is done with a *non-conformity measure* which is based on a conventional machine learning algorithm, called the *underlying algorithm* of the CP. This measure assigns a numerical score α_i to each pair (x_i, y_i) indicating how much it disagrees with all other pairs in (1). In effect it measures the degree of disagreement between the prediction of the underlying algorithm for x_i after being trained on (1) with its actual label y_i; in the case of x_{l+1}, y_{l+1} is assumed to be Y_j.

To convert the non-conformity score $\alpha_{l+1}^{(Y_j)}$ of (x_{l+1}, Y_j) into something informative, CP compares it with all the other non-conformity scores $\alpha_i^{(Y_j)}, i = 1, \ldots, l$. This comparison is performed with the function

$$p((x_1, y_1), \ldots, (x_l, y_l), (x_{l+1}, Y_j)) = \frac{|\{i = 1, \ldots, l+1 : \alpha_i^{(Y_j)} \geq \alpha_{l+1}^{(Y_j)}\}|}{l+1}. \quad (2)$$

The output of this function, which lies between $\frac{1}{l+1}$ and 1, is called the p-value of Y_j, also denoted as $p(Y_j)$, as this is the only unknown part of (1). If the data are independent and identically distributed (i.i.d.), the output $p(y_{l+1})$ for the true classification of x_{l+1} has the property that $\forall \delta \in [0, 1]$ and for all probability distributions P on Z,

$$P^{l+1}\{((x_1, y_1), \ldots, (x_{l+1}, y_{l+1})) : p(y_{l+1}) \leq \delta\} \leq \delta; \quad (3)$$

for a proof see [15]. Therefore all classifications with a p-value under some very low threshold, say 0.05, are highly unlikely to be correct as such sets will only be generated at most 5 % of the time by any i.i.d. process.

Based on the property (3), given a *significance level* δ, or confidence level $1 - \delta$, a CP calculates the p-value of all possible classifications Y_j and outputs the prediction set

$$\{Y_j : p(Y_j) > \delta\}, \quad (4)$$

which has at most δ chance of not containing the true classification of the new unclassified example. In the case where a single prediction is desired (forced prediction) instead of a prediction set, CP predicts the classification with the largest p-value, which is the most likely classification, together with a confidence and a credibility measure for its prediction. The confidence measure is calculated as one minus the second largest p-value, i.e. the significance level at which all but one classifications would have been excluded. This gives an indication of how likely the predicted classification is compared to all other classifications. The credibility measure on the other hand, is the p-value of the predicted classification. A very low credibility measure indicates that the particular instance seems very strange for all possible classifications.

2.1   Inductive Conformal Prediction

The transductive nature of the original CP technique, which means that all computations have to start from scratch for every new test example, makes it too computationally demanding for a mobile phone application. For this reason the proposed approach follows the inductive version of the framework, called Inductive Conformal Prediction (ICP), which only performs one training phase to generate a general rule with which it can then classify new examples with minimal processing.

Specifically, ICP divides the training set (of size l) into the *proper training set* with $m < l$ examples and the *calibration set* with $q := l - m$ examples. It then uses the proper training set for training the underlying algorithm (only once) and the examples in the calibration set for calculating the p-value of each possible classification of the new test example. In effect, after training the underlying algorithm on the proper training set the non-conformity scores $\alpha_{m+1}, \ldots, \alpha_{m+q}$ of the calibration set examples are calculated. Then to calculate the p-value of each possible classification $Y_j \in \{Y_1, \ldots, Y_c\}$ of a new test example x_{l+1}, ICP only needs to calculate the non-conformity score of the pair (x_{l+1}, Y_j) using the already trained underlying algorithm and compare it to the non-conformity scores of the calibration set examples with the function

$$p(Y_j) = \frac{|\{i = m+1, \ldots, m+q, l+1 : \alpha_i \geq \alpha_{l+1}^{(Y_j)}\}|}{q+1}. \tag{5}$$

Notice that the steps that need to be repeated for each test example have almost negligible computational requirements.

3   Proposed Approach

In this study a Multilayer Perceptron (MLP) was used as underlying algorithm of the ICP. The MLP used was a 2-layer fully connected feed-forward network with a sigmoid activation function in all units. It was implemented using the Multilayer Perceptron class of the WEKA data mining software libraries [10]. When given an instance x_i to classify, the trained MLP produces a probabilistic value $\hat{P}(Y_j|x_i)$ for each classification Y_j. In this work the classification task is binary, therefore $Y_j \in \{0,1\}$ and two probabilistic values are produced by the MLP: $\hat{P}(0|x_i)$ and $\hat{P}(1|x_i)$.

The nonconformity measure used for the proposed MLP-ICP was

$$\alpha_i^{Y_j} = 1 - \hat{P}(y_i|x_{m+i}), \quad i = 1, \ldots, q, \tag{6a}$$

$$\alpha_{l+1}^{Y_j} = 1 - \hat{P}(Y_j|x_{l+1}), \tag{6b}$$

where $\hat{P}(y_i|x_i)$ is the output of the MLP for the true classification of x_i and $\hat{P}(Y_j|x_{l+1})$ is the output of the MLP for the assumed class $Y_j \in \{0,1\}$ of x_{l+1}.

The complete process followed by the proposed approach is detailed in Algorithm 1. Lines 1 to 5 correspond to the training phase that needs to be performed only once. This phase trains the MLP on the proper training set (line 1) and calculates the nonconformity score of each calibration example x_{m+i}, $i = 1, \ldots, q$, by inputing it to the trained MLP to obtain the probabilistic outputs $\hat{P}(0|x_{m+i})$ and $\hat{P}(1|x_{m+i})$ (line 3) and using them in (6a) to calculate

---

Algorithm 1. Binary MLP-ICP

Input: proper training set $\{(x_1, y_1), \ldots, (x_m, y_m)\}$,
      calibration set $\{(x_{m+1}, y_{m+1}), \ldots, (x_{m+q}, y_{m+q})\}$,
      test example x_{l+1},
      significance level δ

1   $h \leftarrow$ train the MLP on $\{(x_1, y_1), \ldots, (x_m, y_m)\}$;
2   **for** $i = 1$ **to** q **do**
3      $\{\hat{P}(0|x_{m+i}), \hat{P}(1|x_{m+i})\} \leftarrow h(x_{m+i})$;
4      $\alpha_{m+i} \leftarrow 1 - \hat{P}(y_{m+i}|x_{m+i})$;
5   **end**
6   $\{\hat{P}(0|x_{l+1}), \hat{P}(1|x_{l+1})\} \leftarrow h(x_{l+1})$;
7   **for** $j = 0$ **to** 1 **do**
8      $\alpha_{l+1}^j \leftarrow 1 - \hat{P}(j|x_{l+1})$;
9      $p(j) = \dfrac{|\{i=1,\ldots,q : \alpha_{m+i} \geq \alpha_{l+1}^j\}| + 1}{q+1}$;
10 **end**
   Output:
11   Prediction set $R \leftarrow \left\{Y_j : p(Y_j) > \delta\right\} \cup \left\{\arg\max_{k=1,\ldots,c} p(Y_k)\right\}$.

α_{m+i} (line 4). The testing phase, lines 6 to 11, is the only part that needs to be repeated for every new instance. This phase obtains the probabilistic outputs of the trained MLP for the new instance (line 6) and then for each possible class, it calculates the corresponding nonconformity score with (6b) in line 8 and uses it together with the nonconformity scores of the calibration examples in (5) to calculate the p-value of the new instance belonging to that class (line 9). Finally, in line 11, it outputs the prediction set (4).

4   Data Used

For evaluating the proposed approach a malware dataset[1] created by B. Amos [3] was used. The data was collected using a shell script to automatically record the behavior of Android application files (.apk) by installing and running them on an Android emulator. After each installation the emulator simulated random user interaction with the application using the Android "adb-monkey" tool. The data collected for each application included Binder, Battery, Memory, CPU, Network, and Permission information. The battery related features were removed in our experiments since they all had the same value across all instances, as they were collected through an emulator. Table 1 presents all the features used in our experiments.

    The dataset consisted of 6832 data samples in total, with 5121 benign samples and 1711 malicious samples described by 40 features (after the removal of the battery related data). In our experiments the data samples were divided

---

[1] Available on-line at: https://github.com/VT-Magnum-Research/antimalware.

Table 1. Features used in this work divided into categories.

Category	Features
Binder	Transaction, Reply, Acquire, Release, ActiveNodes, TotalNodes, ActiveRef, TotalRef, ActiveDeath, TotalDeath, ActiveTransaction, TotalTransaction, ActiveTransactionComplete, TotalTransactionComplete, TotalNodesDiff, TotalRefDiff, TotalDeathDiff, TotalTransactionDiff, TotalTransactionCompleteDiff
CPU	User, System, Idle, Other
Memory	Active, Inactive, Mapped, FreePages, AnonPages, FilePages, DirtyPages, WritebackPages
Network	TotalTXPackets, TotalTXBytes, TotalRXPackets, TotalRXBytes, TotalTXPacketsDiff, TotalTXBytesDiff, TotalRXPacketsDiff, TotalRXBytesDiff
Permissions	TotalPermissions

randomly into a training set consisting of 6165 samples (4617 benign and 1548 malicious) and a test set consisting of 667 samples (504 benign and 163 malicious). For MLP-ICP the training set was further divided (also randomly) into the proper training set with 5466 samples (4097 benign and 1369 malicious) and the calibration set with 699 samples (520 benign and 179 malicious). All experiments were repeated 10 times with different random divisions of the data so as to ensure that the obtained results are not dependent on a particular division.

5   Experimental Results

In order to evaluate the proposed approach in the environment it is intended for, both the MLP-ICP and its underlying algorithm were implemented on the Android OS using the WEKA data mining software libraries and all experiments were performed on a LG E400 Android device. Before each experiment all input features were normalized to the range [0,1], based only on the training set of each of the 10 random divisions of the data and the resulting transformation was applied to the test set.

5.1   Accuracy

Our first set of experiments evaluates the proposed technique in terms of the accuracy of forced predictions. That is when the MLP-ICP predicts the most likely classification together with a confidence and credibility measure to that classification. We compare the accuracy of the proposed approach to that of its underlying algorithm. Note that the aim of the proposed approach is not to improve performance in terms of accuracy, but rather to provide probabilistically valid additional information about the likelihood of each prediction being correct,

Table 2. Forced prediction performance of the proposed approach compared to its underlying algorithm. The standard deviation of each value is given in the brackets.

	Accuracy (%)	Mean confidence (%)
MLP-ICP	94.24 (0.0116)	98.83 (0.0020)
Conventional MLP	94.17 (0.0117)	-

which will aid user decision making. Therefore our aim here is to examine if any degree of accuracy is sacrificed by the proposed approach in order to provide this additional information.

Table 2 reports the mean accuracy of the proposed approach and of the conventional MLP over 10 experimental runs with different random divisions of the data. For MLP-ICP we also report the mean of the confidence values produced. The values in brackets are the standard deviation of the corresponding values over the 10 experiments. The reported results show that there is no significant difference in terms of the accuracy of the two approaches; in fact the MLP-ICP gives higher accuracy, but the difference is insignificant. This shows that there is no "price to pay" in terms of accuracy for obtaining the much more informative outputs of the proposed approach. Furthermore, the high mean confidence produced by the proposed approach gives a first indication of the usefulness of its outputs. Finally, the small standard deviation of both accuracy and mean confidence over the 10 experimental runs shows that the presented mean values are a reliable indication of performance.

5.2   Empirical Validity

Next we evaluate the empirical validity of the prediction sets, and indirectly of the p-values and confidence measures, produced by the proposed approach. Figure 1 plots the error percentages of these prediction sets as the significance level changes. The left pane displays the error percentages for all significance levels $\delta \in [0,1]$, while the right pane displays the lower left corner of the first plot, for $\delta \in [0,0.1]$, which is the most important part of the plot as we are generally interested in high confidence levels. The solid line represents the actual error percentages observed, while the dashed line represents the diagonal, i.e. where the error is equal to the required significance level δ. The closeness of the actual errors to the diagonal confirms the empirical validity of the proposed approach, i.e. the errors made are always extremely near the required significance level; the very small deviation of about 0.01 visible in the left pane is due to statistical fluctuations. This confirms empirically the guaranteed validity of the ICP.

5.3   Quality of p-values

To evaluate the quality of the resulting p-values we examine the percentage of prediction sets that contained both classifications at different confidence levels.

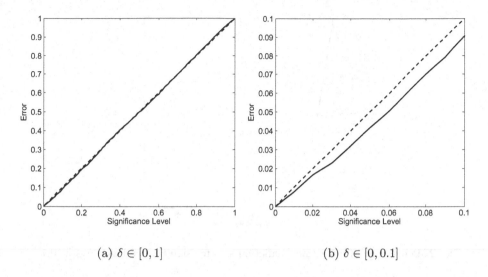

(a) $\delta \in [0, 1]$ (b) $\delta \in [0, 0.1]$

Fig. 1. Empirical validity of the proposed approach. The error percentage of the ICP is plotted with a solid line and the diagonal (exact validity) with a dashed line.

This is the percentage of instances for which the ICP was not able to reject one of the two possible classes at the required confidence level, we call these *uncertain prediction sets*. The lower these values are the more the instances for which the prediction sets are informative at the given confidence level and the higher the quality of the p-values used to compute them.

Figure 2 plots the percentage of uncertain prediction sets for the range of significance levels $\delta \in [0, 0.1]$, i.e. confidence levels between 100% and 90%. The plot shows that the prediction sets produced by the proposed approach are quite informative. With a confidence level as high as 99%, i.e. guaranteeing only 1% of errors, we can be certain in 63.7% of applications on whether they are malware or not. If we decrease the required confidence level by 1% (to 98%) we obtain certain predictions for 80.28% of the cases. Further lowering the confidence level to 95%, which is still rather high, increases the percentage of certain predictions to 95.53%. While for confidence levels 93% and 92% the number of certain predictions rises to more than 99% and becomes 100% from 91% confidence and below. This means that we can have a certain prediction for the large majority of applications with a very small risk, while for applications with uncertain predictions further investigation might be the best option.

The informativeness of the obtained prediction sets also shows the quality of the p-values used for constructing them, and as a result of the confidence measures produced based on the same p-values in the forced prediction case. In this case, the user can decide based on the confidence measure for a given malware prediction and on the risk he/she is willing to take on whether to remove it or not.

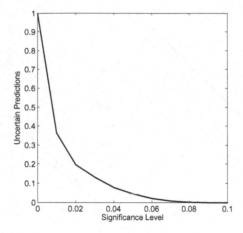

Fig. 2. Percentage of uncertain prediction sets for significance levels $\delta \in [0, 0.1]$.

5.4 Computational Efficiency

Finally, since in a mobile device environment the amount of recourses needed are an important consideration, we examine the execution time of the proposed approach on a LG E400 Android device and compare it to that of its underlying algorithm. The training time for the MLP-ICP was 21 min, while the training time for the conventional MLP was 29 min. The reason for the lower training time of the ICP is the removal of the calibration examples from the proper training set. This works in favour of the proposed approach in terms of computational efficiency, while it has no negative impact on its accuracy as shown in Subsect. 5.1. Although the training time needed seems somewhat long, it should be kept in mind that training only needs to be performed once and this can even be done on a desktop device. The testing time for both methods was 0.005 s, which is very computationally efficient for a mobile device.

6   Conclusions

This work proposed a machine learning approach for Android malware detection, which unlike traditional machine learning based malware detection techniques, produces valid confidence measures in each of its predictions. Such measures enable the user to take informed decisions on whether to remove an application identified as a possible malware, knowing the risk associated with each decision. The proposed approach is based on the Conformal Prediction framework, which produces provably valid confidence measures that have a clear probabilistic interpretation without assuming anything more than i.i.d. data.

Our experimental results on a real Android device have shown that the proposed approach gives high accuracy, which is equal to (or even better than) that of the conventional machine learning approach it is based on. Furthermore, they demonstrate the empirical validity and high informational efficiency of the produced prediction sets and confidence measures. Finally in terms of computational efficiency, the training time needed by our approach is smaller than that of the conventional MLP on which it bases its predictions. While the time needed for classifying a new instance is adequately small for the limited resources of mobile devices.

The particular CP used in this work is based on MLP, which is one of the most popular machine learning techniques. However, the same general approach proposed here can be followed with other high performing machine learning techniques preserving the desirable properties observed in our experiments. The evaluation and performance comparison of CPs based on different conventional machine learning techniques for the particular task is our immediate future plan. Additionally the collection of a dataset from actual mobile devices rather than in an emulated environment in order to study the performance of the proposed approach on more realistic data is another future goal.

References

1. Cabir, smartphone malware (2004). http://www.f-secure.com/v-descs/cabir. shtml. Accessed 12 May 2016
2. G DATA, mobile malware report (threat report: Q4/2015) (2016). https://secure. gd/dl-us-mmwr201504. Accessed 16 May 2016
3. Amos, B., Turner, H., White, J.: Applying machine learning classifiers to dynamic android malware detection at scale. In: Proceedings of the 9th International Wireless Communications and Mobile Computing Conference (IWCMC 2013), pp. 1666–1671. IEEE (2013)
4. Christodorescu, M., Jha, S.: Static analysis of executables to detect malicious patterns. In: Proceedings of the 12th Conference on USENIX Security Symposium, vol. 12, p. 12. USENIX Association (2003)
5. Demertzis, K., Iliadis, L.: SAME: an intelligent anti-malware extension for android ART virtual machine. In: Núñez, M., Nguyen, N.T., Camacho, D., Trawiński, B. (eds.) Computational Collective Intelligence. LNCS. Springer, Switzerland (2015)
6. Demertzis, K., Iliadis, L.: Bio-inspired hybrid intelligent method for detecting android malware. In: Kunifuji, S., Papadopoulos, A.G., Skulimowski, M.A., Kacprzyk, J. (eds.) Knowledge, Information and Creativity Support Systems: Selected Papers from KICSS 2014, pp. 289–304. Springer, Switzerland (2016)
7. Egele, M., Scholte, T., Kirda, E., Kruegel, C.: A survey on automated dynamic malware-analysis techniques and tools. ACM Comput. Surv. 44(2), 6:1–6:42 (2012). http://doi.acm.org/10.1145/2089125.2089126
8. Jacob, G., Debar, H., Filiol, E.: Behavioral detection of malware: from a survey towards an established taxonomy. J. Comput. Virol. 4(3), 251–266 (2008)
9. Griffin, K., Schneider, S., Hu, X., Chiueh, T.: Automatic generation of string signatures for malware detection. In: Kirda, E., Jha, S., Balzarotti, D. (eds.) RAID 2009. LNCS, vol. 5758, pp. 101–120. Springer, Heidelberg (2009)
10. Hall, M., Frank, E., Holmes, G., Pfahringer, B., Reutemann, P., Witten, I.H.: The weka data mining software: an update. SIGKDD Explor. Newsl. 11(1), 10–18 (2009). http://doi.acm.org/10.1145/1656274.1656278

11. Joshua, A., Waziri, O.V., Abdullahi, M.B., Arthur, U.M., Adewale, O.S.: A machine learning approach to anomaly-based detection on android platforms. Int. J. Netw. Secur. Appl. **7**(6), 15–35 (2015)
12. Menahem, E., Shabtai, A., Rokach, L., Elovici, Y.: Improving malware detection by applying multi-inducer ensemble. Comput. Stat. Data Anal. **53**(4), 1483–1494 (2009)
13. Moser, A., Kruegel, C., Kirda, E.: Limits of static analysis for malware detection. In: Proceedings of the 23rd Annual Computer Security Applications Conference, pp. 421–430. IEEE (2007)
14. Moskovitch, R., Elovici, Y., Rokach, L.: Detection of unknown computer worms based on behavioral classification of the host. Comput. Stat. Data Anal. **52**(9), 4544–4566 (2008)
15. Nouretdinov, I., Vovk, V., Vyugin, M.V., Gammerman, A.J.: Pattern recognition and density estimation under the general i.i.d. assumption. In: Helmbold, D.P., Williamson, B. (eds.) COLT 2001 and EuroCOLT 2001. LNCS (LNAI), vol. 2111, pp. 337–353. Springer, Heidelberg (2001)
16. Papadopoulos, H.: Inductive conformal prediction: theory and application to neural networks. In: Fritzsche, P. (ed.) Tools in Artificial Intelligence, Chap. 18, pp. 315–330. InTech, Vienna, Austria (2008). http://www.intechopen.com/download/pdf/pdfs_id/5294
17. Papadopoulos, H., Proedrou, K., Vovk, V., Gammerman, A.J.: Inductive confidence machines for regression. In: Elomaa, T., Mannila, H., Toivonen, H. (eds.) ECML 2002. LNCS (LNAI), vol. 2430, pp. 345–356. Springer, Heidelberg (2002)
18. Rieck, K., Holz, T., Willems, C., Düssel, P., Laskov, P.: Learning and classification of malware behavior. In: Zamboni, D. (ed.) DIMVA 2008. LNCS, vol. 5137, pp. 108–125. Springer, Heidelberg (2008)
19. Sahs, J., Khan, L.: A machine learning approach to android malware detection. In: Proceedings of the 2012 European Intelligence and Security Informatics Conference (EISIC), pp. 141–147. IEEE (2012)
20. Shabtai, A., Moskovitch, R., Elovici, Y., Glezer, C.: Detection of malicious code by applying machine learning classifiers on static features: a state-of-the-art survey. Inf. Secur. Tech. Rep. **14**(1), 16–29 (2009)
21. Vovk, V., Gammerman, A., Shafer, G.: Algorithmic Learning in a Random World. Springer, New York (2005)

Auto Regressive Dynamic Bayesian Network and Its Application in Stock Market Inference

Tiehang Duan[✉]

Department of Computer Science and Engineering,
The State University of New York at Buffalo, Buffalo, NY 14226, USA
tiehangd@buffalo.edu

Abstract. In this paper, auto regression between neighboring observed variables is added to Dynamic Bayesian Network (DBN), forming the Auto Regressive Dynamic Bayesian Network (AR-DBN). The detailed mechanism of AR-DBN is specified and inference method is proposed. We take stock market index inference as example and demonstrate the strength of AR-DBN in latent variable inference tasks. Comprehensive experiments are performed on S&P 500 index. The results show the AR-DBN model is capable to infer the market index and aid the prediction of stock price fluctuation.

Keywords: Auto Regressive Dynamic Bayesian Network (AR-DBN) · Expectation-Maximization (EM) · Kullback–Leibler divergence · Sum-product algorithm · Belief propagation

1 Introduction

Dynamic Bayesian Network (DBN) uses directed graph to model the time dependent relationship in the probabilistic network. The method achieved wide application in gesture recognition [17,20], acoustic recognition [3,22], image segmentation [9] and 3D reconstruction [6]. The temporal evolving feature also makes the model suitable to model the stock market [7]. The classic DBN model assumes the observed variable only depend on latent variables, and we know for the instance of stock market, auto regression widely exists in neighboring observed stock prices due to the momentum of market atmosphere. So we introduce explicit auto regressive dependencies between adjacent observed variables in DBN, forming the Auto Regressive Dynamic Bayesian Network (AR-DBN).

In Sect. 2, we have a brief review of previous work in related fields. In Sect. 3, the structure of the network is formed and probability factors are derived. In Sect. 4, we specify the parameter estimation of AR-DBN and experiments are conducted in Sect. 5. Conclusion is reached in Sect. 6.

2 Related Work

Researchers have been using Dynamic Bayesian Networks(DBN) to model the temporal evolution of stock market and other financial instruments [19]. In 2009,

© IFIP International Federation for Information Processing 2016
Published by Springer International Publishing Switzerland 2016. All Rights Reserved
L. Iliadis and I. Maglogiannis (Eds.): AIAI 2016, IFIP AICT 475, pp. 419–428, 2016.
DOI: 10.1007/978-3-319-44944-9_36

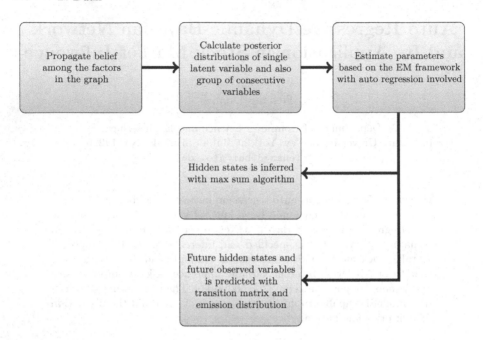

Fig. 1. Overview of work flow for AR-DBN model

Aditya Tayal utilized DBN to analyze the switching of regimes in high frequency stock trading [21]. In 2013, Zheng Li et al. used DBN to explore the dependence structure of elements that influence stock prices [10]. And in 2014, Jangmin O built a price trend model under the DBN framework [7]. Auto regression is an important factor that contribute to the fluctuation of stock prices and has been studied among researchers in financial mathematics [12–14]. Auto regressive relationships among adjacent observed variables is also used in Hidden Markov Models(HMM). In 2009, Matt Shannon et al. illustrated Auto Regressive Hidden Markov Model(AR-HMM) and used it for speech synthesis [18]. In 2010, Chris Barber et al. used AR-HMM to predict short horizon wind with incomplete data [2]. And in 2014, Bing Ai et al. estimated the smart building occupancy with the method [1]. To our knowledge, there haven't been previous investigation of auto regression applied to Dynamic Bayesian Network (DBN), and in this paper, we integrate the auto regressive property into DBN and forms the Auto Regressive Dynamic Bayesian Network (AR-DBN) (Fig. 1).

3 The Dynamic Bayesian Network for Stock Market Inference

The original DBN before integrating auto regression is shown in Fig. 2. For each time slice i, it includes m observed stock price variables $Y_{1i}, ..., Y_{mi}$ and hidden variable X_i. We denote the graph of each slice as $B = (G, \theta)$, where G is the structure of Bayesian Network (BN) for the slice, whose nodes corresponds to the variables and whose edges represent their conditional dependencies, and θ represents the set of parameters encoding the conditional probabilities of each node variable given its parent [11]. The distribution is represented as CPD (conditional probabilistic distribution) [5]. In our case, as the observed variables are continuous, exact inference is achieved with sum product algorithm [4], which iterate between summation of belief in different states for each clique and combining the belief of neighboring cliques. After the end of each iteration, the marginal probability of each variable is inferred based on likelihood of the whole graph [15].

The likelihood of the graph in Fig. 2 is

$$\phi = P(X_1, Y_{11}, ..., Y_{m1}, X_2, Y_{12}, ..., Y_{m2}, ..., X_n, Y_{1n}, ..., Y_{mn}) \tag{1}$$

Define

$$\phi_1 = P(X_1, Y_{11}, ..., Y_{m1})$$
$$\phi_2 = P(X_1, Y_{11}, ..., Y_{m1}, X_2, Y_{12}, ..., Y_{m2})$$
......
$$\phi_n = P(X_1, Y_{11}, ..., Y_{m1}, X_2, Y_{12}, ..., Y_{m2}, ..., X_n, Y_{1n}, ..., Y_{mn})$$
$$\psi_n = P(X_n, Y_{1n}, ..., Y_{mn})$$
$$\psi_{n-1} = P(X_{n-1}, Y_{1(n-1)}, ..., Y_{m(n-1)}, X_n, Y_{1n}, ..., Y_{mn})$$
......
$$\psi_1 = P(X_1, Y_{11}, ..., Y_{m1}, X_2, Y_{12}, ..., Y_{m2}, ..., X_n, Y_{1n}, ..., Y_{mn})$$

also define

$$f(X_{i-1}, X_i) = P(X_i|X_{i-1})$$
$$f(X_i, Y_{1i}, ..., Y_{mi}) = P(Y_{1i}, ..., Y_{mi}|X_i)$$

Based on sum product algorithm, we have

$$\phi_i = \left(\sum_{x_{i-1}} \phi_{i-1} \times f(X_{i-1}, X_i) \right) \times f(X_i, Y_{1i}, ..., Y_{mi}) \tag{2}$$

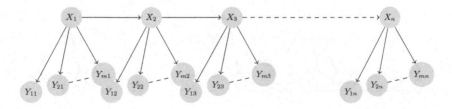

Fig. 2. Structure of Dynamic Bayesian Network (DBN)

$$\psi_i = \left(\sum_{x_{i+1}} \psi_{i+1} \times f(X_i, X_{i+1}) \right) \times f(X_i, Y_{1i}, ..., Y_{mi}) \tag{3}$$

And after the belief completes one round bidirectional propagation through the network, the ϕ_i and ψ_i can be readily used to calculate the posterior distribution of each latent variable X_i

$$P(X_i|Y) \propto \frac{\phi_i \times \psi_i}{f(X_i, Y_{1i}, ..., Y_{mi})} \tag{4}$$

And we can similarly calculate the marginal probability of k consecutive hidden units $X_i, X_{i+1}, ..., X_{i+k}$

$$P(X_i, X_{i+1}, ..., X_{i+k}|Y) \propto \phi_i \times \psi_{i+k} \times f(X_i, X_{i+1}) \times ... \times f(X_{i+k-1}, X_{i+k})$$
$$\times f(X_{i+1}, Y_{1(i+1)}, ..., Y_{m(i+1)}) \times ... \times f(X_{(i+k-1)}, Y_{1(i+k-1)}, ..., Y_{m(i+k-1)}) \tag{5}$$

After the sum product algorithm completes, we can use the marginal probability distribution to estimate the parameters in the network based on EM algorithm.

4   Formulation of Auto Regressive Dynamic Bayesian Network

As mentioned in Sect. 2, due to the ubiquitous auto regressive relationship in stock prices, we can add directed auto regressive edges between neighboring observed variables in the network. The resulting network is shown in Fig. 3. For each observed variable Y_{ki}, with the new assumption, it is not only conditioned on the latent variable, but also influenced directly by the previous observed variable.

For each observed variable, we have

$$Y_t = \beta_1 Y_{t-1} + \beta_2 Y_{t-2} + ... + \beta_k Y_{t-k} + \alpha U_t \tag{6}$$

Where k is the depth of the regression, Y_{t-k} to Y_{t-1} is the previous observed variables and U_t is directly emitted by the latent variable X_t. The coefficients satisfy $\beta_1 + \beta_2 + ... + \beta_k + \alpha = 1$.

Denote $Y_t' = Y_t - \beta_1 Y_{t-1} - \beta_2 Y_{t-2} - ... - \beta_k Y_{t-k}$, we have $U_t = \alpha^{-1} Y_t'$. From which we can estimate the parameters in AR-DBN, including conditional

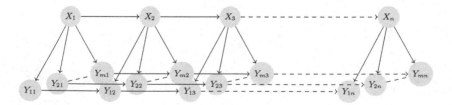

Fig. 3. Structure of Auto Regressive Dynamic Bayesian Network (AR-DBN)

probability distribution $P(X_i|X_{i-1})$, the initial probability distribution $P(X_1)$ and the emission probability $P(Y_{1i}, Y_{2i}, ..., Y_{ki}|X_i)$ which can be modeled as a multi variable Gaussian distribution $N(\boldsymbol{\mu}, \boldsymbol{\sigma})$.

Based on EM algorithm, we estimate the parameters with marginal probabilities from (4) and (5) by maximizing the first term of KL divergence

$$Q(\theta, \theta^{old}) = \sum_X P(X|Y, \theta^{old}) ln P(X, Y|\theta)$$

$$= \sum_X P(X|\alpha^{-1}Y', \theta^{old}) ln \Big[P(X_1|\pi) \prod_{n=2}^N P(X_n|X_{n-1}, A) \prod_{n=1}^N P(\alpha^{-1}Y'_n|X_n, \phi) \Big]$$

$$= \sum_X P(X|\alpha^{-1}Y', \theta^{old}) ln P(X_1|\pi) + \sum_X P(X|\alpha^{-1}Y', \theta^{old}) \Big(\sum_{n=2}^N ln P(X_n|X_{n-1}, A) \Big)$$

$$+ \sum_X P(X|\alpha^{-1}Y', \theta^{old}) \Big(\sum_{n=1}^N ln P(\alpha^{-1}Y'_n|X_n, \phi) \Big)$$

Applying Lagrange method to fulfill the criteria $\sum_{k=1}^K \pi_k = 1$ and $\sum_{j=1}^K A_{ij} = 1$ for each $i \in 1, ..., K$, where K is the number of states for the hidden variables, the parameters are derived by maximizing

$$R = Q(\theta, \theta^{old}) + \lambda_1 (\sum_{k=1}^K \pi_k - 1) + \sum_{i=1}^K \lambda_{2i} (\sum_{j=1}^K A_{ij} - 1) \tag{7}$$

After setting the first order partial derivatives of R with respect to each individual parameter to zero, the explicit expression of the parameters is derived as below

$$\pi_k = \frac{P(X_k|\alpha^{-1}Y')}{\sum_{j=1}^K P(X_j|\alpha^{-1}Y')} \tag{8}$$

$$A_{jk} = \frac{\sum_{n=1}^{N-1} P(X_n = j, X_{n+1} = k|\alpha^{-1}Y')}{\sum_{X_{n+1}=1}^K \sum_{n=1}^{N-1} P(X_n = j, X_{n+1}|\alpha^{-1}Y')} \tag{9}$$

$$\mu_i = \frac{\sum_{n=1}^N P(X_n|\alpha^{-1}Y')Y_{in}}{\sum_{n=1}^N P(X_n|\alpha^{-1}Y')} \tag{10}$$

$$\Sigma_i = \frac{\sum_{n=1}^N P(X_n|\alpha^{-1}Y')(Y_{in} - \mu_i)(Y_{in} - \mu_i)^T}{\sum_{n=1}^N P(X_n|\alpha^{-1}Y')} \tag{11}$$

After the training phase completed, we infer the hidden states that form the highest likelihood path with max sum algorithm [8], similar to DBN. The inferred result for the stock market is shown in the next section.

5 Application in Stock Market

We use the historical $S\&P$ 500 stock price dataset provided by Quantquote [16], covering the period from Jan. 02, 1998 to Aug. 09, 2013. The individual stock price temporal fluctuation forms each observed chain and the hidden states are inferred from the multiple observed chains, as shown in Fig. 3. We randomly pick k individual stocks out of the dataset, where k varies in [2,12], then we infer the hidden states and parameters from the data. The trained CPD is visualized and shown in Fig. 4. From which we can see for the inferring task of 6 individual stocks ($k = 6$), 8 hidden state is an overkill with first 3 hidden state actually not functioning in transition, while 4 hidden states are not enough to represent all different positions of the market. 6 hidden states is the optimized choice for the model.

The AR-DBN outperforms DBN both in the ultimate likelihood achieved and also in the convergence speed, as shown in Fig. 5(a). It reveals that AR-DBN is more suitable and efficient to apply in stock market. The likelihood is also positive correlated with the number of latent states as shown in Fig. 5(b).

The inferred latent states with max sum algorithm is shown in Fig. 6. The latent states plotted in temporal order form the path that produces highest likelihood in the whole network. The absolute value of increase/decrease ratio in $S\&P$ 500 index is discretized into three intervals $0 < ratio < 1\%$, $1\% < ratio < 2\%$

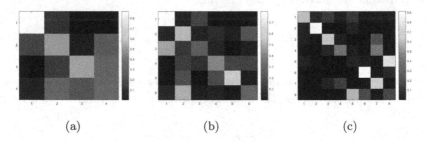

(a)   (b)   (c)

Fig. 4. The visualization of learned CPD. (a) 4 hidden states for each latent variable, (b) 6 hidden states for each latent variable, (c) 8 hidden states for each latent variable.

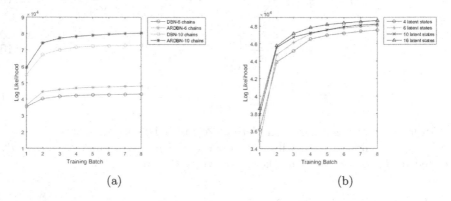

(a)   (b)

Fig. 5. (a) The comparison of log likelihood between DBN and AR-DBN with different number of observed chains. (b) The comparison of log likelihood between different number of latent states in AR-DBN.

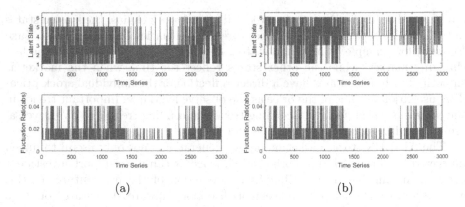

Fig. 6. Trend comparison of inferred latent state with *S&P* 500 fluctuation ratio. (a) Inference with 2 observed individual stock chains. (b) Inferrence with 6 observed individual stock chains.

and $2\% < ratio$, then we compare it with the evolving trend of hidden states in Fig. 6. It can be seen that with more observable chains included, the corresponding relationship between the latent states and the market index is more

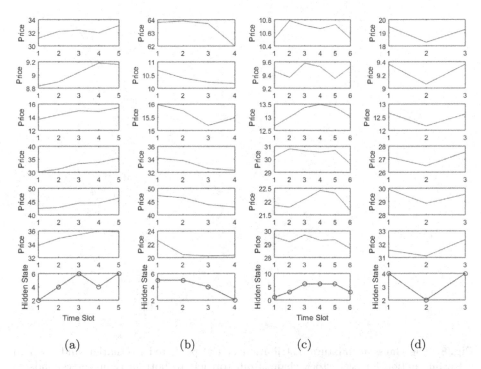

Fig. 7. The micro view of correlation in stock price trends and hidden states. (a) The ascending trend. (b) The descending trend. (c) and (d) The "V" reversal.

obvious and accurate. With 6 observed individual stock chains, the hidden states in the model is capable to precisely capture the fluctuation of market index. Take Fig. 6(b) for example, state 1 and state 2 capture the characteristics of the fast changing market, while the other states reflect the time when the market is smooth. The hidden states have a direct reflection of the individual stock prices if we use absolute value of the price as the observed variable. Important trends in the price movement such as ascending trend, descending trend and "V" reversal are reflected in hidden states as shown in Fig. 7.

The AR-DBN stock market model we generated is not only useful to unveil market rules contained in historical stock price, it can also support investment decision making based on probabilistic prediction of the near future. In this application, we predict price movement direction (upward/downward) of the 6 observed stocks for the first day following the end of chain based on the para-

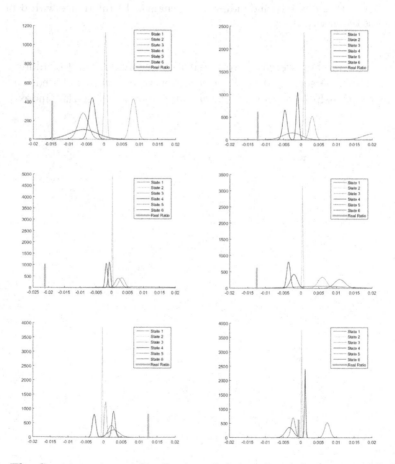

Fig. 8. The Gaussian mixture distribution of the predicted fluctuation and the real price fluctuation for each stock chain, from top left to bottom right corresponds to observed stock price chain 1 to 6.

meters we learned. We multiply the latent states inferred from the max sum algorithm with the conditional probability distribution $P(X_{t+1}|X_t)$ and turns the future stock price Y_{t+1} into a Gaussian mixture distribution conditioned on the probability distribution of future latent states X_{t+1}. The prediction result together with the real fluctuation for each of the stocks is shown in Fig. 8. For 5 out of the 6 stocks, the centroid of the probability distribution are of the same direction (upward/downward) with the real price trend, which shows our model can help with the prediction of fluctuation direction.

6 Conclusion and Future Work

We derived Auto Regressive Dynamic Bayesian Network (AR-DBN) by adding auto regression among the adjacent observed variables in Dynamic Bayesian Network (DBN). We presented a new approach to model the stock market based on the proposed AR-DBN and comprehensive inference tasks were implemented with the model. The results showed the latent variables in the model accurately inferred the market index fluctuation and the stock price trends.

In this paper, we mainly focused on the inference part of the model. There is more work to be done with quantitative prediction of future price fluctuations and also the application of AR-DBN to other temporal analysis domains. We leave this for future work.

References

1. Ai, B., Fan, Z., Gao, R.X.: Occupancy estimation for smart buildings by an auto-regressive hidden Markov model. In: 2014 American Control Conference, pp. 2234–2239, June 2014
2. Barber, C., Bockhorst, J., Roebber, P.: Auto-regressive HMM inference with incomplete data for short-horizon wind forecasting. In: Lafferty, J., Williams, C., Shawe-taylor, J., Zemel, R., Culotta, A. (eds.) Advances in Neural Information Processing Systems, vol. 23, pp. 136–144 (2010). http://books.nips.cc/papers/files/nips23/NIPS2010_1284.pdf
3. Barra-Chicote, R., Fern, F., Lutfi, S., Lucas-Cuesta, J.M., Macias-Guarasa, J., Montero, J.M., San-segundo, R., Pardo, J.M.: Acoustic emotion recognition using dynamic Bayesian networks and multi space distributions. Annu. Conf. Int. Speech Commun. Assoc. 1, 336–339 (2009)
4. Bishop, C.M.: Pattern Recognition and Machine Learning (Information Science and Statistics). Springer, New York (2006)
5. Das, B.: Generating conditional probabilities for Bayesian networks: easing the knowledge acquisition problem. CoRR cs.AI/0411034 (2004)
6. Delage, E., Lee, H., Ng, A.Y.: A dynamic Bayesian network model for autonomous 3D reconstruction from a single indoor image. In: 2006 IEEE Computer Society Conference on Computer Vision and Pattern Recognition (CVPR 2006), vol. 2, pp. 2418–2428 (2006)
7. Jangmin, O., Lee, J.-W., Park, S.-B., Zhang, B.-T.: Stock trading by modelling price trend with dynamic Bayesian networks. In: Yang, Z.R., Yin, H., Everson, R.M. (eds.) IDEAL 2004. LNCS, vol. 3177, pp. 794–799. Springer, Heidelberg (2004)

8. Jordan, M.I., Ghahramani, Z., Jaakkola, T.S., Saul, L.K.: An introduction to variational methods for graphical models. Mach. Learn. **37**(2), 183–233 (1999)
9. Kampa, K., Principe, J.C., Putthividhya, D., Rangarajan, A.: Data-driven tree-structured Bayesian network for image segmentation. In: 2012 IEEE International Conference on Acoustics, Speech and Signal Processing (ICASSP), pp. 2213–2216, March 2012
10. Li, Z., Yang, J., Tan, S.: Systematically discovering dependence structure of global stock markets using dynamic Bayesian network. J. Comput. Inf. Syst. **9**, 7215–7226 (2013)
11. Likforman-Sulem, L., Sigelle, M.: Recognition of degraded characters using dynamic Bayesian networks. Pattern Recogn. **41**(10), 3092–3103 (2008)
12. Marcek, D.: Stock price forcasting: autoregressive modelling and fussy neural network. J. Inf. Control Manage. Syst. **7**, 139–148 (2000)
13. Marcek, D.: Some intelligent approaches to stock price modelling and forecasting. J. Inf. Control Manage. Syst. **2**, 1–6 (2004)
14. Mathew, O.O., Sola, A.F., Oladiran, B.H., Amos, A.A.: Prediction of stock price using autoregressive integrated moving average filter. Glob. J. Sci. Front. Res. Math. Decis. Sci. **13**, 78–88 (2013)
15. Mihajlovic, V., Petkovic, M.: Dynamic Bayesian networks: a state of the art, dMW-project(2001)
16. Quantquote: S&P 500 daily resolution dataset
17. Rett, J., Dias, J.: Gesture recognition using a marionette model and dynamic Bayesian networks (DBNs). In: Campilho, A., Kamel, M.S. (eds.) ICIAR 2006. LNCS, vol. 4142, pp. 69–80. Springer, Heidelberg (2006)
18. Shannon, M., Byrne, W.: A formulation of the autoregressive HMM for speech synthesis. Technical report, Cambridge University Engineering Department, August 2009
19. Sipos, I.R., Levendovszky, J.: Optimizing sparse mean reverting portfolios with AR-HMMs in the presence of secondary effects. Periodica Polytech. Electr. Eng. Comput. Sci. **59**(1), 1–8 (2015)
20. Suk, H.I., Sin, B.K., Lee, S.W.: Hand gesture recognition based on dynamic Bayesian network framework. Pattern Recogn. **43**(9), 3059–3072 (2010)
21. Tayal, A.: Regime switching and technical trading with dynamic Bayesian networks in high-frequency stock markets. Ph.D. thesis, University of Waterloo (2009)
22. Zweig, G., Russell, S.: Speech recognition with dynamic Bayesian networks. In: AAAI-98 Proceedings, American Association for Artificial Intelligence (1998)

A Consumer BCI for Automated Music Evaluation Within a Popular On-Demand Music Streaming Service "Taking Listener's Brainwaves to Extremes"

Fotis Kalaganis[1(✉)], Dimitrios A. Adamos[2,3], and Nikos Laskaris[1,3]

[1] AIIA Lab, Department of Informatics, Aristotle University of Thessaloniki,
54124 Thessaloniki, Greece
kalaganis@csd.auth.gr, laskaris@aiia.csd.auth.gr
[2] School of Music Studies, Aristotle University of Thessaloniki, 54124 Thessaloniki, Greece
dadam@mus.auth.gr, d.adamos@ieee.org
[3] Neuroinformatics GRoup, Aristotle University of Thessaloniki, 54124 Thessaloniki, Greece
http://neuroinformatics.gr

Abstract. We investigated the possibility of a using a machine-learning scheme in conjunction with commercial wearable EEG-devices for translating listener's subjective experience of music into scores that can be used for the automated annotation of music in popular on-demand streaming services.

Based on the established -neuroscientifically sound- concepts of brainwave frequency bands, activation asymmetry index and cross-frequency-coupling (CFC), we introduce a Brain Computer Interface (BCI) system that automatically assigns a rating score to the listened song.

Our research operated in two distinct stages: (i) a generic feature engineering stage, in which features from signal-analytics were ranked and selected based on their ability to associate music induced perturbations in brainwaves with listener's appraisal of music. (ii) a personalization stage, during which the efficiency of extreme learning machines (ELMs) is exploited so as to translate the derived patterns into a listener's score. Encouraging experimental results, from a pragmatic use of the system, are presented.

Keywords: EEG · Music evaluation · Recommendation-systems · Human machine interaction · Spotify

1 Introduction

Until recently, electroencephalography (EEG) was exclusive to doctors' facilities and specialists' workplaces, where trained experts operated expensive devices. The vast majority of the research was dedicated to the diagnosis of epilepsy, sleep disorders, Alzheimer's disease as well as monitoring certain clinical procedures such as anesthesia. By the same token, the application of Brain-Computer Interfaces (BCIs) has so far been confined to neuroprosthetics and for building communication channels for the physically impaired people [1].

© IFIP International Federation for Information Processing 2016
Published by Springer International Publishing Switzerland 2016. All Rights Reserved
L. Iliadis and I. Maglogiannis (Eds.): AIAI 2016, IFIP AICT 475, pp. 429–440, 2016.
DOI: 10.1007/978-3-319-44944-9_37

Recent advances in medical sensors offer commercial EEG headsets at affordable prices. The procedure of recording the electrical activity of the brain via electrodes on the human scalp is pretty simple nowadays and does not require any expertise [2]. The latter can lead to a wide variety of commercial applications in almost any aspect of everyday life. Meanwhile, many researchers are drawn into the exploration of brain patterns that are not related to healthcare [3]. At the same time, the manufacturing of new portable neuroimaging devices favors innovation, as novel applications of non-invasive BCIs are anticipated within real-life environments [4].

Since the beginning of the twenty-first century, the digital revolution has radically affected the music industry and is continuously reforming the business model of music economy [5]. Until now, previously-established channels of music distribution have been replaced and new industry stakeholders have emerged. Among them, on-demand music recommendation and streaming services emerge as the *"disruptive innovators"* [6] of the new digital music ecosystem.

In our previous work [7], we presented our vision for the integration of bio-personalized features of musical aesthetic appreciation into modern music recommendation systems to enhance user's feedback and rating processes. With the current work, we attempted the first step toward implementation with the realization of an automated music evaluation process, that is performed in nearly real-time by decoding aesthetic brain responses during music listening and feeding them back to a contemporary music streaming service (i.e. Spotify) as listener's ratings. We aimed for a flexible BCI system that could easily adapt to new users and, after a brief training session, would reliably predict the listener's ratings about the listened songs.

To facilitate convenience and friendliness of the user's experience, the recording of brain activity was implemented using a modern commercial wireless EEG device. Thus, we first adapted our approach to the device capabilities, investigating which combinations of available brainwave descriptors and electrode sites are reliably and consistently reflecting the listeners' evaluation about the music being played. Then we synthesized these descriptors into a composite biomarker, common for all listeners. Finally, we examined different learning machines that could incorporate this biomarker and generalize efficiently from a very small set of training examples (paired biomarker-patterns and ratings).

The proposed music-evaluation BCI relies on standard ELMs for translating a particular set of readily-computable signal descriptors, as extracted from our 4-channel wearable EEG device, into a single numbered score expressing the appraisal of music within the range [1–5]. It can be personalized with very limited amount of training and runs with negligible amount of delay. In all our experimentations the passive listening paradigm was followed, since this is closer to the real life situation where someone enjoys listening to music.

The preliminary results, reported in this paper, include evidence about the existence of a robust set of brain activity characteristics that reliably reflect a listener's appraisal. Moreover, the effectiveness of ELMs in the particular regression task is established, by comparing its performance with alternative learning machines. Overall, the outcomes of this work are very encouraging for conducting experiments about music perception in real-life situations and embedding brain signal analysis within the contemporary technological universe.

The remaining paper is structured as follows. Section 2 serves as an introduction to EEG and its role in describing and understanding the effects of music. Section 3 outlines the essential tools that were employed during data analysis. Section 4 describes the experimental setup and the adopted methodology for analyzing EEG data. Section 5 is devoted to the presentation of results, while the last section includes a short discussion about the limitations of this study and its future perspectives.

2   Electroencephalography and Music Perception Studies

Electroencephalogram is a recording of the electrical changes occurring in the brain, produced by placing electrodes on the scalp and monitoring the developed electrical fields. EEG reflects mainly the summation of excitatory and inhibitory postsynaptic potentials at the dendrites of ensembles of neurons with parallel geometric orientation. While the electrical field produced by distinct neurons is too weak to be recorded with surface EEG electrode, as neural action gets to be synchronous crosswise a huge number of neurons, the electrical fields created by individual neurons aggregate, resulting to effects measurable outside the skull [8].

The EEG brain signals, also known as *brainwaves*, are traditionally decomposed (by means of band-pass filtering or a suitable transform) and examined within particular frequency bands, which are denoted via greek letters and in order of increasing central frequency are defined as follows: δ (0.5–4) Hz, θ (4–8) Hz, α (8–13) Hz, β (13–30) Hz, γ (>30) Hz. EEG is widely recognized as an invaluable neuroimaging technique with high temporal resolution. Considering the dynamic nature of music, EEG appears as the ideal technique to study the interaction of music, as a continuously delivered auditory stimulus, with the brainwaves. For more than two decades neuroscientists study the relationship between listening to music and brain activity from the perspective of induced emotions [9–11]. More recently, a few studies appeared which shared the goal of uncovering patterns, lurked in brainwaves, that correspond to subjective aesthetic pleasure caused by music [12, 13].

Regarding music perception, the literature has reported a wide spectrum of changes in the ongoing brain activity. This includes a significant increase of power in β-band over posterior brain regions [14]. An increase in γ band, which was confined to subjects with musical training [15], an asymmetrical activation pattern reflecting induced emotions [16] and an increase of frontal midline θ power when contrasting pleasant with unpleasant musical sounds [17].

Regarding the particular task of decoding the subjective evaluation of music from the recorded brain activity, the role of higher-frequency brainwaves has been identified [18], and in particular the importance of γ-band brainwaves recorded over forebrain has been reported [19]. More recently, a relevant CFC biomarker based on the concept of nested oscillations in the brain was introduced for the assessment of spontaneous aesthetic brain responses during music listening [7]. The reported experimental results indicated that the interactions between β and γ oscillations, as reflected in the brainwaves recorded over the left prefrontal cortex, are crucial for estimating the subjective aesthetic appreciation of a piece of music.

3   Methods

This section presents briefly the methodological elements employed in the realization of the proposed framework. More specifically, instantaneous signal energy, activation asymmetry index and a CFC estimator were used to derive brainwave descriptors (i.e. features reflecting neural activity associated with music listening). The importance of each descriptor (or combination of descriptors) was evaluated using *Distance Correlation*. ELMs, an important class of artificial neural network (ANNs), were exploited to convert the derived patterns into scores that represented the listener's appraisal.

3.1   Signal Descriptors

Brainwaves are often characterized by their prominent frequency and their (signal) energy content. Here, we adopted a quasi-instantaneous parameterization of Brainwaves content, by means of Hilbert transform. The signal from each sensor x(t), was first filtered within the range corresponding to the frequency-band of a brain rhythm (like δ-rhythm) and the envelope of the filtered activity $\alpha_{rhythm}(t)$ was considered as representing the momentary strength of the associated oscillatory activity. Apart from the amplitude of each brain rhythm, its relative contribution was also derived by normalizing with the total signal strength (summed from all brain rhythms).

In neuroscience research, activation refers to the change in EEG activity in response to a stimulus and is of great interest to investigate differences in the way the two hemispheres are activated [20]. To this end, an activation asymmetry index was formed by combining measurements of activation strength from two symmetrically located sensors, i.e. $AI(t) = {}^{left}\alpha_{rhythm}(t) - {}^{right}\alpha_{rhythm}(t)$. The normalized version of this index was also employed as an additional alternative descriptor.

A third brainwaves' descriptor was based on the CFC concept, which refers to the functional interactions between distinct brain rhythms [21]. A particular estimator was employed [22] that quantified the dependence of amplitude variations of a high-frequency brain rhythm on the instantaneous phase of a lower-frequency rhythm (a phenomenon known as phase-amplitude coupling (PAC)). This estimator operated on each sensor separately and used to investigate all the possible PAC couplings among the defined brain rhythms.

It is important to notice here, that the included descriptors were selected so as to cover different neural mechanisms and share a common algorithmic framework. Their implementation -and mainly their integration- within a unifying system did not induce time delays unreasonable for the purposes of our real-time application.

3.2 Distance Correlation

In statistics and in probability theory, distance correlation is a measure of statistical dependence between two random variables or two random vectors of arbitrary, not necessarily equal, dimension. Distance Correlation, denoted by R, generalizes the idea of correlation and holds the important property that R(X, Y) = 0 if and only if X and Y are independent. R index satisfies $0 \leq R \leq 1$ and, contrary to Pearson's correlation coefficient, is suitable for revealing non-linear relationships [23]. In this work, it was the core mechanism for identifying neural correlates of subjective music evaluation, by detecting associations between the results of signal descriptors (or combinations of them) and the listener's scores.

3.3 Learning Machines

Machine learning deals with the development and implementation of algorithms that can build models able to generalize a particular function from a set of given examples. Regression is a supervised learning task that machine learning can handle with efficiency of similar, or even higher, level than the standard statistical techniques, and -mainly- without imposing hypotheses. In this work, the decoding of subjective music evaluation was cast as a (nonlinear) regression problem. A model was then sought (i.e. learned from the experimental data) that would perform the mapping of patterns derived from the brain activity descriptors to the subjective music evaluation.

ELMs appeared as a suitable choice due to their documented ability to handle efficiently difficult tasks without demanding extensive training sessions [24]. They are feedforward ANNs with a single layer of hidden nodes, where the weights connecting inputs to hidden nodes are randomly assigned and never updated [25].

4   Implementation and Experiments

Our experimentations evolved in two different directions. First, a set of experiments were run so as to use the experimental data for establishing the brainwave pattern that would best reflect the music appraisal of an individual and train the learning machine from music pieces of known subjective evaluation. Next, additional experiments were run that implemented the real-time scoring by the trained ELM-machine so as to justify the proposed BCI in a more naturalistic setting. In both cases, the musical pieces were delivered through a popular on-demand music streaming service (Spotify) that facilitates the registration of the listener's feedback ('like'-'dislike') to adapt the musical content to his taste and make suggestions about new titles.

4.1 Participants

All 5 participants were healthy students of AUTH Informatics department. Their average age was 23 years and music listening was among their daily habits. They signed an informed consent after the experimental procedures had been explained to them.

4.2 Data Acquisition

Having in mind the user-friendliness of the proposed scheme, we adopted a modern commercial dry-sensor wireless device (i.e. Interaxon's Muse device - http://www.choosemuse.com) in our implementations. This "gadget" offers a 4-channel EEG signal, with a topological arrangement that can be seen in Fig. 1. The signals are digitized at the sampling frequency of 220 Hz. Data are transmitted under OSC, which is a protocol for communication among computers, sound synthesizers, and other multimedia devices that is optimized for modern networking technology [26].

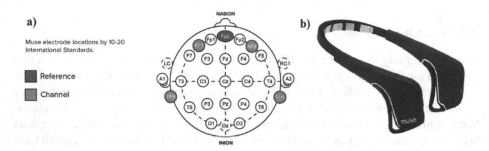

Fig. 1. Topological arrangement of Electrodes (A) and Interaxon's Muse headset (B).

4.3 Experimental Procedure

Prior to placing the headset, subjects sat in a comfortable armchair and volume of speakers was set to a desirable level. They were advised to refrain from body and head movements and enjoy the music experience. Recording was divided into sessions of 30 min duration. The music streaming service was operated in radio mode, hence randomly selected songs (from the genre of their preference), were delivered to each participant while his/her brain activity was registered. Among the songs there were advertisements. That part of recordings was isolated from the rest. The recording procedure was integrated, in MATLAB, together with all necessary information from the streaming audio signal (i.e. song id, time stamps for the beginning and termination of each song). Participants evaluated each of the listened song, using as score one of the integers $\{1, 2, 3, 4, 5\}$, during a separate session just after the end of the recording. These scores, together with the associate patterns extracted from the recorded brain signals, comprised our training set. The overall procedure was repeated for every subject, in order to build an independent testing set.

4.4 Data Analysis

The preprocessing of signal included 50 Hz component removal (by a built-in notch filter in *MuseIO* - the software that connects to and streams data from Muse), DC offset removal, and removal of the signal segments that corresponded to the 5 first second of each song (in order to avoid artificial transients).

The digital processing included (i) band-pass filtering for deriving the brain-waves of standard brain rhythms, (ii) Hilbert Transform for deriving their instantaneous amplitude and phase and (iii) computation of the descriptors described in Sect. 3.1. To increase frequency resolution, we divided the β rhythm into β_{low} (13–20 Hz), and β_{high} (20–30 Hz) sub-bands and derived descriptors separately. Similarly the γ rhythm was divided into γ_{low} (30–49 Hz) and γ_{high} (51–90 Hz). The recorded brain activity was segmented into overlapping windows of fixed duration (that during off-line experimentation had been varied between 30 and 100 s). The overall set of descriptors associated with the segments of brain activity had been used for selecting the best combination in a training-phase (applied collectively to all participants). A particular subset of descriptors (treated as a composite pattern including all the selected features) was extracted from each segment and utilized in an additional training phase (applied individually), during which an ELM-model was tailored to the user. In the testing phase, the subject specific ELM-model was applied to streaming composite patterns (representing segments of brain signals), in order to predict the listener's evaluation.

5 Results

5.1 Selecting Features - the Synthesized Biomarker

Feature selection was based on the Distance Correlation scores R's of all signal descriptors as averaged across all participants (Fig. 2). The descriptors were ranked in descending order (regarding their music evaluation expressiveness) and a *dynamic programming* methodology was applied. Starting with the feature of highest R, we traversed systematically the ranked list for combinations that would eventually maximize the Distance Correlation. This procedure led to a particular combination of 5 features, the **synthesized music appraisal biomarker**, which included the normalized temporal asymmetry index in β_{low} band, relative energy of α band at temporal electrodes, $\gamma_{low} \rightarrow \gamma_{high}$ PAC at FP1 sensor and relative energy of θ band at TP9. The relevance of this descriptor to music evaluation showed a dependence on the length of segment based on which it had been evaluated (using the timecourses from all participants, the averaged trace of Fig. 3 was computed). It can be seen that the effectiveness of the biomarker constantly increases with the duration of the listened music. This kind of investigations can provide indications about how fast the automated music evaluation system could operate.

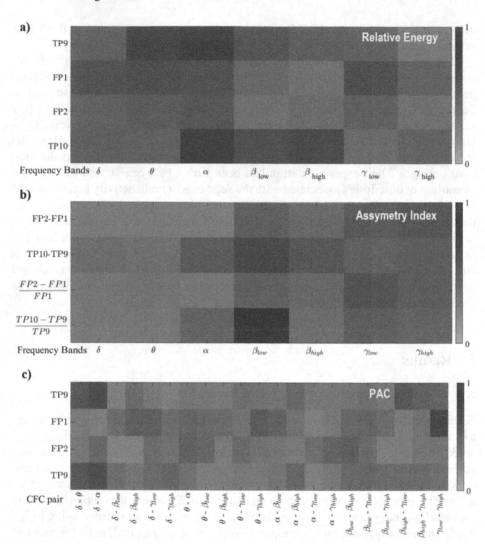

Fig. 2. Distance Correlation R of all employed signal descriptors.

5.2 Designing the ELM

Using the common biomarker, an ELM was designed per listener. The number of neurons in the hidden player, the only parameter to be tuned in ELMs, was selected as the smallest number of neurons such that the training and testing error were converging at an acceptable level, lower than 0.01.

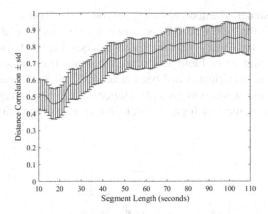

Fig. 3. Distance Correlation R as a function of the segment length for the synthesized biomarker.

5.3 Evaluating the Music-Evaluation BCI

The proposed BCI was evaluated in two ways. During the first stage, the available data (the biomarker patterns of each of the five participants along with the associated ratings) were randomly partitioned in training and testing set, in a 60 %–40 % proportion for cross-validation. An ELM was trained using the former set and its performance was quantified using the latter set. The overall procedure was repeated 10 times, and averaged results are reported. The normalized root mean squared error (RMSE), as defined for regression tasks, was found to be 0.063 ± 0.0093 (mean $\pm$ std across participants). For comparison purposes, we also employed Support Vector Machines (SVMs), which performed slightly inferiorly. Although the difference was marginal, the very short training time was another factor in favor of employing ELMs.

During the second stage, a pilot online BCI was developed in order to realize the testing phase of the ELM in a realistic setting. Two of the previous subjects, participated in an additional recording session during which the already tailored ELMs were providing, based on segments of 90 s, a read-out of the subjective music evaluation. The predicted scores were registered and compared with the ones provided by the listeners just after the experimental session. The normalized RMSE was estimated to be 0.09 and 0.07.

6 Discussion

This paper constitutes the report from a pilot study, during which we attempted to associate the listener's brainwaves with the subjective aesthetic pleasure induced by music. The main novelty of the presented proof-of-concept, is that it was realized based on a modern consumer EEG device and in conjunction with a popular on-demand music streaming service (Spotify). Our results indicate that signals from a restricted number of sensors (located over frontal and temporal brain areas) can be combined in a computable biomarker reflecting the listener's subjective music evaluation. This brainwaves' derivative can be efficiently decoded by regression-ELM and therefore leads to a reliable readout from the listener. The main advantage of the approach is that complies with idea of employing EEG-wearables in daily activities and is readily embedded within the contemporary on-demand music streaming services (see Fig. 4). However, the problem of artifacts (noisy signals of biological origin) has not been addressed yet. For the presented results, the participants had been asked to limit body/head movements and facial expressions and as much as possible. Hence, before employing such a system to naturalistic recordings, methodologies for real-time artifact suppression (as in [27]) have to be incorporated.

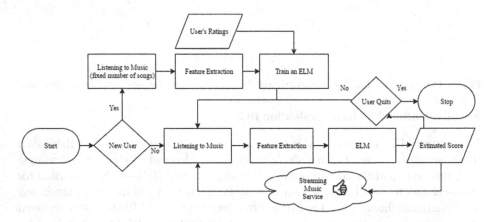

Fig. 4. Flow chart of the proposed music evaluation BCI

Today, we live in the world of *Internet of Things* (IoT) where the interconnectedness among devices has already been anticipated and supportive technologies are now being realized [28]. However, to achieve a seamless integration of technology in people's lives, there is still much room for improvement. In a world beyond IoT, technology would proactively facilitate people's expectations and wearable devices would transparently interface with systems and services. To enable such scenarios would require to move from the IoT to the *Internet of People* (IoP) [29]. People would then participate as first-class citizens relishing the benefits of holistic human-friendly applications. As such, we present a pragmatic use of a consumer BCI that ideally fits in the anticipated forms of the digital music universe and demonstrate a first series of encouraging experimental results.

References

1. Niedermeyer, E., da Silva, F.L.: Electroencephalography: Basic Principles, Clinical Applications, and Related Fields. Lippincott Williams & Wilkins, New York (2005)
2. Casson, A.J., Yates, D., Smith, S., Duncan, J.S., Rodriguez-Villegas, E.: Wearable electroencephalography. IEEE Eng. Med. Biol. Mag. **29**, 44–56 (2010)
3. Ariely, D., Berns, G.S.: Neuromarketing: the hope and hype of neuroimaging in business. Nat. Rev. Neurosci. **11**, 284–292 (2010)

4. He, B., Baxter, B., Edelman, B.J., Cline, C.C., Ye, W.W.: Noninvasive brain-computer interfaces based on sensorimotor rhythms. Proc. IEEE **103**(6), 907–925 (2015)
5. Wikström, P., DeFillippi, R. (eds.): Business Innovation and Disruption in the Music Industry. Edward Elgar Publishing, Cheltenham (2016)
6. Downes, L., Nunes, P.: Big bang disruption. Harvard Bus. Rev. **91**, 44–56 (2013)
7. Adamos, A.D., Dimitriadis, I.S., Laskaris, A.N.: Towards the bio-personalization of music recommendation systems: a single-sensor EEG biomarker of subjective music preference. Inf. Sci. **343–344**, 94–108 (2016)
8. Cohen, M.X.: Analyzing Neural Time Series Data: Theory and Practice. MIT Press, Cambridge (2014)
9. Altenmüller, E.: Cortical DC-potentials as electrophysiological correlates of hemispheric dominance of higher cognitive functions. Int. J. Neurosci. **47**, 1–14 (1989)
10. Petsche, H., Ritcher, P., von Stein, A., Etlinger, S.C., Filz, O.: EEG coherence and musical thinking. Music Percept. Interdisc. J. **11**, 117–151 (1993)
11. Birbaumer, N., Lutzenberger, W., Rau, H., Braun, C., Mayer-Kress, G.: Perception of music and dimensional complexity of brain activity. Int. J. Bifurcat. Chaos **6**, 267 (1996)
12. Hadjidimitriou, S.K., Hadjileontiadis, L.J.: Toward an EEG-based recognition of music liking using time-frequency analysis. IEEE Trans. Biomed. Eng. **59**, 3498–3510 (2013)
13. Schmidt, B., Hanslmayr, S.: Resting frontal EEG alpha-asymmetry predicts the evaluation of affective musical stimuli. Neurosci. Lett. **460**, 237–240 (2009)
14. Nakamura, S., Sadato, N., Oohashi, T., Nishina, E., Fuwamoto, Y., Yonekura, Y.: Analysis of music-brain interaction with simultaneous measurement of regional cerebral blood flow and electroencephalogram beta rhythm in human subjects. Neurosci. Lett. **275**(3), 222–226 (1999)
15. Bhattacharya, J., Petsche, H.: Musicians and the gamma band: a secret affair? Neuroreport **12**(2), 371–374 (2001)
16. Schmidt, A.L., Trainor, L.J.: Frontal brain electrical activity (EEG) distinguishes valence and intensity of musical emotions. Cognit. Emot. **15**(4), 487–500 (2001)
17. Sammler, D., Grigutsch, M., Fritz, T., Koelsch, S.: Music and emotion: electrophysiological correlates of the processing of pleasant and unpleasant music. Psychophysiology **44**(2), 293–304 (2007)
18. Hadjidimitriou, S.K., Hadjileontiadis, L.J.: EEG-based classification of music appraisal responses using time-frequency analysis and familiarity ratings. IEEE Trans. Affect. Comput. **4**, 161–172 (2013)
19. Pan, Y., Guan, C., Yu, J., Ang, K.K., Chan, T.E.: Common frequency pattern for music preference identification using frontal EEG. In: 6th International IEEE/EMBS Conference on Neural Engineering, pp. 505–508 (2013)
20. Coan, A.J., Allen, J.B.J.: Frontal EEG asymmetry as a moderator and mediator of emotion. Biol. Psychol. **67**(1–2), 7–50 (2004)
21. Canolty, T.R., Knight, T.R.: The functional role of cross-frequency coupling. Trends Cognit. Sci. **14**(11), 506–515 (2010)
22. Dimitriadis, S.I., Laskaris, N.A., Bitzidou, M.P., Tarnanas, I., Tsolaki, M.N.: A novel biomarker of amnestic MCI based on dynamic cross-frequency coupling patterns during cognitive brain responses. Front. Neurosci. **9**(350) (2015). doi:10.3389/fnins.2015.00350
23. Szekely, J.G., Rizzo, L.M., Bakirov, K.N.: Measuring and testing dependence by correlation of distances. Ann. Stat. **35**(6), 2769–2794 (2007)
24. Huang, G.B., Zhu, Q.Y., Siew, C.K.: Extreme learning machine: theory and applications. Neurocomputing **70**(1–3), 489–501 (2006)

25. Huang, G.B.: What are extreme learning machines? Filling the gap between Frank Rosenblatt's Dream and John von Neumann's Puzzle. Cogn. Comput. **7**, 263–278 (2015)
26. Wright, M., Freed, A., Momeni, A.: Open sound control: state of the art 2003. In: NIME 2003: Proceedings of the 3rd Conference on New Interfaces for Musical Expression (2003)
27. Akhtar, M.T., Jung, T.P., Makeig, S., Cauwenberghs, G.: Recursive independent component analysis for online blind source separation. In: IEEE Internet Symposium on Circuits and Systems, vol. 6, pp. 2813–2816 (2012)
28. Want, R., Schilit, B. N., Jenson, S.: Enabling the internet of things. Computer (1), pp. 28–35 (2015)
29. Miranda, J., Makitalo, N., Garcia-Alonso, J., Berrocal, J., Mikkonen, T., Canal, C., Murillo, J.M.: From the internet of things to the internet of people. IEEE Internet Comput. **19**(2), 40–47 (2016)

Information Abstraction from Crises Related Tweets Using Recurrent Neural Network

Mehdi Ben Lazreg$^{(\boxtimes)}$, Morten Goodwin, and Ole-Christoffer Granmo

University of Agder, Grimstad, Norway
{mehdi.ben.lazreg,morten.goodwin,ole.granmo}@uia.no
http://www.uia.no

Abstract. Social media has become an important open communication medium during crises. The information shared about a crisis in social media is massive, complex, informal and heterogeneous, which makes extracting useful information a difficult task. This paper presents a first step towards an approach for information extraction from large Twitter data. In brief, we propose a Recurrent Neural Network based model for text generation able to produce a unique text capturing the general consensus of a large collection of twitter messages. The generated text is able to capture information about different crises from tens of thousand of tweets summarized only in a 2000 characters text.

Keywords: Information abstraction · Recurrent neural network · Twitter data · Crisis management

1 Introduction

Social media has become the de facto open crises communication medium [1]. It plays a pivotal role in most crises today, from getting life signs from people affected to communicating with responders [2]. However, processing and extracting useful information and inferring valuable knowledge from such social media messages is difficult for several reasons. The messages are typically brief, informal, and heterogeneous (mix of languages, acronyms, and misspellings) with varying quality, and it may be required to know the context of a message to understand its meaning. Moreover, people also post information on other mundane events, which introduces additional noise into the data.

The state-of-the-art in the area of information discovery using machine learning mostly centres on supervised learning techniques. Those techniques are based on training an algorithm on sets of text from each topic to learn a predictive function, which in turn is used to classify new texts into a previously learnt topic [3]. A limitation of this approach is the scope of the topics: If a new text about an unforeseen topic is presented to the algorithm, such as a new crisis, it will wrongly classify it as one of the existing topics. Another challenge is that crises are diverse and the number of topics discussed in social media during a single

© IFIP International Federation for Information Processing 2016
Published by Springer International Publishing Switzerland 2016. All Rights Reserved
L. Iliadis and I. Maglogiannis (Eds.): AIAI 2016, IFIP AICT 475, pp. 441–452, 2016.
DOI: 10.1007/978-3-319-44944-9_38

crisis is large, dynamic, and changing from crisis to crisis. Moreover, applying a classifier trained on data from previous disasters on the next disaster may not perform well in practise. This can be explained by the fact that the next disaster will typically be more or less unique compared to the previous ones. Accordingly, a loss of accuracy occurs even if the crises have many similarities. Alternatively, unsupervised techniques try to look for co-occurrences of terms in the text as a metric of similarity [4] or infer the word distribution of a set of words the text contains and use it for document clustering [5]. Moreover, different methods based on graph theory has been used to extract information from a document. As an example, a graph based ranking model for document processing was adopted to extract key words from a text document [6]. In addition, a stochastic graph based method has also been employed to extract the most important sentence in a text document [7].

Recurrent Neural Networks and its long-short term memory variant have emerged as an efficient model in a variety of application involving sequential data [8]. This includes handwriting recognition [9], speech recognition [10], and video analysis [11]. As an example, RNN was trained on Wikipedia articles for text generation with great success. The power of recurrent neural network comes from their high dimensional hidden state with non-linear dynamics which has the ability to remember and process past input information [12]. The goal of this paper is to make a model that summarises and reproduces content from massive Twitter streams. The model is based on recurrent neural network to predict the next character in a stream of text. The approach allows to generate a text that compresses the information contained in the text that the network has been trained on.

This paper is organised as follow. Section 2 gives an overview of the state-of-the-art twitter analysis in crises situations. Section 3 introduces recurrent neural network and illustrates its basic features. Section 4 proposes a recurrent network based model for topic discovery in crisis related twitter data. Section 5 presents tests and results of the model. Finally, Sect. 6 concludes and provides pointers to further work.

2   Twitter Analysis for Crisis Situations

There is no doubt that valuable, high throughput data is produced on social media only seconds after a crisis occurs [1]. To cope with the complexity of the social media data, and extract information from crisis related messages, machine learning techniques have been applied [2]. Two main approaches were investigated: supervised and unsupervised learning.

In a supervised approach, the goal is to classify a social media message as part of one particular crisis event. To achieve this, the algorithm learns a predictive function so that it can classify any new unknown message as part of one of the categories of crises. A number of approaches have been investigated including Naïve Bayes, Support Vector Machine (SVM) [13], Random Forests [14], and Logistic Regression [15]. Further, some research focus on only analysing tweets containing certain keywords [2]. In this way, the tags can replace manual labelling for training. As an example, SVM was used to classify tweets related to earthquakes and landslides [3,16]. In a supervised approach, labels are necessary for training the classifiers, but they might be highly difficult to obtain especially in the case of multi-language messages or context knowledge [2]. To address this problem, unsupervised learning techniques are used.

Unsupervised methods are used to identify patterns in unlabelled data. They are most useful when the information seekers do not know specifically what information to look for in the data –which is the case in many crises situations. An example is grouping tweets into stories (clusters of tweets) after a keyword filter [17]. This method reduces the number of social media messages to be handled by humans since it groups equivalent messages together. Another application using unsupervised learning identifies events related to public and safety using a spatio-temporal clustering approach [18]. In addition to strictly clustering elements into groups, soft clusters have been used to allow items to simultaneously belong to several clusters with variant degrees. In this approach, the tweets similarity is based on words they contain and the length of the tweets [19]. The approach was applied on the Indonesia earthquake (2009) data and detected different aspects related to the crisis (relief, deaths, missing persons, and so on).

3 Artificial Neural Networks

An Artificial Neural Network (ANN) is a machine learning algorithm developed to mimic the human brain and reach its information processing capabilities [20]. An ANN is a network of processing units (analogous to neurons) joined by weighted connections (analogous to synapse). The network is activated by giving an input to some or all of the units. This activation is then spread throughout the hidden layers of the network until it reaches the output layer (see Fig. 1).

Many varieties of ANN exist all with different sets of properties [20]. One major distinction is between networks where the connections are acyclic called Feedforwad Neural Networks (FNN) and networks where the connections form cycles called Recurrent Neural Networks (RNN). RNN is most suited for tasks that involve sequential input such as speech and language. A RNN processes an input sequence one element at a time, and maintains information about the history of the past elements in the sequence. This ability makes it suitable for learning patterns to form text since a text is a series of correlated characters. RNN is successfully used to predict the next word in a sequence of semantically related words [8]. It also has some success in predicting the next character in a sequence of characters which is used to generate text, and in machine translation [8].

3.1 Recurrent Neural Networks

A Recurrent neural network (RNN) is an ANN where the connections between neurons are allowed to form a circle (see Fig. 1) [20]. As Fig. 1 shows, the connections between units on the same layer allow mapping the history of previous inputs to the output vectors. For each unit k in the RNN, the activation a_k of that unit depends on the inputs $\{x_1, x_2, ..., x_n\}$ of the unit and the weights $\{w_1, w_2, ..., w_n\}$ of their respective connections as shown in Eq. 1.

$$a_k = f(\sum_{i=1}^{n} w_i x_i + b_k) \tag{1}$$

The most widely used activation functions are sigmoid, hyperbolic tangent (in this case the unit is called a logistic unit) and linear functions (in this case the unit is called linear unit) [20]. b_k is a bias term that represents the expected mean value of the activation when all the inputs are zeroes.

During the training phase of an RNN, the aim is to update the weights so that for a given input, the output produced minimises a loss function that measures the similarity between the output of the network and the desired output [20]. The training of a RNN goes through three major steps [20]:

1. Initialise the weights w_i to a generally small value (in the range $[-0.1, 0.1]$).
2. Forward pass: Computes the activations a_k of all the unit in the RNN.
3. Backward pass: Updates the weights of the network in a manner that minimises the loss function between the output of the RNN and the desired output. This is performed using gradient descent. Backpropagation is used to efficiently compute the gradient and update the weights.

The three steps are repeated until a minimum of the loss function is reached. Note that the solution converged to may actually represent a local minimum.

3.2   Long-Short Term Memory

The main benefit of a RNN is its ability to use the input at previous time steps to produce an output. Nevertheless, in a standard RNN the range of past inputs that can influence output is quite limited because gradients can either decay or blow up exponentially as feedback cycles around the network recurrent connections [20]. This problem is known as the vanishing/exploding gradient problem [21]. To address this problem, a Long-Short Term Memory (LSTM) architecture was proposed [9]. In a LSTM architecture, the unit in the hidden layer of a RNN is replaced by a block (see Fig. 2) analogous to a memory block. The block is composed of:

- A linear unit "c_t": resenting the state of the block a time t.
- A logistic input unit "i_t": analogous to a write gate that updates the value in "c_t" when on (outputs a value close to 1)
- A logistic output unit "o_t": analog to a read gate that retrieves the value in "c_t" when on.
- A logistic forget unit "f_t": analog to a keep gate that maintains the value in "c_t" when on.

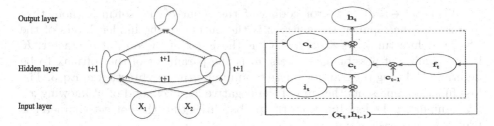

Fig. 1. Example of a RNN.   **Fig. 2.** A block of LSTM architecture.

The state of units "i_t", "o_t" and "f_t" are updated based on the input x_t, the output of the block at the previous time step h_{t-1}, and the output of other blocks in the LSTM network. This architecture is proved successful at a range of tasks that require long range memory including text generation, speech recognition and handwriting recognition [8].

4   Approach

Our approach aims at learning patterns from crises related tweets, and later generate a compressed text deducing the main topic (see Fig. 3). This means that the model learns characteristics of tweets, understands the context and is able to reproduce similar tweets. Note that this is very different from reciting tweets in that the model has learnt concepts. It does not copy Twitter text. It is designed to capture the main concept even with a noisy set of tweets i.e. a collection of tweets about different topics. However, the model does not aim at presenting a comprehensive assessment of the crisis at this point. It only present fragments of the crisis present in the generated text. To achieve this goal, we train a character based LSTM architecture on crises related tweets.

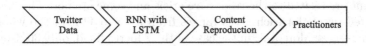

Fig. 3. A hight level overview of the proposed model.

The used LSTM architecture contains multiple hidden layers, each containing multiple LSTM blocks presented in Fig. 2. The input of the network is a vector representing the current character x_t where the character can be a letter or a special character. The output node of the network h_t is a softmax distribution over characters [20]. The softmax function produces an output in the $[0, 1]$ interval that represents the probability of the next character given the input of the node. For a training set $\{(x^1, y^1), (x^2, y^2), .., (x^n, y^n)\}$, the softmax distribution is defined by Eq. 2.

$$p(y^i = k | x^i) = \frac{\exp(\theta_i x^i)}{\sum_{j=1}^{K} \exp(\theta_j x^i)} \tag{2}$$

Where $x^i \in \mathbb{N}^K$ is vector coding of the input of the softmax (note that in our case the input of the softmax is the output of the hidden layers of the LSTM architecture), $y^i = \{1, 2, .., K\}$ is the index of the output character, K is the number of possible characters. θ_i is the parameter of the softmax to be determined during the training phase to minimize the loss function in Eq. 3. The loss function represents the sum of the negative log-likelihood of y^i knowing x^i. By minimizing the loss function, the probabilities that the correct character is predicted approaches one.

$$J(\theta) = -\frac{1}{n}\sum_{i=1}^{n}\sum_{j=1}^{K} 1\{y^i = j\}\log(p(y^i = j|x^i)) \tag{3}$$

$$1\{y^i = j\} = 1 \text{ if } y^i = j \text{ ; } 0 \text{ otherwise} \tag{4}$$

The hypothesis is that the generated text should provide a description of the tweets the network has been trained on. Hence, a practitioner can easily get an overview of the underlying topics of the tweets.

5   Tests and Results

The model described in the previous section was used to learn patterns from Twitter data related to several crises and then generate a unique text containing information present in the initial tweets.

5.1   The Data

The crises data used to train our model is provided by *CrisisLex* [22] a platform for collecting and filtering communications during a crisis. Table 1 refers to more than fifty four thousand tweets about several crises. The data is a mix of tweets where some are related to the respective crisis and some are not. The percentage of unrelated tweets for each crisis ranges from 38 % to 44 % of the whole set of Twitter messages. Moreover, in the case of the Alberta flood, only 30 % out of the related tweets gave concrete useful information about the crisis. The remaining tweets include other mundane topics. The percentage of informative tweets out of the related tweets goes up to a maximum 48 % in the Queensland flood data.

Table 1. Crises related twitter data

Crisis	Year	Number of tweets	Percentage of unrelated tweets
Boston bombing	2013	11012	40 %
Texas explosion	2013	11006	44 %
Alberta flood	2013	11036	44 %
Hurricane Sandy	2012	10008	38 %
Queensland flood	2011	11233	43 %

5.2 Experiments

The network was trained on a Twitter data collect from several crises (see Sect. 5.1). Table 2 summarises the empirical results of the model tested with different setups. The performance of the model can be measured with the training loss indicating the difference between the predicted and true value during the training period (Eq. 3), and the validation loss indicating the difference between the predicted and true value over a validation data (additional data over which the model is applied after training).

Table 2. Experiments results

Parameter	Value	Training loss	Validation loss
Architecture	RNN	1.97	1.81
	LSTM	1.48	1.50
Number of nodes	128	1.81	1.65
	256	1,48	1.50
	512	1.17	1.47
Number of layers	1	1.53	1.57
	2	1.48	1.50
	3	1.56	1.49
Dropout	0	0.80	1.90
	0.50	1.48	1.50
	0.90	2.00	1.90
Batch size	50	1.48	1.61
	100	1.48	1.50
	500	1.66	1.62
	1000	1.85	1.79
Sequence length	30	1.52	1.48
	50	1.48	1.50
	140	1.49	1.51

The first conducted experiment was to show the improvement in training and validation loss the LSTM brings over a simple RNN. Table 2 shows that the validation loss drops from 1.81 for a simple RNN to 1.5 for LSTM. Similarly, by showing the evolution of the training loss over the amount of train data for RNN and LSTM architectures, Fig. 4 illustrates the margin in training loss between LSTM and RNN even for small amount of training data. The training loss ends at a value of 1.97 for RNN and 1.48 for LSTM.

The LSTM architecture possesses different parameters that can be tuned to improve the model's ability to learn patterns from the training set and predict the next character in a sequence. An important parameter is the number of units (or nodes) in the network. Table 2 shows that increasing the number of nodes in the network improves the validation loss from 1.65 with 128 nodes to 1.47 with 512 nodes. Figure 5 displays the same trend: A 512 units network reaches a training loss of 1.17 while the loss for 128 units network remains at 1.81. However, with 512 nodes the validation loss is significantly higher then training loss strongly indicating that the network is over-fitting the data. Over-fitting will cause the generated text later on to be a copy of the tweets existing in the training set which will present no value added. Moreover, using 512 nodes will require more processing time for little gain in validation loss over 256 nodes (1.5 for 256 units and 1.47 for 512 units). An equilibrium is reached at 265 units where the validation loss 1.5 is slightly higher then the training loss 1.48.

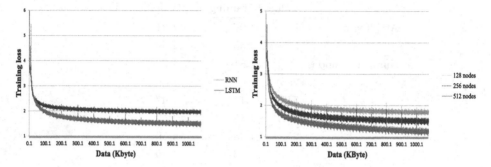

Fig. 4. Training loss for RNN and LSTM architectures.

Fig. 5. Training loss for different number of nodes.

Another parameter of LSTM is the number of hidden layers in the network. Table 2 suggests that increasing the number of hidden layers slightly improves the validation loss from 1.57 for one layer to 1.5 and 1.49 for 2 and 3 layers respectively. Nevertheless, the validation loss barely changes between 2 and 3 hidden layer. Likewise, the training loss is not greatly influenced by the number of layers going from 1.53 to 1.48 and 1.56 for 1, 2 and 3 layers respectively.

The dropout intends to avoid over-fitting by dropping out each node in the network with a certain probability at each training step [23]. Table 2 shows that 0 dropout improves the training loss to 0.8 compared to 2 for a dropout of 0.9. Nevertheless, the validation loss remains high, which again indicates over-fitting. In contrast, a high dropout causes both losses to be high (training loss of 2 and a validation loss of 1.9) which suggest under-fitting. Under-fitting means that the model cannot capture data patterns and fails to fit the data well enough.

The batch size determines how many examples the model looks at before making a weight update [24]. Lower batch sizes should intuitively improve the validation and training loss. However, the change is no longer compelling after reaching an batch size of 100. As Table 2 shows, the training loss goes from 1.85 to 1.48 for a change of batch size from 1000 to 100 and stays at 1.48 for a batch size of 50. Similarly, the validation loss goes from 1.79 to 1.5 for a change of batch size from 1000 to 100 but then increases to 1.61 for a batch size of 50.

The sequence length is the maximum number of characters that remains in the network memory to perform a prediction. Our experiments tested 3 values: the most frequent length of tweets (30 characters), the average length of tweets (50 characters) and the maximum length of tweets (140 characters). The results presented in Table 2 show an improved validation loss with shorter sequence moving from 1.51 to 1.48 for sequence lengths of 140 and 30 respectively. The model starts degrading in fitting the training data for shorter sequence shown by an increase of training loss from 1.49 to 1.52. The best combination of validation and training loss for 50 sequences.

We will in the remainder of the paper apply the configuration and parameters that provide the best performance above. The best setup consists of an LSTM architecture with 2 hidden layers and 256 hidden nodes, which represents approximately 400 thousand parameters to train. We used a dropout of 0.5, a batch size of 100, and the network keeps a memory of last 50 characters to use in its predictions.

5.3   Results

The model was successful in generating a 2000 character text for each crisis. A sample of the generated text is presented in Table 3. The aim is to generate tweets which are concise, explanatory, and extract the main topic of the big twitter data. The generated text is unique and only generated by the model. This means that all generated text is different and is not contained in the training data. Hence, from a structural point of view, the model was able to learn the basic component of a tweet: RT (retweets), hashtags, the "@" to address a specific person and the hypertext links. It was also able to predict an open bracket following two points, :(, for a sad smiley face in the first hurricane Sandy related text.

From a content point of view, the text contains misspelling and the sentences are unstructured. However, they clearly present valuable information that exists in the training data. An example is the Boston bombing, the first tweets clearly indicates the event of a bomb at the Boston marathon and presents a name: "Jeff Bauran". The name is actually a misspelling of "Jeff Bauman", a witness who identified one of the attackers [25]. The second text indicates that the FBI released a video about the Bombing. What actually happened, and present in the training data, is that the FBI released pictures and videos of the attackers [25]. An arrest was also made as the training data suggest and this was also captured by the model in the third tweet.

For the Texas explosion in West Fertilizer Company, the model was able to capture the number 60 surrounded by "killed" and "injured" in the first tweet. Actually, in the training data, this number appears sometimes as the number of killed in the explosion and other times as the number of injured. It is worth noticing that the number of deaths declared by the authorities was 14 [26]. However, this number was not part of the training Twitter data. This is an example of misleading information that can be present in the generated text cause by people spreading rumors through Twitter. The second tweet is unrelated to the crises and handles mundane topic present in the training data.

Table 3. Generated text

Crisis	Generated tweets
Boston bombing	1. Jeff Bauran, was bomb at #Boston Marathon and rew for the Boston marathon. My thoughts and prayers go out to everyone affected by today.
	2. RT @CSDiggren: Wook, suspect in Boston Marathon bombings: FBI releases video of the #BostonMarathon bombing: Any are peosled of sole. Same name gun...
	3. RT @SportsCenter: BREAKING: An arrest has been made in the Boston Marathon bombings, CNN reports.
Texas explosion	1. ?earcharliess killed 60 injured) http://t.co/jaKhPJfNND
	2. He's saying we won't knight.
	3. Eeproike for ouf the texas #PrayForWest PHOTO: Fertilizer plant explosion httexas. I gagrings! Thoughts To fertilizer plant explosionor the mesting, #WestTXClr4MwWUp
Alberta flood	1. Now video all the stay safe. They work to help the flood victims to help. Everyone pock the floods to my content for calgary floods: RT @nenshi:
	2. This pic of an awesome firefighter in Mission needs to go viral. Please share! #abflood http://t.co/x1Y
	3. #job #hiring http://t.co/MDEIzdl5lN
Hurricane Sandy	1. Nice in plays of Hurricane Sandy election to some going hurricane sandy. Tomorrow :(
	2. School destroyed http://t.co/nIsP2NJK
Queensland flood	1. Queensland's flood crisis deepens - The Australian: Sydney Morning
	2. Photos: Flood water rises in Australia: http://t.co/eFEOl

The same applies to the third tweet about Alberta flood. Moreover, the generated tweets about the Texas explosion do not explicitly state the nature of the crises. This might be due to the fact that the training data related the crisis present the highest percentage of unrelated tweets which influence the generated text. Therefore, we tried to remove the unrelated tweets from the train data related to the Texas explosion, retrain the model and generate a new text. The last tweet about the Texas explosion represents a sample of what we obtain. Even though the nature of the crisis is explicitly stated in the tweet, the tweet contains much more misspellings and unstructured. This is caused by reducing the training data (eliminating unrelated tweets), the model had not enough data to capture the structure and learn words.

For the remaining crises in the data, the tweets indicate the type of crisis, its location and provide some update on it status like the school destroyed during hurricane sandy, and the deepening of the Queensland flood. Note that some tweets are related to the crises but do not provide value added information about its status, like the firefighters mission in the Alberta flood second tweet. Nevertheless, these tweets present a significant chunk of the training data (see Sect. 5.1) linked to the crisis but not presenting useful information about it.

When we tried to generate the text anew, the information present in the new text was similar to previously discussed text. It is also worth noticing that some valuable information was present in the training data that was not extracted by the generated text which could be an area of further improvement in addition to automatically displaying the text in a manner that improves situational awareness further.

6 Conclusion

During a crises, valuable and substantial information about the crisis is shared on social media. However, the complexity and heterogeneity of tweeted text render extraction of useful information a difficult task for machine learning techniques. This papers presents a first step towards an approach for information extraction from large Twitter data collections. We propose training an LSTM architecture on crises related tweets and then use the trained model to generate a novel text. The model is able to capture valuable information from tens of thousands of tweets summarized only in an adjustable 2000 character text. Work is still to be made to filter useful crises related information and present it automatically in a more intuitive manner. Another future direction is to adopt a diversification mechanism that uses scores to rank the generated text based on their similarities and the information they bring.

References

1. Zielinski, A., Middleton, S.E., Tokarchuk, L.N., Wang, X.: Social media text mining and network analysis for decision support in natural crisis management. In: Proceedings of the ISCRAM. Baden-Baden, Germany (2013)
2. Imran, M., Castillo, C., Diaz, F., Vieweg, S.: Processing social media messages in mass emergency: a survey. ACM Comput. Surv. (CSUR) **47**(4), 67 (2015)
3. Sakaki, T., Okazaki, M., Matsuo, Y.: Earthquake shakes twitter users: real-time event detection by social sensors. In: Proceedings of the 19th International Conference on World Wide Web (2010)
4. Dumais, S.T.: Latent semantic analysis. Ann. Rev. Inf. Sci. Technol. **38**(1), 188–230 (2004)
5. Blei, D.M., Ng, A.Y., Jordan, M.I.: Latent dirichlet allocation. J. Mach. Learn. Res. **3**, 993–1022 (2003)
6. Mihalcea, R., Tarau, P.: Textrank: bringing order into texts. In: Conference on Empirical Methods in Natural Language Processing (2004)
7. Erkan, G., Radev, D.R.: Lexrank: graph-based lexical centrality as salience in text summarization. J. Artif. Intell. Res. **22**, 457–479 (2004)
8. LeCun, Y., Bengio, Y., Hinton, G.: Deep learning. Nature **521**(7553), 436–444 (2015)

9. Graves, A., Liwicki, M., Fernández, S., Bertolami, R., Bunke, H., Schmidhuber, J.: A novel connectionist system for unconstrained handwriting recognition. IEEE Trans. Pattern Anal. Mach. Intell. **31**(5), 855–868 (2009)

10. Farahat, M., Halavati, R.: Noise robust speech recognition using deep belief networks. Int. J. Comput. Intell. Appl. **15**(1), 1650005 (2016)

11. Kahou, S., Michalski, V., Konda, K., Memisevic, R., Pal, C.: Recurrent neural networks for emotion recognition in video. In: ICMI 2015 - Proceedings of the 2015 ACM International Conference on Multimodal Interaction (2015)

12. Karpathy, A., Johnson, J., Li, F.F.: Visualizing and understanding recurrent networks. arXiv preprint arXiv:1506.02078 (2015)

13. Yin, J., Lampert, A., Cameron, M., Robinson, B., Power, R.: Using social media to enhance emergency situation awareness. IEEE Intell. Syst. **27**(6), 52–59 (2012)

14. Imran, M., Castillo, C., Lucas, J., Meier, P., Vieweg, S.: Aidr: artificial intelligence for disaster response. In: Proceedings of the Companion Publication of the 23rd International Conference on World Wide Web Companion (2014)

15. Ashktorab, Z., Brown, C., Nandi, M., Culotta, A.: Tweedr: mining twitter to inform disaster response. In: Proceedings of ISCRAM (2014)

16. Musaev, A., Wang, D., Pu, C.: Litmus: landslide detection by integrating multiple sources. In: 11th International Conference Information Systems for Crisis Response and Management (ISCRAM) (2014)

17. Rogstadius, J., Vukovic, M., Teixeira, C., Kostakos, V., Karapanos, E., Laredo, J.A.: Crisistracker: crowdsourced social media curation for disaster awareness. IBM J. Res. Dev. **57**(5), 1–13 (2013)

18. Berlingerio, M., Calabrese, F., Di Lorenzo, G., Dong, X., Gkoufas, Y., Mavroeidis, D.: Safercity: a system for detecting and analyzing incidents from social media. In: 2013 IEEE 13th International Conference on Data Mining Workshops (ICDMW). IEEE (2013)

19. Kireyev, K., Palen, L., Anderson, K.: Applications of topics models to analysis of disaster-related twitterdata. In: NIPS Workshop on Applications for Topic Models: Text and Beyond, vol. 1. Whistler, Canada (2009)

20. Graves, A.: Supervised sequence labelling. In: Graves, A. (ed.) Supervised Sequence Labelling with Recurrent Neural Networks. SCI, vol. 385. Springer, Heidelberg (2012)

21. Pascanu, R., Mikolov, T., Bengio, Y.: On the difficulty of training recurrent neural networks. arXiv preprint arXiv:1211.5063 (2012)

22. Olteanu, A., Castillo, C., Diaz, F., Vieweg, S.: Crisislex: a lexicon for collecting and filtering microblogged communications in crises. In: ICWSM (2014)

23. Srivastava, N., Hinton, G., Krizhevsky, A., Sutskever, I., Salakhutdinov, R.: Dropout: a simple way to prevent neural networks from overfitting. J. Mach. Learn. Res. **15**(1), 1929–1958 (2014)

24. LeCun, Y.A., Bottou, L., Orr, G.B., Müller, K.-R.: Efficient backprop. In: Orr, G.B., Müller, K.-R. (eds.) NIPS-WS 1996. LNCS, vol. 1524, pp. 9–48. Springer, Heidelberg (1998)

25. Fbi, B.: Updates on investigation into multiple explosions in boston. (2013). https://www.fbi.gov/news/updates-on-investigation-into-multiple-explosions-in-boston

26. Gillam, C., MacLaggan, C.: Ammonium nitrate stores exploded at texas plant: state agency (2013). http://www.reuters.com/article/us-usa-explosion-texas-idUSBRE9460Gp.20130507

Mining Humanistic Data Workshop (MHDW)

A Scalable Grid Computing Framework
for Extensible Phylogenetic Profile Construction

Emmanouil Stergiadis, Athanassios M. Kintsakis, Fotis E. Psomopoulos[(✉)],
and Pericles A. Mitkas

Department of Electrical and Computer Engineering,
Aristotle University of Thessaloniki, Thessaloniki, Greece
fpsom@issel.ee.auth.gr

Abstract. Current research in Life Sciences without doubt has been
established as a Big Data discipline. Beyond the expected domain-specific
requirements, this perspective has put scalability as one of the most crucial aspects of any state-of-the-art bioinformatics framework. Sequence
alignment and construction of phylogenetic profiles are common tasks
evident in a wide range of life science analyses as, given an arbitrary
big volume of genomes, they can provide useful insights on the functionality and relationships of the involved entities. This process is often a
computational bottleneck in existing solutions, due to its inherent complexity. Our proposed distributed framework manages to perform both
tasks with significant speed-up by employing Grid Computing resources
provided by EGI in an efficient and optimal manner. The overall workflow
is both fully automated, thus making it user friendly, and fully detached
from the end-users terminal, since all computations take place on Grid
worker nodes.

1 Introduction

Over the last decade, the amount of available data in the life sciences domain has
increased exponentially and is expected to keep growing at an ever accelerating
pace. This significant increase in data acquisition leads to a pressing need for
scalable methods that can be employed to interpret them; a scaling that cannot
be met by traditional systems as they cannot provide the necessary computational power and network throughput required. Several efforts are evident in
recent literature towards developing new, distributed methods for a number of
bioinformatics workflows through the use of HPC systems and paradigms, such
as MapReduce [6]. However most, if not all, of these efforts necessitate the setup
of a rather complex computing system, as well as the expertise to manage and
update an independent software project, since most implementations radically
differ from their vanilla counterparts. This is in stark contrast with the situation for most life science researchers who lack the expertise needed to use and
manage those systems. As a result, and despite the overall advantages of these
frameworks, their ultimate use is fairly limited.

© IFIP International Federation for Information Processing 2016
Published by Springer International Publishing Switzerland 2016. All Rights Reserved
L. Iliadis and I. Maglogiannis (Eds.): AIAI 2016, IFIP AICT 475, pp. 455–462, 2016.
DOI: 10.1007/978-3-319-44944-9_39

In order to overcome these issues, while at the same time providing the much needed computing power for complex analyses, we developed a bioinformatics framework on top of a Grid architecture that is able to perform common comparative genomics workflows at a massive scale using EGI resources. Special care has been taken to make the framework as automated as possible, increasing the user friendly factor in order to further facilitate wider use by the scientific community. Moreover, every major submodule in the framework utilizes the latest vanilla version available to the community, in order to ensure that the framework can always stay up-to-date through automatically consuming updates of its individual vanilla parts.

The rest of this paper is structured as follows; Sect. 2 provides an overview of the concepts and technologies used throughout this work. Section 3 outlines the proposed framework, with particular focus on the requirements driving the current implementation. Section 4 establishes the technical aspects of the implementation, including the different modes of operations and the expected input and output. In Sect. 5, the efficiency of the framework is validated for all supported execution modes and finally, Sect. 6 provides some insights towards future steps in this direction.

2   Background

2.1   Comparative Genomics

BLAST Algorithm. BLAST has become the industry and research standard algorithm for sequence alignment. An open source implementation is provided by the NCBI organization which produces all alignments, as well as a number of parameters defining the significance of the alignment. A comprehensive list of the available parameters, as well as an in-depth description of the algorithm can be found in the online manual of the application. Our framework relies in particular on one of these parameters, namely the e-value e for each match. The e-value refers to the number of alignments expected to happen by chance. Therefore, a low e-value indicates a high statistical significance of an alignment.

Phylogenetic Profiles. Although the BLAST algorithm has been designed to readily identify and quantify sequence similarity, it will inherently miss any information that is not directly tied to this aspect. In particular, given a pair of protein sequences that may exhibit a high functional correlation (e.g. similar domains, active sites etc.) while at the same time do not expose any significant similarity in their composition, BLAST will fail to detect their interesting relationship. An alternative way to detect and quantify such relations, is to observe the joint presence or absence of sequences across a common set of genomes, for example protein sequences which only appear in the same family of genomes are highly likely to be related to that familys distinct functionality.

Phylogenetic profiles are vectors that characterize each sequence. Specifically, they indicate the sequence's homologs with every genome found in a given dataset. Each element in the vector corresponds to presence (denoted as the number 1) or absence (denoted as the number 0) of any homologue of the protein sequence under study in the respective genome. Beyond the standard (binary) phylogenetic profiles, an extended version may be constructed by replacing the presence/absence attribute by the actual number of homologs between that sequence and a target genome.

An even stronger evolutionary link between two sequences can be inferred through the use of best bidirectional hits (BBH). By definition, the best hit of a given sequence seq_A derived from genome A to a target genome B, is the sequence seq_B in genome B that represents the best match, i.e. a homologue scoring lower e-value e than all other hits produced. This particular match is also bidirectional if the seq_A is also the best hit for seq_B. A bidirectional best hit represents a very strong similarity between two sequences, and is considered evidence that the genes may be orthologs arising from a common ancestor [6]. A graphical representation of the two cases of phlyogenetic profiles is show in Fig. 1.

Genome-1	Genome-2	Genome-3
$seq_1 1$	$seq_2 1$	$seq_3 1$
	$seq_2 2$	

Hypothetical Genomes, the subscript refers to the sequence's genome.

Query	Genome 1	Genome 2	Genome 3
$seq_1 1$	X	2	1
$seq_2 1$	1	X	1

Extended phylogenetic profiles

Query	Database	Identity Score
$seq_1 1$	$seq_2 1$	89.9
$seq_1 1$	$seq_2 2$	73.2
$seq_1 1$	$seq_3 1$	61.1
$seq_2 1$	$seq_1 1$	92.8
$seq_2 1$	$seq_3 1$	88.1

Sample BLAST output for 2 query seqs

Query	Genome 1	Genome 2	Genome 3
$seq_1 1$	X	1	1
$seq_2 1$	1	X	1

Binary phylogenetic profiles

Genome 1	Genome 2
$seq_1 1$	$seq_2 1$

BBH detected from sample BLAST output

Fig. 1. Graphical overview of the general hit and bidirectional hit profiles, given a set of example sequences.

Grid Computing.

A computer Grid is a system that coordinates resources that are not subject to centralized control using standard, open, general-purpose protocols and interfaces to deliver non-trivial quality of service [3]. A grid computing architecture can bring massive processing power to bear on a problem, as SETI (Search for ExtraTerrestrial Intelligence) and other similar projects have shown [1,2]. An additional abstraction layer, called middleware, makes creating and controlling grids easier.

The European Grid Infrastructure (EGI) is the result of pioneering work that has, over the last decade, built a collaborative production infrastructure of uniform services through the federation of national resource providers that supports multidisciplinary science across Europe and around the world. Our framework is deployed on the HellasGrid infrastructure, part of EGI, which offers high performance computing services to Greek universities and research institutes.

HellasGrid is also the biggest infrastructure for Grid computing in the area of South-East Europe.

3   Framework Description

We designed and implemented a framework capable of performing sequence alignment and phylogenetic profiling in a time-efficient and user-friendly manner, by utilizing EGI resources. These processes were both scaled to resources and automated due to the fact that they comprise an essential part in a plethora of biological analysis pipelines including, but not limited to: species identification, DNA mapping and domain localization [5,7]. The overall design and ultimate implementation of our framework has been dictated by the following requirements:

- Optimal efficiency
- Fully automated and robust against system failures
- Reusable output for building on top of previous results
- Reuse of existing, optimized and tested submodules

Workflow efficiency was maximized through an optimal use of EGI resources, minimizing queue delays as well as potential load failures that lead to resubmission of computational jobs [8]. The workflow does not require any interaction from the user after the initial submission phase; instead everything runs automatically in the background thus releasing the end-users' machine and allowing for better utilization of the local resources until the analysis is finished. Possible system errors are also handled automatically, with the possible exception of fatal errors. The final output is presented in a simple, widely accepted format that can be either consumed as a final output or used as an intermediate step to the next steps of any analysis pipeline. Finally, instead of reinventing the wheel, our framework utilizes well documented and maintained vanilla implementations of its submodules when available in order to accommodate both updates as well as support any further development of the framework through targeted extensions.

3.1   Program Flow

As outlined earlier in the requirements of the framework, the workflow is fully automated. A Grid job is submitted for each file in the query directory corresponding to a genome to be examined, meaning that the number of submitted jobs is equal to the number of genomes we are interested in. Job submission is in reverse order to the size of the query file, thus ensuring that the most computationally expensive jobs are submitted first. Given that the total time needed for our analysis is equal to the worst time of its parallel counterparts, minimizing the delay of the lengthier jobs results in a notable improvement in efficiency, namely 20–30 h for large queries as shown in the experiments performed (Sect. 5). In order to minimize the uploading time from the user's UI, as well as be able to handle an arbitrarily large database file, we exploit the Grid's storage services. Specifically, we upload the database only once to a Storage Element (SE), and provide each worker node with its qualified name. The worker node is thus able to retrieve the database directly downloading from the SE. Finally, each job runs BLAST to align its specific genome file with the database (which is common for all jobs).

The alignments produced by BLAST are then used to construct the phylogenetic profile of any type requested and for each sequence in the query genome. Since the computation at each node is completely independent of other nodes, no blocking occurs which qualifies the procedure as "embarrassingly parallel". Finally, the output of every job is returned to the users terminal and combined in a way that facilitates visualization as well as post processing. An instance of a job handler script is launched in the background for each of those jobs in order to monitor their state and resubmit them in case of failure caused by stability or load issues that the Grid might face at a given time (Fig. 2).

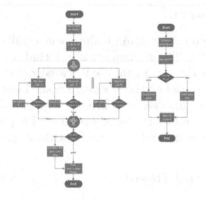

Fig. 2. (A) Diagram of the workflow from the users point of view (B) Diagram of the workflow taking place at each WN.

4   Technical Overview

4.1   Modes of Operation

The driving use case for the proposed framework is to provide users with the means to run an independent analysis. Specifically, the user specifies both the sequences under study as well as the set of genomes against which the sequences are going to be aligned and consequently profiled. Beyond this information and during the initialization phase, additional parameters can be configured such as the type of phylogenetic profiles to be constructed (BBH or plain, extended or binary), the type of the input sequence data type (nucleotides or aminoacid) etc.

Moreover, our framework provides support for building upon previous results; due to the frequent appearance of such analyses in typical biological workflows, the framework provides the ability to combine new results (i.e. new genomes) with any previously computed results. Therefore, and instead of running the whole process from scratch, the framework can expand the previous query without recalculating all involved alignments and phylogenetic profiles' elements.

4.2   Expected Input

The input for every analysis comprises of the database file, including the sequences of every genome in our dataset, and a set of files corresponding to the genomes under study. The former files will be referred to as query files and should reside in a common directory, namely the query directory. It is important to note that both the query files and the database file must comply to the FASTA specification. The user must also supply a text file mapping each genome found in his database with a unique identifier. Finally, the user fills in a properties file in which the exact mode of operation and its parameters are configured.

4.3   Framework Output

The output is comprised of the BLAST alignment results and the phylogenetic profile for each sequence. The framework also includes a library of usual post processing utilities that the user might use to visualize the output. These include (a) filters that are able to isolate the most significant lines of the output, such as alignments with a top e-value score or profiles with specific matches, as well as (b) reducers to collapse profiles at the genome level, i.e. present the phylogenetic profile of a whole genome instead of a profile at the sequence level.

5   Experiments and Results

The framework has been implemented within a standard UNIX environment in mind, i.e. any Debian based version of Linux. This restriction has been placed as a requirement of the underlying Grid infrastructure. The basic workflow is comprised of scripts to exploit the machines native calls, whereas complementary functionalities are designed in an object oriented manner using the Java programming language. In order to validate our framework for scalability we employed a test case, performing an all-vs-all analysis, since this is the most computationally demanding mode of operation. The total of 30 genomes used comprised a BLAST database including 544,538 sequences.

Our experiment involves the creation of BBH profiles for a total of 30 Genomes (all-vs-all operation). It can be seen in Fig. 3 that the total time for each job is comprised of 3 parts. First, there is the submission delay. As explained before, we attempt to minimize the number potential job failures, by inserting a 1-hour delay between consecutive submissions. We take special care to ensure that bigger jobs are scheduled first, thus minimizing the total time for the lengthier jobs. Since the total query time equals that of its most demanding job, this method results in a huge efficiency boost as shown above. The green part corresponds to the scheduling time, which is usually minimal. However, there exist situations where a job fails and has to be resubmitted multiple times, resulting

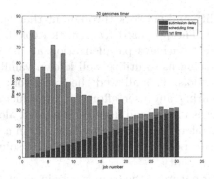

Fig. 3. Execution times in a 30 genome all-vs-all BBH profile construction query.

in extended waiting time such as the case of job number 17. Lastly, we can see the running time on each WN. This demonstrates significant variance, caused not only by input anomalies, but also by the heterogeneity of hardware resources employed by the Grid infrastructure.

Finally, the produced phylogenetic profiles can be readily used for any further analysis, including functional correlations, pangenome approaches and evolutionary studies. Figure 4 provides some preliminary results through the use of BBH phylogenetic profiles.

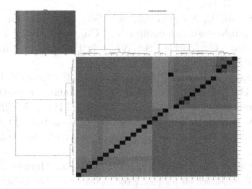

Fig. 4. Visual representation of the BBH profiles produced by the all-vs-all comparison of 30 genomes. Each cell corresponds to the number of BBH that have been identified in the respective pair of genomes. Moreover, the matrix has been hierarchically clustered in order to produce evolutionary meaningful clusters of genomes, as evident by the highly intra-correlated groups (bottom left, middle and top right).

6   Discussion and Conclusions

The ever increasing biological big data that need to be processed and interpreted in almost every modern bioinformatics workflow, require extreme computational power, storage capability and network throughput that only large-scale distributed systems can offer. The proposed framework exploits the native nature of the Grid infrastructure to achieve parallelization at the data layer instead of the software layer, allowing us to utilize well known vanilla implementations of comparative genomics tools. This allowed the framework to achieve significant speed-up without requiring any user effort, or introducing additional maintenance and setup concerns for the end user. This aspect, along with the fact that the main functionality offered poses a significant role in a wide range of diverse biological workflows, can potentially provide an easy solution to researcher in the wider scientific community.

Finally, the modular nature of the frameworks inner workings not only allows for the automatic updating of its submodules, but can also facilitate the addition of new features, such as a user friendly GUI, through integration with existing visualization platforms like Galaxy [4].

Acknowledgements. This work used the European Grid Infrastructure (EGI) through the National Grid Infrastructure - HellasGrid.

References

1. Anderson, D.P.: Boinc: A system for public-resource computing and storage. In: Proceedings of the Fifth IEEE/ACM International Workshop on Grid Computing, pp. 4–10. IEEE (2004)
2. Anderson, D.P., Cobb, J., Korpela, E., Lebofsky, M., Werthimer, D.: Seti@ home: an experiment in public-resource computing. Commun. ACM **45**(11), 56–61 (2002)
3. Foster, I., Kesselman, C., Tuecke, S.: What is the grid?-A three point checklist. Gridtoday (6), July 2002
4. Giardine, B., Riemer, C., Hardison, R.C., Burhans, R., Elnitski, L., Shah, P., Zhang, Y., Blankenberg, D., Albert, I., Taylor, J., et al.: Galaxy: a platform for interactive large-scale genome analysis. Genome Res. **15**(10), 1451–1455 (2005)
5. Marcotte, E.M., Xenarios, I., Van der Bliek, A.M., Eisenberg, D.: Localizing proteins in the cell from their phylogenetic profiles. Proc. Nat. Acad. Sci. **97**(22), 12115–12120 (2000)
6. Overbeek, R., Fonstein, M., Dsouza, M., Pusch, G.D., Maltsev, N.: The use of gene clusters to infer functional coupling. Proc. National Acad. Sci. **96**(6), 2896–2901 (1999)
7. Pellegrini, M.: Using phylogenetic profiles to predict functional relationships. Bacterial Molecular Networks: Methods and Protocols, pp. 167–177 (2012)
8. Vrousgou, O.T., Psomopoulos, F.E., Mitkas, P.A.: A grid-enabled modular framework for efficient sequence analysis workflows. In: Iliadis, L., et al. (eds.) EANN 2015. CCIS, vol. 517, pp. 47–56. Springer, Heidelberg (2015). doi:10.1007/978-3-319-23983-5_5

Community Detection of Screenplay Characters

Christos Makris and Pantelis Vikatos[✉]

Computer Engineering and Informatics Department,
University of Patras, Patras, Greece
{makri,vikatos}@ceid.upatras.gr

Abstract. In this paper, we present a model for automatic community detection of the characters by parsing movie screenplays. In our procedure it is proposed a classification model to predict the casting by categorize each line of the script in a character or not and the co-occurrence of the characters in the same scene constitutes the link between two characters in the social network. We use an existing modularity based community detection algorithm for cutting the created graph in communities. The innovation of our methodology is contained in the extraction of the casting of the screenplay from Wikipedia pages in order to train and build an efficient classifier to identify the characters in a screenplay. The proposed methodology for extracting automatically the social network and communities of screenplay characters can be probably used for enhancing movie recommendations.

Keywords: Community detection · Machine learning · Social network analytics

1 Introduction

The automatic extraction of social network from sources such as unstructured text has gained the interest of researchers as is presented in several studies [1–6]. In some studies such as in [2,5] the extraction of social networks is by parsing literature text which the characters of the book are the nodes and the dialogue between a pair of character constitutes a link. This procedure is expanded also to movie screenplays such as in [3,4,6] which can be seen as unstructured literary works which contain interactions between characters that could be presented as social network. In this paper we introduce an innovative scheme to extract communities of the characters involving in the same scene by only parsing the script of a film using an efficient classification model to predict the characters in a screenplay.

The main challenges is this research is the lack of pre-annotated dataset which can be used as a training and test set of a model. Also most of the screenplays are not well-structured so as to declare a universal rule of regular expression for automatic detection of significant information.

© IFIP International Federation for Information Processing 2016
Published by Springer International Publishing Switzerland 2016. All Rights Reserved
L. Iliadis and I. Maglogiannis (Eds.): AIAI 2016, IFIP AICT 475, pp. 463–470, 2016.
DOI: 10.1007/978-3-319-44944-9_40

An important aspect of our work is exploiting by envisaging a similarity metric between movie screenplays that should be based on the structure of their social network derived from the co-occurrence of characters in scenes that could be useful for movie recommendations.

2   Related Work-Motivation

In [1] is presented a first approach in this field mapping out texts according to geography, social connections and other variables.

A study to extract social networks from nineteenth-century British novels and serials is presented in [2] which the networks have been constructed by dialogue interactions.

A similar study is presented in [5] related to social event detection and social network extraction from a literary text and particularly to the book Alice in Wonderland.

An expansion also to movies' screenplay as a source of as unstructured literary works is presented in [3–6]. In [3] is proposed the extraction of social network by parsing screenplay in order to investigate communities, hidden semantic information and innovations to automation of story segmentation. Another study in [4] is focused in character interaction and networks between characters from plays and movies.

In [6] it is presented a formalization of the task of parsing movies' screenplays as well as a extraction of social network of all characters having a dialogue with each other in a scene with links.

3   Methodology-Model Overview

In the following methodology the automatic detection of communities using movie's screenplay is described. In Fig. 1 the whole system architecture and the separated modules is depicted. In the following sections the detailed description of the each part of proposed model is included and in the Sect. 5 the experimental procedure and the results are presented.

3.1   Crawling IMSDB Webisite

We collect a set of movie screenplays via crawling the Internet Movie Script Database (IMSDB) website[1] which contains a list of movies and the sceenplays for most of them. In studies [6,8] it is mentioned that scripts are constituted by 5 elements:

1. the scene boundary which describes if the scene is to take place inside or outside using the tags *INT.* and *EXT.* respectively and the name of the location.

---

[1] IMSDB website: http://www.imsdb.com/.

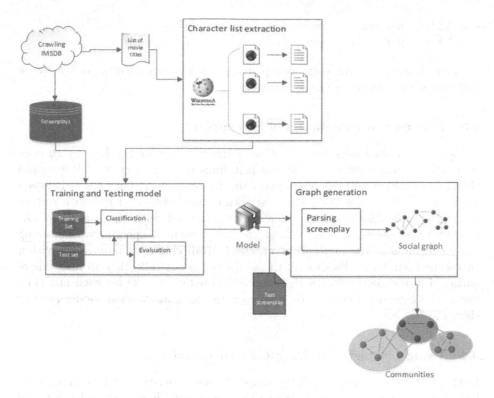

Fig. 1. System architecture

2. Scene description which is below the scene boundary and declares detailed information about the scene.
3. Character name which is the role that is involved in the scene.
4. Dialogue that the active role will act.
5. Meta-data for special information in the script.

In our case it is necessary to recognize the scene boundary element in the script and split the scenes in order to identify the characters between them. Also an initial test about *EXT./INT.* tags in the text should be occurred in order to check remove unstructured screenplays from our data collection.

3.2 Scrapping Wikipedia Information

In study [6] it is mentioned that regular expressions is not the appropriate way to identify and annotate the character names from the scripts. We propose a procedure using Wikipedia website[2] as a source to extract the roles in the movie. Wikipedia site uses a well-structured information about the cast in a certain movie. We created a log file with the specific urls for all movie titles and we scraped the html code. The information is between the following html tags:

---

[2] Wikipedia website: https://www.wikipedia.org/.

$- < h2 >< span\ class = "mw - headline"\ id = "Cast" >$
$- < h2 >< span\ class = "mw - headline"\ id = "Production" >$

The characters form each Wikipedia page are extracted in order to be matched with the ones in the screenplay.

3.3   Parsing and Annotation of Screenplays

We annotate each line of the script with the tag "C" for the line which only contains the character name and the rest lines with the tag "O". It is noted that in the well-structured screenplays the name of each character is between scene boundaries, after the scene description and before dialogue in direct or indirect speech. Also character names are capitalized, with an optional (V.O.) or (O.S.) information for "Voice Over" or "Off-screen." respectively. Examining the correctness of the annotation we check if all character names are within two scene boundaries. Parsing all the scripts from our collection all lines where gathered in a super-text with the proposed annotation C/O for each line. The tag is the category that will be predicted by the classification model as it is shown in Fig. 1.

3.4   Using Linguistics to Create Feature Vector

In this section we aggregate all the scripts to one super-text and we argue each line of the text will be transformed to a vector with linguistic and emotional features. We used LIWC [7] as a tool to extract linguistic characteristics. As a result, we created a vector of 80 characteristics. The linguistic and emotional analysis of each line will provide the appropriate type of features in order to train efficiently the classification model.

3.5   Training the Classification Model

As it is mentioned our scope is to identify the characters in the screenplay. A classification model has been used and trained for this purpose. The first step is the separation of our dataset in train and test set as it using K-Fold Cross-Validation. The main advantage is that all instances in the dataset are eventually used for both training and testing. In Sect. 5 the procedure for the separation of the dataset in training and test set as well as the performance of the classifier is described in detail.

4   Graph Generation and Community Detection

The social graph represents the co-occurrence of characters in the same scene. More specifically the nodes of the graph constitute the characters and the link the co-occurrence of the pair of nodes in the same scene. Each link contains an attribute weight related to the frequency of the co-occurrence in the whole screenplay. We parse the screenplay in order to recognize the scenes. According to the structure of the screenplay the scene are between the $INT.$ and $EXT.$ tags. The communities in the social graph are detected by a well-used community detection approach [9] using modularity optimization as algorithmic progresses. The density of edges inside communities and outside communities is related to a modularity in a scale value between -1 and 1. Heuristic algorithms are used in order to eliminate checking all possible iterations of the nodes into groups while optimizing the modularity value. Using the Louvain Method [9, 10] of community detection, first small communities are detected by optimizing modularity locally and then each small community is clustered into one node and the procedure is repeated.

5   Implementation and Results

We implemented the crawler of the IMSDB website in python 2.7. The total number of files that we gathered was 1112 screenplays. In Sect. 3.1 it is noted that a well-structured screenplay contains 5 elements and thus we checked each file to overview it. Firstly it was checked the existence of the element $EXT./INT.$ for the scene boundary and 972 screenplays had this element in their text. Also we checked the existence of characters' name inside the scene boundaries and the number of files that passed this test was 501 which means that 45 % of the initial screenplays were well-structured and appropriated to be introduced in our methodology as it is shown in Table 1.

Table 1. Instances after preprocessing

Preprocessing step	Number of screenplays	(% of dataset)
$EXT./INT.$ occurrence	972	87 %
Existence of characters between $EXT./INT.$	501	45 %

Based on the films' name a list was created with the urls that link films with Wikipedia pages. The scrapping of html code for discovering characters' name is described in detail in Sect. 3.2 using regular expression to identify the appropriate tags. We annotated each line of the screenplays with labels C/O for character and other respectively and we gather all lines for the screenplays in a super-text with total size 220 MB. A vector with 80 linguistic and emotional characteristics for each line was created with LIWC [7] software forming a dataset with the 86 % and 14 % of instances in label C and O respectively as it is presented in Table 2.

We developed a classifier to predict each label using Weka[3] environment. We used the J48 decision tree classifier in order to predict the label in the test instances. The classifier from Weka is used with the default settings. We

---

[3] Weka toolkit: http://www.cs.waikato.ac.nz/ml/weka/.

Table 2. Distribution of labels

Label	Number of instances	(% in dataset)
Other Line	2859281	86 %
Character	465464	14 %

separated the dataset to training and test set, using K-Fold Cross-Validation (K = 10 Fold). We evaluated the classifier based on F-measure which is the harmonic mean of precision and accuracy of the classification and our classification model achieved 99.3 %.

We worked in two case studies for "X-Men" and "The Lord of the Rings: The Fellowship of the Ring" (LOR) to present the produced social networks and communities in our methodology. In Sect. 4 it is described the procedure for social graph generation and the modularity optimization algorithm for community detection. According to the predicted character list which was derived from our classifier the pattern matching in the screenplays led to the information of Table 3.

Table 3. Parsing screenplay and attributes extraction

Screenplay attributes	X-men	LOR
# Characters	28	21
# Scenes	185	111
# Characters' co-occurrence	192	134
Mean Characters per Scene	2	3

In Fig. 2 is depicted the social networks of co-occurrence characters in screenplay scenes. Line thickness express the weight in the link between two characters (Fig. 3). In the social networks we calculate node centralities as it is shown in Table 4.

Table 4. Social network centralities

Social network centralities	X-men	LOR
Degree Centrality	0.342	0.638
Betweenness Centrality	0.024	0.019
Closeness Centrality	0.582	0.755

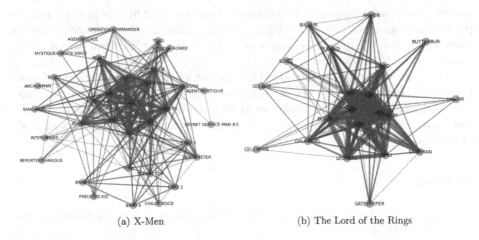

(a) X-Men   (b) The Lord of the Rings

Fig. 2. Social Networks of co-occurrence characters in screenplay scenes

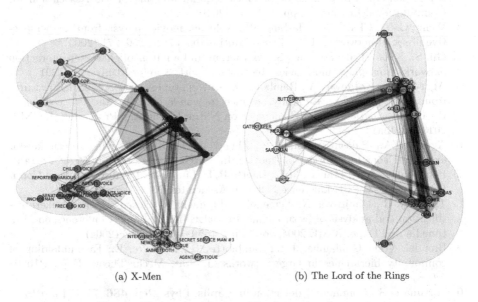

(a) X-Men   (b) The Lord of the Rings

Fig. 3. Characters communities in screenplay scenes

6 Conclusions and Future Work

In this paper there is the detailed description of the methodology for extracting social network and communities of the characters involving in the same scene by only parsing the movie screenplay. The proposed methodology is based on the performance of an efficient classifier for character recognition.

We deal with the main challenge of the lack of pre-annotated dataset which can be used as a training and test set of the classifier adopting the non use of universal regular expressions for character recognition which is not the appropriate way based on previous studies. On the other hand we introduce as an external source Wikipedia webpages in order to extract by html scraping the appropriate information for screenplays.

As future work, we are interested in examining the temporal information of the scenes and discovering the fluctuation of the sentiment between the characters' dialogues. Also this model could be used as a subsidiary factor for the recommendation of similar movies based on the structure and attributes analysis of their social network.

References

1. Moretti, F.: Graphs, Maps, Trees: Abstract Models for a Literary History. Verso, London (2005)
2. Elson, D.K., Dames, N., McKeown, K.R.: Extracting social networks from literary fiction. In: Proceedings of the 48th Annual Meeting of the Association for Computational Linguistics, pp. 138–147 (2010)
3. Weng, C.-Y., Chu, W.-T., Ja-Ling, W.: RoleNet: movie analysis from the perspective of social networks. IEEE Trans. Multimedia 11(2), 256–271 (2009)
4. Gil, S., Kuenzel, L., Caroline, S.: Extraction and analysis of character interaction networks from plays and movies. Technical report, Stanford University (2011)
5. Agarwal, A., Kotalwar, A., Rambow, O.: Automatic extraction of social networks from literary text: a case study on alice in wonderland. In: the Proceedings of the 6th International Joint Conference on Natural Language Processing (IJCNLP 2013) (2013a)
6. Agarwal, A., Balasubramanian, S., Zheng, J., Dash, S.: Parsing screenplaysfor extracting social networks from movies. In: EACLCLFL 2014, pp. 50–58 (2014b)
7. Pennebaker, J.W., Francis, M.E., Booth, R.J.: Linguistic Inquiry and Word Count (LIWC): LIWC2001. Lawrence Erlbaum Associates, New Jersey (2001)
8. Turetsky, R., Dimitrova, N.: Screenplay alignment for closed-system speakeridentification and analysis of feature films. In: IEEE International Conference on Multimedia and Expo, ICME 2004, vol. 3, pp. 1659–1662. IEEE (2004)
9. Blondel, V.D., Guillaume, J.-L., Lambiotte, R., Lefebvre, E.: Fast unfolding of community hierarchies in large networks. J. Stat. Mech: Theory Exp., P10008 (2008)
10. Fortunato, S.: Community detection in graphs. Phys. Rep. 486, 75–174 (2010)

Customer Behaviour Analysis
for Recommendation of Supermarket Ware

Stavros Anastasios Iakovou, Andreas Kanavos[✉], and Athanasios Tsakalidis

Computer Engineering and Informatics Department,
University of Patras, Patras, Greece
{iakovou,kanavos,tsak}@ceid.upatras.gr

Abstract. In this paper, we present a prediction model based on the behaviour of each customer using data mining techniques. The proposed model utilizes a supermarket database and an additional database from Amazon Company, both containing information about customers' purchases. Subsequently, our model analyzes these data in order to classify customers as well as products; whereas being trained and validated with real data. This model is targeted towards classifying customers according to their consuming behaviour and consequently propose new products more likely to be purchased by them. The corresponding prediction model is intended to be utilized as a tool for marketers so as to provide an analytically targeted and specified consumer behavior.

Keywords: Supervised learning · Data analytics · Customer behaviour · Knowledge extraction · Personalization · Recommendation system

1 Introduction

During the last years, more and more companies store their data into large data-centers so as to initially analyze them and to further understand how their consumers behave. Every day, a large amount of information is accessed and processed by companies in order to get a deeper knowledge about their products' sales and consumers' purchases. From small shops to large enterprises, owners try to record information that probably contains useful data regarding consumers.

In addition, the rapid development of technology provides high quality network services. A large percentage of users utilize the Internet for information of each field. For this reason, companies try to take advantage of this situation by creating systems that store information about users who entered their site in order to provide them with personalized promotions. Companies concentrate on their desired information and personal transactions. Businesses provide their customers with cards so they can record every buying detail. This procedure has led to a huge amount of data and search methods for data processing.

Historically, several analysts have been involved in the collection and processing of data. In modern times, the data volume is so huge that it requires the use

© IFIP International Federation for Information Processing 2016
Published by Springer International Publishing Switzerland 2016. All Rights Reserved
L. Iliadis and I. Maglogiannis (Eds.): AIAI 2016, IFIP AICT 475, pp. 471–480, 2016.
DOI: 10.1007/978-3-319-44944-9_41

of specific methods so as to enable analysts to export correct conclusions. Due to the increased volume for automatic data analysis, methods use complex tools; along with the help of modern technologies, data collection can be now considered as a simple process. Analyzing a dataset is a key aspect to understanding how customers think and behave during each specific period of the year. There are many classification and clustering methods which can be successfully used by analysts to aid them broach in consumers' mind. More specifically, supervised machine learning techniques are utilized in the present manuscript in the specific field of supermarket.

Besides the development of web technologies, the abundance of social networks has created a huge number of reviews on products and services, as well as opinions on events and individuals. Concretely, consumers are used to being informed by other users' reviews in order to carry out a purchase of a product, service, etc. One other major benefit is that businesses are really interested in the awareness of the opinions and reviews concerning all of their products or services and thus appropriately modify their promotion along with their further development. As a previous work on opinion clustering emerging in reviews, one can consider the setup presented in [5].

Furthermore, the emotional attachment of ample customers to a brand name is a topic of interest in recent years in the marketing literature; it is defined as the degree of passion that a customer feels for the brand [13]. One of the main reasons for examining emotional brand attachment is that an emotionally attached person is highly probable to be loyal and pay for a product or service [15]. In [6], authors infer details on the love bond between users and a brand name being considered as a dynamic ever evolving relationship. More concretely, users that have demonstrated emotional connection to the brand through their tweets are considered. Thus, the aim is to find those users that are engaged and rank their emotional terms accordingly.

In this paper, we present a work on modeling and predicting customer behavior using information concerning supermarket ware. More specifically, we propose a new method for product recommendation by analyzing the purchases of each customer. With the use of category of the current dataset, we were able to classify the aforementioned data and subsequently create clusters. According to viral marketing [9], clients influence each other by commenting on specific fields of e-shops. Practically, this method appears in real life when people communicate in real time and affect each other on the products they buy. The aim of this model is to analyze every purchase and propose new products for each customer. More to the point, we want to perform the following steps in the below mentioned order: firstly the analysis of the sales rate is utilized, then the distances of each customer from the corresponding supermarket is clustered and finally the prediction of new products that are more likely to be purchased from each customer separately is implemented.

The remainder of the paper is structured as follows: Sect. 2 presents the related work. Section 3 presents our model, while in Sect. 4 we utilize our experiments. Moreover, Sect. 5 presents the evaluation experiments conducted and the results gathered. Ultimately, Sect. 6 presents conclusions and draws directions for future work.

2 Related Work

In recent years, a large percentage of companies maintain an electronic sales transaction system aiming at creating a convenient and reliable environment for their customers. On the other hand, retailers are able to gather significant information for the corresponding customers. Moreover, since the number of data is significantly increasing, more and more researchers have developed efficient methods as well as rule algorithms for market basket analysis [2]. Researchers have also developed applications for optimal product selection on supermarket data; one of them being the "Profset model". Using cross-selling potential, this model selects the most interesting products from a variety of ware. Additionally, Li et al. [10] analyzed and designed a model of E-supermarket shopping recommender. Lu et al. [12] developed a personalized recommendation system for government to business e-services.

Furthermore, according to [8], researchers have invented a new recommendation system where supermarket customers were able to get new products. This recommendation system was first presented as a part of "SmartPad", which was a system that allowed customers to prepare their shopping list in advance. In this system, matching products and clustering methods are used so as to provide new products to less frequent customers. The analysis of this method showed 1.8 % increase in the income of the supermarket.

In addition, since consumers started to resist to traditional methods of marketing, it was necessary for companies to invent a new advertising method based on alternative strategies. Leskovec et al. [9] presented an analysis of a person-to-person recommendation network which included 4 million people along with 16 million recommendations. This model illustrated how effective the recommendation network for both sender and receiver was. Despite the fact that average recommendation networks are not very effective in increasing purchases, this model had successfully managed it.

Another significant model was presented by Dickson and Sawyer [4], where they created a model for a grocery shop so as to analyze how customers respond to price and other point-of-purchase information. Moreover, according to [1], consumers know exactly the price of the product they purchase; the results of this work showed that customers did not pay attention to the price of the products they buy. More specifically, they did not know if the price was reduced since it was on special offer.

In addition, Yao et al. [16] created a recommendation system targeted towards supermarket products for consumers since supermarkets are too big to find the desirable product; using RFID technology with mobile agents, they constructed a mobile-purchasing system. Furthermore, Kim et al. [7] presented another recommendation system based on the past actions of individuals, where they provided their system to an Internet shopping mall in Korea.

In point of fact, in [3], authors showed a new method on personalized recommendation in order to get further effectiveness and quality since collaborative methods presented limitations such as sparsity. Regarding Amazon.com, they used for each customer many attributes, including item views and subject interests, since they wanted to create an effective recommendation system. This view is echoed throughout [11], where authors analyzed and compared traditional collaborative filtering, cluster models and search-based methods. Also, Weng and Liu [17] analyzed customers' purchases according to product features and as a result managed to recommend products that are more likely to fit with customers' preferences.

Finally, authors in [14] analyzed the product range effect in purchase data. Since market society is affected by two factors (e.g. rationality and diversity in the price system), consumers try to minimize their spending and maximize the number of products they purchase. So, researchers invented an analytic framework based on big customers' transaction data. They observed that customers did not always choose the closest supermarket and then they tried to answer why consumers buy specific products.

3   Model Overview

In our model, we want to predict whether a customer will purchase a product or not based on a supermarket ware dataset using data analytics and machine learning algorithms. We can classify this problem as a classification one, since the opinion class consists of specific options. Furthermore, we have gathered the reviews of Amazon Company and in particular the reviews of each customer, in order to analyze the affection of person-to-person influence in each product's market.

The overall architecture of the proposed system is depicted in Fig. 1 while the proposed modules and sub-modules of our model are modulated in the following steps.

3.1   Customer Metrics Calculation

From the supermarket dataset, we randomly sampled 10000 records regarding customers' purchases, containing information about sales over a vast period of 4 years. More specifically, the implementation of our method goes as follows: initially, we sample the customers as well as the products. Subsequently, the clustering of the products based on the sales rate takes place. Then, we cluster our customers related on the distance of their houses from the supermarket. Furthermore, a recommendation model, with new products separately proposed to each customer based on their consumer behavior, is utilized. Furthermore, we sampled the customers of Amazon Company and then using the rates of the reviews, we came up with the fraction of the satisfied customers.

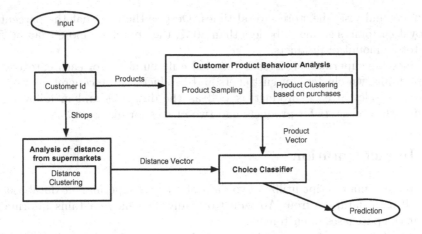

Fig. 1. Supermarket model

The training set of the supermarket data contains 8 features, as presented in following Table 1, including customer ID, the category of the product, the product ID, the shop, the number of items purchased, the distance of each supermarket, the price of the product as well as the choice.

3.2 Decision Analysis

In this subsection, the choice analysis based on classification and clustering tools is described. This method gathers eight basic features of the supermarket database and eleven different methods of classification in order to further analyze our dataset, as shown in the next section. In [8], researchers used clustering to find customers with similar spending history. Furthermore, as [18] indicate, the loyalty of customers to a supermarket is measured in different ways. Specifically, a person is considered as loyal to a specific supermarket if they purchase specific

Table 1. Training set features

Features	Description
Customer ID	The ID of the customer
Product category	The category of the product
Product ID	The ID of the product
Shop	The shop where the customer makes the purchase
Number of items	How many products he purchased
Distance cluster	The cluster of the distance
Product price	The price of the product
Choice	Whether the customer purchases the product or not

products and visit the store several times. Despite the fact that the percentage of loyal customers seems to be less than 30 %, they purchase more than 50 % of the total amount of products.

Since the supermarket dataset included only numbers for each category, we created our own clusters for customers and products. More concretely, we measured the sales of each product as well as the distances and in following we created three clusters for products and two classes for distances.

4   Implementation

The present manuscript utilizes two datasets, e.g. a supermarket database [14] as well as a database from Amazon Company [9] which contains information about the purchases of customers.

Initially, we based our experiments on the supermarket database [14] and we extracted the data using C# language so as to calculate customer metrics. We have implemented an application with which we have measured all the purchases of the customers. In following, a sample of the customers, so as to further analyze the corresponding dataset, was collected. The final dataset consists of 10000 randomly selected purchases with all the information from the supermarket dataset as previously mentioned.

The prediction of any new purchase is based on the assumption that customers are affected by each other. Consumers communicate every day and exchange reviews for products. On the other hand, since their budget is tight, they select products that correspond better to their needs. Therefore, a model that recommends new products to every customer from the supermarket they mostly prefer, is proposed.

By analyzing the prediction model, information about consumers' behavior is extracted. We measured the total amount of products that customers purchased and then categorized them accordingly. Several classifiers are trained using the dataset of vectors. We separated the dataset and we used 10-Fold Cross-Validation to evaluate training set and test set. The classifiers that were chosen, are evaluated using TP (True Positive) rate, FP (False Positive) rate, Precision, Recall, as well as F-Measure metrics. We chose classifiers from five categories of Weka library[1] including "bayes", "functions", "lazy", "rules" and "trees". The classifiers from Weka are used with their default settings and the results are introduced in following in Table 3.

Additionally, we evaluated a model using the results of our experiments on Amazon Company [9] since we wanted to measure the number of contented customers regarding five product categories, namely music, book, dvd, video and toy. In Table 2, we show the number of delighted and on the other hand, the number of not satisfied customers.

Next Fig. 2 illustrates the amount of customers who are satisfied with products of every category. We can observe that the number of satisfied customers

---

[1] Weka toolkit: http://www.cs.waikato.ac.nz/ml/weka/.

Table 2. Measurement of satisfaction of customers

Product category	Satisfied customers	Not satisfied customers
Music	80149	15377
Book	235680	68152
DVD	41597	16264
Video	38903	13718
Toy	1	1

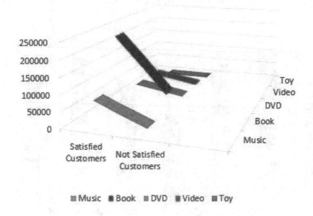

Fig. 2. Customers Reviews

is much bigger than the one of not satisfied ones in four out of five categories (regarding category entitled toy, the number is equal to 1 for both category of customers). With these results, one can easily figure out that Amazon customers are loyal to the corresponding company and prefer purchasing products from the abovementioned categories.

5   Evaluation

The reported values in the charts for the classification models are recorded as AdaBoost, J48, JRip, Multilayer Perceptron, PART, REPTree, RotationForest and SMO. The results for each classifier for their several values are illustrated in Table 3. Depicted in bold, are the selected best classifiers for each value. What is more, Fig. 3 depicts the values of F-Measure for each classifier.

We can observe that J48 achieves the highest score in every category except FP rate. Subsequently, REPTree follows with almost 89 % TP rate and F-Measure, whereas JRip has value of F-Measure equal to 84 %. In addition, concerning F-Measure metric, the other algorithms range from 42 % of Multilayer

Table 3. Classification for predicting

Classifiers	TP rate	FP rate	Precision	Recall	F-Measure
AdaBoost	0.605	0.484	0.598	0.605	0.564
J48	**0.922**	0.095	**0.924**	**0.922**	**0.921**
JRip	0.847	0.187	0.854	0.847	0.843
Multilayer perceptron	0.596	**0.539**	0.66	0.596	0.482
PART	0.748	0.325	0.783	0.748	0.728
REPTree	0.892	0.119	0.892	0.892	0.891
RotationForest	0.785	0.285	0.83	0.785	0.768
SMO	0.574	0.574	0.33	0.574	0.419

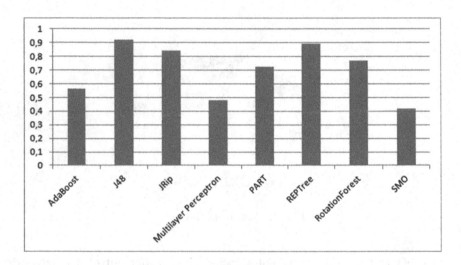

Fig. 3. Customers classification

Perceptron to 77 % of Rotation Forest. Moreover, we see that almost all classifiers achieve a TP rate value of above 60 %, while the percentages for FP rate are relatively smaller. Precision and Recall metrics have almost the same values for each classifier, ranging from 60 % to 92 %.

6   Conclusions and Future Work

In our work we present a methodology to model and predict the purchases of a supermarket using machine learning techniques. More specifically, two datasets are utilized; a supermarket database as well as a database from Amazon Company which contains information about the purchases of customers. Given the analysis of the dataset from Amazon Company, a model that predicts new products for every customer based on the category and the supermarket they prefer is created. We also examine the influence of person-to-person communication, where we found that customers are greatly influenced by other customer reviews.

As future work, we plan to create a platform using the recommendation network. Customers will have the opportunity to choose among many options on new products with lower prices. Next, we can take into consideration more features of the supermarket dataset, in order to improve classification accuracy. In conclusion, we could use a survey to have further insights and get an alternative verification of user's engagement.

References

1. Allen, J.W., Harrell, G.D., Hutt, M.D.: Price Awareness Study. The Food Marketing Institute, Grocer, Washington, D.C. (1976)
2. Brijs, T., Goethals, B., Swinnen, G., Vanhoof, K., Wets, G.: A data mining framework for optimal product selection in retail supermarket data: the generalized PROFSET model. In: Proceedings of the Sixth ACM SIGKDD International Conference on Knowledge Discovery and Data Mining, pp. 300–304 (2000)
3. Cho, Y.H., Kim, J.K., Kim, S.H.: A personalized recommender system based on web usage mining and decision tree induction. Expert Syst. Appl. **23**(3), 329–342 (2002)
4. Dickson, P.R., Sawyer, A.G.: The price knowledge and search of supermarket shoppers. J. Mark. **54**, 24–53 (1990)
5. Gourgaris, P., Kanavos, A., Makris, C., Perrakis, G.: Review-based entity-ranking refinement. In: WEBIST, pp. 402–410 (2015)
6. Kanavos, A., Kafeza, E., Makris, C.: Can we rank emotions? a brand love ranking system for emotional terms. In: IEEE International Congress on Big Data, pp. 71–78 (2015)
7. Kim, J.K., Cho, Y.H., Kim, W.J., Kim, J.R., Suh, J.H.: A personalized recommendation procedure for Internet shopping support. Electron. Commer. Res. Appl. **1**(3–4), 301–313 (2002)
8. Lawrence, R.D., Almasi, G.S., Kotlyar, V., Viveros, M.S., Duri, S.S.: Personalization of supermarket product recommendations. In: Applications of Data Mining to Electronic Commerce, pp. 11–32 (2001)
9. Lescovec, J., Adamic, L.A., Huberman, B.A.: The dynamics of viral marketing. ACM Trans. Web (TWEB) **1**(1), 5 (2007)
10. Li, Y., Meiyun, Z., Yang, B.: Analysis and design of e-supermarket shopping recommender system. In: ICEC 2005 Proceedings of the 7th International Conference on Electronic Commerce, pp. 777–779 (2005)
11. Linden, G., Smith, B., York, J.: Amazon.com recommendations: item-to-item collaborative filtering. IEEE Internet Comput. **7**(1), 76–80 (2003)
12. Lu, J., Shambour, Q., Xu, Y., Lin, Q., Zhag, G.: BizSeeker: a hybrid semantic recommendation system for personalized government-to-business e-services. Internet Res. **20**(3), 342–365 (2010)
13. Malar, L., Krohmer, H., Hoyer, W.D., Nyffeneggeri, B.: Emotional brand attachment and brand personality: the relative importance of the actual and the ideal self. J. Mark. **75**, 35–52 (2011)
14. Pennacchioli, D., Coscia, M., Rinzivillo, S., Pedreschi, D., Giannotti, F.: Explaining the product range effect in purchase data. In: IEEE International Conference on Big Data, pp. 648–656 (2013)

15. Thomson, M., MacInnis, D.J., Park, C.W.: The ties that bind: measuring the strength of consumers' emotional attachments to brands. J. Consum. Psychol. **15**(1), 77–91 (2005)
16. Yao, C., Tsui, H., Lee, C.: Intelligent product recommendation mechanism based on mobile agents. In: 4th International Conference on New Trends in Information Science and Service Science, pp. 323–328 (2010)
17. Weng, S.S., Liu, M.J.: Feature-based recommendations for one-to-one marketing. Expert Syst. Appl. **26**(4), 493–508 (2004)
18. West, C., MacDonald, S., Lingras, P., Adams, G.: Relationship between product based loyalty and clustering based on supermarket visit and spending patterns. Int. J. Comput. Sci. Appl. **2**(2), 85–100 (2005)

Dealing with High Dimensional Sentiment Data Using Gradient Boosting Machines

Vasileios Athanasiou[(⊠)] and Manolis Maragoudakis

Department of Information and Communication Systems Engineering,
University of the Aegean, Palama 2 Street, 83200 Karlovasi, Samos, Greece
{icsdm15041, mmarag}@aegean.gr

Abstract. One of the most common classification tasks that applies on textual information is sentiment analysis, i.e. the prediction of the sentiment of a given document. With the vast use of social media and internet applications such as e-commerce, e-tourism and e-government, numerous comments and opinions are broadcasted per day, thus an automatic way of analyzing them is of great importance. The present paper focuses on sentiment analysis for Greek texts, obtained from Web 2.0 platforms. Greek is a language that lacks an in-depth availability of natural language processing tools in the sense that most of them are not publicly available. The novelty of the article is that instead of utilizing preprocessing tools such as Part-of-Speech taggers, text stemmers and polar-word lexica, it incorporates the translation of the Greek token as provided by the Google Translator® API. Since automatic translation of Greek sentences often results in poor translations where the meaning of the original sentence is severely deteriorated, the translation of each token individually is almost 100 % correct. However, taking the translation of every Greek token poses a significant issue to the outcome of the classification process for practically any classifier, therefore, we introduce the use of a powerful ensemble algorithm that is highly customizable to the particular needs of the application, such as being learned with respect to different loss functions and thus dealing with a large number of dimensions. This algorithm is called Gradient Boosting Machines and experimental results support our claim that it surpasses other, well-known machine learning techniques with a significant improvement for our task.

Keywords: Gradient Boosting Machines · Sentiment analysis · High-dimensional data · Modern Greek

1 Introduction

Throughout recent years, we have witnessed a remarkable rise of the Internet and the World Wide Web that has altered the way they communicate, seek for relevant information, and work on the most fundamental level. The arrival of Web 2.0 technologies has resulted in a vast popularity and ubiquity of web resources built around the ideas of social media and user-generated content (e.g., Facebook, Twitter, and LinkedIn). In the beginning such concepts were part of more traditional web resources such as online retailers (e.g., Amazon) or electronic media, in order to having their products or articles enhanced with the users' opinions. Therefore, the task of relevant

© IFIP International Federation for Information Processing 2016
Published by Springer International Publishing Switzerland 2016. All Rights Reserved
L. Iliadis and I. Maglogiannis (Eds.): AIAI 2016, IFIP AICT 475, pp. 481–489, 2016.
DOI: 10.1007/978-3-319-44944-9_42

information from the vast amounts of human communication information over the Internet is of utmost importance for robust sentiment analysis modules. In fact, the origin of opinionated data has caused the development of Web Opinion Mining (WOM) [1], as a new concept in Web Intelligence. WOM deals with the issue of extracting, analyzing and aggregating web data about opinions. The analysis of users' opinions is significant because through them it is possible to determine how people feel about a product or service and know how it was received by the market. In general, traditional sentiment analysis mining techniques apply to social media content as well, however, there are certain factors that make Web 2.0 data more complicated and difficult to be parsed. An interesting study about the identification of such factors was made by Maynard et al. [2], in which they exposed important features that pose certain difficulties to traditional approaches when dealing with social media streams. Modern Greek language is posing additional obstacles to sentiment analysis since the majority of preprocessing tools for Greek such as POS tagger, stemmer and polarity lexica are not freely available.

In the present paper, we deal with modeling a sentiment analyzer for Modern Greek based on a simple, yet novel idea. We bypass the need for extensive preprocessing tools and utilize a freely available translation API that is provided by Google®, in order to augment the feature set of the training data. However, since automatic translation of sentences often suffer from poor performance, mainly due to the large degree of ambiguity, we decided to translate each Greek token individually, a process that rarely makes mistakes. The resulted feature set was of course double the original size which also poses certain difficulties to the majority of classification algorithms. Hence, we experimented with an ensemble classification algorithm named as Gradient Boosting Machines (GBM) which is theoretically proven of being able to cope with large number of features [3]. According to the referenced study, "a possible explanation why boosting performs well in the presence of high-dimensional data is that it does variable selection (assuming the base learner does variable selection) and it assigns variable amount of degrees of freedom to the selected predictor variables or terms". We experimented with numerous well-known algorithms using the initial Greek-only feature set as well as the enhanced translated one and also some basic feature reduction techniques such as Principal Component Analysis [4] and found that GBM are superior to any other implementation.

2   Related Work

The nature as well as the popularity of feedback-oriented content in domains such as news, social events, services and products is something of a duty to the human urge to post what they feel and think online. As soon as the most common type of message on Twitter is about 'me now' [5], it is evident that users talk often about their own feelings and opinions. Bollen et al. [6] stated that users express both their own mood in tweets about themselves and more generally in messages about other subjects. Another study [7] estimates that approximately 19 % of microblog messages mention a brand name and from those that do, around 20 % contain sentiment.

There are numerous challenges in applying typical sentiment analysis and opinion mining techniques to social media. Short texts in social media, known as micro-posts, are, perhaps, the most challenging text type for opinion mining, given the fact that they do not contain much contextual information and take much implicit knowledge into account. Ambiguity is a common phenomenon since one cannot easily make use of co-reference information: unlike the situations of blog posts and comments, tweets do not typically follow a conversation thread, and appear much more in isolation from other tweets. They also contain much more language variation, tend to be less structured than longer posts, contain unorthodox forms of writing such as emoticons, abbreviations and hashtags, which can form an important part of the meaning. Typically, they contain extensive use of irony and sarcasm, which are particularly difficult for a machine to identify. On the contrary, their shortness can also be beneficial in focusing the topics more explicitly: it is very uncommon for a single tweet to be related to more than one topic, which can therefore contribute in disambiguation.

The research of [8] presents a wide-ranging and detailed review of traditional automatic sentiment detection techniques, including many sub-components, which we shall not repeat here. In general, sentiment detection techniques can be roughly divided into lexicon-based methods (e.g. [9]) and machine learning methods, e.g. [10]. Lexicon-based methods rely on a sentiment lexicon, a collection of known and pre-compiled sentiment terms. Machine learning approaches make use of shallow syntactic and/or linguistic features [11, 12], and hybrid approaches are also very common, with sentiment lexicons playing a key role in the majority of methods, e.g. [13]. Decision Trees, Naïve Bayes and SVM (Support Vector Machine) have been applied from the supervised side [13, 14]. On the unsupervised side there is a pattern-logic classification according to a lexicon. Combination of the above can characterized as semi-supervised [13]. In "On Mining Opinions From Social Media" [15] the authors confirm that Naïve Bayes outperforms a large number of other methods. Techniques that work based on rules normally use sentiment dictionaries plus rules [16].

3 Gradient Boosting Machines

Generally, the boosting methods work by adding sequentially new models and in every iteration every weak base-learner is re-trained [3]. The target is to reduce the total error of the model. The GBM works by constructing new base-learners which are correlated with the negative gradient of the loss function. The loss function can be set from the user (Gaussian L2 loss function, Binomial etc.), this is the main reason that GBM is highly customizable. Suppose there is a dataset $(x, y)_{i=1}^{N}$ where $x = (x_1, x_2 ... x_d)$ are the input variables and y the corresponding labels. The relation between x and y is unknown, this means it is needed to reconstruct this relation in a way to minimize a loss function.

$$\hat{f}(x) = y, \hat{f}(x) = arg \, \frac{min}{f(x)} \, \Psi(y, f(x))$$

Where $f(x)$ indicates the dependence between x, y, $\hat{f}(x)$ is the estimation function and $\Psi(y, f(x))$ is the loss function. The same problem expressed in terms of expectations concludes to

$$\hat{f}(x) = arg \, \underset{f(x)}{min} \quad \overset{expected \ loss}{E_x[E_y(\Psi[y, f(x)])|x]} \atop expectation \ over \ the \ whole \ dataset$$

where $E_y(\Psi[y, f(x])$ is the response variable depended from x. Significant notation is that y can became from not only one distribution provides the ability for more loss functions Ψ.

In machine learning, there are cases that the data set produce a vector with very high dimensionality, for example in sentiment analysis. That means the rows of the document term matrix have very low density because the appearance of the words in every observation is very rare. GBM can to overcome this difficulty because is able to create sparse models, additionally can work with different types of base learners. The ability of use different loss functions combining the ability to use different type of learners provides great regularization capabilities. This has impact to avoid overfitting and achieve generalization to our model. Apart from the above advantages GBM algorithm has drawbacks such as memory consumption and evaluation speed. The first one has impact to the other. GBM works by doing iterations which stored in memory, easily in hard cases for the algorithm like intrusion detection systems the number if iteration can be tens of thousands. This problem appears to all ensemble algorithms.

4 Experimental Setup and Results

4.1 Data Collection

The corpus of data retrieved from online newspaper articles. The types of articles came from financial domain, political, society and sports in plain text type and Greek language. All articles have been selected under the condition to be able classified as positive or negative only. The corpus is balanced and two datasets have been derived from it. The first consists from the articles as gathered in Greek language (Single Dataset), the second one from the article in Greeks plus the translation in English (Mixed Dataset).

4.2 Data Preprocessing

The data included URLs, which have been automatically removed using an HTML tag identifier. As mentioned before, due to lack of availability of Greek preprocessing tools, only some base linguistic tools have been incorporated. The preprocessing phase consists of the following steps:

- Tokenization: To extract only the words, all stop-words have been removed, all letters lowercased.

- Translation of each Greek token using the Google® Translation API.
- Creation of a document-term matrix that contains n rows, where n equals the number of comments (n was 520 for our case, having 180 positive and 340 negative ones) and m rows where m corresponds to the size of the Greek vocabulary, almost double when applied the translation, since some tokens shared the same translation. In each cell of the document-term table, the value of the tf-idf weight of each term is contained, given by the following formula:

$$tf - idf(term) = frequency(term) \cdot log\left(\frac{m}{N(term)} + 1\right),$$ where m is the total

number of terms in the collection and N (term) is a function that returns the number of documents the term appears in.
- Dimensionality reduction using the Principal Component Analysis (PCA): This step applied only in cases we used Naïve Bayes, Decision Trees and Support Vector Machines (SVM) and not in GBM since it could cope with the large number of attributes.

PCA reduces the dimensionality by performing transformation of possibly correlated variables to a fully new dataset with linearly uncorrelated variables, called principal components. Every principal component has variance starting from the first one which has the largest value and this means the item has the biggest impact in dataset comparing the other principal components. All the principal components are the eigenvectors of the covariance matrix, that means are orthogonal.

4.3 Evaluation Criteria and Performance

For the experiments have been used three measures accuracy, precision and recall as described below:

$$\text{Accuracy:} \frac{T_p + T_n}{T_p + T_n + F_p + F_n}, \text{Precision:} \frac{T_p}{T_p + F_p}, \text{Recall:} \frac{T_p}{T_p + F_n},$$

where T_p, T_n (true positive, true negative) are the correct positive and correct negative predictions respectively and F_p, F_n (false positive, false negative) are the wrongly positive and wrongly negative predictions respectively. In order to evaluate the performance for every classifier we used the 10-fold cross validation approach. Figure 1 illustrates the methodology in 3 sub-processes. The first two are preprocess the data. The first one produce the Single dataset. The second one produce the translated dataset, adding the Single dataset with translated dataset produces the Mixed dataset. The third sub-process applies the classifier to the input dataset.

4.4 Classification Benchmark

In order to examine the performance of the proposed GBM method and evaluate it against other, well-known classifiers that have previously applied in sentiment analysis tasks with success, as explained in the related work section, the following classifiers

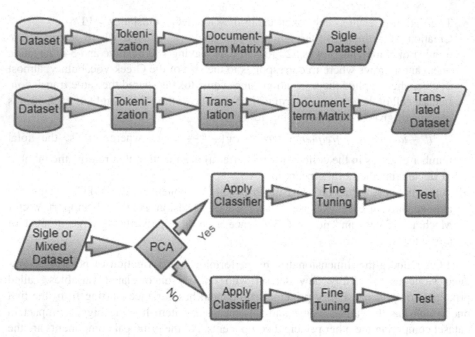

Fig. 1. Methodology flowchart

that have been additionally incorporated: Decision Trees, Naïve Bayes, SVM, Gradient Boosting Machine and Deep Neural Network Learning. In the case of Decision Trees, Naïve Bayes and SVM we have experimented using two different data representation techniques, namely the original tf-idf vector representation of the document-term matrix and the reduced dataset of the PCA transformation using the first 5 principal components.

4.5   Experimental Results

As previously explained, we incorporated two different datasets based on the same set of annotated documents. The first set, called as the **Single** dataset contained only the Greek tokens. The second set, called as the **Mixed** set, contained both Greek tokens and their translation. Finally each of the aforementioned datasets was also undertaken the PCA data dimensionality technique. For reasons of space, we provide detailed outcomes for the Mixed dataset and discuss what happens when considering the other option. As regards to the Decision Tree classifier without performing PCA technique in the Mixed Dataset the outcome is depicted on Table 1.

Table 1. Decision Trees performance on the Mixed Dataset.

Accuracy 60.62 %	Positive	Negative
Precision	42.00 %	64.07 %
Recall	17.80 %	85.64 %

When applied PCA on data the performance has been decreased by 10 %. In addition, when Decision Trees tried on the Single dataset the performance was decreased by 6 %. The reason of this is that the Mixed dataset is has more examples thus our model trained better. For Naïve Bayes classifier without performing PCA on the Mixed Dataset the outcome is depicted on Table 2.

Table 2. Naïve Bayes performance on the Mixed dataset.

Accuracy 81.25 %	Positive	Negative
Precision	82.80 %	81.94 %
Recall	65.25 %	92.08 %

When applied PCA on data the performance in accuracy has been decreased more than 35 %. In addition when Naïve Bayes tried on the Single dataset the performance was decreased by 5 %. For SVM classifier without performing PCA technique on the Mixed dataset the outcome is depicted on Table 3.

Table 3. SVM performance on the Mixed dataset

Accuracy 71.11 %	Positive	Negative
Precision	60.15 %	78.84 %
Recall	66.67 %	73.76 %

When applied PCA on data the performance in accuracy has been decreased by 20 %. In addition when SVM tried on the Single dataset the performance was decreased by 5 %. For the GBM classifier in the Mixed Dataset the outcome is depicted on Table 4.

Table 4. GBM performance on the Mixed dataset

Accuracy 87.26 %	Positive	Negative
Precision	83.19 %	89.66 %
Recall	82.50 %	90.10 %

In Single dataset the performance of GBM in accuracy is 7 % lower. For Deep Learning classifier in Mixed Dataset the outcome is depicted on Table 5.

Table 5. Deep Learning performance on the Mixed dataset

Accuracy 74.69 %	Positive	Negative
Precision	62.34 %	85.88 %
Recall	80.00 %	71.57 %

In Single dataset the performance in accuracy of Deep Learning is 5 % lower. The above tables shows that GBM classifier has the best accuracy, precision and recall comparing with the all of other classifiers we have used. The unique exception is in Naïve Bayes classifier recall in negatives in which the Naïve Bayes is 1 % higher. The comparison of GBM and Naïve Bayes it seems clear in the next diagrams which Illustrates the performance of GBM, Naïve Bayes and Naïve Bayes after preprocessing the corpus with PCA. These diagrams are separated by the performance in Positives on the left and Negatives observations on the right (Fig. 2).

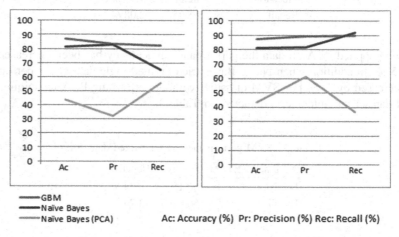

Fig. 2. Performance for GBM, Naïve Bayes and Naïve Bayes with PCA

In both of the cases GBM method slightly takes precedence over Naïve Bayes. In addition GBM algorithm seems to have stable in performance in both cases positives and Negatives on the Recall measure. That means Naïve Bayes in false negatives performed differently. The GBA classifier performs well even in cases we have two languages the Mixed Dataset. The difference is 7 %, in case the Naïve Bayes classifier the difference is at the same level 6 %.

5   Conclusions

This work dealt with the issue of performing sentiment analysis on Greek texts, obtained from Web 2.0 platforms using limited NLP resources. The novelty lied on the fact that instead of utilizing a manual polarity lexicon for Greek, which would have inadequate impact on social media linguistic style, we proposed a method that takes the translation of each token as additional input features. Even though this process may seem to bear additional effort and complexity to the majority of classification algorithms, the use of GBM, a robust boosting approach that can cope with high dimensional data appeared to be beneficial for the task at hand, outperforming a family of well-known methods for sentiment analysis. In the future we can try this methodology on imbalanced dataset to see the performance. On this benchmark we can include more classifiers like Recurrent Neural Networks.

References

1. Taylor, E.M., Rodríguez, O.C., Velásquez, J.D., Ghosh, G., Banerjee, S.: Web opinion mining and sentimental analysis. In: Velásquez, J.D., Palade, V., Jain, L.C. (eds.) Advanced Techniques in Web Intelligence-2. SCI, vol. 452, pp. 105–126. Springer, Heidelberg (2013)
2. Maynard, D., Bontcheva, K., Rout, D.: Challenges in developing opinion mining tools for social media. In: Proceedings of @NLP can u tag #usergeneratedcontent? Workshop at LREC 2012, Turkey (2012)
3. Natekin, A., Knoll, A.: Gradient boosting machines, a tutorial. Front. Neurorobotics 7, 21 (2011)
4. Shlens, J.: A tutorial on principal component analysis, derivation, discussion and singular value decomposition, 25 March 2003
5. Naaman, M., Boase, J., Lai, C.: Is it really about me? message content in social awareness streams. In: Proceedings of the 2010 ACM Conference on Computer Supported Cooperative Work, pp. 189–192. ACM (2010)
6. Bollen, J., Pepe, A., Mao, H.: Modeling public mood and emotion: twitter sentiment and socio-economic phenomena (2009). http://arxiv.org/abs/0911.1583
7. Jansen, B.J., Zhang, M., Sobel, K., Chowdury, A.: Twitter power: tweets as electronic word of mouth. JASIST 60(11), 2169–2188 (2009)
8. Pang, B., Lee, L.: Opinion mining and sentiment analysis. Found. Trends Inf. Retrieval 2(1–2), 1–135 (2008)
9. Taboada, M., Brooke, J., Tofiloski, M., Voll, K., Stede, M.: Lexicon-based methods for sentiment analysis. Comput. Linguist. 1, 1–41 (2011)
10. Boiy, E., Moens, M.-F.: A machine learning approach to sentiment analysis in multilingual web texts. Inf. Retrieval 12(5), 526–558 (2009)
11. Go, A., Bhayani, R., Huang, L.: Twitter sentiment classification using distant supervision. Technical report CS224N Project Report, Stanford University (2009)
12. Pak, A., Paroubek, P.: Twitter based system: using twitter for disambiguating sentiment ambiguous adjectives. In: Proceedings of the 5th International Workshop on Semantic Evaluation, pp. 436–439 (2010)
13. Zhang, L., Ghosh, R., Dekhil, M., Hsu, M., Liu, B.: Combining Lexicon-based and Learning-based Methods for Twitter Sentiment Analysis
14. Bengio, Y.: Learning deep architectures for AI. Found. Trends Mach. Learn. 2(1), 1–127 (2009)
15. Politopoulou, V., Maragoudakis, M.: On mining opinions from social media. In: Iliadis, L., Papadopoulos, H., Jayne, C. (eds.) EANN 2013, Part I. CCIS, vol. 383, pp. 474–484. Springer, Heidelberg (2013)
16. Maynard, D., Funk, A.: Automatic detection of political opinions in tweets. In: García-Castro, R., Fensel, D., Antoniou, G. (eds.) ESWC 2011. LNCS, vol. 7117, pp. 88–99. Springer, Heidelberg (2012)

Discovering Areas of Interest
Using a Semantic Geo-Clustering Approach

Evaggelos Spyrou[1,2]([✉]), Apostolos Psallas[2], Vasileios Charalampidis[2],
and Phivos Mylonas[3]

[1] Institute of Informatics and Telecommunications,
National Center for Scientific Research - "Demokritos", Athens, Greece
espyrou@iit.demokritos.gr
[2] Department of Computer Engineering,
Technological Educational Institute of Central Greece, Lamia, Greece
{apsallas,vcharalabidis}@teilam.gr
[3] Department of Informatics, Ionian University Corfu, Corfu, Greece
fmylonas@ionio.gr

Abstract. Living in the era of social networking, coupled together with great advances in digital multimedia user-generated content, motivated us to focus our research work on humanistic data generated by such activities towards new, more efficient ways of extracting semantically meaningful information in the process. More specifically, the herein proposed approach aims to extract areas of interest in urban areas, utilizing the increasing socially-generated knowledge from social networks. A part of the area of interest is selected, then split into "tiles" and processed with an iterative merging approach whose goal is to extract larger, "homogeneous" areas which are of special (e.g., tourist) interest. In this work generated areas of interest focus on interesting points from the humanistic point of view, thus covering in general main touristic attractions and places of interest. In order to achieve our goals, we exploit two types of metadata, namely location-based information (geo-tags) geo-tags and simple user-generated tags.

Keywords: Areas of interest · Semantics · Geo-clustering · Flickr

1 Introduction

The recent great emerge of social media and social activities, together with advances in digital multimedia user-generated content, shifted research interest to unprecedented domains, like the ones related to the acquisition of information and analysis of the online "footsteps" or presence of users. The latter is often used to produce semi-automatic knowledge about users' whereabouts, interests or even recommend them additional, semantically related information towards covering their information needs. In this framework photographs accompanied by useful metadata information, like tags and/or geo-tags, are considered to be

© IFIP International Federation for Information Processing 2016
Published by Springer International Publishing Switzerland 2016. All Rights Reserved
L. Iliadis and I. Maglogiannis (Eds.): AIAI 2016, IFIP AICT 475, pp. 490–498, 2016.
DOI: 10.1007/978-3-319-44944-9_43

the ideal source of information for the discovery of meaningful, popular trends with respect to users' behavior. More specifically, location-based info mined from such geo-tagged images offers a great opportunity to analyze users' preferences in their daily lives and complement the knowledge of their social activities through the utilization of associated tags. In this paper we present a novel approach that exploits both tags and geo-tags, towards the discovery of areas of interest.

Still, the above interpretation would be insufficient, in case it ignored the underlying semantics. By introducing a semantic geo-clustering approach we provide a novel analysis framework to merge meaningful areas of interest. In this manner we simultaneously take into consideration both location-based information in the sense of user transitions and user-location relations by incorporating respective semantic knowledge. The proposed method attempts to improve related supervised clustering approaches, by adding the inherent semantics of user tagged images derived from Flickr social network in order to enhance the precise establishment of the analysis classes. More specifically, the herein presented approach is evaluated using a large Flickr dataset consisting of approx. 80K geo-tagged images taken in Athens, Greece.

The rest of the paper is organized as follows: In next Sect. 2 we present relevant research efforts that also exploit user-generated geodata and metadata from Flickr. The proposed method is presented in Sect. 3. Then, in Sect. 4 we present early experimental results, along with the dataset used. Finally, we draw our conclusions and discuss plans for future work in Sect. 5.

2   Related Work

The motivation of this work is to ultimately "discover" large and somehow "homogeneous" areas of interest, by merging small geographic "tiles", based on sets of tags that have been added spontaneously by Flickr users. We feel that this work in novel and to the best of our knowledge there does not exist one to be compared with in terms of the produced results. However and since the aforementioned areas of interest are mainly constituted by tourist attractions (since tags have been harvested by touristic photos), it is related to research activities that aim to provide recommendations of places and/or trends using information directly from geo-tagged photos of Flickr. In general, metadata extracted from Flickr, with or without the aid of visual information have been extensively used in the literature for various research goals. A survey may be found in [8].

Since tags form the most "primitive" type of used-generated knowledge, they have been used in many research efforts. Tags, date information and geo-tags have been exploited by Chen and Roy [3], who used temporal and location distributions and photo visual similarity to extract mainly periodic events. Discovering trends for tourist attractions was the goal of Van Canneyt et al. [9], whose recommendation system adopted a probabilistic approach, ranking places of interest according to their popularity and user–related temporal information. Data clustering on geo-tagged photos was also the goal of Kisilevich et al. [5] who aimed to determine urban areas of interest by analyzed spatial and temporal distributions of metatdata so as to identify events and ranked places of interest. Cao et al. [2], proposed a tourism recommendation system. They used mean shift clustering and built a set of representative images and tags for each cluster. Ahern et al. [1] analyzed tags that have been collected from geo-tagged photos of a specific area, and upon a TF-IDF-based approach extracted a set of the most representative ones. Similarly, Serdyukov et al. [6] aimed to predict the location photos were taken, relying solely on textual tags.

3   Geo-Clustering Algorithm

In this Section we shall present in detail the proposed algorithm. It first divides a large region into square "tiles" (sub–regions), of small, fixed size, then adopts a graph-based representation to model connectivity of neighboring tiles, each described by a set of tags. Tiles are merged and upon an iterative process, a set of larger areas is determined within the initial region. At the following we will use "tile" and "sub-region" interchangeably.

3.1   Notation and Definitions

We first select a region from the urban area of interest. Then, we divide this region into sub-regions. Many approaches have been proposed for this, e.g., in our previous work [7] we used equally-sized, round, overlapping regions. However, this would imply that overlapped tiles would share descriptions (tags), which is not a desired property in the context of this work. Thus, herein we adopt a simpler square grid-based approach, since we focus on the description and merging of sub-regions, each having an empirically set, fixed width, W_R.

Now, let R denote a given region containing a set of photos P. Let also $R_{i,j}$ denote its tiles, each containing a set of photos $P_{i,j}$, thus $\bigcup_R P_{i,j} = P$, a set of tags $T_{i,j}$, containing all tags from photos in $P_{i,j}$ and a subset $D_{i,j}$ of $T_{i,j}$, which constitutes the tag-based region description. In the aforementioned grid, i and j denote the corresponding line and column.

Since we have adopted a square grid, the most intuitive approach is to use 4–connectivity, to define the set of the initial neighboring tiles $N_{i,j} = \{R_{i,j}^{\text{up}}, R_{i,j}^{\text{right}}, R_{i,j}^{\text{down}}, R_{i,j}^{\text{left}}\}$. Obviously, $R_{i,j}^{\text{up}} = R_{i-1,j}$, $R_{i,j}^{\text{right}} = R_{i,j+1}$, $R_{i,j}^{\text{down}} = R_{i+1,j}$ and $R_{i,j}^{\text{left}} = R_{i,j-1}$. Of course, when tiles are merged, the set of neighbors of the resulting sub–region is the intersection of neighbors of the initial tiles.

3.2   Region Description

For each tile, we exploit $T_{i,j}$ to create its semantic representation $D_{i,j}$. We expect that among the user-generated tags, we shall encounter some that describe it by means of locality (e.g., *Thiseio*) or landmark(s) (e.g., *Acropolis*). Even though users tend to add "personal" tags (e.g., a name), we expect that a subset of the most "popular" tags (i.e., selected by the majority of users) will be able to describe a tile in a discriminable way. Thus, for $R_{i,j}$ the region description $D_{i,j}^L$ is the set of the L most "popular" tags, where popularity is measured in terms of the number of users that have used a specific tag within it.

Algorithm 1. Semantic Geo–Clustering

Input: Set of tiles R, set of Descriptions D, Similarity thres. S, Description length L
Output: Final Set of merged regions R

1: **function** GEO_CLUSTERING(R, S, L)
2:     **for each** $R_{i,j} \in R$ **do**
3:         merge_flag $\leftarrow$ TRUE
4:         best_match_N $\leftarrow \emptyset$
5:         best_match $\leftarrow S$
6:         **while** merge_flag = TRUE **do**
7:             **for each** $N \in N_{i,j}$ **do**
8:                 **if** JACCARD($D_{i,j}, D_{i,j}^{N}, L$) $> S$ && JACCARD($D_{i,j}, D_{i,j}^{N}, L$) >best_match **then**
9:                     best_match $\leftarrow$ JACCARD($D_{i,j}, D_{i,j}^{N}, L$)
10:                     best_match_N $\leftarrow N_{ij}$
11:                 **end if**
12:             **end for**
13:             **if** best_match_N$\neq \emptyset$ **then**
14:                 $R_{ij} \leftarrow$ MERGE(R_{ij},best_match_N)
15:                 merge_flag$\leftarrow$ TRUE
16:             **else**
17:                 merge_flag$\leftarrow$ FALSE
18:             **end if**
19:         **end while**
20:     **end for**
21: **end function**

3.3 Region Merging

One of the challenges when comparing two sets is to select an appropriate (dis)similarity measure. Herein we use the Jaccard distance [4], which consists a well-known measure for comparing the similarity and diversity of sample sets. Jaccard similarity $J(A, B)$ between two sets A, B is given by

$$J(A, B) = \frac{|A \cap B|}{|A \cup B|} = \frac{|A \cap B|}{|A| + |B| - |A \cap B|} , \tag{1}$$

where in our case A, B are the sets of tags representing two tiles, extracted using the methodology described in Sect. 3.2. Using the aforementioned notation, tiles R_A, R_B with descriptions D_A, D_B are merged when (a) they are neighbors and (b) $J(D_A^L, D_B^L) > S$, where $S \in [0, 1]$ is a user–defined similarity threshold.

The merging process starts from $R_{1,1}$ and continues horizontally. The distance to all its neighbors is checked. It is merged with the tile whose similarity is the max among all those whose similarity is greater than S, if any. For a new tile, its description is calculated based on the union of the sets of tags and the process continues by checking the similarities to its neighbors. In case there does not exist a neighbor with similarity greater than S, the process continues with the next unmerged tile. A graphical example of the tile merging process is illustrated in Fig. 1. Semantic geo-clustering, Jaccard similarity and region merging are presented in pseudocode in Algorithms 1, 2 and 3, respectively.

4 Experiments

For the experimental evaluation of our approach we used an urban image dataset which consists of a total of $79,465$ photos collected from the center of the

494   E. Spyrou et al.

Algorithm 2. Jaccard Similarity

Input: Tile Description D_1, Tile Description D_2, Tile Description Length L
Output: Jaccard Similarity Measure J_S of Descriptions D_1, D_2

1: **function** JACCARD(D_1, D_2, L)
2:     $D_1 \leftarrow$ SORT(D_1) // sort tags based on number of users
3:     $D_1 \leftarrow$ TRIM(D_1, L) // keep first L tags
4:     $D_2 \leftarrow$ SORT(D_2)
5:     $D_2 \leftarrow$ TRIM(D_2, L)
6:     $J_S \leftarrow |D_1 \cap D_2| / (|D_1| + |D_2| - |D_1 \cap D_2|)$
7:     **return** J_S
8: **end function**

Algorithm 3. Region Merging

Input: Sub–region R_1, Sub–region R_2
Output: Merged sub region R_{new}

1: **function** MERGE(R_1, R_2)
2:     CREATE(R_{new}) // Create an "empty" new region
3:     $D_{new} \leftarrow$ DESCRIPTION(T_1, T_2) // Create description of new region, based on tags of R_1, R_2
4:     $N_{new} \leftarrow N_1 + N_2 - N_1 \cap N_2$ // Create set of neighbors of new region
5:     **for each** $N \in N_{new}$ **do**
6:         **if** $N.N_{ij} = R_1$ **or** $N.N_{ij} = R_2$ **then**
7:             $N.N_{ij} \leftarrow R_{new}$
8:         **end if**
9:     **end for**
10:    **return** R_{new}
11: **end function**

---

city of Athens, Greece. All these photos are geo-tagged, dated between January 2004–December 2015 and collected from Flickr using its public API[1]. More specifically we queried Flickr for a region covering what is in general considered to be the center of the Athens, (i.e., where the main touristic attractions are located) and retrieved all geo-tagged photos. This rectangular area is equal to $7.7 \, km^2$. Its Northern-Western and Southern-Eastern points have coordinates $(37.9836, 23.7153), (37.9643, 23.7541)$ respectively.

The collected photos have been captured by 5038 users of various nationalities, thus they contain tags of different languages. Although the majority of these tags is in English, we used the Google Translate API[2], in order to translate

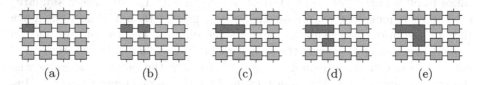

(a)   (b)   (c)   (d)   (e)

Fig. 1. Merging process: For a given tile (a), at a given step, one of its neighbors is considered as a candidate for merging (b). Their similarity is above the given threshold S, thus they are merged (c). At another step, a neighbor of the new tile is considered as a candidate for merging (d). Their similarity is above S, thus, they are merged (e).

---

[1] https://www.flickr.com/services/api/.
[2] https://cloud.google.com/translate/.

non-English tags (leaving English ones unchanged). This way, tags which would otherwise act as "noise", became of use. Additionally, we also created a manual stoplist, whose goal was to remove non–relevant (to our goals) tags. For example, many cameras and smartphones automatically add brand, model and settings; also tags such as *Greece* or *Hellas* or even *Athens* are both common and spread to the whole area, thus do not provide any useful information, while also tend to be amongst the most popular. In Fig. 2 we illustrate the sets of tags extracted from the tiles that correspond to the Panathenaic (Kallimarmaro) Stadium[3].

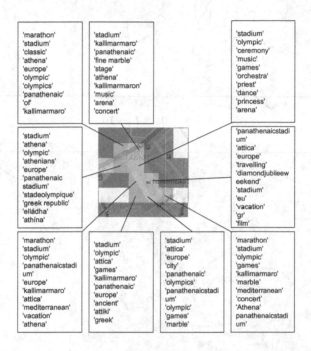

Fig. 2. Sets of tags extracted from merged tiles that correspond to the Panathenaic (Kallimarmaro) Stadium.

The entire region was divided into 770 square regions, of fixed width equal to 100 m. The similarity threshold S and the number L of tags to consider were empirically selected. More specifically, after considering several try-and-error rounds where the algorithm failed to produce satisfactory results, either by merging the majority of tiles into a single region, or by performing only a few merges, resulting to small and non-meaningful areas, we identified their optimal values. Furthermore, for the sake of fair evaluation, we constructed an empirical "ground truth" set of areas upon discussion with residents (which were not involved in this work). In the following we depict two early results of our algorithm compared to the constructed ground truth in Fig. 3, where

---

[3] https://en.wikipedia.org/wiki/Panathenaic_Stadium.

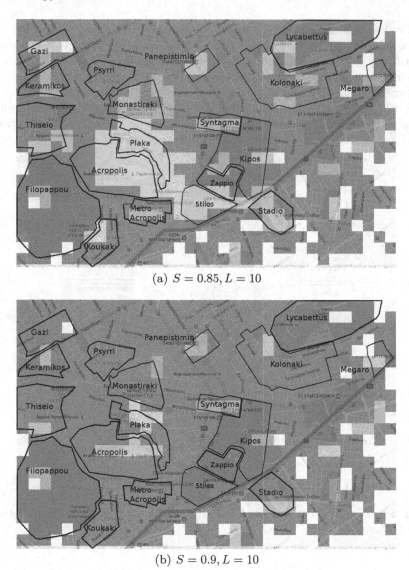

(a) $S = 0.85, L = 10$

(b) $S = 0.9, L = 10$

Fig. 3. Results compared to an empirically constructed ground truth set of areas.

we may observe that even a small increase in S may lead to a significantly improved qualitative result. Also, the result of Fig. 3(a) is quite close to the ground truth. We should note that at this early stage of research it is pointless to provide a more quantitative result, e.g., by measuring the coverage of the ground truth. Finally, it is also worth mentioning that the algorithm merges also some additional regions, irrespectively to the ground truth ones; the latter still produce a meaningful merging result, however not of touristic interest, like the ones on the bottom right of Fig. 3(a).

5   Conclusions and Discussion

In this paper we presented an approach which aimed to extract areas of interest in urban areas, using socially-generated knowledge from Flickr. We selected a part of the city of Athens, split it into "tiles" and proceeded with an iterative merging approach whose goal was to extract larger, "homogeneous" areas of the city which are of tourist interest. We showed that our approach succeeds when its parameters are selected appropriately. The generated areas cover in general the main touristic attractions and places of interest of the city center, and their boundaries often are close to those that a local resident may determine if asked. We should emphasize that we used almost "raw" tags, i.e., we did not use any "intelligent" technique to select/filter tags based on their relevance to the problem at hand, apart from the process described in Sect. 4.

In future work we plan to further improve the proposed algorithm and apply it into larger urban areas (e.g. the whole city of Athens and other major European cities). We also aim to perform an exhaustive evaluation, regarding the sizes of tiles, sets of tags to consider and similarity threshold and investigate whether a set of parameters shows satisfactory performance in more cities. We also wish to evaluate the results by real life users. A possible way is to consider the output of our system as a set of tourist recommendations and focus on user satisfaction. However, such an evaluation has been shown to be a difficult and expensive task, which may involve empirical issues in the process. Thus it should not also involve local residents but also visitors in both the construction of a "ground truth" and the assessment of the system's output.

References

1. Ahern, S., Naaman, M., Nair, R., Yang, J.H.-I.: World explorer: visualizing aggregate data from unstructured text in geo-referenced collections. In: Proceedings of the ACM/IEEE-CS JCDL (2007)
2. Cao, L., Luo, J., Gallagher, A., Jin, X., Han, J., Huang, T.S.: A worldwide tourism recommendation system based on geotagged web photos. In: Proceedings of IEEE ICASSP (2010)
3. Chen, L., Roy, A.: Event detection from Flickr data through wavelet-based spatial analysis. In: Proceedings of ACM CIKM (2009)
4. Duda, R.O., Hart, P.E., Stork, D.G.: Pattern Classification, 2nd edn. Wiley, New York (2001)
5. Kisilevich, S., Keim, D., Andrienko, N., Andrienko, G.: Towards acquisition of semantics of places and events by multi-perspective analysis of geotagged photo collections. In: Moore, A., Drecki, I. (eds.) Geospatial Visualisation. Lecture Notes in Geoinformation and Cartography, pp. 211–233, Springer, Heidelberg (2013)
6. Serdyukov, P., Murdock, V., Van Zwol, R.: Placing Flickr photos on a map. In: Proceedings of ACM SIGIR (2009)
7. Spyrou, E., Mylonas, P.: Analyzing Flickr metadata to extract location-based information and semantically organize its photo content. Neurocomputing **172**, 114–133 (2016). Elsevier

8. Spyrou, E., Mylonas, P.: A survey on Flickr multimedia research challenges. Eng. Appl. Artif. Intell. **51**, 71–91 (2016)
9. Van Canneyt, S., Schockaert, S., Van Laere, O., Bart, B.D.: Time-dependent recommendation of tourist attractions using Flickr. In: Proceedings of BNAIC (2011)

Diversifying the Legal Order

Marios Koniaris[1]([✉]), Ioannis Anagnostopoulos[2], and Yannis Vassiliou[1]

[1] KDBS Lab, School of ECE,
National Technical University of Athens, Athens, Greece
mkoniari@dblab.ece.ntua.gr
[2] Department of Computer Science and Biomedical Informatics,
University of Thessaly, Lamia, Greece

Abstract. "Public legal information from all countries and international institutions is part of the common heritage of humanity. Maximizing access to this information promotes justice and the rule of law." In accordance with the aforementioned declaration on Free Access to Law by Legal information institutes of the world (http://www.worldlii.org/worldlii/declaration/), a plethora of legal information is available through the Internet, while the provision of legal information has never before been easier. Given that law is accessed by a much wider group of people, the majority of whom are not legally trained or qualified, diversification techniques, should be employed in the context of legal information retrieval, as to increase user satisfaction. We address diversification of results in legal search by adopting several state of the art methods from the web search domain. We provide an exhaustive evaluation of the methods, using a standard data set from the Common Law domain that we subjectively annotated with relevance judgments for this purpose. Our results reveal that users receive broader insights across the results they get from a legal information retrieval system.

1 Introduction

Nowadays, as a consequence of many open data initiatives, more and more publicly available portals and datasets provide legal resources to citizens, researchers and legislation stakeholders. Thus, legal data that was previously available only on a specialized audience and in "closed" format is now freely available on the internet. Portals as the EUR-Lex[1], the European Union's database of regulations, the on-line version of the United States Code[2], United Kingdom[3], and the Australian[4], just to mention a few, serve as an endpoint to access millions of regulations, legislation, judicial cases, or administrative decisions. Such portals allow for multiple search facilities, as to assist users to find the information

---

[1] http://eur-lex.europa.eu/.
[2] http://uscode.house.gov/.
[3] http://www.legislation.gov.uk/.
[4] https://www.comlaw.gov.au/.

© IFIP International Federation for Information Processing 2016
Published by Springer International Publishing Switzerland 2016. All Rights Reserved
L. Iliadis and I. Maglogiannis (Eds.): AIAI 2016, IFIP AICT 475, pp. 499–509, 2016.
DOI: 10.1007/978-3-319-44944-9_44

they need. For instance the user can perform simple search operations or utilize predefined classificatory criteria (e.g. year, legal basis, subject matter) to find relevant to his/her information needs legal documents.

At the same time, however, the amount of Open Legal Data makes it difficult, both for legal professionals or the citizens to find relevant and useful legal resources. For example, it is extremely difficult to search for a relevant case law, by using boolean queries or the references contained in the judgment. Consider, for example, a patent lawyer who want to find patents as reference case and submits a user query to retrieve information. A diverse result, i.e. a result containing several claims, heterogeneous statutory requirements and conventions -varying in the numbers of inventors and other characteristics- is intuitively more informative than a set of homogeneous results that contain only patents with similar features. In this paper, we propose a novel way to efficiently and effectively handle similar challenges when seeking information in the legal domain.

Diversification is a method of improving user satisfaction by increasing the variety of information shown to user. As a consequence, the number of redundant items in a search result list should decrease, while the likelihood that a user will be satisfied with any of the displayed results should increase. There has been extensive work on query results diversification (see Sect. 2), where the key idea is to select a small set of results that are sufficiently dissimilar, according to an appropriate similarity metric.

Diversification techniques in legal information systems can be helpful not only for citizens but also for law issuers and other legal stakeholders in companies and large organizations. Having a big picture of diversified results, issuers can choose or properly adapt the legal regime that better fits their firms and capital needs, thus helping them operate more efficiently. In addition, such techniques can also help lawmakers, since deep understanding of legal diversification promotes evolution to better and fairer legal regulations for the society [3].

The objective of this paper is to define and evaluate the potential of results diversification in the field of legal information retrieval. To this end, we adopt various methods from the literature that are introduced for search result diversification [MMR [5], Max-Sum [12], Max-Min [12] and MonoObjective [12]]. We evaluate the performance of the above methods on a legal corpus subjectively annotated with relevance judgments, using metrics employed in TREC Diversity Tasks. To the best of our knowledge none of these methods were employed in the context of diversification in legal information retrieval and evaluated using diversity-aware evaluation metrics.

Our findings reveal that, diversification methods, employed in the context of legal IR, demonstrate notable improvements in terms of enriching search results with otherwise hidden aspects of the legal query space. Furthermore our qualitative analysis can provide helpful insights for legal IR systems, wishing to balance between reinforcing relevant documents, result set similarity, or sampling the information space around the query, result set diversity.

The remainder of this paper is organized as follows: Sect. 2 reviews previous work in query result diversification and in the field of legal text retrieval.

Section 3 introduces the concepts of search diversification and presents diversification algorithms, while Sect. 4 describes our experimental results and discuss their significance. Finally, we draw our conclusions and future work aspects in Sect. 5.

2   Related Work

In this section, we firstly present related work on query result diversification and then we focus on same issues in legal text retrieval techniques.

In order to satisfy a wide range of users, query results diversification has attracted a lot of attention in the field of text mining. The published literature on search result diversification is reviewed in [8]. The maximal marginal relevance criterion (MMR), presented in [5], is one of the earliest works on diversification and aims at maximizing relevance while minimizing similarity to higher ranked documents. Search results are re-ranked as the combination of two metrics, one measuring the similarity among documents and the other the similarity between documents and the query. In [12] a set of diversification axioms is introduced and it is proven that it is not possible for a diversification algorithm to satisfy all of them. Additionally, since there is no single objective function that is suitable for every application domain, the authors propose three diversification objectives, which we adopt in our work. These objectives differ in the level where the diversity is calculated, e.g. whether it is calculated per separate document or on the average of the currently selected documents.

In another approach, researchers utilized explicit knowledge as to diversify search results. [18] proposed a diversification framework, where the different aspects of a given query are represented in terms of sub-queries and documents are ranked based on their relevance to each sub-query. [1] propose a diversification objective that tries to maximize the likelihood of finding a relevant document in the top-k positions given the categorical information of the queries and documents. [14] organizes user intents in a hierarchical structure and proposes a diversification framework to explicitly leverage the hierarchical intent. The key difference between these works and the ones utilized in this paper is that we do not rely on external knowledge e.g. taxonomy, query logs to generate diverse results. Queries are rarely known in advance, thus probabilistic methods to compute external information are not only expensive to compute, but also have a specialized domain of applicability. Instead, we evaluate methods that rely only on implicit knowledge of the legal corpus utilized and on computed values, using similarity (relevance) and diversity functions (e.g., tf-idf cosine similarity) in the data domain.

In respect to legal text retrieval that traditionally relies on external knowledge sources, such as thesauri and classification schemes, various techniques are presented in [17]. Several supervised learning methods that have been proposed to classify sources of law according to legal concepts can be found in [4,13,15]. Legal document summarization techniques that scope to make the content of the legal documents, notably cases, more easily accessible are described in [9,10,16].

Finally, a similar approach with our work is described in [2], where the authors utilize information retrieval approaches to determine which sections within a bill tend to be outliers. However, our work differs in a sense that we maximize the diversify of the result set, rather than detect section outliers within a specific bill.

3   Legal Document Ranking Using Diversification

Here, we firstly provide an overview of general diversification processes focusing in the problem we address. Then, we define the ranking features and describe the diversification algorithms employed in this work.

3.1   Diversification Overview

Initially, the user submits his/her query as a way to express an information need and receives relevant documents. Diversification aims at finding a subset of those documents that maximize an objective function that quantifies the diversity of documents in S. More specifically, the problem is formalized as follows:

Definition 1 (Legal document diversification). *Let q be a user query and N a set of documents relevant to the user query. Find a subset $S \subseteq N$ of documents that maximize an objective function f that quantifies the diversity of documents in S.*

$$S = \underset{\substack{|S| = k \\ S \subseteq N}}{\operatorname{argmax}} f(N) \tag{1}$$

Typicaly, diversification techniques measure diversity in terms of content, where textual similarity between items is used in order to quantify information similarity. In the Vector Space model, each document u can be represented as a term vector $U = (is_{w1u}, is_{w2u}, ..., is_{wmu})^T$, where $w_1, w_2, ..., w_m$ are all the available terms, and is can be any popular indexing schema e.g. $tf, tf-idf, logtf-idf$. Queries are represented in the same manner as documents.

- **Document Similarity.** Various well-known functions from the literature (e.g. Jaccard, cosine similarity etc.) can be employed at computing the similarity of legal documents. In this work, we choose cosine similarity as a similarity measure, thus the similarity between documents u and v, with term vectors U and V is:

$$sim(u, v) = \cos(u, v) = \frac{U \cdot V}{\| U \| \| V \|} \tag{2}$$

- **Document Distance.** The distance of two documents is

$$d(u, v) = 1 - sim(u, v) \tag{3}$$

- **Query Document Similarity.** The relevance of a query q to a given document u can be assigned as the initial ranking score obtained from the IR system, or calculated using the similarity measure e.g. cosine similarity on the corresponding term vectors

$$r(q, u) = \cos(q, u) \tag{4}$$

3.2   Diversification Heuristics

Diversification methods usually retrieve a set of documents based on their relevance scores, and then re-rank the documents so that the top-ranked documents are diversified to cover more query subtopics. Since the problem of finding an optimum set of diversified documents is NP-hard, a greedy algorithm is often used to iteratively select the diversified set S. Let N the document set, $u, v \in N$, $r(q, u)$ the relevance of u to the query q, $d(u, v)$ the distance of u and v, $S \subseteq N$ with $|S| = k$ the number of documents to be collected and $\lambda \in [0..1]$ a parameter used for setting trade-off between relevance and similarity. In this paper, we focus on the following representative diversification methods:

- **MMR:** Maximal Marginal Relevance [5], a greedy method to combine query relevance and information novelty, iteratively constructs the result set S by selecting documents that maximizes the following objective function

$$f_{MMR}(u, q) = (1 - \lambda)\ r(u, q) + \lambda \sum_{v \in S} d(u, v) \qquad (5)$$

MMR incrementally computes the standard relevance-ranked list when the parameter $\lambda = 0$, and computes a maximal diversity ranking among the documents in N when $\lambda = 1$. For intermediate values of $\lambda \in [0..1]$, a linear combination of both criteria is optimized. The set S is usually initialized with the document that has the highest relevance to the query. Since the selection of the first element has a high impact on the quality of the result, MMR often fails to achieve optimum results.
- **MaxSum:** The Max-sum diversification objective function [12] aims at maximizing the sum of the relevance and diversity in the final result set. This is achieved by a greedy approximation algorithm that selects a pair of documents that maximizes Eq. 6 in each iteration.

$$f_{MAXSUM}(u, v, q) = (1 - \lambda)\ (r(u, q) + r(v, q)) + 2\lambda\ d(u, v) \qquad (6)$$

where (u, v) is a pair of documents, since this objective considers document pairs for insertion. When $|S|$ is odd, in the final phase of the algorithm an arbitrary element in N is chosen to be inserted in the result set S.
- **MaxMin:** The Max-Min diversification objective function [12] aims at maximizing the minimum relevance and dissimilarity of the selected set. This is achieved by a greedy approximation algorithm that select a document that maximizes Eq. 7 in each iteration.

$$f_{MAXMIN}(u, q) = (1 - \lambda)\ r(u, q) + \lambda \min_{v \in S} d(u, v) \qquad (7)$$

where $\min_{v \in S} d(u, v)$ is the minimum distance of u to the already selected documents in S.
- **MonoObjective:** MonoObjective [12] combines the relevance and the similarity values into a single value for each document. It is defined as:

$$f_{MONO}(u, q) = r(u, q) + \frac{\lambda}{|N| - 1} \sum_{v \in N} d(u, v) \qquad (8)$$

4   Experimental Setup

In this section, we describe the legal corpus we use, the set of query topics and the respective methodology for subjectively annotating with relevance judgments for each query, as well as the metrics employed for the evaluation assessment. Finally, we provide the results along with a short discussion.

4.1   Legal Corpus

Our corpus contains 3.890 Australian legal cases from the Federal Court of Australia[5]. The cases were originally downloaded from AustLII[6] and were used in [11] to experiment with automatic summarization and citation analysis. The legal corpus contains all cases from the Federal Court of Australia spanning from 2006 up to 2009. From the cases, we extracted all needed text for our diversification framework. Our index was built using standard stop word removal and porter stemming, with log based $tf - idf$ indexing technique, resulting in a total of 3.890 documents, 9.782.911 terms and 53.791 unique terms.

Table 1 summarizes testing parameters and their corresponding ranges. To obtain the candidate set N, for each query sample we keep the $top - n$ elements using cosine similarity and a log based $tf - idf$ indexing schema. Our experimental studies are performed in a two-fold strategy: (i) qualitative analysis in terms of diversification and precision of each employed method with respect to the optimal result set and (ii) scalability analysis of diversification methods when increasing the query parameters.

Table 1. Parameters tested in the experiments

Parameter	Range
Tradeoff l values	0.1, 0.2, 0.3, 0.4, 0.5, 0.6, 0.7, 0.8, 0.9
Candidate set size n = $|N|$	100
Result set size k = $|S|$	5, 10, 20
# of sample queries	298

4.2   Evaluation Metrics

We evaluate diversification methods using metrics employed in TREC Diversity Tasks[7]. In particular we report:

---

[5] http://www.fedcourt.gov.au.
[6] http://www.austlii.edu.au.
[7] http://trec.nist.gov/data/web10.html.

- **a-nDCG:** a-Normalized Discounted Cumulative Gain [7] metric quantifies the amount of unique aspects of the query q that are covered by the $top-k$ ranked documents. We use $a = 0.5$, as typical in TREC evaluation.
- **ERR-IA:** Expected Reciprocal Rank - Intent Aware [6] is based on interdependent ranking. The contribution of each document is based on the relevance of documents ranked above it. The discount function is therefore not just dependent on the rank but also on the relevance of previously ranked documents.
- **S-Recall:** Subtopic-Recall [19] quantifies the amount of unique aspects of the query q that are covered by the $top-k$ ranked documents

4.3   Relevance Judgements

As mentioned above, the evaluation of diversification requires a data corpus, a set of query topics and a set of relevance judgments, preferably made by human assessors for each query. In the absence of a standard dataset and since it was not feasible to involve legal experts in this study, we have employed an subjective way to annotate our corpus with relevance judgments for each query. To this end, we employed the following method:

User Profiles/Queries. We used the West Law Digest Topics[8] as candidate user queries. In other words, each topic was issued as candidate query to our retrieval system. Outlier queries, whether too specific/rare or too general, where removed using the interquartile range, below or above values $Q1$ and $Q3$, sequentially in terms of number of hits in the result set and score distribution for the hits, demanding in parallel a minimum cover of $min|N|$ results. In total, we kept 289 queries. Table 2 provides a sample of the topics we further consider as user queries.

Table 2. West Law Digest Topics as user queries

1:	Abandoned and lost property	3:	Abortion and birth control
24:	Aliens immigration and citizenship	88:	Compromise and settlement
291:	Privileged communications and confidentiality	363:	Threats stalking and harassment

Query assessments and ground-truth. For each topic/query we kept the $top-n$ results. An LDA topic model, using an open source implementation[9], was trained on the $top-n$ results for each query. Based on the resulting topic

---

[8] The West American Digest System is a taxonomy of identifying points of law from reported cases and organizing them by topic and key number. It is used to organize the entire body of American law.

[9] http://mallet.cs.umass.edu/.

distribution and with an acceptance threshold of 20 %, we can infer whether a document is relevant for an aspect. We have made available our complete dataset, ground-truth data, queries and relevance assessments in standard qrel format, as to enhance collaboration and contribution in respect to diversification issues in legal IR[10].

4.4 Results

As a baseline to compare diversification methods, we consider the simple ranking produced by cosine similarity and log based tf-idf indexing schema. The interpolation parameter $\lambda \in [0..1]$ is tuned in 0.1 steps separately for each method. Results are presented with fixed parameter $n = |N|$. Note that each of the diversification variations, is applied in combination with each of the diversification algorithms and for each user query.

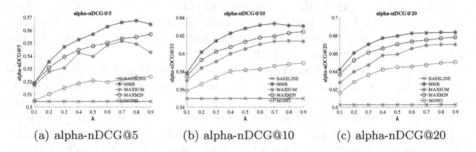

(a) alpha-nDCG@5   (b) alpha-nDCG@10   (c) alpha-nDCG@20

Fig. 1. alpha-nDCG at various levels @5, @10, @20 for baseline, MMR, MAXSUM, MAXMIN and MONO methods [Best viewed in color]

Figure 1 shows the a-nDCG of each method for different values of λ. Interestingly, all methods (MMR, MaxSum, MaxMin and Mono) outperformed the baseline ranking, while as λ increases, preference to diversity also increases for all methods. The trending behavior of MMR, MaxMin, and MaxSum is very similar especially at levels @10, and @20, while at level @5 MaxMin and Max-Sum presented nearly identical a-nDCG values in many λ values (e.g. 0.1, 0.2, 0.4, 0.6, 0.7). Finally, MMR constantly achieves better results in respect to the rest methods, while MaxMin and MaxSum follow. MONO despite the fact that performs better than the baseline in all λ values, still always presents the lower performance when compared to MMR, MaxMin, and MaxSum.

Figure 2 depicts the ERR-IA plots for each method in respect to different values of λ, while similarly Fig. 3 shows the Subtopic-Recall plots. It is clear that all of the approaches (MMR, MaxSum, MaxMin and Mono) tend to perform better than the selected baseline ranking method. Moreover, as λ increases, preference to diversity as well as Subtopic-Recall accuracy increases for all tested

---

[10] http://www.dbnet.ntua.gr/~mkoniari/LegalDiv.

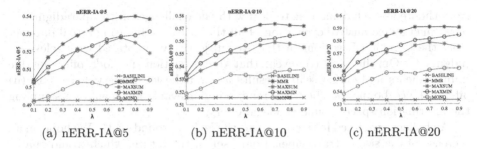

(a) nERR-IA@5   (b) nERR-IA@10   (c) nERR-IA@20

Fig. 2. nERR-IA at various levels @5, @10, @20 for baseline, MMR, MAXSUM, MAXMIN and MONO methods. [Best viewed in color]

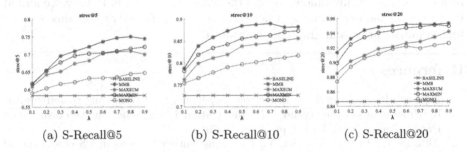

(a) S-Recall@5   (b) S-Recall@10   (c) S-Recall@20

Fig. 3. SubTopic Recall at various levels @5, @10, @20 for baseline, MMR, MAXSUM, MAXMIN and MONO methods. [Best viewed in color]

methods. We noticed a Similar trending behavior with the one discussed for Fig. 1. We also observed that MaxMin tends to perform better than MaxSum. There were few cases where both methods presented nearly similar performance especially in lower recall levels (e.g. for nERR-IA@5 when λ equals to 0.1, 0.4, 0.6, 0.7, and for S-Recall@5 when λ equals to 0.1, 0.2, 0.6, 0.7, 0.8). Once again, MONO presents the lower performance when compared to MMR, MaxMin, and MaxSum for both nERR-IA and S-Recall metric for all λ values applied.

In summary, among all the results, we note that the trends in the graphs look very similar. Clearly enough, the utilized diversification methods statistically significantly[11] outperform the baseline method, offering legislation stakeholders broader insights in respect to their information needs. Furthermore trends across the evaluation metric graphs, highlight balance boundaries for legal IR systems between reinforcing relevant documents or sampling the information space around the legal query.

5   Conclusions

In this paper, we studied the novel problem of diversifying legal search results. We adopted and compared the performance of several state of the art methods

---

[11] Statistical significance with the paired two-sided t-test ($p - value < 0.05$).

from the web search domain as to deal with the challenges in this paradigm. We performed an exhaustive evaluation of all the methods, by using a real data set from the Common Law domain that we subjectively annotated with relevance judgments. Our findings (i) reveal that diversification methods offer notable improvements and enrich search results around the legal query space and (ii) offer balance boundaries between reinforcing relevant documents or sampling the information space around the legal query.

A challenge we faced in this work was the lack of ground-truth. We hope on an increase of the size of truth-labeled data set in the future, which would enable us to draw further conclusions about the diversification techniques. We also plan to incorporate additional features in our legal search result diversification framework, specifically tailored across the legislation domain. Finally, we aim at investigating the performance of heuristics provided for other domains, e.g. for text summarization and graph diversification.

References

1. Agrawal, R., Gollapudi, S., Halverson, A., Ieong, S.: Diversifying search results. In: Proceedings of the Second ACM International Conference on Web Search and Data Mining - WSDM 2009, pp. 5–14 (2009)
2. Aktolga, E., Ros, I., Assogba, Y.: Detecting outlier sections in US congressional legislation. In: Proceedings of the 34th International ACM SIGIR Conference on Research and Development in Information Retrieval - SIGIR 2011, pp. 235–244 (2011)
3. Alces, K.A.: Legal diversification. Columbia Law Rev. **113**, 1977–2038 (2013)
4. Biagioli, C., Francesconi, E., Passerini, A., Montemagni, S., Soria, C.: Automatic semantics extraction in law documents. In: Proceedings of the 10th International Conference on Artificial Intelligence and Law - ICAIL 2005 (2005)
5. Carbonell, J., Goldstein, J.: The use of MMR, diversity-based reranking for reordering documents and producing summaries. In: Proceedings of the 21st Annual International ACM SIGIR Conference on Research and Development in Information Retrieval - SIGIR 1998, pp. 335–336 (1998)
6. Chapelle, O., Metlzer, D., Zhang, Y., Grinspan, P.: Expected reciprocal rank for graded relevance. In: Proceedings of the 18th ACM Conference on Information and Knowledge Management - CIKM 2009, pp. 621–630 (2009)
7. Clarke, C.L.A., Kolla, M., Cormack, G.V., Vechtomova, O., Ashkan, A., Büttcher, S., MacKinnon, I.: Novelty and diversity in information retrieval evaluation. In: Proceedings of the 31st Annual International ACM SIGIR Conference on Research and Development in Information Retrieval - SIGIR 2008 (2008)
8. Drosou, M., Pitoura, E.: Search result diversification. ACM SIGMOD Rec. **39**(1), 41 (2010)
9. Farzindar, A., Lapalme, G.: Legal text summarization by exploration of the thematic structures and argumentative roles. In: Text Summarization Branches Out Workshop Held in Conjunction with ACL, pp. 27–34 (2004)
10. Farzindar, A., Lapalme, G.: LetSum, an automatic legal text summarizing system. In: Legal Knowledge and Information Systems, JURIX 2004, pp. 11–18 (2004)
11. Galgani, F., Compton, P., Hoffmann, A.: Combining different summarization techniques for legal text. In: Proceedings of the Workshop on Innovative Hybrid Approaches to the Processing of Textual Data, pp. 115–123 (2012)

12. Gollapudi, S., Sharma, A.: An axiomatic approach for result diversification. In: Proceedings of the 18th International Conference on World wide web - WWW 2009, pp. 381–390 (2009)
13. Grabmair, M., Ashley, K.D., Chen, R., Sureshkumar, P., Wang, C., Nyberg, E., Walker, V.R.: Introducing luima. In: Proceedings of the 15th International Conference on Artificial Intelligence and Law - ICAIL 2015 (2015)
14. Hu, S., Dou, Z., Wang, X., Sakai, T., Wen, J.R.: Search result diversification based on hierarchical intents. In: Proceedings of the 24th ACM International on Conference on Information and Knowledge Management - CIKM 2015, pp. 63–72 (2015)
15. Loza Mencía, E., Fürnkranz, J.: Efficient pairwise multilabel classification for large-scale problems in the legal domain. In: Daelemans, W., Goethals, B., Morik, K. (eds.) ECML PKDD 2008, Part II. LNCS (LNAI), vol. 5212, pp. 50–65. Springer, Heidelberg (2008)
16. Moens, M.F.: Summarizing court decisions. Inf. Process. Manage. **43**(6), 1748–1764 (2007)
17. Moens, M.: Innovative techniques for legal text retrieval. Artif. Intell. Law **9**(1), 29–57 (2001)
18. Santos, R.L., Macdonald, C., Ounis, I.: Exploiting query reformulations for web search result diversification. In: Proceedings of the 19th International Conference on World Wide Web - WWW 2010, pp. 881–890 (2010)
19. Zhai, C.X., Cohen, W.W., Lafferty, J.: Beyond independent relevance. In: Proceedings of the 26th Annual International ACM SIGIR Conference on Research and Development in Informaion Retrieval - SIGIR 2003 (2003)

Efficient Computation of Clustered-Clumps in Degenerate Strings

Costas S. Iliopoulos, Ritu Kundu, and Manal Mohamed[⊠]

Department of Informatics, King's College London, London WC2R 2LS, UK
{costas.iliopoulos,ritu.kundu,manal.mohamed}@kcl.ac.uk

Abstract. Given a finite set of patterns, a clustered-clump is a maximal overlapping set of occurrences of such patterns. Several solutions have been presented for identifying clustered-clumps based on statistical, probabilistic, and most recently, formal language theory techniques. Here, motivated by applications in molecular biology and computer vision, we present efficient algorithms, using String Algorithm techniques, to identify clustered-clumps in a given text. The proposed algorithms compute in $\mathcal{O}(n + m)$ time the occurrences of all clustered-clumps for a given set of degenerate patterns $\tilde{\mathcal{P}}$ and/or degenerate text $\tilde{T}$ of total lengths m and n, respectively; such that the total number of non-solid symbols in $\tilde{\mathcal{P}}$ and $\tilde{T}$ is bounded by a fixed positive integer d.

Keywords: Conservative degenerate string · Pattern · Overlapping occurrences · Clustered-clump

1 Introduction

The ability to identify and compute various repeated patterns in strings is known to play a central role in many aspects of computer science fields including data compression, computer vision, computer-assisted music analysis and molecular biology. One of the most fundamental questions arising in such studies is to locate the *regions/windows* of overlapping occurrences of patterns in a given longer string named as text. This question is of particular interest in molecular biology, e.g. finding patterns with unexpectedly high or low frequencies and gene recognition.

In this paper, we consider a recently studied problem of computing clumps in a given text [2,10]. In particular, given a finite set of patterns $\mathcal{P}$, we compute all factors in a given text T such that each factor is composed of the maximal overlapping occurrences of patterns from $\mathcal{P}$; these will be referred to as clustered-clumps hereafter. Such findings may be utilised, for example, for gene prediction, that is to find genes within a genome, based on the occurrences of specific DNA sequence motifs before or after them. Examples of such motifs include gene

---

This research is partially supported by The Leverhulme Trust.

L. Iliadis and I. Maglogiannis (Eds.): AIAI 2016, IFIP AICT 475, pp. 510–519, 2016.
DOI: 10.1007/978-3-319-44944-9_45

promoters; start and stop codons; and poly(A) tails. An example of overlapping motifs, specifically, recognition sites to which proteins bind, is presented in [6].

In molecular biology (where sequences are considered as stings over fixed size alphabet Σ) if the specific nature of biological data is to be accommodated, it is required to allow some positions in the sequence to contain, instead of a single letter from Σ, a subset of Σ. Such *degenerate (indeterminate)* symbols can be interpreted as information that the exact letter at the given position is not known, but is suspected to be one of the specified letters.

Other than the aforementioned applications in genomics, identification of clustered-clumps in *degenerate* data also finds applications in areas such as computer vision or image processing. One such application can be the matching and retrieval of roughly aligned images containing the same scene, except nuances of some regions, and allowing for transformations like shifting, scaling, rotation etc.

Several solutions have been presented for identifying clustered-clumps based on statistical, probabilistic, and most recently, formal language theory techniques [2,10]. To the best of our knowledge, no solution heretofore explores the problem accounting for *degeneracy* in data. Here, we present efficient algorithms, using String Algorithm techniques, to identify clustered-clumps in a given text. Our solution considers *degenerate* strings arising from the nature of real data. The proposed algorithms compute in $\mathcal{O}(n + m)$ time the occurrences of all clustered-clumps for a given set of degenerate patterns $\tilde{\mathcal{P}}$ and/or degenerate text $\tilde{T}$ of total lengths m and n, respectively; such that the total number of non-solid symbols in $\tilde{\mathcal{P}}$ and $\tilde{T}$ is bounded by a fixed positive integer d.

The rest of the paper is organised in the following format: The next section introduces the vocabulary and the notions that will be used throughout paper. The algorithmic tools and data-structures required to build the solutions have been described in Sect. 3. Section 4 formally defines the problem and its variations along with presenting and analysing the algorithms. Finally, Sect. 5 concludes the paper.

2   Terminology and Technical Background

We begin with basic definitions and notation. We think of a *string* X of *length* n as an array $X[1 .. n]$, where every $X[i]$, $1 \leq i \leq n$, is a *letter* drawn from some fixed *alphabet* Σ of size $|\Sigma| = \mathcal{O}(1)$. The *empty string* is denoted by ε. The set of all strings over Σ (including the empty string ε) is denoted by Σ^*. A string Y is a *factor* of a string X if there exist two strings U and V, such that $X = UYV$. Hence, we say that there is an *occurrence* of Y in X, or, simply, that Y *occurs in* X. Consider the strings X, Y, U, and V, such that $X = UYV$. If $U = \varepsilon$, then Y is a *prefix* of X. If $V = \varepsilon$, then Y is a *suffix* of X.

A *degenerate symbol* $\tilde{\sigma}$ over an alphabet Σ is a non-empty subset of Σ, i.e., $\tilde{\sigma} \subseteq \Sigma$ and $\tilde{\sigma} \neq \emptyset$. $|\tilde{\sigma}|$ denotes the size of the set and we have $1 \leq |\tilde{\sigma}| \leq |\Sigma|$. A *degenerate string* is built over the potential $2^{|\Sigma|} - 1$ non-empty sets of letters belonging to Σ. In other words, a degenerate string $\tilde{X} = \tilde{X}[1 .. n]$, is a string such that every $\tilde{X}[i]$ is a degenerate symbol, $1 \leq i \leq n$. For example, $\tilde{X} = \{a,b\}\{a\}\{c\}\{b,c\}\{a\}\{a,b,c\}$ is a degenerate string of length 6 over $\Sigma = \{a,b,c\}$. If $|\tilde{X}[i]| = 1$, that is, $\tilde{X}[i]$ is a single letter of Σ, then we say that $\tilde{X}[i]$ is a *solid* symbol and i is a *solid position*. Otherwise, $\tilde{X}[i]$ and i are said to be a *non-solid symbol* and a *non-solid position*, respectively. For convenience, we often write $\tilde{X}[i] = \sigma$ ($\sigma \in \Sigma$), instead of $\tilde{X}[i] = \{\sigma\}$ in case of solid symbols. Consequently, the degenerate string $\tilde{X}$ mentioned previously will be written as $\tilde{X} = \{a,b\}ac\{b,c\}a\{a,b,c\}$. A string containing only solid symbols will be called a *solid string*. A *conservative degenerate string* is a degenerate string where its number of non-solid symbols is upper-bounded by a fixed positive constant.

For degenerate strings, the notion of symbol equality is extended to single-symbol *match* between two degenerate symbols in the following way. Two degenerate symbols $\tilde{\alpha}_1$ and $\tilde{\alpha}_2$ are said to *match* (represented as $\tilde{\alpha}_1 \approx \tilde{\alpha}_2$) if $\tilde{\alpha}_1 \cap \tilde{\alpha}_2 \neq \emptyset$. Extending this notion to degenerate strings, we say that two degenerate strings $\tilde{X}$ and $\tilde{Y}$ *match* (denoted as $\tilde{X} \approx \tilde{Y}$) if $|\tilde{X}| = |\tilde{Y}|$ and $\tilde{X}[i] \approx \tilde{Y}[i]$, for $i = 1, \cdots, |\tilde{X}|$. Note that the relation $\approx$ is not transitive. A degenerate string $\tilde{Y}$ is said to *occur* at position i in another degenerate (resp. solid) string $\tilde{X}$ (resp. X) if $\tilde{Y} \approx \tilde{X}[i..i+|\tilde{Y}|-1]$ (resp. $\tilde{Y} \approx X[i..i+|\tilde{Y}|-1]$). Note that for a fixed-sized alphabet, the matching relation can be implemented in $\mathcal{O}(1)$ time if degenerate symbols are represented by bit-vectors of size $|\Sigma|$.

A set of strings $\mathcal{P} = \{P_1, \cdots, P_r\}$ is *reduced* if no P_i is factor of a P_j with $i \neq j$. For instance, the set $\{aa,aba\}$ is reduced whereas the sets $\{aa,aab\}$, $\{aa,baa\}$, and $\{aa,baab\}$ are non-reduced.

In [1], a *clustered-clump* of a given reduced set of strings $\mathcal{P} = \{P_1, \cdots, P_r\}$, where each P_i of length at least 2, is defined as follows:

Definition 1 (Clustered-Clump). *A clustered-clump of a given reduced set of strings $\mathcal{P} = \{P_1, \cdots, P_r\}$ is a string W such that any two consecutive positions in W are covered by the same occurrence in W of a string $P \in \mathcal{P}$. The position i of the string W is covered by a string P if $P = W[j..j+|P|-1]$ for some $j \in \{1, \cdots, |W| - |P| + 1\}$ and $j \leq i \leq j + |P| - 1$. More formally, W is a clustered-clump for the set $\mathcal{P}$ such that*

$$\forall i \in \{1, \cdots, |W|\} \ \exists P \in \mathcal{P}, \exists j \in Pos_W(P) \text{ such that } j \leq i \leq j + |P| - 1,$$

where $Pos_W(P)$ is the set of positions of occurrences of P in W.

For a given text (string) T, a factor W is a clustered-clump if it is *maximal* in the sense that there exists no occurrence of the set $\mathcal{P}$ in T that overlaps W without being a factor of it.

Example 1. Consider the set $\mathcal{P} = \{aba, bba\}$ and the text $T = bbbabababababb$ bbabaababb, we have the following clumps underlined:

```
b  b b a b a b a b a b a b b  b b b a b a a b a b b
1        5            10             15          20
```

Notice that the factor ababa at position 6 is not a clustered-clump since it is not maximal. Also, the factor bbabaaba at position 15 does not form a single clustered-clump, because its two-letter factor aa is not covered by an occurrence of either aba or bba.

3   Algorithmic Tools

In the following we present two fundamental data structures supporting a wide variety of string matching algorithms. Both data structures are used in the proposed algorithms presented in Sect. 4.

Suffix Tree:

The *suffix tree* $\mathcal{S}(X)$ of a non-empty string X of length n is a compact trie representing all the suffixes of X such that $\mathcal{S}(X)$ has n leaves, labelled from 1 to n. Additionally, each edge is labelled with a factor of X. For any $i, 1 \leq i \leq n$, the concatenation of the edges' labels on the path from the root of $\mathcal{S}(X)$ to leaf i is precisely the suffix $X[i \mathbin{..} n]$. For any two suffixes $U = X[i \mathbin{..} n]$ and $V = X[j \mathbin{..} n]$ of X, if W is the longest common prefix of U and V, then the path in $\mathcal{S}(X)$ corresponding to W is the same for U and V. For a general introduction of suffix trees, see [3].

The construction of the suffix tree $\mathcal{S}(X)$ of the input string X takes $\mathcal{O}(n)$ time and space, for string over a fixed-sized alphabet [8,11,12]. Once the suffix tree of a given string (called text) has been constructed, it can be used to support queries that return the occurrences of a given string (called pattern) in time linear in the length of the pattern.

Aho-Corasick Automaton:

The *Aho-Corasick automaton* of a set of strings $\mathcal{P}$, denoted $\mathcal{A}(\mathcal{P})$, is the minimal partial deterministic finite automaton accepting the set of all strings having a string of $\mathcal{P}$ as a suffix (see [5, Sect. 7.1] for more description and for efficient construction); an example is given in Fig. 1. This data structure has an *initial* state, denoted s_0, and a *transition function* represented by the edges in the figure. A state is marked as terminal if the string it represents is in the set $\mathcal{P}$; note that all the leaves are terminal states. Let 'goto' denote the transition function, then the *suffix-link*, represented by the dotted line, is defined as follows: For a given non empty string X such that $s_i = \mathrm{goto}(s_0, X)$, the suffix-link of state s_i points at $s_j = \mathrm{goto}(s_0, X')$, where X' is the longest suffix of X such that $s_i \neq s_j$.

The construction of the suffix automaton $\mathcal{A}(\mathcal{P})$ together with the suffix-links can be done in linear time and space [3] independent of the alphabet size. Note that the transition function can be implemented in $\mathcal{O}(1)$ time for a fixed size alphabet.

Once the automaton $\mathcal{A}(\mathcal{P})$ has been constructed, searching a text T for occurrences of the patterns in $\mathcal{P}$ can be realized in time linear in the length of T; such a problem is known as the dictionary matching problem. The matching involves the Aho-Corasick automaton scanning the text, reading every letter exactly once. If the automaton is in state s_i and reads letter α of the text, it moves to state $s_j = \text{goto}(s_i, \alpha)$ if defined, otherwise, it moves to the nearest s_k such that $s_k = \text{goto}(s_j, \alpha)$ is defined and s_j is the state identified by the following of suffix-links starting from s_i. Also, if the automaton encounters a terminal state, it outputs an occurrence(s) of one or more patterns. Note that if $\mathcal{P}$ is reduced then at most one pattern from $\mathcal{P}$ occurs at each position of the text. In the rest of the paper, we assume that $\mathcal{P}$ is a reduced set.

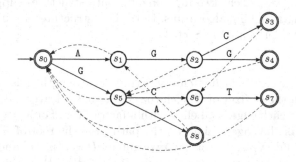

Fig. 1. Aho-Corasick automaton of set $\mathcal{P} = \{\text{AGC}, \text{AGG}, \text{GCT}, \text{GA}\}$.

4   Clustered-Clumps Algorithms

The Clustered-Clump problem is formally defined as follows:

FINDING CLUSTERED-CLUMPS
Input: A text T of length n and a set of patterns $\mathcal{P} = \{P_1, \cdots, P_r\}$, such that $m = \sum_{1 \leq i \leq r} |P_i|$.
Output: All clustered-clumps in T.

While the above problem can efficiently be solved using any standard dictionary matching algorithm, extending the definition to include degenerate strings makes the problem more interesting, challenging, and useful for practical applications. In the following, we reformulate the definition to generate three variations of the problem - the patterns are degenerate (but the text is solid), the text is degenerate (but the patterns are solid), and both the text and the patterns are degenerate. Further, we assume that the number of non-solid symbols in either the text or the set of patterns is bounded by a given constant, i.e. we will deal with conservative degenerate strings.

4.1   Problem 1: Solid Text and Degenerate Patterns

PROBLEM 1: FINDING CLUSTERED-CLUMPS IN SOLID TEXT GIVEN DEGEN-
ERATE PATTERNS
Input: A text T of length n, a set of conservative degenerate patterns
$\tilde{\mathcal{P}} = \{\tilde{P}_1, \cdots, \tilde{P}_r\}$, and integers d and m, such that total number of non-
solid symbols in $\tilde{\mathcal{P}} \leq d$, and $m = \sum_{1 \leq i \leq r} |\tilde{P}_i|$.
Output: All clustered-clumps in T.

Example 2. Consider the text $T = $ TTCGACTAACATAACGAAGCTAATCTTAAC and the
set of degenerate patterns $\tilde{\mathcal{P}} = \{\text{AC}\{\text{T},\text{G}\}\text{AA}\{\text{C},\text{G}\}\{\text{A},\text{C},\text{G}\}\text{TAA}, \text{AT}\{\text{C},\text{G}\}\text{TT}\}$, a
clustered-clump occurring at position 5 has been shown below:

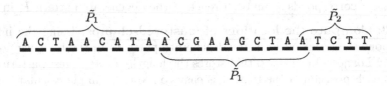

The solution we propose for this problem is based on the idea used in [9].
Each degenerate pattern $\tilde{P}_i \in \tilde{\mathcal{P}}$ can be seen as consisting of solid subpatterns,
interspersed with *non-solid regions*. Let $P_{i,j}$ be a solid subpattern of $\tilde{P}_i$, $1 \leq i \leq$
r, $1 \leq j \leq sub(i)$, where $sub(i)$ denotes the number of solid subpatterns in $\tilde{P}_i$.
Additionally, let $\Re^i_{j-1,j}$ represents a *non-solid region* between the subpatterns
$P_{i,j-1}$ and $P_{i,j}$. In other words, a pattern $\tilde{P}_i$ can be viewed as:

$$\tilde{P}_i = P_{i,1} \ \ \Re^i_{1,2} \ \ P_{i,2} \ \ \Re^i_{2,3} \ \ P_{i,3} \ \ .. \ \ P_{i,sub(i)}$$

Note that if the pattern $\tilde{P}_i$ ends with a non-solid symbol(s), then the last
non-solid region is represented as $\Re^i_{sub(i),\infty}$.

The following steps outline our solution:

Step 1: Split: In this step, each degenerate pattern in $\tilde{\mathcal{P}}$ is split into its component subpatterns; we call the set of all these solid subpatterns so obtained
$\mathcal{P}$. Effectively, we are breaking every degenerate pattern into subpatterns by
chopping out *non-solid regions* so that each of the subpatterns is solid.

Example 3. Suppose $\tilde{\mathcal{P}} = \{\text{AC}\{\text{T},\text{G}\}\text{AA}\{\text{C},\text{G}\}\{\text{A},\text{C},\text{G}\}\text{TAA}, \text{AT}\{\text{C},\text{G}\}\text{TT}\}$ as in
Example 2. Here $r = 2$. Splitting $\tilde{P}_1$ results in $P_{1,1} = \text{AC}$, $P_{1,2} = \text{AA}$, $P_{1,3} = \text{TAA}$,
while splitting $\tilde{P}_2$ results in $P_{2,1} = \text{AT}$ and $P_{2,2} = \text{TT}$. Then $\mathcal{P} = \{\text{AC}, \text{AA}, \text{TAA},$
$\text{AT}, \text{TT}\}$, $sub(1) = 3$, and $sub(2) = 2$. Here, $\Re^1_{1,2} = \{\text{T},\text{G}\}$, $\Re^1_{2,3} = \{\text{C},\text{G}\}\{\text{A},\text{C},\text{G}\}$,
and $\Re^2_{1,2} = \{\text{C},\text{G}\}$.

Step 2: Find occurrences of solid subpatterns in T: We next build the
Aho-Corasick automaton of the set $\mathcal{P}$; denoted by $\mathcal{A}(\mathcal{P})$. Using the automaton,
we compute all the occurrences of the solid subpatterns in the text T.

The occurrences of the subpatterns of the set $\mathcal{P}$ are maintained using a boolean matrix $Valid$ of size $|\mathcal{P}| \times n$ such that we can test in constant time whether or not a specific solid subpattern occurs at a given text position. If an occurrence of $P_{i,j}$ for which $\Re^i_{j-1,j}$ exists is found at a position (say k), then we need to check:

1. Whether $P_{i,j-1}$ (if $j > 1$) occurs in the corresponding position (i.e. $k - (|P_{i,j-1}| + |\Re^i_{j-1,j}|)$ in T.
2. Whether the non-solid symbols in $\Re^i_{j-1,j}$ match the corresponding positions in T. If $j = sub(i)$, then the non-solid region $\Re^i_{sub(i),\infty}$ is also tested for matching.

If both conditions are true, then an occurrence of $P_{i,j}$ is marked **true** in the matrix $Valid$. Notice that proceeding in this way, an occurrence marked **true** for $P_{i,sub(i)}$ corresponds to an occurrence of the degenerate pattern $\tilde{P}_i$ in T.

Step 3: Compute the locations of clustered-clumps: Using the information about the occurrences of the degenerate patterns in the text, we populate an array $LongestOcc$ of size n that stores the length of the longest pattern occurring at each position of the text. It is easy to see that simple calculations done in a single scan of this array can report the positions of all the clustered-clumps in T; see Function 1 below for more details.

Function 1 ComputeClusteredClumps($LongestOcc, n$)

    input : $LongestOcc$ is the array storing the length of the longest pattern
           occurring at each position of the text
           integer n represents length of the text.
    output: A set of all pairs (i, l) such that a maximal clustered clump of
           length l occurs at position i in the text.

    $\mathcal{R} \leftarrow \Phi$;
    $start \leftarrow last \leftarrow 1$;
    for $u \leftarrow 1$ **to** n **do**   // Scan $LongestOcc$
        if $LongestOcc[u] = 0$ **then**
            if $last < u$ **then**
                $start \leftarrow last \leftarrow u + 1$;
        else if $u + LongestOcc[u] - 1 > last$ **then**
            $last \leftarrow u + LongestOcc[u] - 1$;
        if $u = last$ **then**
            if $last - start > 0$ **then**
                Add $(start, last - start + 1)$ to $\mathcal{R}$;
            $start \leftarrow last \leftarrow last + 1$;
    return $\mathcal{R}$;

Running Time Analysis: Computing both $\mathcal{P}$ and $\mathcal{A}(\mathcal{P})$ takes $\mathcal{O}(m)$ time, while $\mathcal{O}(n)$ time is required for finding all the occurrences of all the solid subpatterns using the Aho-Corasick automaton. Subsequent symbol-by-symbol matching of *non-solid regions* is bounded by $\mathcal{O}(d)$ in the worst case for each position in the text, implying that overall $\mathcal{O}(dn)$ time is required. Computations in the last step can be done in $\mathcal{O}(n)$ time. Thus, the solution finds all the clustered-clumps in the text in $\mathcal{O}(n + m)$ time for constant d.

4.2   Problem 2: Degenerate Text and Solid Patterns

PROBLEM 2: FINDING CLUSTERED-CLUMPS IN DEGENERATE TEXT GIVEN SOLID PATTERNS

Input: A conservative degenerate text $\tilde{T}$ of length n, a set of patterns $\mathcal{P} = \{P_1, \cdots, P_r\}$, and integers d and m, such that the total number of non-solid symbols in $\tilde{T} \leq d$, and $m = \sum_{1 \leq i \leq r} |P_i|$.

Output: All clustered-clumps in $\tilde{T}$.

Example 4. Consider the text $T = \text{CATTA}\{A,G\}\text{GAGC}\{T,G\}\text{CTTTA}$ and the set of patterns $\tilde{\mathcal{P}} = \{\text{AGC}, \text{AGG}, \text{GCT}, \text{GA}\}$, a clustered-clump occurring at position 5 has been shown below:

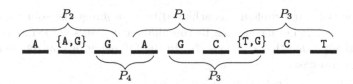

Our solution for Problem 2 is built using the recently developed algorithm described in [4] that solves, in linear time, pattern matching problem in conservative degenerate text (applying an adapted version of Landau and Vishkin's [7] algorithm for approximate pattern matching). Please refer to [4] for full details, but for the sake of completeness, a brief description has been provided in the following steps:

Step1: Substitute: In this step, each of the non-solid symbols occurring in the given degenerate text is replaced by a unique symbol which is not present in Σ. Let Λ be the set of these unique symbols i.e. $\Lambda = \{\lambda_i\}$ such that $0 < i \leq d$. It is to be noted that the text, T_λ, obtained by such a substitution will be a solid string. For example, if $T = \text{CATTA}\{A,G\}\text{GAGC}\{T,G\}\text{CTTTA}$ as in Example 4 then $T_\lambda = \text{CATTA}\lambda_1\text{GAGC}\lambda_2\text{CTTTA}$; here $\Lambda = \{\lambda_1, \lambda_2\}$.

Step 2: Find occurrences of solid patterns in T_λ: We concatenate the text T_λ and the patterns as follows:

$$\overline{T} = T_\lambda P_1 \#_1 .. P_r \#_r,$$

where each delimiting symbol $\#_i$, $1 \leq i \leq r$ is a unique symbol that is not present in $\Sigma \cup \Lambda$. Next, the suffix tree $\mathcal{S}(\overline{T})$ of $\overline{T}$ is constructed. As detailed in [4], checking whether a pattern P_i occurs at a certain position in T is realized by at most d longest common ancestor (LCA) queries on $\mathcal{S}(\overline{T})$.

Step 3: Compute the positions of clustered-clumps: We proceed in similar fashion as in Step 3 of Problem 1, reporting the positions of all clustered-clumps in $\tilde{T}$ using Function 1.

Running Time Analysis: The substitution and the concatenation steps to obtain T_λ and $\overline{T}$, and later constructing the suffix tree $\mathcal{S}(\overline{T})$ can be performed in $\mathcal{O}(n + m)$ time. Finding all the occurrences of all the patterns in $\mathcal{P}$ takes $\mathcal{O}(dn)$ [4]. Thus, for constant d, the solution computes all the clustered-clumps in the text in $\mathcal{O}(n + m)$ time.

4.3   Case 3: Degenerate Text and Degenerate Patterns

PROBLEM 3: FINDING CLUSTERED-CLUMPS IN DEGENERATE TEXT GIVEN DEGENERATE PATTERNS

Input: A conservative degenerate text $\tilde{T}$ of length n, a set of conservative degenerate patterns $\tilde{\mathcal{P}} = \{\tilde{P}_1, \cdots, \tilde{P}_r\}$, and integers d and m, such that total number of non-solid symbols in both $\tilde{T}$ and $\tilde{\mathcal{P}} \leq d$ and $m = \sum_{1 \leq i \leq r} |\tilde{P}_i|$.

Output: All clustered-clumps in $\tilde{T}$.

The solution for Problem 3 is achieved by combining the solutions for both Problems 1 and 2. Let d_T and $d_{\mathcal{P}}$ be the number of non-solid symbols in $\tilde{T}$ and $\tilde{\mathcal{P}}$, respectively, such that $d = d_T + d_{\mathcal{P}}$. In the following, we outline our solution for this general case:

Step1: Substitute: Replace each non-solid symbol in $\tilde{T}$ with a unique symbol $\lambda_i; 0 < i \leq d_T$, to obtain a solid text T_λ.

Step 2: Split: Split each of the degenerate patterns in $\tilde{\mathcal{P}}$ into its component subpatterns to obtain the set of solid subpatterns $\mathcal{P}$.

Step 3: Find occurrences of solid subpatterns: Construct the suffix tree $(\mathcal{S}(\overline{T}))$ for $\overline{T}$ (string obtained by appending T_λ with all the subpatterns in $\mathcal{P}$ delimited by unique symbols $\#_i$, $1 \leq i \leq d_{\mathcal{P}} + r$). Find all the occurrences of each of the subpatterns in $\mathcal{P}$ by LCA queries on $\mathcal{S}(\overline{T})$ (as in Step 2 of Problem 2). Similar to Step 2 in Problem 1, maintain a boolean matrix *Valid* such that an occurrence marked **true** for $P_{i,sub(i)}$ corresponds to an occurrence of the degenerate pattern $\tilde{P}_i$ in $\tilde{T}$.

Step 4: Compute the locations of clustered-clumps: Use Function 1 to report the positions of all clustered-clumps in $\tilde{T}$.

Running Time Analysis: As the solution makes use of the steps of the solutions for Problems 1 and 2, both of which have been shown to be bounded by $\mathcal{O}(n + m)$ time, thus, Problem 3 can overall be solved in $\mathcal{O}(n + m)$ time.

5   Conclusion

In this paper, we studied the problem of identifying clustered-clumps in conservative degenerate strings and presented $\mathcal{O}(n+m)$-time algorithms that compute the occurrences of all clustered-clumps for a given set of degenerate patterns $\tilde{\mathcal{P}}$ or/and a degenerate text $\tilde{T}$ of total lengths m and n, respectively; such that the total number of non-solid symbols in $\tilde{\mathcal{P}}$ and $\tilde{T}$ is bounded by a given constant d. The presented algorithms are promising for applications in genomics and computer vision. We intend to conduct larger-scale experiments, using genomic as well as digitized-images datasets. Furthermore, other domains that involve web-mining applications may find the presented solutions interesting and beneficial.

References

1. Bassino, F., Clément, J., Fayolle, J., Nicodème, P.: Constructions for clumps statistics. CoRR abs/0804.3671 (2008)
2. Boeva, V., Clément, J., Régnier, M., Vandenbogaert, M.: Assessing the significance of sets of words. In: Apostolico, A., Crochemore, M., Park, K. (eds.) CPM 2005. LNCS, vol. 3537, pp. 358–370. Springer, Heidelberg (2005)
3. Crochemore, M., Hancart, C., Lecroq, T.: Algorithms on Strings. Cambridge University Press, Cambridge (2007). p. 392
4. Crochemore, M., Iliopoulos, C.S., Kundu, R., Mohamed, M., Vayani, F.: Linear algorithm for conservative degenerate pattern matching. Eng. Appl. Artif. Intell. **51**, 109–114 (2016)
5. Crochemore, M., Rytter, W.: Text Algorithms. Oxford University Press, Oxford (1994)
6. Kvietikova, I., Wenger, R.H., Marti, H.H., Gassmann, M.: The transcription factors ATF-1 and CREB-1 bind constitutively to the hypoxia-inducible factor-1 (HIF-1) DNA recognition site. Nucleic Acids Res. **23**(22), 4542–4550 (1995)
7. Landau, G.M., Vishkin, U.: Fast parallel and serial approximate string matching. J. Algorithms **10**(2), 157–169 (1989)
8. McCreight, E.M.: A space-economical suffix tree construction algorithm. J. ACM (JACM) **23**(2), 262–272 (1976)
9. Rahman, M.S., Iliopoulos, C.S.: Pattern matching algorithms with don't cares. In: van Leeuwen, J., Italiano, G.F., van der Hoek, W., Meinel, C., Sack, H., Plasil, F., Bielikova, M. (eds.) SOFSEM 2007, pp. 116–126. Institute of Computer Science AS CR, Prague (2007)
10. Régnier, M.: A unified approach to word statistics. In: Proceedings of the Second Annual International Conference on Computational Molecular Biology, RECOMB, USA, pp. 207–213. ACM, New York (1998)
11. Ukkonen, E.: On-line construction of suffix trees. Algorithmica **14**(3), 249–260 (1995)
12. Weiner, P.: Linear pattern matching algorithms. In: Proceedings of the 14th IEEE Annual Symposium on Switching and Automata Theory, pp. 1–11. Institute of Electrical Electronics Engineer (1973)

Learning and Blending Harmonies in the Context of a Melodic Harmonisation Assistant

Maximos Kaliakatsos-Papakostas[1], Dimos Makris[2], Asterios Zacharakis[1],
Costas Tsougras[1(✉)], and Emilios Cambouropoulos[1]

[1] Department of Music Studies, Aristotle University of Thessaloniki,
54124 Thessaloniki, Greece
{maxk,aszachar,tsougras,emilios}@mus.auth.gr
[2] Department of Informatics Studies, Ionian University, 49100 Corfu, Greece
c12makr@ionio.gr

Abstract. How can harmony in diverse idioms be represented in a machine learning system and how can learned harmonic descriptions of two musical idioms be blended to create new ones? This paper presents a creative melodic harmonisation assistant that employs statistical learning to learn harmonies from human annotated data in practically any style, blends the harmonies of two user-selected idioms and harmonises user-input melodies. To this end, the category theory algorithmic framework for conceptual blending is utilised for blending chord transition of the input idioms, to generate an extended harmonic idiom that incorporates a creative combination of the two input ones with additional harmonic material. The results indicate that by learning from the annotated data, the presented harmoniser is able to express the harmonic character of diverse idioms in a creative manner, while the blended harmonies extrapolate the two input idioms, creating novel harmonic concepts.

1 Introduction

Machine learning allows a machine to acquire knowledge from data forming concrete conceptual spaces, while conceptual blending [10] between two input spaces allows new spaces to be constructed expressed as new structural relations or even new elements, creating new and potentially unforeseen output [27]. In music, harmony is an characteristic and well-circumscribed element of an idiom that can be learned from human annotated musical data using techniques such as Hidden Markov Models, N-grams, probabilistic grammars, inductive logic programming (see [6,7,12,15,17,20–23,25,26] among others). In the context of computational creativity in music, a challenging task tackled in the Concept Invention Theory (COINVENT) [3,18,24] project is to blend different/diverse input harmonic idioms learned from data to create new idioms that are creative supersets of the input ones.

The paper at hand briefly presents the extension of a melodic harmonisation assistant (introduced in [15]) that learns harmonic idioms by statistical learning on human data, for inventing new harmonic spaces by blending transitions between chords. The blended transitions are created by combining the features characterising pairs of transitions belonging to two idioms (expressed as sets of potentially learned transitions) according to an amalgam-based algorithm [5,9] that implements the theory presented in [10] for conceptual blending, through the categorical-based methodology presented in [11]. The transitions are then used in an extended harmonic space that accommodates the two initial harmonic spaces, linked with the new blended transitions.

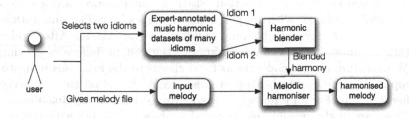

Fig. 1. System overview.

2   Statistical Learning of Harmonies from Human Annotated Datasets

Before blending harmonies, the system learns different aspects of harmony through annotated training data, while it produces new harmonisations according to guidelines provided in a melody input file given by the user. The system learns the available chord types within diverse dataset (according to their root notes) based on the *General Chord Type* (GCT) [2] representation, which can be used not only to represent but also to describe meaningful relations between harmonic labels [16] – even in non-tonal music idioms [1,14]. The training data include musical scores from many idioms (from Modal harmonisations in the Middle Ages to harmonisations of popular music and jazz in the 20th century), with expert annotations. Specifically, the notes of harmonic manually annotated reductions are regarded for the harmonic learning process, where only the most important harmonic notes are included; additional annotated layers of information are given regarding the tonality and the metric structure of each piece. Accordingly, the format of the user melody input file includes indications of several desired attributes that the resulting harmonisation should have.

After the system is trained, it is able to harmonise a user-given melody that, in this stage, includes manual annotations about harmonic rhythm, harmonically important notes, key and phrase structure. Additionally, the user has the freedom to choose specific chords at desired locations (constraint chords), forcing the system creatively to produce chord sequences that comply with the user-provided constraints, therefore allowing the user to 'manually' increase the interestingness of the produced output.

The cHMM [17] algorithm is used for modelling/learning chord progression probabilities for a given idiom. Then statistical information from the user-defined melody is combined with the chord progression model to generate chord progressions that best represent the idiom. Additionally the algorithm offers the possibility for prior determination of intermediate 'checkpoint' chords [4]). The fixed intermediate chords on the one hand help towards preserving some essence of higher level harmonic structure through the imposition of intermediate and final cadences, while on the other hand allow interactivity by enabling the user to place desired chord at any position. Statistics for cadences are learned during the training process, where expert annotated files including annotations for phrase endings are given as training material to the system. After collecting the statistics about cadences from all idioms, the system, before employing the cHMM algorithm, assigns cadences as fixed chords to the locations indicated by user input. The cadence to be imported is selected based on three criteria: (a) whether it is a final or an intermediate cadence; (b) the cadence likelihood (how often it occurs in the training pieces); and (c) how well it fits with the melody notes that are harmonised by the cadence's chords. Direct human intervention allows the user of the system to specify a harmonic 'spinal chord' of anchor chords that are afterwards connected by chord sequences that give stylistic reference to a learned idiom.

Regarding voice leading, experimental evaluation of methodologies that utilise statistical machine learning techniques demonstrated that an efficient way to harmonise a melody is to add the bass line first [26]. The presented harmoniser, having defined the optimal sequence of GCT chords, uses a modular methodology for determining the bass voice leading presented in [19], which utilises independently trained modules that include (a) a hidden Markov model (HMM) deciding for the bass contour (hidden states), given the melody contour (observations); (b) distributions on the distance between the bass and the melody voice; and (c) statistics regarding the inversions of the chords in the given chord sequence.

The bass voice motion provides abstract information about the motion of the bass, however, assigning actual pitches for a given set of GCT chords requires additional information: *inversions* and *melody-to-bass* distance distributions are also learned from data. The inversions of a chord play an important role in determining how eligible is each chord's pitch class to be a bass note, while the melody-to-bass distance captures statistics about the pitch height region that the bass voice is allowed to move according to the melody. After obtaining the exact bass pitch, the exact voicing layout, i.e. exact pitches for all chord notes, for each GCT chord is defined. To this end, a simple statistical model is utilised that finds the best combination of the intermediate voices for every chord according to some simple criteria. These criteria include proximity to a *pitch-attractor*, evenness of *neighbouring notes distances* and inner voice *movement distances between chords.* These criteria form an aggregate wighted sum that defines the optimal setting for all the intermediate notes (between the bass and the melody) in every GCT chord.

3   Blending Learned Harmonies

In the presented system, the harmony of an idiom is represented by first order Markov matrices, which include one respective row and column for each chord in the examined idiom. The probability value in the i-th row and the j-th column exhibits the probability of the i-th chord going to the j-th—the probabilities of each row sum to one. Figure 2(a) illustrates a grayscale interpretation of the transition in a set of major-mode Bach chorales. An important question is: *Given two input idioms as chord transition matrices, how would a blended idiom be expressed in terms of a transition matrix?* The idea examined in the present system is to create an *extended* transition matrix that includes new transitions that allow moving across chords of the initial idioms by potentially using new chords. The examined methodology uses *transition blending* to create new transitions that incorporate blended characteristics for creating a smooth 'morphing' harmonic effect when moving from chords of one space to chords of the other. An abstract illustration of an extended matrix is given in Fig. 2(b).

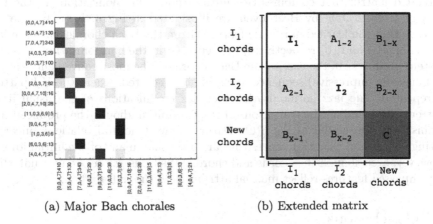

(a) Major Bach chorales   (b) Extended matrix

Fig. 2. Graphical description of (a) the transition matrix in a set of major-mode Bach chorales and (b) an *extended* matrix that includes transition probabilities of two initial idioms – like the ones depicted in (a) – and of several new transitions generated through transition blending.

In an extended matrix (Fig. 2), when using transitions in I_i only chords of the i-th idiom are used, while (blended) transitions in A_{i-j} create direct jumps from chords of the i-th to chords of the j-th idiom. Transitions in B_{i-x} constitute harmonic motions from a chord of idiom i to a newly created chord by blending. For moving from idiom i to idiom j using one external chord c_x that was produced by blending, a chain of two consecutive transitions is needed ($\mathtt{B_{i-x-j}}$): $c_i \rightarrow c_x$ followed by a transition $c_x \rightarrow c_j$, where c_i in idiom i and c_j in idiom j respectively. Transitions in $\mathtt{C}$ are disregarded since they incorporate pairs of chords that exist outside the i-th and j-th idioms.

Based on this analysis of the extended matrix, a methodology is proposed for using blends between transitions in $\mathtt{I_1}$ and $\mathtt{I_2}$. Thereby, transitions in $\mathtt{I_1}$ are blended with ones in $\mathtt{I_2}$ and a number of the best blends is stored for further investigation, creating a pool of best blends. A greater number of blends in the pool of best blends introduces a larger number of possible commuting paths in $\mathtt{A_{i-j}}$ or in $\mathtt{B_{i-x-j}}$. *Transition blending* is performed through *amalgam-based* conceptual blending that has already been applied to invent chord cadences [8,28]; in this setting, cadences are considered as special cases of chord transitions—pairs of chords, where the first chord is followed by the second one—that are described by means of features such as the roots or types of the involved chords, or intervals between voice motions, among others. For more information on transition blending the reader is referred to [13].

4   Harmonisation Examples

To briefly demonstrate the effect that transition blending has on forming the extended matrix that combines two initial idioms, harmonisations of the first part of '*Ode to Joy*' by Beethoven are illustrated in Fig. 3. Initially, one can observe that the learned harmonic features from the Bach chorales and the jazz sets (Fig. 3(a) and (b) respectively) are reflected in the harmonisations that the system produces. In the case of the Bach chorales, the most convenient (yet not so musically impressive) sequence of chords is generated, where the V-I pattern is repeated. The jazz harmonisation includes modifications of the usual ii-V-I pattern. By blending the transitions of the two initial idioms, the produced harmonisation (illustrated in Fig. 3(c)) features new structural relations, incorporating chords and chord sequences that are not usual in any of the initial idioms. However, even though the chords and chord transitions *per se* are unusual, their encompassed features reflect musical attributes of the initial idioms.

5   Conclusions

This paper describes a melodic harmonisation system which receives as inputs a melody file and two harmonic idioms and produces a melodic harmonisation with the blended harmony of two input idioms. To this end, diverse harmonic datasets were compiled and annotated by experts, while the harmonic description of each idiom was based on chords extracted with the General Chord Type (GCT) algorithm, statistical learning of chord progressions, cadences and chord voicing over these data. Blending of harmonies is performed through blending chord transitions (one chord leading to another) from the input idioms using an algorithmic realisation of conceptual blending based on category theory. Chord

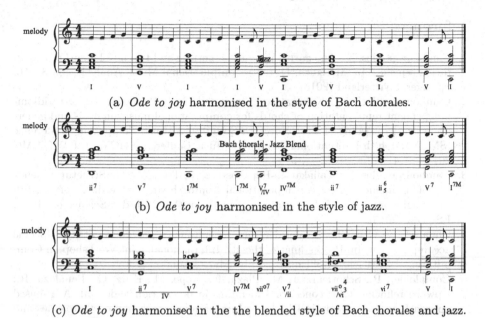

(a) *Ode to joy* harmonised in the style of Bach chorales.

(b) *Ode to joy* harmonised in the style of jazz.

(c) *Ode to joy* harmonised in the the blended style of Bach chorales and jazz.

Fig. 3. Beethoven's *Ode to joy* theme harmonised by the system with learned idioms and their blend.

transition blending combines features between pairs of transitions from the two input harmonic idioms, producing new transitions that potentially include new chords for both idioms and incorporate blended features. These new blended transitions act as connection points between the two input harmonic idioms, generating the *extended idiom* that is a blended harmonic superset of the two input ones.

A thorough experimental process that evaluates the usefulness of the produced harmonisations in real-world applications (e.g. when the system is used as an assistant for composers) is underway. Initial experimental results indicate that the blended melodic harmonisations are more interesting than the ones produced by using each input idiom separately. Additional experimental processes are expected to provide insights into whether the blended harmonic spaces are perceived as alterations of one of the input spaces (one-sided blends), balanced blends or radically new harmonic idioms, as well as into the role of the user-defined melody in the process of using blended or non-blended harmonising idioms.

Acknowledgments. This work is supported by the COINVENT project (FET-Open grant number: 611553).

References

1. Cambouropoulos, E.: The harmonic musical surface and two novel chord representation schemes. In: Meredith, D. (ed.) Computational Music Analysis, pp. 31–56. Springer, Switzerland (2015)
2. Cambouropoulos, E., Kaliakatsos-Papakostas, M., Tsougras, C.: An idiom-independent representation of chords for computational music analysis and generation. In: Proceeding of the Joint 11th Sound and Music Computing Conference (SMC) and 40th International Computer Music Conference (ICMC), ICMC-SMC 2014 (2014)
3. Cambouropoulos, E., Kaliakatsos-Papakostas, M., Tsougras, C.: Structural blending of harmonic spaces: a computational approach. In: Proceedings of the 9th Triennial Conference of the European Society for the Cognitive Science of Music, ESCOM (2015)
4. Chuan, C.H., Chew, E.: A hybrid system for automatic generation of style-specific accompaniment. In: Proceedings of the 4th International Joint Workshop on Computational Creativity. Goldsmiths, University of London (2007)
5. Confalonier, R., Schorlemmer, M., Plaza, E., Eppe, M., Kutz, O., Peñaloza, R.: Upward refinement for conceptual blending in description logic - an ASP-based approach and case study in EL^{++}. In: International Workshop on Ontologies and Logic Programming for Query Answering, International Joint Conference on Artificial Intelligence, IJCAI 2015 (2015)
6. Conklin, D.: Representation and discovery of vertical patterns in music. In: Anagnostopoulou, C., Ferrand, M., Smaill, A. (eds.) ICMAI 2002. LNCS (LNAI), vol. 2445, pp. 32–42. Springer, Heidelberg (2002)
7. Dixon, S., Mauch, M., Anglade, A.: Probabilistic and logic-based modelling of harmony. In: Ystad, S., Aramaki, M., Kronland-Martinet, R., Jensen, K. (eds.) CMMR 2010. LNCS, vol. 6684, pp. 1–19. Springer, Heidelberg (2011)
8. Eppe, M., Confalonier, R., Maclean, E., Kaliakatsos-Papakostas, M., Cambouropoulos, E., Schorlemmer, M., Codescu, M., Kühnberger, K.U.: Computational invention of cadences and chord progressions by conceptual chord-blending. In: International Joint Conference on Artificial Intelligence, IJCAI 2015. p. Submitted (2015)
9. Eppe, M., Maclean, E., Confalonieri, R., Kutz, O., Schorlemmer, M., Plaza, E.: ASP, amalgamation, and the conceptual blending workflow. In: Calimeri, F., Ianni, G., Truszczynski, M. (eds.) LPNMR 2015. LNCS, vol. 9345, pp. 309–316. Springer, Heidelberg (2015)
10. Fauconnier, G., Turner, M.: The Way We Think: Conceptual Blending and the Mind's Hidden Complexities. Basic Books, New York (2003). Reprint edition
11. Goguen, J.: Mathematical models of cognitive space and time. In: Andler, D., Ogawa, Y., Okada, M., Watanabe, S. (eds.) Reasoning and Cognition, Interdisciplinary Conference Series on Reasoning Studies, vol. 2. Keio University Press (2006)
12. Granroth-Wilding, M., Steedman, M.: A robust parser-interpreter for jazz chord sequences. J. New Music Res. 1–20 (2014)
13. Kaliakatsos-Papakostas, M., Confalonier, R., Corneli, J., Zacharakis, A., Cambouropoulos, E.: An argument-based creative assistant for harmonic blending. In: Proceedings of the 7th International Conference on Computational Creativity (ICCC) (2016, accepted)

14. Kaliakatsos-Papakostas, M., Katsiavalos, A., Tsougras, C., Cambouropoulos, E.: Harmony in the polyphonic songs of epirus: representation, statistical analysis and generation. In: 4th International Workshop on Folk Music Analysis, FMA 2014, June 2014

15. Kaliakatsos-Papakostas, M., Makris, D., Tsougras, C., Cambouropoulos, E.: Learning and creating novel harmonies in diverse musical idioms: an adaptive modular melodic harmonisation system. J. Creative Music Syst. 1(1) (2016, in press)

16. Kaliakatsos-Papakostas, M., Zacharakis, A., Tsougras, C., Cambouropoulos, E.: Evaluating the General Chord Type representation in tonal music and organising GCT chord labels in functional chord categories. In: Proceedings of the 4th International Conference on Music Information Retrieval, ISMIR 2015, Malaga, Spain (2015)

17. Kaliakatsos-Papakostas, M., Cambouropoulos, E.: Probabilistic harmonisation with fixed intermediate chord constraints. In: Proceeding of the Joint 11th Sound and Music Computing Conference (SMC) and 40th International Computer Music Conference (ICMC), ICMC-SMC 2014 (2014)

18. Kaliakatsos-Papakostas, M., Cambouropoulos, E., Kühnberger, K.U., Kutz, O., Smaill, A.: Concept invention and music: creating novel harmonies via conceptual blending. In: Proceedings of the 9th Conference on Interdisciplinary Musicology (CIM2014), CIM 2014, December 2014

19. Makris, D., Kaliakatsos-Papakostas, M., Cambouropoulos, E.: Probabilistic modular bass voice leading in melodic harmonisation. In: Proceedings of the 4th International Conference on Music Information Retrieval (ISMIR 2015). Malaga, Spain (2015)

20. Pérez-Sancho, C., Rizo, D., Inesta, J.M.: Genre classification using chords and stochastic language models. Connection Sci. 21(2-3), 145-159 (2009)

21. Raphael, C., Stoddard, J.: Functional harmonic analysis using probabilistic models. Comput. Music J. 28(3), 45-52 (2004)

22. Rohrmeier, M.: Towards a generative syntax of tonal harmony. J. Math. Music 5(1), 35-53 (2011)

23. Scholz, R., Vincent, E., Bimbot, F.: Robust modeling of musical chord sequences using probabilistic n-grams. In: IEEE 2009 International Conference on Acoustics, Speech and Signal Processing, ICASSP 2009, pp. 53-56. IEEE (2009)

24. Schorlemmer, M., Smaill, A., Kühnberger, K.U., Kutz, O., Colton, S., Cambouropoulos, E., Pease, A.: Coinvent: towards a computational concept invention theory. In: 5th International Conference on Computational Creativity, ICCC 2014, June 2014

25. Steedman, M.: The blues and the abstract truth: music and mental models. In: Mental Models in Cognitive Science, pp. 305-318 (1996)

26. Whorley, R.P., Wiggins, G.A., Rhodes, C., Pearce, M.T.: Multiple viewpoint systems: Time complexity and the construction of domains for complex musical viewpoints in the harmonisation problem. J. New Music Res. 42(3), 237-266 (2013)

27. Wiggins, G.A., Pearce, M.T., Müllensiefen, D.: Computational modeling of music cognition and musical creativity. In: The Oxford Handbook of Computer Music. OUP, USA (2009)

28. Zacharakis, A., Kaliakatsos-Papakostas, M., Cambouropoulos, E.: Conceptual blending in music cadences: a formal model and subjective evaluation. In: ISMIR, Malaga (2015)

Lyrics Mining for Music Meta-Data Estimation

Hasan Oğul[⊠] and Başar Kırmacı

Department of Computer Engineering,
Baskent University, Ankara, Turkey
hogul@baskent.edu.tr

Abstract. Music meta-data comprise a number of structured attributes that provide descriptive annotations such as singer, author, genre and date of a song deposited in a digital library. While they provide a crucial knowledge to represent music entry in current information retrieval and recommendation systems applications, they suffer from two limitations in practice. First, they may contain missing or wrong attributes due to incomplete submissions. Second, available attributes may not suffice to characterize the music entry for the objective of the retrieval or recommendation task being considered. Here, we offer an automated way of estimating the meta-data of a song using its lyrics content. We focus on attributing the author, genre and release date of songs solely based on the lyrics information. To this end, we introduce a complete text classification framework which takes raw lyrics data as input and report estimated meta-data attributes. The performance of the system is evaluated based on its retrieval ability on a large dataset of Turkish songs, which was gathered in this study and made publicly available. The results promote the use of such technique as a complementary tool in organizing music repositories and implementing music information retrieval systems.

Keywords: Music information retrieval · Song classification · Author attribution

1 Introduction

We have witnessed a drastic shift in promotion trends of music industry in response to changes in the way of people's music listening habits in recent years. Collective online shops and libraries have been more popular compared to individual album records. Consequently, the amount of music data available online has increased dramatically over the past few years. There is a pressing need to develop intelligent tools which will improve the usability of these data such that users can access, enjoy and communicate the content in a more effective and flexible way. This need has been further emphasized by the contribution of mobile devices in accessing online music content. The research effort into music information retrieval and recommendation systems has been therefore raised significantly to respond these challenges in the last decade (Orio 2006; Schedl et al. 2014).

Music is a multimodal object: it comprises an audio signal, song lyrics and other textual annotations that provide descriptive information about the entry, including e.g. song's singer, composer, author, genre, release date and social data. The latter is usually

© IFIP International Federation for Information Processing 2016
Published by Springer International Publishing Switzerland 2016. All Rights Reserved
L. Iliadis and I. Maglogiannis (Eds.): AIAI 2016, IFIP AICT 475, pp. 528–539, 2016.
DOI: 10.1007/978-3-319-44944-9_47

referred as meta-data and constitutes a brief yet helpful representation of music entry for accessing, searching or organizing the content in online media. Music information retrieval is a fruitful effort in more enjoyable access of music content in digital libraries. It usually requires an abstraction technique to represent the music objects and a model of comparing them to retrieve relevant music entries in available repositories. This abstraction task for music objects has been usually approached with meta-data or audio content (Moutselakis and Karakos 2009; Debaecker eta al. 2011). There are two major practical limitations with using meta-data approach. First, some attributes in the collection might be missing or incorrectly entered by database administrator or submitting user. Second, available attributes may not suffice to characterize the music entry for the objective of the task being considered. For example, if existing meta-data structure does not provide any attribute about the release date of a song, it is useless when a user wishes to retrieve similar songs from a particular period of time. Employing audio content is another widespread approach in extracting features for abstract representation of music (Costa et al. 2004; Sarkar and Saha 2015). Even though audio is the core component of a music entry, it has several drawbacks to serve as a feature generator to encode the music object. It contains a variety of signal layers corresponding to for example each instrumental sound, voice of a singer and background noise. Therefore, mining an audio content is a difficult task. There is indeed no well-established way of feature representation for a song from its audio content (Schedl et al. 2014).

Although general music perception is largely characterized by melodic and acoustic components comprised by audio content, the overall perception of non-instrumental song in fact can be elucidated only by considering all modalities including its lyrics. In spite of its great potential to characterize several cognitive concepts, research efforts on lyric-based music information retrieval and classification are very few. It was hypothesized that lyrics may contain lexical components that emphasize a certain mood and as such can be used to recognize the underlying mood (van Zaannen and Kanters 2010). In fact, the words such as "happy", "angry", "smile" and "dead" do not have to be spelled with a strong emotional voice or melody. In this respect, lyrics were used to classify the songs based on several mood categories such as "happy", "sad", "depressed" and "desire" (Hu et al. 2009). In some studies, a similar attempt was referred as lyric-based song sentiment classification, which seeks to assign songs appropriate sentiment labels such as light-hearted and heavy-hearted (Xia et al. 2008). Genre of a song was shown to be predictable by lyrics content (Mayer et al. 2008; Fell and Sporleder 2014). A feasibility study was presented to recognize the genre of a song from its lyrics written in Nordic language (Adriano et al. 2014), which is, to our knowledge, the only study that considers lyrics-based song classification in a language other than English.

In this study, we focus on estimating a set of meta-data attributes, i.e. author, genre and release date, solely from song lyrics to address the aforementioned challenges. There are a few attempts for genre or mood classification from lyrics in the literature. However, this study, to our knowledge, is the first attempt to predict the author and release date of a song from its lyrics, where some preliminary results have been presented in Kirmaci and Ogul (2015). We offer a number of novel features that are believed to be representative for song lyrics. We present the results of a comprehensive analysis conducted on a large dataset of Turkish songs, which was also collected in this study, on a rigorous experimental setup. Experimental results suggest the use of proposed technique as a complementary tool in music information retrieval applications.

2 Approach

2.1 Classification Framework

The task is to assign unknown lyrics content into known classes of different attributes. Here, we consider three meta-data attributes for a song: author, genre and release date. For each attribute, classes are restricted to a set of known labels. In a supervised approach, the framework learns a model to distinguish between classes for an attribute. This is achieved by feeding a learning classifier by a set of numeric values that represent the content of song objects in the training collection. Having a trained model, the prediction stage involves feeding the learned model with the representation of query song, which is encoded in the same way. The result is an estimation of class among the restricted label set of that attribute. There are two issues that should be dealt with in this framework: (1) which classification model will be learned, and (2) which features will be used to feed that classifier. Pertaining to its prior success in similar applications, we opted to use here a Naive Bayes classifier with a multinomial assumption in the distribution of data (Kibriya 2004).

Naive Bayes is a supervised classification technique based on Bayesian statistics. Bayesian statistics approach assumes an underlying probabilistic model and it allows user to capture uncertainty about the model in a principled way by determining probabilities of the outcomes. Parameter estimation for naive Bayes models is done using maximum likelihood. In Multinomial Naïve Bayes (MNB), a multinomial probability distribution of data is assumed.

Given a set of variables, $X = \{x_1, x_2, ..., x_d\}$, it is desired to construct the posterior probability for the class C_j among a set of possible outcomes $C = \{c_1, c_2, ... c_d\}$. In our case, X is the vector of features derived from the lyrics content and C is the set of classes present in the meta-data attribute being studied. Using Bayes' rule:

$$p(C_j | x_1, x_2, ..., x_d) = p(x_1, x_2, ..., x_d | C_j) p(C_j)$$

where $p(Cj | x_1, x_2, ..., x_d)$ is the posterior probability of class membership, i.e., the probability that the song X belongs to C_j. Since Naive Bayes assumes that the conditional probabilities of the independent variables are statistically independent we can decompose the likelihood to a product of terms:

$$p(X | C_j) = \prod_{k=1}^{d} p(x_k | C_j)$$

and rewrite the posterior as:

$$p(C_j | X) = p(C_j) \prod_{k=1}^{d} p(x_k | C_j)$$

Using Bayes' rule above, we label a new song X with a class label C_j that achieves the highest posterior probability. Although the assumption that the variables are independent is not always accurate, it can significantly simplify the classification task, since it allows the class conditional densities $p(x_k|C_j)$ to be calculated separately for each variable, i.e., it reduces a multidimensional task to a number of one-dimensional ones. Furthermore, the assumption does not largely affect the posterior probabilities, thus, it leaves the classification task unaffected.

2.2 Lyrics Features

A major concern in text classification is how to select the numerical features to be derived from lyrics text to represent the category that we seek to assign the song being queried. In fact, same representation is needed for all samples in a collection which will be used to train a supervised model. Since a song lyrics has a different characteristic than text content of a scientific article, a blog entry or a magazine news, typical feature representations used in previous text classification applications formerly discussed needs to be re-evaluated for the objective of song classification by text. Since lyrics may exhibit certain structures due to its specific parts such as chorus, bridge and verse, other features might be more relevant in retrieval studies.

In this study, we consider five feature sets applicable to song lyrics: bag-of-words, word N-grams, character N-grams, global text statistics and line length statistics. Simply concatenating some of these feature sets can provide fusion of several characteristics which might be useful in associating lyrics content with meta-data attributes.

Bag-of-words. Bag-of-words method is a classical yet powerful method for numerical representation of text objects. In this method, each unique word appeared in any document in the collection is considered as a separate feature. To encode a text entry, the information about the presence of each word is quantified to and used to fill out the feature vector indexed by these words. A simple way to quantify a word is to use a Boolean model, which only identifies whether the given word appears at least only once in the current document. Even if it is sometimes useful, a more sophisticated method has been used based on weighting each word and associating it with its frequency in the document. The method is called as term frequency-inverse document frequency (tf-idf), which considers both the frequency of terms in current documents and its occurrence statistics in other documents in the collection. Although using tf-idf has been shown to be more explanatory in pairwise comparison of documents in many cases, we opt to use only term frequency in our framework. The reason behind this choice is the assumption that the machine learning classifier that we use can already handle the statistics of each feature in other samples of the collection. That is, for example, a term will have no effect in separating hyper-plane of the trained model if it is present with a near frequency in all documents of the collection. Another crucial choice in implementing a bag-of-word method is whether to use a stemming for words or not. Stemming implies the preference of the use of original words without considering suffixes and reduces the size of feature space. We hypothesize here the using the frequencies of word roots as separate features but supporting this representation with the use of suffixes will present a more descriptive representation of author preferences. Hence, we opt to apply stemming in our bag-of-words representation. We implement word stemming by relevant tool in Zemberek, which is an open-source NLP library developed for Turkish language (Akin and Akin 2007).

Word N-grams. A word N-gram is the frequency of a phrase having N consecutive words. Concatenating the frequency of N-grams for all possible such phrases creates a word N-gram model for text representation. It is commonly used in general text classification applications and author attribution studies. Although it usually provides a complementary representation, it exponentially increases the size of feature space in terms of the value of N. Moreover, use of a large value of N makes the feature vector to have a too sparse content and pervert the learning model. Enlarging the dimension of feature vector with a sparse content will indeed increase the need for the number of samples in training data. Therefore, we determine to use only 2-grams to represent the preference of word phrase uses.

Character N-grams. Use of character frequencies in the form of N-grams (Cavnar and Trenkle 1994) has been shown to be helpful in several contexts (Iliev et al. 2014). A character N-gram refers to the presence of a string with a length of N characters in a text object. Similar to word N-gram model, a feature set of character N-grams contains the frequency of all possible strings having N characters from a finite alphabet. Character N-gram model does not entail that all such strings correspond to meaningful words in the language. Rather than providing a semantic value, it promises to infer lexical, grammatical or orthographic preferences without any linguistic background. In our case, these features may indirectly help in capturing the preferences in the use of similar rhymes in distinct genres or authors, since their frequency will be high. Another implicit value that character N-grams model maintains is the ability to measure the suffix composition in the text. This is particularly useful for agglutinating languages such as Turkish. In our framework, we built a feature set comprising character 2-grams, 3-grams and 4-grams since longer strings will be resulted with sparse and high dimensional vector content which is not usually desirable as discussed for word N-grams.

Global Text Statistics. In addition to vocabulary and language preferences, some other global indicators might be descriptive in representing lyrics content. In our framework, we consider four features in this set: total word count, total character count, average word length and number of unique words in the text. The first two features measure the tendency of telling something with long or short phrases. Third feature is related to the preference of using sophisticated and unordinary words, which are usually longer than daily conversation terms. Last feature is about the enrichment of vocabulary use.

Line Length Statistics. While previous feature sets are application to other text classification studies, we offer some new features only relevant to song lyrics. Lyrics have a certain characteristic of poetical design over a rhythm and melody. This leads to particular layout of lines and stanzas over the entire text, which is not observed in plain texts. Our observation is that the characteristics of this design exhibit a fronting diversity over authors, genres and even the periods of time when the song is released.

Therefore, we include three novel features in this set: average line length, standard deviation of line length and difference between lengths of longest and shortest lines.

3 Results

3.1 Data

We gathered 1048 Turkish song lyrics of 12 authors who have distinct styles. Each author often writes lyrics for the songs composed in one of three different genres: pop, rock and arabesk. The genres were selected based on their popularity in Turkey (Angi 2013). The authors included in the dataset are listed in Table 1 with the number of their song lyrics in the collection. In the resulting set, the number of authors in each genre is equal, and the number of samples is very close to each other in all categories. When assigning the songs into classes based on their release dates, we concerned with the issue of sociocultural changes in the country which are relevant to corresponding dates. This consideration resulted with the following three classes for the songs: released before 1993, released between 1994 and 2006, released after 2007. The number of samples in each class then became 293, 562, 193 respectively. The year 1993 is the date that Turkish pop music had a great explosion and highly popularized in comparison to arabesk music in especially urban populations. This resulted with a free-style content in both melody and lyrics compared to more conservative and artistic trends in older productions. In mid of 2000 s, several rock groups appeared in the country, which significantly affect the listening styles of new generation. The popularity of rock music affected the lyric styles in general. Instead of using words and phrases from daily conversations, songwriters were being enforced to creative linguistic designs in their lyrics.

Table 1. Data set content.

Author	Genre	Number of song lyrics
Sezen Aksu	Pop	124
Serdar Ortaç	Pop	134
Yaşar	Pop	49
Mustafa Sandal	Pop	58
Teoman	Rock	68
Haluk Levent	Rock	57
Barış Manço	Rock	76
Şebnem Ferah	Rock	65
Selami Şahin	Arabesk	92
Yıldız Tilbe	Arabesk	57
Ferdi Tayfur	Arabesk	177
Hakan Altun	Arabesk	91

3.2   Experiment Setup

We conducted classification experiments independently per meta-data category in a ten-fold cross-validation setup. In this setup, the dataset is divided into ten equal partitions such that each partition has a balanced number of samples from all categories. Each sample is then predicted using the classification models trained by other nine partitions which do not have the query sample. All samples are guaranteed to go through a prediction stage after repetition of same experiments ten times with a different training set in each. Proposed classification and feature selection methods were compiled with Weka, an open source data mining tool (Hall et al. 2009). In addition to traditional metrics such as recall, precision and F-measure, we determine the classification performance using ROC (Receiver Operating Characteristic) score. The ROC score is a metric often used to evaluate the performance of information retrieval systems. A score is computed for each category separately by the area under the corresponding ROC curve. Average of these scores is reported as a general ROC score. We used the ROC score to determine the ability of the feature representation schemes to infer the relevance of two songs in terms of meta-data attributes defined.

3.3   Empirical Results

We first report the effect of each feature set in meta-data estimation performance. To this end, we compiled MNB to predict author, genre, and release date with a variety of feature set combinations. Pertaining to the ROC scores obtained from cross-validation experiments, the combination of bag-of-words and character N-grams achieved the greatest performance in author prediction (Table 2). In genre and release date prediction, addition of global statistics and line length features improved the overall performance compared with the other combinations of feature sets. According to the results, word-N-grams are not descriptive enough for the song meta-data attributes studied as they can not provide an improvement in ROC scores when combined with bag-of-words features. For that reason, they were not incorporated into further feature set combinations.

Table 2. ROC scores of MNB classifier with varying feature sets with no stemming and no feature selection.

Feature sets					ROC		
Bag-of-words	Word N-grams	Character N-grams	Global text statistics	Line length statistics	Author	Genre	Date
√					0,839	0,822	0,650
√	√				0,811	0,820	0,621
√		√			0,862	0,865	0,665
√		√	√		0,856	0,868	0,664
√		√	√	√	0,860	0,873	0,674

To discern the effect of word stemming, we repeated the experiments with the last configuration in Table 2 but now using the words on which a stemming was not applied. Table 3 shows that stemming can promote the estimation performance to some degree. According to the results, author and date predictions can be improved by approximately 2 % when stemming is applied, however, genre prediction is not affected by this treatment. This slight improvement in performance indicates that language-specific information, when available, can improve the descriptive ability of representation scheme in retrieval task. In fact, since Turkish is an agglutinating language, stemming can be considered as an informative attempt to characterize the writing style of a Turkish author. On the other hand, positive contribution of character N-grams in prediction performance argues that use of suffixes is also an importance indicator in writing style. This justifies our hypothesis that using the frequencies of word roots as separate features but supporting this representation with the use of suffixes will present a more descriptive representation of author preferences.

Table 3. Effect of stemming in ROC score with best feature set and no feature selection

Category	Without stemming	With stemming
Author	0,841	0,860
Genre	0,875	0,873
Release date	0,654	0,674

Since the machine learning classifier used for training and prediction can sometimes significantly affect the general performance, we compared the results of multinomial Naive Bayes with some other methods. To justify the assumption of multinomial distribution of data, we repeated the experiments with Gaussian Naive Bayes. We also compiled two different versions of Support Vector Machine (SVM) classifier. SVM is a popular learning method, which has shown to be superior to other algorithms in several classification contexts. A key component in the success of SVM is the kernel function that maps the current data into a higher dimension to search for a hyperplane that can linearly separate the training samples. We opted to use 'linear' and 'radial basis function (RBF)' kernels with their default parameters in our comparative experiments. Table 4 illustrates that multinomial assumption is a reasonable decision since Gaussian NB performed very badly in predicting all attributes. Furthermore, multinomial NB significantly outperformed both SVM-based classifiers in all categorization schemes.

Table 4. Comparing classifiers in terms of ROC score

Category	Naive Bayes (Gaussian)	SVM (RBF)	SVM (Linear)	Naive Bayes (Multinomial)
Author	0,679	0,579	0,736	0,860
Genre	0,687	0,678	0,770	0,873
Release date	0,578	0,514	0,610	0,674

Overall performance of proposed framework is discerned in Table 5. The table demonstrates the recall, precision, F-measure and average ROC score evaluations for the best classifier configuration for each meta-data attribute. These results suggest that the model for genre classification from lyrics can be used in practical applications as such, pertaining to high scores in both recall and precision. Recall and precision scores for release date prediction are around 70 %. As the release date is actually a continuous variable, but here it was enforced to have categorical labels using artificial boundaries, the prediction results for these labels can be considered to be promising. A relatively high ROC score justifies this argument. According to the results, direct prediction of author from lyrics is not reliable enough when we base our assessment on recall and precision scores. On the other hand, a reasonable high ROC achieved in the experiments indicates that the model can successfully rank the candidate songs in relevant to their authors. This suggests that the knowledge contained in feature representation model can provide a good complement to music audio and other modalities in an information retrieval setup when author is an issue of relevance.

Table 5. Estimation performance of NMB with best classification setup

Category	Recall	Precision	F-measure	ROC
Author	0,49	0,60	0,44	0,862
Genre	0,82	0,82	0,82	0,925
Release date	0,72	0,71	0,71	0,818

We further analyzed the results of best configuration for individual class labels in corresponding meta-data attributes. The results in Table 6 indicate that the authors writing in the same genre category have similar preferences. For example, most confusing author for SA was SO, where both write songs in pop category. This observation was same for all pop music authors. While FT had the highest recall, other authors in Arabesk category were mostly confused with this author.

Table 6. Confusion matrix for author prediction with best classification setup

Predicted actual	SA	SO	Y	MS	T	HL	BM	ŞF	SŞ	YT	FT	HA
Sezen Aksu	78	20	0	0	2	0	5	4	2	0	12	1
Serdar Ortaç	1	110	0	0	1	0	0	1	0	0	19	2
Yaşar	17	10	1	0	1	0	1	0	0	0	18	1
M. Sandal	10	25	0	1	1	0	1	3	0	0	15	2
Teoman	9	7	0	0	47	0	0	3	0	0	2	0
Haluk Levent	17	6	0	0	2	8	4	2	2	0	15	1
Barış Manço	13	9	1	0	1	1	37	2	1	0	11	0
Ş. Ferah	16	10	0	0	5	0	0	30	1	0	3	0
Selami Şahin	6	10	0	0	0	0	1	1	20	0	50	4
Yıldız Tilbe	9	28	0	0	0	1	0	0	1	3	15	0
Ferdi Tayfur	4	9	1	0	0	0	2	0	2	1	158	0
Hakan Altun	3	21	0	0	2	0	0	0	0	0	43	22

According to the confusion matrix for genre classification (Table 7), arabesk is the most distinguishable category. The recall for arabesk classification is computed as 87.5 % while it is 76.7 % and 78.6 for pop and rock respectively. It is obvious that the lyrics in arabesk songs have special characteristics due to their dispirited emotion, while pop and rock songs may have versatile spiritual characteristics. On the other hand, highest precision was achieved for rock classification with 88.2 %. It can also be inferred from the confusion matrix that pop and arabesk songs are usually confused with each other but not with rock songs. This result might be attributable to the fact that rock authors tend toward using an original and more sophisticated vocabulary to create a narrative or legendary effect in their lyrics rather than putting toward a poetic sound. While this tendency differentiates their song from the other, dissimilarity of authors within same category makes them unrecognizable.

Table 7. Confusion matrix for genre prediction with best classification setup

Predicted actual	Pop	Rock	Arabesk
Pop	280	23	62
Rock	31	209	26
Arabesk	47	5	365

Evaluating the release date prediction performance is not an easy task since the determined boundaries for categorical labeling is intuitive. However, the confusion matrix in Table 8 clearly indicates that most recognizable category is the interval between 1993 and 2006. We can attribute this result to the free-style writing preferences of pop music authors in conjunction with the growth of pop music in popularity in young population in those years. This trend created an unrestricted grammatical structure and simpler vocabulary borrowed from daily conservations.

Table 8. Confusion matrix for release date prediction with best classification setup

Predicted actual	1972–1993	1994–2006	2007–2014
1972–1993	219	65	9
1994–2006	103	410	49
2007–2014	21	71	101

Table 9 lists top features selected by chi-square metric. According to the list, the most informative features for all attribute categories are those relevant to line length statistics. Global text statistics such as number of words, number of characters and number of unique words, stand for the second most contributing feature set. Among bag-of-words, the words "aşk" (the act of loving), "sevgi" (the act of liking), "vücut" (body), "kalp" (heart), "barış" (peace), "ümit" (hope), "bir" (a or one) and "yıl" (year) are the most representative. It seems that, in spite of their wide use, the frequencies of some common words such as "aşk", "sevgi" and "ümit" are effective in determining the writing preferences. Most informative character N-grams are "lar", "ik", "dim", "dü",

Table 9. Most descriptive features for each category

Author	Genre	Release date
Difference between longest and shortest line lengths	Number of characters	Difference between longest and shortest line lengths
Number of words	Difference between longest and shortest line lengths	Average line length
Number of characters	Number of unique words	Word frequency of "Aşk"
Number of unique words	Word frequency of "Bir"	Average line length
Word frequency of "Barış"	N-gram frequency of "he"	Word frequency of "Sevgi"
Word frequency of "Bir"	N-gram frequency of "da"	N-gram frequency of "urb"
Average line length	Average line length	Word frequency of "Kalp"
Word frequency of "Vücut"	N-gram frequency of "nda"	Word frequency of "Ümit"
N-gram frequency of "lar"	Word frequency of "Gönül	N-gram frequency of "ik"
Word frequency of "Aşk"	N-gram frequency of "lar"	N-gram frequency of "dim"
N-gram frequency of "her"	N-gram frequency of "dü"	Word frequency of "Barış"
N-gram frequency of "da"	N-gram frequency of "miş"	Word frequency of "Yıl"

"miş" and "da", which refer to suffix terms for making plural, past tense for first person plural, past tense for first person singular, past tense for third person singular, past perfect tense for third person singular, and stating location (e.g. "at", "in" or "on" in English) respectively. This indicates that character N-grams can essentially capture the distribution of certain suffixes, thus, determine the grammatical preferences in writing.

4   Conclusions

Analyzing lyrics data by quantitative techniques is a promising task to provide an effective means of organizing, accessing and manipulating music content in digital libraries. In this article, we present a comprehensive study to infer valuable knowledge from lyrics content using intelligent data analysis techniques. The contribution of the study is fourfold.

First, a complete framework is introduced for music libraries to estimate three different music meta-data attributes, i.e. author, genre, and release date, using only song lyrics content. To the best of our knowledge, this is the first attempt for predicting author and release date of a song from its lyrics.

Second, a number of relevant feature sets are evaluated and incorporated into a single feature encoding scheme to represent song lyrics. The results have shown that some particular features proposed in this study, such as the difference between the longest and shortest line lengths can help in predicting certain attributes. This new representation scheme can be used in several applications such as music classification, information retrieval and recommendation systems.

Third contribution is the large data set containing carefully annotated Turkish song lyrics, which is gathered and made publicly available in this study. The dataset is public accessible at www.baskent.edu.tr/~hogul/lyrics. We anticipate that researchers in information retrieval, NLP and music analysis communities will benefit from this data set in their future experimental studies.

Finally, lyrics classification task in general is addressed for the first time in the context of Turkish language. We have shown that incorporation of some language-specific pre-processing steps, such as stemming, can improve the classification and retrieval ability. Experimental results suggest the use of proposed models as complementary tools in music information retrieval applications not only for Turkish lyrics but also for songs from other languages.

References

Adriano, A., Rodrigo, M., Ribeiro, R.P., Silla, C.N.: Nordic music genre classification using song lyrics. In: Métais, E., Roche, M., Teisseire, M. (eds.) NLDB 2014. LNCS, vol. 8455, pp. 89–100. Springer, Heidelberg (2014)

Akın, M.D., Akın, A.A.: Türk dilleri için açık kaynaklı doğal dil işleme kütüphanesi: ZEMBEREK. Elektrik Mühendisliği **431**, 38 (2007)

Angi, C.: Müzik Kavramı ve Türkiye'de Dinlenen Bazı Müzik Türleri. Idil **2**, 59–81 (2013)

Cavnar, W.B., Trenkle J.M.: N-gram-based text categorization. In: 3rd Annual Symposium on Document Analysis and Information Retrieval, pp. 161–175 (1994)

Costa, C.H.L., Valle, J.D., Koerich A.L.: Automatic classification of audio data. In: IEEE International Conference on Systems, Man and Cybernetics (2004)

Debaecker, J., Widad, M.H.: Music indexing and retrieval: evaluating the social production of music metadata and its use. Facets of Knowledge Organization, 353–363 (2011)

Fell, M., Sporleder, C.: Lyrics-based analysis and classification of music. COLING **2014**, 620–631 (2014)

Hall, M., Frank, E., Holmes, G., Pfahringer, B., Reutemann, P., Witten, I.H.: The WEKA data mining software: an update. SIGKDD Explor. **11**, 10–18 (2009)

Iliev, R., Dehghani, M., Sagi, E.: Automated text analysis in psychology: methods, applications, and future developments. Lang. Cogn. **7**, 265–290 (2014)

Kibriya, A.M., Frank, E., Pfahringer, B., Holmes, G.: Multinomial naive bayes for text categorization revisited. In: Webb, G.I., Yu, X. (eds.) AI 2004. LNCS (LNAI), vol. 3339, pp. 488–499. Springer, Heidelberg (2004)

Kırmacı. B., Oğul, B.: Evaluating text features for lyrics-based songwriter prediction. In: IEEE 19th International Conference on Intelligent Engineering Systems (2015)

Mayer, R., Neumayer, R., Rauber, A.: Rhyme and style features for musical genre classification by song lyrics. In: 9th International Conference on Music Information Retrieval, pp. 337–342 (2008)

Moutselakis, E.V., Karakos, A.S.: Semantic web multimedia metadata retrieval: a music approach. In: 13th Panhellenic Conference on Informatics (2009)

Orio, N.: Music retrieval: a tutorial and review. Found. Trends Inf. Retrieval **1**, 1–90 (2006)

Sarkar, R., Saha, S.K.: Music genre classification using EMD and pitch based feature. In: Eighth International Conference on Advances in Pattern Recognition (2015)

Schedl, M., Gómez, E., Urbano, J.: Music information retrieval: recent developments and applications. Found. Trends Inf. Retrieval **8**, 127–261 (2014)

van Zaannen, M., Kanters, P.: Automatic mood classification using TF*IDF based on lyrics. In: 11th International Society for Music Information Retrieval Conference (2010)

Xia, Y., Wang, L., Wong, K.F., Xu, M.: Sentiment vector space model for lyric-based song sentiment classification. In: 46th Annual Meeting ACL HLT, pp. 133–136 (2008)

Mining Domain-Specific Design Patterns

Vassiliki Gkantouna[1]($\boxtimes$), Giannis Tzimas[2], Basil Tampakas[2], and John Tsaknakis[2]

[1] Department of Computer Engineering and Informatics, University of Patras, Patras, Greece
gkantoun@ceid.upatras.gr
[2] Computer and Informatics Engineering Department,
Technological Educational Institute of Western Greece, Patras, Greece
{tzimas,tampakas,jtsaknakis}@teiwest.gr

Abstract. Most catalogues of web design patterns contain patterns of general purpose, making it difficult for developers to properly apply them. This has led to the advent of domain-specific design patterns, encapsulating design experience which is in alignment with the natural constraints of a particular domain. Towards this end, we have developed a methodology which when applied on a collection of websites in a particular domain, leads to the automated identification of domain-specific design patterns. At the level of a single website, the methodology analyzes its conceptual model in terms of the incorporated recurrent patterns and evaluates their consistent use. The identified design patterns are stored in a central repository. By applying the methodology on a set of websites of the same application domain, we can populate a repository containing all the design patterns identified within the various websites designs, categorized towards various aspects such as the domain functionalities they perform. In this work, we focus on the domain of educational websites and present our preliminary results.

Keywords: Domain-specific · Design pattern · Web application · CMS · Design quality · Web design

1 Introduction

Web design patterns support developers facing the intrinsic complexity of web application development by providing them with proven solutions to recurring design problems that can be reused in different contexts where the correspondence problem arises. However, due to the fact that most design patterns are of general purpose (i.e., too abstract and divorced from the context in which sites are being developed), developers often find it difficult to properly use them. This fact has led to the advent of domain-specific design patterns, encapsulating design experience which is in alignment with the natural constraints of a particular application domain. They offer developers solutions to common design problems that arise particularly in a specific application domain, enabling them to produce quality designs.

Towards this end, we have developed a methodology which when applied on a collection of websites in a particular domain facilitates the automated identification of domain-specific design patterns. In the case of a single website, the methodology

© IFIP International Federation for Information Processing 2016
Published by Springer International Publishing Switzerland 2016. All Rights Reserved
L. Iliadis and I. Maglogiannis (Eds.): AIAI 2016, IFIP AICT 475, pp. 540–551, 2016.
DOI: 10.1007/978-3-319-44944-9_48

analyzes and inspects its conceptual model in terms of the incorporated design fragments that are used repeatedly for supporting the various functionalities within the application's context. At the conceptual level, we consider these fragments as recurrent patterns occurring in the application model, consisting of a configuration of front-end interface components that interrelate each other and interact with the end-user to achieve certain behavior or functionality. To be able to inspect the consistent use of these patterns, we also consider pattern variants. More specifically, we consider that a pattern consists of a core specification, i.e., an invariant composition of front-end design elements that characterizes the pattern and by a number of pattern variants which extend the core specification with all the valid modalities in which the pattern composition can start (starting variants) or terminate (termination variants). The proposed methodology automatically extracts the conceptual model of a web application and subsequently performs a pattern-based analysis of it in order to identify the occurrences of all the incorporated recurrent patterns. Then, to verify that the identified patterns actually perform certain functionality, we additionally inspect their occurrences to examine whether the recurrence of the design elements at the hypertext design goes with a recurrence at the data level, i.e. the content they deliver to the end-users. This is done by utilizing a semantic similarity measurement technique. Finally, we calculate evaluation metrics on them, revealing whether these patterns are used consistently throughout the entire application design and store them on a central patterns repository.

To be able to mine domain-specific design patterns, the methodology must be applied on a collection of websites that belong to the same application domain. By applying the methodology on the websites, we can detect and store the design patterns used in the various websites designs in one single central repository. This way, we can identify the most typical design patterns that are commonly used by developers in this specific application domain for achieving certain application behavior or communication effects such as the checkout process in e-commerce applications. Furthermore, we can also categorize the identified patterns towards various aspect such as the application functionality they perform. In the context of this work, we focus on the domain of educational websites and present a number of design patterns that we have identified for this domain. These patterns can assist developers produce more consistent and predictable educational website designs, meeting users' expectations about the system and improving the ease of use of the application.

In order to automate the process of analyzing the design of a web application, we have to narrow down the methodology's scope to the domain of Content Management Systems (CMSs), since they provide a common base of source code which can be systematically processed. The remaining of this paper is organized as follows: Sect. 2 provides an overview of the related work. Section 3 presents in detail the methodology for analyzing and evaluating the conceptual model of a Web application, while Sect. 4 presents some examples of domain-specific design patterns for educational websites. Finally, Sect. 5 discusses conclusions and future work.

2 Related Work

Research in the field of domain-specific design patterns is not mature. In [7], authors present an ontology-based approach that allows designers and domain experts to enrich their domain-specific patterns with semantic annotations using their domain concepts and terms. The representation framework keeps the pattern template and the domain specific knowledge separate, since each domain depending on its needs uses more or less fields to describe its patterns and the later ontology changes according to the domain knowledge that the pattern captures. In [8], authors present an approach of using design patterns at PIM (platform independent model) level, which combines the benefits of model-driven software engineering and design patterns directly for the domain expert. In [9, 10], authors present an integrated pattern definition process that supports both the top-down and bottom-up design approaches. It is distinguished from existing methods in two ways: (i) it is based on UML-profile language that visually distinguishes between the fundamental and variable elements of a pattern, and (ii) it defines the different steps that must be taken to obtain a domain-specific design pattern as well as a precise set of unification rules that identifies the commonalities and differences between applications in the domain. In all the aforementioned works, the domain-specific design patterns are manually devised by human experts after analyzing the designs of successful applications. The key difference of our approach is that we provide a methodology for the automated identification of the recurrent design patterns lying in the designs of the websites in a specific application domain. We can detect even patterns which may be hidden in a particular instantiation of a design problem in a specific website design, making it hard even for experienced designers to recognize them and come up with reusable design examples. By applying the methodology on their websites, developers can compare their designs, the design patterns they have used for supporting certain application functionalities, and see how their design choices differentiate from the most common design patterns used in an application domain.

3 Methodology Overview

In this section, before presenting the design patterns that we have identified for the educational websites domain, we present a brief overview of the methodology that we have used to detect them. At the level of a single website, the application of the methodology results in the automated identification and evaluation (towards consistency) of all the recurrent design patterns lying within its conceptual model. It is comprised by three main phases (Fig. 1). First, we extract the conceptual model of the website, at hypertext level, which is then submitted to a pattern-based analysis with the aim of (i) identifying the occurrences of all the recurrent patterns lying within it, (ii) verifying which of them can actually be considered as design patterns (i.e. support the realization of a certain functionality). Then, we calculate a set of evaluation metrics to assess if they are used consistently throughout the application model. Finally, the identified design patterns are stored in a repository available at [1]. A detailed description of the methodology can be found in [2].

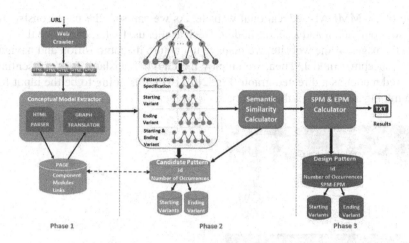

Fig. 1. The tree main phases of the methodology.

3.1 Phase 1: Extracting the Conceptual Model

At the hypertext level, the conceptual model of a web application specifies its composition and navigation, i.e. the organization of its front-end interfaces in terms of pages, made of design elements which are linked to support the user's navigation and interaction. Thus, the main task for automatically extracting the conceptual model of a website is to identify the organization of the front-end design elements that compose the hypertext of its HTML pages.

In the context of a Joomla!-based website, the hypertext of its pages is composed by assembling a number of predefined structural and navigational design elements which are called components and modules. A page is composed by one component, specifying the organization of content in its main part, and by a set of modules, specifying the organization of content in the peripheral positions. There is a variety of categories for components and modules, each one providing various types for interacting with the system (such as forms, confirmation buttons, etc.) and supporting alternative ways of arranging the content delivered to the end-users (e.g., blogs, lists, etc.). The content that they publish is extracted from the tables of the website's underlying database. To specify the hypertext organization for all the pages of a website, we have developed a set of tools as depicted in the first phase of Fig. 1.

Given the URL of a Joomla!-based website, the Web Crawler crawls its pages which are then parsed by the Conceptual Model Extractor to identify their organization as a set of components and modules. In the HTML code of a page, components and modules can be found as < div > elements having an HTML class attribute value (i.e. < div class = "value" >) which characterizes them and specifies the exact type of the component-module they represent (the complete list with these values is available in [1]). Thus, by parsing the HTML code of a page and locating the occurrences of these characteristic values within it, we can recover the page's organization as a set of Joomla! design elements. For example, Fig. 2(a) presents the Joomla! design elements identified within a page of the MMLAB[1] educational website. As we can see, the page consists of the "Article" component and a set of modules such as menus, footer, etc. Once this is done for all the pages of the website, we manage to capture its composition and navigation, i.e. its conceptual model. Then, we employ the Graph Translator for representing the recovered model as a directed graph (Fig. 2(b)), which is going to be the input for the graph mining algorithm of the next phase.

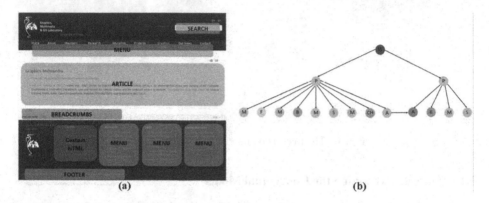

Fig. 2. (a) The organization of a page in terms of Joomla! design elements (component and modules). (b) The equivalent graph representation of the page's hypertext.

3.2   Phase 2: Identification of the Recurrent Design Patterns

After the extraction of the application's conceptual model, the next step is to inspect and analyze it in terms of the recurrent patterns (i.e., design fragments) that support the various application's functionalities for achieving certain purposes. These patterns are configurations of Joomla! components and modules which when located in a particular layout, they may serve a certain application purpose. To this end, we perform a pattern-based analysis of the recovered model to identify and evaluate the occurrences of all the incorporated recurrent patterns (Fig. 1, Phase 2).

Mining the Recurrent Patterns. To identify the recurrent patterns (their core speci-fications along with their starting and termination variants), we have used an approach which is based on the subgraph isomorphism problem, synopsized into finding whether the isomorphic image of a subgraph exists in a larger graph. As presented in detail in [2], the identification of the isomorphic subgraphs within a graph is an alternative way to obtain the identification of the incorporated recurrent patterns. Based on this, we applied the gSpan algorithm [4] on the conceptual model graph G, which identifies the occur-rences of all the recurrent patterns lying in the conceptual model, by locating all the isomorphic subgraphs images within it. In this way, the graph G is analyzed in terms of

---

[1] Available at http://mmlab.ceid.upatras.gr/en/graphicsmultimedia/.

its frequent subgraphs (representing recurrent patterns). In this way, we manage to identify the core specifications as well as the starting and ending variants of the recurrent subgraph-pattern.

Design Patterns Selection. In order to verify that the identified recurrent patterns can actually specify design patterns, i.e. they perform a certain functionality, we additionally inspect their occurrences on the website to examine whether the recurrence of the design elements occurring at the hypertext design goes with a recurrence at the data level, i.e., the content they deliver to the end-users. To achieve this, we have used a semantic similarity measurement technique [5] among the patterns occurrences. The content published by the various components and modules is extracted from the tables of the website's database. Thus, to examine for a given recurrent pattern whether there is a coexisting recurrence at the data level, we have to examine from which database tables the corresponding Joomla! components and modules (that make up the pattern) among its occurrences extract content. If they publish content from the same tables, then there is a high possibility of identifying a reusable design pattern for implementing a certain task. However, in this work we assume that we do not have access to the database of a website, since this is the common scenario in real-life websites. Based on this, we attempt to examine if there is a recurrence at the data level, by computing the semantic closeness of the content published by the pattern's corresponding Joomla! design elements among its occurrences. The rationale behind this is that the contents of the pages that come from the same database's table usually have a very close semantic relation. So, if the pattern's occurrences are semantically close, we can assume that they could probably derive content from the same database tables, and infer that the pattern is used for certain purpose, i.e., to provide certain information object to end-users.

To compute the semantic closeness of the published content between two occurrences of a pattern, we have defined two metrics, the "SemSimScore" and the "AverageSemSimScore". On the grounds that the main content of a page, published by components, is indicative of the page's semantics, the "SemSimScore" metric addresses the semantic similarity measurement of the content published by the Joomla! components occurring in a pattern. This is why there are empty cells for the "SemSimScore" computations in Table 1, when it comes to measure semantic similarity among content published by modules. Given two contents, the "SemSimScore" metric determines how similar the meaning of two contents is. The higher the score, the more similar the meaning of the two contents is, increasing the possibility that there is also a recurrence at the content displayed by the components of a pattern between its occurrences. Then, the "AverageSemSimScore" computes the average value of the individual "SemSimScore" values between the pattern occurrences. In Table 1, we can see an example of a pattern occurring on the MMLAB website. To verify that this pattern is actually used in the website for supporting a functionality, we inspect its occurrences to examine if there is also a recurrence at the content that the pattern's components deliver to the end-users. In Table 1, we can see three occurrences of the pattern Occ.1, Occ.2 and Occ.3. By comparing the semantic similarity of the content published by the pattern's corresponding components for Occ.1 and Occ.2, they have an AverageSemSimScore of 85 % which means that they are semantically very close. Similarly, the AverageSemSimScore for Occ.1 and Occ.3 is 0.04 % implying that the content published in these two occurrences is irrelevant. As a result, we can assume that the pattern in Table 1 is used for supporting the user's navigation among the various categories of a specific information object, i.e., the various fellowships categories. In this way, we can obtain a safe estimation about the recurrence at the data level among the occurrences of the identified patterns. We compute the AverageSemSimScore metric for all the occurrences of the identified patterns (core specification and variants) and we select and store in a "Candidate Patterns Repository" only the ones having an AverageSemSiScore over 70 %.

Table 1. Measuring semantic similarity among pattern's occurrences.

Pattern	Menu [Module]	Category list [Component]	Article [Component]
Occ.1	Top menu	Postgraduate fellowships list	Fellowship description
Occ.2	Top menu	Undergraduate fellowships list	Fellowship description
SemSimScore Occ.1-2		90 %	80 %
AverageSemSimScore Occl.-Occ.2			85 %
Occ.3	Top menu	R&D projects	Project description
SemSimScore Occ.1-3		5 %	2 %
AverageSemSimScore Occl.-Occ.3			0.04 %

3.3 Evaluation of Pattern Variants Consistent Use

In this final step, we focus on evaluating the consistent use of the identified patterns for determining their impact on the overall application's quality. Patterns which are used consistently throughout the conceptual model of an application result in high design quality, facilitating end-users identify typical sequences of interactions with the system for performing common tasks. This results in foreseeable navigation behavior, and thus high design quality. On the other hand, patterns which are not used consistently may cause serious design inconsistencies. To this end, we calculate some metrics to evaluate whether the patterns stored in the "Candidate Patterns Repository" are used consistently throughout the conceptual model. These metrics are called Start-Point Metric (SPM) and End-Point Metric (EPM) and intuitively they compute the statistical variance of the occurrences of the starting and termination variants of a pattern throughout the application model. SPM is defined as (EPM is defined in an analogous way):

$$SPM = \sigma^2 / \sigma_{BC}^2 \qquad (1)$$

σ^2 is the statistical variance of the N starting variants occurrences which is calculated according to the formula (2):

$$\sigma^2 = \frac{1}{N} \sum_{i=0}^{N} \left(p_i - \frac{1}{N} \right)^2 \qquad (2)$$

where p_i is the percentage of occurrences for the i-th pattern variant. σ_{BC}^2 is instead the best case variance and it is calculated by the formula (2) assuming that only one variant has been coherently used throughout the application. More details about the metrics definition can be found in [2]. We have also specified a measurement scale which defines a mapping between the numerical results obtained through the calculus method of the SPM-EPM metrics and a set of (predefined) meaningful and discrete values, expressing different consistency levels (Table 2).

Table 2. The measurement scale for the EPM and SPM metrics.

EPM-SPM range	Measurement scale value
$0 \leq \text{SPM} < 0.2$	Insufficient
$0.2 \leq \text{SPM} < 0.4$	Weak
$0.4 \leq \text{SPM} < 0.6$	Discrete
$0.6 \leq \text{SPM} < 0.8$	Good
$0.8 \leq \text{SPM} \leq 1$	Optimum

We compute these metrics for all the occurrences of all the patterns' starting and ending variants and store the results in the "Design Patterns Repository", which are also provided in a TXT file ("Results"). By applying the methodology on a website, developers can gain important information regarding its design quality and particularly about the various design fragments that they have chosen to use for realizing the various tasks in the application's context. On one side, the methodology can detect effective reusable design solutions consistently used throughout the application for supporting certain functionality. Such reusable solutions facilitate the discovery of new interaction design patterns for the CMS domain. On the other side, the methodology can also detect recurrent design constructs indicating design inconsistencies lowering the application's quality.

4   Domain-Specific Design Patterns - a Case Study: The Educational Websites Domain

In the context of this work, we focused on the domain of educational web-sites and we applied the previously described methodology on a collection of such type of websites. This way, we have identified a set of design patterns for this specific domain, stored in a central repository available at [1], a number of which are presented in the current section. Our dataset is composed by a number of educational websites that have been developed by our team, such as the MMLAB website, and a set of websites from educational websites category of the official Joomla! Community Showcase website catalogue [6]. In the following sections, we provide a short description of this domain and a number of the design patterns we have identified for this area.

4.1 Domain Description

College and university websites are part of institutions' identity, having an important role in their marketing practices. They provide useful information to prospective students and they have a strong impact on forming user's opinion during the college search process. Thus, educational websites must be designed in such a way, that it enables end-users to recognize typical interaction patterns with the system in order to quickly find the information that they are looking for. In the context of educational websites, the common types of end-users are the prospective students, the current under-graduate and postgraduate students, and the alumni. Furthermore, the common types of information objects that can be found in such websites include: courses catalogues, degree programs for graduate diplomas, master and doctoral degrees, research and publications, admissions and financial aid, faculty, students' life, news and events, etc.

4.2 Design Patterns for Educational Websites

In this section, we present some of the design patterns that we have identified within the various website designs. We have classified these patterns into three main categories, as listed below.

Layout: This category includes patterns that provide developers with solutions for standard page types within the context of educational websites, in terms of the design components that specify the page's layout. These page types include:

Homepage: specifies a template for the design of the website's homepage. For example, Fig. 3 depicts one of the most prevalent design patterns for the design of the homepage that we found in our websites dataset. It consists of a welcome message based on the single article component (usually having a link to the About us page) and a number of modules such as banners which provide quick links usually to undergraduate and postgraduate information pages. Additionally, they use the most read module for direct access to the available educational programs pages and the latest articles module for publishing their news.

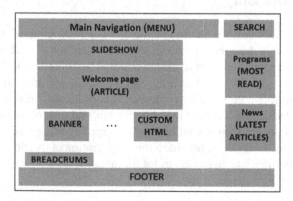

Fig. 3. Design patterns for the design of the Homepage.

Browsing Information Object's Categories Hierarchy: specifies a template for supporting the user's navigation among the various categories and subcategories of a specific information object. In, Fig. 4, we can see two of the most common patterns for browsing the categories hierarchy of typical object information types.

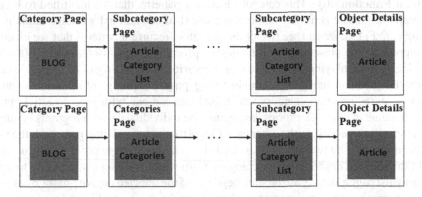

Fig. 4. The patterns for browsing the categories hierarchy of an information object type.

The top pattern is mostly used to browse the categories hierarchy of the courses whereas the other pattern is mainly used for browsing the educational programs, since the module Article Categories component provides a short description of each published list item (i.e. for each educational program). Similarly, we have also identify additional layout patterns such as the Search Results Page, the About Institution Page, the Information Object List Page, the Information Object Details Page, etc. The description of these patterns is available in [1]. Generally, these patterns support the consistent use of standard page structures, composed by a specific combination of Joomla! components and modules, used as templates for assisting designers in the predictability of user navigation. After all, consistency across pages and page design elements enforces a coherent design style improving the ease of use of a website while at the same time the users can enjoy a more pleasant user experience by reducing the number of unexpected variations in the page layout.

Navigation: This category includes patterns (such as navigation bars, menus, pagination, module tabs) for supporting user's navigation for locating content or accomplishing certain task. An example can be the use of a "fat" footer (Fig. 5) for providing users with a mechanism for quick access to specific sections of the website, bypassing the navigational structure. An example can be found on the website of the Graduate School of Arts and Sciences[2] (GSAS) in which the footer contains links to pages frequently used by users.

---

[2] Available at https://www.gsas.harvard.edu/.

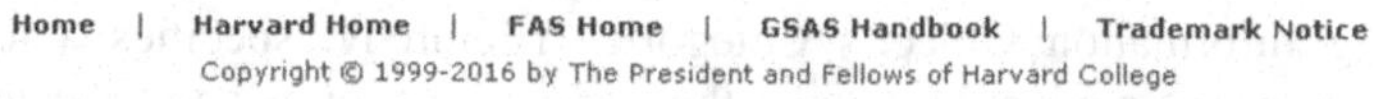

Fig. 5. An example of a fat footer.

Domain Functionality: This category includes patterns that we identified to be used for supporting basic activities within the context of educational websites domain. For example, in Fig. 6, we can see two of the prevalent recurrent patterns that we identified for supporting the presentation of the degree programs to the users. In Fig. 6(a), there is a blog page displaying an overview of the various degree programs categories from which the user can navigate to another blog page providing information about the selected degree program category which includes a menu with links to article pages, corresponding to its subcategories, presenting the individual degree programs belonging to this specific category. In the other case Fig. 6(b), the second blog page with the menu has been replaced by two other pages, an article categories page providing information about the selected degree program category from which the user can navigate to article category list page enlisting all the subcategories of the selected degree program category. These are two pat-terns for supporting the user navigation to the hierarchy of categories and subcategories of the various degree programs.

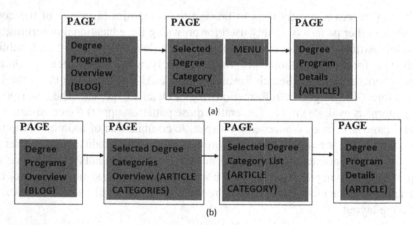

Fig. 6. An example of a domain functionality design pattern.

Miscellaneous: This category includes patterns that we have not found their main category yet. In this case, we have detected certain configurations of Joomla! component and modules occurring frequently in the hypertext design of educational websites but we can-not find a clear semantic connection with the domain's concepts.

The benefits of using the above domain-specific design patterns will be increased reuse in future designs. Furthermore, in this way, we can derive website design practices and design guidelines (i.e., the prevalent design patterns) for the domain of educational websites, and also identify commonalities and differences in the design of such kind of applications.

5   Conclusion and Future Work

In this paper, we have illustrated a model-driven approach for mining domain-specific design patterns. We focus on the domain of educational websites and present our preliminary results, i.e. the design patterns which we identified for this particular domain. Most of the work presented here can be applied also to web applications built by using other CMSs with slight straightforward modifications. By applying the methodology on a website, developers can gain important information regarding its design quality and submit the patterns identified in their designs to the central repository. In the future, we plan to explore other application domains and thus, come up with useful design guidelines for building successful Joomla!-based websites.

References

1. CMS Patterns (2015). http://alkistis.ceid.upatras.gr/research/modeling/patterns/. Accessed 22 Feb. 2016
2. Gkantouna, V., Tsakalidis, A., Tzimas, G.: Mining Interaction Patterns in the Design of Web Applications for Improving User Experience, ACM Hypertext (2016)
3. Yan, X., Han, J.: gSpan: Graph-based substructure pattern mining. In: ICDM 2002, p. 721, Washington, DC, USA. IEEE Computer Society (2002)
4. Philippsen, M.: ParSeMiS - the Parallel and Sequential Mining Suite (2011). https://www2.cs.fau.de/EN/research/zold/ParSeMiS/index.html. Accessed 22 Feb. 2016
5. Simpson, T., Dao, T.: WordNet-based semantic similarity measurement (2010). http://www.codeproject.com/Articles/11835/WordNet-based-semantic-similarity-measurement. Accessed 22 Feb. 2016
6. Joomla! Community Showcase website. http://community.joomla.org/showcase/sites/educational.html, Accessed 22 Feb. 2016
7. Montero, S., Díaz, P., Aedo, I.: A semantic representation for domain-specific patterns. In: Wiil, U.K. (ed.) MIS 2004. LNCS, vol. 3511, pp. 129–140. Springer, Heidelberg (2005)
8. Herzner, W., Csertan, G., Balogh, A.: Design Patterns for Domain-specific Application Modelling. http://static.inf.mit.bme.hu/decoscd/papers/DECOS_paper_Design_Patterns_for_Domain_specific_Application.pdf
9. Rekhis, S., Bouassida, N., Duvallet, C., Bouaziz, R., Sadeg, B.: A process to derive domain-specific patterns: application to the real time domain. Adv. Databases Inf. Syst. **51**, 475–489 (2010)
10. Rekhisa, S., Bouassidaa, N., Bouaziza, R., Duvalletb, C., Sadegb, B.: A new method for constructing and reusing domain specific design patterns: Application to RT domain. J. King Saud Univ. Comput. Inf. Sci. (2016)

Modelling Cadence Perception Via Musical Parameter Tuning to Perceptual Data

Maximos Kaliakatsos-Papakostas[✉], Asterios Zacharakis,
Costas Tsougras, and Emilios Cambouropoulos

Department of Music Studies, Aristotle University of Thessaloniki,
54124 Thessaloniki, Greece
{maxk,aszachar,tsougras,emilios}@mus.auth.gr

Abstract. Conceptual blending when used as a creative tool combines the features of two input spaces, generating new blended spaces that share the common structure of the inputs, as well as different combinations of their non-common parts. In the case of music, conceptual blending has been employed creatively, among others, in generating new cadences (pairs of chords that conclude musical phrases). Given a specific set of input cadences together with their blends, this paper addresses the following question: are some musical features of cadences more salient than others in defining perceived relations between input and blended cadences? To this end, behavioural data from a pairwise dissimilarity listening test using input and blended cadences as stimuli were collected, thus allowing the construction of a 'ground-truth' human-based perceptual space of cadences. Afterwards, the salience of each cadence feature was adjusted through the Differential Evolution (DE) algorithm, providing a system-perceived space of cadences that optimally matched the ground-truth space. Results in a specific example of cadence blending indicated that some features were distinguishably more salient than others. This pilot study was a first step towards building self-aware blending systems and revealed that the salience of features in conceptual blending is an essential part for producing perceptually relevant blends.

1 Introduction

The cognitive theory of conceptual blending by [9] has been extensively used in linguistics, music composition [21], music cognition [1,2] and other domains mainly as an analytical tool for explaining the cognitive processes that humans undergo when engaged in creative acts. In computational creativity, conceptual blending has been modelled by Goguen [10] as a generative mechanism, according to which two *input spaces* are blended to generate novel *blended spaces*, using tools of *category theory*. A computational framework that extends Goguen's approach has been developed in the context of the COncept INVENtion Theory[1]

---

[1] http://www.coinvent-project.eu.

© IFIP International Federation for Information Processing 2016
Published by Springer International Publishing Switzerland 2016. All Rights Reserved
L. Iliadis and I. Maglogiannis (Eds.): AIAI 2016, IFIP AICT 475, pp. 552–561, 2016.
DOI: 10.1007/978-3-319-44944-9_49

(COINVENT) project [17] based on the notion of *amalgams* [8,16]. Following this framework, systems have been developed that blend features of two input musical cadences [7,19] (pair of ending chords) or chord transitions [13], producing novel blended ones that incorporate meaningful characteristics of the input ones.

The paper at hand presents a pilot study for defining the salient features of cadences in the context of a cadence blending system, based on the data collected from perceptual experiments. The system-produced blended cadences incorporate combinations of features from manually-made input cadences, while the importance of each feature differs according to its perceptual salience. The salience of features is obtained by applying the Differential Evolution (DE) algorithm for optimally matching (in terms of pairwise dissimilarity) the system-perceived cadence relations with the perceptual space extracted by the experiment with humans. This study is a first step towards increasing self-awareness in a creative system that produces cadences through conceptual blending.

2   A Formal Description of Cadences for Generative Conceptual Blending

In this paper a cadence is considered as a special case of a transition (a chord following another), but with the second chord is fixed. Therefore, when blending two input cadences, the characteristics of the penultimate chords of the inputs are combined to produce new penultimate blended chords that are paired with the fixed final chord to constitute the blended cadence. For instance, the case of blending, e.g., the perfect with the Phrygian cadences is described by the transitions I_1: G7 → Cm and I_2: B♭m → Cm respectively, while a blend of these inputs is the tritone substitution cadence, C♯7 → Cm [7,19]. The perceptual characteristics of the penultimate chord that are considered for describing a cadence are the following:

1. *fcRoot*: Root of the first chord.
2. *fcType*: Type of the first chord as presented by the GCT.
3. *fcPCs*: Pitch classes of the penultimate chord.
4. *rDiff*: Root difference for the transition.
5. *DIC0*: Existence of fixed pitch class.
6. *DIC1*: Existence of upward semitone movement in pitch classes.
7. *DIC-1*: Existence of downward semitone movement in pitch classes.
8. *DIC*: The compete DIC vector of the chord transition.
9. *asc*: Existence of ascending semitone to the tonic.
10. *desc*: Existence of descending semitone to the tonic.
11. *semi*: Existence of semitone movement towards the tonic.

For computing the root and type in a consistent manner for all utilised chords, the General Chord Type (GCT) [5,14] has been employed, which allows the rearrangement of the notes of a harmonic simultaneity such that abstract types of chords along with their root may be derived. The GCT algorithm finds the maximal subset that forms the base upon which the chord type is built, while the lowest note of the base is the root of the chord; the potentially remaining notes are assembling the extension of the GCT, which is the set of notes that would not be a part of the maximally consonant subset. For example, the GCT representation of the first degree (I) chord in a major scale is [0, [0 4 7]], where 0 indicates the root note in relation to the scale (0 is the scale as first degree) and [0 4 7] is the chord's type (4 indicates a major third and 7 a perfect fifth). Accordingly, a V7 chord is denoted by [7, [0 4 7], [10]], where 10 is the extension (minor seventh), which cannot be included in the base considering that the tritone and minor seventh intervals are dissonant.

The aforementioned properties 1–2 describe the first chord of the cadence and the first two properties (chord root and type) are extracted from the GCT algorithm, considering that all the examined cadences are in the key of C minor. Property 4, the difference between the chord roots is an integer between -5 and 6, indicating the pitch class difference between the roots of the first and the second chords of the cadence. Property 5 captures the existence of a common note between the two chords, while properties 6 and 7 indicate the existence of a semitone movement (upward and downward respectively) in any pitch class of the cadence transition. These properties actually indicate if there is a 0, 1 or −1 in the Directional Interval Class (DIC) [6], flagging whether there are small pitch class voice leading movements (repeating notes or semitone movements) in the cadence. Property 8 incorporates the entire DIC vector of the transition/cadence. Properties 9 to 11 are used to highlight the importance of whether there is a semitone movement (property 11) to the tonic from the first to the second chord of the cadence as well as whether this movement is ascending (property 9) or descending (property 10); these properties reflect the importance of the leading note (upwards or, even, downwards).

Table 1 illustrates a blending example, where the tritone substitution cadence is created from the perfect and the phrygian cadences. This blend incorporates properties from both input spaces with a good balance, i.e. many properties that are common in both input spaces, while new properties have also been added through completion. Specifically, this blend includes five properties of input 1, four properties of input 2, three common properties and four new properties that were not present in any input space. The properties of the blended space come from either input space, or are completed by logical deduction through axioms describing cadences, as indicated in the parentheses next to each respective property.

3   Approximating the Importance of Properties According to the Perceptual Pairwise Distances of Cadences

As discussed in the introduction, one desirable property for a creative system is the ability to self-evaluate its products [12]. In this respect, the cadence blending system should be able to make accurate predictions of how the blends are perceived in relation to the inputs. To this end, a vector containing differences

Table 1. Example of the tritone substitution cadence invention, by blending the perfect and the phrygian cadences.

Property name	Input 1 (Perfect)	Input 2 (Phrygian)	Possible blend
fcRoot	7	10	1
fcType	[0 4 7 10]	[0 3 7]	[0 4 7 10]
fcPCs	$\{7, 11, 2, 5\}$	$\{10, 1, 5\}$	$\{11, 1, 5, 8\}$
rDiff	5	2	1
DIC0	1	0	0
DIC1	1	0	1
DIC-1	1	1	1
DIC	$[1, 2, 0, 2, 0, 1, 2, 1, 0, 1, 2, 0]$	$[1, 0, 1, 1, 1, 0, 0, 3, 0, 0, 1, 1]$	$[2, 1, 0, 1, 2, 0, 1, 2, 0, 2, 0, 1]$
asc	1	0	1
desc	0	1	1
semi	1	1	1

on the utilised music properties, denoted by P_i^C. Since it is assumed that not all properties are of equal importance in deciding the distance between pairs of cadences, each property (P_i^C) is assumed to have a *weight* of importance, denoted by w_i. The overall distance between cadences can be then calculated by summing the weight values of the properties that are different in these cadences. Specifically, the distance between two cadences X and Y is calculated by:

$$D(X, Y) = \sum_{i=1}^{11} w_i \, f_i, \tag{1}$$

where f_i is a function related to how distance is measured for each property, as analysed in Eqs. 2, 3 and 4.

Properties with indexes 1, 2, 4, 5, 6, 7, 9, 10, 11 have a binary f_i function similar to the Kronecker delta function:

$$f_i = \begin{cases} 1, & \text{if } P_i^X \neq P_i^Y, \\ 0, & \text{if } P_i^X = P_i^Y \end{cases}, \text{ for } i \in \{1, 2, 4, 5, 6, 7, 9, 10, 11\}. \tag{2}$$

Equation 2 indicates that these properties need to be equal in both cadences in order not to be penalised by the respective w_i values. The function for property 3 is computing the number of common over the number of total pitch classes in the first chords of two cadences. Specifically,

$$f_3 = \frac{N(\cup(P_3^X, P_3^Y)) - N(\cap(P_3^X, P_3^Y))}{N(\cup(P_3^X, P_3^Y))}, \tag{3}$$

where $N(\cap(P_3^X, P_3^Y))$ and $N(\cup(P_3^X, P_3^Y))$ is the number of elements in the intersection and union of the pitch class sets. Equation 3 indicates that there is a proportional penalty to w_i for pitch classes that are not common in the first chord of two cadences. Finally, DIC information (property 8) is measured according to the correlation of the DIC vectors of the cadences under examination. Weaker correlations are penalised proportionally with regards to w_8, according to the following equation:

$$f_8 = (1 - \mathrm{corr}(P_8^X, P_8^Y))/2. \tag{4}$$

Correlation between DIC vectors conveys harmonic meaning at some extend, as indicated by the genre categorisation results based on DIC correlation reported in [4].

By calculating the distances between all pairs of the examined cadences according to Eq. 1, a dissimilarity matrix that represents the pairwise differences among the nine cadences is constructed. This dissimilarity matrix is subsequently analysed through non-metric weighted MDS and results in a spatial configuration of the cadences that from now on will be called the 'algorithmic space'. Therefore, in order to define the contribution (i.e. weight value) of each parameter on deciding the overall distance, a differential evolution (DE) algorithm [18] was used to optimise the fit between pairwise distances in the perceptual space (used as ground truth) and the respective ones in the algorithmic space. An overview of the optimisation process is schematised in Fig. 1.

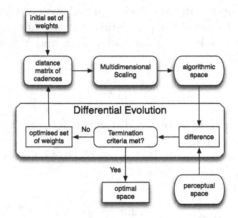

Fig. 1. The optimisation through the Differential Evolution.

The difference between the perceptual and algorithmic spaces is quantified through a fitness function that is estimated by taking the average of two similarity metrics, namely the m^2 statistic for Procrustes analysis [11] and Tucker's congruence coefficient [3]. For a detailed discussion on the application of these metrics to comparison between timbre spaces please see [20].

4   Perceptual Experiment

A pairwise dissimilarity listening test was deemed appropriate to act as a ground truth for modelling how a set of musical cadences is perceived by listeners, as the dissimilarity matrices it produces allow Multidimensional Scaling (MDS) analysis to generate geometric configurations that represent the relationships between percepts.

Twenty listeners (age range = 18–44, mean age 24.9, 10 male) participated to the listening experiment. Participants were students in the Department of Music Studies at the Aristotle University of Thessaloniki. All of them reported normal hearing and long term music practice (16.5 years on average, ranging from 5 to 35).

Participants were asked to compare all the pairs among 9 cadences using the free magnitude estimation method. Therefore, they rated the perceptual distances of 45 pairs (same pairs included) by freely typing in a number of their choice to represent dissimilarity of each pair (i.e., an unbounded scale) with 0 indicating a same pair. Each stimulus lasted around 4 s and interstimulus interval was set at 0.5 s. Listeners became familiar with the range of cadences under study during an initial presentation of the stimulus set (random order). This was followed by a brief training stage where listeners rated the distance between four selected pairs of cadences. For the main part of the experiment, participants were allowed to listen to each pair of cadences as many times as needed prior to submitting their dissimilarity rating. The pairs were presented in random order and participants were advised to retain a consistent rating strategy throughout the experiment. In total, the listening test sessions, including instructions and breaks, lasted around thirty minutes for most of the participants.

The stimulus set consisted of the two input cadences (the perfect and Phrygian) together with seven blended cadences. The selection of cadences was made manually after evaluating their blending elements so as to attain a theoretically valid, maximally diverse corpus. All cadences were assumed to be in C minor tonality/modality, consisted of two chords and the final chord was kept constant (C minor), thus variation between the stimuli resulted from altering the penultimate chords. The nine cadential pairs of chords are described from a music-theoretical perspective in the following list:

1. Perfect authentic cadence, featuring the full V7 dominant chord that resolves to the i tonic chord without 5th, in order to achieve correct voice leading.
2. Phrygian cadence, with the $\flat$vii chord in first inversion resolving to the i tonic chord.
3. Tritone substitution progression, with the $\flat$II7$^{\flat}$ chord (German-type augmented-6th chord) leading to the tonic.
4. Backdoor progression, with the $\flat$VII7 chord in first inversion, in order to achieve maximum voice-leading uniformity.
5. Contrapuntal-type tonal cadence, with the viio6 resolving to the minor tonic.
6. Plagal-type cadence, with the iio6/5 progressing to the tonic.
7. Minor-dominant to minor-tonic progression, utilising chords from the natural minor scale (Aeolian mode).
8. Altered dominant-7th chord to minor-tonic progression, with the dominant in second inversion and with its 5th lowered (French-type augmented 6th chord).
9. Half-diminished 'dominant'-7th chord to minor-tonic progression.

5   Results

Before proceeding to the main body of the analysis for the dissimilarity data
we examined the internal consistency of the dissimilarity ratings. Cronbach's
alpha was .94 indicating high inter-participant reliability. In the main body of
the analysis, the dissimilarity ratings within each linguistic group were analysed
through non-metric (ordinal) MDS with dimension weighting (INDSCAL within
SPSS PROXSCAL algorithm) [15]. A two-dimensional solution was deemed
optimal for data representation as the improvement of measures-of-fit when
adding a third dimension was minimal. Figure 2a and shows the configuration of
the cadences within this 2-D space.

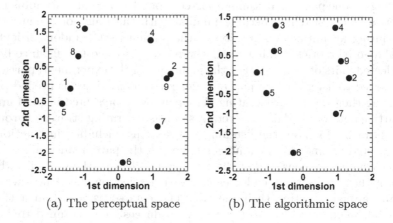

(a) The perceptual space   (b) The algorithmic space

Fig. 2. (a) The 2 dimensional perceptual space of the nine cadences. The perfect and
the Phrygian cadences (No. 1 & 2) are positioned far away from each other on the 1st
dimension. (b) The optimised algorithmic cadence space that resulted from modelling
the perceptual space through optimal weighting of the musical parameters.

The optimisation process presented in Sect. 3 produced combinations of
weights for each cadential feature that minimise the differences between the perceptual and the algorithmic spaces, providing an optimised modelling of cadence
perception. The ideal combination should offer the highest possible fit (quantified
by the similarity metrics) with the perceptual space.

Several optimisation simulations were run with different setups for the DE
algorithm (concerning population members, number of generations etc.). Table 2
shows a property weight configuration that provided a satisfactory modelling of
the perceptual space (population members = 50, number of iterations = 30).
This configuration featured an $m^2 = .027$, a congruence coefficient = .994 and
an RV modified coefficient = .966 indicating an excellent fit between the perceptual and algorithmic configurations. Had no optimisation taken place (i.e.,
all parameters assigned importance equal to 1), the similarity metrics between
the two configurations would become: $m^2 = .361$, congruence coefficient = .942

and an RV modified coefficient = .523, representing a serious divergence of the algorithmic space in relation to the perceptual. Figure 2b shows the optimised 2-dimensional configuration of the algorithmic space. As expected, based on the similarity measures reported above, the two spaces (perceptual and algorithmic) are very closely related since the obtained algorithmic space maintains the majority of the perceptual relationships between cadences.

Table 2. Optimal weights of the musical properties for modelling cadence relationships. A combination of four prominent parameters (in bold) and two weaker ones (in italics) achieved an excellent model of the perceptual space.

Weights (w_i values)										
fcRoot	fcType	fcPCs	rDiff	DIC0	DIC1	DIC-1	DIC	asc	desc	semi
.020	**.790**	**.751**	.000	*.311*	.014	*.270*	**.760**	**.731**	.001	.003

Based on the above, it can be concluded that the penultimate chord types, their pitch classes, the information provided by the DIC vector and the presence or absence of a leading note account for the way listeners perceived the relationships of cadences within this particular set. It should be also noted that according to the DE simulations there are two additional properties of 'moderate' importance: those that examine whether there is at least one fixed (*DIC0*) or one descending by one semitone (*DIC-1*) pitch class.

6   Conclusions

This paper presents a first pilot study towards increasing the self-evaluation ability of a creative cadence blending algorithm, by utilising data from perceptual experiments. The listening experiment incorporated two cadences (the perfect and the phrygian) as a starting point along with seven system-produced blends. The blending algorithm combines characteristics of the input spaces, generating several blends that include different combinations of characteristics from the input cadences. Aim of this paper is to identify whether any of the cadence characteristics are perceptually more salient in defining pairwise cadence similarities. The Differential Evolution (DE) algorithm was employed in order to fine-tune the salience weight of every cadence property, so that the relative placement of cadences obtained with system-based metrics optimally matches the user-obtained perceptual space.

The dissimilarity rating experiment revealed a categorical perception of cadences reflected by positioning on the 1st MDS dimension and clearly dictated by the existence of an upward semitone movement to the tonic (leading note) in the left-hand cadences in comparison to the lack of a leading note in the right-hand cadences. This fact is also evident by the high weight of the *asc* value shown in Table 2. Two major clusters of cadences were formed based on this differentiation together with one outlier (the plagal cadence) that featured neither and upward semitone nor an upward tone to the tonic but a duplication of the tonic. It is also shown that both the intra and inter-cluster relations could be adequately modelled mainly through four salient musical properties, namely the penultimate chord type, its pitch classes, the DIC vector of the cadence and the existence of the leading note.

This being a pilot study, the generalisation of these findings for a wider range of cadences as well as a detailed mapping of musical properties to perceptual dimensions is a necessary step that is left for future work. For instance, initial findings indicate that the differentiation of cadences along the 2nd dimension could be explained by the inherent dissonance of the penultimate chords (as expressed by the MIR Toolbox roughness calculation) together with their distances from the final chord in Lerdahl's Tonal Pitch Space. Identification of such complementary measures could help towards increasing self-awareness of a cadence blending system, according to various diverse aspects of its creative products.

Acknowledgments. This work is funded by the Concept Invention Theory (COIN-VENT) project. The project COINVENT acknowledges the financial support of the Future and Emerging Technologies (FET) programme within the Seventh Framework Programme for Research of the European Commission, under FET-Open grant number: 611553.

References

1. Antovic, M.: Musical metaphors in serbian and romani children: an empirical study. Metaphor Symbol **24**(3), 184–202 (2009)
2. Antovic, M.: Musical metaphor revisited: Primitives, universals and conceptual blending. Universals and Conceptual Blending (2011)
3. Borg, I., Groenen, P.J.F.: Modern Multidimensional Scaling: Theory and Applications, 2nd edn. Springer, New York (2005)
4. Cambouropoulos, E.: A directional interval class representation of chord transitions. In: Proceedings of the Joint Conference ICMPC-ESCOM (12th International Conference for Music Perception and Cognition, & 8th Conference of the European Society for the Cognitive Sciences of Music) (2012)
5. Cambouropoulos, E., Kaliakatsos-Papakostas, M., Tsougras, C.: An idiom-independent representation of chords for computational music analysis and generation. In: Proceeding of the Joint 11th Sound and Music Computing Conference (SMC) and 40th International Computer Music Conference (ICMC) (2014)
6. Cambouropoulos, E., Katsiavalos, A., Tsougras, C.: Idiom-independent harmonic pattern recognition based on a novel chord transition representation. In: Proceedings of the 3rd International Workshop on Folk Music Analysis (FMA) (2013)
7. Eppe, M., Confalonier, R., Maclean, E., Kaliakatsos-Papakostas, M., Cambouropoulos, E., Schorlemmer, M., Codescu, M., Kühnberger, K.U.: Computational invention of cadences and chord progressions by conceptual chord-blending. In: International Joint Conference on Artificial Intelligence (IJCAI) (2015)

8. Eppe, M., Maclean, E., Confalonieri, R., Kutz, O., Schorlemmer, M., Plaza, E.: ASP, amalgamation, and the conceptual blending workflow. In: Calimeri, F., Ianni, G., Truszczynski, M. (eds.) LPNMR 2015. LNCS, vol. 9345, pp. 309–316. Springer, Heidelberg (2015)
9. Fauconnier, G., Turner, M.: The Way We Think: Conceptual Blending and The Mind's Hidden Complexities. Basic Books, New York (2003)
10. Goguen, J.: Mathematical models of cognitive space and time. In: Andler, D., Ogawa, Y., Okada, M., Watanabe, S. (eds.) Reasoning and Cognition, vol. 2. Keio University Press (2006)
11. Gower, J.C., Dijksterhuis, G.B.: Procrustes Problems. Oxford University Press, Oxford (2004)
12. Jordanous, A.K.: Evaluating computational creativity: a standardised procedure for evaluating creative systems and its application. Ph.D. thesis, University of Sussex (2013)
13. Kaliakatsos-Papakostas, M., Confalonier, R., Corneli, J., Zacharakis, A., Cambouropoulos, E.: An argument-based creative assistant for harmonic blending. In: Proceedings of the 7th International Conference on Computational Creativity (ICCC) (2016)
14. Kaliakatsos-Papakostas, M., Zacharakis, A., Tsougras, C., Cambouropoulos, E.: Evaluating the General Chord Type representation in tonal music and organising GCT chord labels in functional chord categories. In: ISMIR. Malaga, Spain (2015)
15. Meulman, J.J., Heiser, W.J.: PASW Categories 18. SPSS Inc., Chicago (2008). Chapter 3
16. Ontañón, S., Plaza, E.: Amalgams: a formal approach for combining multiple case solutions. In: Bichindaritz, I., Montani, S. (eds.) ICCBR 2010. LNCS, vol. 6176, pp. 257–271. Springer, Heidelberg (2010)
17. Schorlemmer, M., Smaill, A., Kühnberger, K.U., Kutz, O., Colton, S., Cambouropoulos, E., Pease, A.: Coinvent: towards a computational concept invention theory. In: 5th International Conference on Computational Creativity (ICCC) (2014)
18. Storn, R., Price, K.: Differential evolution - a simple and efficient adaptive scheme for global optimization over continuous spaces. J. Global Optim. **11**, 341–359 (1997)
19. Zacharakis, A., Kaliakatsos-Papakostas, M., Cambouropoulos, E.: Conceptual blending in music cadences: a formal model and subjective evaluation. In: ISMIR. Malaga (2015)
20. Zacharakis, A., Pastiadis, K., Reiss, J.D.: An interlanguage unification of musical timbre: bridging semantic, perceptual and acoustic dimensions. Music Percept. **32**(4), 394–412 (2015)
21. Zbikowski, L.M.: Conceptualizing Music: Cognitive Structure, Theory, and Analysis. Oxford University Press, New York (2002)

Musical Track Popularity Mining Dataset

Ioannis Karydis[1]([⊠]), Aggelos Gkiokas[2], and Vassilis Katsouros[2]

[1] Department of Informatics, Ionian University, 49132 Kerkyra, Greece
karydis@ionio.gr
[2] Institute for Language and Speech Processing, Athena RIC, 15125 Athens, Greece
{agkiokas,vsk}@ilsp.gr

Abstract. Music Information Research requires access to real musical content in order to test efficiency and effectiveness of its methods as well as to compare developed methodologies on common data. Existing datasets do not address the research direction of musical track popularity that has recently received considerate attention. Existing sources of musical popularity do not provide easily manageable data and no standardised dataset exists. Accordingly, in this paper we present the Track Popularity Dataset (TPD) that provides different sources of popularity definition ranging from 2004 to 2014, a mapping between different track/author/album identification spaces that allows use of all different sources, information on the remaining, non popular, tracks of an album with a popular track, contextual similarity between tracks and ready for MIR use extracted features for both popular and non-popular audio tracks.

1 Introduction

One of the most important requirements of Music Information Research (MIR) is access to pertinent musical content. The experimentation on this content mostly aims on the testing of the efficiency and effectiveness of the MIR methods, while providing reference for comparison of new and existing methods in order to show progress. In rare cases, the use of synthetic data can be helpful to the aforementioned use of data in MIR experiments, though music, being highly an artistic form of expression, does not always adhere to a set of deterministic rules that researchers could rely on in order to avoid the requirement for access to real musical content.

Accordingly, in MIR, as in most areas of scientific research, the collection, distribution and use of datasets is of great importance, despite the litany of legal issues [6] that may arise from such practices. Music data for the purposes of MIR usually refer to audio files of recorded performed musical pieces, symbolic representation of a piece, lyrics, metadata as well as contextual to the piece information mainly collected through social networks pertaining to the users' perception of or activities on the pieces. Thus, following the need for such content

© IFIP International Federation for Information Processing 2016
Published by Springer International Publishing Switzerland 2016. All Rights Reserved
L. Iliadis and I. Maglogiannis (Eds.): AIAI 2016, IFIP AICT 475, pp. 562–572, 2016.
DOI: 10.1007/978-3-319-44944-9_50

exchange and its intended use, MIR datasets additionally include commonly used derivative transformations of all the aforementioned musical information in order to avoid legal implications as well as to spare users of time and resources required for these to be produced.

Numerous datasets exist in MIR [9] that cover a broad area of the domain, though none is immediately applicable for knowledge extraction from the popularity that musical pieces receive. The process of track popularity prediction prior to or during the initial period of a track's release has long been a requirement of the musical industry. Interestingly enough, the gains of such a prediction go far beyond the obvious benefits of allowing musical labels to identify financially interesting clients, as the whole ecosystem (artists and listeners) also profits. Despite the aforementioned benefits, it was only after the commercial application by Polyphonic HMI[1] that the issue gained significant attention as a research direction, as early as 2005 [4].

1.1 Motivation and Contribution

Existing commonly used services, such as Spotify[2], Billboard[3] and Last.fm[4] that provide popularity of musical content do not offer easily manageable data. Spotify's localised charts, although provided an Application Programming Interface (API), have temporarily according to the service's community helpdesk, ceased to function as of approximately March 2015 and are still offline. Last.fm's localised charts do not offer an API, though Last.fm does indeed provide the aggregated number of listeners and playcounts for all available tracks. Billboard's Hot 100 Chart does not offer an API but does provide the most long termed archives, dating back to August 9th, 1958.

To add to the difficulties of collecting track popularity information, the aforementioned services utilise their respective track identification space making collective use of multiple popularity sources rather difficult. Moreover, collecting just the tracks that exceed the popularity threshold, research cannot deal integrally with the separation of hits from non-hits as no information on non-hits is available, since the collected information only contains the degree of popularity. Finally, having access to the content of the audio files of the popularity chart is, among other parameters, very important in the selection of the track's representative features that will lead to high quality predictions.

To address these requirements, we introduce the Track Popularity Dataset (TPD), a collection of track popularity data for the purposes of MIR, containing:

1. different sources of popularity definition ranging from 2004 to 2014,
2. information on the remaining, non popular, tracks of an album with a popular track,

---

[1] http://polyphonichmi.blogspot.gr/p/about-company.html.
[2] https://charts.spotify.com.
[3] http://www.billboard.com/charts/hot-100.
[4] http://www.last.fm/charts.

3. a mapping between different track/author/album identification spaces that allows use of all different sources,
4. contextual similarity information between all popular tracks,
5. ready for MIR use extracted features for popular & non-popular audio tracks,

The rest of the paper is organised as follows: Sect. 2 presents background information on Hit Song Science and related work, while Sect. 3 discusses the proposed dataset, its creation processes as well as a detailed analysis of its content. Next, Sect. 4 details future directions concerning the dataset that could ameliorate is usability and further support MIR research. Finally the paper is concluded in Sect. 5.

2 Background and Related Research

2.1 Hit Song Science

Hit Song Science (HSS) [13] refers to the MIR direction aiming in predicting the popularity of musical tracks, as presented in top-charts. A number of scenarios' parameters exist as to the prediction's prerequisites, such as the little or no availability of early popularity information, the granularity of popularity definition, the type of input sources representing the musical tracks and many others.

Similarly, under the auspices of HSS numerous research tasks also take place: popularity pattern modelling, binary (hit/non-hit) or otherwise granulated popularity classification, tracks' future position on the popularity chart prediction given current position, popularity correlation to other activities (i.e. twitter posts, music search/download in peer-to-peer networks, etc.), prediction of the popular track subset of an album and many more.

The ability to predict the popularity of musical tracks is of great importance to all parties involved in the musical content lifecycle. Creators can work reversely the process of HSS and focus on characteristics that make their songs more probable to be popular in addition to customised characteristics of listeners, markets or distribution channels. The music industry, aiming at maximum profit, could benefit by selecting the most promising of the works for publication as well as, given that popularity predictions can be attributed to specific profile candidate consumers, modify accordingly its marketing plans. Finally, music consumers indirectly increase enjoyment of listening by receiving music that the distribution channels have either selected to fit their profile or that is in general more probable to be of high popularity and thus more probable widely liked.

It is widely claimed that the breadth of characteristics that lead to the popularity of a musical piece exceeds the *per se* track's content i.e., the audio and lyrics. Factors such as artist preferential attachment [1], society and culture [4], the changing musical tastes leading to evolving popularity pattern [11], psychological parameters on the reasons for preferring a track and listening exposure to tracks [12], the video clip of the track [3] just to name a few, play also an important role.

Nevertheless, existing research in the area agree that, beyond the very hard to measure characteristics, quantifiable qualities of musical tracks that contribute to a track's popularity do exist. Accordingly, the burden remains with the transformation representations of musical tracks that need to adhere to the track's popularity pertinent attributes.

2.2   Existing Research

In the first work on the area, Dhanaraj and Logan [4], utilise SVM and boosting classifiers on both acoustic and lyric information for the purposes of hit songs' separation from non-hits. Their aim is to determine if such a task is feasible or if hit song science claims are to be deemed as impossible, arriving after experimentation at the former.

Chon et al. [3] research for meaningful patterns within musical data while also attempting to predict both how long an album will stay in chart as well as a new album's position in chart on a certain week in the future using with the first few weeks' sales data. The results presented therein indicate interesting correlations.

Pachet and Roy, in [13], and Pachet, in [12], describe a large scale experiment aiming at the validation of current state-of-the-art methods' capabilities to predict the popularity of musical titles based on acoustic and/or contextual features. Both these works suggest that the commonly used features for music analysis are not informative enough to offer judgment on notions related to subjective aesthetics.

In [1], Bischoff et al. propose the music pieces' success prediction by exploitation of social interactions and annotations leaving out content characteristics of the musical tracks. Thus, their method relies on data mined from the Last.fm[5] music social network and the relationship between tracks, artists and albums while reaching promisingly improved results.

In a differentiated scenario, the work of Koenigstein et al. [8] compares peer-to-peer file sharing information on songs to their popularity, as described on the Billboard[6] charts. Their work indicates popularity trends of songs on the Billboard having a strong correlation to their respective popularity on peer-to-peer network. Based on this result, Koenigstein et al. propose a methodology that utilises the aforementioned correlation in order to to predict a songs' success on the popularity charts offering a 2–3 weeks prior to chart announcement prediction with high accuracy.

Following the work [13], Ni et al. [11] on a slightly alternated research question argue the feasibility of popularity prediction, "given a relevant feature set". Based on that work, the website "Score a hit"[7] was also created.

The work of Kim et al. [7] proposes the collection of users' music listening behaviour from Twitter, based on music-related hashtags, for the purposes of pre-

---

[5] http://www.last.fm/.
[6] http://www.billboard.com/.
[7] http://www.scoreahit.com.

Table 1. Existing HSS research dataset details.

Research	content-based audio	lyrics	subjective contextual	objective contextual	objective metadata	P2P queries	album sales data	Dataset size	Hit definition	Top charts time span
[4]	✓	✓						1700 songs	Billboard top 1	Jan 1956 - Apr 2004
[3]							✓	291 albums	Billboard top 1-25	Sep 2002 - Jun 2006
[12], [13]	✓		✓		✓			32000 songs	HiFind popularity label: low, medium, high	?
[1]		✓						50555 songs	Billboard top 1, 3, 5, 10, 20, 30, 40, 50	Aug 1958 - Apr 2008
[8]						✓		185598176 p2p queries, 200 songs	Billboard top 10, 20, 20, 30, 40, 50, 100	Jan 2007 - Jul 2007
[11]	✓							5000 songs	Billboard top 5	1962-2011
[14]		✓						6815 songs	Billboard top 15, 25, 35	2008-2013
[7]				✓	✓			1806438 tweets, 168 songs	Billboard top 10, 20, 30, 40, 50	Nov 2013 - Jan 2014

dicting popularity rankings. The results reported show high correlation between users' music listening behaviour on Twitter and general music popularity trend.

Singhi and Brown [14] propose a hit detection model based Bayesian networks on solely lyrics' features.

Finally, Burgoyne et al. [2] present a close to the theme of musical track popularity work studying musical content's "catchiness", or the "long term musical salience" of a piece. Despite the broader scope of the musical popularity prediction task, the correlation of catchiness to popularity is evident although most probably one directional, since numerous less-memorable top-chart tracks do exist.

The aforementioned existing research, with the exception of [12,13], have utilised different datasets to perform experimentation. The diversity of the utilised datasets in terms of size vary greatly as shown in Table 1.

3 The Dataset

The TPD is a collection of information revolving around the notion of track popularity. Its aim is to provide an easy to use collection of information for the purposes of track popularity data mining research tasks. In this Section we detail the creation process and content of the TPD.

3.1 Creation Process

In order to create the TPD, we separated the potential information sources into three distinct categories: the popularity sources, the metadata/content sources and the contextual similarity source.

The selection of the popularity periods was made based on the availability of both popularity information from the sources and access to the tracks' content. Thus, from popularity sources Last.fm and Spotify we collected all available popularity charts at the time of collection, that is from 17 September 2006 up to 28 December 2014 and 28 April 2013 up to 18 January 2015, respectively. From popularity source Billboard we collected the last 10 years, ranging from 03 January 2004 up to 24 January 2015.

Following the collection of the popular tracks from the popularity sources, we utilised the metadata/content sources Apple[8], Spotify[9], 7digital[10] in order to identify and get information for albums of the collected popular tracks and then to gather information on remaining, non-popular, tracks of each album.

Access to the content of the collection's tracks was based on the metadata/content sources' (Apple, Spotify and 7digital) 30 second previews clips as all three web services provide an API for the purposes of searching and streaming the audio clips. The collected files were converted to an appropriate format in order to undergo feature extraction.

While performing the above mentioned information collection processes, it was confirmed that multiple identification spaces do indeed exist for all track/author/album entities. Accordingly, and in order to facilitate the interoperability of the collected information, we performed exact match searches in all sources producing thus a mapping between different track/author/album identification spaces. As not all sources engulf information on all collected data, the mapping is not complete, but nevertheless, far from sparse (~55% of the matrix cells contain values). Content for the mapping was collected from both popularity sources and metadata/content sources.

To enrich further the TPD, we additionally included contextual information as to the similarity of the collection's tracks based on Last.fm's API *track.getSimilar* method that provides similarity between tracks, based on listening data.

Finally, for each track of the TPD, three feature-sets extracted directly from the audio content are included in matlab variable MAT-files. The first feature-set, *feature-set A*, is based on jAudio [10] and contains only single overall average and standard deviation values performed on all values of the features over all windows with window size 512 samples and 0 % overlap between successive windows. The second feature-set, *feature-set B*, was created with MIRToolbox offering per window feature extraction with window size 1024 samples and 50 % overlap between successive windows. The two feature-sets provide different levels of detail on the audio content in order to suit a broad range of applications. The third feature-set, *feature-set C* is based on the periodicity function of the tempo estimation method presented in [5].

3.2   The Content

The TPD contains 23.385 tracks of which, 9.193 are designated as popular by appearing in any of the popularity sources charts, while 14.192 are tracks that

---

[8] https://www.apple.com/itunes/affiliates/resources/documentation/itunes-store-web-service-search-api.html.

[9] https://developer.spotify.com.

[10] http://developer.7digital.com/.

appear in one of the 1.843 albums of the popular tracks and are not designated as popular by any of the popularity sources. The popularity ratings records, contain the position of a track for a specific week, collected from Billboard are 57.800, while for Last.fm and Spotify are 43.300 and 6.500, respectively. Of the popular tracks, 1,5 % are designated in all three sources of popularity, 5,9 % in two sources and 92,6 % in just one source. The discrepancy in proportions is due to the range of available data by the popularity sources. As far as the contextual similarity based on Last.fm's API *track.getSimilar* method is concerned, 78 % of the popular tracks of the dataset have a degree of contextual similarity to other popular tracks of the dataset. As not all tracks' audio files were possible to be found, the TPD contains audio derived features for ∼74% of the tracks.

Of the three feature-sets included in the TPD described in Sect. 3.1, *feature-set A* is meant as a small, less detailed feature set for fast and simple research applications. The features included in *feature-set A* are: overall standard deviation & overall average of spectral centroid (dimension: 1), spectral rolloff point (dim: 1), spectral flux (dim: 1), compactness (dim: 1), spectral variability (dim: 1), root mean square (dim: 1), fraction of low energy windows (dim: 1), zero crossings (dim: 1), strongest beat (dim: 1), beat sum (dim: 1), strength of strongest beat (dim: 1), strongest frequency via zero crossings (dim: 1), strongest frequency via spectral centroid (dim: 1), strongest frequency via FFT maximum (dim: 1), MFCCs (dim: 13), LPCs (dim: 10), method of moments (dim: 5), partial based spectral centroid (dim: 1), partial based spectral flux (dim: 1), peak based spectral smoothness (dim: 1), relative difference function (dim: 1), area method of moments (dim: 10). The second feature-set, *feature-set B*, contains windowed MFCCs (dim: 13), rolloff (dim: 1), brightness (dim: 1), flux (dim: 1), zero crossings (dim: 1), inharmonicity (dim: 1), centroid (dim: 1), spread (dim: 1), skewness (dim: 1), kurtosis (dim: 1), flatness (dim: 1), entropy (dim: 1). The third feature-set, *feature-set C*, contains 276 target tempi. For each target tempo this feature-set contains eight energy bands and one chroma (dim: 9).

In order to provide an aggregated glimpse of the popularity records of the dataset by contrasting the popularity sources, Fig. 1 shows the normalised probability density of next week's rank increase/decrease (position change) given

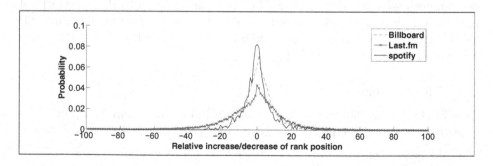

Fig. 1. The normalised probability density of position change.

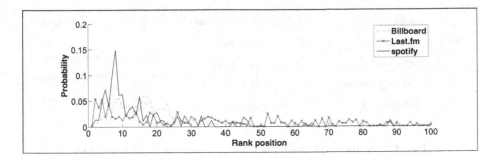

Fig. 2. Popularity chart entry position probability density.

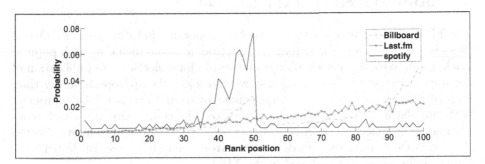

Fig. 3. Popularity chart leave position probability density.

current position for all three sources of popularity. Moreover, Figs. 2 and 3 show the probability density of rank position when entering and leaving respectively the top-100 popularity chart for all three sources of popularity.

3.3 Format and Usage

The dataset is divided into two separate parts: part A includes the relations/metadata of the tracks and their popularity while part B contains the files of the three feature-sets.

The first part is in the form of a relational database, the compact schema of which is shown in Fig. 4. The archive of part A contains the SQL statements that will create the TPD database and tables and subsequently load all the information into the tables of an existing MySQL installation. Moreover, the contents of the first part are also provided in CSV format, in order to support fast use of the data and alleviate the necessity for a relational database. The second part consists of compressed archives of bz2 type that contain the feature-sets in a one file with features per track manner.

The complete TPD can be downloaded from http://mir.ilsp.gr/track_popularity.html.

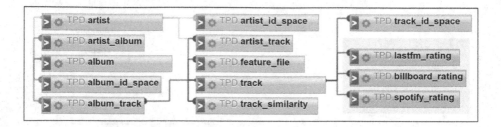

Fig. 4. Schema for the metadata and the popularity of the tracks.

4   Future Direction of the Dataset

The TPD is not without issues that can be ameliorated in future versions. One of these issues pertains to the automatic selection of album including each popular track: as more than one such albums may exist (hit collections, re-publication of the same artist, etc.), there is no easy way to select the appropriate other than manual filtering. Moreover, the requirement of having access to the content of both popular and non-popular tracks elevated the complexity and timely conclusion of the collection process, which in order to remain within limits affected the size of the popularity records collected from the only source, Billboard, containing information not included in the TPD.

Some of the future actions that would greatly ameliorate the TPD are:

API. A documented API for the purposes of accessing from a single point, aggregated, integrated and fully up-to-date popularity information.

Automated updates. The design and implementation of a fully automated collection and integration web-based service that will update the dataset by harvesting the sources using event-driven or periodical triggers.

Popularity sources. The addition of more popularity sources mostly oriented to social networks, such as twitter based hash-tags (e.g. *#nowplaying* with mention of track's metadata) as well as directly collecting tracks' airtime from e-radios using common protocols (e.g. *Shoutcast, Icecast*, etc.).

5   Conclusion

This work introduces the Track Popularity Dataset. The dataset is, to the best of the authors' knowledge, the first complete attempt to create an integrated dataset for the purposes of mining information from musical track popularity. It includes three different sources of popularity definition with records ranging from 2004 to 2014, a mapping between different track/author/album identification spaces, information for the remaining, non popular, tracks of an album with popular track(s), contextual similarity between tracks and ready for MIR use extracted features.

Despite the inherent difficulty of popularity prediction prior to or during the initial period of a track's release, such a process has long been a requirement of the musical industry, while interestingly enough, the gains of such a prediction also profit artists and listeners. Thus, the availability of datasets that will allow music information researchers to experiment and compare their methods would greatly support the advancement of the research direction.

Future directions of the dataset include its manual filtering in order to enhance its content, the creation of an API for the dissemination of the dataset's information, an automated collection of up-to-date popularity information process and the expansion of the sources by addition of social networks and e-radios.

References

1. Bischoff, K., Firan, C.S., Georgescu, M., Nejdl, W., Paiu, R.: Social knowledge-driven music hit prediction. In: Huang, R., Yang, Q., Pei, J., Gama, J., Meng, X., Li, X. (eds.) ADMA 2009. LNCS, vol. 5678, pp. 43–54. Springer, Heidelberg (2009)
2. Burgoyne, J.A., Bountouridis, D., Balen, J.V., Honing, H.: Hooked: a game for discovering what makes music catchy. In: International Society for Music Information Retrieval Conference, pp. 245–250 (2013)
3. Chon, S.H., Slaney, M., Berger, J.: Predicting success from music sales data: a statistical and adaptive approach. In: ACM Workshop on Audio and Music Computing Multimedia, pp. 83–88 (2006)
4. Dhanaraj, R., Logan, B.: Automatic prediction of hit songs. In: International Society for Music Information Retrieval Conference, pp. 488–491 (2005)
5. Gkiokas, A., Katsouros, V., Carayannis, G., Stajylakis, T.: Music tempo estimation and beat tracking by applying source separation and metrical relations. In: 2012 IEEE International Conference on Acoustics, Speech and Signal Processing (ICASSP), pp. 421–424 (2012)
6. Karydi, D., Karydis, I., Deliyannis, I.: Legal issues in using musical content from itunes and youtube for music information retrieval. In: International Conference on Information Law (2012)
7. Kim, Y., Suh, B., Lee, K.: #nowplaying the future billboard: mining music listening behaviors of twitter users for hit song prediction. In: International Workshop on Social Media Retrieval and Analysis, pp. 51–56 (2014)
8. Koenigstein, N., Shavitt, Y., Zilberman, N.: Predicting billboard success using data-mining in p2p networks. In: International Symposium on Multimedia, pp. 466–470 (2009)
9. Makris, D., Kermanidis, K.L., Karydis, I.: The Greek audio dataset. In: Iliadis, L., Maglogiannis, I., Papadopoulos, H., Sioutas, S., Makris, C. (eds.) Artificial Intelligence Applications and Innovations. IFIP AICT, vol. 437, pp. 165–173. Springer, Heidelberg (2014)
10. McEnnis, D., McKay, C., Fujinaga, I., Depalle, P.: jAudio: a feature extraction library. In: International Society for Music Information Retrieval Conference, pp. 600–603 (2005)
11. Ni, Y., Santos-Rodriguez, R., McVicar, M., De Bie, T.: Hit song science once again a science? In: International Workshop on Machine Learning and Music. ACM (2011)
12. Pachet, F.: Hit song science. In: Tao, T.O. (ed.) Music Data Mining, chap. 10, pp. 305–326. Chapman & Hall/CRC Press (2011)

13. Pachet, F., Roy, P.: Hit song science is not yet a science. In: International Society for Music Information Retrieval Conference, pp. 355–360 (2008)
14. Singhi, A., Brown., D.G.: Hit song detection using lyric features alone. In: International Society for Music Information Retrieval Conference (2014)

On the Computational Prediction
of miRNA Promoters

Charalampos Michail[1], Aigli Korfiati[1,2(✉)],
Konstantinos Theofilatos[2], Spiros Likothanassis[1],
and Seferina Mavroudi[3]

[1] Department of Computer Engineering and Informatics,
University of Patras, Patra, Greece
{cmichail,korfiati,likothan}@ceid.upatras.gr
[2] InSyBio Ltd., London, UK
k.theofilatos@insybio.com
[3] Department of Social Work,
Technological Institute of Western Greece, Patra, Greece
mavroudi@teiwest.gr

Abstract. MicroRNAs transcription regulation is an open topic in molecular biology and the identification of the promoters of microRNAs would give us relevant insights on cellular regulatory mechanisms. In the present study, we introduce a new computational methodology for the prediction of microRNA promoters, which is based on the hybrid combination of an adaptive genetic algorithm with a nu-Support Vector Regression (nu-SVR) classifier. This methodology uses genetic algorithms to locate the optimal features set and to optimize the parameters of the nu-SVR classifier. The main advantage of the proposed solution is that it systematically studies and calculates a vast number of features that can be used for promoters prediction including frequency-based properties, regulatory elements and epigenetic features. The proposed method also handles efficiently the issues of over-fitting, feature selection, convergence and class imbalance. Experimental results give accuracy over 87 % in the miRNA promoter prediction.

Keywords: miRNA promoters · Classification · Computational prediction · Feature selection · Transcription start sites

1 Introduction

One of the current trends in molecular biology is studying the various types of short and long non-coding RNAs (ncRNAs) [1]. MicroRNAs (miRNAs) are the most thoroughly characterized subclass of short RNAs in the recent literature [2]. miRNAs are short (21–23 nt) and single stranded endogenous RNA molecules. They regulate protein coding genes by binding to the 3' untranslated regions (3' UTRs) of their target mRNAs. This binding event causes translational repression of the target gene or stimulates rapid degradation of the target transcript [3]. miRNAs are involved in diverse biological processes, including development, differentiation, apoptosis, cell proliferation, and disease [3]. A growing number of studies indicate that miRNAs play

© IFIP International Federation for Information Processing 2016
Published by Springer International Publishing Switzerland 2016. All Rights Reserved
L. Iliadis and I. Maglogiannis (Eds.): AIAI 2016, IFIP AICT 475, pp. 573–583, 2016.
DOI: 10.1007/978-3-319-44944-9_51

crucial roles in human disease development, progression, prognosis, diagnosis and evaluation of treatment response [4] and miRNAs have been linked to cancer, neurodegenerative and cardiovascular diseases.

Many algorithms are able to predict miRNA genes and their targets, but their transcription regulation is still under investigation [5]. It is generally believed that intragenic (intronic, exonic) miRNAs (located in introns or exons of protein coding genes) are co-transcribed with their host genes [6], but literature has indicated that intragenic miRNA genes may be transcribed by their own promoter [7, 8]. Intergenic miRNAs (located between protein coding genes) are independent transcription units, with their own transcriptional regulatory elements [9]. For intergenic miRNAs the distances between transcription initiation sites (TSSs) and miRNA-coding regions dramatically vary, ranging from a few hundred bases to 30-kb upstream and the nature of the primary transcript of intergenic miRNAs and promoter organization are largely unknown.

Transcription initiation is a key step in the regulation of gene expression. During this process, transcription factors bind promoter region of a gene in a sequence-specific manner and recruit the RNA polymerase to form an active initiation complex around the transcription start site (TSS) [10]. The promoter is commonly referred to as the region upstream of a gene that contains the information permitting the proper activation or repression of the gene that it controls [11]. The promoter region is divided into three parts:

- the core-promoter is 100 bp long, surrounds the TSS and contains binding sites for RNA polymerase II (Pol II) and general transcription factors;
- the proximal promoter is several hundred base pairs long upstream the core promoter and contains several regulatory elements;
- the distal promoter is up to thousands of base pairs long upstream of the TSS and contains additional regulatory elements called enhancers and silencers.

As it contains primary information to control gene transcription, it is a fundamental step to identify the core-promoter in study of gene expression patterns and constructing gene transcription networks.

In the present study, a new computational methodology for the prediction of miRNA promoters is introduced and it is based on the hybrid combination of an adaptive genetic algorithm with a nu-SVR classifier. This methodology uses genetic algorithms to locate the optimal features set and optimize the parameters of the nu-SVR classifier. The main advantage of the proposed solution is that it systematically studies the different features that can be used for miRNA promoters prediction. The simple method and script provided can be used to calculate effectively most of the features that have been correlated with promoter attributes without the need for combining different tools. In terms of classification performance, the main advantages of the proposed method are that it handles efficiently the issues of over-fitting, feature selection, convergence and class imbalance. Experimental results give accuracy over 87 % in the miRNA promoter prediction.

The rest of the article is organized as follows: Sect. 2 describes the existing methods for the prediction of protein coding gene and miRNA promoters, Sect. 3 analyzes the proposed methodology and the relevant datasets and features which were used, Sect. 4 presents the experimental results and Sect. 5 concludes the paper and discusses interesting future research directions.

2 Promoter Prediction Methods

The promoter of a gene is a significant region for its transcription initiation and thus, identifying miRNA promoters would give us insights on their regulatory mechanism. A common practice in miRNA promoters identification is to first apply a promoter prediction method to predict their promoters, and then to verify the predictions by wet lab experiments. Developing the promoter identification algorithm is a very challenging problem [12]. A number of computational methods for predicting promoters of protein-coding genes have been developed, however their performances are far from satisfactory, because our understanding of the transcription process is incomplete.

Literature indicates that there are features of the promoter regions that differentiate them from other parts in the genome. These features include TATA-box, GC-box, CAAT-box, and Inr [10]. Others features include the CpG islands close to the TSS, binding sites of typical transcription factors, chromatin modifications and statistical properties of the core and proximal promoter. The similarities between orthologous promoters and information from mRNA transcripts have also been used to identify promoters [11]. Some well-known promoter prediction programs are CoreBoost_HM [10], McPromoter [13] and EP3 [11].

Concerning miRNA promoter prediction, initial approaches trained classifiers on protein coding genes promoters and applied them to identify miRNA promoters [9, 12, 14, 15]. These techniques provided the first indications of miRNA transcription start site (TSS) positions on a genome-wide scale. However, they were not built based on the promoters of microRNA genes and they exhibit high false-positive rates. Additionally, although miRNA promoters present several similarities with RNA Pol II promoters, this is mainly true for intergenic miRNAs, as too little is still known about intragenic miRNA promoters. For these reasons, a supervised method trained on protein-coding genes is not the optimal choice for identifying miRNA promoters [16].

Other studies for miRNA promoter prediction are based on experimental data, such as cap analysis of gene expression (CAGE) data, RNA Pol II data or histone modification data. CAGE tags were used to identify miRNA TSSs by considering its possibility to capture the 5' cap. MiRStart [15], PROmiRNA [16] and the method of Saini et al. [9] are representative examples. However, there exist uncapped pre-miRNAs that can't be captured by CAGE technology. The methods of Wang et al. [10], Zhou et al. [12] and Corcoran et al. [7] are based on RNA Pol II data. However, these studies were limited to small amount of miRNAs due to the insufficiency of Pol II data. Chromatin signature based methods use histone mark profiles, such as H3K4me3 [17, 18] or nucleosome positioning patterns [8] in specific cell lines to annotate miRNA promoters de novo. A recent method [19] combined data of H3K4me3 and DNase I hypersensitive sites (DHSs) with conservation and sequence features to identify cell-specific TSSs. Although histone mark-based methods have good results, they have been designed for specific cell lines. Additionally, due to the nature of ChIP-seq experiments, chromatin-based methods represent a valuable strategy for detecting intergenic and host gene miRNA promoters, but they might lack the sensitivity required to identify intronic promoters [16].

3   Materials and Methods

3.1   Datasets

Exclusively experimentally validated miRNA TSSs have been used in order to construct a positive dataset of high quality. For each TSS, the [−1000, 1000] bp region around it has been used as the promoter region. Even though a rather smaller region, [−250, 50] or [−450, 50] bp around the TSS, is usually used in other methods, we have included 1000 bases downstream the TSS, since in [20] it is suggested that downstream elements also regulate transcription. The experimentally validated TSSs have been downloaded from the miRT [21] database. From the total of 670 TSSs, we have selected only the 306 that are related to the gene assembly hg19. With these TSSs we have then queried the UCSC DAS server [22] in order to extract the promoter sequences.

For the construction of the negative dataset, i.e. a set of sequences that do not contain miRNA promoters, a pool of 1224 sequences was formed: the four (two upstream and two downstream) non-overlapping consecutive segments immediately upstream and downstream of the positive dataset, as suggested in [23].

Since we wanted to preserve a 1:1 ratio between the positive and the negative dataset, 306 sequences from the negative pool of 1224 are selected at random in every execution of the proposed method. This rate has been maintained because an imbalanced distribution could affect the performance of the classifier.

3.2   Features

Representative features are essential in order to train the proposed model for efficiently distinguishing miRNA promoters from a negative data set. Features from several broad categories are presented in the literature [23–25] and since the proposed method is able to handle large numbers of features and to extract the optimal subset of them, we have used them all as inputs. They include (i) frequency-based properties of the promoters such as k-mers, word commonality, skew, palindromes; (ii) regulatory elements such as CpG islands, repetitive elements; (iii) epigenetic features such as chromatin states. The features employed are summarized in Table 1. For the features calculation we have implemented a Python script, which is freely available at: https://github.com/bioinfoceid/miRNAPromoters. Some features are calculated over the whole examined sequences, while others on sliding windows. When sliding windows are employed, their size is 100 bp and their step is 50 bp, thus resulting in 39 windows for an examined sequence of 2000 bp.

Concerning the k-mers, we have calculated the frequencies of mono-, di-, tri-consecutive nucleotides (4, 16, 64 frequencies, respectively) in the upstream [−1000, −1] and downstream [+1, +1000] regions relative to TSSs separately. This generates 2* (4 + 16 + 64) = 168 features.

Table 1. Number of features per category

Feature category	Number of features
K-mers	2*(4 + 16 + 64) = 168
Observed/expected ratio di- and tri-nucleotides	16 + 64 = 80
Word commonality	39
AT- and CG-skews	2*39 = 78
Palindromes	39
CpG islands	2*39 = 78
Repetitive elements	39
Chromatin states	7*15 = 105
Total	626

The observed/expected ratio of di-nucleotides is

$$\frac{Obs}{Exp}b_i b_j = \frac{\#(b_i b_j)}{\#b_i * \#b_j} * N$$

and of tri-nucleotides is

$$\frac{Obs}{Exp}b_i b_j b_k = \frac{\#(b_i b_j b_k)}{\#b_i * \#b_j * \#b_k} * N^2$$

where b_i, b_j, b_k are the nucleotides A, C, G, T and $\#b_i, \#b_j, \#b_k, \#(b_i b_j), \#(b_i b_j b_k)$, are numbers of mono-, di- and tri-nucleotides and N is the total number of nucleotides in the examined sequence. The calculation of these ratios yields 16 features for the di-nucleotides and 64 for the tri-nucleotides.

For the word commonality feature category we have downloaded 1000 random sequences of 1000 bp from the gene assembly hg19 and then counted the frequency of all possible hexamers. The frequency of each hexamer has been normalized so that the least common hexamer has the score 0 and the most common one has the score 1. Finally, we have calculated a score in each sliding window of the examined sequence by adding the score of all hexamers occurring in that window, resulting in 39 features.

The AT-skew:

$$ATskew = \frac{\#A - \#T}{\#A + \#T}$$

and CG-skew:

$$CGskew = \frac{\#C - \#G}{\#C + \#G}$$

where $\#A, \#T, \#C, \#G$ represent the number of A, T, C and G have been calculated over each of the 39 sliding windows, thus producing 78 features.

In each sliding window, we have calculated the number of nucleotides overlapping with any palindrome of length six or more, taking into account that a sequence is considered a palindrome if it is equal to its reverse complement. This generates 39 features.

For the CpG islands features category we have calculated the following two metrics in each of the sliding windows:

$$M1 = \frac{\#CG}{\#C + \#G} * window\_length$$

$$M2 = \frac{\#C + \#G}{window\_length} * 100$$

This results in 2*39 = 78 features.

In each sliding window, we have calculated the number of nucleotides overlapping with any repetitive element, producing 39 features.

For each examined sequence we have calculated the percentage of the total number of positions overlapping with each of 15 different chromatin states in each of 7 cell types (GM12878, H1-hESC, HMEC, HSMM, HUVEC, NHEK, and NHLF) [26]. The coordinates of the states have been downloaded from the UCSC Genome Browser. This produces 7*15 = 105 features.

3.3   Proposed Methodology

The proposed method is an embedded classification method that combines an adaptive GA with a nu-SVR classifier. It is inspired by EnsembleGASVR, a method suggested in [27] for classifying missense single nucleotide polymorphisms. In principle, SVR classifiers present high classification performance and low complexity. The nu parameter of a nu-SVR classifier allows for the tuning of the number of the support vectors in the resulting classification model. GAs are stochastic meta-heuristic optimization algorithms. One advantage of GAs is their ability to explore efficiently large search spaces and identify possible solutions, without getting trapped in local optima, while at the same time locating near-to-optimal solutions. In the proposed method, the adaptive GA is used to identify the best feature subset and to tune the nu-SVR parameters.

The produced hybrid algorithm mainly consists of the iterative application of the evaluation, selection, crossover and mutation steps in a population of candidate solutions (chromosomes) which are initially randomly generated. Binary encoding has been used to represent each chromosome. Specifically, a 680-bit string is used where 626 bits encode features and 54 bits encode parameters. The parameters are (i) classifier parameters C (20 bits) and (ii) nu (10 bits); (iii) radial basis kernel bandwidth *gamma* (14 bits); and (iv) classification *threshold* (10 bits).

A rank-based roulette wheel selection method controls the selection of the best candidates in each GA generation. This selection mechanism is preferred compared with the single roulette wheel selection to raise the selection pressure toward better solutions when all solutions of the population present similar fitness values. Elitism is used to force the best solution of each population to be selected at least once in the next generation. This selection mechanism has been suggested in [27].

The evaluation of each chromosome in the population is performed according to the following fitness function:

$$Fitness = a * Accuracy + b * GeometricMean - c * 10^2 * MSE - d$$

$$* \frac{1}{626} Features - e * \frac{1}{408} * SupportVectors$$

where *Accuracy* is the nu-SVR's accuracy, *GeometricMean* is the geometric mean of sensitivity and specificity, *MSE* is the mean square of errors, *Features* is the size of the selected features subset and *SupportVectors* is the number of support vectors included in the trained nu-SVR model.

The ranges of the examined variables in the proposed fitness function are $Accuracy \in [0, 1]$, $GeometricMean \in [0, 1]$, $MSE \in [0, 0.01]$, $Features \in [1, 626]$, where 626 represents the maximum number of features that can be selected by this method, and $SupportVectors \in [1, 408]$, where 408 represents the number of the training samples. $MSE, Features$ and $SupportVectors$ are multiplied by constants, as shown in the equation, to normalize their values in the range $[0, 1]$. The constants $a = 0.5$, $b = 0.5$, $c = 0.01$, $d = 0.005$ and $e = 0.001$ are user-defined weights assigned without experimentation and selected so as to reflect the priorities of each goal. More specifically, the classification accuracy and the geometric mean are the most significant. Then the MSE of the classifier follows. The number of selected features follows next and the number of support vectors is the least significant objective. To avoid over-fitting problems, we did not attempt to optimize these values, as suggested in [27]. The weights of the goals have been set so as to achieve high classification performance and simultaneously generate a simple and effective model.

Then the differentiation operators, crossover and mutation are applied to the top-ranked candidate solutions to create a new population. The crossover operator applies 2-point crossover to obtain a new offspring from two parents. The crossover rate is constant and set to 0.9, in order to leave some part of the population to survive unchanged to the next generation. This property is essential when good solutions emerge in early stages of the algorithm, as proposed in [27].

The mutation operator is very important to avoid local optima and explore a wide area of the search space. In the first generations it is preferable to explore a wider search space (exploration), while in the last generations it is preferable to search locally near the most promising areas of the search space (exploitation). To balance the tradeoff between the exploration and exploitation, the proposed method uses an adaptive mutation probability starting with a high value, 0.2, and gradually decreasing. The mutation rate is computed according to the following equation:

$$P_m(n) = 0.2 - n * \frac{0.2 - \frac{1}{P\_S}}{MAX\_G}$$

where n is the current generation, P_S is the size of the population and MAX_G is the maximum generation specified by the termination criterion. The mean similarity of every chromosome with the best chromosome of the population is measured at every generation. If the mean similarity is greater than 90 % then the mutation rate is increased by a factor of $\frac{0.2-\frac{1}{P\_S}}{MAX\_G}$ instead of being decreased to avoid stagnation, i.e. getting trapped to local optima

The size of the population is set to 80 chromosomes and the termination criterion is 250 generations.

4   Experimental Results

In order to evaluate the performance of our method for predicting miRNA promoters against other sequences, we have performed 10 5-fold external cross validation experiments and then we took the average in order to better assess the performance. In each fold, the $\frac{2}{3}$ of the samples were used to train the SVM model and $\frac{1}{3}$ of the samples were used as validation samples to measure the performance and calculate the fitness values. Table 2 summarizes the results which have been achieved by the proposed method. It presents the average values of all 10 5-fold external cross validation experiments for the classification metrics: accuracy, specificity, sensitivity and geometric mean. The last column is the average value for the 10 5-fold cross validation experiments of the number of the selected features which are used as inputs in our method.

Table 2. Metrics

Accuracy	Sensitivity	Specificity	Geometric mean	Number of features
0.8786	0.8447	0.9126	0.8780	307

The proposed methodology achieves on average accuracy 87.86 %, sensitivity 84.47 %, specificity 91.26 % and geometric mean 87.8 %. The average of selected features is 307 out of the total 626. These results are comparable with those of the better performing methods in the literature. The recent method of Hua et al. [19] presents 84 % sensitivity and 91.3 % precision and the miRStart method [15] presents sensitivity of 90.36 %, specificity of 90.05 %, accuracy of 90.21 % and precision of 90.08 %. The accuracy, sensitivity, specificity, precision and Matthews Correlation Coefficient (MCC) of [25] are 92.00 %, 91.56 %, 92.15 %, 79.74 % and 0.80, respectively.

5   Conclusion

The proposed approach for the prediction of miRNA promoters is a computational methodology based on the hybrid combination of an adaptive genetic algorithm with a nu-SVR classifier. The adaptive genetic algorithm is responsible for locating the optimal features set and optimizing the parameters of the nu-SVR classifier. The main advantage of the proposed solution is that it systematically studies a vast number of features that can be used for miRNA promoters prediction. They include (i) frequency-based properties of the promoters such as k-mers, word commonality, skew, palindromes; (ii) regulatory elements such as CpG islands, repetitive elements; (iii) epigenetic features such as chromatin states. The provided script can be used to calculate effectively most of the features that have been correlated with promoter attributes without the need for combining different tools. The proposed script is of general usage as it can be used to structurally, sequentially and epigenetically annotate candidate promoters not only for miRNAs but also for protein coding genes and other non-coding RNA categories. In terms of classification performance, the proposed method handles efficiently the issues of over-fitting, feature selection, convergence and class imbalance. Experimental results give accuracy over 87 %, sensitivity over 84 % and specificity over 91 % in the miRNA promoter prediction.

Our future research plans involve a more extensive study on the calculated features in order to gain insight on miRNA promoters characteristics. Additionally, in order to better handle the class imbalance issue, we plan to employ the Synthetic Minority Over-sampling Technique (SMOTE) [28]. Finally, we plan to compare the proposed solution with other existing solutions in the same datasets in order to gain more fair comparative results and to better validate the performance of the proposed solution.

Acknowledgement. Insybio participates in NBG Business Seeds Program by NBG.

References

1. Stefani, G., Slack, F.J.: Small non-coding RNAs in animal development. Nat. Rev. Mol. Cell Biol. **9**(3), 219–230 (2008)
2. Krol, J., Loedige, I., Filipowicz, W.: The widespread regulation of microRNA biogenesis, function and decay. Nature Rev. Genet. **11**(9), 597–610 (2010)
3. Bartel, D.P.: MicroRNAs: target recognition and regulatory functions. Cell **136**(2), 215–233 (2009)
4. Jiang, Q., Wang, Y., Hao, Y., Juan, L., Teng, M., Zhang, X., Liu, Y., et al.: miR2Disease: a manually curated database for microRNA deregulation in human disease. Nucleic Acids Res. **37**(suppl 1), D98–D104 (2009)
5. Kleftogiannis, D., Korfiati, A., Theofilatos, K., Likothanassis, S., Tsakalidis, A., Mavroudi, S.: Where we stand, where we are moving: surveying computational techniques for identifying miRNA genes and uncovering their regulatory role. J. Biomed. Inform. **46**(3), 563–573 (2013)
6. Rodriguez, A., Griffiths-Jones, S., Ashurst, J.L., Bradley, A.: Identification of mammalian microRNA host genes and transcription units. Genome Res. **14**(10a), 1902–1910 (2004)
7. Corcoran, D.L., Pandit, K.V., Gordon, B., Bhattacharjee, A., Kaminski, N., Benos, P.V.: Features of mammalian microRNA promoters emerge from polymerase II chromatin immuno precipitation data. PLoS ONE **4**(4), e5279 (2009)

8. Ozsolak, F., Poling, L.L., Wang, Z., Liu, H., Liu, X.S., Roeder, R.G., Fisher, D.E., et al.: Chromatin structure analyses identify miRNA promoters. Genes Dev. **22**(22), 3172–3183 (2008)

9. Saini, H.K., Griffiths-Jones, S., Enright, A.J.: Genomic analysis of human microRNA transcripts. Proc. Nat. Acad. Sci. **104**(45), 17719–17724 (2007)

10. Wang, X., Xuan, Z., Zhao, X., Li, Y., Zhang, M.Q.: High-resolution human core-promoter prediction with CoreBoost_HM. Genome Res. **19**(2), 266–275 (2009)

11. Abeel, T., Saeys, Y., Bonnet, E., Rouzé, P., Van de Peer, Y.: Generic eukaryotic core promoter prediction using structural features of DNA. Genome Res. **18**(2), 310–323 (2008)

12. Zhou, X., Ruan, J., Wang, G., Zhang, W.: Characterization and identification of microRNA core promoters in four model species. PLoS Comput. Biol. **3**(3), e37 (2007)

13. Ohler, U., Niemann, H., Liao, G.C., Rubin, G.M.: Joint modeling of DNA sequence and physical properties to improve eukaryotic promoter recognition. Bioinformatics **17**(suppl 1), S199–S206 (2001)

14. Monteys, A.M., Spengler, R.M., Wan, J., Tecedor, L., Lennox, K.A., Xing, Y., Davidson, B. L.: Structure and activity of putative intronic miRNA promoters. RNA **16**(3), 495–505 (2010)

15. Chien, C.H., Sun, Y.M., Chang, W.C., Chiang-Hsieh, P.Y., Lee, T.Y., Tsai, W.C., Huang, H.D., et al.: Identifying transcriptional start sites of human microRNAs based on high-throughput sequencing data. Nucleic Acids Res. **39**(21), 9345–9356 (2011)

16. Marsico, A., Huska, M.R., Lasserre, J., Hu, H., Vucicevic, D., Musahl, A., Vingron, M., et al.: PROmiRNA: a new miRNA promoter recognition method uncovers the complex regulation of intronic miRNAs. Genome Biol. **14**(8), R84 (2013)

17. Barski, A., Jothi, R., Cuddapah, S., Cui, K., Roh, T.Y., Schones, D.E., Zhao, K.: Chromatin poises miRNA-and protein-coding genes for expression. Genome Res. **19**(10), 1742–1751 (2009)

18. Marson, A., Levine, S.S., Cole, M.F., Frampton, G.M., Brambrink, T., Johnstone, S., Calabrese, J.M., et al.: Connecting microRNA genes to the core transcriptional regulatory circuitry of embryonic stem cells. Cell **134**(3), 521–533 (2008)

19. Hua, X., Chen, L., Wang, J., Li, J., Wingender, E.: Identifying cell-specific microRNA transcriptional start sites. Bioinformatics, btw171 (2006)

20. Maston, G.A., Evans, S.K., Green, M.R.: Transcriptional regulatory elements in the human genome. Ann. Rev. Genomics Hum. Genet. **7**, 29–59 (2006)

21. Bhattacharyya, M., Das, M., Bandyopadhyay, S.: miRT: a database of validated transcription start sites of human microRNAs. Genomics, Proteomics Bioinform. **10**(5), 310–316 (2012)

22. Karolchik, D., Hinrichs, A.S., Furey, T.S., Roskin, K.M., Sugnet, C.W., Haussler, D., Kent, W.J.: The UCSC table browser data retrieval tool. Nucleic Acids Res. **32**(suppl. 1), D493–D496 (2004)

23. Zhao, X., Xuan, Z., Zhang, M.Q.: Boosting with stumps for predicting transcription start sites. Genome Biol. **8**(2), R17 (2007)

24. Alam, T., Medvedeva, Y.A., Jia, H., Brown, J.B., Lipovich, L., Bajic, V.B.: Promoter analysis reveals globally differential regulation of human long non-coding RNA and protein-coding genes. PLoS ONE **9**(10), e109443 (2014)

25. Bhattacharyya, M., Feuerbach, L., Bhadra, T., Lengauer, T., Bandyopadhyay, S.: MicroRNA transcription start site prediction with multi-objective feature selection. Stat. Appl. Genet. Mol. Biol. **11**(1), 1–25 (2012)

26. Ernst, J., Kheradpour, P., Mikkelsen, T.S., Shoresh, N., Ward, L.D., Epstein, C.B., Ku, M., et al.: Mapping and analysis of chromatin state dynamics in nine human cell types. Nature **473**(7345), 43–49 (2011)

27. Rapakoulia, T., Theofilatos, K., Kleftogiannis, D., Likothanasis, S., Tsakalidis, A., Mavroudi, S.: EnsembleGASVR: a novel ensemble method for classifying missense single nucleotide polymorphisms. Bioinformatics **30**(16), 2324–2333 (2014)
28. Chawla, N.V., Bowyer, K.W., Hall, L.O., Kegelmeyer, W.P.: SMOTE: synthetic minority over-sampling technique. J. Artif. Intell. Res. **16**, 321–357 (2002)

New Methods and Tools for Big Data Wokshop (MT4BD)

Building Multi-occupancy Analysis and Visualization Through Data Intensive Processing

Dimosthenis Ioannidis[1,2], Pantelis Tropios[1], Stelios Krinidis[1(✉)],
Dimitris Tzovaras[1], and Spiridon Likothanassis[2]

[1] Information Technologies Institute, Centre for Research and Technology
Hellas, 6th Km Charilaou-Thermi, 57001 Thermi-Thessaloniki, Greece
{djoannid,ptropios,krinidis,
Dimitrios.Tzovaras}@iti.gr
[2] Computer Engineering and Informatics,
University of Patras, Rio, Patras, Greece
likothan@ceid.upatras.gr

Abstract. A novel Building Multi-occupancy Analysis & Visualization through Data Intensive Processing techniques is going to be presented in this paper. Building occupancy monitoring plays an important role in increasing energy efficiency and provides useful semantic information about the usage of different spaces and building performance generally. In this paper the occupancy extraction subsystem is constituted by a collection of depth image cameras and a multi-sensorial cloud (utilizing big data from various sensor types) in order to extract the occupancy per space. Furthermore, a number of novel visual analytics techniques allow the end-users to process big data in different temporal resolutions in a compact and comprehensive way taking into account properties of human cognition and perception, assisting them to detect patterns that may be difficult to be detected otherwise. The proposed building occupancy analysis system has been tested and applied to various spaces of CERTH premises with different characteristics in a real-life testbed environment.

Keywords: Big data analysis · Building occupancy · Occupancy extraction · Human presence · Building occupancy visualization

1 Introduction

Knowing the true occupancy, the presence or the actual number of occupants of a building at any given time is fundamental for the effective management of various building operation functions ranging from security concerns to energy savings targets, especially in complex buildings with different internal kind of use [1–4]. The accurate definition of occupancy is the amount of people per building's spaces at any given time. Furthermore the influence that the occupants' actions have in the indoor environment [5], including those related to their business processes can also be added to the definition. Occupant's locations within the building varies throughout the day, therefore it is difficult to characterize the number of people that occupy a particular space

L. Iliadis and I. Maglogiannis (Eds.): AIAI 2016, IFIP AICT 475, pp. 587–599, 2016.
DOI: 10.1007/978-3-319-44944-9_52

and for what duration because human behavior is considered stochastic in nature [6]. Due to the random nature of individuals' behavior and challenges accessing accurate data, current studies include the creation of deterministic schedules where a standard workday profile is the same for the whole workweek and both weekend days have the same profile [7].

There are numerous techniques to detect space occupancy and even track their movements, which can be found in the literature. These techniques range from user surveys, interviews or walkthrough inspections [8–13] to a more or less complex deployment of sensors within the area of study [1, 2, 4, 14, 15]. The sensors used to this purpose are of various kinds and in general present lack of accuracy. In most cases, a combination of different sensors types is preferred to achieve better results [4]. Measurement of occupancy is more commonly undertaken in residential environments rather than offices or commercial buildings [14]. The most commonly used are:

- **CO_2 (Carbon dioxide) sensors** are often deployed in commercial buildings to obtain CO_2 data that are used to automatically modulate rates of outdoor air supply. Furthermore, CO_2 sensors show very slow response to the change of the occupancy [16]. Sometimes, more than 15 min is necessary so as to indicate a change to the occupancy of the space [1]. An additional drawback is that the CO_2 measurements are highly influenced by the ventilation system and the open doors windows, etc.;
- **Passive infrared (PIR) sensors** are commonly used for non-individualized occupancy detection. PIR sensors suffer from two main limitations: (i) they only give information about whether a room is occupied or not providing no indication about the exact number of people and (ii) they often do not detect stationary occupants, leading in false negative signals [14]. To overcome these limitations, they are often coupled with other sensors;
- **Video Cameras:** Video imaging typically uses small cameras mounted overhead of a doorway and video analytics to count and differentiate between people entering and exiting a building or space. The video analytics creates two lines, similar to the infrared using two beams and detects motion as to when the line is broken. It is not unusual to find the people counting capability as an add-on module of a video surveillance system. However, video cameras can heavily raise privacy and ethical issues and are sensitive to lighting levels. Video cameras if improperly installed and configured show substantial errors [1];
- **RFID System** uses wireless radio communication technology to determine the location of occupants who carry special tags. Depending on the layout of the receivers the zones can overlap and detect occupants going from one zone to another while they are not moving [17].

It is difficult to determine the actual number of occupants in a predefined space and their patterns of movement with current sensing techniques [14], even more, low-cost and non-intrusive environmental sensors to measure occupancy in commercial buildings are not fully explored [14]. In this paper the results of new occupancy extraction system [18] by means of depth cameras are presented; the developed system offers data anonymity and privacy preserving. Depth-image cameras (such as Microsoft Kinect) are used to extract occupancy (exact number of people, location and track) within a space or zone through the analysis of the depth-images collected. They are usually installed near ceiling to cover the examined area as much as possible but can also be installed on entrances/exits if zoning isn't required. They are more suitable for closed spaces but can also be used for open spaces and they need proper calibration in order to provide proper results. The equipment needed (number of cameras, cable extensions etc.) and the topology used depends on the current application, space layout and limitations (e.g. wooden separators between offices), cameras limitations (maximum distance, depth image limitations), overlapping FOV (Field of View) and number/location of entrances (which should be inside FOV). For example, 2-3 cameras would be adequate for a one-door closed space of 50 m^2 given that there are not many obstacles. Real-time depth-image analysis is demanding as it requires high computational power in order to detect the presence of an occupant and extract his/her location. Moreover, occupancy detection based on depth-image cameras is sensitive to changes in image background and requires time consuming setup and calibration for defining the monitored area and tracking the detected occupants in the monitored space. On the contrary, depth-image cameras provide quite high accuracy with relatively low cost and have the capability to exhibit totally transparent properties (i.e. no colour images are recorded for human detection). Also, they are not sensitive to the lighting levels of the environment, although they will not work in direct sunlight conditions since they use infrared (IR) radiation.

Real-time estimation of the number of occupants in a building's space is a challenging task. A main challenge is to determine the method of processing the input received by the multiple sensor types. There are two main occupancy extraction approaches from a sensor fusion model: the **rule-based** approach and the **probabilistic model** approach.

The use of a rule-based system results in logical inference from sensor data [21]. According to this approach, a set of rules for a set of installed sensors are defined and applied. Rules are defined by a domain expert and knowledge about sensor characteristics is required. The set of applied rules usually depends on the combination of sensors that are used. Two studies where this method is applied for occupancy detection is [22, 23].

The probabilistic model approach views occupancy extraction as a classification problem. A probabilistic model is created by training a selected classifier and the target is to infer the occupancy class based on the input from the various sensors. Different models have been examined such as support vector machines, neural networks, hidden Markov models, agent-based models, decision trees etc. [24, 25]. A training phase must be performed in advance in order to learn the parameters and be able to start the occupancy extraction process. On the contrary, a training phase is not mandatory for the rule-based approach.

The rest of the paper is organized as follows. Section 2 briefly presents the system used to extract the building measurements (data extraction), while the data analysis process is presented in Sect. 3. Finally the conclusions are drawn in Sect. 4.

2   System Overview and Data Extraction

2.1   Occupancy Extraction Approach

The subsystem utilized for the occupancy extraction is constituted by a collection of depth image cameras and a multi-sensorial cloud (various types of simple sensors) in order to extract the occupancy per space. The system is able to monitor multi-space environments and it has been built based on a client-server architecture.

The proposed occupancy extraction system is a real-time system since it is able to detect and track people and visualize the results in real-time and in high accuracy and details. Since **depth image cameras** are utilized in order to extract the occupancy of the building, issues like camera calibration, overlapping areas, error propagation, etc. have to be dealt with [18]. Furthermore the depth image cameras provide only depth information in order to take into account all legal and ethical issues regarding individual privacy and provide anonymity. Finally depth information is immutable to luminance and shadow changes [18].

As aforementioned various types of simple sensors are also utilized in order to detect human activities and collect occupancy data which have been analysed by S. Zikos et al. [30] using a Conditional Random Field approach. **Double-Beam Sensors** [27] are established in specific locations, where a semantic event may occur. These locations are the doors of the building, as well as the doors of all building spaces. Moreover two **Pressure Mats Sensors** [28] are placed next to each other, separated by a small space in order to detect movement direction and **PIR Motion Sensors** [29] which were already installed through the Alarm system installation. When movement is detected, an activation event is sent by the sensor and after a specified period (configured to a few seconds) of no movement detection, a deactivation event is sent. Finally **CO_2 Sensors** are established in some spaces which measure the CO_2 concentration of the air and can be very useful when combined with other sensors mentioned above, since it can provide information on occupancy density.

2.2   Building Installation

The proposed occupancy extraction system can operate in any type of building. An indicative example of the physical installation of each sensor in CERTH premises is depicted in Fig. 1. The test bed consists of eight (8) main areas with different usage (offices, corridors, rest area, meeting room and kitchen). The most remarkable spaces of the building are the Developers' office which is 56.7 m^2, the corridor and the rest area which are 81.5 m^2, the Meeting Room (26.4 m^2) while the kitchen is 33.7 m^2. The majority of the sensors for occupancy extraction have been installed at the developers' office, a characteristic area of the building, since it is a core element for testing the real-time occupancy extraction system. In total six (6) Kinect Cameras were used as depth image sensors to provide occupancy information to a sub-space level, while three (3) Pressure Mats (×2), eight (8) Active Infrared Beams (×2) per door, ten (10) PIR sensors and three (3) CO_2 sensors covered the area at a space level.

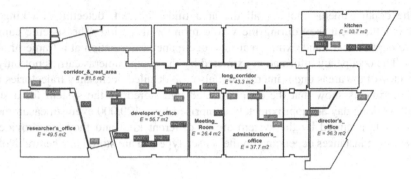

Fig. 1. Physical configuration of sensors installed in CERTH premises

The number and the type of the sensor cloud installed in this building are shown in Table 1.

Table 1. Sensors used in CERTH premises Test Bed (Fig. 1)

Type	Measurement	Period	Qty
Depth cameras	Occupancy flows	20 fps	6
CO_2	Carbon dioxide	15 min	3
PIR sensors	Occupancy Density & Presence	15 min	10
Beams	Movement direction	15 min	8
Pressure mats	Movement direction	15 min	3

2.3   Data Acquisition

The data acquisition is performed utilizing the sensor cloud and the system that has been described in Sect. 2.1. The VGA resolution of the infrared depth cameras, i.e. the pixel size, determines the point scaling of the depth data on the XY plane (perpendicular to camera axis). Since each depth image contains a constant 320×240 pixels the point density will decrease with increasing distance of the object surface from the sensor. Considering the point density as the number of points per unit area, while the number of points remains constant the area is proportional to the square distance from the sensor. Therefore, the point density is inversely proportional to the square distance from the sensor. The depth resolution is determined by the number of bits per pixel used to store the disparity measurements. The specific cameras disparity measurements are stored as 11-bit integers. Therefore, a disparity image contains 2048 levels of disparity. Since depth is inversely proportional to disparity the resolution of depth is also inversely related to the levels of disparity. That is, the depth resolution is not constant and decreases with increasing distance to the sensor. For instance, at a range of 2 meters one level of disparity corresponds to 1 cm depth resolution, whereas at 5 meters one disparity level corresponds to about 7 cm depth resolution. Furthermore, they have an angular field of view of 57° horizontally and 43° vertically.

The depth cameras monitor all the area under interest, detecting, tracking and extracting the occupancy during the whole monitoring period. The specific cameras capture depth images and extract real-time occupancy information at a frame of about 20 fps. The occupancy information extracted by the depth cameras carries not only the occupancy of the areas under interest, but also the detailed occupancy trajectories in it. The experiments show that the data extracted and stored by the system for a single normal working day are approximately comprised by > 120.000 events/measurements.

The data acquired by the system are of different kind and they are provided at different time instances depending on the sensor type and are stored in a central NoSQL database.

3 Data Analysis

The assessment of the building performance towards occupant's comfort and energy savings has been set as a main concern nowadays by using the required and equivalent software. Building Occupancy Extraction and more specifically occupants' trajectories are really important for building performance, occupants' work efficiency, building usage and is directly related to occupants' comfort, therefore occupancy statistics are depicted per space, as well as the number of transitions from one space to another. Based on these meaningful information, one can extract Key Performance Indicators (KPIs) related to the building occupancy.

The basic building occupancy related KPIs are: (a) average work efficiency and (b) average building usage. The **average work efficiency KPI** provides a measurement of the work efficiency and it is defined as:

$$avgWE = \frac{totalActivityHours}{totalOccupancyHours} * 100 \qquad (1)$$

where *totalActivityHours* is the overall hours that the occupants are involved in any activity in a specific space and *totalOccupancyHours* is the overall hours that the building is occupied. The **average building usage KPI** provides the usage of the building and it is defined as:

$$avgBU = \frac{totalOccupancyHours}{Time * N_{spaces}} * 100 \qquad (2)$$

where *totalOccupancyHours* is the overall hours that the building is occupied, *Time* is the overall is the duration (in years) of the monitoring activity and N_{spaces} is the total number of building spaces.

Finally, the basic KPIs related to occupant's comfort are: (a) average overcrowding factor, (b) average Predicted Mean Vote (PMV), and (c) average Predicted Dissatisfied (PPD) [31]. The **average overcrowding factor** is defined as:

$$avgOF = \frac{1}{N_{spaces}} \sum_{i \in SP} \frac{\frac{\sum_{j \in Nocc_i} occHours_{i,j}}{Time \cap occupancyHours_i}}{cap_i} 100 \tag{3}$$

where N_{spaces} is the number of spaces in the building under interest, SP is the set of all spaces, cap_i is the capacity of space i, $Time$ is the duration (in hours) of the monitoring activity, $occupancyHours_i$ represents the hours that the space i is occupied, $occHours_i$ represents the hours that the space i is occupied by occupant j, and N_{occ} is the number of occupants at space i. The **average Predicted Mean Vote (PMV) KPI** [31] is a thermal comfort model, which is defined as:

$$avgPMV = \left(0.303e^{-0.036M} + 0.028\right)L \tag{4}$$

$$
\begin{aligned}
L = q_{met,heat} \quad & - 3.96e^{-8} f_{cl} [(t_{cl} + 273)^4 - (t_r + 273)^4] \\
& - f_{cl} h_c (t_{cl} - t_a) \\
& - 3.05(5.73 - 0.007 q_{met,heat} - p_a) \\
& - 0.42(q_{met.heat} - 58.15) \\
& - 0.0173M(5.87 - p_a) \\
& - 0.0014M(34 - t_a)
\end{aligned}
\tag{5}
$$

where M is the rate of metabolic rate (W/m^2), $q_{met,heat} = M\text{-}w$ is the metabolic heat loss, the difference between the metabolic generation converted to work (e.g., lifting, running), w is the external work (W/m^2), f_{cl} is ratio of clothed surface area to DuBois surface area (Acl/AD), h_c is the convection heat transfer coefficient $(Btu/h\ m^2\ °C)$, t_{cl} is the average surface temperature of clothing $(°C)$, t_a is the air temperature $(°C)$, t_r is the mean radiant temperature $(°C)$, p_a is the vapour pressure of air [kPa]. Since, all these parameters are not available (e.g. the ratio of clothed surface area of a human), some default values that have been used:

$$
\begin{aligned}
M &= 115 \\
f_{cl} &= 1.15 \\
h_c &= 4.69 \\
t_{cl} &= 30.2 \\
h_c &= 0.7 \\
w &= 0
\end{aligned}
\tag{6}
$$

The **average Predicted Percentage Dissatisfied (PPD) KPI** [31] predicts the percentage of occupants that will be dissatisfied with the thermal conditions, which is defined as:

$$avgPPD = 100 - 95 * e^{-0.3353*avgPMV^4 - 0.2179*avgPMV^2} \tag{7}$$

All the above mentioned KPIs are calculated during the measurement extraction procedure for a 4 month testing period of CERTH's premises. The collected data are over 1.15 billion of information, which is analyzed as shown in Table 2.

Table 2. Events produced for a 4 month testing period of time

Type of event	Number of events/data
Space occupancy events	316.921
Occupancy trajectories	41.089
Occupancy trajectory points	1.166.397.820
Total	**1.166.755.830**

In order to handle and process the big amount of data, which can be a very difficult and time consuming task due to size and diverse data types, a visual analytic intuitive user friendly application has been developed which presents a set of visualization techniques that facilitate users to perceive readily the data extracted. The data visualization application that was developed uses a coarse-to-fine approach to visualize information. Coarse-to-fine approaches are becoming more widespread as statistical problems grow into larger and significant domains. The coarse-to-fine approach minimizes the loss of accuracy, while executes the process at successively finer granularities. In accordance with the coarse-to-fine approach, the user can observe the key performance indicators for the building, which was described above in a kiviat diagram (Fig. 2).

The space occupancy data can be displayed per day, week or month in relation to user preferences, thus an extension of the **Clock Map** was utilized proposed in [19], with the addition of a 3rd dimension encoded in the radius using concentric cycles. Figure 3 illustrates the space occupancy in Clock-view form, where each building space is represented by different colour. More specifically, orange colour indicates the

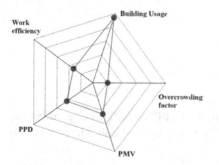

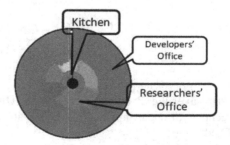

Fig. 2. Kiviat Diagram: Top level visualization displaying key performance indicators for the building

Fig. 3. Occupancy Clock-view for three different building spaces (for one working hour) (Color figure online)

kitchen; the dark green denotes the researcher's office and the developer's office is represented by light green colour. Respectively the radius of each building space denotes the portion of the total building occupancy and occupants per building space, where the intensity of the colour denotes the relation of the space occupancy with the corresponding completeness space occupancy.

Except from the Clock Map, **occupancy heat maps** were developed to further detect human presence and their trajectories. The system tracks occupants' movements and turns this information into heat maps as shown in Fig. 5. Heat maps are a popular and intuitive visualization technique for encoding quantitative values derived from gaze points or fixations at corresponding image or surface locations. Heat maps enabled us to gain additional insights into temporal and spatial patterns present in the data. More specifically, the occupants' trajectories through the space are depicted in Fig. 5 where the colours on the floor correspond to foot traffic during a particular time period. Pink areas are hotspots that lots of occupants walked through, while the small splotch of blue in some regions indicates lower traffic congestion. Moreover, an alternative heat map view is depicted in Fig. 6. It is two-dimensional graphical representation of building occupancy data where the different values are shown as colours. The intuitive nature of the colour scale, as it relates to the temperature minimizes the amount of learning necessary to understand it. From experience we know that red is warmer than orange, yellow or light green. As it is observed the Developer's office (intense red colour) is the most congested space of CERTH premises while distinct spaces such as the meeting room, director's office and long corridor are areas with the least occupants (intense green), since they are not permanently occupied. This can be easily also observed in Kiviat Diagram (Fig. 4).

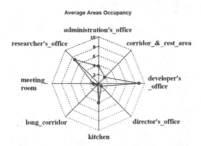

Fig. 4. Average Occupancy per space in Kiviat diagram (Color figure online)

Fig. 5. Occupancy heat map to further detect occupants' trajectories

The distribution of the occupants among the different spaces in the CERTH pilot building is shown in Fig. 8 for a typical week. All spaces are occupied during the working days, while during the weekends only partial part of the building (e.g. developers, kitchen) is used for a small portion of time (Fig. 7).

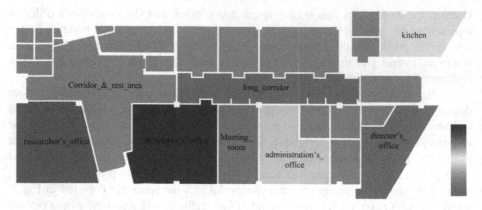

Fig. 6. Building occupancy heat map

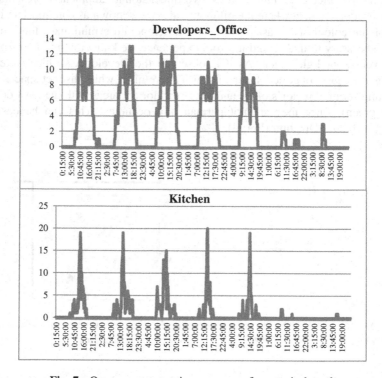

Fig. 7. Occupancy extraction per space for a typical week

Finally, all experimental results have been performed on an Intel I7 (8 cores with 3.5 GHz) workstation with 16 GB RAM under Windows 7 without any particular code optimization. All visualizations can be generated in near real time, e.g. less than a second, for 4-month data.

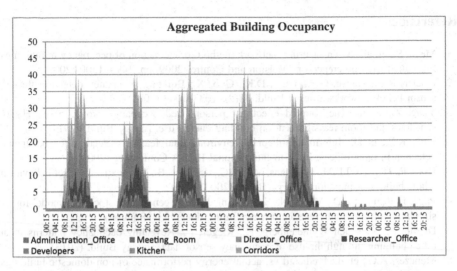

Fig. 8. Occupancy distribution among building spaces for a typical week

4   Conclusions

To respond to the need for improved detection of building occupants resolved in space and time, we developed a universal building occupancy big data analysis system in which information collected by a network of multiple and low cost privacy preventing depth sensors and various types of sensors including, but not limited to, CO_2, double beam sensors, pressure mats sensors and PIR Motion sensors for detecting occupants' movement, trajectories and direction.

Finally it must be pointed out that along with the building occupancy extraction system, visual analytics techniques have been developed in order to visualize the collected big data through a large number of visualizations, allowing the user to evaluate the performance of a building from a building occupancy point of view using an intuitive graphical user interface. Also the proposed system is designed in such a way that can deal with the big data produced by the sensor network over a long period of time.

In future, further experimentation in other buildings is planned covering a larger time period, in order to better explore the potential outcomes. Furthermore, it is planned to extend the system in order to combine information from various other building elements except from occupancy, such as environmental conditions and energy consumption of equipment devices, HVACs or lights. This will allow a much more assiduous evaluation of the building performance.

Acknowledgement. This work has been partially supported by the European Commission through the project HORIZON 2020-RESEARCH & INNOVATION ACTIONS (RIA)-696129-GREENSOUL.

References

1. Meyn, S., et. al.: A sensor-utility-network method for estimation of occupancy in buildings. In: 48th IEEE Conference on Decision and Control 2009, pp. 1494–1500 (2009)
2. Dodier, R.H., Henze, G.P., Tiller, D.K., Guo, X.: Building occupancy detection through sensor belief networks. Energy Build. **38**(9), 1033–1043 (2006)
3. Yang, Z., et al.: The coupled effects of personalized occupancy profile based HVAC schedules and room reassignment on building energy use. Energy Build. (2014)
4. Lam, K.P., et al.: Information-theoretic environmental features selection for occupancy detection in open offices. In: 11th International IBPSA Conference (2009)
5. Hoes, P., Hensen, J.L.M., Loomans, M.G.L.C., de Vries, B., Bourgeois, D.: User behavior in whole building simulation. Energy Build. (2009)
6. Page, J., Robinson, D., Morel, N., Scartezzini, J.-L.: A generalised stochastic model for the simulation of occupant presence. Energy Build. **40**(2), 83–98 (2008)
7. Duarte, C., Van Den Wymelenberg, K., Rieger, C.: Revealing occupancy patterns in an office building through the use of occupancy sensor data. Energy Build. (2013)
8. Menezes, A.C., et al.: Predicted vs. actual energy performance of non-domestic buildings: using post-occupancy evaluation data to reduce the performance gap. Appl. Energy (2012)
9. (Annie) Egan, A.M.: Occupancy of Australian office buildings: how accurate are typical assumptions used in energy performance simulation and what is the impact of inaccuracy. ASHRAE Trans. **118**(1), 217–224 (2012)
10. Eguaras-Martinez, M., et al.: Simulation and evaluation of building information modeling in a real pilot site. Appl. Energy **114**, 475–484 (2014)
11. Caucheteux, A., et al.: Occupancy measurement in building: a literature review, application on an energy efficiency research demonstrated building. Int. J. Metrol. (2013)
12. Erickson, V.L., Carreira-Perpiñán, M.Á., Cerpa, A.E.: Occupancy modeling and prediction for building energy management. ACM Trans. Sens. Networks **10**(3), 1–28 (2014)
13. Ke, M., et al.: Analysis of building energy consumption parameters and energy savings measurement and verification by applying eQUEST software. Energy Build. **61**, 100–107 (2013)
14. Dong, B., et al.: An information technology enabled sustainability test-bed (ITEST) for occupancy detection through an environmental sensing network. Energy Build. (2010)
15. Kuutti, J., et al.: Real Time Building Zone Occupancy Detection and Activity Visualization Utilizing a Visitor Counting Sensor Network, February 2014
16. Mahdavi, A.: Patterns and implications of user control actions in buildings. Indoor Built Environ. **18**(5), 440–446 (2009)
17. Erickson, V.L., et al.: Occupancy based demand response HVAC control strategy. In: Proceedings of the 2nd ACM Workshop on Embedded Systems for EE in Building (2010)
18. Tabak, V.: User Simulation of Space Utilisation (2009)
19. Krinidis, S., et al.: A Robust and real-time multi-space occupancy extraction system exploiting privacy-preserving sensors. In: 6th International Symposium on Communications, Control and Signal Processing (ISCCSP 2014) (2014)
20. Kintzel, C., Fuchs, J., Mansmann, F.: Monitoring large IP spaces with clock-view. In: 8th International Symposium on Visualization for Cyber Security, ACM, New York (2011)
21. Stellmach, S., Nache, L., Dachselt, T.: 3D attentional maps: aggregated gaze visualizations in three dimensional virtual environments. In: Proceedings of the International Conference on Advanced Visual Interfaces, pp. 345–348. ACM, New York (2010)

22. Nguyen, T.A., Aiello, M.: Energy intelligent buildings based on user activity: a survey. Energy Build. **56**, 244–257 (2013). http://dx.doi.org/10.1016/j.enbuild.2012.09.005, ISSN 0378-7788

23. Agarwal, Y., Balaji, B., Gupta, R., Lyles, J., Wei, M., Weng, T.: Occupancy-driven energy management for smart building automation. In: Proceedings of the 2nd ACM Workshop on Embedded Sensing Systems for Energy-Efficiency in Building, BuildSys 2010, pp. 1–6. ACM, New York (2010)

24. Nguyen, T.A., Aiello, M.: Beyond indoor presence monitoring with simple sensors. In: Proceedings of the 2nd International Conference on Pervasive and Embedded Computing and Communication Systems (PECCS), pp. 5–14 (2012)

25. Hailemariam, E., et al.: Real-time occupancy detection using decision trees with multiple sensor types. In: Proceedings of the 2011 Symposium on Simulation for Architecture and Urban Design. Society for Computer Simulation International (2011)

26. Yang, Z., et al.: A systematic approach to occupancy modelling in ambient sensor–rich buildings. Simulation (2013). doi:10.1177/0037549713489918

27. Kuutti, J., Saarikko, P., Sepponen, R.E.: Real time building zone occupancy detection and activity visualization utilizing a visitor counting sensor network. In: 11th International Conference on Remote Engineering and Virtual Instrumentation (REV), IEEE, Polytechnic of Porto (ISEP) in Porto, Portugal, 26-28 February, 2014, pp. 219-224 (2014)

28. Ekwevugbe, T., Brown, N., Fan, D.: A design model for building occupancy detection using sensor fusion. In: 6th IEEE International Conference on Digital Ecosystems and Technologies (DEST), Campione d'Italia, 18-20 June 2012, pp. 1–6 (2012)

29. Wahl, F., Milenkovic, M., Amft, O.: A distributed PIR-based approach for estimating people count in office environments. In: Proceedings of the IEEE 15th International Conference on Computational Science and Engineering (CSE 2012), IEEE, Washington, DC (2012)

30. Zikos, S., et al.: Conditional Random fields-based approach for real-time building occupancy estimation with multi-sensory networks. Autom. Constr. **68** (2016)

31. Fanger, P.O.: Analysis and Applications in Environmental Engineering. McGraw-Hill Book Company, New York (1970)

Discovering the Discriminating Power in Patient Test Features Using Visual Analytics: A Case Study in Parkinson's Disease

Panagiotis Moschonas[1(✉)], Elias Kalamaras[1], Stavros Papadopoulos[1],
Anastasios Drosou[1], Konstantinos Votis[1], Sevasti Bostantjopoulou[2],
Zoe Katsarou[2], Charalambos Papaxanthis[3],
Vassilia Hatzitaki[4], and Dimitrios Tzovaras[1]

[1] Information Technologies Institute, Centre for Research and Technology Hellas,
Thessaloniki, Greece
{moschona,kalamar,spap,drosou,kvotis,Dimitrios.Tzovaras}@iti.gr
[2] Department of Neurology, Aristotle University of Thessaloniki,
Thessaloniki, Greece
[3] Universit de Bourgogne, UFR STAPS, Campus Universitaire, Dijon, France
[4] Department of Physical Education and Sports Sciences,
Aristotle University of Thessaloniki, Thessaloniki, Greece

Abstract. This paper presents a novel methodology for selecting the most representative features for identifying the presence of the Parkinson's Disease (PD). The proposed methodology is based on interactive visual analytic based on multi-objective optimisation. The implemented tool processes and visualises the information extracted via performing a typical line-tracking test using a tablet device. Such output information includes several modalities, such as position, velocity, dynamics, etc. Preliminary results depict that the implemented visual analytics technique has a very high potential in discriminating the PD patients from healthy individuals and thus, it can be used for the identification of the best feature type which is representative of the disease presence.

Keywords: Parkinson's disease · Visual analytics · Multi-objective optimisation · Feature discrimination power

1 Introduction

Parkinson's disease (PD) is a degenerative disorder of the central nervous system that is mainly affecting the motor system. Its most obvious symptoms are movement-related, including shaking, rigidity, slowness of movement and difficulty with walking and gait [8]. Tracking in a correct way the progress of the

© IFIP International Federation for Information Processing 2016
Published by Springer International Publishing Switzerland 2016. All Rights Reserved
L. Iliadis and I. Maglogiannis (Eds.): AIAI 2016, IFIP AICT 475, pp. 600–610, 2016.
DOI: 10.1007/978-3-319-44944-9_53

disease is crucial for the quality of patient's life, thus the need of an objective and consistent way of measuring the patient motor dexterity is essential.

There are several established scales for measuring the progress of the disease. The Unified Parkinson's Disease Rating Scale [16] is the most commonly used metric for clinical study and is used as a severity rating method. An older scaling method known as the Hoehn and Yahr scale [8], and a similar scale known as the Modified Hoehn and Yahr scale, have also been commonly used. The later defines five basic stages of progression. These scale measurement are derived from qualitative questionnaires asked by doctors, thus they enclose the danger of subjectivity in their results: different clinical examinations may produce different answers. Additionally, the repetition period of such solutions are in the class of several months.

Electronic devices from various domains have been also used for measuring the PD progress. The current market offers a variety of wristbands capable of measuring the tremor [6]. However, these are used for complimentary assessment to the typical clinical examination. Moreover, their effectiveness is high when the tremor is present to the patient arm, but their efficiency is not guaranteed in early stages of the disease. An increasing set of smartphone and tablet PD applications has arisen in the market. Despite most of them are used for scheduling, there is a small subset which makes use of tests for tracking the progress of the disease. In most of these tests, the user has to use his/her hand to track a line or a shape and the software extract metrics representative of the current state. Factors such as velocity, target deviation, reaction time, minimum jerk, are recorded and compared to the ones measured from previous states. The results are displayed in curves, allowing the user to track the PD progress.

However, even providing the tests, there is still the matter of how the program is going to correlate the captured information with the disease presence on the individual. Comparing the velocity or deviation profiles per day measurements is not enough as it could not be representative of the specific patient symptoms [18]. For such reasons, proper visualisation capable of identifying which of the extracted features encapsulate and discriminate better the examined person status is needed. In this paper we apply a novel multi-objective visualisation for identifying which of the measured quantities are best for discriminating the presence of the Parkinson's Disease.

1.1 Relevant Work

Information visualization concerns the use of interactive computer graphics to get insight into large amounts of abstract data, such as multivariate, hierarchical and network data [23]. Conventional visualization techniques, such as bar-charts, pie-charts, and line-charts, are useful for the depiction of information of a higher level, but fail to depict large and complex data sets in detail. Therefore, and in order to enable the easy extraction of patterns, trends and outliers, a variety of novel visualization methods are continuously being developed for specific applications, e.g. [1, 4, 22].

Their ability to provide insight into large amounts of complex data make visual analytics techniques especially useful for healthcare and biomedical applications. This has resulted in numerous existing methods and tools which utilize various visualization types and user interaction levels [10,21]. Recent works have employed visual properties such as color and position [15,24] or animation [17], in order to visually encode patient information, group patients with similar characteristics together and discriminate between different events. Visual queries have also been utilized in combination with pattern mining and interactive visualization, in order to explore large datasets more efficiently.

The continuous development in the field of information visualization has led to a variety of new methods and techniques that enable people to understand the phenomena behind large amounts of data, and which increasingly find their way to Health and real-life applications. However, and due to their specialized nature, many of these methods have limited use outside their initially targeted application. For the majority of methods, data are considered to be homogenous and in most cases just one or two types of data are supported. In addition, data are assumed to come from a single source and to be clean and exact, whereas noisy, polluted, uncertain, and missing data are rarely dealt with. Scalability is another common issue, since it is typically limited to thousands of elements. Therefore, if one of the above assumptions is violated, standard methods from information visualization fall short, and a need for new representations arises.

Since several attributes are usually available for each patient, such as age, speed of task completion, acceleration, etc., they can be combined to provide better insights in the data. The combination of multiple sources of information for classification, clustering, visualization, etc. has generally been handled by multimodal fusion methods [2]. A first category of fusion methods simultaneously combine characteristics of all modalities, e.g. through weighted sums. In [20], a graph is constructed for each of multiple attributes and the graph Laplacians are then fused. Graph-based techniques are also used in [13], which employs Multiple Kernel Learning [9] for fusion. A second category are methods that utilize information of one modality to assist learning in another modality in an iterative manner, such as [3,14].

A different principle is followed by works such as [11,12], where visualization of each modality is formulated as an optimisation problem and then multi-objective optimisation techniques [5,7] are utilized to simultaneously optimize all objectives and resulting in a set of Pareto-optimal solutions, instead of one. Although such methods demonstrate the effectiveness of multimodal fusion, for a wide range of applications, they have not been adopted for visualization of healthcare data, where the combination of multiple modalities may reveal important patient behavioural characteristics that can assist decision making. In this respect, this paper presents an application of the techniques presented in [11,12], in the task of clustering patients according to multiple attributes.

2 Applied Methodology

A set of features is first extracted by performing a simple test on a tablet device. During the test a vertical line appears at random position, while the subject has to track it as fast as possible with a stylus pen. After the test, a set of features is extracted and passed into the visual analytics engine.

The extracted features are the following: (a) deviation in x axis from target, i.e. position in pixels, (b) velocity, (c) acceleration, and (d) simulated muscle activation, extracted using methodology described in [19]. These features are normalised in the time component. This is achieved by splitting the whole recording into events which start when the target line changes its position. Then a representative signal of each feature type is extracted by averaging the event signals. The resulted signals are fed to the Multi-objective visual analytics engine in order to analyse the discriminating power of each of the feature types.

2.1 Multi-objective Visualisation

The feature extraction procedure results in feature vectors describing various characteristics of a patient, such as age, speed of task completion, acceleration. Each of these types of types of features determines a specific type of similarities and dissimilarities among the users. An adequate visualisation of the patient data should be able to visualize these similarities. A common and straightforward visualisation scheme is to consider each patient as a point on the screen and using the relative position of the patients to denote similarity, by placing similar patients close to each other and dissimilar ones away from each other. However, the presence of multiple notions of similarity, due to the multiple types of features extracted from a single patient, renders the problem of visualizing similarities non-trivial, since many types of similarity must be visualized simultaneously, or some kind of compromise among the various feature types should be considered. In this respect, the multi-objective visualisation method of [11] is used hereby, which exactly considers multiple notions of similarity. The method is briefly presented in this section. The method proceeds by first considering each feature type separately, formulating visualisation as a single-objective optimisation problem, and then combining multiple objectives in a multi-objective optimisation setting.

Visualisation of a Single Feature Type: Let $\mathcal{O} = \{O_1, O_2, \ldots, O_N\}$ denote a set of N patients. Each patient O_i is itself a set of M feature vectors:

$$O_i = \{\mathbf{o}_{i,1}, \mathbf{o}_{i,2}, \ldots, \mathbf{o}_{i,M}\}, \ \mathbf{o}_{i,m} \in \mathbb{R}^{D_m}.$$

Each feature vector $\mathbf{o}_{i,m}$ is a vector of dimension D_m, which is specific for the m-th feature type.

Considering just a single feature type m, a distance function

$$d_m : \mathbb{R}^{D_m} \times \mathbb{R}^{D_m} \to \mathbb{R}$$

is used to calculate the distance between the feature vectors of two patients. Hereby, the L1 distance measure is used:

$$d_m(\mathbf{o}_{i,m}, \mathbf{o}_{j,m}) = \sum_{k=1}^{D_m} |o_{i,m,k} - o_{j,m,k}|,$$

where $o_{i,m,k}$ is the k-th component of vector $\mathbf{o}_{i,m}$. The distance measure is used to construct a complete graph $G_m(\mathcal{O}, E_m)$, $E_m \subseteq \mathcal{O} \times \mathcal{O}$, of the patients, where each patient $O_i \in \mathcal{O}$ is represented by a vertex and there is an edge $\{O_i, O_j\} \in E_m$ between every pair of patients, weighted by the corresponding distance between them. As a further step, the minimum spanning tree (MST), $T_m(\mathcal{O}, E'_m)$, $E'_m \subset E_m$, of the graph is computed, in order to reduce the number of edges, while keeping the similarity information.

The MST is then used to guide the positioning of the patient vertices on the 2-dimensional screen. Let $\mathbf{p}_i = (x_i, y_i)^T$, $x_i, y_i \in \mathbb{R}$, be the coordinates of patient i on the screen. Let also $\mathbf{P} = (\mathbf{p}_1, \mathbf{p}_2, \ldots, \mathbf{p}_N)^T$ be the matrix collecting all the 2D points as its rows. Then, the placement $\mathbf{P}$ of the N points is calculated by minimizing the following objective function:

$$J_m(\mathbf{P}) = \sum_{i=1}^{N} \sum_{j=1, j \neq i}^{N} \frac{q^2}{||\mathbf{p}_i - \mathbf{p}_j||} + \sum_{i,j:\ (O_i, O_j) \in E'_m} k||\mathbf{p}_i - \mathbf{p}_j||^2.$$

Intuitively, this function evaluates the potential energy of the tree, if each vertex is considered as a charged particle and the edges as springs attached to pairs of them. The first term of the sum is the energy of the charges repelling one another, following Coulomb's law. The second term is the energy of the attracting springs, according to Hooke's law. If such a system of charges and springs is let to act freely, the particles will tend to repel one another, while the springs will keep together particles connected by edges, thus unfolding the tree structure. The lowest energy position is one where the tree has been completely unfolded

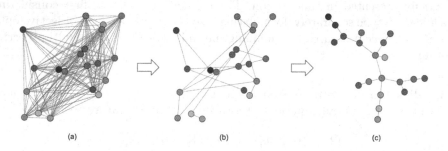

(a)　　　　　　　　　(b)　　　　　　　　　(c)

Fig. 1. Example of the potential objective minimization. (a) The points and their similarities form a complete graph. The graph's vertices are placed at random positions. (b) The minimum spanning tree of the graph is calculated. (c) By minimizing the potential objective, the vertices are moved so that the tree's structure is apparent. (Color figure online)

and its structure is easily visible. In this position, the patients with similar characteristics will be put close to each other, while dissimilar patients will be put away from each other. An example of the overall procedure is illustrated in Fig. 1, where the color of each point is used to visually denote the example characteristics of each point. After the minimization of the potential objective, in Fig. 1(c), the structure of the tree is apparent and similar colors are placed close to each other.

Handling Multiple Feature Types: The above procedure holds for a specific feature type m. If a different feature type is used, different distances will be calculated among the points, resulting in a different tree, a different objective function and a different ultimate positioning. The multiple feature types result in a set of objective functions $\mathcal{J}(\mathbf{P})$, instead of a single one:

$$\mathcal{J}(\mathbf{P}) = \{J_1(\mathbf{P}), J_2(\mathbf{P})), \ldots, J_M(\mathbf{P})\},$$

so that the goal is to minimize them all simultaneously:

$$\mathbf{P}_{\text{opt}} = \arg\min_{\mathbf{P}} \mathcal{J}(\mathbf{P}).$$

The multiple objective J_m are conflicting, since the optimal solution for one of them is not optimal for another, thus a single solution cannot in general be reached. Such problems of conflicting objectives are handled by multi-objective optimisation techniques, which result in a set of optimal trade-offs among the objectives, namely the *Pareto set*. Multi-objective optimisation is based on the notion of *Pareto dominance* among the feasible solutions. A solution dominates another one if it has a smaller value for at least one objective and there is no objective for which it has a larger value. Between two solution, the dominant one is preferred, since it is impartially better than the other one, with respect to all objective functions, without sacrificing any of them. Formally, a solution $\mathbf{P}_1$ dominates another solution $\mathbf{P}_2$, if

$$J_m(\mathbf{P}_1) \leq J_m(\mathbf{P}_2), \ \forall m \in \{1, \ldots, M\}, \text{ and}$$

$$\exists k \in \{1, \ldots, M\} : J_k(\mathbf{P}_1) < J_k(\mathbf{P}_2).$$

If two solutions mutually do not dominate each other, they are said to be *incomparable*, since there can be no impartial judgment as to which is better than the other. The goal of multi-objective optimisation is to compute the set of solutions that dominate all other feasible solutions but are mutually incomparable. This set is called the Pareto set and consists a set of optimal trade-offs among the multiple objectives. In Fig. 2, the feasible and optimal solutions for an example problem of two objectives are depicted.

The gray-shaded area represents the set of all feasible solutions, while the bold border in the lower left of the feasible area represents the solutions of the Pareto set. In such diagrams in the space of the objective function values, the

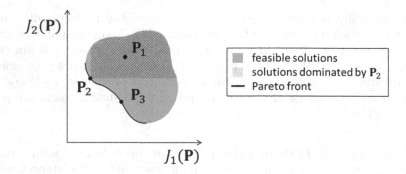

Fig. 2. Example Pareto diagram illustrating the Pareto front for a problem of two objectives, J_1 and J_2. The gray area represents the set of all feasible solutions, while the bold border is the Pareto front. Solution $\mathbf{P}_2$ dominates $\mathbf{P}_1$, as well as all solutions within the hatched area. Solutions $\mathbf{P}_2$ and $\mathbf{P}_3$ are incomparable.

solutions corresponding to the Pareto set are called the *Pareto front*. Three example solutions are shown. Solution $\mathbf{P}_2$ dominates $\mathbf{P}_1$ since both objectives have smaller values at $\mathbf{P}_2$. Similarly, $\mathbf{P}_2$ dominates all solutions within the hatched area. On the other hand, solutions $\mathbf{P}_2$ and $\mathbf{P}_3$ are incomparable, since none dominates the other. All the solutions of the Pareto front are mutually incomparable, since decreasing one objective leads to increasing the other.

Computing the Pareto front, which is the goal of multi-objective optimisation, means presenting the decision maker with a minimal set of optimal trade-offs, from which to select. In this paper, this set of solutions corresponds to different trade offs among the various medical and simulated features describing the patients. By selecting among the trade-offs, through an application interface, the doctor or any decision maker can put more focus on various characteristics and thus view different kinds of relationships and groupings among the data.

3   Preliminary Results

A total of 42 PD patients of various ages (mean: 63, stdev: 8.1) and 10 healthy/control subjects of matching ages (mean: 64.9, stdev: 7.75) were measured. All PD subjects were tested with an on-medication treatment state. The subjects performed the test with both hands and the extracted features were averaged, thus resulting into one feature-set per subject. The proposed algorithm computes the visualisation output of the 52 instances almost instantly ($t < 0.05\,\mathrm{s}$). The interactive environment allows its user to re-adjust the objective weights and updates the dots position without any lag. In terms of scalability the algorithm performs at $O(N \log N)$, where N is the number of nodes. Experiments in the application domain of mobile network security have shown that a dataset of 4800 points needs less than $3\,\mathrm{s}$ for the output solution [12]. The aforementioned experiments were performed using a 8-threaded Intel Core i7 processor running at $4\,\mathrm{GHz}$.

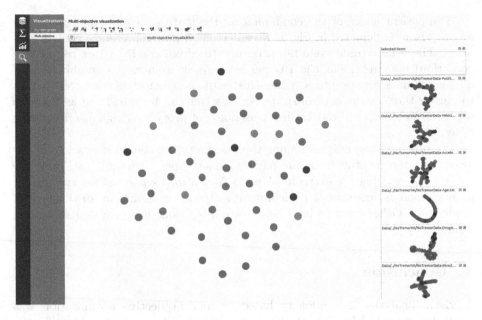

Fig. 3. Multi-objective result of all the feature type combined. The gathering of the green dots representing the control subjects in the lower part is not concentrated. (Color figure online)

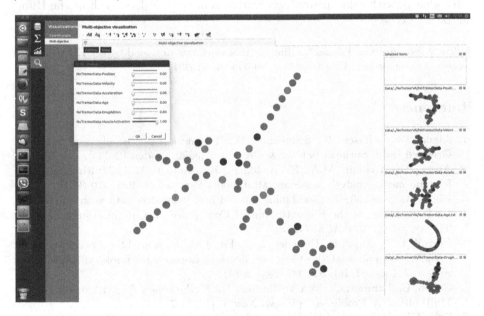

Fig. 4. Testing the impact of each feature: increasing the weight of the "Muscle Activation" feature results into a robust patient-healthy separation. (Color figure online)

The patient space of the combining all the features, using equal weighting among them, is depicted in Fig. 3. Green coloured dots represent a control subject, while yellow (mild condition), orange (medium condition) or red (severe condition) dots represent the PD patients. As it is shown, a combination of all the features results into a sparse cluster of the green dots. However, it fails to cluster three of the control subjects. This fact can be translated as that the selected combination of the feature types does not provide enough discriminating power.

By adjusting the weights among the feature types, the most crucial one for separating the healthy from the patient cohort may be found. In Fig. 4, the feature representing the extracted Muscle Activations separates the two groups in much more robust way, thus, defining it as the most significant for this type of application. Other features have been tested, but none has provided such good discrepancy.

4   Conclusion

A visual analytics methodology based on multi-objective optimisation was applied to a typical PD line tracking test output and managed to identify successfully the most crucial feature able to separate the healthy from the patient groups. In the future work, there are plans of investigating the potential of the methodology with other neurodegenerative genetic disorders, such as the Huntington's Disease.

Acknowledgement. Partial funding for this work was provided by the EU co-funded project NoTremor (EC FP7 Grant agreement no 610391).

References

1. Aigner, W., Miksch, S., Schumann, H., Tominski, C.: Visualization of Time-Oriented Data. Springer Science & Business Media, London (2011)
2. Atrey, P.K., Hossain, M.A., El Saddik, A., Kankanhalli, M.S.: Multimodal fusion for multimedia analysis: a survey. Multimedia Syst. **16**(6), 345–379 (2010)
3. Blum, A., Mitchell, T.: Combining labeled and unlabeled data with co-training. In: Proceedings of the Eleventh Annual Conference on Computational Learning Theory, pp. 92–100. ACM (1998)
4. Carroll, L.N., Au, A.P., Detwiler, L.T., Fu, T.C., Painter, I.S., Abernethy, N.F.: Visualization and analytics tools for infectious disease epidemiology: a systematic review. J. Biomed. Inf. **51**, 287–298 (2014)
5. Coello, C., Lamont, G., Van Veldhuizen, D.: Evolutionary Algorithms for Solving Multi-objective Problems. Springer, New York (2007)
6. Coy, J.A., Mehrkens, J.H., Roppenecker, D.B., Lueth, T.C.: Finding the center of parkinson's disease. a novel measurement device for quantifying motor symptoms during DBS-surgery. In: 2014 IEEE International Conference on Robotics and Biomimetics (ROBIO), pp. 1691–1696. IEEE (2014)
7. Ehrgott, M.: Multicriteria Optimization, vol. 2. Springer, Heidelberg (2005)

8. Goetz, C.G., Poewe, W., Rascol, O., Sampaio, C., Stebbins, G.T., Counsell, C., Giladi, N., Holloway, R.G., Moore, C.C., Wenning, G.K., et al.: Movement disorder society task force report on the Hoehn and Yahr staging scale: status and recommendations the movement disorder society task force on rating scales for parkinson's disease. Mov. Disord. **19**(9), 1020–1028 (2004)
9. Gönen, M., Alpaydın, E.: Multiple kernel learning algorithms. J. Mach. Learn. Res. **12**, 2211–2268 (2011)
10. Holzinger, A., Jurisica, I.: Knowledge discovery and data mining in biomedical informatics: the future is in integrative, interactive machine learning solutions. In: Holzinger, A., Jurisica, I. (eds.) Interactive Knowledge Discovery and Data Mining in Biomedical Informatics. LNCS, vol. 8401, pp. 1–18. Springer, Heidelberg (2014)
11. Kalamaras, I., Drosou, A., Tzovaras, D.: A multi-objective clustering approach for the detection of abnormal behaviors in mobile networks. In: 2015 IEEE International Conference on Communication Workshop (ICCW), pp. 1491–1496. IEEE (2015)
12. Kalamaras, I., Papadopoulos, S., Drosou, A., Tzovaras, D.: MoVA: a visual analytics tool providing insight in the big mobile network data. In: Chbeir, R., Manolopoulos, Y., Maglogiannis, I., Alhajj, R. (eds.) AIAI 2015. IFIP AICT, vol. 458, pp. 383–396. Springer, Heidelberg (2015). doi:10.1007/978-3-319-23868-5_27
13. Lin, Y.Y., Liu, T.L., Fuh, C.S.: Multiple kernel learning for dimensionality reduction. IEEE Trans. Pattern Anal. Mach. Intell. **33**(6), 1147–1160 (2011)
14. Nigam, K., Ghani, R.: Analyzing the effectiveness and applicability of co-training. In: Proceedings of the Ninth International Conference on Information and Knowledge Management, pp. 86–93. ACM (2000)
15. Ordóñez, P., DesJardins, M., Lombardi, M., Lehmann, C.U., Fackler, J.: An animated multivariate visualization for physiological and clinical data in the ICU. In: Proceedings of the 1st ACM International Health Informatics Symposium, pp. 771–779. ACM (2010)
16. Ramaker, C., Marinus, J., Stiggelbout, A.M., van Hilten, B.J.: Systematic evaluation of rating scales for impairment and disability in parkinson's disease. Mov. Disord. **17**(5), 867–876 (2002)
17. Rind, A., Aigner, W., Miksch, S., Wiltner, S., Pohl, M., Drexler, F., Neubauer, B., Suchy, N.: Visually exploring multivariate trends in patient cohorts using animated scatter plots. In: Robertson, M.M. (ed.) EHAWC 2011 and HCII 2011. LNCS, vol. 6779, pp. 139–148. Springer, Heidelberg (2011)
18. Snow, B.: Objective measures for the progression of parkinsons disease. J. Neurolo. Neurosurg. Psychiatry **74**(3), 287–288 (2003)
19. Stanev, D., Moschonas, P., Votis, K., Tzovaras, D., Moustakas, K.: Simulation and visual analysis of neuromusculoskeletal models and data. In: Chbeir, R., Manolopoulos, Y., Maglogiannis, I., Alhajj, R. (eds.) AIAI 2015. IFIP AICT, vol. 458, pp. 411–420. Springer, Heidelberg (2015). doi:10.1007/978-3-319-23868-5_29
20. Tong, H., He, J., Li, M., Zhang, C., Ma, W.Y.: Graph based multi-modality learning. In: Proceedings of the 13th Annual ACM International Conference on Multimedia, pp. 862–871. ACM (2005)
21. Turkay, C., Jeanquartier, F., Holzinger, A., Hauser, H.: On computationally-enhanced visual analysis of heterogeneous data and its application in biomedical informatics. In: Holzinger, A., Jurisica, I. (eds.) Interactive Knowledge Discovery and Data Mining in Biomedical Informatics. LNCS, vol. 8401, pp. 117–140. Springer, Heidelberg (2014)

22. Von Landesberger, T., Kuijper, A., Schreck, T., Kohlhammer, J., van Wijk, J.J., Fekete, J.D., Fellner, D.W.: Visual analysis of large graphs: state-of-the-art and future research challenges. In: Computer Graphics Forum, vol. 30, pp. 1719–1749. Wiley (2011)
23. Ward, M.O., Grinstein, G., Keim, D.: Interactive Data Visualization: Foundations, Techniques, and Applications. CRC Press, Natick (2010)
24. Wongsuphasawat, K., Guerra Gómez, J.A., Plaisant, C., Wang, T.D., Taieb-Maimon, M., Shneiderman, B.: LifeFlow: visualizing an overview of event sequences. In: Proceedings of the SIGCHI Conference on Human Factors in Computing Systems, pp. 1747–1756. ACM (2011)

ERMIS: Extracting Knowledge from Unstructured Big Data for Supporting Business Decision Making

Christos Alexakos[✉], Konstantinos Arvanitis, Andreas Papalambrou,
Thomas Amorgianiotis, George Raptis, and Nikolaos Zervos

Industrial Systems Institute, ATHENA Research and Innovation Centre,
Patras Science Park Building, Platani, 265 04 Rio, Patras, Greece
{alexakos,nzervos}@isi.gr, konstantinos.arvanitis@gmail.com,
andreas@papalambrou.gr, amorgianio@ceid.upatras.gr,
george.raptis@ymail.com

Abstract. Business managers support that decisions based on data analysis are better decisions. Nowadays, in the era of digital information, the accessible information sources are increasing rapidly, especially on the Internet. Also, the most critical information for business decisions is hidden in a large amount of unstructured data. Thus, Big Data analytics has become the cornerstone of modern Business Analytics providing insights for accurate decision making. ERMIS (Extensible pRoduct Monitoring by Indexing Social sources) system is able to aggregate unstructured and semi-structured data from different sources, process them and extracting knowledge by semantically annotating only the useful information. ERMIS Knowledge Base that is created from this process is a tool for supporting business decision making about a product.

Keywords: Big Data · Business Analytics · Ontologies · Data driven decision making · Knowledge extraction

1 Introduction

During the last decade the term Big Data has the pride of place in both academia and industry on the area of Business Analytics [1]. Big Data analytics methodologies and technologies permit the processing of large amount of data for providing accurate insights for a business [2]. Especially today, the century of digital information, businesses can collect useful data from various sources such as Internet Sites, blogs, social media and IoT infrastructure, even from their own information systems. This not only means the need for processing of large volumes of data but also the necessity for the analysts to face the fact that these data volumes are increasing in high rates [3]. Since, the industry sector supports that decisions based on data analysis are better decisions, the utilization of Big Data analytics enables managers to conclude to decisions based on evidence rather than intuition [4].

Another immense challenge of Business Big Data analytics is the extraction of useful knowledge from the millions of unstructured data existing in various sources on the

Published by Springer International Publishing Switzerland 2016. All Rights Reserved
L. Iliadis and I. Maglogiannis (Eds.): AIAI 2016, IFIP AICT 475, pp. 611–622, 2016.
DOI: 10.1007/978-3-319-44944-9_54

Internet. This problem is entitled as Variety and it is one of the three major challenges in Big Data, called the three Vs of Big Data, alongside with Velocity and Volume [5]. Big data can be characterized in three types: (a) structured, (b) semi-structured and (c) unstructured. Structured data is provided in an already tagged and easily sorted format. Unstructured data is random thus it is difficult to be processed. Semi-structured data has separated data elements but they do not conform to fixed fields [6]. When analysts undertake problems of Business Intelligence that require multi-domain and multi-source knowledge such as the processing of crowd opinions about a product or spot consumer's behavior, the information data sources are usually web sites, posts in forums or social networks, blogs, reviews and news on portals [7]. In such cases the really useful information is "hidden" in unstructured data among other non-critical information [8].

The main scope of this article is to introduce an intelligent integrated system – called ERMIS (Extensible pRoduct Monitoring by Indexing Social sources) - which allows the digital mediation to the optimum decision making, via the intelligent process of large volume of structured and semi-unstructured data derived from various Internet sources. The specialized technological aim is the development of an intelligent layer that will integrate useful information from various sources in a unified Knowledge Base. The users of this system are both the consumers as well as the decision making persons of a business. Consumers can get information about a specific knowledge on a product by setting a query in natural language. The business managers can monitor the information for a specific product or service, by studying the extracted knowledge from a large amount of sources on the Internet.

ERMIS allows users to compose a query related to a product in natural language (i.e. "I want to buy DELL X345 laptop", "The monitor in my iPhone 6 has been broken"). The system processes the query and exports the main semantics that characterized it. It tries to recognize the product, the producer and the purpose of the query (i.e. if an article refers to damage or an intentional buy). In the background, the system collects data from the Internet (social media, news portals, e-shops, etc.), it processes them, annotates them based on an integrated ontology and feeds them to a unified Knowledge Base. Knowledge Base keeps only the useful information based on the semantics of the ERMIS integrated ontology. The extraction of knowledge related to a user's query comes with the semantic inference to the axioms stored in the Knowledge Base. The extracted knowledge is presented to the user through a user-friendly web interface. ERMIS system is designed and implemented for processing text data in Greek language, a most challenging effort due to the diversity and variety of the grammar and syntax of Greek language. Nevertheless, the system can be easily adapted for the English language and for any other language.

The proposed approach is presented in detail in the rest of the paper. In Sect. 2 some significant technologies and related work in the area of Big Data analytics based decision making are presented. Section 3 describes the basic functional components of the proposed architecture and Sect. 4 presents in detail the ontological model of the ERMIS Knowledge Base. Section 5 describes the data collection, processing and semantic annotation, while Sect. 6 refers to the knowledge extraction and presentation to the users. Finally, Sect. 7 concludes the paper.

2   Big Data and Data-driven Decision Making

A lot of decisions in the industry are based on the analysis of data. This practice is called Data-driven decision making (DDD). Decision makers can choose between two practices: in the first, more traditional, the managers based their decisions on their experience and their intuition, in the second, the managers take advantage of the analysis of business-related data in order to interpret the market trends. The second one is based on DDD techniques and it is supported by various Data Analytics tools. As DDD is not an all-or-nothing practice and it can be easily combined with the practices based on manager's experience, it is gaining the confidence of industry in the last decade [9]. A study by Brynjolfsson, Hitt and Kim [10] shows that one standard deviation higher on the DDD scale is associated with a 4–6 % increase in productivity and also affects higher return on assets, return on equity, asset utilization, and market value.

Big data technologies, especially from the side of data engineering, permit analysts to process large volume of data which leads them to more accurate decisions. It is remarkable, that a study presented by Tambe [11] shows that utilization of Big Data technologies correlate with productivity growth that can reach 1–3 % higher productivity for one standard deviation higher utilization of big data. Specifically, one standard deviation higher than the average business.

In the last years a lot of Big Data Analytics tools have been proposed by the key players in data analytics. IBM big data platform [12] and SAS Big Data Insights[1] are some paradigms of platforms that provide Big Data engineering techniques to their customers for creating analysis processes and reports. Also, most of the cloud providers have services for Big Data analysis such as Microsoft's Azure HDInsight[2] and Amazon Web Services Big Data platform[3]. Regarding the academia, the proposed DDD approaches are mainly driven to solve current problems as higher accuracy in data mining and data visualization in various fields. Visual analytics for Big Data is a big challenge as they provide users a friendly way for analysing their data. The use of visual analytics is the center of many research works such as network bandwidth evaluation for security vulnerabilities detection [13] and human muscles movement and forces simulation for diagnosis purposes [14]. Also, the proposed BIG [15] is a Multi Agent System for collecting data, unstructured text processing and decision making. BIG text process in based on keyword extraction and it does not support natural language querying.

3   ERMIS Architecture

The ERMIS system targets two user models. Everyone can create a Public User account, which allows one to make queries and receive answers. Both the queries and the answers will be visible to every user and only a single pending query is permitted for each public

---

[1] http://www.sas.com/en_us/insights/big-data.html.

[2] https://azure.microsoft.com/en-us/services/hdinsight/.

[3] https://aws.amazon.com/big-data/.

user account. Specific users can also create Industry User accounts, which allows the Industry User to perform multiple concurrent queries and keep the query and results private and linked to their account. In addition, Industry Users have access to Monitoring queries (queries which continue to provide answers while they remain active), notifications when their queries are answered or new data become available on a Monitoring query and the ability to generate statistical reports based on their queries. In addition, Industry User accounts are provided with access to a REST service API to allow integrating the ERMIS system with their internal software.

The ERMIS system is composed of two main subsystems the Front-End and the Back-End, each with specific component modules as presented in Fig. 1.

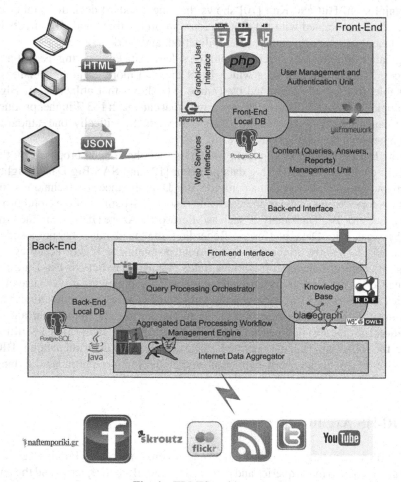

Fig. 1. ERMIS architecture

The **Front-End** components are a UI component, a user management component, a query management component and a Web Service component. The UI component is responsible for providing a web-based graphical User interface that allows interaction with the ERMIS functionality. The user management component is responsible for user authentication and authorization for the whole ERMIS system. The query management component is responsible for accepting and handling the users' queries, the system's answers to them and all the various monitoring data and reports, by scheduling tasks in the backend subsystem and handling the back-end's answers. The Web Service component provides a REST API that makes the ERMIS functionality available to third party systems.

The **Back-End** components are databases for structured and unstructured data, the Knowledge Base, a Query Processing Orchestrator and the Aggregated Data Processing Workflow Management. The databases are used to keep track of data such as reports and queries, as well as metadata for the various information sources and structured data related entities in the Knowledge base. The Knowledge Base contains the knowledge gathered through external sources, expressed in axioms composed in following RDF format and according to the OWL ontologies that are described in the next chapter. The Query Processing Orchestrator is a Multi-Agent System and it is responsible for scheduling all tasks related to query processing, such as analysis to extract relevant terms and periodic checks to update monitoring queries. It compiles the reports and answers that are then made available to the Front-End. The data processing workflow provides constant monitoring of select RSS feeds and other data sources, such as site-specific information APIs and links located through processing of the various feeds. It utilizes syntactic and lexicographic analysis of data from information sources and extracts relevant metadata and terms to store in the Knowledge base.

The Front-End is installed in a web server (NGINX[4]) and is developed on top of Yii[5] PHP framework. The provided screens to the users are empowered with HTML5, CSS and JavaScript and designed following responsive design patterns utilizing the Bootstrap[6] framework. The Back-End is developed in Java, as a set of interoperating software modules. To provide for future extension and alternative implementations, Apache UIMA (Unstructured Information Management Architecture)[7] is used to create and manage the analysis engines used for natural language processing for both queries and unstructured data downloaded from other information sources. Source specific modules are used to take advantage of site-specific APIs (such as Skroutz[8], Twitter[9] and YouTube[10]) that provide structured or semi-structured data. It uses PostgreSQL to store structured data and BlazeGraph[11] to provide the Knowledge Base functionality. Blazegraph is a high performance graph database [16] platform that supports RDF/SPARQL with scalable solutions including embedded, High Availability, scale-out, and GPU-acceleration.

---

[4] http://nginx.org/.

[5] http://www.yiiframework.com/.

[6] http://getbootstrap.com/.

[7] https://uima.apache.org/.

[8] http://developer.skroutz.gr/.

[9] https://dev.twitter.com/rest/public.

[10] https://developers.google.com/youtube/.

[11] https://www.blazegraph.com/.

4   ERMIS Knowledge Base

ERMIS's knowledge management is accompanied by a Knowledge Base where collected information related to products is stored in RDF graph format. The structure of the graph along with the rules defining the interrelationships are based on an integrated ontology model represented in OWL. This model defines the basic system ontology and a group of secondary ontologies and taxonomies focusing on expanding the main ontology in order to define axioms and entities which add and interrelate information gathered from multiple internet sources.

4.1   ERMIS Ontology

ERMIS ontology is the main ontology of the Knowledge Base. It describes the concepts of the Product, Document (information source) and Query. The concepts of ERMIS ontology are depicted in Fig. 2.

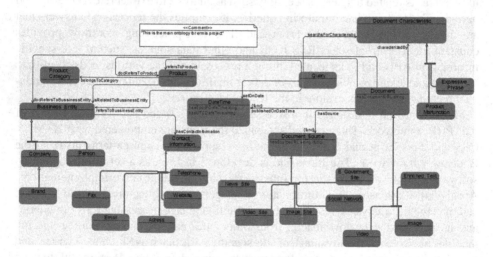

Fig. 2. ERMIS ontology

The primary classes of the ontology are:

- *Product:* This class defines a product, and is interrelated with other classes of the system. The relationship *isRelatedToBussinessEntity* is used to correlate products with business entities, such as suppliers and manufacturers, and the relationship *belongsToCategory* is used to classify the products into categories.
- *Bussiness_Entity:* This class defines all the business entities which are interrelated to a product, through its subclasses such as Person and Company. The contact information of each business entity, such as address, telephone and email address, is defined by the class *Contact_Information* and the relationship *hasContactInformation*.

- *Document:* This class defines the entities providing information related to products, collected from multiple internet sources. Each instance could be an enriched text or a multimedia object, such as image or video, as it is defined in the corresponding subclasses. Multimedia could be part of a document and this axiom is represented in the *includedInDocument* relationship. Documents' characteristics include expressive phrases, article polarity and referred product malfunction. Hence, each document refers to a product and/or is interrelated to a business entity. It is collected from different types of internet sources, such as news websites, image or video galleries, e-government services and social networks.
- *Query:* This class represents the queries set by the ERMIS users. When a query is set, a search for related documents starts based on characteristics, referred product and referred business entity.
- *DateTime:* This class defines the date and time of an action and is used when a document is published or when a query is set.

4.2 Integrated Ontology

The integrated ontology ErmisIO is a generic ontology created to combine the afore-mentioned ontologies of the system. All the required ontologies were imported in ErmisIO and different namespaces were used to distinguish them. The additional ontologies except from ERMIS ontology are:

- Dublin Core from Protégé for the description digital objects[12].
- GPT Ontology derived from Google Product Taxonomy used at Google Merchant[13].
- SIOC (Semantically-Interlinked Online Communities)[14] ontology for social media content
- Twitter Engineering Ontology for describing twitter content [17].

In particular, for the ontology DUBLIN CORE the annotation properties *dc:title, dc:creator, dc:format, dc:language* and *dc:description* were added. To support the ontology Google Product Taxonomy the class *ermisio:GooglePT_Entity*, a subclass of the *ermisonto:Product_Category* class, was created. The *ermisio:GooglePT_Entity* class is the superclass of all the classes on the first level of the Google Product Taxonomy ontology, and hence the taxonomy is provided as *ermisonto:Product_Category*. For the ontology SIOC the class *ermisio:SIOC_Entity* was created, which is the superclass of all the classes on the first level of the SIOC. The *sioc:Post* class is the subclass of *ermisonto:Product*, which classifies a post as document as it is defined in the ERMIS ontology. Similarly, for the Twitter Engineering ontology, the *twtronbto:Tweet* class is the subclass of *ermisonto:Product*. In the same way, other domain ontologies that describe structured data from an internet source can be integrated in ERMIS Integrated Ontology.

---

[12] http://dublincore.org/.
[13] https://support.google.com/merchants/answer/1705911?hl=en.
[14] http://rdfs.org/sioc/spec/.

4.3   WordNet Ontology

WordNet [18] is a lexical database of English words, which groups the words into synonyms sets, called synsets. Each synset represents a distinct lexical meaning, provides short definitions and is connected to various lexical and semantical relationships. It was created in 1985 by G.A. Miller [19] who was inspired by artificial intelligence experiments trying to understand the human semantic memory. Its main purpose was to provide a combination of dictionary and thesaurus features to support the automatic text analysis in interfacial intelligence applications.

Balkanet[15] expanded the number of European languages developed by Euro-WordNet[16]. The Greek WordNet was established by the Databases Lab (DBLab) of the University of Patras along with the participation of the University of Athens [20]. The biggest ambition of BalkaNet is the semantic connection of the words of each language in order to create a multilingual semantic network.

The ERMIS project requires words recognition in a gathered text and their correlation with a product malfunction or deficiency, along with the polarity of the information (positive, negative, or neutral). For these purposes, the Greek WordNet schema was expanded in order to annotate each synset with any related information. A brief ontology was created for this purpose and is presented in Fig. 3 and its primary classes are the Malfunction and Polarity classes.

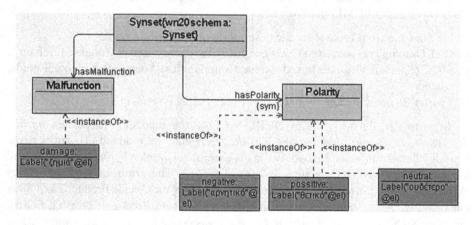

Fig. 3. WordNet extension for ERMIS project

5   Process Unstructured Data Towards to Semantic Annotation

Unstructured information found on the Internet typically exists in texts that can be formatted (containing heading, bold text etc.) or not. One of the goals of the ERMIS system is to process these texts and correlate them semantically to the ErmisIO ontology.

---

[15] http://www.dblab.upatras.gr/balkanet.
[16] http://www.illc.uva.nl/EuroWordNet/.

In order for this process to take place, every document fetched from the Internet (web page, forums posts etc.) is processed as depicted in Fig. 4. In more detail, the stages are as follows:

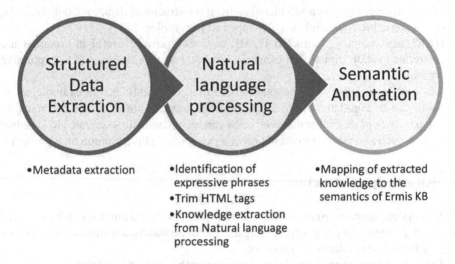

•Metadata extraction   •Identification of   •Mapping of extracted
                              expressive phrases   knowledge to the
                            •Trim HTML tags   semantics of Ermis KB
                            •Knowledge extraction
                              from Natural language
                              processing

Fig. 4. Process unstructured data towards to semantic annotation

Structured data extraction: Each retrieved document is typically accompanied by metadata such as dates, authors, titles etc. For example, Twitter tweets can provide time and date, user and HTML links. HTML documents typically contain metadata information in the HEAD section. These metadata can be directly correlated with the Dublin-Core annotation properties.

Natural language processing (NLP): NLP involves the processing and understanding of natural language (in our case, in written form). In order to perform the NLP, the Apache UIMA framework is used. UIMA uses Analysis Engines to annotate documents of unstructured information and provide metadata. The following steps are needed from the retrieval of a document to the semantical analysis of the text.

- *HTML cleanup:* This involves the removal of HTML tags leaving only plain text. However, important information that could assist in the semantic analysis (such as bold, italics etc.) is kept in the form of UIMA annotations.
- *Grammatical and syntactic processing:* This involves the recognition of grammar and syntax tokens as well as meaningful tokens in the written text. First, basic grammatical processing in the form of sentence and word annotation takes place. Then words are recognized regarding their Parts of Speech and are also stemmed in order to correlate all possible grammatical forms to a specific semantic token. Finally, Named entities are also recognized.

Semantic Annotation: This stage involves the mapping of the above annotated properties to the semantics of ERMIS knowledge base. The mapping involves the following specific stages.

- *Metadata mapping.* Extracted metadata from the structured data, for example dates, are converted to RDF triplets and mapped to the ontology.
- *HTML tag mapping.* Extracted HTML tags that provide useful information are converted to RDF triplets. For example, bold (< b>) tags are mapped to expressive phrases.
- *NLP results mapping.* NLP annotations are mapped to entities. For example, some nouns can be correlated to Product Categories, named entities can be correlated to companies or places, other nouns or verbs can be semantically correlated to Wordnet synsets such as a product malfunction or a positive/negative opinion on a product.

6   Knowledge Extraction

ERMIS system supports users to compose their query in natural language. When a query is set to the system, the latter is processing it in the same way as described in the previous chapter for free text information sources.

From the query the annotator is trying to extract the product, producer, product category, keywords, the type of the question (question for damage or general question) and the damage of the product if it exists. With this information the appropriate SPARQL queries are composed for getting information from the Knowledge Base. The results are evaluated in a rank system of 1 to 3 grade (match, high match and great match) according to their closeness to query's initial information. Finally, the results are presented to the user in a timeline from the most recent to the oldest one. Each answer is marked with

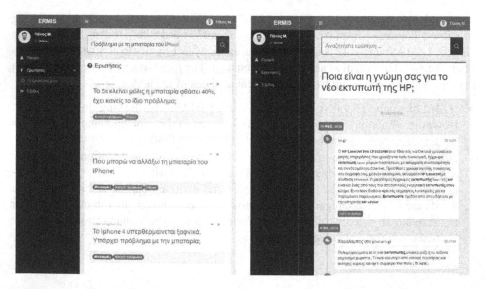

Fig. 5. Presentation of extracted knowledge to the user

different icon and color according to its ranking. Figure 5 depicts two screenshots from the ERMIS system, one with the list of user's queries and one with the timeline of the answers for a specific query.

7 Conclusion

The adaptation of Data-driven decision making to modern businesses has been proved to provide significant added value to business growth. When the problem comes to the analysis of large volume of Big Data the challenges range from data engineering tasks to data explanation and knowledge extraction. ERMIS (Extensible pRoduct Monitoring by Indexing Social sources) system offers an integrated solution for collecting, processing, semantically annotating structured and unstructured data accumulated from various sources over the Internet. The main scope of ERMIS is to elaborate the natural language processing techniques for both providing users a simple way to set their queries and extracting valuable knowledge which is hidden in unstructured data sources such as web sites, posts in forums or social networks, blogs, reviews and news on portals. ERMIS system introduces an intelligent layer that integrates useful information from various sources in a unified Knowledge Base. The users of ERMIS system, which are both the consumers and the industry decision makers, are able to compose a query related to a product in natural language and view the inferred knowledge in a user-friendly web interface.

Although ERMIS supports state-of-art algorithms for Natural Language Processing, it can be extended with methods for extracting more knowledge from text processing such as sentiment analysis. Also, during the evaluation, we spotted a lot of mismatching values caused by the complexity of Greek language. In a future version of ERMIS system, we are expecting to resolve such issues. Furthermore, ERMIS is going to be extended to more information sources with both new data processing annotators and domain definition ontologies. Finally, the adaptation of new languages such as English and French is in the ERMIS' future roadmap.

References

1. Chen, H., Chiang, R.H.L., Storey, V.C.: Business intelligence and analytics: from big data to big impact. MIS Q. **36**(4), 1165–1188 (2012)
2. Wixom, B., et al.: The current state of business intelligence in academia: the arrival of big data. Commun. Assoc. Inf. Syst. **34**(1), 1 (2014)
3. Gantz, J., Reinsel, D.: The digital universe in 2020: big data, bigger digital shadows, and biggest growth in the far east. In: IDC iView: IDC Analyze the Future, pp. 1–16 (2012)
4. McAfee, A., Brynjolfsson, E., Davenport, T.H., Patil, D.J., Barton, D.: Big data. Manage. Revolution Harvard Bus. Rev. **90**(10), 61–67 (2012)
5. Singh, S., Singh, N.: Big data analytics. In: International Conference on Communication, Information and Computing Technology, Mumbai, India, October 2011. IEEE (2012)
6. Sagiroglu, S., Sinanc, D., Big data: a review. In: 2013 International Conference on Collaboration Technologies and Systems (CTS), pp. 42–47. IEEE (2013)

7. Kambatla, K., Kollias, G., Kumar, V., Grama, A.: Trends in big data analytics. J. Parallel Distrib. Comput. **74**(7), 2561–2573 (2014)
8. boyd, d., Crawford, K.: Six provocations for big data (21 September 2011). In: A Decade in Internet Time: Symposium on the Dynamics of the Internet and Society, September 2011. SSRN: http://dx.doi.org/10.2139/ssrn.1926431
9. Provost, F., Fawcett, T.: Data science and its relationship to big data and data-driven decision making. Big Data **1**(1), 51–59 (2013)
10. Brynjolfsson, E., Hitt, L.M., Kim, H.H.: Strength in numbers: how does data-driven decision making affect firm performance? Working paper, SSRN working paper (2011). SSRN: http://ssrn.com/abstract=1819486
11. Tambe, P.: Big data know-how and business value. Working paper. NYU Stern School of Business, NY, New York (2012)
12. Zikopoulos, P., Parasuraman, K., Deutsch, T., Giles, J., Corrigan, D.: Harness The Power of Big Data. The IBM Big Data Platform. McGraw Hill Professional, New York (2012)
13. Kalamaras, I., Papadopoulos, S., Drosou, A., Tzovaras, D.: MoVA: a visual analytics tool providing insight in the big mobile network data. In: Chbeir, R., Manolopoulos, Y., Maglogiannis, I., Alhajj, R. (eds.) AIAI 2015. IFIP AICT, vol. 458, pp. 383–396. Springer, Heidelberg (2015). doi:10.1007/978-3-319-23868-5_27
14. Stanev, D., Moschonas, P., Votis, K., Tzovaras, D., Moustakas, K.: Simulation and visual analysis of neuromusculoskeletal models and data. In: Chbeir, R., Manolopoulos, Y., Maglogiannis, I., Alhajj, R. (eds.) AIAI 2015. IFIP AICT, vol. 458, pp. 411–420. Springer, Heidelberg (2015). doi:10.1007/978-3-319-23868-5_29
15. Lesser, V., Horling, B., Klassner, F., Raja, A., Wagner, T., Zhang, S.X.Q.: BIG: an agent for resource-bounded information gathering and decision making. Artif. Intell. **118**(1–2), 197–244 (2000)
16. Sikos, L.F.: Graph databases. In: Mastering Structured Data on the Semantic Web, pp. 145–172. Apress, Berkeley (2015)
17. Allen, T.: Twitter ontology, Report, National Center for Ontological Research (2013). http://ncor.buffalo.edu/2013/IE500/Reports/Travis-Allen-Twitter-Ontology.docx. Accessed
18. Miller, G.A.: WordNet: a lexical database for English. Commun. ACM **38**(11), 39–41 (1995)
19. Miller, G.A.: Wordnet: a dictionary browser' in information in data. In: Proceedings of the First Conference of the UW Centre for the New Oxford Dictionary. University of Waterloo, Waterloo, Canada (1985)
20. Stamou, S., Nenadic, G., Christodoulakis, D.: Exploring Bal-kanet shared ontology for multilingual conceptual indexing. In: LREC, European Language Resources Association (2004)

Superclusteroid 2.0: A Web Tool for Processing Big Biological Networks

Maria Tserirzoglou-Thoma[1], Konstantinos Theofilatos[2(✉)], Eleni Tsitsouli[1],
Georgios Panges-Tserres[1], Christos Alexakos[2], Charalampos Moschopoulos[1],
Georgios Alexopoulos[1], Konstantinos Giannoulis[1], Spiros Likothanassis[1(✉)],
and Seferina Mavroudi[3]

[1] Department of Computer Engineering and Informatics, University of Patras, Patras, Greece
{thom,tsitsoue,panges,mosxopul,alexopo,giannul,
likothan}@ceid.upatras.gr
[2] InSyBio Ltd., 109 Uxbridge Road, London, UK
{k.theofilatos,c.alexakos}@insybio.com
[3] Department of Social Work, Technological Institute of Western Greece, Patras, Greece
mavroudi@teiwest.gr

Abstract. Biological networks have been the most prevalent model to analyze the complexity of cellular mechanisms. The expansion of the existing knowledge on known intracellular players such as genes, RNA molecules and proteins as long as the continued study on their interactions has increased lately the ability to construct big biological networks of increased complexity. Many web tools have been introduced in the last decade but they are incomplete, as they do not provide all required features for a full research study neither they can handle the big and complex nature of these networks and the increased needs of researchers for fast and uninterrupted analysis. In the present paper, the new version of the Superclusteroid tool is presented which includes among others new visualization features, network comparison tools and new clustering algorithms. Moreover, a new strategy is proposed to deal with the necessity of handling effectively the increased work load of the tool as long as to improve the speed in the two most time consuming steps: network visualization and network clustering.

Keywords: Web tool · Biological networks · Protein-protein interaction networks · Network clustering · Network visualization

1 Introduction

Network visualization is a fundamental method that helps scientists in understanding biological networks and important properties in underlying biochemical processes. Molecules such as DNA, RNA, proteins, metabolites and interactions between them are related to highly important biological networks. Whenever such molecules are connected by physical interactions, they form molecular interaction networks that are generally classified by the nature of the compounds involved. Many biological networks have been characterized in detail: Protein-Protein Interaction (PPI), Gene co-expression

© IFIP International Federation for Information Processing 2016
Published by Springer International Publishing Switzerland 2016. All Rights Reserved
L. Iliadis and I. Maglogiannis (Eds.): AIAI 2016, IFIP AICT 475, pp. 623–633, 2016.
DOI: 10.1007/978-3-319-44944-9_55

networks (Transcript-Transcript association networks), Gene regulatory networks (DNA-protein interaction networks), protein phosphorylation, metabolic interactions, and genetic interaction networks [1].

The PPIs represent the interaction between proteins: e.g. the formulation of protein complexes and the activation of one protein by another protein. These interactions are essential to almost every process in a cell, thus understanding of them is crucial. Such a network can be defined as an un-directed graph G = (V, E) .where V is the set of proteins represented as nodes and E.the set of interactions represented as edges. Graphs of a whole cell PPIs are complex and difficult to be generated. For this reason bioinformatics tools have been developed to simplify the difficult task of visualization, such as Cytoscape [2] which is an open-source software commonly used. Large scale identification of PPIs generates hundreds of thousands interactions, which were collected together in specialized biological databases, such as BIND and DIP, that are continuously updated in order to provide complete interactomes [3].

The Gene Co-expression networks represent the pairs of genes which show a similar expression pattern across samples, since the transcript levels of two co-expressed genes rise and fall together across samples. Gene co-expression networks are of biological interest since co-expressed genes are controlled by the same transcriptional regulatory program, functionally related, or members of the same pathway or protein complex. These networks can be defined as un-directed graphs where each node corresponds to a gene and a pair of nodes is connected with an edge when there is a significant co-expression relationship between them. Modules or the highly connected subgraphs in gene co-expression networks correspond to clusters of genes that have a similar function or involve in a common biological process which causes many interactions among themselves. Gene co-expression networks are usually constructed using datasets generated by high throughput gene expression profiling technologies such as Micro-array or RNA-Seq. [4].

The Gene Regulatory networks (DNA-Protein interaction networks) represent the DNA segments in a cell which interact with each other indirectly, through their RNA and protein expression products and with other substances in the cell to govern the gene expression levels of mRNA and proteins. The modeling techniques for such networks involves the use of Coupled Ordinary Differential Equations, Boolean networks, Bayesian networks, Graphical Gaussian models, Stochastic Gene network et al. The Gene Regulatory networks can be defined as graphs in which the un-directed edge connects two genes, representing a biochemical process such as a reaction, transformation, interaction, activation or inhibition [5].

The problem of analyzing biological networks is a big data problem for two reasons. First, it is related to intensive time-consuming analysis of big datasets as for example a PPI network can implicate more than 20000 proteins and more than 200000 interactions. Second, this analysis is not an one off procedure. New versions of biological networks are being available daily as new interactions are being studied and thus the tool's workload is extremely big. Additionally, if we consider the vast number of different organisms as well as the enormous number of biological conditions we can easily understand the "big data" nature of the problem.

In the present paper, we introduce a new version of the Superclusteroid tool. Super-clusteroid is a web based tool which enables the analysis of various types of biological networks including the aforementioned categories with the main constraint being that only undirected networks can be used. This new version is differentiated from the previous one by including a new more advanced algorithm for biological network clustering, providing new features for the analysis and using a new mechanism to handle high workload of very large networks.

2   Existing Tools for the Analysis of Biological Networks

Large scale biological studies produce huge amounts of data that reveal various layers of molecular interaction networks. As we saw graphs have been used to represent, study and integrate such biological networks, which leads in large-scale analyses. Specialized tools are required to extract and compare information obtained from multiple data sources, and apply various statistical parameters treatments to describe and understand networks properties. Following are the basic web-based or standalone tools for analyzing biological networks, which are based on graphs and their visualization.

NETAL [6], is a new graph-based method for global alignment of protein-protein interaction networks. It uses a greedy method, based on the alignment scoring matrix, which is derived from both biological and topological information of input networks to find the best global network alignment.

NeAT (Network Analysis tool) [7], is another tool which provides a user-friendly web access to a collection of modular tools for the analysis of networks (graphs) and clusters (e.g. microarray clusters, functional classes, etc.). This tool is designed to cope with large datasets and provide a flexible toolbox for analyzing biological networks stored in various databases (protein interactions, regulation and metabolism) or obtained from high-throughput experiments (two-hybrid, mass-spectrometry and microarrays). The web interface interconnects the programs in predefined analysis flows, enabling to address a series of questions about networks of interest.

GraphWeb [8], is a public web server for biological network analysis and module discovery. It provides methods to integrate heterogeneous and multispecies data for constructing directed and undirected, weighted and un-weighted networks, to discover network modules using a variety of algorithms and topological filters and interpret modules using functional knowledge of the Gene Ontology and pathways, as well as regulatory features such as binding motifs and microRNA targets.

Giba [9], is a clustering tool that offers the ability to detect important protein modules such as protein complexes. GIBA implements a two-steps strategy, where in the first one the whole protein – protein interaction graph is divided into clusters and in the second step these clusters are filtered and only the ones considered important are kept.

jClust [10] is an application which provides access to a set of widely used clustering and clique finding algorithms. The toolbox allows a range of filtering procedures to be applied and is combined with an advanced implementation of the Medusa interactive visualization module. These implemented algorithms are k-Means, Affinity Propagation, Bron–Kerbosch, MULIC, Restricted neighborhood search cluster algorithm,

Markov clustering and Spectral clustering, while the supported filtering procedures are haircut, outside–inside, best neighbors and density control operations. The tool provides a powerful tool for data analysis and information extraction.

Cluto [11] is a software package for clustering low- and high-dimensional datasets and for analyzing the characteristics of the various clusters. Cluto is well-suited for clustering data sets arising in many diverse application areas including information retrieval, customer purchasing transactions, web, GIS, science, and biology. Cluto's distribution consists of both stand-alone programs and a library via which an application program can access directly the various clustering and analysis algorithms implemented in Cluto.

VisANT [12], is an application for integrating biomolecular interaction data into a cohesive, graphical interface. This software features a multi-tiered architecture for data flexibility, separating back-end modules for data retrieval from a front-end visualization and analysis package. This system is integrated with standard databases for organized annotation, including GenBank, KEGG and SwissProt. It provides a general tool for mining and visualizing such data in the context of sequence, pathway, structure, and associated annotations. Interaction and predicted association data can be combined, overlaid, manipulated and analyzed using a variety of built-in functions.

Most of the aforementioned tools cover only a subset of the analysis related to biological networks, while most of them are not PPI network specific and some of them include obsolete algorithmic solutions. Superclusteroid 1.0 [13] is a web tool dedicated to data processing of protein-protein interaction networks which was initially introduced to cover all these caveats. The tool is implemented in the GNU/Linux environment and is written in Perl. It supports various input file formats and provides the following services. First, clustering, by choosing one of the available clustering algorithms. Second, PPI network visualization, such as the original network or other DOT files and also the algorithms results can be automatically visualized or can be downloaded for later use. Third, protein cluster function prediction in which the user, by choosing a specific protein, may continue with the analysis by implementing the Majority Vote Prediction Algorithm (MVPA) [14] or the Hypergeometric Distribution Prediction Algorithm (HDPA) [15].

3   Superclusteroid 2.0: A Web Tool for Processing Big Protein-Protein Interaction Networks

3.1   Superclusteroid 1.0

The Superclusteroid 1.0 [13] uploads and manipulates input of PPI data, which can have one the following formats: tab-delimited text files, adjacency matrices in text files, DOT files-using the DOT network description languages and SIF files, a popular tab-delimited text file mostly used in Cytoscape.

It offers a selection of widely used clustering algorithms to process the input data. These algorithms are the MCL-Markov Cluster [15], the Restricted Neighbourhood Search Clustering Algorithm–RNSC [16], the Highly Connected Subgraphs Algorithm-HCS [17] and the SideS, a variation of HCS which uses a statistical model to express the statistical significance of a cluster [18]. The resulting files are tab-delimited data with two columns, one for the name of the cluster and one for the protein belonging to that cluster.

The results from the clustering algorithm can be automatically visualized or can be downloaded for later use. Also the original network or other DOT files can be viewed by choosing the "visualize" tab. In either case, a java applet named "ZGRViewer" [19] is used to support the "fdp" and "twopi" GraphViz/DOT tools for spring model and radial layouts respectively.

The analysis of the network data can be continued, since the user can choose a specific protein and implement the Majority Vote Prediction Algorithm (MVPA) [14] or the Hypergeometric Distribution Prediction Algorithm (HDPA) [15]. Both methods are applied only on PPI data with Uniprot IDs and for the S. cerevisiae organism.

A help page is available on the tool which contains explicit instructions describing its services and a comprehensive list of the web services available, along with their description and the access URL for each of them. Additionally, the web tool provides demo data to help the user to understand its functionality.

Despite the success of the first version of Superclusteroid tool it lacked some extra functionalities in order to enable its users to realize the whole analysis of PPI networks without the need of referring to external tools. Moreover, new algorithms have emerged for some of the tasks of analyzing PPI networks, such as new clustering algorithms for overlapping clusters, since the initial launching of Superclusteroid 1.0 and some of them were needed to be incorporated in it. Finally, the successful adoption of Superclusteroid tool from the scientific community (more than 200 unique users/visitors per month) raised some performance issues and more sophisticated solutions were required to manage work load and improve the efficiency of the tool. For all these reasons, a new version of Superclusteroid was essential and the new features incorporated in it are described in the next section.

3.2   New Features Incorporated in Superclusteroid 2.0

EEMC. The Evolutionary Enhanced Markov Clustering (EEMC) [20] is a hybrid combination of an adaptive evolutionary algorithm and a state-of-the-art clustering algorithm. It is based on the MCL algorithm [15] which is one of the most commonly used methods in clustering PPI graphs in order to predict protein complexes. Although it has some strong limitations, with the most important one being its restriction to assign each protein to only one protein complex. To overcome the above MCL's problem, Moschopoulos et al. (2008) [21], proposed the Enhanced Markov Clustering (EMC) method which is an improvement of the MCL algorithm. In specific, it deploys the MCL algorithm to make an initial clustering and then it improves it by applying 4 different filtering methods: density filter, haircut, best neighbor and cutting edge operators. The last method requires tuning of their parameters and it is not able to function on weighted graphs.

The EEMC is a fully unsupervised method, which is a combination of an adaptive Genetic Algorithm and an extension of the EMC method in which, the filtering methods of the EMC algorithm were adjusted to enable handling of weighted PPI graphs. The

Genetic Algorithm was used to optimize on parallel the inflation rate and the parameters of the filtering methods of EMC algorithm.

The EEMC starts with the creation of the initial population ie. the chromosomes with binary representation. Then the adjusted filters from the EMC algorithm are applied on the chromosomes. These filtering methods are density filter, haircut, best neighbor and cutting edge operators. The next step is the evaluations of the chromosomes, with an evolutionary framework using unsupervised fitness value which will produce scaled fitness, to assign high values for high performance clustering. The selection operator is then taking place on the algorithm. The roulette wheel selection assigns probabilities of selection in each chromosome proportional to its performance. The variation operators used are the two-point crossover and the binary mutation. A dynamic control of the mutation parameter is used, to estimate the variation which is applied in the mutation probability for each iteration of the EEMC. Finally the termination criteria of the algorithm are a combination of the maximum number of generations to be reached and a convergence criterion.

New Functions of Version 2.0. Superclusteroid 2.0 includes a set of new features which enable users to conduct all required biological network analysis without the need to use other tools.

To begin with, a new way of cluster visualization is proposed, using the Cytoscape program [22], a program for analyzing and visualizing network data. Specifically, the users can now not only to see the nodes of a cluster but also their interactions and their weights in the cases of weighted networks. This can enable a more detailed analysis of the cluster and its connectivity.

Furthermore, with the new version of Superclusteroid, the users have the ability to compare and evaluate clusters, with a collection of human core protein complexes from the CORUM database [23] or with a dataset which the user chooses to upload. The uploaded datasets should be in a tab delimited format, with every row representing a cluster and including its nodes separated with tabs. The calculated metrics include the standard metrics of sensitivity, positive predictive value, arithmetic accuracy, geometric accuracy, separation [24].

Another newly introduced feature is the linkage of clustering results with the he Gene Set Enrichment Analysis (GSEA) [25] tool. With this feature the users are able to evaluate and characterize the importance of gene sets, i.e., gene groups with a common biological function, chromosomal location or setting.

Moreover, the pipeline has been completed with a network comparison option which allow the comparison of similar biological networks which have either be constructed under different biological conditions or refer to different organisms. This analysis, additionally to the standard network metrics calculation, such as clustering coefficient, also offers a graph with the degree distributions of the networks under comparison and allow the biomarker discovery by locating the nodes and edges which have been differentiated between the under comparison networks.

Some other features of secondary significance which have been included in the current version of Superclusteroid tools for the meta-analysis of protein clustering results such as a haircut filter to discard nodes with low connectivity, a neighborhood analysis tool to locate nodes being connected with a specific node of interest and cluster function prediction tool which is based on the HDPA algorithm but is only applicable on *Homo Sapiens* and *Saccharomyces cerevisiae*.

The most intensive tasks for the analysis of biological networks are network clustering and network visualization. In order to satisfy the increased demands of Superclusteroid's users for these two types of analysis, Superclusteroid 2.0 introduce a queuing mechanism presented in Fig. 1. In specific, the Superclusteroid web server, using iteratively the min cut algorithm splits the initial network to sub-networks and provides them to the virtual infrastructure dedicated to clustering and visualization analysis. When the analysis is conducted then it is provided to Superclusteroid's web server which undertakes to assemble the results and present them to the user. File sharing between web server and virtual infrastructure is done via a Network File System (NFS) memory unit. RabbitMQ server [26] is utilized in order to handle efficiently the work load of the Virtual Infrastructure while all required scripts are written in Python programming language version 2.7.

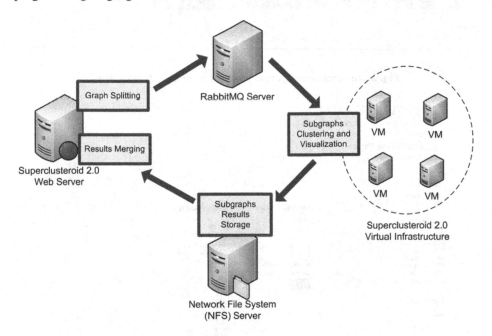

Fig. 1. Proposed Superclusteroid 2.0 architecture for handling effectively increased work load on time intensive biological network analysis tasks

4 Results

Superclusteroid 2.0 has been tested using as input the PPI graph of the Yeast organism as described in [27]. The dataset contains 1430 proteins and 26531 interactions between them. Experimental results [20] have proved the superiority of the EEMC algorithms in all examined clustering metrics. Figures 2, 3 and 4 depict some screenshots of the analysis.

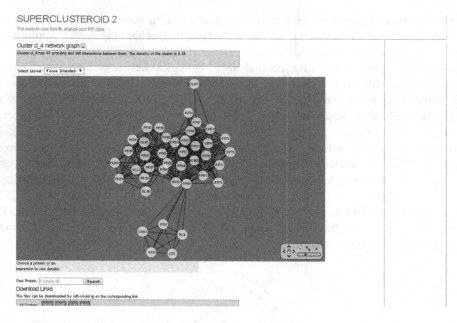

Fig. 2. Superclusteroid's new protein function prediction view.

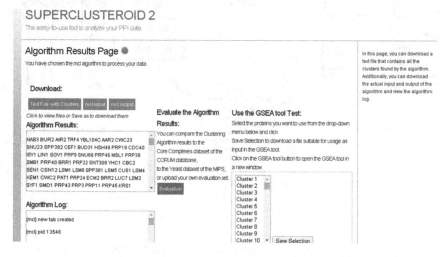

Fig. 3. Newly designed view for showing clustering results and allowing algorithms evaluation and GSEA analysis.

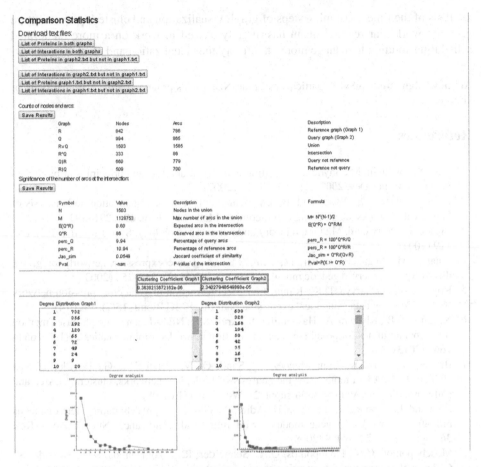

Fig. 4. Superclusteroid's Biological Network comparison module. This figure shows the comparison of two yeast PPI networks. This analysis allows for a network comparison using general network metrics such as clustering coefficient as well as a comparison using the node degree distribution figures and a more detailed analysis on the union, intersection and differences between the two examined networks.

5 Discussion

In this paper we have presented a new version of the Superclusteroid web tool. This new version extends significantly the previous one in order to provide to the users all the necessary tools to perform a full analysis and meta-analysis on undirected weighted biological networks. Moreover, a new more elaborated network clustering solution was added to allow the discovery of overlapping clusters. This clustering solution was proved to overcome the constraints of previous clustering solutions achieving higher clustering metrics.

Finally, in order to handle the increased traffic of the web tool, we have implemented and incorporated in it an architecture which faces effectively the work load on the analysis of the time-consuming steps of graph visualization and clustering. As a future research work, our research team has already started to work on a more elaborated scheduling solution to manage more effectively the visualization and clustering tasks.

Acknowledgments. InSyBio participates in the NBG seeds program by the National Bank of Greece.

References

1. Zhu, X., Gerstein, M., Snyder, M.: Getting connected: analysis and principles of biological networks. Genes Dev. **2007**(21), 1010–1024 (2007)
2. Kohl, M., Wiese, S., Warscheid, B.: Cytoscape: software for visualization and analysis of biological networks. data mining in proteomics. Meth. Mol. Biol. **696**, 291–303 (2011)
3. Mason, O., Verwoerd, M.: Graph theory and networks in biology. Syst. Biol. IET **1**(2), 89–119 (2007)
4. Stuart, J.M., Segal, E., Koller, D., Kim, S.K.: A gene-coexpression network for global discovery of conserved genetic modules. Science **302**(5643), 249–255 (2003)
5. Roy, S., Bhattacharyya, D.K., Kalita, J.K.: Reconstruction of gene co-expression network from microarray data using local expression patterns. BMC Bioinf. **15**(7), 1 (2014)
6. Neyshabur, B., Khadem, A., Hashemifar, S., Arab, S.S.: NETAL: a new graph-based method for global alignment of protein–protein interaction networks. Bioinformatics **29**(13), 1654–1662 (2013)
7. Brohée, S., Faust, K., Lima-Mendez, G., Sand, O., Vanderstocken, G., Deville, Y., van Helden, J.: NeAT: a toolbox for the analysis of biological networks, clusters, classes and pathways. Nucleic Acids Res. **36**(suppl. 2), W444–W451 (2008)
8. Reimand, J., Tooming, L., Peterson, H., Adler, P., Vilo, J.: GraphWeb: mining heterogeneous biological networks for gene modules with functional significance. Nucleic Acids Res. **36**(suppl. 2), W452–W459 (2008)
9. Moschopoulos, C.N., Pavlopoulos, G.A., Schneider, R., Likothanassis, S.D., Kossida, S.: GIBA: a clustering tool for detecting protein complexes. BMC Bioinf. **10**(6), 1 (2009)
10. Pavlopoulos, G.A., Moschopoulos, C.N., Hooper, S.D., Schneider, R., Kossida, S.: jClust: a clustering and visualization toolbox. Bioinformatics **25**(15), 1994–1996 (2009)
11. Zhao, Y., Karypis, G.: Data clustering in life sciences. Mol. Biotechnol. **31**(1), 55–80 (2005)
12. Hu, Z., Mellor, J., Wu, J., DeLisi, C.: VisANT: an online visualization and analysis tool for biological interaction data. BMC Bioinf. **5**(1), 1 (2004)
13. Ropodi, A., Sakkos, N., Moschopoulos, C., Magklaras, G., Kossida, S.: Superclusteroid: a web tool dedicated to data processing of protein-protein interaction networks. EMBnet. J. **17**(2), 10 (2011)
14. Bu, D., Zhao, Y., Cai, L., Xue, H., Zhu, X., Lu, H., Li, G.: Topological structure analysis of the protein–protein interaction network in budding yeast. Nucleic Acids Res. **31**(9), 2443–2450 (2003)
15. Enright, A.J., Van Dongen, S., Ouzounis, C.A.: An efficient algorithm for large-scale detection of protein families. Nucleic Acids Res. **30**(7), 1575–1584 (2002)
16. King, A.D.: Graph clustering with restricted neighbourhood search (Doctoral dissertation, University of Toronto) (2004)

17. Hartuv, E., Schmitt, A., Lange, J., Meier-Ewert, S., Lehrach, H., Shamir, R.: An algorithm for clustering cDNAs for gene expression analysis. In: Proceedings of the Third Annual International Conference on Computational Molecular Biology, pp. 188–197. ACM, April 1999
18. Koyutürk, M., Szpankowski, W., Grama, A.: Assessing significance of connectivity and conservation in protein interaction networks. J. Comput. Biol. **14**(6), 747–764 (2007)
19. Pietriga, E.: Zgrviewer-a 2.5 D graph visualizer for the DOT language (2005)
20. Theofilatos, K., Pavlopoulou, N., Papasavvas, C., Likothanassis, S., Dimitrakopoulos, C., Georgopoulos, E., Mavroudi, S.: Predicting protein complexes from weighted protein–protein interaction graphs with a novel unsupervised methodology: evolutionary enhanced Markov clustering. Artif. Intell. Med. **63**(3), 181–189 (2015)
21. Moschopoulos, C.N., Pavlopoulos, G.A., Likothanassis, S.D., Kossida, S.: An enhanced Markov clustering method for detecting protein complexes. In: 8th IEEE International Conference on BioInformatics and BioEngineering, BIBE 2008, pp. 1–6. IEEE, October 2008
22. Shannon, P., Markiel, A., Ozier, O., Baliga, N.S., Wang, J.T., Ramage, D., Ideker, T.: Cytoscape: a software environment for integrated models of biomolecular interaction networks. Genome Res. **13**(11), 2498–2504 (2003)
23. Ruepp, A., Weagele, B.: CORUM: the comprehensive resource of mammalian protein complexes—2009. Nucleic Acids Res. **38**(Database issue), D497–D501 (2006)
24. Brohee, S., Van Helden, J.: Evaluation of clustering algorithms for protein-protein interaction networks. BMC Bioinf. **7**(1), 1 (2006)
25. Lander, E.S., et al.: Gene set enrichment analysis: a knowledge-based approach for interpreting genome-wide expression profiles. PNAS 2005 **7**(1), 15545–15550 (2005)
26. Russell, J., Cohn, R.: Rabbitmq. Book on Demand (2012)
27. Gavin, A.C., Aloy, P., Grandi, P., Krause, R., Boesche, M., Marzioch, M., Edelmann, A.: Proteome survey reveals modularity of the yeast cell machinery. Nature **440**(7084), 631–636 (2006)

Systematic Mapping Study on Performance Scalability in Big Data on Cloud Using VM and Container

Cansu Gokhan[1], Ziya Karakaya[2(✉)], and Ali Yazici[2]

[1] Institute of Natural and Applied Sciences, Atilim University, Ankara, Turkey
cansugokhann@gmail.com
[2] Faculty of Engineering, Atilim University, Ankara, Turkey
{ziya.karakaya,ali.yazici}@atilim.edu.tr

Abstract. In recent years, big data and cloud computing have gained importance in IT and business. These two technologies are becoming complementing in a way that the former requires large amount of storage and computation power, which are the key enabler technologies of Big Data; the latter, cloud computing, brings the opportunity to scale on-demand computation power and provides massive quantities of storage space. Until recently, the only technique used in computation resource utilization was based on the hypervisor, which is used to create the virtual machine. Nowadays, another technique, which claims better resource utilization, called "container" is becoming popular. This technique is otherwise known as "lightweight virtualization" since it creates completely isolated virtual environments on top of underlying operating systems. The main objective of this study is to clarify the research area concerned with performance issues using VM and container in big data on cloud, and to give a direction for future research.

1 Introduction

Big data applications continue to receive an ever-increasing amount of attention, thus they become a dominant class of applications deployed over virtualized environments [1]. On the other hand, the resource utilization feature of cloud computing is mostly based on virtualization techniques, which is the common way to run different services on the cloud [2]. By combining these two, most of the big data on cloud environments are using hypervisor to provision the virtual machines. In this technique, the VMs have their own operating systems which run on the virtual hardware resources provided by hypervisor [3]. Although it is proven to be a very useful technique in resource utilization, still there is an inherent overhead because of the hypervisor [1].

In recent years, containers, which are also called "lightweight virtualization", are gaining popularity due to their ability to offer superior performance because they do not have their own operating systems [2]. Instead, they use the OS kernel underlined with the host machine and they work similar to a regular

© IFIP International Federation for Information Processing 2016
Published by Springer International Publishing Switzerland 2016. All Rights Reserved
L. Iliadis and I. Maglogiannis (Eds.): AIAI 2016, IFIP AICT 475, pp. 634–641, 2016.
DOI: 10.1007/978-3-319-44944-9_56

application and are completely isolated from each other as well as from the underlying system. This technique receives its popularity mostly in Linux OS virtualization, since it uses the features provided by Linux OS kernel itself, such as "cgroup", "namespace", etc., in order to completely isolate each container from the rest.

In this study, along with the other research questions, the main purpose of investigation was to identify if there is a gap in the literature and to what extend those techniques are being studied by using experimental approach.

There are three different databases used in this study to search for relevant papers. The authors found 308 papers that appeared to be relevant. After applying the inclusion and exclusion criteria, there were only 62 papers containing significant information either directly or indirectly related with the research questions.

The remainder of this paper is structured as follows: the related works on performance and comparison of VM vs. containers are discussed in Sect. 2. Section 3 presents our methodology and research questions investigated. In Sect. 4, the results are summarized and the findings together with discussion is given. Finally, the limitations of this study is appear in Sect. 5.

2   Related Work

Readers requiring in-depth information about the technologies and techniques used in virtual machines and containers together with their relationships to hypervisor and underlying operating systems are offered to read the white paper published by Intel [4].

There are many studies in the literature about big data on cloud environment. One of the Systematic Mapping (SM) studies conducted on this subject is the work of Ibrahim and his colleagues [5]. They have analysed the scalability issues of storage, but not the scalability issues of performance within the cloud. They have proposed a classification for big data, a conceptual view of big data, and a cloud services model. This model was compared with several representative big data cloud platforms. They have discussed the background of Hadoop technology and its core component, namely MapReduce, and Hadoop Distributed File System (HDFS).

Yanzhang focuses on the scalability performance issues of Hadoop Virtual Cluster with cost consideration [6]. They compared the scalability performance with respect to scale-up and scale-out methods under different workloads. Hadoop benchmarks and real parallel machine learning algorithms were used to evaluate the scalability performance. Their experimental results showed that the scale-up method outperformed the scale-out method for CPU-bound applications, and the opposite for I/O-bound applications. They also noted that disk and network I/O are the main bottlenecks of cloud platform due to shared resource contention and interference.

The most comprehensive work on performance comparison of virtual machines and Linux containers has been done by Felter and colleagues [3].

Their goal was to isolate and understand the overhead introduced by virtual machines (specifically KVM) and containers (specifically Docker) relative to non-virtualized Linux on Cloud. They have concluded that both VM and containers are mature technologies, and that both have negligible performance overheads with respect to CPU and Memory performance. Nevertheless, they warned about the use of these technologies in case of I/O intensive works, which is the case in Big Data Application on Cloud.

Yang et al. have discussed the impact of virtual machine on Hadoop [7]. They describe the effect of different virtualization technologies such as KVM, Xen and OpenVZ on MapReduce environment. Also, they evaluate performance and stability of HDFS (Hadoop Distributed File System) on KVM, Xen and OpenVZ. Besides this, Pedro et al. have presented the performance of KVM and OpenVZ using micro benchmarks for disk and CPU [8].

There are many other papers in the literature that have studied the performance scalability of Big Data Applications. However, only a few of them have focused on performance scalability comparison of Container vs. VM technologies.

3   Research Methodology

A systematic map study was performed to obtain the current research map on the performance scalability issues in big data on cloud. The guidelines proposed by Peterson and colleagues [9] is followed in this study. The mapping study was conducted in three main stages, namely planning, execution and result.

3.1   Systematic Mapping Plan

In the planning stage, we defined research questions, search strategy, screening of papers for inclusion and exclusion, classification of papers and data extraction.

Research Questions. The following research questions were identified as relevant to purpose:

1. *To what extend are the published papers on the performance scalability issues in big data on cloud are based on experimental study?*
2. *What is the percentage of the mostly studied technologies in big data performance scalability issues on cloud environment?*
3. *Which is the most investigated hypervisor in big data performance scalability issues?*
4. *What types of containers are being studied in big data on cloud?*
5. *Which components of the resources are mostly investigated for performance impact on big data analysis?*
6. *How frequent is the dominating technology being studied in the last five years as a tool in big data on cloud?*

Table 1. Selected databases

Database	Location
IEEE explore	http://ieeexplore.ieee.org/
Science direct	http://www.sciencedirect.com/
ACM digital library	http://dl.acm.org/

Search Strategy. The selected databases for the study are shown in Table 1 in order to identify potentially relevant conference articles and journal publications.

The following keywords were used in order to perform the search for the study: Big data, Cloud Computing, Performance, Scalability, Container, Virtual Machine, Comparision of VM vs. Containers. Search strings were applied to check keyword, title, and abstract fields in order to perform the automatic search in the selected digital libraries.

These strings are given as follows:

[("Big Data") AND ("Cloud Computing") AND (performance OR Scalability) AND (Container OR VM OR "Virtual Machine")]

Inclusion and Exclusion Criteria. The aim of this process is to identify the most relevant studies for the mapping study. According to the research questions, the inclusion and exclusion criteria given in Table 2 were applied to the selected papers.

Table 2. Inclusion and Exclusion criteria

Inclusion criteria	
1	Studies addressing performance scalability issues in big data on cloud.
2	Journal and/or conference papers.
3	Studies that describe virtual machine and container types in the big data on cloud.
4	Primary or secondary studies.
Exclusion criteria	
1	Studies not accessible in full text.
2	Studies that do not address the performance scalability issues in big data on cloud.
3	Studies not presented in English.
4	Prefaces, slides, panels, editorials or tutorials.
5	Studies that do not answer the research questions

Classification of Papers. The present work classified the papers according to properties and categories listed in Table 3.

Table 3. Classification scheme

Properties	Categories
Research approach	Theory, survey, review, experimental
Year	Years between 2009 and 2016
Article title	Name of the article
Container type	Docker, OpenVZ, LXC, Linux-VServer
VM type	Xen, KVM, Vmware (ESX, ESXi), Others
Technology	MapReduce, Hadoop, Spark, Storm, FLink
Component of hardware resource	CPU, Disk I/O, Network speed, Memory (RAM), # of VM/Container

Data Extraction. In order to extract data from the selected studies, we designed a data extraction Excel table. Each selected paper appears as a record item in this file. The data extraction table consist of article name, year of publication, technology, component of hardware resource, VM types and container types. Then, the data that is specifically related to research questions were extracted from each study.

3.2 Execution of Systematic Mapping

At the execution stage, we conducted a systematic mapping study according to the plan stated in the previous section. The search string was modified for the different syntax as of databases according to the search criteria, and we have found 308 papers as candidate studies from all the selected sources. The title, abstract, and keywords were analysed, and then, some of the articles were eliminated by applying the exclusion criteria. In case of uncertainties as to inclusion of some papers, the introduction and conclusion sections of these articles were also taken into consideration. As a result of eliminating unrelated articles, 62 relevant research papers were selected.[1]

3.3 Results of Mapping

RQ1 Objective: The main objective of answering this question is to identify the proportion of experimental researches already done when compared to others.
RQ1 Results: Considering the performed study, 60 of 62 papers were based on experimental studies making them a majority. Figure 1(a) shows the number of experimental and non-experimental studies with respect to publication years.
RQ2 Objective: The main objective of answering this question is to identify to what extend the mostly studied technology is dominating the research area.

---

[1] List of articles: http://bit.ly/1Ux6H5M.

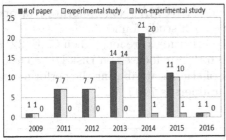

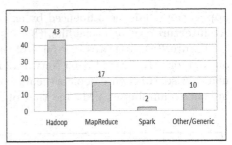

(a) Annual Publication count for each of frameworks

(b) Distribution of Technology

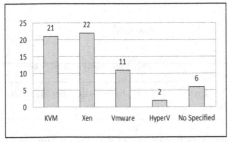

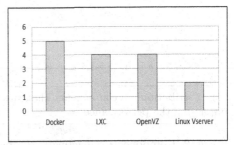

(c) Distribution of VM's

(d) Distribution of Containers

Fig. 1. Distribution of Technology and VM's

RQ2 Results: The papers were categorized as follows: Hadoop, MapReduce and Spark. If a paper's study could not be defined with a specific technology, it is shown as the category called "Other/Generic".

According to the results given in Fig. 1(b), we can conclude that the majority of these studies are based on MapReduce, since Hadoop is itself based on the MapReduce technique.

RQ3 Objective: The main objective of this question is to identify the most widely used virtual machine types for big data performance scalability issues.

RQ3 Results: According to Fig. 1(c), KVM (Kernel based Virtual Machine) and Xen are found to be the mostly investigated VM type in Big Data performance issues. This is probably because of their open source nature.

RQ4 Objective: Our aim with this question is to find the dominating container technology being studied.

RQ4 Results: As shown in Fig. 1(d), Docker is the most commonly studied container type in Big Data performance scalability issues, while LXC and OpenVZ are equally distributed.

RQ5 Objective: The aim of this question is to determine most commonly studied hardware component, to show its performance impact.

RQ5 Results: The most investigated and resulted components are CPU, Disk I/O, Memory and Network Speed. The findings are consistent with the needs of big data applications on cloud. Although the performance of memory intensive application could be influenced by existence of NUMA (Non-uniform Memeory Architecture), or by cache hits, there is no such study found.

Figure 2(a), indicates that 56 % of the total studies are focused on CPU, 32 % on Memory (RAM), 44 % on Network and 69 % are focused on Disk I/O components. Also, this figure depicts the number of studies attributing the performance impact to number of VMs and that of containers is 7 (11 %).

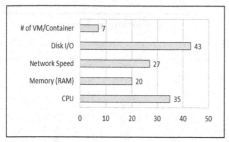 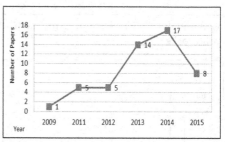

(a) Distribution of components that are impacting the performance

(b) Frequency of the studies on MapReduce

Fig. 2. Components impacting performance and frequency of studies on Hadoop

RQ6 Objective: The aim of this question is to determine the frequency of research on the subject within the last five years.

RQ6 Results: The result shows that 50 papers out of 62 are about the MapReduce based tool. According to this Fig. 2(b), it is easily seen that the number of papers which used MapReduce as a tool increased suddenly in 2013, and the frequency of papers has decreases after 2014.

4   Conclusion

In this paper, we used systematic mapping technique, to obtain a current research map on performance scalability issues in big data on cloud. The guidelines proposed by Peterson and colleagues [9] is followed. Of the 308 papers found in three popular databases according to our search string, 62 papers were found to be related either directly or indirectly with the research subject and questions. The papers are then analysed with respect to the research questions.

The following conclusions are deduced from the analysis of those related papers; (i) considering the performed study, 61 of 62 papers were based on experimental studies achieving a majority of the papers; (ii) the mostly studied technologies are Hadoop and MapReduce; (iii) KVM and Xen are the dominant hypervisor technology used in these studies; (iv) Docker is the mostly studied container type, and LXC and OpenVZ are the technologies that are used equally; and (v) CPU and Disk I/O are the two issues that are mostly handled when comparing these technologies. It is better to state that KVM and Xen are opensource, and Spark is relatively newer than Hadoop.

The most important finding of this research is that there are only a handful academic papers which compare the performance scalability of hypervisor-based virtual machines vs. containers for the big data applications on cloud. On the other hand, although there are many researches about the comparison of either VM vs. "bare metal" (physical) on cloud, only three of them compares the performance in Big Data on Cloud. This was the second gap found during this study.

Another most important implication of this study shows that there is a lack of empirical study conducted on the other popular Big Data analysis frameworks, such as Spark, Storm, FLink, etc.

5   Limitations

In this study, only three databases are searched for relevant papers. These technologies have gained popularity over the past few years, and journal/conference papers may contain more than those of found in these databases. We believe that other databases should also be considered in such analysis, and we plan to further analyse those papers.

References

1. Mytilinis, I., Tsoumakos, D., Kantere, V., Nanos, A., Koziris, N.: I/O performance modeling for big data applications over cloud infrastructures. In: Proceeding of 2015 IEEE International Conference on Cloud Engineering, pp. 201–206 (2015)
2. Morabito, R., Kjllman, J., Komu, M.: Hypervisors vs. lightweight virtualization: a performance comparison. In: Proceeding of 2015 IEEE International Conference on Cloud Engineering, pp. 386–393 (2015)
3. Felter, W., Ferreira, A., Rajamony, R., Rubio, J.: An updated performance comparison of virtual machines and linux containers. In: 2015 IEEE International Symposium on Performance Analysis of Systems and Software (ISPASS). IEEE (2015)
4. Intel White Paper: Linux Containers Streamline Virtualization and Complement Hypervisor-Based Virtual Machines (2014). http://www.intel.com/content/dam/www/public/us/en/documents/white-papers/linux-containers-hypervisor-based-vms-paper.pdf
5. Abaker, I., Hashem, T., Yaqoob, I., Anuar, N.B., Mokhtar, S., Gani, A., Khan, S.U.: The rise of 'big data' on cloud computing: review and open research issues. J. Inf. Syst. **47**, 98–115 (2015)
6. He, Y., Jiang, X., Wu, Z., Ye, K., Chen, Z.: Scalability analysis and improvement of hadoop virtual cluster with cost consideration. In: 2014 IEEE 7th International Conference on Proceeding of the Cloud Computing (CLOUD), pp. 594–601 (2014)
7. Yang, Y., Long, X., Dou, X., Wen, C.: Impacts of virtualization technologies on hadoop. In: 2013 Third International Conference Proceeding of Intelligent System Design and Engineering Applications (ISDEA), pp. 846–849 (2013)
8. Vasconcelos, P.R.M., de Araújo Freitas, G.A.: Performance analysis of hadoop MapReduce on an OpenNebula cloud with KVM and OpenVZ virtualizations. In: Proceeding 9th International Conference for Internet Technology and Secured Transactions, ICITST (2014)
9. Kai, P., Vakkalanka, S., Kuzniarz, L.: Guidelines for conducting systematic mapping studies in software engineering: an update. Inf. Softw. Technol. **64**, 1–18 (2015)

Ineffective Efforts in ICU Assisted Ventilation: Feature Extraction and Analysis Platform

Achilleas Chytas[1,2], Katerina Vaporidi[3], Dimitris Babalis[3],
Dimitris Georgopoulos[3], Nicos Maglaveras[1,2],
and Ioanna Chouvarda[1,2(✉)]

[1] Lab of Computing and Medical Informatics,
PB 323 Medical School, AUTH, 54124 Thessaloniki, Greece
ioannach@auth.gr
[2] Institute of Applied Biosciences, CERTH, 57001 Thessaloniki, Greece
[3] Department of Intensive Care Medicine, School of Medicine,
University of Crete, 71500 Heraklion, Greece

Abstract. Intensive Care Unit (ICU) is a challenging environment, requiring continuous monitoring and treatment adaptations, raising the need for tools and platforms to support medical decisions. In this context, the focus of this work is in supporting clinicians in managing assisted ventilation intervention (AVI). In AVI the need for patient-ventilator coupling exists. Attention may be required in cases when patient's effort doesn't trigger the ventilator at all, and the assisted ventilation event is lost, i.e. when an ineffective effort (IE) event takes place. A high exposure to IEs has been related to adverse clinical outcomes. The purpose of this work is to create new features that complement the already existing IE index in terms of estimating the adverse effects of ventilation exposure. A series of tools varying from raw data handling to the creation of predictive models are created and implemented in a custom platform, utilizing open-source software.

1 Introduction

In an Intensive Care Unit (ICU) context, patient bio-signals are continuously monitored and displayed towards recognizing alerting events. These recordings of physiologic waveforms along with data coming from laboratory examinations and patient interaction (medications, medical procedures) can be difficult to be interpreted especially in an environment as demanding as the ICU. Thus there is a need for analysing and displaying the data in an easy-to-understand manner. Real-time analysis of patients' bio-signals can be used to detect conditions that precede medical complications using both domain expert knowledge and knowledge obtained by automated procedures [1].

The ventilation dy-synchronization is a known issue; a prime example are the incidences of ineffective triggering [2]. Ineffective triggering, (IT) is common, but factors affecting IT vary considerably and can be contributed to patient condition and factors related to the ventilation system. As a result, the IT frequency of appearances and overall distribution varies as well.

© IFIP International Federation for Information Processing 2016
Published by Springer International Publishing Switzerland 2016. All Rights Reserved
L. Iliadis and I. Maglogiannis (Eds.): AIAI 2016, IFIP AICT 475, pp. 642–650, 2016.
DOI: 10.1007/978-3-319-44944-9_57

In the medical literature, the total time spent under ineffective effort has been proposed as an index related to adverse clinical events. However, these events are sometimes not equally distributed in time. Ineffective triggering of the ventilator is frequent but highly variable among patients and during the course of mechanical support for each patient. As previously reported [3], most patients have small (5 min) periods with high intensity of ineffective synchronization, i.e. ineffective efforts (IEs). It is still an open issue whether the cumulative effect of IE exposure during the total time in ventilation or the temporal patterns of IEs relate to patient deterioration.

The AEGLE [4] project was created by the need to provide the tools and the necessary data for physicians and researchers to explore new data and answer research questions. The datasets and tools described in this paper will enable clinicians to explore these new features and gain new insight in the related phenomena. In this work we propose a set of features that describe the morphology of IE event time-series and complement the IE index that might be proved helpful in better describing patients and estimating their hospital prognosis, and a web-based platform that is used to perform the analysis and evaluate the outcomes.

2 The AEGLE Project Approach

The design and implementation of the 1st AEGLE platform prototype is currently underway, parallel to that, analytic tools such as the ones presented in this paper, are also being designed. Having the data providers, domain experts and analytics developers in different sites can be a hindrance in the design of novel analytics, since direct communication and information flow is of outmost importance especially in the early stages.

The solution was to expose directly to the domain experts the in-development analytics via a web-based platform, so that a feedback loop between the involved parties will be created having as a result a more efficient designing phase.

For the purpose of developing the PVI analytics we chose R, an open source and powerful scripting language. For the last few years it has been among the most common used software packages in scholarly articles, and in 2015 it became the 2nd most used, while also being in fourth position in terms of usage growth in the same domain [5]. That makes R the most popular free analytic software, having as an added bonus its' high flexibility in that it provides the users the ability to create custom analytics. At the same time, it benefits greatly from a large and active community that contributes to analytics and visualization libraries, and frameworks.

3 Methodology

3.1 Data Collection

For the purposes of this study recordings from 108 patients for multiple days were obtained from the ICU clinic of the University Hospital of Heraklion, Crete (PAGNI) using an experimental protocol; Patient-Ventilator Interaction (PVI) Monitor [6], as well as a selection of fields from the hospital Electronic Health Records (EHR). The raw data produced is approximately 12 Mb per patient per day as an average.

The data providers uploaded pseudo-anonymized data as files on a secure location. Later a more appropriate solution for datasets containing time-series was used by utilizing the NoSQL Apache Cassandra, although at this point it is implemented over workstation level resources.

3.2 Preprocessing

The main problem with the dataset that hinders a frequency based analysis is that the recordings are event driven, and thus the time difference between consecutives recordings varies, and in several cases significant. We applied a pre-processing tool that first utilizes general data cleaning methods and case specific rules (e.g. no breaths or attempts in breathing for 3 min is considered an artefact), and afterwards resamples the data to a fixed sampling rate of 30 s [7, 8].

Ineffective effort for more than 10 % of total breath count is believed to cause problems and prolong hospitalization time [2, 9]. We focused on the examination of periods in which the patient was experiencing serious troubles in breathing. We call such incidents IE events [7].

3.3 Feature Extraction for IE

A common used feature for the patient-ventilator interaction is the IE index.

$$\frac{Ineffective\ Efforts}{Ineffective\ Efforts\ +\ Breaths}. \tag{1}$$

It is defined as the sum of IE to the combined sum of IE and breaths over a course of period, be it the totality of the recording, or smaller time segments (e.g. 1 h to 5 min). It gives us a general idea on how the patient faired over a period of time. It is reported that it IEs can lead to extended ventilation time, prolonged hospitalization and that it has an impact on mortality for patients with high IE index (>10 %) [2, 9]. A question that arises is whether or not a patient that might have a low overall IE index over the course of hours, has increased health risks because he is subject to short periods of intense IT activity. In such cases, even a much lower IE index might be related to patient deterioration. Also, a single feature might not be enough to sufficient describe the complexity of a signal distribution.

In order to address this issue, this paper suggests a set of indices that better describe the IE signal morphology. They can be divided into two categories:

1. Indices that are calculated based on IE signal morphology and are independent from the IE even definition (Table 1).
2. Indices that are calculated based on the IE event definition (Table 2), and thus can vary depending on the researcher's input.

Table 1. Indices of 1st category

Name and description	Equation
Power: The sum of all IE, this is a helping index used to describe others	$Pow_{tot} = \sum IE$
Mean: The average of the IE. Also descriptive index for others	$Pow_m = \frac{1}{n}\sum IE$
Density: The percentage of samples in which the patient is experiencing at least one IE.	$Density = 1/N \sum_{i=1}^{N} \begin{cases} 1, & IE_i > 0 \\ 0, & IE_i == 0 \end{cases}$, where $i = 1,\ldots,N$ the samples of the IE signal
Mean to Density: Utilizing the previous indices we define this new one. For two patients having the same Mean score, the patient with the lowest Density score (same percentage of IE but more concentrated) will end up with a higher Mean to Density score, thus differentiating between those two.	$M2D = \frac{Pow_m}{Density}$
Area over x: This index is based on Density and represents the percentage of samples where the patient is experiencing at least x amount of IE.	$AreaOver_x = \frac{1}{N}\sum_{i=1}^{N} \begin{cases} 1, IE_i > x \\ 0, IE_i == x \end{cases}$
Variation: Coefficient of variation for the IE signal.	$VarCof = 100\frac{SD(IE)}{Pow_m}$
IE distance: Median time distance between consecutives IE	
Max clean Area: Percentage of the maximum IE free time period in reference to the total recording period.	

3.4   Exposing ICU Analytics via Web-Based Platform

In order to make the developed tools accessible to physicians a custom web based ICU platform was created. The main goal was to present the analytic tools in a simplified way, hiding functionality from the user when the steps were already predefined while giving them the ability to parameterize tools when required. The platform supports two major functionalities (Fig. 1).

The first functionality is a module regarding data pre-processing and processing. Currently only a PVI dataset module is in place. The physicians can upload to the database pseudo-anonymized raw data as extracted by the PVI monitor, then query raw datasets to be pre-processed as previously defined. Afterwards, the user can select either of the two analysis pipelines currently in place, the feature extraction and/or the correlation between the ventilator recorded signals and their phase delay with each-other estimation, based on wavelet coherence [7]. The wavelet coherence analysis explores the time-series correlation with each other and produces a set of features that are expected to give insight regarding the physiological phenomena that take place at the vicinity of an IE event, either preceding it (potential causes) or following it

Table 2. Indices of 2nd category

Name and Description	Equation
Event Power: The sum of IE that belong to event period, this is also an index with limited predictive capabilities that is used to better describe other indices.	$Pow_e = \sum_{n=1}^{n} Pow_{e_i}$ **where n the number of events**
Event Duration: The total duration of all IE events combined.	$Dur_e = \sum_{i=1}^{n} Dur_{e_i,}$ *where n the number of events*
Median Event Power: The median power of the IE events	$Pow_{e_m} = Pow_e/n$
Median Event duration: The median duration of the event	$Dur_{e_m} = Dur_e/n$
Event distance: Median time distance between consecutives IE events, for recordings that did not have an event this value was the total recording time, for recordings that had a single event this value is	$\frac{Dur_{tot}}{2} - Dur_e$
Area concentrating power (x%): In this case the IE event, the percentage of time in which the x % percentage of power is concentrated. This requires to (a) Locate the i = 1,...,n events, with Pow_{ei}, Dur_{ei}, (b) Sort them by descending density di, (c) Sum the m more dense events till we reach the desired x%, the index is calculated as the sum of their duration to the total recording duration. In most cases the desired x percentage is either not reached or is surpassed, so the index is scaled to the desired value.	$d_i = Pow_{e_i}/Dur_{e_i},$ **Local event density** $ACPow_x = \frac{1}{Dur_{tot}} \sum_{i=1}^{k} Dur_{e_i},$ $ACPow_x = \frac{ACPow_i * x'}{x},$
Concentrated density (x%): In an effort to differentiate between patients that have low AreaConPowx because they have several IE and most of them are concentrated in small places, or they have almost zero IE.	$ConDensity_x = AreaConPow_x/Density$

(potential consequences). On each case, the user has the option to experiment using different thresholds regarding the IE events, thus affecting the outcome of the analysis.

The features extracted by the ventilator dataset are combined with a segment of the patients' clinical data that was retrieved offline from the hospital database.

The second functionality is to provide a set of analytics and exploratory visualization, offering a set of commonly used analytics. The platform is designed agnostic of the data type it is provided, although it requires them to be on a tabular format. The user can apply data cleaning methods (out of bound, missing values, etc.) either automatically or manually by the use of UI elements. There, the physicians can run the statistical analytic functions they select among the available ones and evaluate the

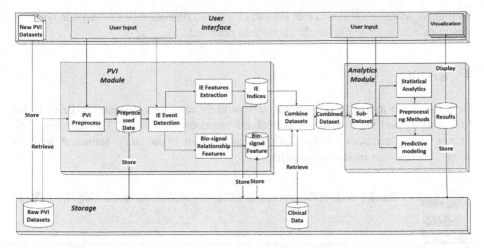

Fig. 1. Overview of platform architecture

clinical outcomes they choose. For this segment, R packages that implement well known and established algorithms were utilized.

3.5   Web-Based Platform Implementation

The web based platform was developed with Shiny, an R framework focused on creating web applications. An important part of the platform was the data visualization. In order to convey the information in a meaningful way based on the user ever-changing needs, we decided to focus on interactive visualization tools. Thus, we examined tools that are available as R packages. As it turned out there wasn't a single package that could cover the entirety of our needs, so we choose the packages described in Table 3.

Specifically, for the Google charts, a custom function was created for the visualization of an entire column based data frame, as individual columns and the relation with each other (Fig. 1). For an N column data frame, a single mega-chart was created that consists of $N \times N$ individual charts. Each sub-chart $C_{ij}(i, j1, \ldots, N)$ depicts the relationship between the variables residing in column i and column j of the data frame,

Table 3. Visualization Packages

Visualization packages	Description
GoogleVis	R interface to Google Charts API, a great collection of charts, but the time-series libraries were somewhat lacking [10]
Dygraphs	R interface to the 'dygraphs' JavaScript charting library. It focus on depicting time-series on highly costumizable charts [11]
D3Heatmap	A single type of chart (Heatmap) that is highly customizable [12]

with appropriate visualization that depends on the combination of the variable types. In the cases where $i == j$, a single variable is depicted.

4   Results

Both the platform and the algorithms are currently used by physicians at PAGNI, the functionality and the interface was evolved based on their input. The following two figures show instances of the platform running an exploratory statistical analysis (Fig. 2) and a data processing pipeline.

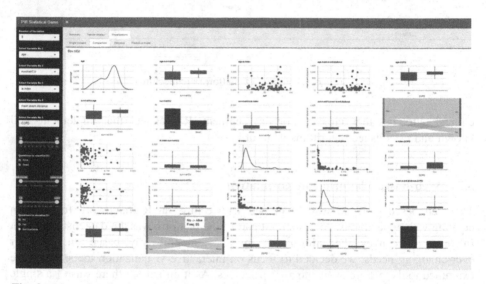

Fig. 2. Mass visualization of 5 variable and their combinations (3 numerical and 2 categorical). Red colored charts represent the diagonal (a single-variable chart) (Color figure online).

Based on analysis enabled by the platform, the optimal threshold for defining respiratory events that relate to adverse outcomes is found to be 10 IE per minute for a time span of at least 3 min. On the current state of this research, performing multivariate analysis, adjusting for age and severity while setting the significance levels at .05 had as results that:

- three indices are found to be related to ICU mortality
- four indices are found to be related to hospital mortality
- two indices are found to be related to the number of days' spent ventilation.

5 Discussion

As the data collection phase proceeds, there will be an increase to the amount of data available (Volume), by including additional equipment recordings such as ICU monitors (recording bio-signals such us Electrocardiogram with a much higher sampling rate than the PVI) and also by increasing the amount of patients whose data are recorded. Provided that the research questions answered by the AEGLE platform yields useful results regarding patient clinical outcome, our goal is to support streaming data and combine them with the acquired knowledge in order to build a decision support system that will require constant processing (the wavelet-based analysis previously described requires 2 to 10 s on workstation level resources depending on IE event duration) of live streaming data (Velocity). Either way, we are gradually stepping into Big Data territory and performance will be an issue.

In order to address this, and as the AEGLE project moves towards the integrated platform, we are shifting our focus on the analytic integration and optimisation, so we can take advantage of the distributed storage and distributed processing that the combination of RHadoop and SparkR frameworks provide.

The knowledge acquired from this analysis, powered by the platform, can generate alarms thresholds that might be eventually applied on a clinical level.

References

1. Blount, M., et al.: Real-time analysis for intensive care: development and deployment of the artemis analytic system. IEEE Eng. Med. Biol. Mag. **29**(2), 110–118 (2010)
2. Georgopoulos, D.: Ineffective efforts during mechanical ventilation: the brain wants, the machine declines. Intensive Care Med. **38**(5), 738–740 (2012)
3. Babalis, D., Lilitsis, E., Akoumianaki, E., et al: Incidence of ineffective efforts in mechanically ventilated critically ill patients. In: ESICM 2011 (2011)
4. Soudris, D., et al.: AEGLE: a big bio-data analytics framework for integrated health-care services. In: 2015 International Conference on Embedded Computer Systems: Architectures, Modeling, and Simulation (SAMOS), SAMOS 2015, pp. 246–253 (2015). doi:10.1109/SAMOS.2015.7363682
5. Muenchen, R.: r4stats - 6/8/16: R Passes SAS in Scholarly Use. http://r4stats.com/2016/06/08/r-passes-sas-in-scholarly-use-finally
6. Younes, M., Brochard, L., Grasso, S., et al.: A method for monitoring and improving patient: ventilator interaction. Intensive Care Med. **33**(8), 1337–1346 (2007)
7. Chytas, A., et al.: Ineffective efforts in ICU assisted ventilation: exploring causalities via multiscale analysis. In: 2015 37th Annual International Conference of the IEEE on Engineering in Medicine and Biology Society (EMBC). IEEE (2015)
8. Chouvarda, I., Babalis, D., Papaioannou, V., et al.: Multiparametric modeling of the ineffective efforts in assisted ventilation within an ICU. Med. Biol. Eng. Comput. **54**(2), 441–451 (2016)
9. De Wit, M., et al.: Ineffective triggering predicts increased duration of mechanical ventilation. Crit. Care Med. **37**(10), 2740–2745 (2009)

10. Gesmann, M., de Castillo, D.: Using the Google visualisation API with R. R J. **3**(2), 40–44 (2011)
11. Vanderkam, D., Allaire, J.J.: Dygraphs: Interface to Dygraphs Interactive Time Series Charting Library. R package version 0.5 (2015)
12. Cheng, J., Galili, T.: d3heatmap: Interactive Heat Maps Using 'htmlwidgets' and 'D3.js'. R package version 0.6.1. (2015)

5G – Putting Intelligence to the Network Edge (5G-PINE)

Security Analysis of Mobile Edge Computing in Virtualized Small Cell Networks

Vassilios Vassilakis[1], Ioannis P. Chochliouros[2(✉)], Anastasia S. Spiliopoulou[2],
Evangelos Sfakianakis[2], Maria Belesioti[2], Nikolaos Bompetsis[2], Mick Wilson[3],
Charles Turyagyenda[3], and Athanassios Dardamanis[4]

[1] School of Computing and Engineering, University of West London, London, W5 5RF, UK
vasileios.vasilakis@uwl.ac.uk
[2] Hellenic Telecommunications Organization (OTE) S.A., 99, Kifissias Avenue,
151 24 Athens, Greece
{ichochliouros,esfak,mbelesioti,nbompetsis}@oteresearch.gr,
aspiliopoul@ote.gr
[3] Fujitsu Laboratories of Europe Ltd., Hayes Park Central, Hayes End Road,
Hayes, Middlesex, UB4 8FE, UK
{Mick.Wilson,Charles.Turyagyenda}@uk.fujitsu.com
[4] SmartNET S.A., 2, Lakonias Street, Agios Dimitrios, 173 42 Attica, Greece
ADardamanis@smartnet.gr

Abstract. Based upon the context of Mobile Edge Computing (MEC) actual research and within the innovative scope of the *SESAME* EU-funded research project, we propose and assess a framework for security analysis applied in virtualised Small Cell Networks, with the aim of further extending MEC in the broader 5G environment. More specifically, by applying the fundamental concepts of the SESAME original architecture that aims at providing enhanced multi-tenant MEC services through Small Cells coordination and virtualization, we focus on a realistic 5G-*oriented* scenario enabling the provision of large multi-tenant enterprise services by using MEC. Then we evaluate several security issues by using a formal methodology, known as the *Secure Tropos*.

Keywords: 5G · Mobile Edge Computing (MEC) · Network Functions Virtualization (NFV) · Security · Software Defined Networking (SDN) · Small Cell (SC) · Virtual Network Function (VNF)

1 Introduction

In the recent years we are witnessing a widespread use of end user devices with advanced capabilities, such as smart-phones and tablet computers, and the emergence of new services and communication technologies. Modern devices implicate for powerful multimedia capabilities and they are increasingly penetrating the global e-communications market, thus creating new demands on broadband (wireless or mobile) access. The challenge becomes greater as devices are also expected to actively communicate with a multiplicity of equipment (such as sensors, smart meters, actuators, etc.) within a fully converged framework of heterogeneous (underlying) network infrastructure(s). This

© IFIP International Federation for Information Processing 2016
Published by Springer International Publishing Switzerland 2016. All Rights Reserved
L. Iliadis and I. Maglogiannis (Eds.): AIAI 2016, IFIP AICT 475, pp. 653–665, 2016.
DOI: 10.1007/978-3-319-44944-9_58

results to the emergence of new data services and/or related applications that can drastically "reshape" the network usage and all associated demands; these are also "key success factors" in order to realize an effective mobile broadband experience for the benefit of our modern societies and economies. This new evolved ecosystem, *however*, imposes very strict requirements on the network architecture and its functionality. Enabling low end-to-end (E2E) latency and supporting a large number of connections at the fitting level, is not possible to be accomplished in current Long-Term Evolution (LTE) networks. In fact, the fundamental limitations of current approaches lie in their centralized mobility management and data forwarding, as well as in insufficient support for multiple co-existing Radio Access Technologies (RATs) [1] and for suitable adaptability to new architectural schemes. Today, a large variety of RATs and heterogeneous wireless networks have been successfully deployed and used. However, under the current architectural framework, it is not easy to integrate -*or to "enable"*- a way of a suitable coordination of these technologies. Despite the fact that the coverage of such wireless and cellular networks has increased by deploying more Base Stations (BSs) and Access Points (APs), the Quality-of-Experience (QoE) of End-Users (EUs) does not increase, *accordingly*. For example, the current architectural approach does not enable a Mobile User (MU) selecting the "best available network" in a dynamic and efficient way. It also does not enable simultaneous and coordinated use of radio resources, from different RATs. This leads to highly inefficient use of hardware resources (wireless infrastructure) and spectrum, which is worsened even more with almost uncontrollable inter-RAT interference [2]. In this paper, we build a novel architecture, proposed for next-generation cellular networks. This architecture benefits from the recent advances in Software Defined Networking (SDN) [3] and Network Function Virtualization (NFV) [4], which are natively integrated into the new and novel architecture. Traditionally, SDN and NFV although not dependent on each other, are seen as "*closely related*" and as "complementary" concepts [5]. This integration enables good scalability in terms of supporting a large number of connections as well as heavy mobility scenarios. Also, the introduction of new services and applications becomes much easier. Decoupling control and data planes, and abstracting network functions from the underlying physical infrastructure, brings much greater flexibility to efficiently utilize radio and computing resources both in the Radio Access Network (RAN) [6] as well as in the Mobile Core Network (MCN). Furthermore, the new approach enables the incorporation of Mobile Edge Computing (MEC) services in an easy and straightforward way.

MEC, also known as "*Fog computing*", is a novel concept that extends the services, typically provided by the Cloud, to the network edge [7, 8]. In case of 5G wireless networks, by the term "edge" we usually mean the RAN and some part of the Cloud services is provided by cognitive BSs. The provided services may include storage, computing, data, and application services. The available MEC infrastructure allows applications to run closer to the end user. This is expected to reduce the E2E network latency and to reduce the backhaul capacity requirements. Moreover, it enables better QoE of fast moving EUs, facilitates highly-interactive real-time applications, and even the emergence of novel applications, such as the *Tactile Internet* [9]. In this work, we focus on Small Cell (SC) BSs, which include both physical BSs as well as BSs that are virtualized via NFV and SDN technologies. Our architectural assumptions are based upon the *SESAME architecture*, which derives from an ongoing European 5G-PPP funded research project that aims at providing enhanced multi-tenant MEC services though Small Cells coordination and virtualization [10]. However, our analysis can be easily extended to alternative network architectures and even in the cases of Macro-Cells or combinations of Macro- and Small-Cells. Thus, in the present work we perform analysis of MEC when the latter is applied upon a selective and realistic 5G scenario, enabling large multi-tenant enterprise services, from the security and privacy viewpoint.

2   Previous Relevant Works

In this section we review the most important and recent works on security and privacy for MEC. The fact that MEC is still at its infancy explains the very limited number of relevant works. These works mainly just touch the security and privacy implications of MEC and no adequate solutions have been proposed to address all the challenges, especially when considering the interaction of MEC with other technologies, such as SDN, and NFV, within the 5G networks context. In [11], a number of security and privacy challenges of MEC have been discussed. The considered security threats are mainly in the context of a cloud-*enabled* IoT (Internet of Things) environment. The study makes a classification of the available security technologies according to the involved network elements, such as technologies to secure a fog node (i.e., the MEC server) and an IoT node, as well as techniques to protect the communication. Next, two threats on the existing security mechanisms have been described, namely the *man-in-the-middle (MitM) attack* and *malicious fog node problem*. Finally, a number of high-level suggestions have been proposed to address the security concerns, such as intrusion detection; malicious node detection; data protection; and secure data management. In [12], the security issues of MEC have been discussed in the context of smart grids, smart traffic lights, wireless sensor networks, and SDN. The focus of this study is the MitM attack and, *in particular*, the stealthy features of this attack that could be addressed by examining the Customer Premises Unit (CPU) and memory consumption of the fog node. This also addresses the assessment of authentication and authorization techniques for connecting the fog with the cloud. The applicability of existing techniques, such as signature- and anomaly-*based* intrusion detection has been studied.

In [13], the challenges of MEC with respect to digital forensics have been discussed. This work mainly considers sensors and various types of smart objects that require connectivity to the cloud and to each other. The focus of this work is to study processes and events that would allow reconstructing past activity for providing digital evidence. Various existing solutions, such as Virtual Machine (VM) introspection and Trusted Platform Module (TPM), have been discussed and analysed. This paper also makes a distinction between the techniques that can be applied in both fog and cloud, and between those that are only applicable in one of them. In [14], the existing data protection techniques have been studied with respect to their suitability in MEC. The conferred data theft attacks include both external intrusion as well as insider attacks. The paper has proposed a novel approach for data protection, using offensive decoy technology.

According to this approach, the data access is initially monitored to detect any abnormal access patterns. Next, when unauthorized access in suspected, large amounts of decoy information is returned to the attacker. Experiments in realistic scenarios indicate that such kind of approach could provide sufficient levels of data protection in MEC environments. In [15], a number of research and security challenges towards realization of MEC have been identified and analysed. One important conclusion drawn is that the MEC paradigm would need to develop security and privacy solutions to explicitly consider coexistence of trusted nodes with malicious ones in distributed edge settings. This will require the enforcement of secure and redundant routing, and trust topologies. Another implication of shifting the computation from the cloud to the edge is that the concentration of information is prevented in comparison to the centralised cloud computing approach. Hence, novel techniques are required to deal with fragmented information that is distributed over a potentially large/heterogeneous set of edge nodes. We observe that the existing works on security analysis of MEC mainly consider M2M-*like* scenarios while lacking of a formal methodological analysis approach and/or of security/privacy study in MEC, related to other coexisting technologies. In this work, we are trying to "fill" this gap.

3   SESAME-*Based* Essential Architecture

In this section, we describe the cellular network architecture developed in the context of the SESAME project [4]. In the following, this architecture is referred to as the "*SESAME architecture*". One of its key elements is the incorporation of MEC concepts at the RAN level, i.e. by enhancing the BSs with MEC servers. Other important characteristic of the architecture is the support of multi-tenancy feature through cellular infrastructure virtualization and NFV. Below we describe the involved actors and their inter-relations (as schematic representation is also given in Fig. 1); afterwards, we describe the functional architecture and its essential elements.

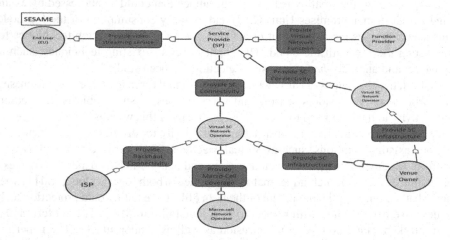

Fig. 1. Actors and their relationships

We distinguish the following essential definitions: **(i)** *End User (EU)*: It can be a mobile device (such as a smart-phone or a laptop) that consumes communication serv ices via the cellular network; **(ii)** *Infrastructure Owner (IO)*: This is the owner of the cellular infrastructure, such as SCs and macro BS. An IO could be, *for example*, a *Venue Owner (VO)* (such as mall, stadium, enterprise or municipality) or the traditional network operator; **(iii)** *IT Equipment Vendor (ITEV)*: It is a legal entity/company that develops, manufactures, and/or sells IT equipment, such as BSs and servers; **(iv)** *Small Cell Network Operator (SCNO)*: It is a legal entity/company that possesses the equipment so as to provide radio communications services and provides radio access to end users locally, by using SCs; **(v)** *Virtual Small Cell Network Operator (VSCNO)*: It is a legal entity/company that does not possess the equipment but lease it from another one, so as to provide radio communications services and deliver services to EUs; **(vi)** *Macro- Cell Network Operator (MCNO)*: It is a legal entity/company that possesses the equipment so as to provide radio communications services and provides radio access to EUs in wide areas at the macro cell level; **(vii)** *Backhaul Provider (BP)*: A legal entity/company that provides the backhaul connection (either wired or wireless) of the Small Cells and Macro Cells. This could be an Internet Service Provider (ISP) or the traditional *Mobile Network Operator (MNO)*; **(viii)** *Service Provider (SP)*: This is a legal entity that produces, controls and distributes services over the MNO/VMNO. *(This could include, for example, the traditional Over-the-Top (OTT) players)*; **(ix)** *Virtual Function Provider (VFP)*: This is a legal entity/company that supplies virtual network functions and other appliances, such as gateways, proxies, firewalls and transcoders. In this way, the need for the customer to acquire, install, and maintain specialised hardware is essentially eliminated, *and*; **(x)** *Spectrum Owner (SO)*: This is a legal entity/company that owes a particular piece of spectrum in a given geographical area. Nowadays, the SO is essentially the MNO who leases the spectrum from the relevant national authority. However, it is envisioned that in the future an independent player may owe the spectrum and lease it to an operator (such as MCNO, SCNO, VSCNO).

As shown in Fig. 1, the EU is dependent on the SP for receiving one or more services (such as the video streaming service). To provide that, the SP depends on the SCNO or the VSCNO who provide the SC connectivity, and also on the VFP who provides the required (virtual) network functions. Both, SCNO and VSCNO are dependent on the IO who owes the SC infrastructure. Finally, the VSCNO is also dependent on the BP (e.g., an ISP) who provides backhaul connectivity as well as on the MCNO who provides the macro-cell connectivity. We describe, *in brief*, the SESAME functional architecture, which is also illustrated in Fig. 2. Firstly, we provide the basic component definitions and afterwards we describe *"how these components interact with each other"*. In fact, we identify the following fundamental components: **(i)** *MEC server*: It is specialised hardware that is placed inside the SC and provides processing power, memory and storage capabilities, and networking resources; **(ii)** *Cloud Enabled Small Cell (CESC)*: This is the SC device which has been enriched with a MEC server; **(iii)** *CESC cluster*: A group of CESCs that are collocated, able to exchange information and properly coordinated; as a trivial case, a CESC cluster could comprise one CESC; **(iv)** *Light Data Center (Light DC)*: It is a cluster of MEC servers. In particular, the Light DC is a logical entity consisting of a set of distributed MEC servers of the same CESC cluster;

(v) *Virtual Infrastructure Manager (VIM)*: This is an entity responsible for management of the virtual hardware (i.e., VMs) and networking resources of a single Light DC; in particular, the VIM manages the lifecycle, provision, placement, and operation of VMs. The VIM is also responsible for the allocation of Virtual Network Functions (VNFs) over the hardware it manages and offers functionalities to control virtual networks across VNF instances and associate storage to them. The VIM offers an aggregated view of compute, network and storage resources of the Light DC; **(vi) *CESC Manager (CESCM)*:** The architectural component in charge of managing and orchestrating the cloud environment of the Light DC; it can simultaneously manage multiple clusters, a cluster or a single CESC. The CESC Manager also manages the radio access and "*self-x*" functionalities, e.g., self-optimising, self-healing and self-configuring of the Small Cells contained in each CESC cluster, in order to guarantee the service continuity and the required performance of services.

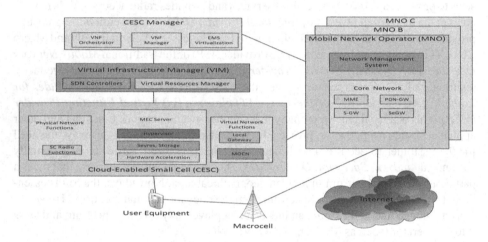

Fig. 2. SESAME functional architecture

The CESCM orchestrates services and, *consequently*, manages the VIM to compose them with virtual resources. A CESCM is actually a functionality that will be "mapped" on to the distributed physical elements. As mentioned before, one important feature of this architecture is the distributed set of MEC servers which can logically "be grouped into clusters", thus effectively forming a Light DC at the network edge. Clusters are able to communicate with each other as well as with the mobile core network (i.e., Evolved Packet Core (EPC) in the LTE terminology). The distributed deployment of MEC servers facilitates flexible and dynamic allocation of resources in cases of flash crowd events and fast EU mobility.

4   Security and Privacy Considerations

Network and system security is a very critical issue because the SESAME system is expected to support both customer enterprises and end users, who cannot tolerate financial losses or data privacy violations and, *therefore*, they seek the highest possible security guarantees. In the present section, the considered SESAME scenario and functional components are evaluated by using a formal methodology known as the *Secure Tropos (SecTro)* [16]. Our goal is to identify, model and analyse security issues from the early stages of system design and software development as well as to model and analyse threats and vulnerabilities in existing software and protocols that will be used in the SESAME system. We aim at preventing a wide range of attacks, such as control hijacking, reverse engineering, malware injection, eavesdropping, *just to name a few*. At the same time, the SESAME concepts can provide invaluable opportunities of developing solutions for attack prevention, management & recovery.

Initially, the physical security of CESC infrastructure and hardware integrity has to be ensured. Hence, appropriate security controls (such as in [17]) should be deployed by the CESC infrastructure owner, to prevent hardware tampering. Likewise, it is important to consider attacks that are initiated from the cloud side. This is particularly relevant in scenarios where multiple enterprises using private clouds are hosted. Especially in the multi-tenant environment of SESAME, the adversary *per se* could be a legitimate tenant interacting with network entities by using valid credentials and having privileged *access* to virtualised resources. Also, the emerging *Bring Your Own Device (BYOD)* trend [18] in many enterprises constitutes many conventional security solutions incapable of protecting the private network; for example, a Trojan horse, that infected an employee's device, can bypass the security of the corporate firewall. Hence, the cloud provider must ensure the physical security of the cloud infrastructure and of the data centres. This can be done, *e.g.*, by following the recommendations from the *Cloud Security Alliance* [19]. Moreover, the selection of suitable cloud provider can be based on formal methodologies to ensure that the security and privacy requirements are properly met [20]. This effectively means that services offered by cloud providers who do not meet the specified requirements and have not implemented the mandatory security controls, could so be restricted or even could be blocked. To ensure confidentiality and integrity of the User Equipment (UE) data, cryptographic security controls must be in place. This implicates that any adopted *Public-Key scheme* that enables the encryption of the communications among CESC, UE and the cloud, must be sufficiently secure. Cryptographic and privacy protection techniques are particularly important in cases where an EU receives service from multiple service or network providers, due to mobility or QoE considerations.

An important category of attacks could potentially "target" the management system (for example, if initiated inside virtualised environments and aims at taking control of the Hypervisor shown in Fig. 2). Also, the NFV Orchestrator is an attractive "attack target" due to being in the "middle" of the system model architecture; the same can be for other components of the management layer, such as the VNF Manager. Also, impersonation by the adversary of one of the VNFs or the MEC server when communicating with the management layer could be a potential threat. Considering again the virtualised environment, both host and guest Operating Systems (OSs) may be targeted, and to alleviate the impact of such an attack, adequate isolation must be enforced between guest VMs, as well as between the host and guest VMs. The adversary could attempt to break the isolation by exploiting, *e.g.*, some flaws of the used virtualisation platform [12].

Therefore, appropriate choice of the virtualisation platform that meets security and privacy requirements is of major importance.

In some cases, to launch an attack against a component, the adversary requires that this component has specific exploitable configuration or runs specific software. For example, a precondition for a *Denial-of-Service (DoS) attack* can be specific configuration of the CESCM with regard to the allocation of resources to tenants. Yet, some flaws in the resource allocation algorithm can allow the adversary to prevent a tenant from accessing its portion of virtual resources. The introduction of the MEC paradigm has also implications on the E2E security in 5G networks. A potential solution to deal with this problem is to facilitate the network slicing concept, according to which each application or network flow "gets" its own slide of the network. This allows the end-to-end security to be enforced within each slice by each application individually and any security breaches would not affect other applications. As security will be a fundamental enabling factor of future 5G networks, we are concerned with identifying and mitigating security threats and vulnerabilities against a broad range of targets at the intersection of MEC with *"Small Cells-as a-Service"* (SCaaS), SDN, and NFV. These can have crucial effect on legal and regulatory frameworks as well as on decisions of businesses, governments and end-users.

4.1   Scenario: *Enabling Large Multi-tenant Enterprise Services by Using MEC*

To further emphasize, we consider an SCNO who is providing a radio interface to a number of distinct mobile operators, virtual mobile network operators (VMNOs) and VSCNOs. The SCNO may transmit by using licensed or unlicensed spectrum over the air interface. In addition to the provision of radio coverage in the business centre and orchestration of multi-tenancy, the SCNO offers a platform for MEC for low latency and compute intensive applications/services. The MOs, VMNOs and VSCNOs provide both in-house and third party services from OTT players or the SPs. The offered services can include *inter-alia*: multi-person real-time video-conferencing, virtual presence 360° video communications with meetings using virtual presence glasses/devices, and assisted reality to actively inform users of ambient interests such as danger warnings to support people with disabilities and improve interactions with their surroundings. The EUs can benefit from fast and cost-effective access to a wide variety of innovative services from third party players. MOs, VMNOs and VSCNOs can benefit from extra market share. VOs can benefit from having a single set of radio and IT equipment installed on the premises, instead of multiple installations from multiple network operators. The CESC is made up of: hardware resources, virtualisation layer, VNFs, and an Element Management System (EMS). The virtualisation layer abstracts the hardware resources and decouples the VNF software from the underlying hardware. A VNF is a virtualisation of a network function in a legacy non-virtualised network. The EMS performs management of one or more VNFs. A cluster of CESCs is managed by the CESCM that constitutes of: VIM, VNF manager and the network functions virtualisation orchestrator (NFVO). The VIM manages the interaction of a VNF with the compute, storage and network resources under its specific authority. The VNF manager is responsible for VNF lifecycle management. The orchestrator is in charge of orchestration, of management NFV infrastructure and software resources and of realising network services.

In Fig. 3 we demonstrate *how this scenario can be supported by the specific SESAME system*. In particular, we see a CESC infrastructure provider who owns, deploys and maintains the network of CESCs inside the premises where different enterprises are hosted. The CESC provider has a *Service Level Agreement (SLA)* with each customer enterprise and SLAs are to enable enterprise users to a number of services offered by the CESC network; the SLA shall cover the target performance metrics for any service (or service category) required by each enterprise, supporting different tenants' requirements. Such sort of services can be categorised in data services and real-time services: these can include, *inter-alia*, Internet access for enterprise users, web browsing, file sharing, electronic mail service, voice communications and video conferencing. The deployment of MEC servers with high processing capabilities can enable close-to-zero latency and enhanced QoE of the enterprise users (i.e., an enhanced handling of the media flows and, *consequently*, an optimal QoE). In addition to the computing resources, MEC servers can provide storage resources and support content caching at the network edge. The reality is that different hosted enterprises may have different traffic patterns which may fluctuate greatly, depending on the time of the day or on special occasions, such as popular events. This leads to the requirement of a "flexible" system which can be scaled up and down, *on demand*. For example, most enterprises may need a higher capacity and higher quality of service (QoS) during the office hours, while a security firm providing security to the building would need a low capacity and the same service quality throughout the day. The main issues may arise from possible service disruptions and from the dynamicity of the enterprise activity. The service quality levels can be dynamic (time variant) as well. In some instants, the total capacity and the number of connected devices for a certain enterprise could rise significantly. This can be an event like an Annual General Meeting or a conference/exhibition organised by the enterprise. This extra capacity/connections may not need the same QoS and may not access the internal enterprise data, so may not need the same level of security. The main requirements are for the available capacity to be rapidly scaled up and other virtual network(s) created mainly for open access. Also in some cases, certain enterprises may downsize their operations or move out of the premise, which requires scaling down. This kind of scalability and flexibility needs to be incorporated into the design of particular use cases for this representative scenario. The enterprise scenario shown in Fig. 3 will leverage on SESAME features such as intrinsic support of multi-tenancy by enabling multiple SC operators since Small Cells operators to provide network services and connectivity over the network owned by a single CESC infrastructure provider. Furthermore, the SESAME system allows native incorporation of self-organizing network techniques, which can be adapted to network behaviour and can optimize service delivery to the enterprise users. In any case, the high level of network security as demanded by the enterprise customers will be an inherent feature of the respective SESAME solutions. We present the actors involved in the scenario, their corresponding goals as well as their dependencies. We identify four major actors involved in the scenario, namely CESC infrastructure provider, Virtual SCNO, ISP and enterprise. The enterprise depends on the SC operator which provides the wireless connectivity. The SC operator requires backhaul connectivity and access to external networks, such as Internet. This can be provided by an ISP. Finally, the SC operator aiming to provide its services to multiple enterprises depends on the CESC infra-structure which is owned and maintained by the CESC provider.

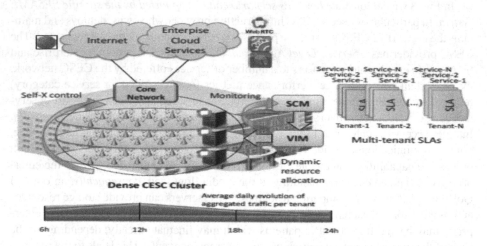

Fig. 3. Scenario: enterprise services in multi-tenant large businesses

In Figs. 4 and 5 we present the *Security Components View* for two main actors of this scenario: the *CESC provider* and the *virtual SCNO*. The security component view of the CESC provider, depicted in Fig. 4, contains two "resources" that need to be protected: the *Hypervisor* and the *Tenant's Data*. A resource in the *Secure Tropos terminology* could be a physical or an informational entity, and in the *SecTro* tool is depicted as a yellow, rectangular box. A resource is required to achieve a specific "goal" of an actor (the CESC provider in this example). A goal represents an actor's strategic interests. In this example, we consider two primary goals (depicted as green ovals): *operating the CESC infrastructure* and *enabling multi-tenancy*. Both these goals require the Hypervisor as a primary resource. Also, to enable multi-tenancy, the Tenant's Data resource has to be created. A goal could be restricted by a "security constraint" (depicted as a red octagon). In this example, the CESC infrastructure operation is restricted by the requirement to *protect the control plane*, whereas the multi-tenancy goal is restricted by the requirement to *prevent unauthorized access to another tenant's VM*. Various security constraints must satisfy a number of "security objectives" (depicted as blue hexagons). In this example, the security constraints are satisfied by the two objectives: *Protect the Control Plane* and *Prevent Access to another Tenant's VM*. These objectives are implemented by using a number of "security mechanisms" (green hexagons), such as VM isolation, Data Encryption, and Server Replication. We also consider a number of "threats" (depicted as pentagons) that impact some of the resources. In this example, the Hypervisor can be impacted by the two threats: *Control Hijacking* and *Denial of Service*. The Tenant's Data resource can be impacted by the *Eavesdropping* threat. The security component view of the Virtual SCNO, depicted in Fig. 5, contains three resources that need to be protected: the *Radio Resources*, the *Radio Spectrum* and the *EMS*.

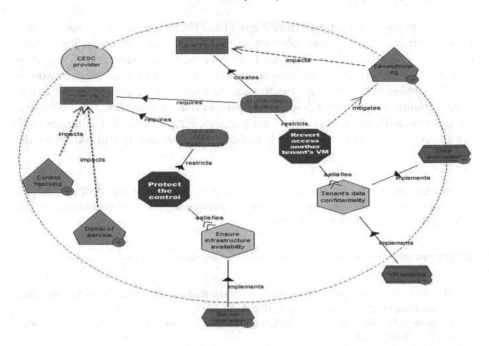

Fig. 4. Security components view for the CESC provider (Color figure online)

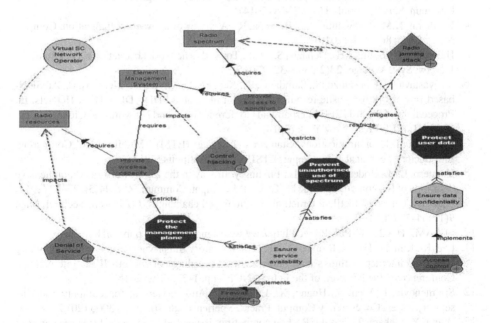

Fig. 5. Security components view for the virtual SC network operator (SCNO) (Color figure online)

In this example, the actor's primary goals (that require the above resources) are to provide wireless capacity and spectrum to the tenants. The corresponding security constraints that restrict these goals are to *protect the management plane,* to *prevent unauthorized access to the wireless spectrum* and to *protect user data.* These constraints must be satisfied by two security objectives: *Ensure service availability* and *ensure data confidentiality.* The corresponding security mechanisms to implement these objectives are using firewalls and access control mechanisms. Finally, a number of threats could impact the considered resources, such as *DoS, control hijacking* and *radio jamming attacks.*

Acknowledgments. This work has been performed in the scope of the *SESAME* European Research Project and has been supported by the Commission of the European Communities (*5G-PPP/H2020, Grant Agreement No. 671596*).

References

1. Demestichas, P., Georgakopoulos, A., et al.: 5G on the horizon: key challenges for the radio-access network. IEEE Veh. Technol. Mag. **8**(3), 47–53 (2013)
2. Andrews, J.G.: Seven ways that HetNets are a cellular paradigm shift. IEEE Commun. Mag. **51**(3), 136–144 (2013)
3. Nunes, B.A.A., Mendonça, M., Ngyen, X.-N., Obraczka, K., Turletti, T.: A survey of software-defined networking: past, present, and future of programmable networks. IEEE Commun. Surv. Tutorials **16**(3), 1–18 (2014)
4. Mosharaf, N.M., Chowdhury, K., Boutaba, R.: A survey of network virtualization. Comput. Netw. **54**(5), 862–876 (2010)
5. Haleplidis, E., Salim, J.H., Denazis, S., et al.: Towards a network abstraction model for SDN. J. Netw. Syst. Manage. **23**(2), 309–327 (2015)
6. Shrivastava, R., Constanzo, S., Samdanis, K., Xenakis, D., Grace, D., Merakos, L.: An SDN-based framework for elastic resource sharing in integrated FDD/TDD LTE-A HetNets. In: Proceedings of the 3rd IEEE International Conference on Cloud Networking (CloudNet), pp. 126–131. IEEE Press, New York (2014)
7. European Telecommunications Standards Institute (ETSI): Mobile-Edge Computing, Introductory Technical White Paper. ETSI, Sophia-Antipolis (2014)
8. Vaquero, L.M., Rodero-Merino, L.: Finding your way in the fog: towards a comprehensive definition of fog computing. ACM SIGCOMM Comput. Commun. Rev. **44**(5), 27–32 (2014)
9. Fetweiss, G.P.: The tactile internet: applications and challenges. IEEE Veh. Technol. Mag. **9**(1), 64–70 (2014)
10. SESAME H2020 5G-PPP Project. http://www.sesame-h2020-5g-ppp.eu/Home.aspx
11. Lee, K., Kim, D., Ha, D., Rajput, U., Oh, H.: On security and privacy issues of fog computing supported internet of things environment. In: Proceedings of the 6th IEEE International Conference on the Network of the Future (NOF), pp. 1–3. IEEE (2015)
12. Stojmenovic, I., Wen, S., Huang, X., Luan, T.H.: An overview of fog computing and its security issues. Concurrency Comput. Pract. Experience **28**(10), 2991–3005 (2015)
13. Wang, Y., Uehara, T., Sasaki, R.: Fog computing: issues and challenges. In: Proceedings of the 39th IEEE Annual COMPSAC Conference, pp. 53–59. IEEE (2015)

14. Stolfo, S.J., Salem, M.B., Keromytis, A.D.: Fog computing: mitigating insider data theft attacks in the cloud. In: Proceedings of the IEEE Symposium on Security and Privacy Workshops (SPW), pp. 125–128. IEEE (2012)
15. Lopez, P.G., Montresor, A., Epema, D., et al.: Edge-centric computing: vision and challenges. ACM SIGCOMM Comput. Commun. Rev. **45**(5), 37–42 (2015)
16. Mouratidis, H., Giorgini, P.: Secure tropos: a security-oriented extension of the tropos methodology. Int. J. Softw. Eng. Knowl. Eng. **17**(2), 285–309 (2007)
17. Skorobogatov, S.: Physical attacks and tamper resistance. In: Tehranipoor, M., Wang, C. (eds.) Introduction to Hardware Security and Trust, pp. 143–173. Springer, New York (2011)
18. Buetnner, R.: Towards a new personal information technology acceptance model: conceptualization and empirical evidence from a bring your own device datataset. In: Proceedings of the 21st AMCIS. AIS, Fajardo, Puerto-Rico (2015)
19. Cloud Security Alliance (CSA). Security guidance for critical areas of focus in cloud computing v2.1. CSA (2009)
20. Mouratidis, H., Islam, S., Kalloniatis, C., Gritzalis, S.: A framework to support selection of cloud providers based on security and privacy requirements. J. Syst. Softw. **86**(9), 2276–2293 (2013)

A Model for an Innovative 5G-*Oriented* Architecture, Based on Small Cells Coordination for Multi-tenancy and Edge Services

Ioannis P. Chochliouros[1(✉)], Ioannis Giannoulakis[2], Tassos Kourtis[2], Maria Belesioti[1], Evangelos Sfakianakis[1], Anastasia S. Spiliopoulou[3], Nikolaos Bompetsis[1], Emmanouil Kafetzakis[4], Leonardo Goratti[5], and Athanassios Dardamanis[6]

[1] Research Programs Section Fixed, Hellenic Telecommunications Organization (OTE) S.A.,
1, Pelika & Spartis Street, 151 22 Athens, Greece
{ichochliouros,mbelesioti,esfak,nbompetsis}@oteresearch.gr
[2] National Centre for Scientific Research "Demokritos",
Patriarchou Georgiou Street, Aghia Paraskevi, 153 10 Athens, Attica, Greece
{giannoul,kourtis}@iit.demokritos.gr
[3] Hellenic Telecommunications Organization (OTE) S.A., 99 Kifissias Avenue,
151 24 Athens, Greece
aspiliopoul@ote.gr
[4] ORION Innovations Private Company, 56, Ioulianou Street, 104 39 Athens, Greece
mkafetz@orioninnovations.gr
[5] CREATE-NET (Centre for Research and Telecommunication Experimentation
for Networked Communities), Via alla Cascata 56D, Trento, Italy
leonardo.goratti@create-net.org
[6] SmartNET S.A., 2, Lakonias Street, 173 42 Agios Dimitrios, Attica, Greece
ADardamanis@smartnet.gr

Abstract. The "core" aim of the *SESAME* EU-funded research project is to design and develop a novel 5G platform based on the use of Small Cells, featuring multi-tenancy between network operators and also attach to them edge cloud capabilities to be offered to both the network operators and the mobile users. SESAME aims at providing a fresh 5G mobile network architecture so as to support the ambitious goal of small cell virtualization, multitenancy and edge cloud services. In the present work we assess the fundamental SESAME components and their role in the respective systems, while analysing the initial framework of the essential relevant architecture to implement the critical targets of the respective approach. Finally we identify future potential extensions.

Keywords: Fifth generation mobile technology (5G) · Mobile edge computing (MEC) · Multi-tenancy · Network Functions Virtualization (NFV) · Software Defined Networking (SDN) · Small cell (SC) · Virtualised network function (VNF)

© IFIP International Federation for Information Processing 2016
Published by Springer International Publishing Switzerland 2016. All Rights Reserved
L. Iliadis and I. Maglogiannis (Eds.): AIAI 2016, IFIP AICT 475, pp. 666–675, 2016.
DOI: 10.1007/978-3-319-44944-9_59

1 Introduction

Internet and electronic communication networks are fundamental "enablers" of our modern economy as they promote, support and disperse an extended variety of digital applications and enhanced facilities, satisfying both residential and corporate customers' various needs. In particular, the immense penetration and the continuous growth of wireless data services driven by mobile Internet and smart devices has further promoted the investigation effort of the fifth generation mobile technology (5G), which is expected to "fulfill" the demands and business contexts of the year 2020 and beyond. 5G is expected to assist an entirely mobile and fully connected and converged society as well as to endow a diversity of socio-economic transformations in immeasurable ways, many of which are unimagined today, including those for productivity, sustainability and well-being [1]. The claims of a modern society where both mobility and connectivity are among the "core" features are characterized by the incredible growth in connectivity and density/volume of traffic, the required multi-layer densification in enabling this, and the broad range of use cases and business models expected to assist any relevant process [2]. Therefore, in 5G there is a need to "push" the envelope of performance to provide, *where necessary*, much greater throughput, much lower latency, ultra-high reliability, much higher connectivity density, and higher mobility range. This enhanced way of performance is expected to be provided along with the capability to "control" a highly heterogeneous environment with the aim, *among others*, to ensure security and trust, identity and privacy [3]. 5G is to become the fundamental means to realize the full potential of the global "networked society". The new capabilities of 5G span several dimensions, including tremendous flexibility, lower energy requirements and improved energy efficiency, greater capacity, bandwidth, security, reliability and data rates, as well as enhanced indoor coverage lower latency and device costs [4].

Technology advancements of the recent years (e.g., SDN [5], NFV [6, 7], big data, All-IP) can drastically modify the way networks are being constructed and managed. These changes "enable" the development of a highly flexible infrastructure that allows cost-efficient development of networks and of associated services. The 5G architecture should comprise modular network functions that could be further expanded and scaled on demand, to accommodate several use cases in an agile and cost-efficient manner [8]. As a consequence, future networks are becoming highly dynamic and distributed, typically consisting of various multi-operator heterogeneous networks and a huge number of network elements and users' devices/equipment.

In order to efficiently deal with automated network management solutions able to "address" resource utilization, one current trend is to "*move functionality at the network edge*", thus reducing complexity at the network core, leveraging on the increasing computing and processing power of the end-user devices [9]. Network virtualization and Software Defined Networking (SDN) compose a much promising solution for orchestrating the allocation of physical and virtual resources that can be instrumental in this direction. Network Functions Virtualization (NVF) distinguishes logical services from physical resources, also allowing for moving resources to other network locations [10]. This concept is different from the notion of SDN networks which proposes the full decoupling of the network control and data planes, moving the control of the network behavior to third party software running in external dedicated or distributed servers [9]. Delivering services at the edge of the modern network enables service providers to offer services and a user experience that cannot be surpassed in terms of responsiveness and performance.

Future networks will also have to be expanded much more densely than today's networks and will become significantly more heterogeneous than today, especially in terms of: transmit power, antenna configuration, supported frequency bands, transmission bandwidths, directional blindness, multi-hop architecture and duplex arrangements. The radio-network architectures of the nodes are expected to vary from stand-alone base stations (BSs) to systems with different degrees of centralized processing, depending on the sort of the available backhaul technology [1]. One major venue in future 5G networks is dense deployment of Small Cells (SCs) coexisting with micro- and macro-cells as well as other systems (such as WiFi, 4G and 3G), thus comprising a Heterogeneous Network (HetNet) [11]. As the number of mobile devices and data traffic increases following a clear exponential growth, the installation of SCs appears to be a *"suitable and efficient way"* so that to achieve enhanced performance and capacity to both indoor and outdoor places/locations, with significant impact upon all any of the corresponding market-*related* scenarios. Small Cells are radio transmitters [12] whose complexity range from just antennas and radio circuits (i.e. remote radio head) up to a full functioning evolved Node B (eNodeB) base station. Small cells are increasingly recognised by the global telecom operators as *"playing an essential role"* in future broadband networks. They address - quite satisfactorily- [13] many of the key challenges faced by market actors such as: (i) increasing capacity (*which actually appears to be the most critical challenge*); (ii) improving depth of coverage, especially inside buildings; (iii) improving user experience, especially the typical available data rates, *and*; (iv) delivering value added services, especially those enabled by high-precision location information. Future 5G networks will be "denser" to realize the capacity increase offered by the deployment of SCs. In that sort of considered scenario, reducing the overall costs becomes an issue of major importance. One option of realizing this is by enabling an effective "sharing" of the network infrastructure. To this aim, Radio Access Networks (RAN) virtualization techniques are ultimately to provide logical isolated pieces ("slices") of the access infrastructure to individual tenants, so they can operate them as *"if each virtual slice were a single real physical infrastructure"*. A well-coordinated sharing of the access infrastructure yields a higher throughput per area.

The target of the SESAME EU-*funded* H2020 5G-PPP project (*Grant Agreement No. 671596*) [14] is to design and develop a novel 5G platform based on small cells, featuring multi-tenancy between network operators and also attach to them edge cloud capabilities to be offered to both the network operators and the mobile users. Thus, the key innovations proposed by SESAME focus on the novel concepts of a multi-operator (multi-tenancy) enabling framework and also on providing an edge-*based*, virtualised execution environment. SESAME aims at providing a fresh 5G mobile network architecture so as to support the ambitious goal of small cell virtualization, multitenancy and edge cloud services. The present work discusses a model for architecture able to fulfil multi-tenancy purposes within a small cell-*based* concept as developed by the actual SESAME effort, still being in dynamic progress. In Sect. 2 we identify several essential definitions, as well as the fundamental SESAME components and their roles. In Sect. 3 we focus upon the proposed architectural approach and we discuss the corresponding "key features" affecting the entire scope. Finally, in Sect. 4 we summarize and we also propose some potential extensions, in parallel with options for future work in more enhanced environments.

2   The SESAME-*Based* Conceptual Approach

In reference to the fundamental SESAME approach, the Small Cell concept is evolved so that not only be able to provide multi-operator radio access capacity with virtualised Small Cells that can be integrated within the operator (tenant) infrastructures, but also to be capable of providing a virtualised implementation environment for delivering Cloud services at the network's edge. In order to achieve this, *however*, the Small Cell needs to offer mobile-edge computing (MEC) capabilities which, *sequentially*, will allow the virtual or mobile operators to increase the capacity of their own 4G/5G RAN infrastructures or to spread the range of their provided services, while preserving the essential agility to be able to offer these extensions, *on demand*.

In order to achieve the previously mentioned essential aims, some further considerations need to be made on "*how to separate-or combine - the network and the computing resources*", and also, "*which small cell functions should be physical network functions*" and "*which ones should be virtual*". The SESAME approach to this challenge is performed by enhancing the Small Cells with Micro-Servers that are able to deliver virtualized computing and networking resources and by being able to "form" clusters, thus creating a kind of Light Data Centre (DC) at the edge. This Light DC is further complemented by supplementary components that reside either close to the edge or to the backbone, such as the VIM or the CESCM, in order to provide the proper reference points - or the scope - for the whole network (as depicted in Fig. 1). In order to "translate" the concept of that figure, we consider the following definitions that are also fundamental for the wider SESAME-*based* scope:

- *Execution infrastructure, micro-server (μS):* Specific hardware that is placed inside the Small Cell and provides processing power (also can potentially include some memory and storage capabilities).
- *Small Cell Network Operator (SCNO):* A legal entity that provides the physical connection to Virtual Small Cells and CESCs.
- *Virtual Small Cell Network Operator (VSCNO):* This implicates companies/legal entities that do not possess the equipment but lease it (through appropriate Service Level Agreements (SLAs) from another company), so as to provide wireless communications services and deliver services to end users.
- *Cloud Enabled Small Cell (CESC):* The Small Cell device which includes a micro-server in hardware form.
- *Cluster of CESCs:* A group of CESCs that are collocated, exchange information and are properly coordinated. As a trivial case, one CESC can be called CESC cluster.
- *Light Data Centre (Light DC):* The hardware entity composed by the micro-servers of the CESCs forming a cluster (see Fig. 3).

- *CESC Manager (CESCM):* The architectural component in charge of managing and orchestrating the cloud environment of the Light DC, as well as management of small cell functions. It can manage, at the same time, multiple clusters, a cluster or a single CESC.
- *Virtualised Infrastructure Management (VIM):* Manager of the HW and networking resources (i.e., lifecycle, provision, placement and operation) constituting of a cluster of micro-servers, namely the Light DC, and the networking nodes and links (i.e., both virtual and physical).
- *Backhaul Provider (BP):* A legal entity/company that provides the backhaul connection (either wired or wireless) of the Small Cells and Macro Cells. This could be an Internet Service Provider (ISP) or the traditional Mobile Network Operator;
- *"NMS"* stands for the respective *Network Management System* and *"EPC"* for the *Evolved Packet Core* (also known as the *System Architecture Evolution (SAE) Core*) which is amongst the main components of the SAE architecture and will serve as the "equivalent" of GPRS networks[1].

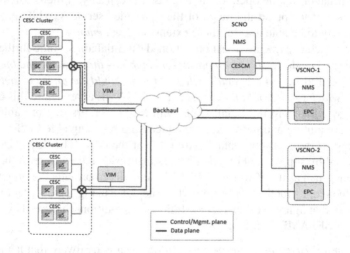

Fig. 1. Scope of SESAME components (physical view)

3   Key Features of the SESAME-*Based* Architecture

The key innovations proposed in the SESAME architecture do emphasize on the novel concepts of virtualising Small Cell networks by leveraging the paradigms of a multi-operator (multi-tenancy) enabling framework coupled with an edge-*based*, virtualised execution environment. SESAME falls in the scope of these two principles and "promotes" the adoption of Small Cell multitenancy - i.e., *multiple network operators*

---

[1] More information about the detailed specific concept of the EPC can be found at: http://www.3gpp.org/technologies/keywords-acronyms/100-the-evolved-packet-core. (Additional supportive information is given in the references therein).

will be able to use the SESAME platform, each one using his own network "slice". Moreover, the principal idea is to "endorse" the deployment of Small Cells with some virtualized functions, with each Small Cell containing also a micro-server through appropriate fronthaul technology. A micro-server is based on a non-x86 architecture[2] using 64-bit ARMv8 technology[3]. Together with the SC, they form the Cloud- Enabled Small Cell (CESC), while a number of CESCs compose the "CESC cluster" capable to provide access to a geographical area with one or more operators. At this point, we illustrate a brief description of the two main technological fields that constitute the core innovative fields of the SESAME framework. This targeted way of "decomposition" has been the "starting point" for building, in the continuity, an accurate framework for the intended SESAME architecture. To that end, the NFV technology is going to be used as a fundamental enabler that will offer a virtualisation platform and "meet" the SESAME requirements, namely NFV-*driven* small cell functions and NFV-*based* network services. The left-hand side of Fig. 2 presents the Management and Orchestration (MANO) framework for the NFV part.

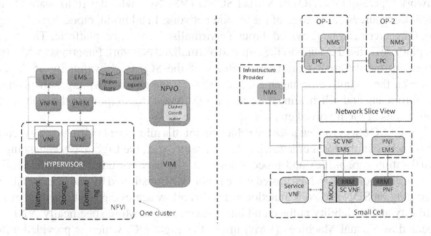

Fig. 2. Scope of SESAME components (physical view)

On the other hand, a Small Cell network capable to support more than one network operator is also envisaged (Fig. 2 – right-hand side). Relevant 3GPP specifications have already added some support for Radio Access Network (RAN) sharing [15]. Although two main architectures are identified, namely *Multi-Operator Core Network (MOCN)*, where the shared RAN is directly connected to each of the multiple operator's core networks, and *Gateway Core Network (GWCN)*, where a shared core network is deployed so that the interconnection of the multiple operator's core networks is done at

---

[2] The *x86* is a family of backward compatible instruction set architectures based on the Intel 8086 CPU and its Intel 8088 variant. More related -and indicative- information for this case can be found, *for example*, at: https://en.wikipedia.org/wiki/X86.

[3] More relevant information about the corresponding technology can be found, *for example*, at: http://www.arm.com/products/processors/armv8-architecture.php.

core network level, the MOCN case has been identified as the exclusive enabler for multitenancy features in SESAME platform. The related infrastructure consists of a number of Small Cells and the corresponding SC network functions such as gateways and management systems. The adopted architecture is based on the current 3GPP framework for network management in RAN sharing scenarios [16, 17]. Assuming LTE (Long Term Evolution) technology as the contextual framework of basis, the interconnection of the SCs of the SCaaS provider to the Evolved Packet Core of the tenant is done through the *S1 interface*, delivering both data (e.g., transfer of end-users traffic) and control (e.g., activation of radio bearers) plane functions.

Based on the required functionalities as well as to the architectural principles that have been mentioned above, it is possible to derive an overall, high-level view of the SESAME system, as the one proposed by Fig. 3, below. To that end, the CESC offers computing, storage and radio resources. Through virtualization, the CESC cluster can be seen - or assessed - as a cloud of resources which can be "sliced" to enable multi-tenancy. Therefore, the CESC cluster becomes a neutral host for mobile Small Cell Network Operators (SCNO) or Virtual SCNO (VSCNO) who desire to share IT- and network-resources at the edge of the mobile network. In addition, cloud-*based* computation resources can be provided through a virtualised execution platform. This execution platform is used to support the required Virtualized Network Functions (VNFs) that implement the different features/capabilities of the SCs (and eventually of the core network), the cognitive management and *"self-x"* operations [18] (e.g. self-planning, self-optimising and self-healing)[4], as well as the computing support for the mobile edge applications of the involved end-users.

The CESC clustering enables the achievement of a micro-scale virtualised execution infrastructure in the form of a distributed data centre (i.e., the Light DC), enhancing the virtualisation capabilities and process power at the network edge. Network Services (NSs) are supported by VNFs hosted in the Light DC -constituted by one or more CESC-, leveraging on SDN and NFV functionalities that allow achieving a satisfactory level of flexibility and scalability at the cloud infrastructure edge. More specifically, VNFs are executed as Virtual Machines (VMs) inside the Light DC, which is provided with a hypervisor (based on the concept of a Kernel-*based* Virtual Machine (KVM)) specifically extended to support carrier grade computing and networking performance.

Over the provided virtualised execution environment (i.e., the Light DC), it is possible to chain different VNFs to meet/fulfil a requested NS by a tenant (i.e., a mobile network operator). Note that, in the context of SESAME, a NS is conceived as a "collection" of VNFs that jointly supports data transmission between User Equipment (UE)

---

[4] Self-Organizing Networks (SON), also referred to as *"self-x"* features, include several techniques for automating the operation of the network, by automatically tuning different network settings; these can be implemented either as physical or as virtual network functions instantiated and executed in the micro server. Typical examples of *self-x* features include: Inter-Cell Interference Coordination (ICIC) to configure the power; time and frequency resources to minimize inter-cell interference; Coverage and Capacity Optimization (CCO) to adjust RF parameters; Automatic Neighbour Relationships (ANR) to manage neighbour lists; Mobility Load Balancing (MLB) to manage traffic loads between cells, *and;* Mobility Robustness Optimisation (MRO) to optimize the operation of handover procedures.

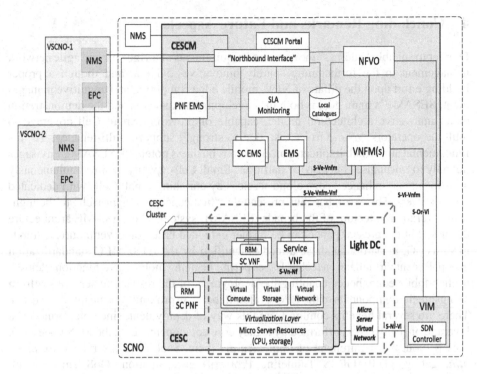

Fig. 3. SESAME overall architecture

and operators' Evolved Packet Core, with the possibility to involve one or several service VNFs in the data path. Therefore, each NS is deployed as a chain of SC VNFs and Service VNFs.

Finally, the CESC Manager (CESCM) is the central service management and orchestration component in the overall architecture figure. Generally speaking, it integrates all the necessary network management elements, traditionally suggested in 3GPP, and the novel recommended functional blocks of NFV MANO [19]. A single instance of CESCM is able to operate over several CESC clusters, each constituting a Light DC, through the use of a dedicated Virtual Infrastructure Manager per cluster. With regard to interfaces, it must be noted that Fig. 3 mostly depicts reference points -*which may contain one or more actual interfaces*- between architectural layers. Each reference point label starts with "S-"to differentiate it from interfaces defined in ETSI NFV ISG documents[5] (and in specific Vi-Vnfm, Or-Vi, Ve-Vnfm, Nf-Vi) – although in several cases the functionality of the reference point will be almost "aligned" to the actual ETSI definitions.

---

[5] More relevant information about the respective ETSI documentation can be found at: http://www.etsi.org/technologies-clusters/technologies/689-network-functions-virtualisation.

4 Concluding Remarks and Future Aspects

In order to suitably "address" the needs and requirements of a vigorous and agile network management in the forthcoming -purely innovative- 5G era, and though a proper building effort upon the pillars of NFV, mobile edge computing and cognitive management, SESAME's main goal is both the development/expansion and the demonstration of an innovative architectural model, capable of providing Small Cell coverage to multiple operators "as a Service", so that to strongly support multi-tenancy. To this fundamental aim which implicates an enormous business potential, SESAME envisages not only to "virtualise" but also to "partition" Small Cell capacity, while simultaneously aims to support enhanced edge cloud services by enriching Small Cells with dedicated micro-servers. This work has thus presented a "first conceptual approach" to the high-level overall architecture of the broader SESAME system. The SESAME architecture discussed in the present paper can be further extended based on several current trends towards 5G, in particular those already identified by the ETSI MEC standardization group. Recent virtualization technologies permit bringing mobile core functions "close" to the mobile edge, hence enabling the deployment of the respective service platform alongside the components of the EPC while still being in proximity to the involved users. There are several benefits coming from this way of deployment, since platform(s) can leverage on many tasks performed by legacy core components (e.g., the PDN Gateway), thus without the requirement of implementing them. Such tasks are related to, *inter-alia*, gating, GTP (GPRS Tunneling Protocol) encapsulation, QoS enforcement, charging, lawful interception and mobility support. The direct influence upon the current SESAME architecture would be to install/deploy the virtual EPC (vEPC) [20], alongside with the virtual small cell and allow the NFV components executing their tasks at the egress point of the core (i.e., the SGi interface[6]), instead of the ingress point (i.e., the S1 interface). In this manner, as aforementioned, NFVs can take actions directly on IP packet flows without the need for implementing GTP encapsulation/decapsulation mechanisms, as currently required in the SESAME architecture.

Acknowledgments. The present work has been performed in the scope of the *SESAME* ("*Small cElls coordinAtion for Multi-tenancy and Edge services*") European Research Project and has been supported by the Commission of the European Communities (*5G-PPP/H2020, Grant Agreement No. 671596*).

References

1. El Hattachi, R., Erfanian, J.: Next Generation Mobile Networks (NGMN) Alliance: 5G White paper, NGMN Alliance Ltd. (2015). https://www.ngmn.org/uploads/media/NGMN_5G_White_Paper_V1_0.pdf

---

[6] More informative data about this interface can be found at: http://lteworld.org/ltefaq/what-are-lte-interfaces.

2. European Commission: Communication on a European Strategy for Key-Enabling Technologies – A Bridge to Growth and Jobs (COM (2012) 341 final, 26.06.2012). European Commission (2012)
3. Zakrzewska, A., Ruepp, S., Berger, M.: Towards converged 5G mobile networks - challenges and current trends. In: Proceedings of the 2014 ITU Kaleidoscope Academic Conference, pp. 39–45. IEEE, June 2014
4. 5G Public Private Partnership (5G-PPP): 5G Vision: The 5G-PPP Infrastructure Private Public Partnership: The Next Generation of Communication Network and Services. European Commission (2015). https://5g-ppp.eu/wp-content/uploads/2015/02/5G-Vision-Brochure-v1.pdf
5. Nadeau, T.D., Gray, K.: SDN: Software Defined Networks, 1st edn. O'Reilly, Sebastopol (2013)
6. Liang, C., Yu, F.-R.: Wireless network virtualization: a survey, some research issues and challenges. IEEE Commun. Surv. Tutorials 17(1), 358–380 (2014)
7. Patouni, E., Merentitis, A., Panagiotopoulos, P., Glentis, A., Alonistioti, N.: Network virtualisation trends: virtually anything is possible by connecting the unconnected. In: Proceedings of the IEEE 2013 SDN Conference for Future Networks and Services (SDN4FNS), pp. 1–7. IEEE (2013)
8. Thompson, J., Ge, X., Wu, H.-C., Irmer, R., et al.: 5G wireless communication systems: prospects and challenges. IEEE Commun. Mag. 52(2), 62–64 (2014)
9. Manzalini, A., Minerva, R., Callegati, F., Cerroni, W., Campi, A.: Clouds of virtual machines in edge networks. IEEE Commun. Mag. 51(7), 63–70 (2013). IEEE
10. European Telecommunications Standards Institute (ETSI): Network Functions Virtualisation - Introductory White paper, ETSI-NFV (2012)
11. Andrews, J.G.: Seven ways that HetNets are a cellular paradigm shift. IEEE Commun. Mag. 51(3), 136–144 (2013)
12. European Commission: 5G: Challenges, Research Priorities, and Recommendations – Joint White paper, European Commission, Strategic Research and Innovation Agenda (2014)
13. Real Wireless Ltd.: An Assessment of the Value of Small Cell Services to Operators (Based on Virgin Media Trials) - Version 3.1. Real Wireless Ltd., October 2012
14. SESAME H2020 5G-PPP Project (Grant Agreement No. 671596). http://www.sesame-h2020-5g-ppp.eu/Home.aspx
15. 3rd Generation Partnership Project (3GPP): 3GPP TS 23.251 v13.1.0 - Network Sharing; Architecture and Functional Description (Release 13). 3GPP, March 2015
16. 3rd Generation Partnership Project (3GPP): 3GPP TS 36.300 v13.2.0 - Evolved Universal Terrestrial Radio Access (E-UTRA) and Evolved Universal Terrestrial Radio Access Network (EUTRAN); Overall Description; Stage 2 Release 13, 3GPP, December 2015
17. 3rd Generation Partnership Project (3GPP): 3GPP TS 32.130 v13.0.0 - Telecommunication Management; Network Sharing; Concepts and Requirements (Release 13), 3GPP, January 2016
18. SESAME H2020 5G-PPP Project: Deliverable D2.4 - Specification of the Infrastructure Virtualisation, Orchestration and Management – First Iteration, April 2016
19. European Telecommunications Standards Institute (ETSI): NFV Management and Orchestration - An Overview, GS NFV-MAN 001 v1.1.1. ETSI (2014)
20. Basta, A., Kellerer, W., Hoffmann, M., Hoffmann, K., Schmidt, E.-D.: A virtual SDN-enabled LTE EPC architecture: a case study for S-/P-Gateways functions. In: Proceedings of the 2013 IEEE SDN Conference for Future Networks and Services (SDN4FNS), pp. 1–7. IEEE, November 2013

Network Architecture and Essential Features for 5G: The SESAME Project Approach

Leonardo Goratti[1], Cristina E. Costa[1(✉)], Jordi Perez-Romano[2],
Oriol Sallent[2], Cristina Ruiz[3], August Betzler[3], Pouria Sayyad Khodashenas[3],
Seiamak Vahid[4], Karim M. Nasr[4], Babangida Abubakar[5], Alan Whitehead[6],
Maria Belesioti[7], and Ioannis Chochliouros[7]

[1] CREATE-NET, Via alla Cascata 56/D, Povo, Italy
{leonardo.goratti,cristina.coasta}@create-net.org
[2] Universitat Politecnica de Catalunya (UPC), Barcelona, Spain
[3] i2CAT, Barcelona, Spain
[4] University of Surrey, Guildford, UK
[5] University of Brighton, Brighton, UK
[6] ip.access, Cambridge, UK
[7] Hellenic Telecommunications Organization, Marousi, Greece

Abstract. The outstanding and continuous growth of the request of mobile broadband Internet access is creating the unprecedented need to rethink most of the design paradigms of the mobile network. Such trend is accompanied by remarkable progresses of miniaturised electronics, together with the proliferation of social services and computation intensive applications such as high definition video. On one hand, the current mobile network is unable to deliver sufficiently high data rates per user in order to support this growth, and a possible solution is provided by the dense deployment of small cell devices. On the other, mobile operators are struggling to lower costs of deployment and maintenance while keeping profitable revenues. This paper aims to provide overview of the solution developed by the 5G-PPP SESAME project. SESAME proposes to leverage on the concept of Small Cell-as-a-service (SCaaS), providing the complete architectural solution to deploy cloud-enabled small cells. The key innovations developed by SESAME include the deployment of computation capabilities at the mobile network edge, and to exploit virtualisation techniques to manage and orchestrate dense small cell scenarios and different use cases.

Keywords: 5G · SCaaS · Mobile edge computing · Self-organising networks · Small cell virtualisation

1 Introduction

In recent years the way mobile users access and consume contents has dramatically changed due to remarkable progresses of miniaturised electronics and the proliferation of portable and user friendly devices including smartphones and tablets [1]. As shown in [2], during the third quarter of 2015, almost 3.4 billion subscriptions were registered

L. Iliadis and I. Maglogiannis (Eds.): AIAI 2016, IFIP AICT 475, pp. 676–685, 2016.
DOI: 10.1007/978-3-319-44944-9_60

worldwide. At the same time new social services and applications have become so popular that mobile connectivity is a preferred way for users. It is evident that with the current pace of growth even LTE-Advanced (LTE-A) will fail to support such high traffic.

Recently the 5G Infrastructure Public Private Partnership (5G-PPP) [3] has started its activities to create the next generation of mobile network. The SESAME project [4] is one amid the nineteen projects which were recently funded. As mentioned above, the emergence of innovative services and increased network availability motivate and drive further mobile user's engagement, creating a loop of increasing expectations and demand [5]. It is worth registering that mobile broadband penetration has indeed risen to 85.5 % in the OECD area, meaning more than four wireless subscriptions for every five inhabitants, while the penetration rate in June 2014 was just 76 % [6]. To cope with this large growth, METIS project [7] has defined a large number of use cases that 5G technology shall serve with superior performance over previous generations of cellular technology.

To meet growing users' demand the solution is offered by denser networks in which small form factor small cell (SC) devices can be deployed in large amounts and operate over licensed spectrum. This trend is also regarded to as network densification. Several approaches and technologies are currently converging in the new 5G mobile network as generally discussed in [8]. In particular, besides new physical layer solutions, Software-Defined Networking (SDN) [9], and Network Functions Virtualisation (NFV) [10, 11] are making their way toward a programmable network solution. Borrowing from cloud computing concepts such as Infrastructure-as-a-Service (IaaS) and more recently XaaS (anything-as-a-Service) have been adopted also in the mobile network domain. These approaches facilitate the decoupling between service provider, infrastructure provider and network provider. As a consequence, besides traditional Mobile Network Operators (MNO), new types of Virtual Network Operators (MVNO) and Over-The-Top (OTT) service providers can find unprecedented opportunities in 5G.

The SESAME project develops the concept of Small Cell-as-a-Service (SCaaS), which leverages on the separation between traditional market roles, with the aim to make resources available through network virtualisation. In this context, SESAME will maximise the opportunities offered by opening small cells to multi-tenancy. This latter concept, in opposition to typical mobile operators which deploy their own network infrastructure in competition with others, encourages both traditional and new market entrants to share the infrastructure. In this case operators can differentiate based on their service offers rather than on network connectivity. To manage the dense SC network, including mitigating interference and assign resources dynamically, the ETSI NFV Industry Specification Group (ISG) has developed the Management and Orchestration (MANO) framework that constitutes a solution for managing virtualised small cells [12]. This flexible and dynamic system allows operators to reduce CAPEX and OPEX as required in the next generation of mobile networks.

1.1 SESAME Enabling Technologies

Small Cells. Small Cells have become pivotal in today's 4G access. Small cells installation is an effective way to achieve greater performance and capacity to both indoor and outdoor places: they provide improved cellular coverage, higher capacity and applications for homes and enterprises, as well as in dense metropolitan and rural areas [13]. Their role is crucial for providing services in specific high traffic places such as office areas, dense urban areas, stadiums, shopping malls, concert venues, and generally, places with (tactic or sporadic) high end-users density [14].

Mobile-Edge Computing. ETSI Mobile-Edge Computing (MEC) offers computing capabilities at the network edge and bring different services near to the mobile subscribers. Providing edge cloud capabilities allows to enable accelerated services, content and application thanks to increased network responsiveness. The approach proposed is to deploy a MEC server between the mobile core and the Radio Access Network (RAN). Typical services which can benefit from mobile-edge computing include Internet-of-Things, augmented reality and data caching.

Virtualisation. Virtualisation of the communication infrastructure, such as core/edge network elements and access points/macrocells, has been extensively studied by several industry and research initiatives up to now. Recently, its applicability to the small cell infrastructure has now started to receive increasing attention. The remainder of this paper is organised as follows. In Sect. 2 we describe the general principles behind the SESAME project. In Sect. 3 we provide detailed description of the SESAME system and conclusions are drawn in Sect. 5.

2 SESAME: Small Cell Coordination for Multi-tenancy and Edge Services

The SESAME project targets innovations around the placement of network intelligence and services in the network edge through NFV and cloud computing. Through the evolution of the SC concept, already mainstream in 4G, SESAME expects to exploit its full potential in challenging highly dense 5G scenarios. SESAME targets providing SCaaS and to consolidate multi-tenancy in communication infrastructures, allowing several operators/service providers to engage in new sharing models, obtaining higher capacity on the access side and exploiting edge computing capabilities.

The key innovations proposed by SESAME focus on the novel concepts of virtualising SC networks by substantially evolving the SC concept under the paradigms of a multi-operator (i.e. multi-tenancy) enabling framework and an edge-based, virtualised execution environment. SESAME leverages on the capability to deliver intelligence directly to the network's edge, in the form of virtual network appliances. The provisioning of multi-operator SC networks is optimized for the most promising scenarios and use cases.

SESAME develops and will demonstrate an innovative architecture, capable of providing SC networks to multiple operators. SESAME fosters the concept of logical partitioning of the SC network in multiple isolated slices, virtualising and partitioning small cells capacity to multiple tenants. SESAME supports enhanced multi-tenant edge cloud services combining SCs with micro-server facilities. The unique characteristics of the SESAME approach allows new SC operators (real estate companies, municipalities, etc.) to enter the value chain deploying access infrastructure in specific high traffic demanding areas, and acting as neutral host providers, offering to existing mobile operators on-demand access to network resources.

2.1 SESAME Design Principles

The SESAME architecture was driven by design principles extracted from the identification and analysis of high-impact use cases and stakeholders' requirements. Each stakeholder involved in the process behaves according to different objectives. MNOs can achieve lower total cost of ownership combining own resources with outsourced access capacity and computing; VMNOs benefit of wireless services without owing the physical infrastructure; Venue Owners as new market entrants can benefit from becoming local network operators; End Users can enjoy personalised services with superior quality of experience.

In order to maximise user experience, the architecture should take into account various *user requirements* such as per-user data rate and latency, robustness and resiliency, mobility, seamless user experience and context-awareness. On the other hand, *network requirements* must be met in order to allow efficient operation and management. The latter include scalability, network capacity, automated system management and configuration, advanced Self-Organizing Network (SON) features, network flexibility, improved coverage, security and flexible spectrum management.

The set of design principles that drives the SESAME architecture is described below. These are derived from both the analysis of use cases and underlying requirements, as well as considering the state-of-the-art in the technological fields relevant to SESAME (e.g. SCs, NFV, cloud computing), and the most prominent standardisation trends. *Principle 1*: The SESAME system is sustainable and reconfigurable; *Principle 2*: SESAME offers an infrastructure shared between operators, transparent and neutral; *Principle 3*: SESAME accelerates the creation of innovative services with superior quality of experience through mobile edge computing; *Principle 4*: SESAME develops a system which is capable of optimising the usage of radio, storage and computing resources.

3 The SESAME System

SESAME proposes a novel 5G platform based on small cells, featuring multi-tenancy and edge cloud capabilities, offered to both network operators and mobile users. In the SESAME system shown in Fig. 1, one key design principle is to support the innovative concept of Virtual Small Cell Network Operators (VSCNO). In particular, VSCNOs use the infrastructure deployed by a SCNO.

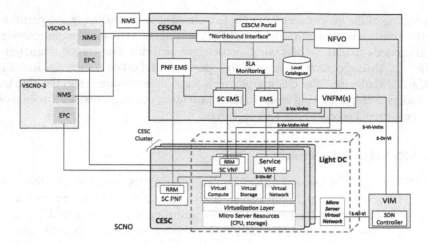

Fig. 1. SESAME system architecture.

The SESAME approach enhances the SC with micro-servers, which together form the Light DC execution environment. Micro-servers in the Light DC are connected whereby a dedicated internal network.

3.1 High-Level Architecture

The SESAME system in Fig. 1 bases its architecture on the concept of the *Cloud-Enabled Small Cell (CESC)*, a new multi-operator enabled SC that integrates a virtualised execution platform (i.e. the Light DC) for deploying VNFs, supporting automated network management and executing novel applications and services inside the access network infrastructure. The Light DC features low-power processors and hardware accelerators for time critical operations and provides a highly manageable clustered edge computing infrastructure. SESAMEs Light DC is based on low-power 64-bit processors supported by hardware accelerators (e.g. GPU, DSPs and FPGAs). In addition to the internal network connecting the micro-servers, a suitable backhaul connection provides connectivity for the SESAME system to external packet data networks.

The CESC *Manager* (CESCM) provides optimized management of the CESC deployment and is a key element of the SESAME architecture. It implements orchestration, NFV management, virtualisation of management views per tenant, Self-x features and radio access management techniques. The *Portal* is used by externals to request resources or apply (re)configuration of parameters. The NorthBound Interface (NBI) is the connecting point between orchestration, Service Level Agreements (SLAs) and VSCNOs. The *SLA Monitoring* module provides inputs to both VSCNOs and infrastructure providers about correct execution of the environment, and it enables the orchestration subsystem to react accordingly to possible changes in the network.

Small cells connect to the operator's domain and in particular to the Evolved Packet Core (EPC) and the Network Management System (NMS). The NMS is responsible for communicating with the typical Element Management System (EMS) in the small cells domain, in charge of Fault, Configuration, Accounting, Performance and Security (FCAPS) for physical and virtual functions, as will be clarified shortly. As shown in Fig. 1, SDN is used to configure the forwarding behaviour of traffic inside the Light DC and the chain of VNFs.

3.2 Essential Features

The essential innovations brought by the SESAME architecture (Fig. 1) delve into virtualising small cells, as well as leveraging on the paradigm of multi-tenancy (i.e. multi-operator) coupled with a virtualised computation environment at the mobile network edge. Multiple network operators will be able to use the SESAME system, each one through its own *slice* of network resources. The idea is to deploy small cells in which functions are virtualised, and each small cell is also connected to a micro-server through appropriate fronthaul technology. A micro-server is based on a non-×86 architecture using 64-bit ARMv8 technology. As anticipated, micro-servers together with the SCs form the CESC, and a number of CESCs form the *CESC cluster*, which can be shared by multiple operators within a geographical area. SESAME will develop important advances also in network management developing a novel edge-computing architecture and deploying Self-x procedures directly to the network edge. To that end, SESAME will support multi-tenancy through Multi-Operator Core Network (MOCN).

3.3 Functional Split in Small Cells

SESAME fully embraces network virtualisation, which allows small cell functions to be decoupled from the physical hardware. Indeed, SC functions can be split between Physical Network Function (SC PNF) and virtual (SC VNF), with one SC VNF which can be connected to multiple SC PNF through the fronthaul. Several studies are currently carried forward by the Small Cell Forum to evaluate the implications of different functional splits between physical and virtual [15]. The fact that one SC VNF can connect to several SC PNF offers several advantages: (1) improved coordination of the radio functions (coordinated scheduling, inter-cell interference coordination, etc.), (2) enhanced scalability of small cell deployments with simplified management, (3) accelerated life-cycle upgrade enabling new features and (4) flexibility to optimally make the workload placement.

3.4 Cloud-Enabled SESAME Environment

SESAME relies on SDN and NFV to use general-purpose computing and storage hardware at the mobile network edge. Along this direction a micro-scale virtualised execution environment proposed by SESAME consists in the Light DC, which offers the cloud execution environment of SESAME. The Light DC is designed to build a clustered infrastructure with high manageability and optimised to reduce power consumption, cabling, space and costs. This execution platform is used to support the required VNFs that implement the different SC features and the cognitive Self-x management operations. In addition, SESAME will develop several service VNFs, which include virtual video transcoding, virtual caching and virtual load balancer, just to name a few. Both SC and service VNFs shall be described by appropriate descriptor files using either TOSCA, JSON or YAML files.

As shown in Fig. 1, the CESCM includes the SESAME MANO amongst other components. The MANO include the NFV Orchestrator (NFVO), the VNF Manager (VNFM) and the Virtualised Infrastructure Manager (VIM) [12]. The SESAME NFVO provides the automated execution environment for VNFs and Network Services (NSs). The NFVO can deploy a new VNF or NS upon receiving the request through the portal or in reaction to the information supplied by SLA Monitoring module through the NBI. The NFVO shall look at the local catalog, solve the VNF placement problem and interact with the VIM to make an effective deployment of Virtual Machines (VMs) over the hardware substrate. Further the VIM is connected to and SDN controller (e.g. Open-Daylight) to make the chain of VNFs. The VNFM is hence in control of the service life-cycle (migration, rescaling, termination etc.).

3.5 Network Slicing Through Virtualisation

A fundamental feature of SESAME will be the virtualisation of small cells and their utilisation and partitioning into logically isolated slices, offered to multiple operators/tenants. A hypervisor software is used to create the Network Functions Virtualization Infrastructure (NFVI) on top of the bare hardware where VMs are executed under the control of the SESAME MANO. The main aspect of this innovation will be the capability to accommodate multiple operators within the same infrastructure, satisfying the SLA and requirements of each operator separately. This significantly reduces the costs of the deployed infrastructure (cost of ownership, maintenance, etc.), since hosted SCs can be treated as an operating resource instead of a capital expenditure. Under this perspective, the creation of neutral host solutions comes to address also the economic viability of investments done by telecommunications operators.

3.6 SLA

The SLA Monitoring module is responsible for gathering information on the use, performance and delivery of network services. It monitors the performance of a tenant network, as well as of the whole SESAME infrastructure. Accordingly, the SLA nego-tiation (encompassing billing issues, accounting and so forth), which lead to liaisons between virtual and physical operators, must be part of an interactive process with the existing support system of the telecommunications operator through appropriate open software. Furthermore, SLA Monitoring will encompass monitoring and analytics as fundamental tools for efficient virtualized network management.

4 Self-organising Network Features

In a multi-tenant scenario like the one considered in SESAME, it should be distinguished between Self-x functions that are tenant-specific (i.e. the configuration of parameters can differ from tenant to tenant) and those that are common to all the tenants. In the following, selected Self-x functions are discussed to illustrate the perspective of tenants and the relation to the SESAME architecture.

4.1 Self-planning Functions

Self-planning is the automation of the decision process to roll out new network nodes in specific areas, identifying adequate configurations and settings of radio parameters, as well as proposing capacity extension for those already deployed (e.g. increasing bandwidth and/or adding new carriers). In SESAME planning of a new cell shall consist of the automatic decision that a new SC has to be deployed in a certain geographical position and the RF planning of such new cell (i.e. transmit power and antenna parameters). The decision is one inherent function to the SCNO, which is responsible to manage the infrastructure to satisfy the capacity demand of different VSCNOs. On the other hand, the spectrum planning function specifies the amount of bandwidth required by a SC (either a new cell or one already deployed), the type of spectrum (e.g. licensed/unlicensed, etc.) and carries out the automatic assignment of the spectrum. In SESAME this function can be tenant-specific if slicing of the resources is implemented assigning separate carriers to each tenant.

4.2 Self-optimisation Functions

Once the network is in operational state, self-optimization includes the set of functions to improve or maintain the network performance in terms of coverage, capacity and service quality by tuning the different network settings. In SESAME the Coverage and Capacity Optimization (CCO) function is used to adjust RF parameters based on coverage and capacity targets. This task is accomplished by the SCNO and it cannot be left to a single tenant since these parameters affect all the tenants sharing the same physical SC.

Automatic Neighbor Relations (ANR) is responsible for automatically building the Neighbor Relation Table (NRT) of each small cell. This is fundamental for mobility purposes because handovers can only be executed between neighbor cells. In SESAME, to manage mobility in a multi-tenant scenario, the NRT of each SC has to include the relation between the SCs of the SCNO and the cells that belong to a specific tenant.

Mobility Load Balancing (MLB) addresses the problem of uneven traffic distribution in mobile networks. The main target of MLB and traffic steering algorithms is to enable overloaded cells to re-direct part of their traffic to neighbouring less loaded cells, hence alleviating congestion problems. This suits particularly well the deployment proposed by SESAME within the CESC cluster, in which the central decision of the CESCM can provide optimisation of the cluster as a whole. The resulting increased network efficiency using MLB postpones the deployment of additional network capacity, in turn reducing costs. This is usually done through range-expansion [16], achieved by either cell coverage parameter adjustments or mobility parameter adjustments. However, in 4G LTE networks, MLB is known to lead to network performance degradation due to the frequency reuse-1 in this technology [17]. SESAME will develop solutions to remedy this inefficiency.

Admission and congestion control are integral parts of any Quality of Service (QoS) mechanism for networks that support different types of traffic. This is the case in SESAME where a variety of different applications are delivered to the tenants of a CESC. SESAME will utilise admission and congestion control algorithms to deliver in effective manner the SLA agreed between VSCNOs and the SCNO. In traditional networks where the infrastructure is owned by a single tenant, admission and congestion control mechanisms utilise resources (e.g. multi-service packet traffic) that this tenant provides to the end-users. SESAME constitutes a more challenging environment since admission and congestion control mechanisms have to take into account the particular features of the tenants to ensure that services which have specific QoS characteristics can be delivered.

4.3 Self-healing Functions

This is the automation of the processes related to fault management and fault correction, usually associated to hardware and/or software problems, in order to keep the network operational while awaiting a more permanent solution to fix it and/or prevent disruptive problems from arising. In SESAME the concept of cell outage can be extended to both the CESC and CESC cluster. Specifically, CESC outage occurs in case of failure of the small cells, the micro-server or VNFs. Cell outage detection is done first collecting information such as alarms, alters, error messages and key performance indicators at both CESC and CESC cluster levels. Afterward, cell outage detection can be done applying different methods, including data mining (e.g. for large small cells deployment) in order to identify possible misbehaviours [18].

5   Conclusion

In this paper we have illustrated the system architecture developed by the 5G-PPP SESAME project, as well features and peculiarities of this system, in the scope of the future 5G mobile network. We have discussed the essential features of the different modules present in the system and we have clearly highlighted the suitability of SESAME to leverage an accrued virtualised network environment. Moreover, we have showed that SESAME incorporates guidelines from standard bodies such as the ETSI NFV ISG. SESAME shall leverage on a cloud-enabled small cell environment at the mobile network edge materialised by the Light DC. Network virtualisation shall allow SESAME to decouple network services from the bare hardware in unprecedented manner, and Self-x features render the system a suitable host for multi-tenant operators.

Acknowledgments.   The research leading to these results has been supported by the EU funded H2020 5G-PPP project SESAME under the grant agreement no. 671596.

References

1. Kim, Y.H., Kim, D.J., Wachter, K.: A study of mobile user engagement (MoEN): engagement motivations, perceived value, satisfaction, and continued engagement intention. Decis. Support Syst. **56**(1), 361–370 (2013)
2. Ericsson Mobility Report: On the Pulse of Networked Society (2015)
3. 5G-PPP: https://5g-ppp.eu/
4. SESAME Project: http://www.sesame-h2020-5g-ppp.eu/
5. Tojib, D., Tsarenko, Y., Sembada, A.Y.: The facilitating role of smart-phones in increasing use of value-added mobile services. New Media Soc. J. **17**, 1220–1240 (2016). doi: 10.1177/1461444814522951
6. OECD Report: OECD Broadband Statistics Update. http://www.oecd.org/internet/broadband-statistics-update.htm. Accessed April 2016
7. Deliverable 1.5: Updated Scenarios, Requirements and KPIs for 5G Mobile and Wireless System with Recommendations for Future Investigations. METSI (2015)
8. Andrews, J.G., et al.: What will 5G be? IEEE J. Sel. Areas Commun. **32**(6), 1065–1082 (2014)
9. Kreutz, D., Ramos, F.M., Verissimo, P., Rothenberg, C.E., Azodolmolky, S., Uhlig, S.: Software-defined networking: a comprehensive survey. Proc. IEEE **103**(1), 14–76 (2014)
10. Chiosi, M., et al.: Network functions virtualisation: an introduction, benefits, enablers, challenges & call for action. In: SDN and OpenFlow World Congress (2012)
11. ETSI GS NFV 001: Network Functions Virtualization (NFV); Use Cases (2013)
12. Network Functions Virtualisation (NFV); Management and Orchestration. ETSI GS NFV-MAN 001 V1.1.1 (2012)
13. Andrews, J.G.: Seven ways that HetNets are a cellular paradigm shift. IEEE Commun. Mag. **51**(3), 136–144 (2013)
14. Osseiran, A., et al.: Scenarios for 5G mobile and wireless communications: the vision of the METIS project. IEEE Commun. Mag. **52**(5), 26–35 (2014)
15. Small Cell Virtualization: Functional Splits and Use Cases. SCF 159.06.02 (2016)
16. Ruiz-Avils, J.M., et al.: Analysis of limitations of mobility load balancing in a live LTE system. IEEE Wireless Commun. Lett. **4**(4) (2015)
17. Mobility Enhancements in HetNets. 3GPP TR 36.839 V11.1.0
18. Wang, W., Zhang, J., Zhang, Q.: Cooperative cell outage detection in self-organising femtocell networks. In: IEEE INFOCOM, pp. 782–790 (2013)

On Learning Mobility Patterns
in Cellular Networks

Juan Sánchez-González[✉], Jordi Pérez-Romero, Ramon Agustí,
and Oriol Sallent

Universitat Politècnica de Catalunya (UPC), Barcelona, Spain
{juansanchez,jorperez,ramon,sallent}@tsc.upc.edu

Abstract. This paper considers the use of clustering techniques to learn the mobility patterns existing in a cellular network. These patterns are materialized in a database of prototype trajectories obtained after having observed multiple trajectories of mobile users. Both K-means and Self-Organizing Maps (SOM) techniques are assessed. Different applicability areas in the context of Self-Organizing Networks (SON) for 5G are discussed and, in particular, a methodology is proposed for predicting the trajectory of a mobile user.

Keywords: Clustering · Cellular networks · Mobility patterns

1 Introduction

The new generation of mobile and wireless systems, known as 5th Generation (5G), intends to provide solutions to the continuously increasing demand for mobile broadband services associated with the massive penetration of wireless equipment while at the same time supporting new use cases associated to customers of new market segments and vertical industries (e.g., e-health, automotive, energy). As a result, the vision of the future 5G Radio Access Network (RAN) corresponds to a highly heterogeneous network with unprecedented requirements in terms of capacity, latency or data rates, as identified in different fora [1, 2]. To cope with this heterogeneity and complexity, the RAN planning and optimization processes can benefit at a large extent from exploiting cognitive capabilities that embrace knowledge and intelligence.

In this direction, legacy systems already started the automation in the planning and optimization processes through Self-Organizing Network (SON) functionalities [3]. In 5G, considering also the advent of big data technologies [4], it is envisioned that SON can be further evolved towards a more proactive approach able to exploit the huge amount of data available by a Mobile Network Operator (MNO) and to incorporate additional dimensions coming from the characterization of end-user experience and end-user behavior [5]. Then, SON can be enhanced through Artificial Intelligence (AI)-based tools, able to smartly process input data from the environment and come up with knowledge that can be formalized in terms of models and/or structured metrics that represent the network behavior. This will allow gaining in-depth and detailed knowledge about the whole 5G ecosystem, understanding hidden patterns, data structures and relationships, and using them for a more efficient network management [6].

© IFIP International Federation for Information Processing 2016
Published by Springer International Publishing Switzerland 2016. All Rights Reserved
L. Iliadis and I. Maglogiannis (Eds.): AIAI 2016, IFIP AICT 475, pp. 686–696, 2016.
DOI: 10.1007/978-3-319-44944-9_61

AI-based SON involves three main stages [6]: (i) the acquisition and pre-processing of input data exploiting the wide variety of available data sources; (ii) the knowledge discovery that smartly processes the input data to come up with exploitable knowledge models that represent the network/user behavior; and (iii) the knowledge exploitation stage that applies the obtained models to drive the decision-making of the SON functions. This paper focuses on the knowledge discovery stage and, in particular, on automatically learning the mobility patterns of the mobile users, trying to identify if the traffic across the cells in a scenario follows specific patterns that can be characterized in terms of prototype trajectories followed by many users.

Different works of the literature have addressed the analysis of trajectories in different contexts such as hurricane trajectories, animal movements, public transportation, etc. Various tools have been considered, such as Self-Organizing Maps (SOM) together with visual analysis [7], density-based clustering [8, 9] or Principal Component Analysis [10]. In wireless networks, [11] proposed a trajectory prediction strategy to deal with routing in mesh sensor networks. It is based on clustering similar trajectories followed by wireless nodes and using them for making predictions of other nodes. However, the concept of trajectory in [11] is defined by the set of nodes that a mobile node would associate with to send or receive data along a path, but not by the geographical locations. Instead, in our work we intend to derive a deeper knowledge about trajectories based on analyzing the geographical coordinates. In turn, [12, 13] address the problem of classifying the trajectory followed by a mobile terminal based on a set of reference trajectories in order to optimize the handover process in LTE. However, while [12, 13] use a simple method for building the set of reference trajectories, based on monitoring certain users with a given probability and adding their trajectories to the set, in our approach we propose the use of clustering techniques, which are more powerful for identifying the most representative trajectories.

In this context, the approach proposed in this paper considers the use of clustering techniques, namely K-means and SOM, to learn the mobility patterns existing in a cellular network. These patterns are materialized in a database of prototype trajectories obtained after having observed multiple trajectories of mobile users. Different applicability areas for these patterns in the context of 5G-SON are discussed and, in particular, a methodology is proposed for predicting the trajectory of a mobile user.

The rest of the paper is organized as follows. Section 2 describes the proposed methodology based on clustering tools for learning mobility patterns. Section 3 discusses the applicability areas and describes the approach for identifying the trajectory of a mobile user. Proposed approach is evaluated in Sect. 4, while Sect. 5 summarizes the concluding remarks.

2 Mobility Pattern Knowledge Discovery

Current cellular networks like 4G already include the capability that the User Equipments (UEs) provide geolocation information, including both geographical coordinates and altitude, as part of the radio measurement reporting processes [14]. Location information can be obtained from UEs in connected mode, who periodically transmit measurement reports to the network. Furthermore, thanks to the use of Minimization of

Drive Tests (MDT) feature [15], UEs in idle mode can log measurements and transmit them later on when the UE enters in connected mode. These capabilities enable MNOs to collect large amounts of data that include valuable knowledge about the spatio-temporal traffic distribution across the cells. This paper proposes a methodology to analyze this data and identify the existing mobility patterns of the UEs.

The approach for learning mobility patterns is graphically illustrated in Fig. 1. It operates on a long-term basis after having observed a large amount of connected and idle mode UEs in different time periods of a certain geographical area and analyzes the collected location information from these UEs to identify the existence of prototype trajectories. As shown in Fig. 1 the first step is the pre-processing, which analyzes consecutive reports for each UE and extracts the geolocation information in order to build a trajectory for this UE. A trajectory is defined here as the concatenation of N coordinates at consecutive time instants $t_1, \ldots, t_N$. Then, assuming for simplicity two-dimensional (2D) coordinates (x, y), the trajectory for the j-th UE is given by the vector of dimension $B = 2N$ denoted as $\mathbf{r}_j = [x_j(t_1), y_j(t_1), \ldots, x_j(t_N), y_j(t_N)]$. The result of the pre-processing task will be a total of J trajectories $\mathbf{r}_j$, $j = 1, \ldots, J$.

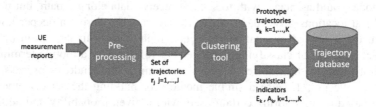

Fig. 1. Procedure for learning mobility patterns

The second step is the clustering, which processes the set of J trajectories by grouping them in K clusters in a way that trajectories of the same cluster are similar among them and different from the trajectories of the rest of the clusters. Two alternative clustering techniques are considered in this work:

- K-means: This strategy belongs to the family of partitioning methods. It groups the J input trajectories in K clusters by trying to maximize the similarity between trajectories of the same cluster and to minimize the similarity between trajectories of different clusters, using the Euclidean distance as a metric of similarity. The process can be summarized as follows (see [16] for further details): (a) The algorithm starts by selecting randomly K out of the J input trajectories. Each of these K trajectories represents an initial cluster. For each cluster k, the algorithm computes the centroid $\mathbf{s}_k$. At this initial stage, where each cluster contains only one trajectory, the centroid $\mathbf{s}_k$ equals the selected trajectory for the k-th cluster. (b) Each of the remaining $J - K$ trajectories is assigned to the cluster to which it is the most similar, based on Euclidean distance between the trajectory and the centroid of each cluster $|\mathbf{r}_j - \mathbf{s}_k|$. Once all the J trajectories have been clustered, the new values of the centroids $\mathbf{s}_k$ are recomputed. In particular, the i-th component of $\mathbf{s}_k$ is the average of the i-th components of all the trajectories belonging to the k-th cluster. (c) Using the new values of the centroids s_k, each of the J trajectories r_j is reassigned to the cluster with lowest distance $|r_j - s_k|$. The new centroids are recomputed and this step is iteratively repeated until convergence (i.e. until there are no changes in the obtained clusters after two consecutive iterations). (d) At the end of the process, each cluster $k = 1, ..., K$ will contain a number of input trajectories N_k and its centroid s_k will be the so-called prototype trajectory that is taken as a representative of all the trajectories belonging to this cluster.

- Self-Organizing Map (SOM): This clustering strategy relies on a neural network model with a total of K neurons and where each neuron is characterized by a B-dimensional weight vector s_k. The process can be summarized as follows (see [17] for details): (a) The weight vectors s_k are initialized. This can be done randomly or through the linear initialization method described in [17]. (b) An iterative unsupervised learning process is used to update the values of the weight vectors s_k of the different neurons according to the Kohonen's algorithm [17] based on the input trajectories r_j. In essence, at iteration t the algorithm identifies, for each trajectory r_j the winning neuron as the one with the lowest Euclidean distance $|r_j - s_k|$. Then, the algorithm updates the weight vector of this winning neuron k as $s_k(t + 1) = s_k(t) + \alpha(t)(r_j - s_k(t))$ where $\alpha(t)$ is a scalar-valued adaptation gain that decreases with successive iterations. A similar update is performed for the weight vectors of the rest of neurons $k' \neq k$ but in this case the adaptation gain $\alpha(t)$ is multiplied by a neighborhood function that decreases with the distance between neurons k' and k. The process is repeated for a certain number of iterations. (c) At the end of the process, all the input trajectories that have neuron k as winning neuron form the k-th cluster. The number of trajectories in the k-th cluster is N_k, and the prototype trajectory of this cluster is the weight vector s_k.

As shown in Fig. 1, the prototype trajectories obtained as a result of the clustering will be stored in the database. In addition, two statistical indicators are also included for each cluster to assess how representative this cluster is:

- Percentage of hits ($A_k = N_k/J$): It is the percentage of input trajectories that belong to the cluster k. The prototype trajectories of clusters with a high value of A_k will be more frequent and representative of the scenario.
- Average squared Euclidean distance of the trajectories in k-th cluster (E_k): It is a metric that captures the degree of similarity between trajectories of the same cluster with respect to the prototype trajectory s_k of the cluster. A high value of E_k reflects a higher dispersion in the cluster, meaning that the prototype trajectory is less representative of the clustered trajectories. It is defined as:

$$E_k = \sum_{j \in \text{Cluster } k} |r_j - s_k|^2 \tag{1}$$

3   Exploitation of Mobility Patterns

It is envisaged that the identification of prototype trajectories as explained in previous section can have applicability for different 5G-SON functions.

For example, prototype trajectories can be used in the context of self-planning to decide appropriate cell locations and antenna settings. For example, if there is a well identified representative trajectory, a sector of a cell site can be pointed in the direction of this trajectory. Typically, this can be the case of a cell site providing coverage over a main street. Despite one could argue that a radio engineer could easily identify such a situation and take such a common sense decision, the interest of the proposed use case remains in the fact that SON involves automatization. That is, self-planning and self-configuration means the capability for the system to automatically identify the trajectories and propose the adequate values for the parameters of a new cell.

Similarly, the learnt mobility patterns can also have applicability in the self-optimization of several functions such as handover, load balancing or admission control. For example, by identifying the trajectory of a UE or group of UEs in relation to a known prototype trajectory it is possible to anticipate the cell that the UEs are heading to and configure these functions so as to avoid call droppings and overload situations. In the following, we focus on proposing a methodology to predict the future positions of a certain UE based on analyzing the actual locations reported by the UE in relation to the learnt prototype trajectories.

3.1   Mobility Prediction

The proposed approach is illustrated in Fig. 2 and is executed on an individual UE basis. The criterion to decide which specific UEs are analyzed is out of the scope of this paper and it will depend on the specific self-optimization function under consideration. For example, the optimization of load balancing may predict the trajectory of UEs that demand a high bit rate in order to anticipate the arrival of these UEs to a cell and take the appropriate actions to ensure there are sufficient resources for these UEs in the cell. Similarly, it is also possible to predict the trajectory of high priority UEs to ensure that they will not experience problems in handovers, etc.

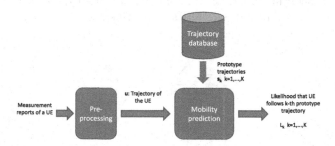

Fig. 2. Exploitation of learnt patterns for predicting the trajectory of a UE

The process of Fig. 2 starts from the measurement reports provided by the UE whose trajectory is being predicted. First, pre-processing stage is carried out to extract the geolocation information and build the trajectory **u** that is currently being observed for this UE. The trajectory **u** is a vector of dimension $C = 2M$ composed by the concatenation of M pairs of coordinates followed by the UE at consecutive time instants $\mathbf{u} = [x(t_1), y(t_1), ..., x(t_M), y(t_M)]$. Without loss of generality, let us consider that the dimension of **u** is lower than the number of elements of the prototype trajectories $\mathbf{s_k}$ (i.e. $C \leq B$). This reflects that, in case that the UE was following a prototype trajectory, the actual location of the UE is somewhere within the prototype trajectory.

The mobility prediction process of Fig. 2 intends to determine the likelihood that the UE is following one of the learnt prototype trajectories. This is done by assessing the similarity between the trajectory **u** followed by the UE and the prototype trajectories $\mathbf{s_k}$ according to the Euclidean distance. Given that $C \leq B$, all the possible portions of C consecutive elements of the vectors $\mathbf{s_k}$ ($k = 1, ..., K$) need to be considered when assessing this similarity. The α-th portion of $\mathbf{s_k}$ is then defined as the vector $[s_k(1 + \alpha), ..., s_k(C + \alpha)]$ with $\alpha = 0, ..., B - C$, where $s_k(i)$ denotes the i-th component of $\mathbf{s_k}$. Then, the squared Euclidean distance between the α-th portion of $\mathbf{s_k}$ and trajectory **u** is computed as:

$$d_{u,k}(\alpha) = \sum_{c=1}^{C} [u(c) - s_k(c + \alpha)]^2 \quad \text{with } \alpha = 0, ..., B - C \tag{2}$$

Then, the similarity between **u** and $\mathbf{s_k}$ is computed as the minimum Euclidean distance between **u** and the possible portions of the prototype trajectory $\mathbf{s_k}$, that is:

$$m_k = \min_{\alpha} d_{u,k}(\alpha) \tag{3}$$

A low value of m_k indicates that the trajectory **u** is very similar to some portion of vector $\mathbf{s_k}$. Then, the likelihood L_k that the UE is following the prototype trajectory $\mathbf{s_k}$ is defined here as:

$$L_k = \frac{1/m_k}{\sum_{k=1}^{K}(1/m_k)} \tag{4}$$

A high value of L_k reflects that the UE is following a trajectory very similar to a portion of $\mathbf{s_k}$. Therefore, $\mathbf{s_k}$ provides information about the positions that the UE may likely follow in the future.

4   Results

This section provides some results to illustrate the performance of the proposed approach. The considered scenario is shown in Fig. 3 and represents an urban area in the intersection between two main streets. The mobility of multiple UEs has been considered including a wide variety of situations as shown Fig. 3a. For example, some

UEs move straight along a street, others move straight and turn right, left or move back. For each kind of trajectory, 100 realizations have been generated by considering UE trajectories that are not perfectly straight but they have lateral movements simulating e.g. cars changing the lane in the road. It is assumed that the distance between two consecutive positions of the trajectory is a random value (simulating that the user speed may be variable). Moreover, 100 realizations of users that move a short distance and stop at a particular position (represented by black arrows in Fig. 3a) have been also generated. Finally, a group of 100 static users (represented by black dots in Fig. 3a) have also been placed randomly in each of the four corners of the scenario. After the preprocessing of the UE measurements, there are a total of J = 2100 trajectories. Each trajectory r_j consists on N = 40 positions.

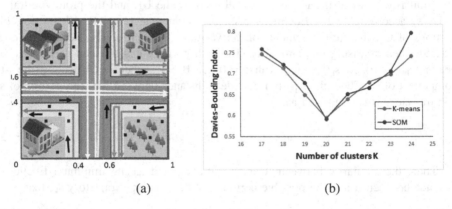

(a)   (b)

Fig. 3. (a) Illustration of the considered scenario. Distances are normalized between 0 and 1. (b) Davies-Bouldin index for different numbers of clusters.

4.1 Clustering Process

The K-means and SOM clustering techniques have been implemented by means of RapidMiner Studio [18]. The K-means algorithm is configured with 1000 runs and a maximum of 100 iterations for each run (i.e. the process explained in Sect. 2 is repeated 1000 times with different initial random selections, and the best result among all runs is kept). In turn, a SOM with one dimension is configured with 10000 iterations, initial adaptation rate equal to 0.1 and final adaptation rate 0.01. The neighborhood function is defined by an initial adaptation radius of 2 and a final adaptation radius equal to 0.01.

First, the impact of the number of clusters K has been analyzed for both K-means and SOM techniques. The Davies-Bouldin index [19] is considered as a relevant metric to assess the quality of the clustering process. This index takes into account how similar are all the trajectories that belong to the same cluster and how different are the prototype trajectories of the different clusters. Low values of the Davies-Bouldin index reflect a better quality of the clustering process. Figure 3b presents the Davies-Bouldin index as a function of the number of clusters for both K-means and SOM methodologies. As shown, for the considered use case, the minimum value of the Davies-Bouldin index is observed with K = 20 clusters for both methodologies. For this case, Fig. 4 illustrates the prototype trajectories s_k obtained by the K-means methodology. The same prototypes are obtained by the SOM methodology with K = 20. The red point marked in each prototype trajectory in Fig. 4 indicates the initial position of the trajectory while the black point indicates its final position (e.g. the prototype trajectory 1 represents a user moving from the left to the right while the prototype trajectory 2 represents a user moving from the right to the left). Note that some shorter prototype trajectories represent users who move on a specific direction and then go back (e.g. prototype trajectory 13 represents to users who move from the left to the right in the scenario, and then go back from the right to the left). Other prototype trajectories, such as prototype 17, represent the centroid of some static users located around this area.

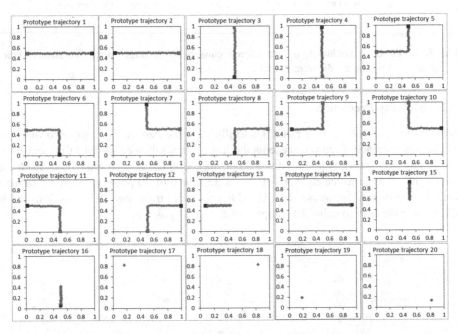

Fig. 4. Prototype trajectories obtained with K-means (K = 20). Horizontal and vertical axes represent normalized distances between 0 and 1.

Figure 5a illustrates the percentage of hits A_k for each cluster with both K-means and SOM, while Fig. 5b represents the average squared Euclidean distance E_k. All the clusters corresponding to long trajectories (i.e. clusters 1 to 12 of Fig. 4) exhibit low E_k. This indicates that these trajectories are well-clustered and their corresponding prototype trajectories are good representatives of the cluster. In turn, clusters 13, 14, 15 and 16 of Fig. 4 include users that move straight and go back, users that move short distances and even some static users. As a consequence, higher percentage of hits A_k and higher values of E_k are observed. Finally, clusters 17, 18, 19 and 20 of Fig. 4 are formed by static users scattered around the four corners of the scenario. These are characterized by high values of E_k, meaning that some static users of these clusters may be located at a relatively high distance of the centroid. A very similar clustering is done by both K-means and SOM methodologies as shown in Fig. 5a and b.

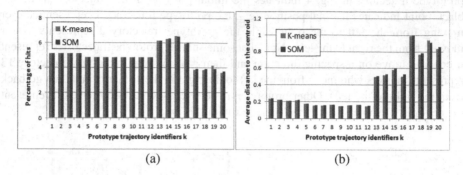

(a)   (b)

Fig. 5. (a) Percentage of hits A_k for the different clusters; (b) Average square Euclidean distance to the centroid E_k for the different clusters.

4.2 Mobility Prediction

This section presents several examples to illustrate the behavior of proposed mobility prediction approach. Figure 6 presents the trajectories followed by four different UEs. UE A has a trajectory that consists of 20 positions representing a movement from the left of the scenario to the right. UE B has a trajectory of 20 positions moving straight and then turning in the intersection. UE C has a shorter trajectory of 10 positions while UE D is static and contains 10 samples of the same position.

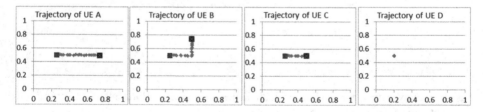

Fig. 6. Example of UEs' trajectories. Horizontal and vertical axes represent normalized distances between 0 and 1.

Figure 7 shows the likelihood L_k that each of the four UEs is following each prototype trajectory. As shown, the likelihood that UE A is following prototype trajectory 1 is almost 100 %. As seen in Fig. 5, this prototype trajectory corresponds to the users that move from the left to the right. Similarly, for UE B there is also a very high likelihood that it is following prototype trajectory 5. For UE C, the likelihood L_1, L_5 and L_6 are similar, because, with the trajectory followed by UE C so far, it may correspond to either trajectories 1, 5 or 6. Finally, trajectory D is not similar to any of

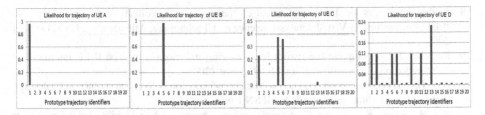

Fig. 7. Likelihood L_k that the UEs are following the learnt prototype trajectories.

the prototypes obtained in the clustering process (see Fig. 4). For this reason, the prediction process provides a very low likelihood for all the clusters.

5 Concluding Remarks

The paper has proposed a methodology for learning mobility patterns in wireless networks based on clustering techniques such as K-means and SOM. Learnt trajectories present applicability in different areas, such as self-planning and self-optimization. In this respect, the paper has proposed a strategy for predicting the mobility of specific users based on the obtained prototype trajectories. Results reflect that both K-means and SOM techniques are able to properly identify the different trajectories existing in the considered scenario.

Acknowledgements. This work has been supported by the EU funded H2020 5G-PPP project SESAME under the grant agreement no 671596 and by the Spanish Research Council and FEDER funds under RAMSES grant (ref. TEC2013-41698-R).

References

1. METIS 2020 Project: http://www.metis2020.com
2. El Hattachi, R., Erfanian, J. (eds.): NGMN 5G White Paper, NGMN Alliance, February 2015
3. Ramiro, J., Hamied, K.: Self-organizing Networks: Self-planning, Self-optimization and Self-healing for GSM, UMTS and LTE. Wiley, Chichester (2012)
4. I, C.-L., Liu, Y., Han, S., Wang, S., Liu, G.: On big data analytics for greener and softer RAN. IEEE Access, August 2015
5. Imran, A., Zoha, A., Abu-Dayya, A.: Challenges in 5G: how to empower SON with Big Data for enabling 5G. IEEE Netw. **28**, 27–33 (2014)
6. Pérez-Romero, J., Sallent, O., Ferrús, R., Agustí, R.: Artificial intelligence-based 5G network capacity planning and operation. In: ISWCS Conference (2015)
7. Schreck, T. et al.: Visual cluster analysis of trajectory data with interactive Kohonen maps. In: IEEE Symposium on Visual Analysis Science and Technology, Columbus, USA, October 2008
8. Lee, J.-G., Han, J., Whang, K.-Y.: Trajectory clustering: a partition-and-group framework. In: SIGMOD, China (2007)

9. Andrienko, G., et al.: Interactive visual clustering of large collections of trajectories. In: IEEE Symposium on Visual Analysis Science and Technology, Atlantic City, USA, October 2009

10. Masciari, E.: A complete framework for clustering trajectories. In: 21st IEEE International Conference on Tools with Artificial Intelligence (2009)

11. Lee, H.J., et al.: Data stashing: energy-efficient information delivery to mobile sinks through trajectory prediction. In: ACM/IEEE IPSN Conference (2010)

12. Sas, B., Spaey, K., Blondia, C.: Classifying users based on their mobility behavior in LTE networks. In: 10th International Conference on Wireless and Mobile Communications (ICWMC) (2014)

13. Sas, B., Spaey, K., Blondia, C.: A SON function for steering users in multi-layer LTE networks based on their mobility behavior. In: VTC Spring Conference (2015)

14. 3GPP TS 36.331 v12.7.0: Radio Resource Control (RRC); Protocol Specification (Release 12), September 2015

15. Hapsari, W.A., Umesh, A., Iwamura, M., Tomala, M., Gyula, B., Sébire, B.: Minimization of drive tests solution in 3GPP. IEEE Commun. Mag. **50**, 28–36 (2012)

16. Han, J., Kamber, M.: Data Mining Concepts and Techniques, 2nd edn. Elsevier, Amsterdam (2006)

17. Kohonen, T.: Essentials of the self-organizing map. Neural Netw. **37**, 52–65 (2013)

18. RapidMiner Studio: http://www.rapidminer.com

19. Davies, D.L., Bouldin, D.W.: A cluster separation measure. IEEE Trans. Pattern Anal. Mach. Intell. **PAM-1**(2), 224–227 (1979). doi:10.1109/TPAMI.1979.4766909

Design of Cognitive Cycles in 5G Networks

Bego Blanco[✉], Jose Oscar Fajardo, and Fidel Liberal

School of Engineering of Bilbao, University of the Basque Country (UPV/EHU),
Alda. Urquijo S/N, 48012 Bilbao, Spain
{bego.blanco,joseoscar.fajardo,fidel.liberal}@ehu.eus

Abstract. Adding cognitive capabilities to the wireless networks makes it possible to leverage the control and management information used in the network operation to infer information about the local state and exploit it to improve the overall performance. This paper deals with the combined use of centralized and distributed cognitive cycles integrated at different planes in 5G networks: an integrated data plane, a unified control plane and a cross-layer management plane. This context-aware cognitive schema acts on the decision making modules depending on the monitored environment to prevent failures, balance the virtualized execution and get a global enhancement in the provision of mobile services. The multi-level cognitive cycle supports the interaction between the edge and the cloud blurring the line that separates two paradigms: centralized radio operation and mobile edge services.

Keywords: 5G · Cognitive cycle · Glocal · Blurring edge

1 Introduction

The challenging requirements of 5G demand an evolution of the existing mobile network architecture. A defining characteristic of 4G networks is the coupling of data and control planes at the network level that interact with a separate service plane. This architecture forces to an adaptive operation: the service layer declares its requirements to the network layer; then the network accommodates the new service together with the other service requests and communicates the service plane the Quality of Service (QoS) levels that can be provided; finally, the service plane rearranges its requirements adapting them to the QoS offer. This iterative process is performed in isolation at both service and network planes without complete information of the other side, leading to non-optimal adaptation of services to network performance and of networks to service requirements.

In order to overcome these drawbacks, an innovative feature of 5G networks against the previous architectures is the clear decoupling of data and control planes and the integration of service management elements in the control plane, while service instantiations are allocated in the data plane. In this context, Software Defined Networking (SDN) provides the abstraction tool needed to split control and data planes in data forwarding. The control plane is now centralized and has the complete picture of the network, even considering different access technologies. Eventually, this plane is capable of driving the different traffic flows along the most suitable paths and, thus,

L. Iliadis and I. Maglogiannis (Eds.): AIAI 2016, IFIP AICT 475, pp. 697–708, 2016.
DOI: 10.1007/978-3-319-44944-9_62

optimizing the use of the available resources in a service/network cross-layered way. At this point, context-aware cognitive techniques can become a useful tool to support this optimization process.

In this context, the data and control plane decoupling of 5G is aligned with the Mobile Edge Computing (MEC) principles. The application of MEC to 5G systems allows the physical separation of the planes, leaving the data plane close to the user in the network edge and uploading the centralized control plane to the cloud servers.

Another evolution step of 5G systems is provided by Network Function Virtualization (NFV) technology. NFV decouples the network functions from proprietary hardware appliances so they can run in software on a data center and can be instantiated in various locations in the network as required. Combining NFV with cloud computing concepts, a centralized orchestrator is responsible for the on-boarding of new network services and VNF packages together with the management of network service lifecycle. Again, at this stage, a second level of cognitive cycle can support the decision making process of the NFV Orchestrator (NFVO).

This paper is organized as follows: Sect. 2 makes a brief introduction of 5G overall architecture and its enabling technologies, i.e. SDN, MEC and NFV. Then, Sect. 3 reviews relevant literature about the integration of cognitive capabilities in network systems. Next, Sect. 4 shows an application proposal of a cloud-based multi-level cognitive cycle to the control and management planes of H2020 SESAME architecture [1]. Finally, Sect. 5 ends summarizing the main conclusions.

2   5G Overview

Although the development of 5G is still in its early steps, the research effort in this area is oriented to achieve to main overall objectives: meet the requirements of the biggest mobile traffic growth ever known in a sustainable way, and provide a consistent end-to-end experience under diverse scenarios with ultra-high data rate, ultra-low latency and massive connections. Based on the analysis of these requirements, the International Telecommunication Union (ITU) [2] proposes the high-level network architecture depicted in Fig. 1.

An innovative feature of 5G systems against previous architectures is the clear separation of control and data plane functions, with open interfaces defined between them in according with SDN principles [3]. The decoupling of hardware and software functions of network elements in all network domains fosters a cost efficient deployment and upgrade possibilities. Other inherent benefits are real-time and on-demand network configuration and automated optimization, flexible and cost efficient network operation, maximization of utilization efficiency of available network resources and dynamic relocation of network resources, fully controlled by the operator. The disengagement of the network control logic and its centralization in an upper level provides a comprehensive view of the network state, so that the Network and Service Orchestration module is capable of configuring and managing the network service (NS) lifecycle (including instantiation, maintenance and termination), while the data plane allocates the NS instances.

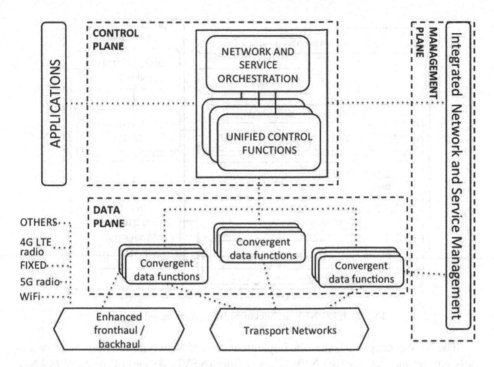

Fig. 1. High level network architecture of 5G based on [2].

This decoupling of control/management planes from the data plane is aligned with the concept of Mobile Edge Computing (MEC). MEC provides IT and cloud computing capabilities within the Radio Access Network (RAN) close to the end user [4]. Traditionally, all data traffic originated at the data centers is forwarded up to the mobile core network and down again to a base station which delivers the content to the mobile devices. In the MEC scenario, cloud servers take some or even all of the tasks originally performed in the data centre and eliminate the need of routing these data flows through the core network. This way, the RAN edge offers a low-latency high-bandwidth service environment as well as direct access to real-time radio network information that can be used by the upper plane to provide context-aware services.

Furthermore, the combination of SDN and MEC techniques over 5G architecture enables the application of Self Organizing Network (SON) principles to replace the classic manual configuration, post deployment optimization, and maintenance in cellular networks with self-configuration, self-optimization, and self-healing functionalities [5]. SON technology constitutes a fundamental change in the way networks are managed, bringing automation and dynamic, predictive resource allocation to the forefront.

Finally, the integration of a NFV framework takes another step forward in the optimization of the system performance. Figure 2 shows the NFV architecture proposed by ETSI ISG NFV [6].

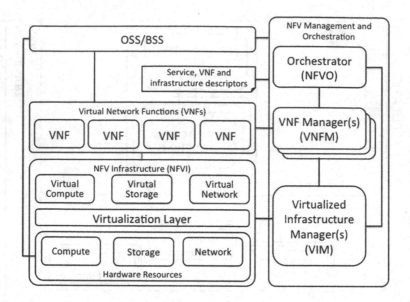

Fig. 2. ETSI NFV architectural framework based on [6].

The NFV concept envisages the implementation of Network Functions as software-only entities that run over the NFV Infrastructure (NFVI). As part of the NFV MANO, the NFV Orchestrator manages the incorporation of the Network Services through the chaining of VNFs and global resource arrangement. The VNF Manager oversees life-cycle management of VNF instances and coordinates the configuration and event reporting between NFVI and E/NMS. Finally, the Virtualized Infrastructure Manager (VIM) controls and manages the NFVI compute, storage, and network resources.

Next section analyzes the existing literature on cognitive cycles in order to insert intelligence into the different decision making elements in the network-service provisioning.

3 Cognitive Networks

The philosophy of a cognitive network is a generalization of the renowned Cognitive Radio model [7]. This concept is proposed to make a more efficient use of the electro-magnetic spectrum in wireless networks. The main idea consists in the variation of transmission and reception parameters according to the observed internal and external factors. Cognitive radio, built over software defined radio, is described as an intelligent wireless communication system that is environment aware and learns from it, adapting to the statistical variations of a set of indicators.

The cognitive network [8] extrapolates the concept of cognitive radio to other levels of the communications reference model, minimizing the dependence on human inter-vention and adapting to the continuous changes of the network conditions and user requirements fast, precise and automatically. The key characteristic of this kind of network is the cognitive process that monitors the current state of the network, plans a future action, makes a decision and acts in consequence. With this objective, a cognitive network must identify the network conditions and predict future situations, constantly adapt to the dynamic state of the network, learn from previous experiences and balance the requirements of all the participants according to service agreements [9].

These characteristics lead to the cognitive cycle proposed in OOPDAL (Observe, Orient, Plan, Decide, Act, Learn) model [10] depicted in Fig. 3.

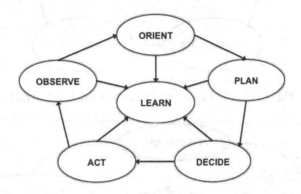

Fig. 3. OOPDAL cognitive cycle model based on [10].

The observation phase collects raw data to feed the orientation phase. The planning phase must identify the goals to achieve and arrange possible actions depending on the current situation to get closer to those goals. The decision phase assesses the available action plan and chooses the most appropriate. The action phase executes the selected plan, which may consist in a reconfiguration of parameters or a replacement of operative modules. Finally, the learning phase receives the feedback from all the previous phases and uses that experience to update the models employed in prevision, orientation, planning and decision processes.

On the other hand, Motorola-IBM proposes a simplified cognitive model within their FOCALE (Foundation-Observation-Comparison-Action-Learn-rEason) architecture for the autonomous network management [11]. Although this framework does not explicitly include a cognitive cycle, the description of its control cycles contains the necessary elements to recognize a cognitive process that involves monitoring, analysis, planning and execution tasks linked by a learning process. While it is a simpler cycle than OOPDAL, it adds the novelty of associating the learning with business goals, something that is not considered in OOPDAL.

CME (Cognitive Management Entity) architecture [12] also suggests a 5 phase cognitive model (as in Fig. 4).

The first phase is perception, responsible for collecting, measure, monitoring, pre-processing and approaching the context or environment information. Then, the analysis phase takes charge of the interpretation and abstraction, the consistence check, the prediction, the information fusion, the reasoning, the model management and update and the learning. The decision state is composed of elements such as the selection of alternatives, its evaluation and optimization and the decision-making. The

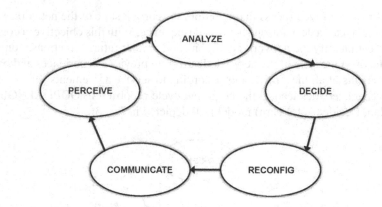

Fig. 4. Cognitive cycle of CME architecture based on [12].

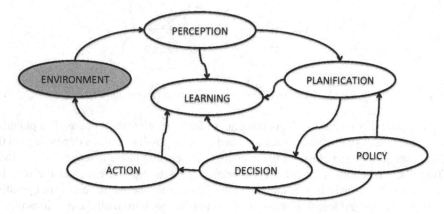

Fig. 5. Fortuna-Mohorcic reference cognitive cycle based on [13].

reconfiguration phase implements the decisions, selects the components and throws warnings and alarms. Finally, the communication phase is responsible for the command and data exchange between the several cognitive motors distributed along the network.

Inspired by the previous models, the cognitive cycle introduced in [13] is composed of the six phases shown in Fig. 5.

In this self-aware model, the nodes of the network have sensors to perceive the information of the environment. The captured observations are employed in the planning, but also to feed the learning module that builds and updates the model that remembers those observations. These models are used in the decision phase to choose the most adequate alternative according to the past experience. The planning module determines the potential strategies to follow based on the stored observations and policies. At last, the actions/reconfigurations related to the taken decision are accomplished. In this model, the learning module is connected to those others from which useful information can be extracted. This way, it can correlate and infer its own knowledge.

Besides the selected model, once the cognitive cycle and the techniques used in each phase are designed, it can be implemented in several ways. The cognitive capability may remain highly centralized or be absolutely distributed according to the design specifications. An example of cognitive cycle decentralization is the reconfigurable node model introduced in [14]. This model splits the phases into two entities (Fig. 6): the cognitive motor, which encompasses reasoning, learning and decision capabilities, and the reconfigurable node, which focuses on observation and action. The reconfigurable node is a structure composed of a set of hardware and software elements that are assembled by the cognitive motor according to the conclusions obtained from the information processing, and can be modified at convenience.

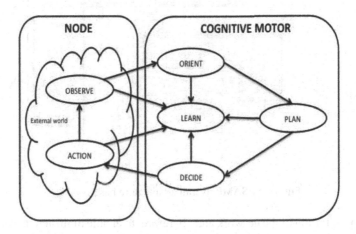

Fig. 6. Reconfigurable node architecture based on [14].

4 Application Example of Cognitive Cycles to SESAME Architecture

The SESAME EU-funded project contributes to the development of 5G focusing on three technological pillars: (i) the use of NFV and MEC to move network intelligence to the edge, (ii) the evolution of the Small Cell concept, and (iii) the promotion of multi-tenancy in communications infrastructures and services.

Based on these structural principles, SESAME suggests the architecture shown in Fig. 7. This architecture proposes the virtualization of small cell networks introducing the concept of Cloud Enabled Small Cell (CESC), a complete small cell that also contains a microserver offering, thus, computing, storage and radio resources. A number of CESCs form a cluster whose virtualized physical resources are shared and controlled by the VIM.

The CESC Manager (CESCM) is the central service management and orchestration component. A single instance of CESCM is able to operate over several CESC clusters, through the use of a dedicated VIM per cluster. Through virtualization, a Small Cell Network Operator (SCNO) can offer a cloud of resources that can be sliced to provide

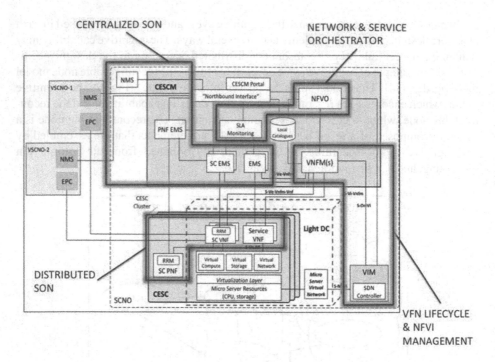

Fig. 7. SESAME overall architecture based on [1].

a logical instantiation of the network and, therefore, to enable multi-tenant services. The virtualized execution platform provided by the Light DC supports the VNF chains that implement the capabilities demanded to serve a requested network service (NS) by a tenant.

In summary, the Network Management System (NMS) arranges the logical network slices which enable SCNOs to provide Network-as-a-Service to the VSCNOs. Network slicing is supported by cloud edge-computing and NFV. The NFV Orchestrator (NFVO) provides management of those NFV services and is responsible for on-boarding of new network services and VNFs packages; NS lifecycle management; global resource management; validation and authorization of network functions virtualization infrastructure resource requests [6]. Finally, the VIM is responsible for controlling and managing the compute, storage, and network resources to allocate the VNFs within the CESC cluster.

At this point, the application of Artificial Intelligence techniques makes sense in those elements of the architecture where decisions are taken. In the context of SESAME platform, the decision-makers should be placed in the control plane (VIM, VNFs and NFVO) and/or the NMS in the management plane, as highlighted in Fig. 7. In particular, the concept of cognitive cycle can be applied to implement learning processes fed by the data generated during system operation.

In principle, the philosophy of MEC combines better with a distributed cognitive model, like the aforementioned OOPDAL, CME or Fortuna-Mohorcic. Such a model observes and collects the local operation data generated in the edge and processes it heuristically to extract new knowledge in the form of enhanced local decision rules. This operation way reinforces the distinctive features of the MEC architecture, i.e. low latency, context awareness, minimized data transit and reduced network congestion.

On the other hand, a centralized cognitive model such as the reconfigurable node can gather all the information about the network operation in the cloud servers and take advantage of higher computational and storage resources. At this point, Big Data techniques can also be useful to process high volume of raw data, in order to achieve optimized decision rules that are later communicated to the edge.

As a result of these considerations, we propose a combination of distributed and centralized model through a cloud-based multi-level cognitive cycle depicted in Fig. 8. The figure represents the four aforementioned decision points of the architecture.

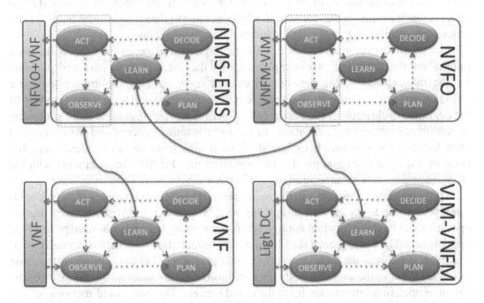

Fig. 8. Cloud-based multi-level cognitive cycle.

At the lower level of this model, on the left, the Distributed SON (D-SON) module applies a simplified version of the OOPDAL cognitive module to manage the VNFs running in the CESC. The NS (VNFs) themselves observe their own environment and self-adapt their operation. On the right side of Fig. 8, the VIM applies the same cognitive cycle to support the arrangement of the physical resources of the Light DC. The indicators that feed the monitoring phase of the cognitive cycle are just those generated in each CESC cluster. In this context, the planning phase analyzes the possible alternatives to allocate and distribute the VNFs in the available resources to later apply a decision rule that will lead to an operation action. All these steps give feedback to the learning module that heuristically analyses the results and infers enhanced decision rules. This way, these actions are entirely performed in the CESC cluster, maintaining all the logic in the edge. In addition, the VNFM can request VNF scaling to the VIM according to pre-established rules, defining, thus, the Distributed NFV Management.

In the upper level, the decision points identified as Centralized SON and Network and Service Orchestration implement a version of the Reconfigurable Node Model. This version consists of the same simplified OOPDAL cycle acting as the cognitive motor and the reconfigurable node that includes in its observation phase the information from the lower levels that is uploaded through the cloud.

In the Network and Service Orchestration, the cognitive cycle supports the tasks of the NVFO aimed at composing service chains (constituted by two or more VNFs located either in one or more CESCs) and managing the deployment of VNFs over the Light DC. The NFVO monitors the performance level of the NSs for all the VSCNOs through the SLA Monitoring, and is able to change the capabilities of the network elements (eventually, VNFs). It must be remarked that the CESC Manager that allocates the NVFO can manage one or more CESC cluster, with the increment of sample data to process that it entails. In this context, this high volume of data may require the Big Data processing techniques that can be implemented in the big cloud servers.

Finally, this structure is repeated in the management plane, at the C-SON decision point. Here, the cognitive motor is implemented in the NMS-EMS, which are responsible for the management of the network slices that serve the NSs. Following the same rationale, the interaction with the lower level is bidirectional. The learning module of the cognitive cycle implemented in the NVFO acts as the reconfigurable node, providing an additional source of observations to feed the learning process and also receiving knowledge updates to enrich its operation with high level decision rules. Thus, the network slicing activity is refined complementing the OSS/BSS requirements with the NFV managing state for an optimized performance.

The proposed variation of the Reconfigurable Node Model supposes that the cognitive cycle at the lower levels can operate independently in a distributed way, but its operation can be uplifted when appropriate incorporating valuable knowledge from the upper levels. The consequence is a blurring-edge effect: the low-level distributed cognitive cycle reinforces the edge-computing concept maintaining the decision logic close to the user, but at the same time, the reconfigurable node model adaptation moves the valuable operation information up to the cloud servers. This way, a bidirectional glocal effect is obtained. First, the cognitive cycle implemented within the distributed decision points can operate just with local information, but the generated knowledge feeds the cognitive motor at the centralized points leading to a global improvement over the performance of the network. In contrast, the knowledge results of the motor reconfigures as well the learning schema of the lower level, providing more accurate decision rules for future samples.

5   Conclusions

The application of cognitive models in the management of networked services has attracted significant effort for long time. This paradigm is especially relevant in mobile networks, where the dynamically variable context of use imposes more stringent requirements to the accurate management. Traditionally, the service and network management planes have been implemented in a rather isolated way, leading to non-optimal adaptation of services to network performance and of networks to service requirements. The evolution of different network softwarization techniques and its introduction into the design of future 5G mobile networks anticipates a revolution in the dynamic management of mobile services.

First, the SDN paradigm splits the control and data planes in data forwarding techniques. In this way, a centralized intelligent element is now capable of determining the most appropriate data paths for the different traffic flows. Additionally, the unification of different radio technologies in the framework of 5G networks introduces new capabilities and challenges in the cognitive management.

Second, the integration of service and network management endows the intelligent element with capabilities to perform consolidated cross-layer decisions. The advancement of mobile edge service technologies also brings new possibilities for an optimized coordination of the network and service data planes.

The use of cognitive cycles in this centralized unified service and network management system anticipates an enhanced performance of mobile services beyond the current capabilities of self organizing networks. However, the nature of end to end network services and the consideration of different network domains require also the inclusion of distributed cognitive cycles that locally work towards the overall enhancement.

The third technological building block analyzed in this paper is the inclusion of NFV techniques in the deployment of future 5G networks. The virtualization of network and service elements breaks the traditional limits of network-service performance maximization in each administrative domain, allowing the network provider to dynamically allocate the hardware resources where they are needed. Therefore, the paper analyses the possibility of building a higher-level cognitive cycle for the adaptive orchestration of virtualized resources, which in turn has an impact on the context information of the local cognitive management processes.

Different cognitive models have been presented and analyzed in the scope of 5G networks and specifically in the framework of dense cloud-enabled small cell deployments. As a result, the most appropriate cognitive schemes are discussed for the different decision making elements in the network-service provisioning.

Acknowledgements. This research received funding from the European Unions H2020 Research and Innovation Action under Grant Agreement No. 671596 (SESAME project).

References

1. H2020-SESAME: SESAME D2.2 v0.2: Overall System Architecture and Interfaces (2016)
2. ITU-T: FG IMT-2020: Report on Standards Gap Analysis (2015)
3. NGMN: 5G White Paper (2015)
4. ETSI: Mobile-Edge Computing - Introductory Technical White Paper (2014)
5. Imran, A., Zoha, A.: Challenges in 5G: how to empower SON with big data for enabling 5G. IEEE Netw. **28**(6), 27–33 (2014)
6. ETSI: ETSI GS NFV-MAN 001 v1.1.1: Network Functions Virtualisation (NFV); Management and Orchestration (2014)

7. Biglieri, E., Goldsmith, A.J., Greenstein, L.J., Mandayam, N.B., Poor, H.V.: Principles of Cognitive Radio. Cambridge University Press, New York (2012)
8. Clark, D.D., Partridge, C., Ramming, J.C., Wroclawski, J.T.: A knowledge plane for the internet. In: Proceedings of 2003 Conference on Applications, Technologies, Architectures, and Protocols for Computer Communications, SIGCOMM 2003. ACM Press, New York (2003)
9. Haigh, K.: AI technologies for tactical edge networks. IEEE J. Sel. Areas Commun. (2011). Keynote Presentation for MobiHoc Workshop on Tactical Mobile Ad Hoc Networking
10. Mitola, J.: Cognitive radio for flexible mobile multimedia communications. In: Proceedings of IEEE International Workshop on Mobile Multimedia Communications (Mo-MuC 1999), pp. 3–10 (1999)
11. Strassner, J., Agoulmine, N., Lehtihet, E.: FOCALE: a novel autonomic networking architecture. In: Proceedings of Latin American Autonomic Computing Symposium, Campo Grande, Brazil (2006)
12. Balamuralidhar, P., Prasad, R.: A context driven architecture for cognitive radio nodes. Wireless Pers. Commun. 45(3), 423–434 (2008)
13. Fortuna, C., Mohorcic, M.: Trends in the development of communication networks: cognitive networks. Comput. Netw. 53(9), 1354–1376 (2009)
14. Sutton, P., Doyle, L.E., Nolan, K.E.: A reconfigurable platform for cognitive networks. In: Proceedings of IEEE 2006 1st International Conference on Cognitive Radio Oriented Wireless Networks and Communications, pp. 1–5 (2006)

Author Index

Printed in the United States
By Bookmasters